Window Settings

Customize a window
TI-82, 10, 440
TI-83, 12, 446–7
TI-85, 14–15, 453–4
Visual Calculus, 427

Obtain a nice window setting
TI-82, 440
TI-83, 447
TI-85, 454
Visual Calculus, 429

Least-Squares Lines

Obtain slope and y-intercept of a least-squares line
TI-82, 37–38, 445
TI-83, 37–38, 452
TI-85, 38, 458
Visual Calculus, 429

Assign the least-squares line to a function; display points and/or line
TI-82, 445
TI-83, 452
TI-85, 458
Visual Calculus, 429

Antiderivatives, Riemann Sums, and Definite Integrals

Graph an antiderivative (or the solution of a differential equation)
TI calculators, 282
Visual Calculus, 431–32

Calculate a Riemann Sum
TI-82, 312, 444
TI-83, 312, 451
TI-85, 312, 457
Visual Calculus, 432

Evaluate a Definite Integral
TI-82, 323, 444
TI-83, 323–4, 451
TI-85, 323–4, 457
Visual Calculus, 433

Functions of Two Variables

Evaluate a function of two variables
TI calculator, 386
Visual Calculus, 433

Evaluate a first partial derivative
TI calculator, 392–3
Visual Calculus, 433

Evaluate a second partial derivative
TI-82 or TI-83, 397
TI-85, 397
Visual Calculus, 433

Applied Calculus

A Graphing Approach

Applied Calculus

A Graphing Approach

David I. Schneider

University of Maryland

David C. Lay

University of Maryland

Prentice Hall
Upper Saddle River, New Jersey 07458

Library of Congress Cataloging-in-Publication Data

SCHNEIDER, DAVID I.
Applied calculus: a graphing approach / David I. Schneider, David C. Lay.
p. cm.
Includes index.
ISBN 0-13-342478-2
1. Calculus. I. Lay, David C. II. Title.
QA303.S36 1997
515—dc20 96-24296
CIP

Acquisition Editor: George Lobell
Editorial Assistant: Gale Epps
Assistant Editor: Audra J. Walsh
Editorial Director: Tim Bozik
Editor-in-Chief: Jerome Grant
Assistant Vice President of Production and Manufacturing: David W. Riccardi
Editorial / Production Supervision: Judy Winthrop
Managing Editor: Linda Mihatov Behrens
Executive Managing Editor: Kathleen Schiaparelli
Manufacturing Buyer: Alan Fischer
Manufacturing Manager: Trudy Pisciotti
Director of Marketing: John Tweeddale
Marketing Assistant: Diana Penha
Creative Director: Paula Maylahn
Art Director: Amy Rosen
Assistant to the Art Director: Rod Hernandez
Cover Design: Jeanette Jacobs
Interior Design: Christine Gehring Wolf
Copyeditor: Kristen Cassereau
Cover photo: "And Justice for All" by Lili LaKich ©1990
Collection: Ace Parking / The Koll Co., San Diego; Photo: Jeff Atherton

Printed in the United States of America

10 9 8 7 6 5 4 3 2 1

ISBN 0-13-342478-2

Prentice-Hall International (UK) Limited, *London*
Prentice-Hall of Australia Pty. Limited, *Sydney*
Prentice-Hall Canada Inc., *Toronto*
Prentice-Hall Hispanoamericana, S. A., *Mexico*
Prentice-Hall of India Private Limited, *New Delhi*
Prentice-Hall of Japan, Inc., *Tokyo*
Simon & Schuster Asia Pte. Ltd., *Singapore*
Editora Prentice-Hall do Brasil, Ltda., *Rio de Janeiro*

CONTENTS

Chapter 3

ALGEBRAIC DIFFERENTIATION AND ITS APPLICATIONS 137

Chapter 4

EXPONENTIAL AND LOGARITHMIC FUNCTIONS 195

Chapter 5

APPLICATIONS OF THE EXPONENTIAL AND LOGARITHMIC FUNCTIONS 235

Chapter 6

ANTIDIFFERENTIATION AND DIFFERENTIAL EQUATIONS 271

Chapter 7

Chapter 8

PREFACE

Applied Calculus: A Graphing Approach is a one-semester, technology-required, reform calculus book for students majoring in business, economics, life sciences, and social sciences. The text places considerable emphasis on translating back and forth among graphs, formulas, numbers, and words. Prime importance is given to understanding the meaning of the first and second derivatives and the integral, especially with regard to applications. Technology is used to gain insights into the fundamental concepts of calculus.

HOW DOES THIS BOOK DIFFER FROM OTHER "APPLIED CALCULUS" BOOKS?

TECHNOLOGY REQUIRED Students are expected to use a graphing calculator or a computer with mathematical software. Individual instructors can determine the balance between technology and symbol manipulation. For instance, students can solve the consumers' surplus problem by evaluating the definite integral with a calculator, with the fundamental theorem of calculus, or with both.

GRAPHICAL EMPHASIS Although students are expected to sketch simple graphs by hand, they will rely on graphing utilities for creating most graphs. The effort saved is put into learning to read and interpret graphs. Also, the text has more graphs than the average applied calculus text.

FUNDAMENTAL CONCEPTS GIVEN PRIORITY OVER SYMBOL MANIPULATION The concept of the derivative is discussed extensively in

Chapter 2, whereas symbolic differentiation is postponed until Chapter 3. Applications of the definite integral precede the fundamental theorem of calculus. Students with weak algebraic skills do get an opportunity to improve their skills; however, the lack of algebraic skills should not hinder their understanding of the important concepts of calculus. Instructors can decide how much algebraic manipulation to require.

The book stays focused on its main objectives—understanding the meaning of the first and second derivatives and the integral, especially within the context of applications. (Graphing tricky functions, mastering every nuance of using graphing calculators, and doing complicated calculations are not objectives.) The book presents in detail the significance of the derivative before it asks the student to calculate derivatives algebraically. Also, it presents the meaning and wide variety of uses of the integral before giving the fundamental theorem of calculus. Key applications are examined several times in the text, each time from a slightly different viewpoint.

WHY DOES THE BOOK REFER SPECIFICALLY TO TI CALCULATORS AND VISUAL CALCULUS?

TI CALCULATORS If the book were written generically to apply to all graphing calculators, *everyone* would have to make adjustments. Since the TI-82, TI-83, and TI-85 calculators are the overwhelming favorite graphing calculators for calculus courses, *nearly all* students will have a book customized to their calculator. Section 1.2 presents the basic tasks to be performed, and other specialized tasks are discussed as needed. Appendices C, D, and E provide straightforward step-by-step summaries of using, respectively, the TI-82, TI-83, and TI-85.

VISUAL CALCULUS This software is the easiest-to-use calculus software available and is customized to the text. (About 500 functions from the text have been be entered into the software.) Appendix A explains how to use Visual Calculus. Colleges adopting the text will receive a free site license. In addition, the student study guide contains a copy of Visual Calculus.

Most students primarily will be using calculators. However, many of them also will have access to a computer. These students can use Visual Calculus to enhance the visualization, and therefore the understanding, of the fundamental concepts of calculus.

HOW DOES THE BOOK MANAGE TO PROVIDE SO MUCH FLEXIBILITY TO THE INSTRUCTOR?

GENEROUS EXERCISE SETS The ample supply of exercises include traditional drill problems, problems that test understanding, and problems requiring technology. Instructors can choose the balance of exercises they feel is most suitable.

ABUNDANCE OF TOPICS The book contains more than a semester's worth of material. Therefore, instructors can tailor the course to the needs and interests of their students.

WHAT SUPPLEMENTS ARE AVAILABLE?

INSTRUCTOR'S SOLUTION MANUAL Contains complete worked solutions to every exercise.

STUDENT STUDY GUIDE Contains detailed explanations and solutions to every sixth exercise. Also, includes helpful hints and strategies for studying that will help students improve their performance in the course.

INSTRUCTOR'S TEST ITEM FILE Contains hundreds of test questions, with many questions that test understanding and/or relate to the use of technology. Graphs appearing in the questions will be available on a diskette as TIF and EPS files, which can be inserted into exams.

WHAT ARE SOME OTHER SPECIAL FEATURES OF THE BOOK?

PRACTICE PROBLEMS FOR EACH SECTION The practice problems are carefully selected exercises located at the end of each section, just before the exercise set. Complete solutions are given following the exercise set. The practice problems often focus on points that are potentially confusing or are likely to be overlooked. We recommend that the reader seriously attempt the practice problems and study their solutions before moving on to the exercises. In effect, the practice problems constitute a built-in workbook.

CONCEPTUAL EXERCISES FOR EACH CHAPTER Each chapter ends with a set of exercises that review the fundamental concepts of the chapter. Many of these exercises require verbalization.

ACKNOWLEDGMENTS

While writing this book, we have received assistance from many persons. And our heartfelt thanks goes out to them all. Especially, we should like to thank the following reviewers, who took the time and energy to share their ideas, preferences, and often their enthusiasm with us. Reviewers: Robert D. Adams, U. of Kansas; John C. Hennessey, Loyola College, Baltimore; Yvette Hester, Texas A&M; Jack Porter, U. of Kansas; Helen Read, U. of Vermont; David C. Royster, U. of North Carolina, Charlotte; and Jon W. Scott, Montgomery College.

Our sincere thanks to Judy Winthrop, who has acted as production editor, and to Kristen Cassereau for her copyediting. We would also like to thank

Amanda Lubell for her contribution of many exercises and ideas throughout the preparation of the book, and Jason Schultz for his careful proofreading and helpful suggestions.

The authors would like to thank the many people at Prentice Hall who have contributed to the success of our books over the years. We appreciate the tremendous efforts of the production, art, manufacturing, and marketing departments.

The authors wish to extend special thanks to our editor George Lobell and typesetter/artist Dennis Kletzing. George's ideas, encouragement, and enthusiasm have nurtured us as we prepared this book. Dennis's considerable skills and congenial manner made for an uncomplicated and pleasant production process.

David I. Schneider
dis@math.umd.edu

David C. Lay

INDEX OF APPLICATIONS

Biology and Medicine

Business and Economics

Ecology

Social Sciences

General Interest

Applied Calculus

A Graphing Approach

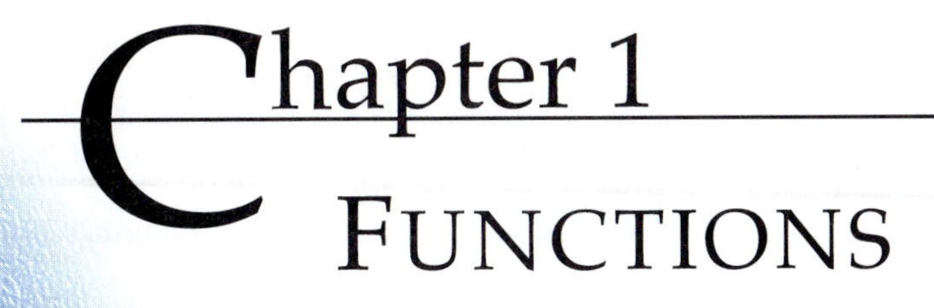

Chapter 1 Functions

Problems in business, management, the social and life sciences, and the physical sciences are often analyzed with the aid of one or more functions. This chapter explores various concepts about functions that will be used throughout the text.

1.1 Functions, Graphs, and Tables

A *function* of a variable x is a rule f that assigns to each value of x a number denoted by $f(x)$, called the *value of the function at* x. [Read "$f(x)$" as "f of x."] The variable x is called the *independent variable*. The set of permissible values of x is called the *domain* of the function. The domain may be explicitly specified as part of the definition of a function, or it may be understood from context. The *range* of a function f is the set of all values $f(x)$ as x varies over the domain.

A function can be defined in different ways, usually by a formula, a verbal description, or a graph. For instance, consider the function f defined for all x by the formula

$$f(x) = 3x - 1.$$

The domain of f is the set of all real numbers. In other words, this function is the rule that multiplies any given real number by 3 and then subtracts 1 from the result. If we specify a value, say $x = 2$, then the value of the function is computed by substituting 2 for x in the formula:

$$f(2) = 3(2) - 1 = 5.$$

This same function can also be defined graphically, as in Figure 1. Note that the point (2, 5) is on the line in Figure 1. In general, a point (x, y) is on the line if and only if y is the value of the function at x, that is, if and only if $y = 3x - 1$.

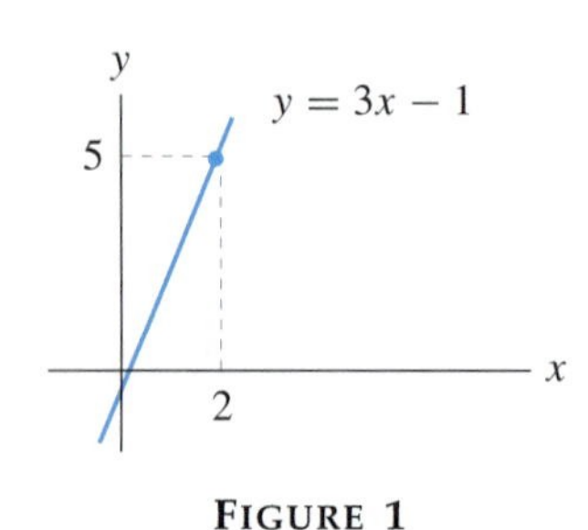

FIGURE 1

EXAMPLE 1 Let f be the function with domain all real numbers and defined by

$$f(x) = x^2 - 2x + 1.$$

(a) Find the value of f at $x = 5$.

(b) Find the value of f at $x = -5$.

(c) Find the value(s) of x at which the value of f is 4.

Solution (a) In symbols, compute $f(5)$. That is, substitute 5 for each occurrence of x in the formula for $f(x)$:

$$f(5) = (5)^2 - 2(5) + 1 = 25 - 10 + 1 = 16.$$

(b) Compute $f(-5)$, that is, substitute -5 for each occurrence of x. Parentheses around each -5 will ensure correct substitution. For instance, x^2 must be replaced by $(-5)^2$, not by -5^2:

$$f(-5) = (-5)^2 - 2(-5) + 1 = 25 + 10 + 1 = 36.$$

(c) To find the value (or values) of x at which the value of the function is 4, solve the equation $f(x) = 4$. That is, solve $x^2 - 2x + 1 = 4$:

$$x^2 - 2x + 1 = 4$$
$$x^2 - 2x - 3 = 0$$
$$(x - 3)(x + 1) = 0.$$

Thus $x - 3 = 0$ or $x + 1 = 0$, that is, $x = 3$ or $x = -1$. It is easy to check that $f(3) = 4$ and $f(-1) = 4$. ■

Note that in part (c) above, a quadratic equation was solved by factoring. Appendix B at the end of this book reviews techniques for solving quadratic equations.

GRAPHICAL REPRESENTATION OF A FUNCTION Functions can be described geometrically on a rectangular xy-coordinate system, as in Figure 1. Each x in the domain of f determines a point $(x, f(x))$ in the xy-plane. The value of the function at x is the y-coordinate of the point. The set of all such points $(x, f(x))$ usually forms a curve in the xy-plane and is called the *graph of the function* f. (Usually, only a part of the graph of a function is actually displayed in a figure.)

The computations described in Example 1 have graphical analogues, as the next example shows.

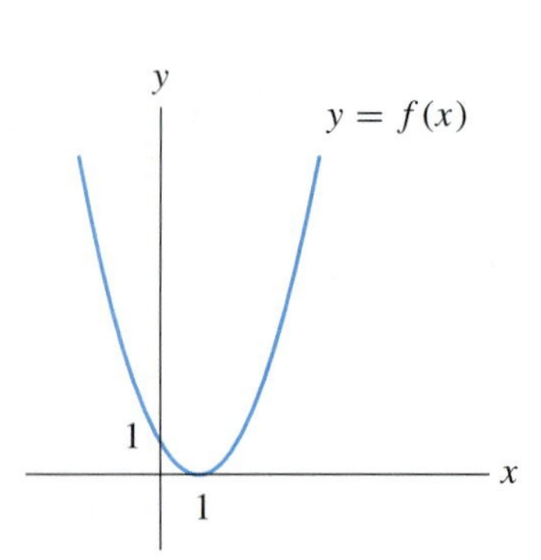

FIGURE 2

EXAMPLE 2 Figure 2 shows the graph of the function $f(x) = x^2 - 2x + 1$. Use the graph to answer the following questions.

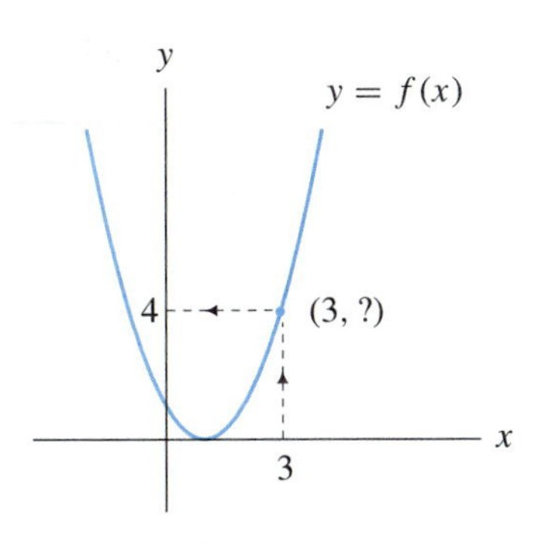

FIGURE 3

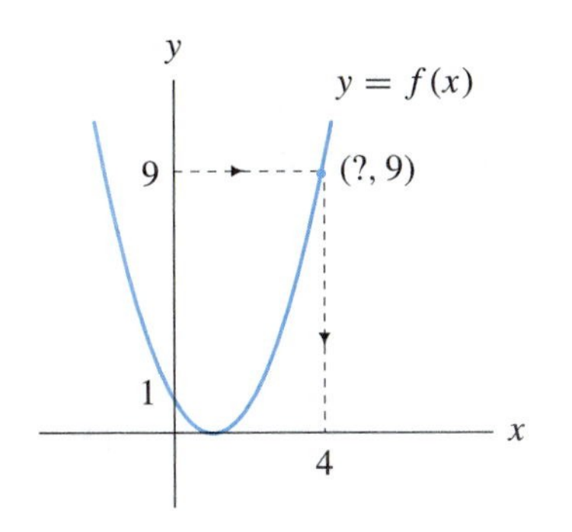

FIGURE 4

(a) What is the value of $f(x)$ when x is 3?

(b) For what x is the value of the function equal to 9?

Solution (a) To find $f(3)$ graphically, start at 3 on the x-axis and move vertically until you reach a point on the graph. The y-coordinate of that point gives the answer. To find it, move from the point horizontally (to the left) and read the value on the y-axis. See Figure 3. In this case, the point is $(3, 4)$ and $f(3) = 4$.

(b) The value of the function is known, not the value of x. In this case, look for a point whose y-coordinate is 9. Begin at 9 on the y-axis and move horizontally until you reach a point on the graph. The x-coordinate of that point gives the answer. To find it, move from the point vertically (down) and read the value on the x-axis. See Figure 4. Actually, there are two answers, because the horizontal line $y = 9$ intersects the graph of f twice. Both $(-2, 9)$ and $(4, 9)$ are on the graph, so the function value is 9 at both $x = -2$ and $x = 4$. ■

EXAMPLE 3 When studying various metabolic processes in a research animal, doctors sometimes add a radioactive isotope of iodine, ^{131}I, to a drug in order to monitor the path of the drug within the body. The iodine decays (decomposes) rapidly and quickly disappears, so it does not pose a health hazard for the animal. The graph in Figure 5 shows the amount of ^{131}I present within an animal at various times after an initial injection of 48 milligrams. Here t (for time) is used on the horizontal axis instead of x and $f(t)$ is the number of milligrams of iodine present after t days.

(a) When will only 10 milligrams of ^{131}I be present?

(b) How much ^{131}I is present 10 days after the initial injection?

Solution (a) The question "when" indicates that the time t is unknown. The 10 milligrams is the value of the function, not the value of t. Look for a point on the graph whose y-coordinate is 10. See Figure 6. The point appears to have a t-coordinate about halfway between 15 and 20, so an approximate answer is $t =$ 17.5 days (after the initial injection).

(b) Now the 10 refers to the time, that is, $t = 10$. The amount of ^{131}I present is the y-coordinate of the point at which $t = 10$. From Figure 7, the y-coordinate appears to be approximately 20. Approximately 20 mg of iodine are present 10 days after the injection. ■

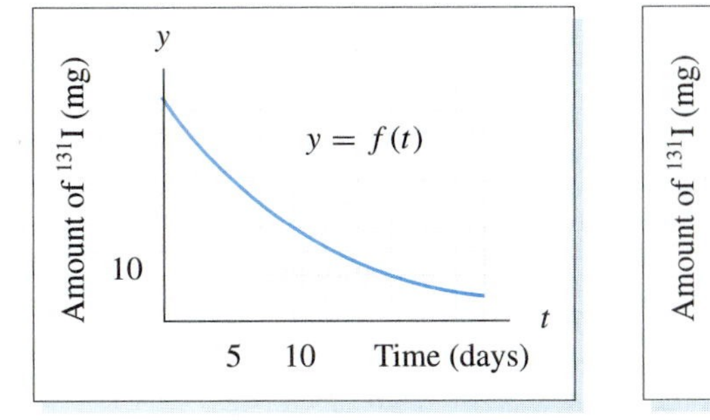

FIGURE 5

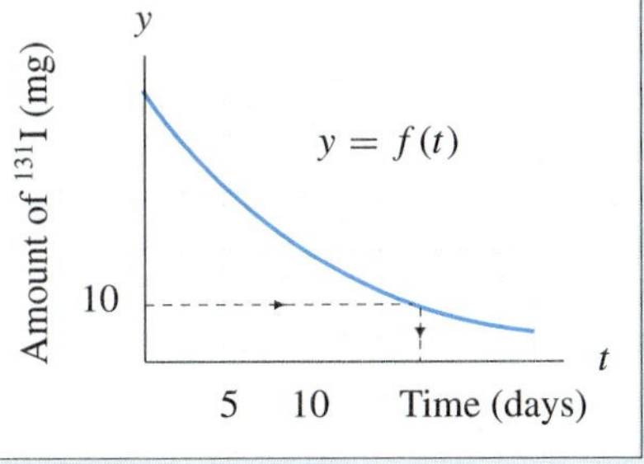

FIGURE 6

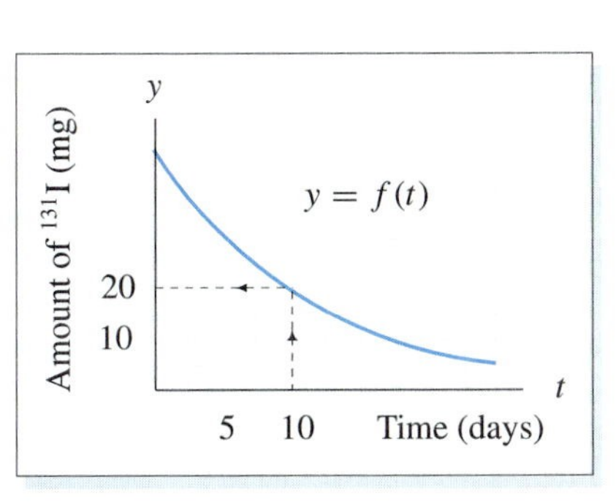

FIGURE 7

Occasionally, the algebraic representation of a function involves more than one formula.

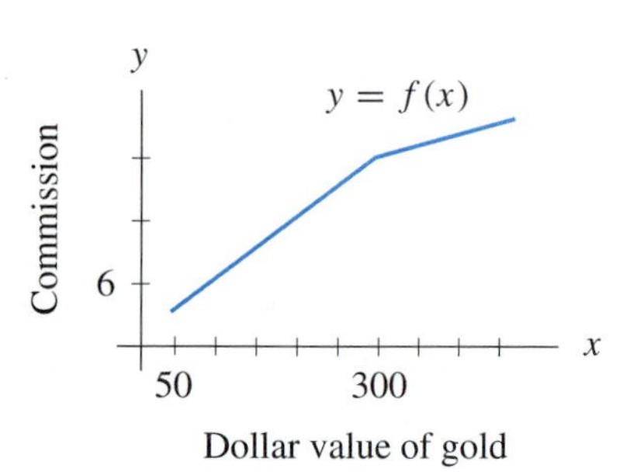

FIGURE 8 Commission charges for gold purchases

EXAMPLE 4 A brokerage firm charges a 6% commission on gold purchases in amounts from \$50 to \$300. (The minimum purchase is \$50.) For purchases exceeding \$300, the firm charges 2% of the amount purchased, plus \$12. Let x denote the amount of gold purchased (in dollars) and let $f(x)$ be the commission charge as a function of x.

(a) Give an algebraic representation of $f(x)$.

(b) Find $f(100)$ and $f(500)$.

(c) Draw the graph of $f(x)$.

Solution (a) The formula for $f(x)$ depends on whether x is in the interval from \$50 to \$300 or x is greater than \$300. We shall write $50 \le x \le 300$ to indicate that x is at least 50 and not more than 300. For such x, the brokerage charge is $.06x$. For x greater than 300, which we write as $x > 300$ or $300 < x$, the brokerage charge is $.02x + 12$.

To show how the two formulas for $f(x)$ fit together, we write

$$f(x) = \begin{cases} .06x & \text{if } 50 \le x \le 300 \\ .02x + 12 & \text{if } 300 < x. \end{cases}$$

(b) Since $x = 100$ satisfies $50 \le x \le 300$, use the first formula for $f(x)$: $f(100) = .06(100) = 6$. Since $x = 500$ satisfies $300 < x$, use the second formula for $f(x)$: $f(500) = .02(500) + 12 = 22$.

(c) The graph of $f(x)$ consists of two line segments, as shown in Figure 8. Graphs involving straight lines will be discussed further in Section 1.4. ■

FUNCTIONS DEFINED BY TABLES Sometimes a function arises in an application as a table that displays corresponding values of two variables, say t and y. The table determines y as a function of t if for each value of t there is only one corresponding value of y in the table.

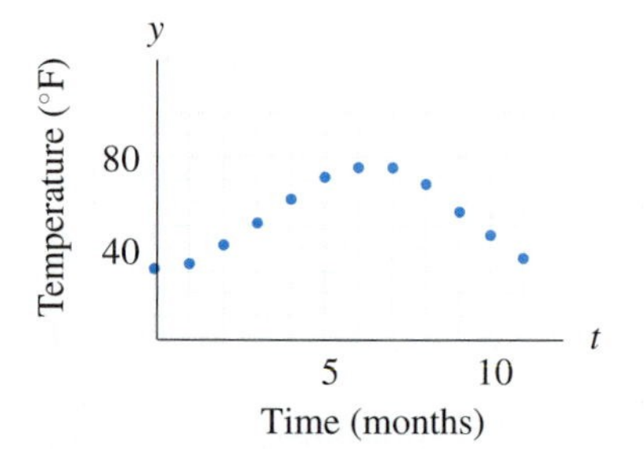

FIGURE 9 Monthly normal mean temperature in Wilmington

EXAMPLE 5 The data in Table 1 is from the World Almanac.* This table determines a function f of a variable t. For instance, let t be the number of months after January and let $f(t)$ be the monthly normal mean temperature, in degrees Fahrenheit. See Table 2. (The months could have been numbered differently, say, from 1 to 12.) The function f in Table 2 can be described graphically as a set of twelve dots, as in Figure 9. ■

Table 1 MONTHLY NORMAL MEAN TEMPERATURE (°F) IN WILMINGTON, DE

Month	Jan	Feb	Mar	Apr	May	Jun	Jul	Aug	Sept	Oct	Nov	Dec
Mean Temp.	31	33	42	52	62	71	76	75	68	56	46	36

* *The World Almanac and Book of Facts*, (New York: Pharos Books, 1993), p. 185.

Table 2 A FUNCTION DEFINED BY TABLE 1

Month	0	1	2	3	4	5	6	7	8	9	10	11
Mean Temp.	31	33	42	52	62	71	76	75	68	56	46	36

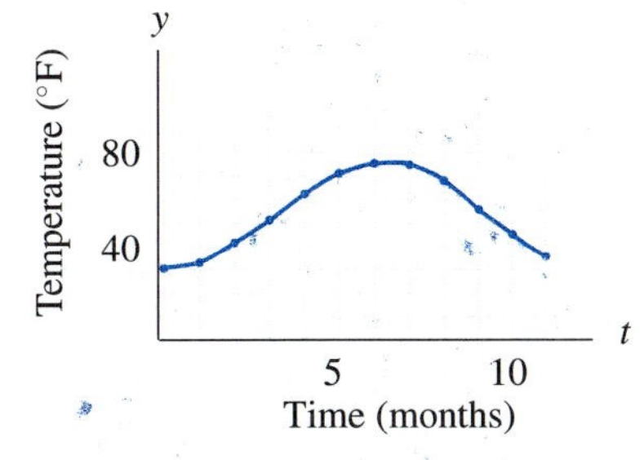

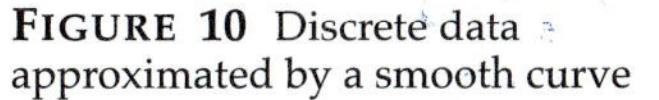
FIGURE 10 Discrete data approximated by a smooth curve

Calculus is used to study functions whose graphs are smooth curves rather than sets of discrete points. So it is often convenient to draw a smooth curve approximately through the points and assume that $f(t)$ is defined for all t in some interval. See Figure 10, for example, where t varies over the interval $0 \leq t \leq 11$.

MORE ABOUT THE DOMAIN OF A FUNCTION When defining a function, it is necessary to specify the domain of the function, which is the set of acceptable values of the variable. In the preceding examples, we explicitly specified the domains of the functions considered. However, throughout the remainder of the text, we will usually mention functions without specifying domains. In such circumstances, we will understand the intended domain to consist of all numbers for which the defining formula(s) makes sense. For example, consider the function

$$f(x) = x^2 - x + 1.$$

The expression on the right may be evaluated for any value of x. So in the absence of any explicit restrictions on x, the domain is understood to consist of all numbers. As a second example, consider the function

$$f(x) = \frac{1}{x}.$$

Here x may be any number except zero. (Division by zero is not permissible.) So the domain intended is the set of nonzero numbers. Similarly, when we write

$$f(x) = \sqrt{x},$$

we understand the domain of $f(x)$ to be the set of all nonnegative numbers, since the square root of a number x is defined if and only if $x \geq 0$.

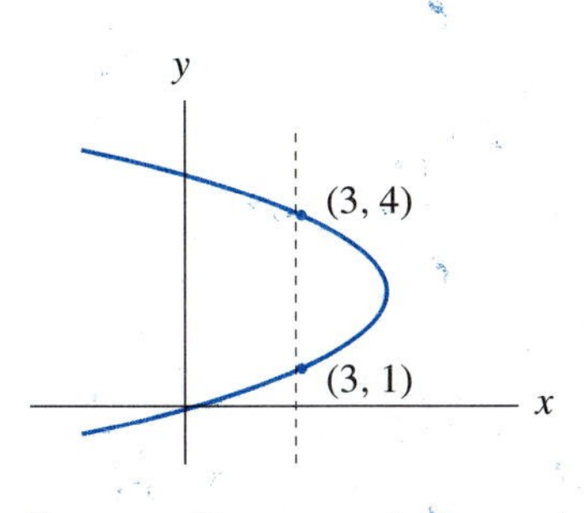

FIGURE 11 A curve that is not the graph of a function

CURVES THAT ARE NOT FUNCTIONS The graphs of the functions shown earlier have the important property that for each x in the domain of the function, there is only one y such that (x, y) is on the graph. This is to be expected, because a function f assigns only one value $f(x)$ to each x in the domain. In contrast, the curve in Figure 11 cannot be the graph of a function because there are values of x, such as $x = 3$, that determine more than one point on the graph.

The following graphical test shows which curves are graphs of functions.

The Vertical Line Test A curve in the xy-plane is the graph of a function if and only if each vertical line cuts or touches the curve at no more than one point.

EXAMPLE 6 Which of the curves in Figure 12 are graphs of functions?

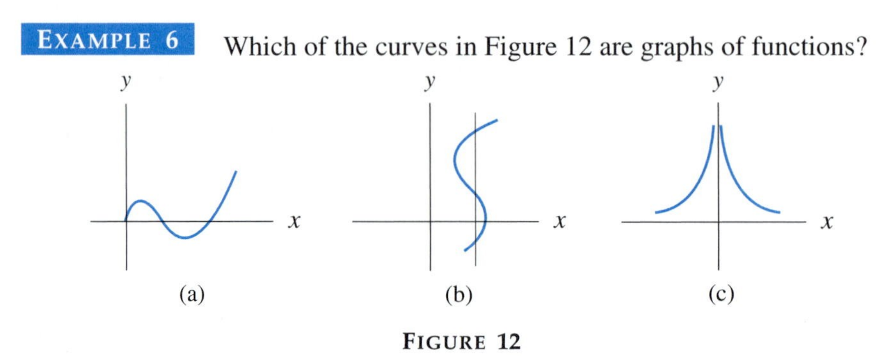

FIGURE 12

Solution The curve in (a) is the graph of a function. It appears that vertical lines to the left of the y-axis do not touch the curve at all. This simply means that the function represented in (a) is defined only for $x \geq 0$. The curve in (b) is *not* the graph of a function, because some vertical lines cut the curve in three places. The curve in (c) is the graph of a function whose domain is all nonzero x. [There is no point on the curve in (c) whose x-coordinate is 0.] ■

OTHER NOTATION FOR FUNCTIONS Suppose that f is a function whose graph is displayed on an xy-coordinate system. The values of $f(x)$ determine the y-coordinates of points on the graph. For this reason, the function is often abbreviated by the letter y, and we find it convenient to speak of "the function $y = f(x)$." For instance, the function $y = 2x^2 + 1$ refers to the function f for which $f(x) = 2x^2 + 1$. The graph of a function f is often called the graph of the equation $y = f(x)$.

Letters other than f can be used to denote functions, and letters other than x, y, and t can be used to denote variables. This is especially common in applied problems in which letters are chosen to suggest the quantities they represent. For instance, the revenue of a company as a function of time might be written as $R(t)$.

PRACTICE PROBLEMS 1.1

1. Rewrite the following commands about a function f, given by a formula with variable x, in verbal form. [Use the symbol f, but not $f(3)$ or $f(x)$.]
   (a) Compute $f(3)$. (b) Solve $f(x) = 5$.
2. Is the point (3, 12) on the graph of the function g given by $g(x) = x^2 + 5x - 10$?

EXERCISES 1.1*

1. Write a formula for the function f that adds three to any number x and then multiplies the result by five.
2. Write a formula for the function h that multiplies the cube of any number t by three and then adds four to the result.

In Exercises 3 and 4, give a verbal description of the function. (*See the statements in Exercises 1 and 2.*)

3. $f(t) = t^3 + 6t$

* A complete solution for every third odd-numbered exercise is given in the Study Guide for this text.

4. $g(x) = 4x - x^2$

In Exercises 5–8, rephrase the verbal statement about a function f of a variable x. Write a complete sentence in a concise form that uses $f()$ in an equation.

5. Find the value of f at $x = -2$.
6. At what value of x is the value of f equal to $\sqrt{3}$?
7. At what value of x is the value of f equal to 20?
8. Find the value of f when x is 300.
9. If $f(x) = x^2 - 3x$, find $f(0)$, $f(5)$, and $f(-7)$.
10. If $f(s) = 9 - 6s + s^2$, find $f(0)$, $f(4)$, and $f(-13)$.
11. If $h(s) = \dfrac{s}{s^2 - 1}$, find $h\left(\frac{1}{2}\right)$ and $h\left(-\frac{3}{2}\right)$.
12. If $g(x) = \dfrac{x^2}{x - 1}$, find $g\left(\frac{3}{2}\right)$ and $g\left(-\frac{1}{2}\right)$.
13. Suppose you are shown the graph of a function f drawn on graph paper. Explain how you would determine the value of $f(5)$. Explain how you would find a solution to the equation $f(x) = 5$. (Write complete sentences.)
14. Suppose you are given the formula for a function. Explain how you would determine the value of $f(5)$. (Write complete sentences.)
15. An office supply firm finds that the number of fax machines sold in year x is given approximately by the function $f(x) = 50 + 4x + \frac{1}{2}x^2$, where $x = 0$ corresponds to 1990.
    (a) What does $f(0)$ represent?
    (b) Find the number of fax machines sold in 1992.
16. When a solution of acetylcholine is introduced into the heart muscle of a frog, it diminishes the force with which the muscle contracts. The data from experiments of A. J. Clark are closely approximated by a function of the form

$$R(x) = \frac{100x}{b + x}, \qquad x \geq 0,$$

where x is the concentration of acetylcholine (in appropriate units), b is a positive constant that depends on the particular frog, and $R(x)$ is the response of the muscle to the acetylcholine, expressed as a percentage of the maximum possible effect of the drug.
    (a) Suppose that $b = 20$. Find the response of the muscle when $x = 60$.
    (b) Determine the value of b if $R(50) = 60$ — that is, if a concentration of $x = 50$ units produces a 60% response.

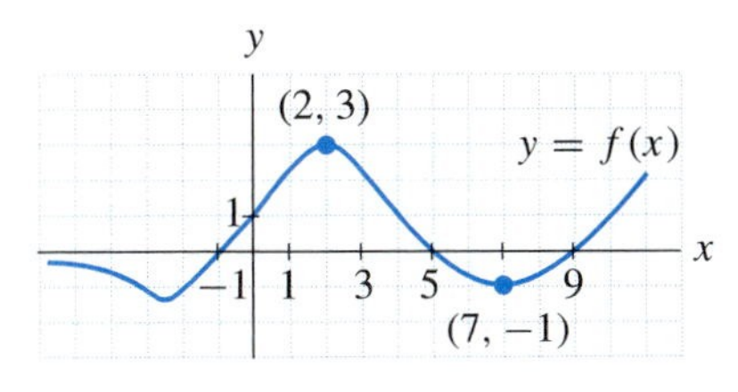

FIGURE 13

Exercises 17–26 relate to the function whose graph is sketched in Figure 13.

17. Find $f(0)$.
18. Find $f(7)$.
19. Find $f(2)$.
20. Find $f(-1)$.
21. Is $f(4)$ positive or negative?
22. Is $f(6)$ positive or negative?
23. Is $f\left(-\frac{1}{2}\right)$ positive or negative?
24. Is $f(1)$ greater than $f(6)$?

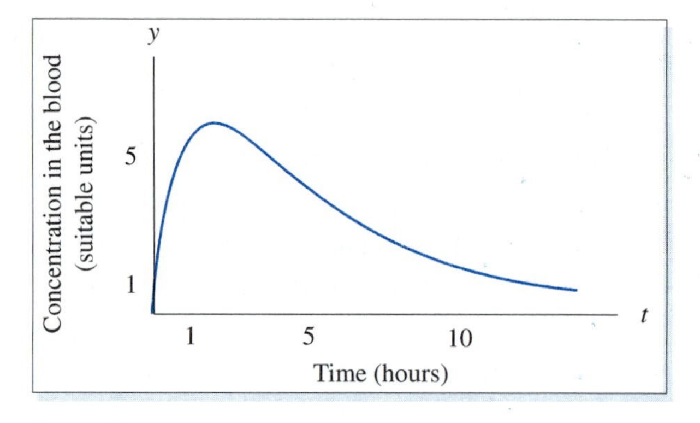

FIGURE 14 Drug time–concentration curve

25. For what values of x is $f(x) = 0$?

26. For what values of x is $f(x) \geq 0$?

Exercises 27–32 relate to Figure 14. When a drug is injected into a person's muscle tissues, the concentration y of the drug in the blood is a function of the time elapsed since the injection. The graph of a typical time–concentration function f is given in Figure 14, where $f(t)$ is the concentration t hours after the injection.

27. What is the concentration of the drug after 12 hours?

28. While the drug concentration is decreasing, when is the concentration 4 units?

29. While the drug concentration is increasing, when is the concentration 5 units?

30. After how many hours does the drug attain its greatest concentration, and what is the greatest concentration?

31. Find $f(5)$ and explain in medical terms what your answer means.

32. Find a solution to $f(t) = 1$ and explain in medical terms what your answer means.

33. Is the point $(3, 12)$ on the graph of the function $f(x) = (x - \frac{1}{2})(x + 2)$?

34. Is the point $(-2, 12)$ on the graph of the function $f(x) = x(5 + x)(4 - x)$?

35. Is the point $\left(\frac{1}{2}, \frac{2}{5}\right)$ on the graph of the function $g(x) = (3x - 1)/(x^2 + 1)$?

36. Is the point $\left(\frac{2}{3}, \frac{5}{3}\right)$ on the graph of the function $g(x) = (x^2 + 4)/(x + 2)$?

37. Suppose that the brokerage firm in Example 4 decides to keep the commission charges unchanged for purchases up to and including \$600 but to charge only 1.5% plus \$15 for gold purchases exceeding \$600. Express the brokerage commission as an algebraic representation of the amount x of gold purchased.

In Exercises 38–40, compute $f(1)$, $f(2)$, and $f(3)$.

38. $f(x) = \begin{cases} \sqrt{x} & \text{for } 0 \leq x < 2 \\ 1 + x & \text{for } 2 \leq x \leq 5 \end{cases}$

39. $f(x) = \begin{cases} 1/x & \text{for } 1 \leq x \leq 2 \\ x^2 & \text{for } 2 < x \end{cases}$

40. $f(x) = \begin{cases} \pi x^2 & \text{for } x < 2 \\ 1 + x & \text{for } 2 \leq x \leq 2.5 \\ 4x & \text{for } 2.5 < x \end{cases}$

In Exercises 41–44, describe the domain of the function.

41. $f(x) = \dfrac{8x}{(x-1)(x-2)}$

42. $f(t) = \dfrac{1}{\sqrt{t}}$

43. $g(x) = \dfrac{1}{\sqrt{3-x}}$

44. $g(x) = \dfrac{4}{x(x+2)}$

In Exercises 45–50, decide which curves are graphs of functions.

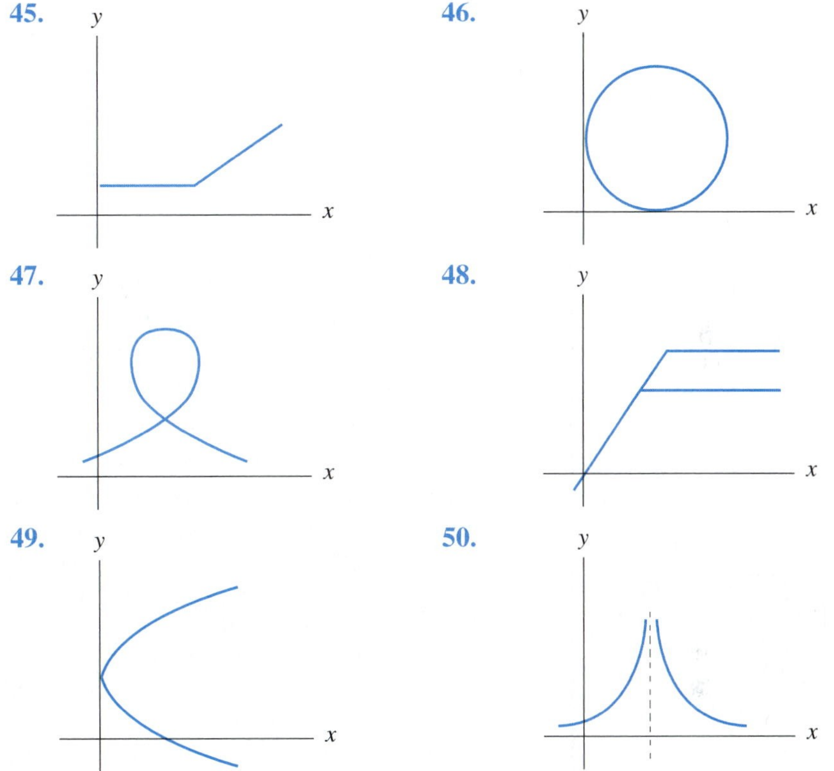

SOLUTIONS TO PRACTICE PROBLEMS 1.1

1. (a) Substitute 3 for each occurrence of x in the formula and carry out the arithmetic.
   (b) Find the value of x at which the function value is 5.
2. No. The point (3, 12) is on the graph of the function g only if $g(3) = 12$. However, $g(3) = 3^2 + 5 \cdot 3 - 10 = 9 + 15 - 10 = 14$.

1.2 GRAPHING UTILITIES

We use the term *graphing utility* to refer to either a graphing calculator or a computer with mathematical software. Because graphing utilities are now readily available and easy to operate, we use them to develop graphical insights that are not obvious through algebra. Most of the understanding of calculus is graphical; the mechanics of calculus are algebraic.

This section provides an introduction to the use of graphic utilities. Throughout this text, specific instructions are given for the TI-82, TI-83, and TI-85 graphing calculators. However, the same results can be produced on many other graphing calculators with slight modifications.

In this text, graphing utilities will be used to carry out computations and to visualize fundamental concepts of calculus. Total mastery of the intricacies of graphing utilities is not one of the objectives of the text. This section discusses the following basic tasks:

1. Graphing functions.
2. Finding the coordinates of points on the graph of a function.
3. Determining where the graph of a function crosses the x-axis.
4. Finding the intersection point of two graphs.

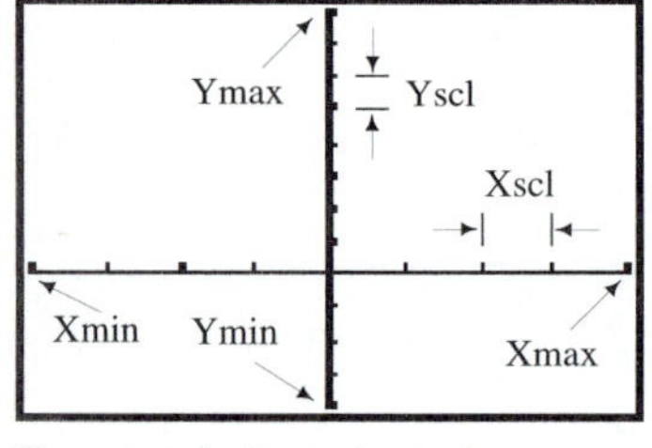

FIGURE 1 Typical window

The following three subsections discuss the basic tasks for the TI-82, TI-83, and TI-85 graphing calculators. If you will be using one of these calculators, read the appropriate subsection and the subsection titled "General Comments on TI Calculators." If you are using a different calculator, skim the TI-82 section and consult your owner's manual to learn how to perform the four basic tasks.

USING THE TI-82 GRAPHING CALCULATOR

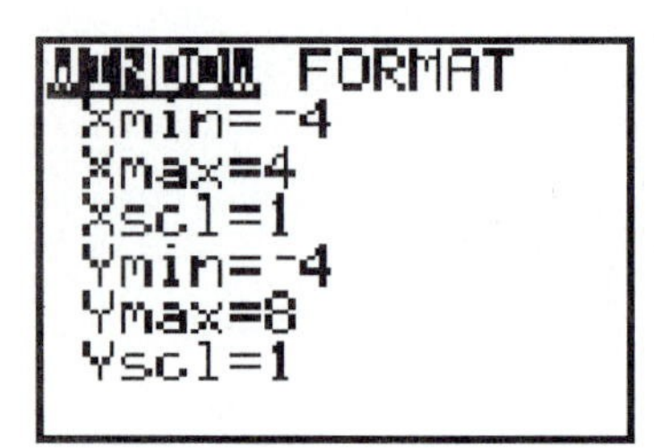

FIGURE 2 TI-82 settings for Figure 1

Functions are graphed in a rectangular window like the one shown in Figure 1. The numbers on the x-axis range from Xmin to Xmax, and the numbers on the y-axis range from Ymin to Ymax. The distances between tick marks are Xscl and Yscl on the x- and y-axis, respectively. To specify these quantities, press [WINDOW] and type in the values of the six variables. Figure 2 gives the settings associated with the window in Figure 1. (*Note*: To enter a negative number, use the [(−)] key on the bottom row of the calculator.)

To specify functions, press the [Y=] key and type expressions next to the function names Y_1, Y_2, ... (*Note*: To erase an expression, use the arrow keys to move the cursor anywhere on the expression and press [CLEAR].) In Figure 3, several functions have been specified. The expressions were produced with the following keystrokes:

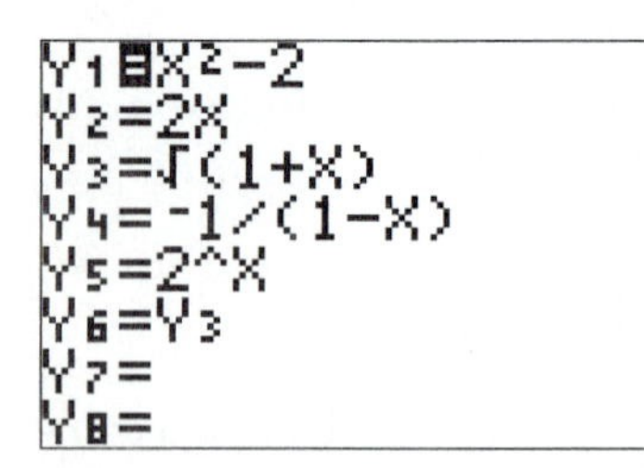

FIGURE 3 TI-82 function declarations

Y_1: [X, T, θ] [x^2] [−] **2**

Y_2: **2** [X, T, θ] (Notice that a multiplication sign is not needed.)

Y_3: [2nd] [√] [(] **1** [+] [X, T, θ] [)]

Y_4: [(−)] **1** [÷] [(] **1** [−] [X, T, θ] [)]

Y_5: **2** [∧] [X, T, θ] (∧ is the symbol for exponentiation.)

Y_6: [2nd] [Y-VARS] **1 3**

Notice that in Figure 3 the equal sign in Y_1 is highlighted, whereas the other equal signs are not highlighted. This highlighting can be toggled by moving the cursor to an equal sign and pressing [ENTER]. Functions with highlighted equal

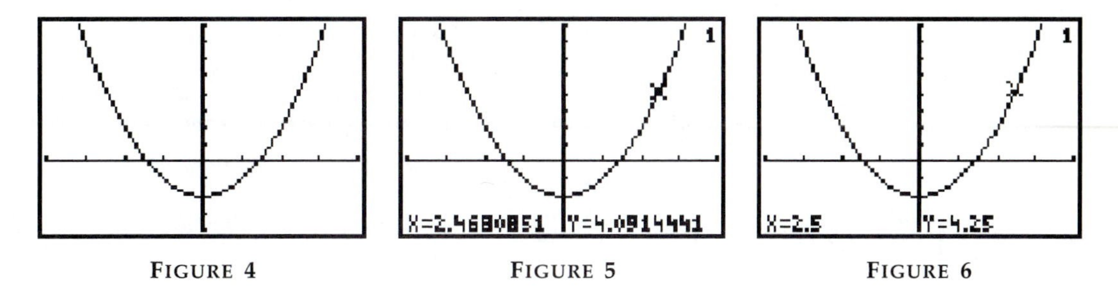

FIGURE 4 FIGURE 5 FIGURE 6

signs are said to be *selected*. Pressing the GRAPH key instructs the calculator to graph all selected functions.

Press GRAPH to obtain Figure 4. Then press TRACE and press ▶, the right-arrow key, 29 times to obtain Figure 5. Each time a right- or left-arrow key is pressed, the trace cursor moves along the curve and the coordinates of the trace cursor are displayed. The small number 1 in the upper-right corner of the window identifies the curve as the graph of Y_1.

To approximate the function value for a specific value of x, move the trace cursor as close as possible to the value of x and read the y-coordinate of the point. For a more precise function value, press 2nd [CALC] **1**, type in a number (such as 2.5), and press ENTER. See Figure 6.

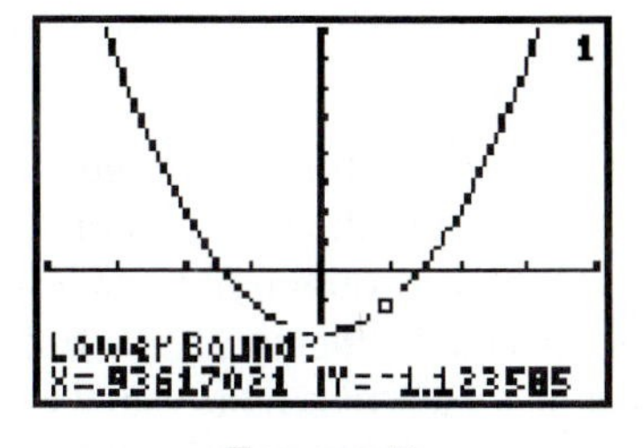

FIGURE 7

A point where the graph crosses the x-axis is called an x-*intercept*. The coordinates of an x-intercept can be approximated by tracing. The x-coordinate of an x-intercept is a root of the equation $Y_1 = 0$. For a more precise value of the root, press 2nd [CALC] **2** and answer the questions. Reply to "Lower Bound?" by moving the cursor to a point whose x-coordinate is less than the root and pressing ENTER. See Figure 7. Reply to "Upper Bound?" by moving the cursor to a point whose x-coordinate is greater than the root and pressing ENTER. See Figure 8. Reply to "Guess?" by moving the cursor near the x-intercept and pressing ENTER. Figure 9 shows the resulting display.

So far, all operations were carried out while looking at the graph of a function. These same operations also can be carried out in the Home screen, which is invoked by pressing 2nd [QUIT]. Figure 10 shows several computations. These computations were produced with the following keystrokes:

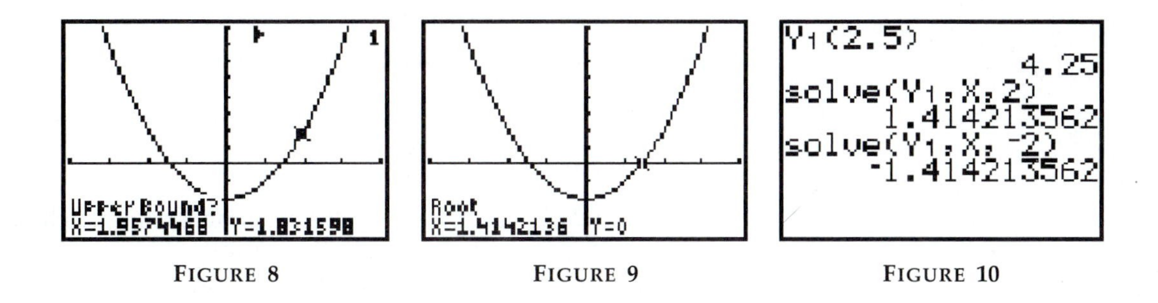

FIGURE 8 FIGURE 9 FIGURE 10

1. 2nd [Y-VARS] **1 1** (**2.5**) ENTER

2. MATH **0** 2nd [Y-VARS] **1 1** , X, T, θ , **2**) ENTER

   (*Note*: The number **2** is a guess for the solution to $\mathrm{Y}_1 = 0$.)

3. 2nd [ENTRY] ◄ ◄ 2nd [INS] (–) ENTER

   (*Note*: Pressing 2nd [ENTRY] repeats the previous instruction. Pressing 2nd [INS] toggles between Insert Mode and Overwrite Mode.)

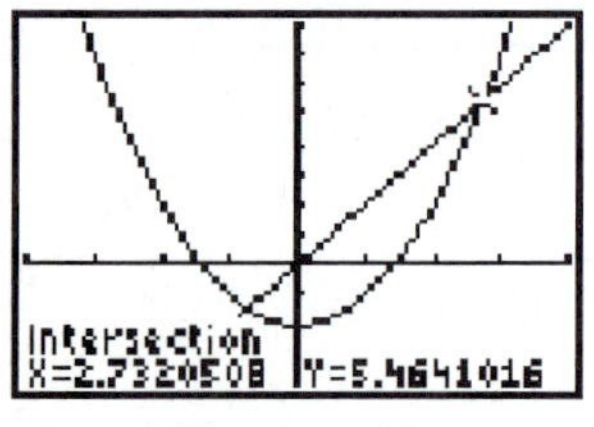

FIGURE 11

Return to the function screen by pressing Y= and then select both Y_1 and Y_2. Press GRAPH to obtain the graphs of the two functions. Press TRACE and then press the right-arrow key several times. Now press the down-arrow key several times. Each time the down-arrow key is pressed, the trace cursor moves from one curve to the other. The small number in the upper-right corner, 1 or 2, identifies the function containing the cursor. Move the cursor as close as possible to the point of intersection of the two curves to approximate the coordinates of the intersection point. For more precise values, press 2nd [CALC] **5** and answer the questions. Reply to "First curve?" by moving the trace cursor to one of the curves and pressing ENTER. Reply to "Second curve?" by moving the trace cursor to the other curve and pressing ENTER. Reply to "Guess?" by moving the trace cursor near the point of intersection and pressing ENTER. Figure 11 shows the resulting display.

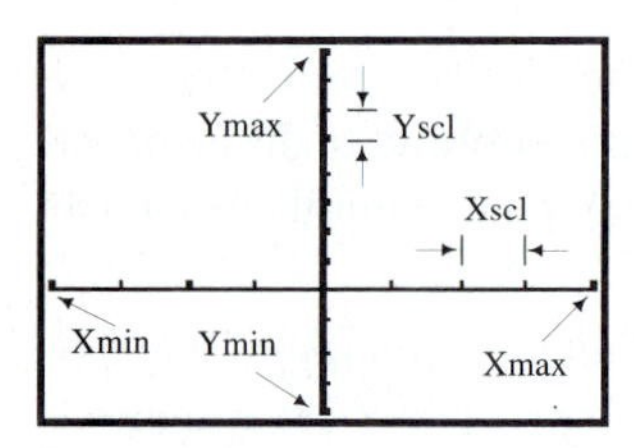

FIGURE 12 Typical window

USING THE TI-83 GRAPHING CALCULATOR

Functions are graphed in a rectangular window like the one shown in Figure 12. The numbers on the x-axis range from Xmin to Xmax, and the numbers on the y-axis range from Ymin to Ymax. The distances between tick marks are Xscl and Yscl on the x- and y-axis, respectively. To specify these quantities, press WINDOW and type in the values of the six variables. Figure 13 gives the settings associated with the window in Figure 12. (*Note*: To enter a negative number, use the (–) key on the bottom row of the calculator.)

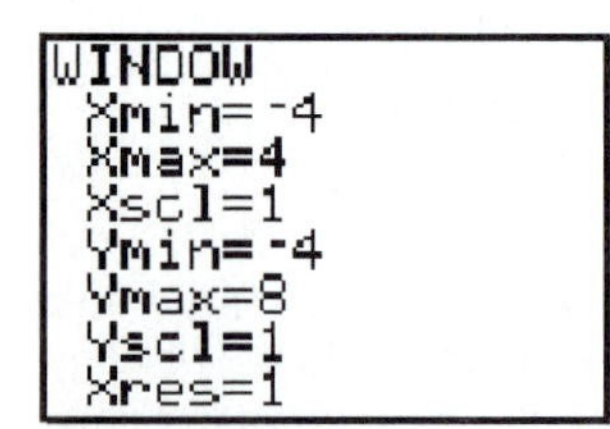

FIGURE 13 TI-83 settings for Figure 12

To specify functions, press the Y= key and type expressions next to the function names Y_1, Y_2, ... (*Note*: To erase an expression, use the arrow keys to move the cursor anywhere on the expression and press CLEAR.) In Figure 14, several functions have been specified. The expressions were produced with the following keystrokes:

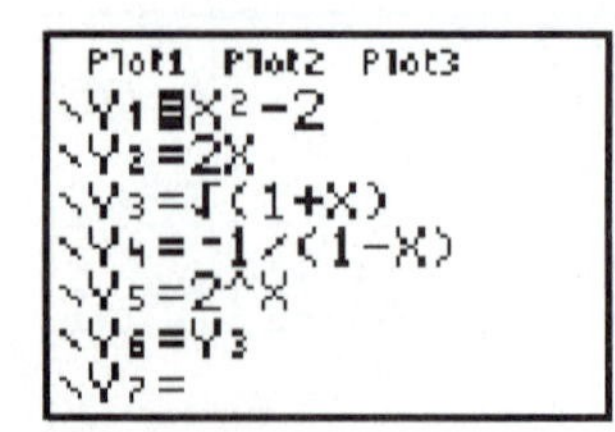

FIGURE 14 TI-83 function declarations

Y_1: X, T, θ, n x^2 – **2**

Y_2: **2** X, T, θ, n (Notice that a multiplication sign is not needed.)

Y_3: 2nd [$\sqrt{\ }$] **1** + X, T, θ, n)

Y_4: (–) **1** ÷ (**1** – X, T, θ, n)

Y_5: **2** ∧ X, T, θ, n (∧ is the symbol for exponentiation.)

Y_6: VARS ► **1 3** (The standard method for entering a function.)

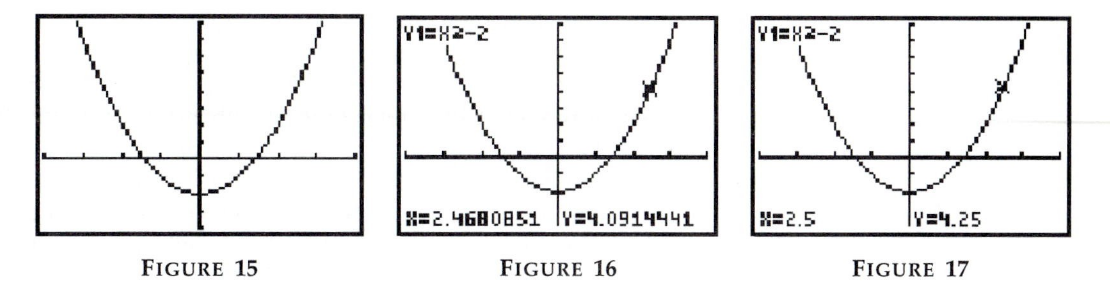

FIGURE 15 FIGURE 16 FIGURE 17

Notice that in Figure 14 the equal sign in Y_1 is highlighted, whereas the other equal signs are not highlighted. This highlighting can be toggled by moving the cursor to an equal sign and pressing [ENTER]. Functions with highlighted equal signs are said to be *selected*. Pressing the [GRAPH] key instructs the calculator to graph all selected functions. (*Note*: The words "Plot1 Plot2 Plot3" on the first row are used for statistical plots. The backslash symbol preceding each function is used to specify one of seven possible styles for the function.)

Press [GRAPH] to obtain Figure 15. Then press [TRACE] and press [▶], the right-arrow key, 29 times to obtain Figure 16. Each time a right- or left-arrow key is pressed, the trace cursor moves along the curve and the coordinates of the trace cursor are displayed.

To approximate the function value for a specific value of x, move the trace cursor as close as possible to the value of x and read the y-coordinate of the point. For a more precise function value while tracing, type in a value for X (such as 2.5) and press [ENTER]. See Figure 17.

A point where the graph crosses the x-axis is called an *x-intercept*. The coordinates of an x-intercept can be approximated by tracing. The x-coordinate of an x-intercept is called a *zero* of the function Y_1 or a *root* of the equation $\mathrm{Y}_1 = 0$. Figure 15 shows that $\mathrm{Y}_1 = 0$ has a zero between 1 and 2 and the value of the zero is about 1.5. For a precise value of the zero, press [2nd][CALC] **2** and answer the questions. Reply to "Left Bound?" by moving the cursor to a point whose x-coordinate is less than the zero and pressing [ENTER]. See Figure 18. Reply to "Right Bound?" by moving the cursor to a point whose x-coordinate is greater than the zero and pressing [ENTER]. See Figure 19. Reply to "Guess?" by moving the cursor near the x-intercept and pressing [ENTER]. Figure 20 shows the resulting display.

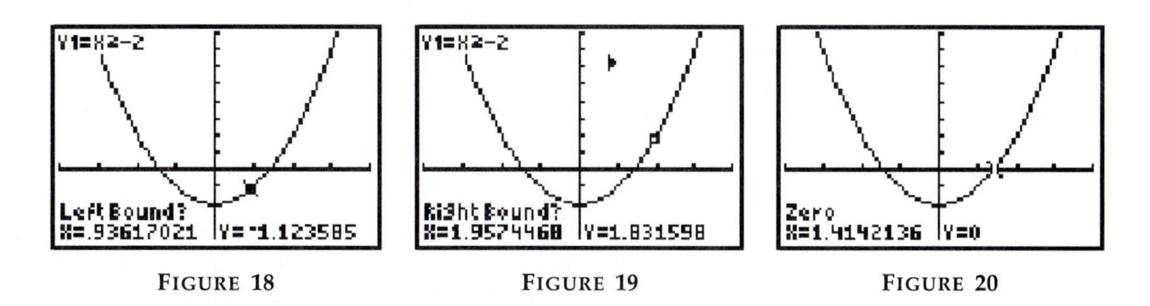

FIGURE 18 FIGURE 19 FIGURE 20

An alternate way to find a zero of a function is to reply to the questions by entering appropriate numbers. For instance, after pressing [2nd] [CALC] **2**, respond to "Left Bound?" by typing in the number 1 and pressing [ENTER], respond to "Right Bound?" by typing in the number 2 and pressing [ENTER], and respond to "Guess?" by typing in the number 1.5 and pressing [ENTER]. The final screen will be identical to Figure 20.

So far, all operations were carried out while looking at the graph of a function. These same operations also can be carried out in the Home screen, which is invoked by pressing [2nd] [QUIT]. The function Y_1 can be evaluated at 4 by entering $Y_1(4)$ and pressing [ENTER]. The zeros of Y_1 can be found by the equation solver with the following steps:

1. Press [MATH] **0** to invoke the solver.
2. Press [▲] [CLEAR] to invoke and clear the equation solver.
3. Enter either Y_1 or $X^2 - 2$ to the right of "eqn:0=".
4. Press [▼] [CLEAR]. The equation to be solved will be on the first line of the screen and the cursor will be just to the right of "X=".
5. Type in a guess for a zero of Y_1 (such as 1) and then press [ALPHA] [SOLVE]. After a little delay, the value you typed in will be replaced by the value of a zero of Y_1. (*Note*: You needn't be concerned with the last two lines displayed.)
6. You can now insert a different guess to the right of "X=" (such as −3) and press [ALPHA] [SOLVE] again to obtain another zero.

Return to the function screen by pressing [Y=] and then select Y_2 so that now both Y_1 and Y_2 are selected. Press [GRAPH] to obtain the graphs of the two functions. Press [TRACE] and then press the right-arrow key several times. Now press the down-arrow key several times. Each time the down-arrow key is pressed, the trace cursor moves from one curve to the other. The identity of the function containing the cursor is given in the upper-left part of the screen. Move the cursor as close as possible to the point of intersection of the two curves to approximate the coordinates of the intersection point. For more precise values, press [2nd] [CALC] **5** and answer the questions. Reply to "First curve?" by moving the trace cursor to one of the curves and pressing [ENTER]. Reply to "Second curve?" by moving the trace cursor to the other curve and pressing [ENTER]. Reply to "Guess?" by either moving the trace cursor near the point of intersection or typing in a number close to the x-coordinate of the point and pressing [ENTER]. Figure 21 shows the resulting display.

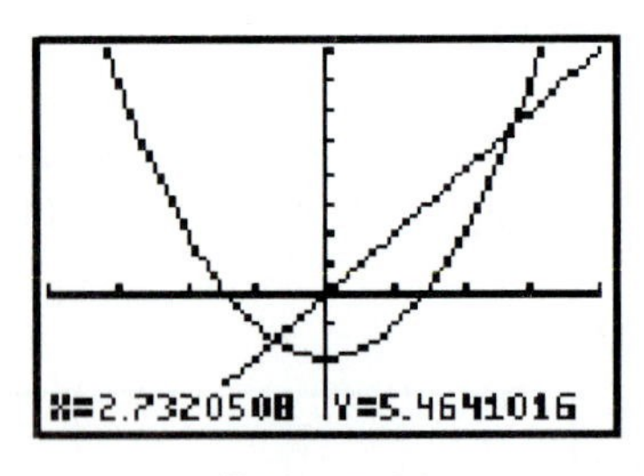

FIGURE 21

USING THE TI-85 GRAPHING CALCULATOR

Functions are graphed in a rectangular window like the one shown in Figure 22. The numbers on the x-axis range from xMin to xMax, and the numbers on the y-axis range from yMin to yMax. The distances between tick marks are xScl and

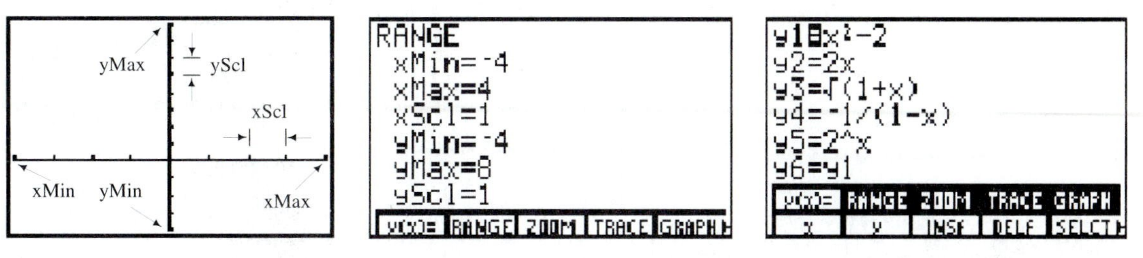

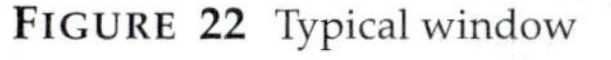

FIGURE 22 Typical window FIGURE 23 Settings for Figure 22 FIGURE 24

yScl on the x- and y-axis, respectively. To specify these quantities, press GRAPH F2 and type in the values of the six variables. Figure 23 gives the settings associated with the window in Figure 22. (*Note*: To enter a negative number, use the (−) key on the bottom row of the calculator.)

To specify functions, press GRAPH F1, type an expression next to the function name y1, press ENTER, and then continue to specify functions as long as desired. (*Note*: To erase an expression, use the arrow keys to move the cursor anywhere on the expression and press CLEAR.) In Figure 24, several functions have been specified. The expressions were produced with the following keystrokes.

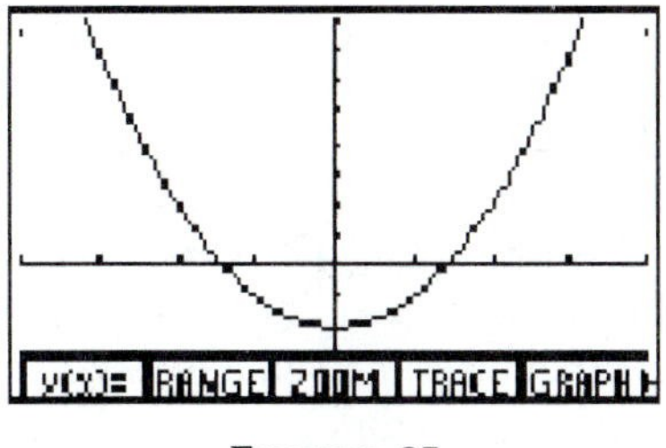

FIGURE 25

y1: x-VAR x^2 − **2**

y2: **2** x-VAR or **2** F1 (Notice that a multiplication sign is not needed.)

y3: 2nd [√] (**1** + x-VAR)

y4: (−) **1** ÷ (**1** − x-VAR)

y5: **2** ∧ x-VAR (∧ is the symbol for exponentiation.)

y6: F2 **1** (*Note*: The function y1 can also be entered with 2nd [VARS] MORE F3 {move arrow to y1} ENTER or 2nd [alpha] **0 1**. These are the standard methods for entering a function in the Home screen.)

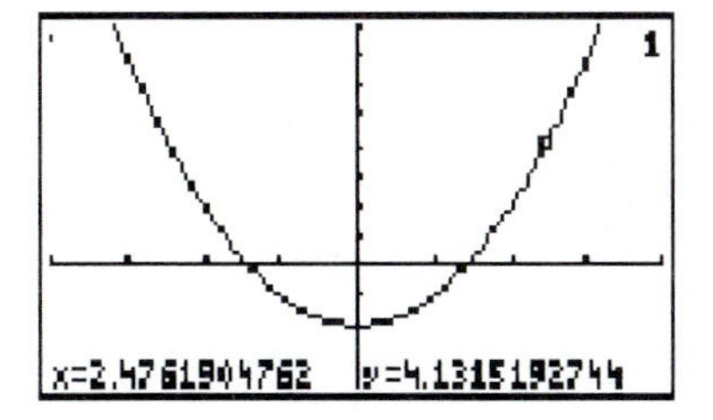

FIGURE 26

Notice that in Figure 24 the equal sign in y1 is highlighted, whereas the other equal signs are not highlighted. This highlighting for a function can be toggled by moving the cursor anywhere on the line containing the function and pressing F5. Functions with highlighted equal signs are said to be *selected*. The GRAPH instruction graphs all selected functions.

Press GRAPH F5 to obtain the graph in Figure 25. Then press F4 to invoke tracing. Press ▶, the right-arrow key, 39 times to obtain Figure 26. Each time a right- or left-arrow key is pressed, the trace cursor moves along the curve and the coordinates of the trace cursor are displayed. (*Note*: The small number 1 in the upper-right corner of the window identifies the curve as the graph of y1.)

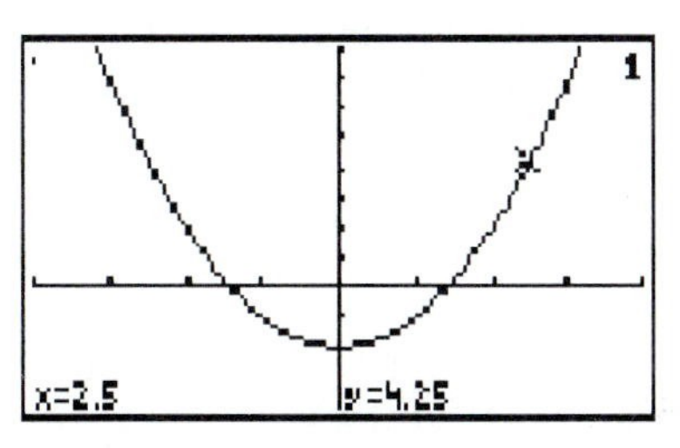

FIGURE 27

To approximate the function value for a specific value of x, move the trace cursor as close as possible to the value of x and read the y-coordinate of the point. For a more precise function evaluation, press GRAPH MORE MORE F1, type in a number (such as 2.5), and press ENTER. See Figure 27.

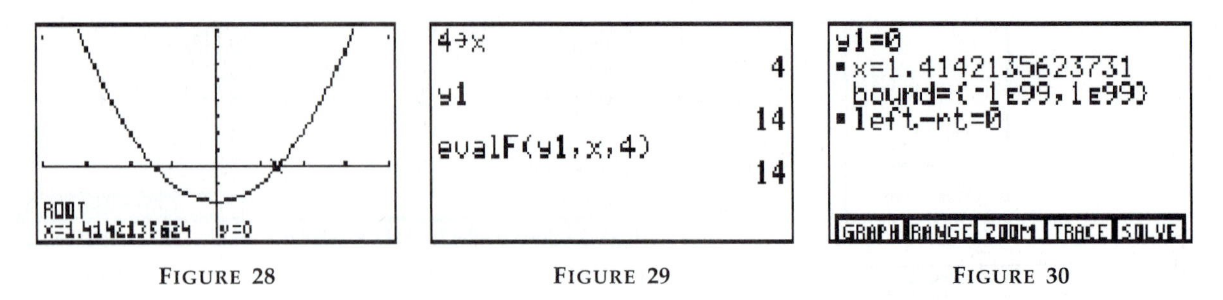

FIGURE 28 FIGURE 29 FIGURE 30

A point where the graph crosses the x-axis is called an *x-intercept*. The x-coordinate of an x-intercept is called a *root* of the equation `y1=0` or a *zero* of the function `y1`. The coordinates of an x-intercept can be approximated by tracing. For a more precise value of a root, press GRAPH MORE F1 F3, move the cursor to a point of the graph close to an x-intercept, and press ENTER. Figure 28 shows the resulting display.

So far, all operations were carried out while looking at the graph of a function. These same operations also can be carried out in the Home screen, which is invoked by pressing 2nd [QUIT]. Figure 29 shows two ways to evaluate `y1` at $x = 4$. The first line assigns the value 4 to the variable x and the third line evaluates the function `y1` at $x = 4$. To display the first line, press **4** STO ▶ x-VAR. To display `"evalF"`, press 2nd [CALC] F1. In general, `evalF`(*function*, *variable*, *number*) assigns the number to *variable* and then uses the value of *variable* to evaluate *function*. (*Note*: The first two instructions of Figure 29 cannot be replaced by `y1(4)`. The instruction `y1(4)` evaluates the function `y1` at the current value of x and then multiplies the function value by 4.)

The following steps solve the equation `y1=0` from the Home screen:

1. Press 2nd [SOLVER] CLEAR.
2. Press F2 ALPHA [=] **0** ENTER.
3. To the right of `"x="` type in a guess for a zero of `y1`, say 2, and then press F5. After a little delay, the value you typed in will be replaced by the value of a zero of `y1`. See Figure 30. (*Note*: You needn't be concerned with the last two lines displayed.)
4. You can now insert a different guess to the right of `"x="` (such as -3) and press F5 again to obtain another zero.

Return to the function screen by pressing GRAPH F1 and then select both `y1` and `y2`. Press GRAPH F5 or 2nd [M5] to obtain the graphs of the two functions. Press F4 and then press the right-arrow key a few times. Now press the down-arrow key a few times. Each time the down-arrow key is pressed, the trace cursor moves from one curve to the other. The small number in the upper-right corner, 1 or 2, identifies the function containing the cursor. Move the cursor as close as possible to the point of intersection of the two curves to approximate the coordinates of the intersection point. For more precise values, press GRAPH MORE F1 MORE F5 to select ISECT. Move the cursor along one of the two curves close to a point of intersection. Press ENTER to select the current curve as one of the two curves. The cursor should now be on the other curve. Press ENTER to select that curve as the second curve and to calculate the coordinates of the point of intersection. See Figure 31. (*Note*: Had more than two curves been visible, you would have used the down-arrow key to select the two you wanted before pressing ENTER.)

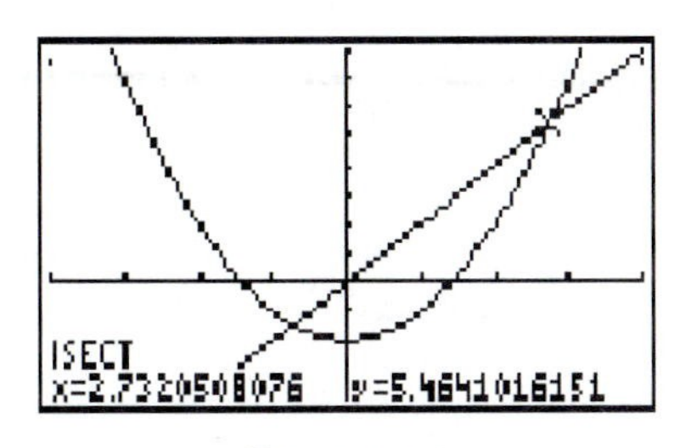

FIGURE 31

General Comments on TI Calculators

1. Colons can be used to place several instructions on a single line. For instance, the line

$$4 \rightarrow \mathrm{X} : \mathrm{Y}_1$$

first assigns the value of 4 to X and then evaluates the function Y_1. (*Note*: Press STO ▶ to obtain the arrow.)

2. We write [Xmin, Xmax] *by* [Ymin, Ymax] to identify the range settings for the viewing window. The TI-82 and TI-83 have three predefined windows, called ZDecimal, ZSquare, and ZStandard. To specify one of these windows, press ZOOM and select item 4, 5, or 6. With the ZDecimal window, $[-4.7, 4.7]$ *by* $[-3.1, 3.1]$, x-values start at 0 and change by .1 when the graph is traced. The ZStandard window is

$$[-10, 10] \text{ by } [-10, 10].$$

Selecting ZSquare changes either the x-range or the y-range so that a unit on the x-axis has the same length as a unit on the y-axis. For instance, after ZSquare has been invoked, the graph of the line $y = x$ actually makes a 45° angle with the x-axis. On the TI-85, these windows are called ZDECM, ZSQR, and ZSTD and are invoked by pressing GRAPH F3 and then making a selection. The ZDECM window is $[-6.3, 6.3]$ *by* $[-3.1, 3.1]$.

3. The ZDecimal window has the nice feature that simple numbers appear as x-coordinates when a graph is traced. This can also be accomplished with a TI-82 or TI-83 by making the difference between xMax and xMin 9.4 or a multiple of 9.4, such as 18.8, 28.2, or 4.7. Some possibilities are

$$(\text{xMin} = 0, \text{xMax} = 18.8), \quad (\text{xMin} = -2.7, \text{xMax} = 6.7),$$

$$(\text{xMin} = 0, \text{xMax} = 94), \quad \text{and} \quad (\text{xMin} = 0, \text{xMax} = 4.7).$$

With a TI-85, the difference between Xmax and Xmin should be a multiple of 12.6.

4. The values of the variables in the viewing window can be changed to zoom in on a portion of a graph or to zoom out to see more of the graph. Zooming can also be done with the zoom menu invoked by pressing ZOOM on the TI-82 or TI-83, and pressing GRAPH F3 on the TI-85. If you are interested in this optional technique, see your guidebook for details.

5. With a TI-82 or TI-83, tables of function values can be produced by pressing 2nd[TABLE]. See Appendix C or D for details.

USING MATHEMATICS COMPUTER SOFTWARE

If you intend to use a "menu-based" computer program, such as Visual Calculus or Derive, read the documentation that explains how to use the menus and how to move about on the screen. Learn how to edit the screen display and how to enter a function for graphing.

If you plan to use a "command-driven" computer program, such as Matlab, Maple, or Mathematica, you need to know how to use the on-line "help" information and to exit the program. Learn the format for entering a few basic commands (such as "plot" or "graph") and read how to use results of previous commands. You might try to graph a function such as x^2. Make sure you can perform the four basic tasks presented at the beginning of this section.

Visual Calculus was designed for this course and has been customized for this text. For instance, all pertinent functions appearing in examples and in odd-numbered exercises are stored in the software and can be accessed without your having to type them. Throughout this text, whenever specific instructions are given for graphing calculators, corresponding Visual Calculus instructions are given in Appendix A. For instance Part I of Appendix A shows how to perform the four basic tasks presented at the beginning of this section.

PRACTICE PROBLEMS 1.2

Write each expression in a form appropriate for a graphing utility.

1. $\dfrac{1}{x+1}$

2. $\dfrac{1}{x}+1$

EXERCISES 1.2

In Exercises 1–6, write each expression in a form appropriate for a graphing utility.

1. $f(x)=\dfrac{1}{2x}$

2. $f(x)=2^{5+x}$

3. $f(x)=\sqrt{1-x^3}$

4. $f(x)=\dfrac{10000}{1+50\left(\frac{1}{3}\right)^x}$

5. $f(x)=(x^2)^x$

6. $f(x)=3x+\dfrac{1}{\sqrt{x^2-2}}$

In Exercises 7–12, graph the function with the given window, evaluate the function at $x=5$, and find all the places where the curve crosses the x-axis.

7. $f(x)=16-x^2$, $[-10, 10]$ *by* $[-10, 20]$

8. $f(x)=x^2-5x+3$, $[-2, 6]$ *by* $[-5, 10]$

9. $f(x)=x^3-14x^2+62x-87$, $[0, 10]$ *by* $[-10, 10]$

10. $f(x)=(2x-9)^3-2x+9$, $[0, 8]$ *by* $[-2, 2]$

11. $f(x)=\dfrac{5x}{2^x}-1$, $[-1, 7]$ *by* $[-5, 3]$

12. $f(x)=\dfrac{x^2}{3^{x-2}}-3$, $[-2, 6]$ *by* $[-4, 10]$

In Exercises 13–18, graph the functions with the given window and find all points of intersection.

13. $f(x) = 2x - 3$, $g(x) = 6 - x$; $[-10, 10]$ *by* $[-10, 10]$
14. $f(x) = 6 - x^2$, $g(x) = x^2 + 2x + 1$; $[-4, 4]$ *by* $[-10, 10]$
15. $f(x) = x^2 - 3x - 4$, $g(x) = x - 1$; $[-2, 6]$ *by* $[-7, 7]$
16. $f(x) = \dfrac{5x}{2^x}$, $g(x) = 2 - \dfrac{(x-3)^2}{2}$; $[-2, 6]$ *by* $[-6, 3]$
17. $f(x) = x^3 + 6x^2 + 9x - 2$, $g(x) = \dfrac{x-6}{2}$; $[-6, 2]$ *by* $[-8, 2]$
18. $f(x) = x^3 - 2x^2$, $g(x) = -(x-1)^2$; $[-1.5, 2.5]$ *by* $[-10, 2]$

In Exercises 19–24, graph the function in each window and choose the best window.

19. $f(x) = x^3 - .5x + 1$;
    $[-2, 2]$ *by* $[-10, 10]$, $[-2, 2]$ *by* $[-1, 3]$, $[-2, 2]$ *by* $[.5, 1.5]$
20. $f(x) = x^4 - x^2 + 1$;
    $[-10, 10]$ *by* $[-10, 10]$, $[-2, 2]$ *by* $[0, 10]$, $[-2, 2]$ *by* $[0, 3]$
21. $f(x) = \left(\dfrac{x}{2}\right)^5 - x^3 - 2x^2 + 3x + 1$;
    $[-10, 10]$ *by* $[-10, 10]$, $[-10, 10]$ *by* $[-100, 2]$, $[-7, 7]$ *by* $[-65, 2]$
22. $f(x) = \dfrac{x}{2^x}$;
    $[-1, 3]$ *by* $[-2, 1]$, $[-1, 10]$ *by* $[-2, 1]$, $[-10, 10]$ *by* $[-10, 1]$
23. $f(x) = \dfrac{1}{x-1}$;
    $[-1, 3]$ *by* $[-2, 2]$, $[-1, 3]$ *by* $[-10, 10]$, $[-10, 10]$ *by* $[-10, 10]$
24. $f(x) = x2^{x^2-1} - x^2$;
    $[-1, 2]$ *by* $[-.5, .5]$, $[-10, 10]$ *by* $[-10, 10]$, $[-3, 3]$ *by* $[-10, 10]$

Use a graphing calculator to answer Exercises 25–36.

25. Graph the function $f(x) = x$ in the window

$$[-100, 100] \text{ by } [-100, 100]$$

    with Xscl = 1 and Yscl = 1. Why do the axes look so thick? How can you correct this situation?

26. Graph the function

$$f(x) = \frac{1}{x-1}$$

    in each of the predefined windows ZStandard, ZSquare, and ZDecimal. Which graph do you prefer?

27. Graph the function $f(x) = \sqrt{9 - x^2}$ in each of the predefined windows ZStandard, ZSquare, and ZDecimal. The graph of $f(x)$ is the top half of the circle of radius 3. Which graph do you prefer?

28. Solve the equation $x^3 - 2x + 4 = 0$ in the Home screen.

29. Solve the equation $4 - 2x + \sqrt{x+1} = 0$ in the Home screen.

30. Solve the equation $3x - 2^x = 2$ in the Home screen.

31. Solve the equation $\frac{20}{x+1} = x^2 - 11$ in the Home screen.

32. Give the sequence of keystrokes to display Y_1 (or y1).

33. Give the sequence of keystrokes to assign the value 23 to the variable X.

34. Give the sequence of keystrokes to select or deselect a function.

35. Give the sequence of keystrokes to completely clear the Home screen.

36. Find a value for xMax between 20 and 30, such that when a function is traced with the window [0, xMax] *by* [0, 30], all x-values have just a single decimal place.

SOLUTIONS TO PRACTICE PROBLEMS 1.2

1. $1/(x + 1)$

2. $(1/x) + 1$

1.3 FUNCTIONS AND GRAPHS IN APPLICATIONS

The key step in solving many applied problems in this text is to construct appropriate functions or equations. Once this is done, the remaining mathematical steps are usually straightforward. This section focuses on representative applied problems and reviews skills needed to set up and analyze functions, equations, and their graphs.

GEOMETRIC SHAPES Many examples and exercises in the text involve dimensions, areas, or volumes of objects similar to those in Figure 1. When a problem involves a plane figure, such as a rectangle or circle, one must distinguish between the *perimeter* and the *area* of the figure. The perimeter of a figure, or "distance around" the figure, is a *length* or *sum of lengths*. Typical units, if specified, are inches, feet, centimeters, meters, and so on. Area involves the *product of two lengths*, and the units are *square* inches, *square* feet, *square* centimeters, and so on.

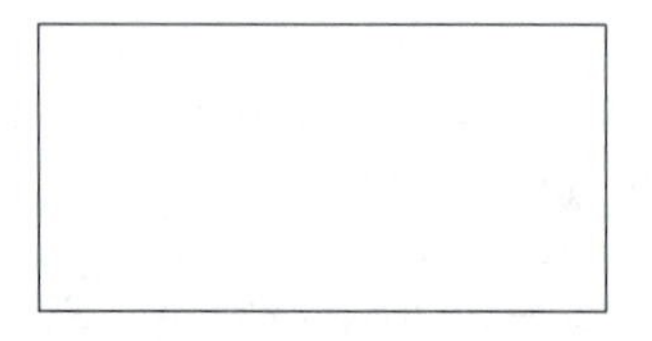

(a) Rectangle

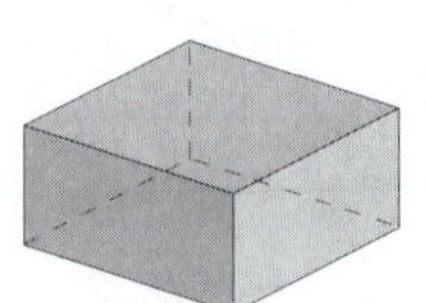

(b) Rectangular box

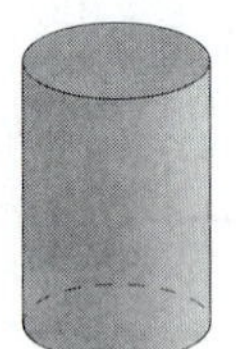

(c) Cylinder

EXAMPLE 1 Suppose the longer side of the rectangle in Figure 1(a) has twice the length of the shorter side, and let x denote the length of the shorter side.

(a) Express the perimeter of the rectangle as a function of x.

(b) Express the area of the rectangle as a function of x.

(c) Suppose the rectangle represents a kitchen countertop to be constructed of a durable material costing $25 per square foot. Write a function $C(x)$ that expresses the cost of the material as a function of x, where lengths are in feet.

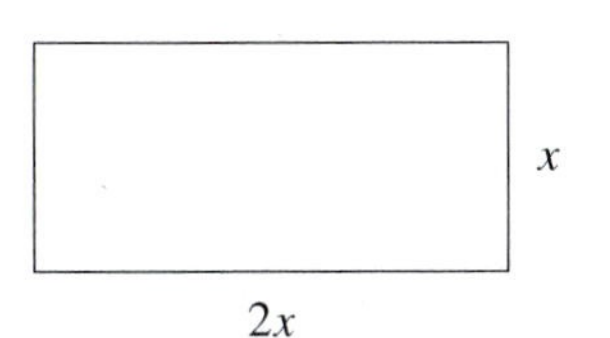

FIGURE 2

Solution (a) The rectangle is shown in Figure 2. The length of the longer side is $2x$. If the perimeter is denoted by P, then P is the sum of the lengths of the four sides of the rectangle, namely, $x + 2x + x + 2x$. That is, $P = 6x$.

(b) The area A of the rectangle is the product of the lengths of two adjacent sides. That is, $A = x \cdot 2x = 2x^2$.

(c) Here the area is measured is square feet. The basic principle for this part is

$$\begin{bmatrix} \text{cost of} \\ \text{materials} \end{bmatrix} = \begin{bmatrix} \text{cost per} \\ \text{square foot} \end{bmatrix} \cdot \begin{bmatrix} \text{number of} \\ \text{square feet} \end{bmatrix}$$

$$\begin{aligned} C(x) \quad &= \quad 25 \quad \cdot \quad 2x^2 \\ &= 50x^2 \quad \text{(dollars).} \end{aligned}$$

When a problem involves a three-dimensional object, such as a box or cylinder, one must distinguish between the *surface area* of the object and the *volume* of the object. Surface area is an area, of course, so it is measured in *square* units. Typically, the surface area is a *sum of areas* (each area is a product of two lengths). The volume of an object is often a *product of three lengths* and is measured in *cubic* units.

EXAMPLE 2 A rectangular box has a square copper base, wooden sides, and a wooden top. The copper costs $21 per square foot and the wood costs $2 per square foot.

(a) Write an expression giving the surface area (that is, the sum of the areas of the four sides, the top, and the bottom of the box) in terms of the dimensions of the box. Also, write an expression giving the volume of the box.

(b) Write an expression giving the total cost of the materials used to make the box in terms of the dimensions.

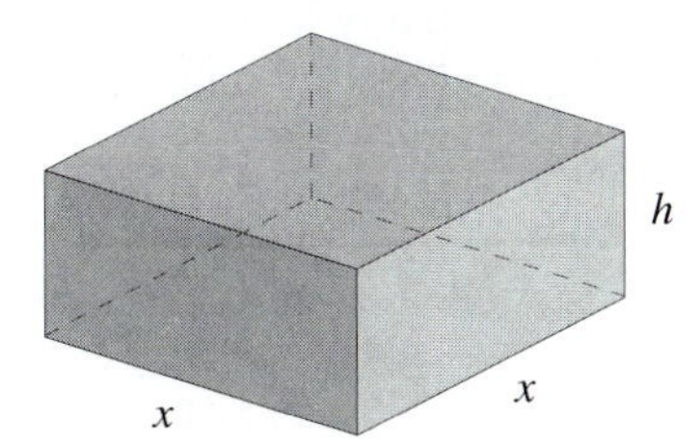

FIGURE 3 Closed box

Solution (a) The first step is to assign letters to the dimensions of the box. Denote the length of one (and therefore every) side of the square base by x, and denote the height of the box by h. See Figure 3.

The top and bottom each has area x^2, and each of the four sides has area xh. Therefore, the surface area is $2x^2 + 4xh$. The volume of the box is the product of the length, width, and height. Because the base is square, the volume is x^2h.

(b) When the various surfaces of the box have different costs per square foot, the cost of each is computed separately:

$$[\text{cost of bottom}] = [\text{cost per sq. ft.}] \cdot [\text{area of bottom}] = 21x^2;$$
$$[\text{cost of top}] = [\text{cost per sq. ft.}] \cdot [\text{area of top}] = 2x^2;$$
$$[\text{cost of one side}] = [\text{cost per sq. ft.}] \cdot [\text{area of one side}] = 2xh.$$

The total cost is

$$\begin{aligned} C &= [\text{cost of bottom}] + [\text{cost of top}] + 4 \cdot [\text{cost of one side}] \\ &= 21x^2 + 2x^2 + 4 \cdot 2xh = 23x^2 + 8xh. \end{aligned}$$

■

BUSINESS PROBLEMS Many applications in the text involve cost, revenue, and profit functions.

EXAMPLE 3 Suppose a toy manufacturer has fixed costs of \$3000 (such as rent, insurance, and business loans) that must be paid no matter how many toys are produced. In addition, there are variable costs of \$2 per toy. At a production level of x toys, the variable costs are $2 \cdot x$ (dollars) and the total cost is

$$C(x) = 3000 + 2x \quad \text{(dollars)}.$$

(a) Find the cost of producing 2000 toys.

(b) What additional cost is incurred if the production level is raised from 2000 toys to 2200 toys?

(c) To answer the question "How many toys may be produced at a cost of \$5000?" should you compute $C(5000)$ or should you solve the equation $C(x) = 5000$?

Solution (a) $C(2000) = 3000 + 2(2000) = 7000$ (dollars).

(b) The total cost when $x = 2200$ is $C(2200) = 3000+2(2200) = 7400$ (dollars). So the *increase* in cost when production is raised from 2000 to 2200 toys is

$$C(2200) - C(2000) = 7400 - 7000 = 400 \quad \text{(dollars)}.$$

(c) This is an important type of question. The phrase "how many toys" implies that the quantity x is unknown. Therefore, the answer is found by solving $C(x) = 5000$ for x:

$$\begin{aligned} 3000 + 2x &= 5000 \\ 2x &= 2000 \\ x &= 1000. \end{aligned}$$

Another way to analyze this problem is to look at the types of units involved. The input x of the cost function is the *quantity* of toys, and the output of the cost function is the *cost*, measured in dollars. Since the question involves 5000 *dollars*, it is the *output* that is specified. The input x is unknown. ■

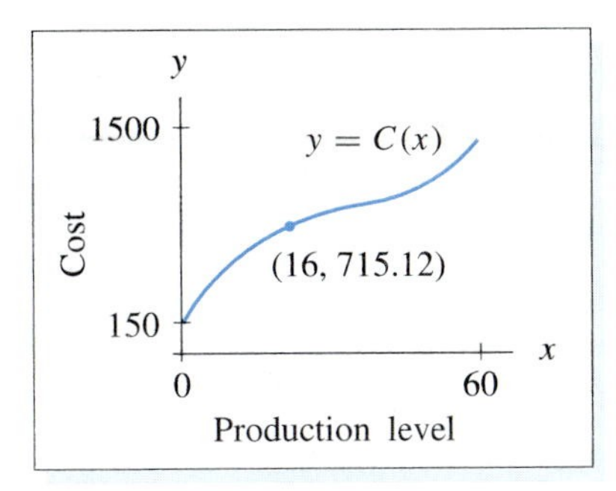

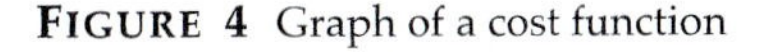

FIGURE 4 Graph of a cost function

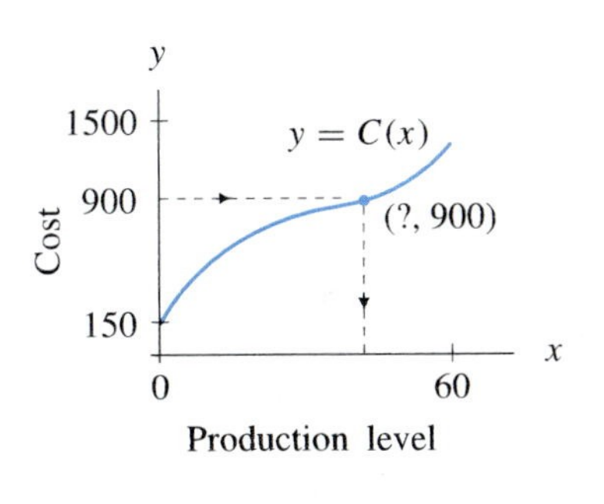

FIGURE 5

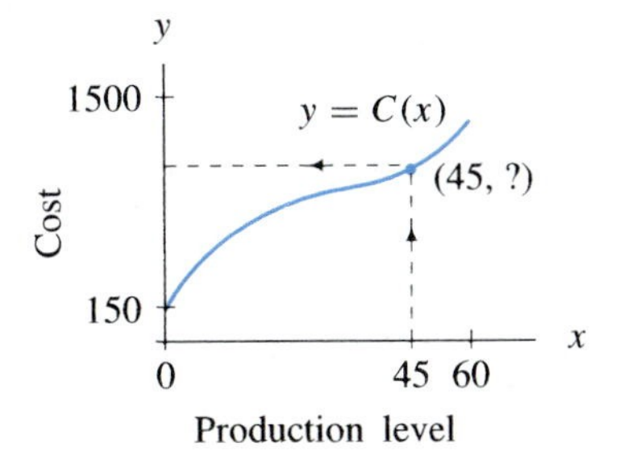

FIGURE 6

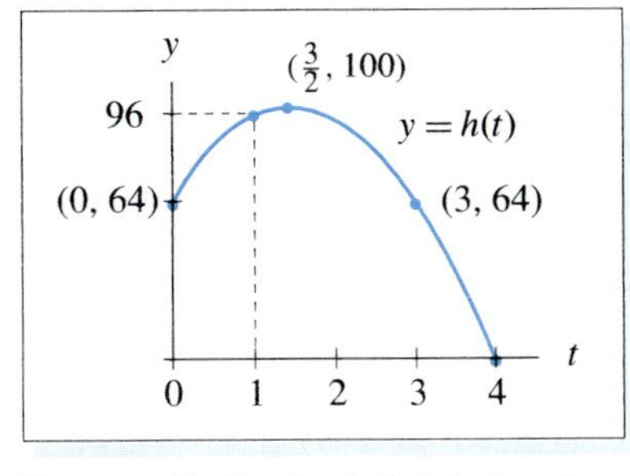

FIGURE 7 Graph of a height function

FUNCTIONS AND GRAPHS When a function arises in an applied problem, the graph of the function provides useful information. Every statement or task involving a function corresponds to a feature or task involving its graph.

EXAMPLE 4 To plan for future growth, a company analyzes production costs for one of its products and estimates that on a daily basis, the total cost (in dollars) of producing x units is given by the function

$$C(x) = 150 + 59x - 1.8x^2 + .02x^3.$$

Suppose the graph of this function is available, either displayed by a graphing utility or perhaps printed on graph paper in a company report. See Figure 4.

(a) The point $(16, 715.12)$ is on the graph. What does this say about the cost function?

(b) Explain how to use a graphing utility to find the level of production that will cost \$900.

(c) Explain graphically how to determine the total cost of producing 45 units.

Solution (a) The fact that $(16, 715.12)$ is on the graph of the cost function C means that $C(16) = 715.12$. That is, if the production level is 16 units per day, then the cost of producing those units is \$715.12.

(b) The graphical problem is to find the x-coordinate of the point on the graph of C whose y-coordinate is 900. The x-coordinate is the solution of $C(x) = 900$. To solve this equation with a graphing utility, graph the two functions $y = C(x)$ and $y = 900$ together and use "intersect" (item 5 in the TI-82 or TI-83 CALC menu), ISECT (in the TI-85 GRAPH/MATH menu), or "Solve" (in the Visual Calculus menu) to determine the point of intersection. To two decimal places, the answer is 39.04, so the production of 39 units will cost about \$900. If the graph is printed on graph paper, start at 900 on the y-axis, move to the right until you reach the graph, and then move down to the x-axis to find an approximate answer. See Figure 5.

(c) The total cost of producing 45 units is $C(45)$. To find $C(45)$ graphically, find the y-coordinate of the point on the graph of C whose x-coordinate is 45. With the TI-82, display the graph of $y = C(x)$ and use "value" (item 1 in the TI-82 CALC menu). With the TI-83, display the graph and then press [TRACE] **45** [ENTER]. With the TI-85, display the graph and then press [MORE] [MORE] [F1] **45** [ENTER]. With Visual Calculus, display the graph of $y = C(x)$ with $0 \le x \le 64$, press the down-arrow key once (to obtain an increment of 1), and press the right-arrow key several times until the value of x is 45. The cost is \$982.50. If the graph is printed on graph paper, start at 45 on the x-axis, move up until you reach the graph, and then move left to the y-axis to find an approximate answer. See Figure 6. ■

EXAMPLE 5 A ball is thrown straight up into the air. The function $h(t)$, the height of the ball (in feet) after t seconds, has the graph shown in Figure 7. (*Note*: This graph is not a picture of the physical path of the ball; the ball is thrown vertically into the air.)

(a) What is the height of the ball after 1 second?

(b) After how many seconds does the ball reach its greatest height, and what is this height?

(c) After how many seconds does the ball hit the ground?

(d) When is the height 64 feet?

Solution (a) Since the point $(1, 96)$ is on the graph of $h(t)$, $h(1) = 96$. Therefore, the height of the ball after 1 second is 96 feet.

(b) The highest point on the graph of the function has coordinates $(\frac{3}{2}, 100)$. Therefore, after $\frac{3}{2}$ seconds the ball achieves its greatest height, 100 feet.

(c) The ball hits the ground when the height is 0. This occurs after 4 seconds.

(d) The height of 64 feet occurs twice, at times $t = 0$ and $t = 3$ seconds. ■

Table 1 on the next page summarizes most of the concepts in Examples 3–5. Although stated here for a profit function, the concepts will arise later for many other types of functions as well. Each statement about the profit is translated into a statement about $f(x)$ and a statement about the graph of $f(x)$. The graph in Figure 8 illustrates each statement.

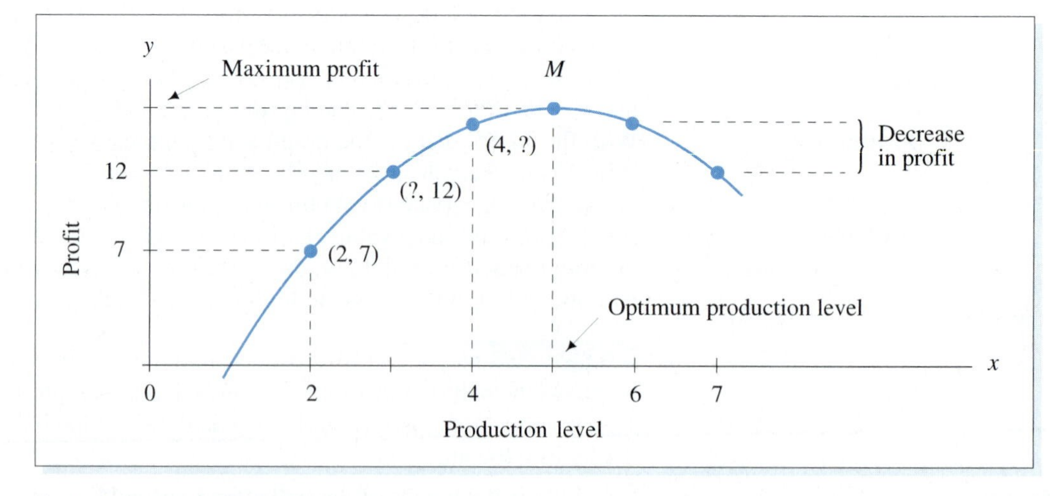

FIGURE 8 Graph of a profit function

Table 1 TRANSLATING AN APPLIED PROBLEM

Assume that $f(x)$ is the profit in dollars at production level x.

Applied Problem	Function	Graph
When production is at 2 units, the profit is $7.	$f(2) = 7$.	The point (2, 7) is on the graph.
Determine the number of units that generates a profit of $12.	Solve $f(x) = 12$ for x.	Find the x-coordinate(s) of the point(s) on the graph whose y-coordinate is 12.
Determine the profit when the production level is 4 units.	Evaluate $f(4)$.	Find the y-coordinate of the point on the graph whose x-coordinate is 4.
Find the production level that maximizes the profit.	Find x such that $f(x)$ is as large as possible.	Find the x-coordinate of the highest point, M, on the graph.
Determine the maximum profit.	Find the maximum value of $f(x)$.	Find the y-coordinate of the highest point on the graph.
Determine the change in profit when the production level is changed from 6 to 7 units.	Find $f(7) - f(6)$.	Determine the difference in heights of the points with x-coordinates 7 and 6.
The profit decreases when the production level is changed from 6 to 7 units.	The function value decreases when x changes from 6 to 7 units.	The point on the graph with x-coordinate 6 is higher than the point with x-coordinate 7.

PRACTICE PROBLEMS 1.3

Consider the cylinder shown in Figure 9.

1. Assign letters to the dimensions of the cylinder.
2. The girth of the cylinder is the circumference of the tinted circle in the figure. Express the girth in terms of the dimensions of the cylinder.
3. What is the area of the bottom (or top) of the cylinder?
4. What is the surface area of the side of the cylinder? (*Hint*: Imagine cutting the side of the cylinder and unrolling the cylinder to form a rectangle.)

FIGURE 9

EXERCISES 1.3

In Exercises 1–6, assign letters to the dimensions of the geometric object.

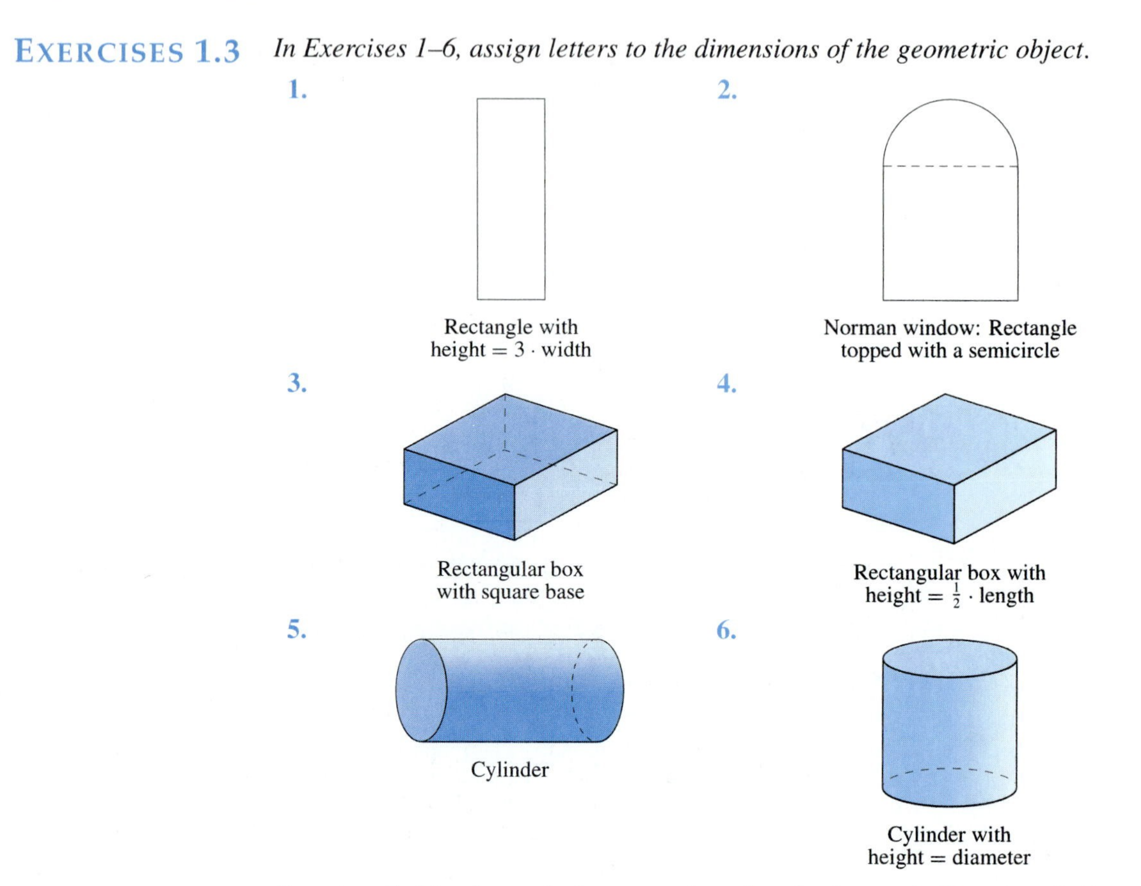

Exercises 7–14 refer to the letters assigned to the figures in Exercises 1–6.

7. Consider the rectangle in Exercise 1. Write an expression for the perimeter. Suppose the area is 25 square feet, and write this fact as an equation.

8. Consider the rectangle in Exercise 1. Write an expression for the area. Write an equation expressing the fact that the perimeter is 30 centimeters.

9. Consider a circle of radius r. Write an expression for the area. Write an equation expressing the fact that the circumference is 15 centimeters.

10. Consider the Norman window of Exercise 2. Write an expression for the perimeter. Write an equation expressing the fact that the area is 2.5 square meters.

11. Consider the rectangular box in Exercise 3, and suppose that it has no top. Write an expression for the volume. Write an equation expressing the fact that the surface area is 65 square inches.

12. Consider the closed rectangular box in Exercise 4. Write an expression for the surface area. Write an equation expressing the fact that the volume is 10 cubic feet.

13. Consider the cylinder of Exercise 5. Write an equation expressing the fact that the volume is 100 cubic inches. Suppose the material to construct the

left end costs \$5 per square inch, the material to construct the right end costs \$6 per square inch, and the material to construct the side costs \$7 per square inch. Write an expression for the total cost of material for the cylinder.

FIGURE 10

14. Consider the cylinder of Exercise 6. Write an equation expressing the fact that the surface area is 30π square inches. Write an equation for the volume.

15. Consider a rectangular corral with a partition down the middle, as shown in Figure 10. Assign letters to the outside dimensions of the corral. Write an equation expressing the fact that 5000 feet of fencing are needed to construct the corral (including the partition). Write an expression for the total area of the corral.

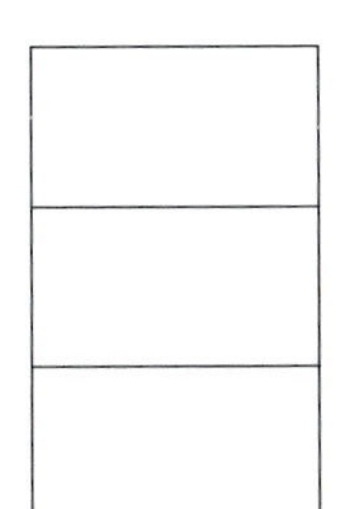

FIGURE 11

16. Consider a rectangular corral with two partitions, as in Figure 11. Assign letters to the outside dimensions of the corral. Write an equation expressing the fact that the corral has a total area of 2500 square feet. Write an expression for the amount of fencing needed to construct the corral (including both partitions).

17. Consider the corral of Exercise 16. Suppose the fencing for the boundary of the corral costs \$10 per foot and the fencing for the inner partitions costs \$8 per foot. Write an expression for the total cost of the fencing.

18. Consider the rectangular box of Exercise 3. Assume the box has no top, the material needed to construct the base costs \$5 per square foot, and the material needed to construct the sides costs \$4 per square foot. Write an equation expressing the fact that the total cost of materials is \$150. (Use the dimensions assigned in Exercise 3.)

19. Suppose the rectangle in Exercise 1 has a perimeter of 40 cm. Find the area of the rectangle.

20. Suppose the cylinder in Exercise 6 has a volume of 54π cubic inches. Find the surface area of the cylinder.

21. A specialty shop prints custom slogans and designs on T-shirts. The shop's total cost at a daily sales level of x T-shirts is $C(x) = 73 + 4x$ dollars.
    (a) At what sales level will the cost be \$225?
    (b) If the sales level is at 40 T-shirts, how much will the cost rise if the sales level changes to 50 T-shirts?

22. A dry cleaning shop estimates that its daily cost for cleaning x suits is $40 + 2.50x$ dollars.
    (a) What is the shop's cost on a day when it cleans 28 suits?
    (b) How many suits did the shop clean on a day when its costs were \$125?
    (c) Suppose the shop cleans about 30 suits on a typical day. What is the additional cost of cleaning just one more suit per day?

23. A frozen yogurt stand makes a profit of $P(x) = .40x - 80$ dollars when selling x scoops of yogurt per day.
    (a) Find the break-even sales level, that is, the level at which $P(x) = 0$.
    (b) What sales level generates a daily profit of \$30?

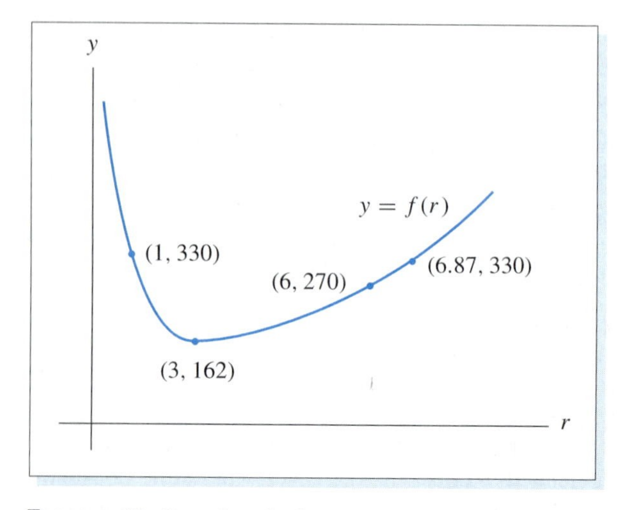

FIGURE 12 Cost of a cylinder

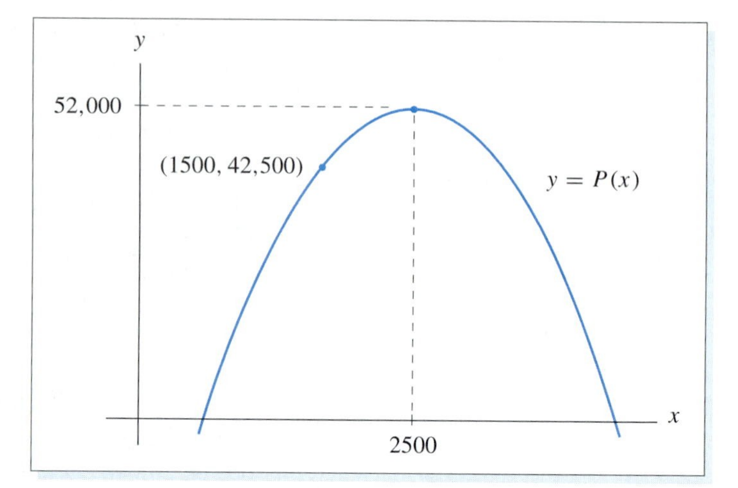

FIGURE 13 A profit function

(c) How many more scoops of yogurt will have to be sold to raise the daily profit from \$30 to \$40?

24. A cellular telephone company estimates that if it has x thousand subscribers, then its monthly profit is $P(x)$ thousand dollars, where $P(x) = 12x - 200$.

(a) How many subscribers are needed for a monthly profit of 160 thousand dollars?

(b) How many new subscribers would be needed to raise the monthly profit from 160 to 166 thousand dollars?

Exercises 25–30 refer to a function $f(r)$ that gives the cost (in cents) of constructing a 100 cubic inch cylinder of radius r inches. The graph of f is shown in Figure 12.

25. What is the cost of constructing a cylinder of radius 6 in?

26. For what value(s) of r is the cost 330 cents?

27. Interpret the fact that the point (3, 162) is on the graph of the function.

28. Interpret the fact that the point (3, 162) is the lowest point on the graph of the function. What does this say in terms of cost versus radius?

29. What is the additional cost of increasing the radius from 3 inches to 6 inches?

30. How much is saved by increasing the radius from 1 inch to 3 inches?

Exercises 31–34 refer to the profit function in Figure 13.

31. The point (2500, 52,500) is the highest point on the graph of the function. What does this say in terms of profit versus quantity?

32. The point (1500, 42,500) is on the graph of the function. Restate this fact in terms of the function $P(x)$.

33. Translate the task "solve $P(x) = 30{,}000$" into a task involving the graph of the function.

34. Translate the task "find $P(2000)$" into a task involving the graph.

A ball is thrown straight up into the air. The function $h(t)$ gives the height of the ball (in feet) after t seconds. In Exercises 35–40, translate the task into both a statement involving the function and a statement involving the graph of the function.

35. Find the height of the ball after 3 seconds.

36. Find the time at which the ball attains its greatest height.

37. Find the greatest height attained by the ball.

38. Determine when the ball will hit the ground.

39. Determine when the height of the ball is 100 feet.

40. Find the height of the ball when it is first released.

In the remaining exercises, use a graphing utility to obtain the graphs of the functions.

41. The daily cost (in dollars) of producing x units of a certain product is given by the function

$$C(x) = 225 + 36.5x - .9x^2 + .01x^3.$$

Graph $y = C(x)$ in the window $[0, 70]$ *by* $[0, 800]$.

(a) Find the cost of producing 50 units.

(b) What is the additional cost of increasing the production to a level of 51 units?

(c) At what production level will the cost be \$510?

42. Refer to the cost function in Exercise 41, and suppose the daily cost is running at about \$800.

(a) What is the current production level?

(b) By how many units can the production level be increased and yet keep the daily cost at or below \$850?

43. A store estimates that the total revenue from the sale of x bicycles per year is $250x - .2x^2$ dollars. Graph this function in the window

$$[200, 500] \text{ by } [42000, 75000].$$

Currently, the annual bicycle revenue is about \$63,000.

(a) Estimate the current sales level of bicycles.

(b) If the sales level were to decrease by 50 bicycles, by how much would the revenue fall?

44. Refer to the revenue function in Exercise 43. Suppose the store expects to sell 400 bicycles next year.

(a) What total revenue should the store expect to receive from its bicycle sales?

(b) The store believes that if it spends \$5000 in advertising, it can raise the total sales from 400 to 450 bicycles next year. Should it spend the \$5000? Why or why not?

SOLUTIONS TO PRACTICE PROBLEMS 1.3

1. Let r be the radius of circular base and let h be the height of the cylinder.
2. Girth $= 2\pi r$ (the circumference of the circle).
3. Area of the bottom $= \pi r^2$.
4. The cylinder is a rolled-up rectangle of height h and base $2\pi r$ (the circumference of the circle). The area is $2\pi rh$. See Figure 14.

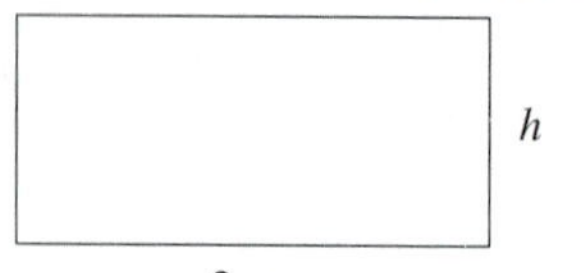

FIGURE 14 Unrolled side of a cylinder

1.4 LINEAR FUNCTIONS

Of all the functions used in calculus, linear functions are the simplest and the most important. Linear functions arise often in applications, and they will be the key to describing the behavior of the more complicated functions studied in Chapter 2.

EXAMPLE 1 Some years ago, a company that manufactures titanium dioxide was dumping sulfuric acid into the Mississippi River. When the U.S. Environmental Protection Agency (EPA) discovered this, it fined the company $125,000, plus $1000 per day until the company complied with federal water pollution regulations. (Nowadays, the fine would be more severe.) Express the total fine as a function of the number x of days the company continued to violate the federal regulations.

Solution The variable fine for x days of pollution, at $1000 per day, is $1000x$ dollars. So the total fine is given by the function

$$f(x) = 1000x + 125{,}000.$$

The graph of this function is the line shown in Figure 1. ■

The function in Example 1 is a linear function because it can be written in the form $f(x) = mx + b$, or simply

$$y = mx + b \tag{1}$$

for some numbers m and b. The graph of equation (1) is a nonvertical line. See Figure 2(a). Any nonvertical line has an equation of the form (1) and is the graph

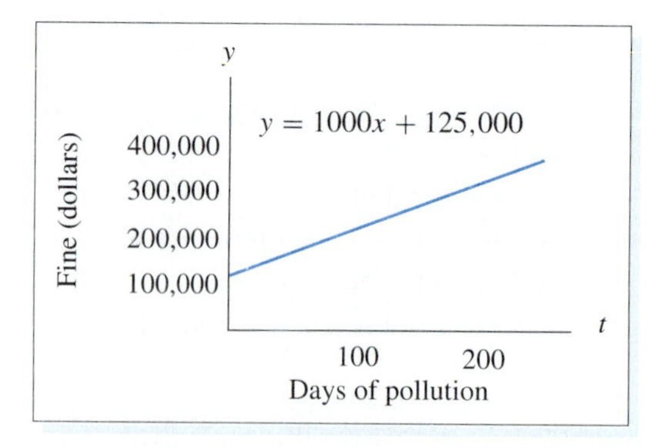

FIGURE 1 Fine for water pollution

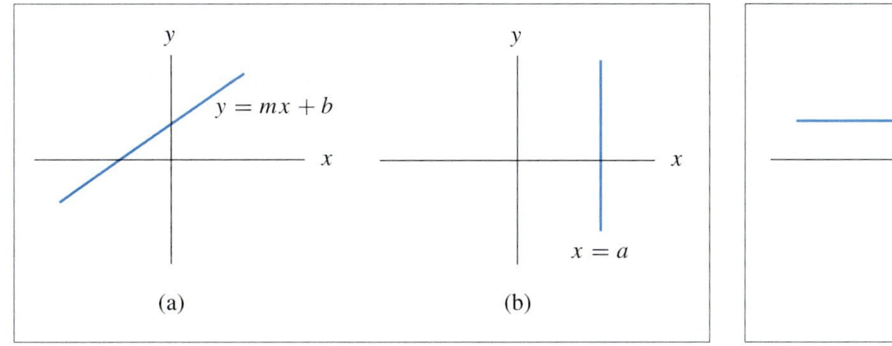

FIGURE 2

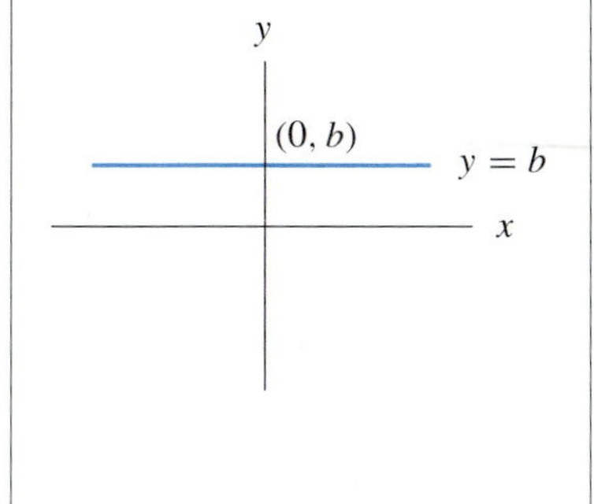

FIGURE 3 Graph of the constant function $f(x) = b$

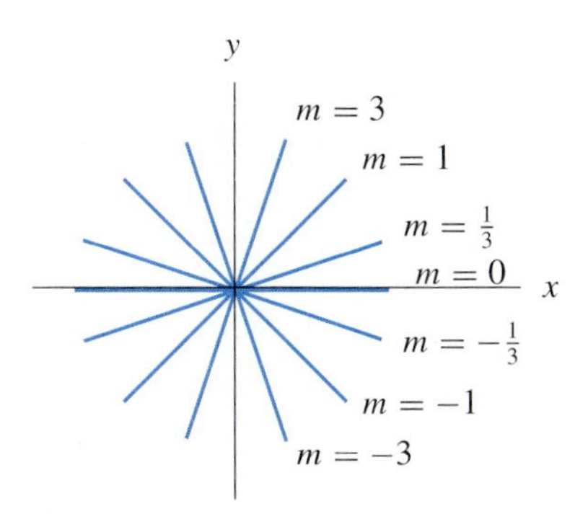

FIGURE 4 Lines with slope m

of a linear function. A vertical line has an equation of the form $x = a$, as in Figure 2(b). A vertical line is not the graph of a function, because the vertical line test is violated.

An important special case of a linear function occurs when the value m is zero, that is, when $f(x) = b$ for some number b. In this case, f is called a *constant function*, since it assigns the same number b to every value of x. Its graph is the horizontal line whose equation is $y = b$. See Figure 3.

THE SLOPE OF A LINE The number m in the equation $y = mx + b$ is called the *slope* of the line and it determines the direction of the line. The point $(0, b)$ is the *y-intercept* of the line, because $y = b$ when x is zero.

The slope of a line is related to the steepness of the line. See Figure 4. Imagine walking along a line from left to right. On lines with positive slope, the path is uphill; the more positive the slope, the steeper the ascent. On lines with negative slope, the path is downhill; the more negative the slope, the steeper the descent.

Figure 5 shows a family of lines with the same slope but differing y-intercepts. The lines are parallel.

The slope and y-intercept of a line often have important interpretations, as the next two examples illustrate.

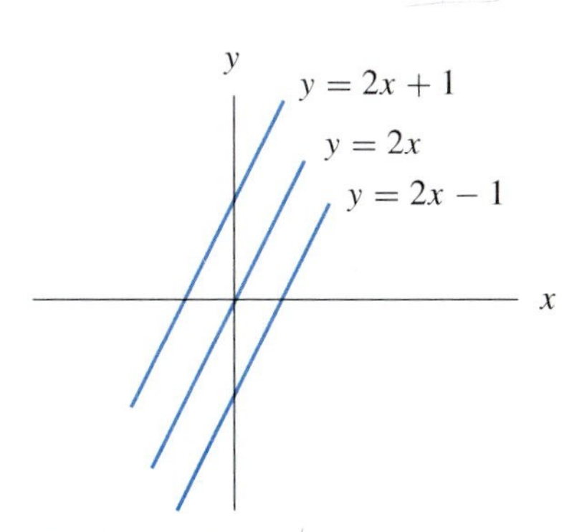

FIGURE 5 Lines with the same slope

EXAMPLE 2 How are the slope and y-intercept of the line in Figure 1 related to the fine the company in Example 1 must pay?

Solution The y-intercept of the line in Figure 1 is (0, 125,000). The company pays only \$125,000 if it stops polluting the river the same day it receives the EPA notice. The slope of the line corresponds to how fast the company's fine rises if the pollution does not stop. Examine the accompanying table of fines.

Days of pollution	Fine the company must pay
$x = 5$	$y = 1000(5) + 125{,}000 = 130{,}000$
$x = 6$	$y = 1000(6) + 125{,}000 = 131{,}000$
$x = 7$	$y = 1000(7) + 125{,}000 = 132{,}000$

The company's fine increases at a constant rate of \$1000 per day for each day the pollution continues. Geometrically, the y-coordinates of points on the line in Figure 1 increase by 1000 for each unit increase in x. ■

EXAMPLE 3 An apartment complex has a storage tank to hold its heating oil. The tank was filled on January 1, but no more deliveries of oil will be made until some time in March. Let t denote the number of days after January 1 and let y denote the number of gallons of fuel oil in the tank. Current records of the apartment complex show that y and t are related approximately by the equation

$$y = 30{,}000 - 400t. \tag{2}$$

What interpretation can be given to the y-intercept and slope of this line? (See Figure 6.)

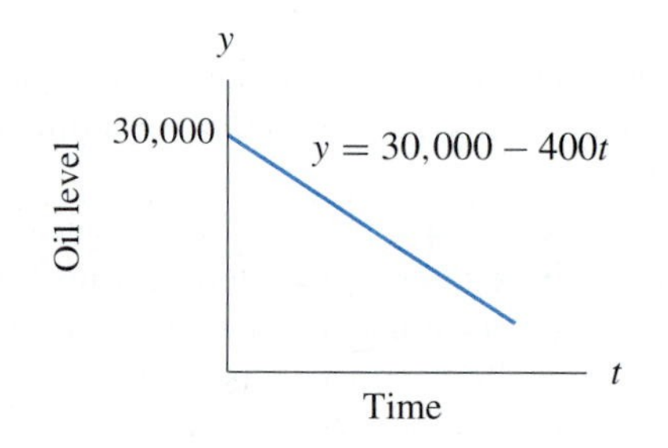

FIGURE 6 Amount of heating oil in a tank

Solution The y-intercept is $(0, 30{,}000)$. This value of y corresponds to $t = 0$, so there were 30,000 gallons of oil in the tank on January 1. Let us examine how fast the oil is removed from the tank.

Days after January 1	Gallons of oil in the tank
$t = 0$	$y = 30{,}000 - 400(0) = 30{,}000$
$t = 1$	$y = 30{,}000 - 400(1) = 29{,}600$
$t = 2$	$y = 30{,}000 - 400(2) = 29{,}200$
$t = 3$	$y = 30{,}000 - 400(3) = 28{,}800$
⋮	⋮

The oil level in the tank drops by 400 gallons each day; that is, the oil is being used at the rate of 400 gallons per day. The slope of the line is -400. Thus the slope gives the rate at which the level of oil in the tank is changing. The negative sign on the -400 indicates that the oil level is decreasing rather than increasing. ■

Examples 2 and 3 illustrate the characteristic property of a linear function f; the function values $f(x)$ change at a constant rate with respect to x. This fact is one of the key ideas needed for the study of calculus.

PROPERTIES OF THE SLOPE OF A LINE Let us now examine several useful properties of the slope of a straight line. At the end of the section we shall explain why these properties are valid.

Slope Property 1 Suppose that we start at a point on a line of slope m and move one unit to the right. Then we must move m units in the y-direction in order to return to the line.

Starting point
m
1
m positive

1
m
Starting point
m negative

Starting point
1
m zero

Slope Property 2 We can compute the slope of a line by knowing two points on the line. If (x_1, y_1) and (x_2, y_2) are on the line, then the slope of the line is $\dfrac{y_2 - y_1}{x_2 - x_1}$.

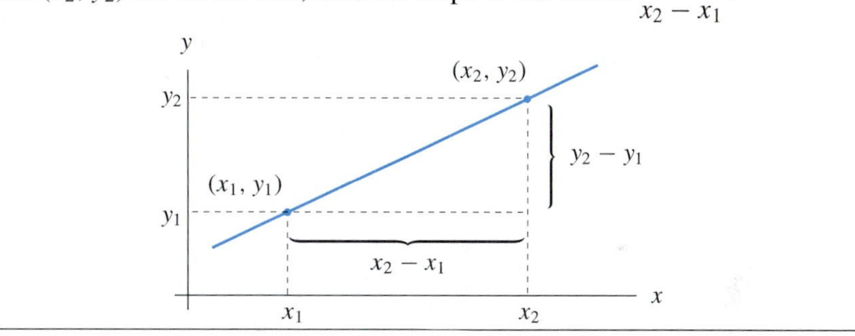

As we move from (x_1, y_1) to (x_2, y_2), the change in the y-coordinates is $y_2 - y_1$ and the change in the x-coordinates is $x_2 - x_1$. Thus the slope of the line is simply the ratio of the change in y to the change in x. This interpretation of the slope is also illustrated by the diagrams given for Slope Property 1. The slope of a line equals the change in y per unit change in x. We say that the slope gives the *rate of change of y with respect to x*.

CALCULATIONS INVOLVING SLOPE OF A LINE

EXAMPLE 4 Find the slope and the y-intercept of the line whose equation is $2x + 3y = 6$.

Solution We solve for y in terms of x:

$$3y = -2x + 6$$

$$y = -\frac{2}{3}x + 2.$$

The slope is $-\frac{2}{3}$ and the y-intercept is $(0, 2)$. ■

Slope Property 3 The equation of a line can be obtained if we know the slope and one point on the line. If the slope is m and if (x_1, y_1) is on the line, then an equation of the line is

$$y - y_1 = m(x - x_1).$$

This equation is called the *point-slope form* of the equation of the line.

Slope Property 4 Distinct lines of the same slope are parallel. Conversely, if two lines are parallel, they have the same slope.

EXAMPLE 5 Sketch the graph of the line

(a) passing through $(2, -1)$ with slope 3,

(b) passing through $(2, 3)$ with slope $-\frac{1}{2}$.

Solution We use Slope Property 1. (See Figure 7.) In each case, we begin at the given point, move one unit to the right, and then move m units in the y-direction (upward for positive m, downward for negative m). The new point reached will also be on the line. Draw the straight line through these two points. ■

EXAMPLE 6 Find the slope of the line passing through the points $(6, -2)$ and $(9, 4)$.

Solution We apply Slope Property 2 with $(x_1, y_1) = (6, -2)$ and $(x_2, y_2) = (9, 4)$.

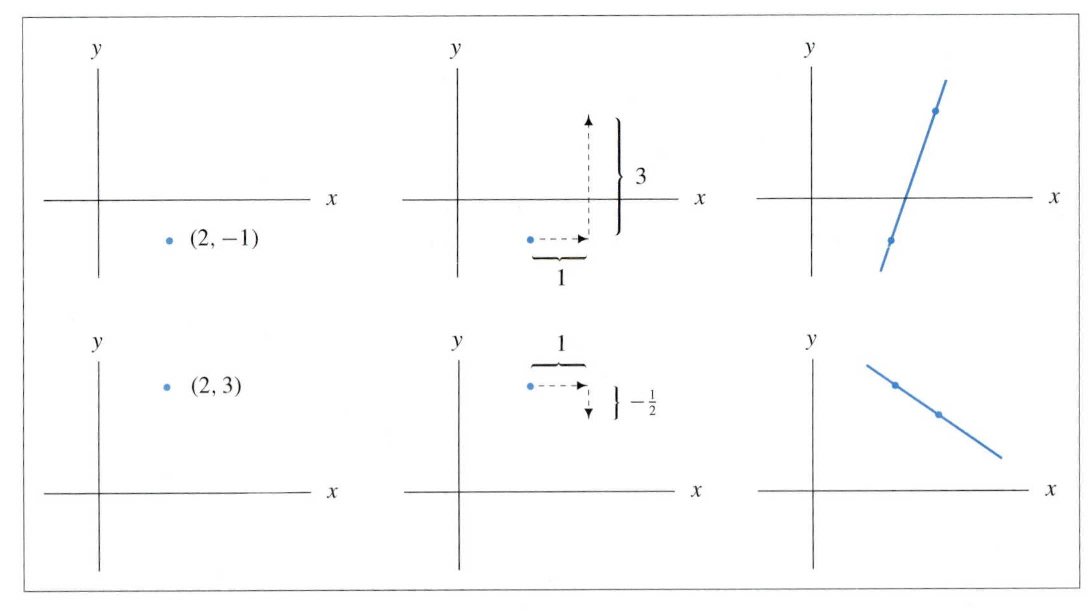

FIGURE 7

Then

$$\frac{y_2 - y_1}{x_2 - x_1} = \frac{4 - (-2)}{9 - 6} = \frac{6}{3} = 2.$$

Thus the slope is 2. [We would have reached the same answer if we had let $(x_1, y_1) = (9, 4)$ and $(x_2, y_2) = (6, -2)$.] The slope is just the difference of the y-coordinates divided by the difference of the x-coordinates, with each difference formed in the same order. ■

EXAMPLE 7 Find an equation of the line passing through $(-1, 2)$ with slope 3.

Solution We let $(x_1, y_1) = (-1, 2)$ and $m = 3$, and use Slope Property 3. An equation of the line is

$$y - 2 = 3[x - (-1)] \quad \text{or} \quad y - 2 = 3(x + 1).$$

If desired, this equation can be put into the form $y = mx + b$:

$$\begin{aligned} y - 2 &= 3(x + 1) = 3x + 3 \\ y &= 3x + 5. \end{aligned}$$

■

EXAMPLE 8 Find an equation of the line passing through the points $(1, -2)$ and $(2, -3)$.

Solution By Slope Property 2, the slope of the line is

$$\frac{-3 - (-2)}{2 - 1} = \frac{-3 + 2}{1} = -1.$$

Since $(1, -2)$ is on the line, we can use Slope Property 3 to get an equation of the line:

$$\begin{aligned} y - (-2) &= (-1)(x - 1) \\ y + 2 &= -x + 1 \\ y &= -x - 1. \end{aligned}$$

■

EXAMPLE 9 Find an equation of the line passing through $(5, 3)$ parallel to the line $2x + 5y = 7$.

Solution We first find the slope of the line $2x + 5y = 7$:

$$\begin{aligned} 2x + 5y &= 7 \\ 5y &= 7 - 2x \\ y &= -\frac{2}{5}x + \frac{7}{5}. \end{aligned}$$

The slope of this line is $-\frac{2}{5}$. By Slope Property 4, any line parallel to this line will also have slope $-\frac{2}{5}$. Using the given point $(5, 3)$ and Slope Property 3, we get the desired equation:

$$y - 3 = -\frac{2}{5}(x - 5).$$

This equation can also be written as $y = -\frac{2}{5}x + 5$. ■

THE METHOD OF LEAST SQUARES

Modern people compile graphs of literally thousands of different quantities: the purchasing value of the dollar as a function of time, the pressure of a fixed volume of air as a function of temperature, the average income of people as a function of their years of formal education, or the incidence of strokes as a function of blood pressure. The observed points on such graphs tend to be irregularly distributed due to the complicated nature of the phenomena underlying them as well as to errors made in observation (for example, a given procedure for measuring average income may not count certain groups). In spite of the imperfect nature of the data, we are often faced with the problem of making assessments and predictions based on them. Roughly speaking, this problem amounts to filtering the sources of errors in the data and isolating the basic underlying trend. Frequently, on the basis of a suspicion or a working hypothesis, we may suspect that the underlying trend is linear—that is, the data should lie on a straight line. But which straight line? This is the problem that the *method of least squares* attempts to answer. To be more specific, let us consider the following problem:

> **PROBLEM** Given observed data points $(x_1, y_1), (x_2, y_2), \ldots, (x_N, y_N)$ on a graph, find the straight line that "best" fits these points.

In order to completely understand the statement of the problem being considered, we must define what it means for a line to "best" fit a set of points. If (x_i, y_i) is one of our observed points, then we will measure how far it is from a given line $y = ax + b$ by the vertical distance, E_i, from the point to the line. (See Figure 8.)

Statisticians prefer to work with the square of the vertical distance E_i. The total error in approximating the data points $(x_1, y_1), \ldots, (x_N, y_N)$ by the line $y = ax + b$ is usually measured by the sum E of the squares of the vertical distances from the points to the line,

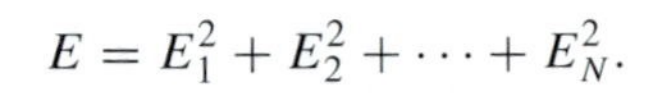

$$E = E_1^2 + E_2^2 + \cdots + E_N^2.$$

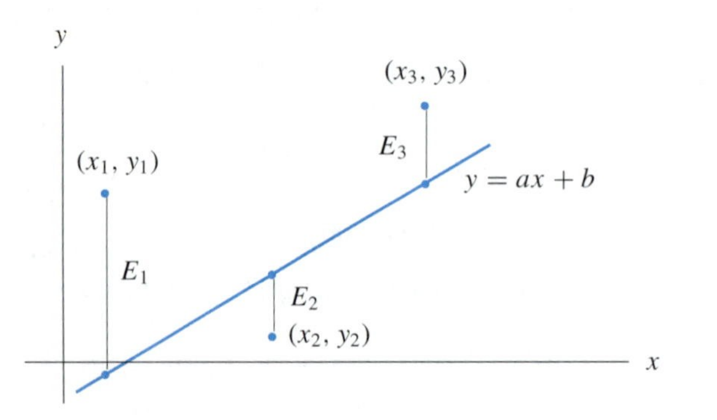

FIGURE 8 Fitting a line to data points

E is called the least-squares error of the observed points with respect to the line. If all the observed points lie on the line $y = ax+b$, then all the E_i are zero and the error E is zero. If a given observed point is far away from the line, the corresponding E_i^2 is large and hence makes a large contribution to the error E. In general, we cannot expect to find a line $y = ax+b$ that fits the observed points so well that the error E is zero. Actually, this situation will occur only if the observed points lie on a straight line. However, we can rephrase our original problem as follows:

> **PROBLEM** Given observed data points $(x_1, y_1), (x_2, y_2), \ldots, (x_N, y_N)$, find the straight line $y = ax + b$ for which the error E is as small as possible.

This line, called the *least-squares line* or the *regression line*, can be found with TI calculators and most software packages, including Visual Calculus.

OBTAINING THE LEAST-SQUARES LINES WITH A TI-82 OR TI-83 The following steps find the straight line that minimizes the least-squares error for the points $(1, 4)$, $(2, 5)$, $(3, 8)$:

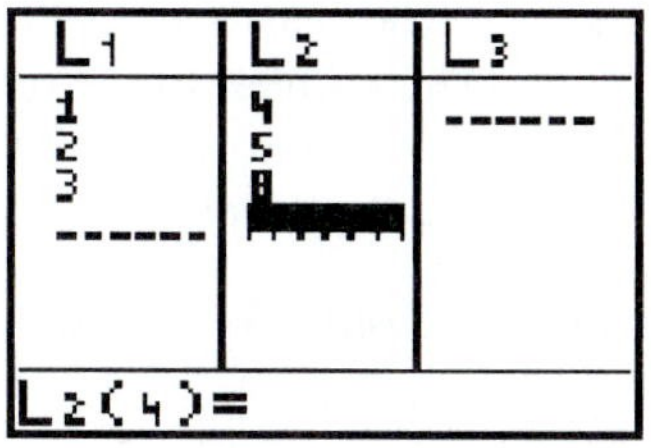

FIGURE 9

1. Press [STAT] **1** to obtain a table used for entering the data.
2. If there is no data in the columns labeled L_1 and L_2, proceed to step 4.
3. Move the cursor up to L_1 and press [CLEAR] [ENTER] to delete all data in L_1's column. Move the cursor right and up to L_2 and press [CLEAR] [ENTER] to delete all data in L_2's column.
4. If necessary, move the cursor left to the first blank row of the L_1 column. Press **1** [ENTER] **2** [ENTER] **3** [ENTER] to place the x-coordinates of the three points into the L_1 column.
5. Move the cursor right to the L_2 column and press **4** [ENTER] **5** [ENTER] **8** [ENTER] to place the y-coordinates of the three points into the L_2 column. The screen should now appear as in Figure 9.
6. Press [STAT] [▶], and press the number for `LinReg(ax+b)`.
7. Press [ENTER]. The TI-82 screen should now appear as in Figure 10. The TI-83 screen might be missing the last line. (*Note*: The number r is called the correlation coefficient. If the absolute value of r is close to 1, then the straight line is a good fit to the points.) The least-squares line is $y = 2x + \frac{5}{3}$. (*Note*: $\frac{5}{3} \approx 1.666666667$.)
8. If desired, the linear function can be assigned to `Y1` with the following steps:
   (a) Press [Y=] [CLEAR] to erase the current expression in `Y1`.
   (b) Press [VARS] **5** [▶] [▶] and the number for `RegEQ` to assign the linear function to `Y1`.

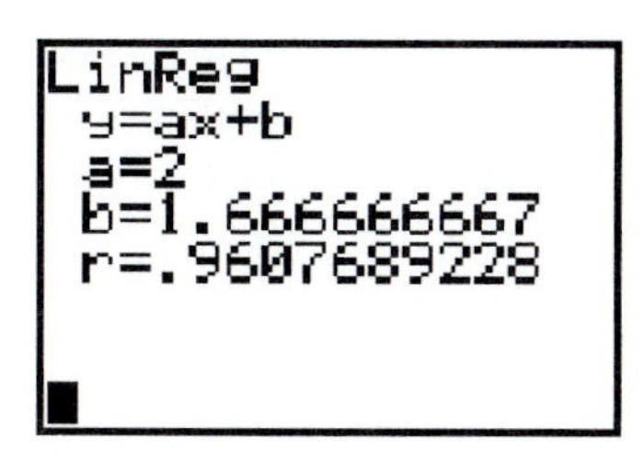

FIGURE 10

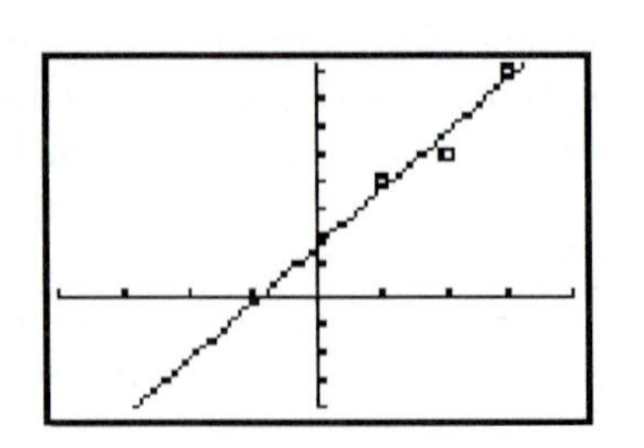

FIGURE 11

9. The original three points can be easily plotted along with the least-squares line. Assume that the linear function has been assigned to `Y1` and all other functions have been cleared or deselected. Set the window to $[-4, 4]$ *by* $[-4, 8]$ and then press 2nd [STAT PLOT] ENTER ENTER GRAPH to see the display in Figure 11. (*Note 1*: To turn off the point-plotting feature, press 2nd [STAT PLOT] ENTER ▶ ENTER. *Note 2*: With the TI-83, the point-plotting feature can be toggled from the function-declaration screen by moving the cursor to the word "`PLOT1`" on the top line and pressing ENTER.)

OBTAINING THE LEAST-SQUARES LINE WITH A TI-85 The following steps find the straight line that minimizes the least-squares error for the points $(1, 4)$, $(2, 5)$, $(3, 8)$.

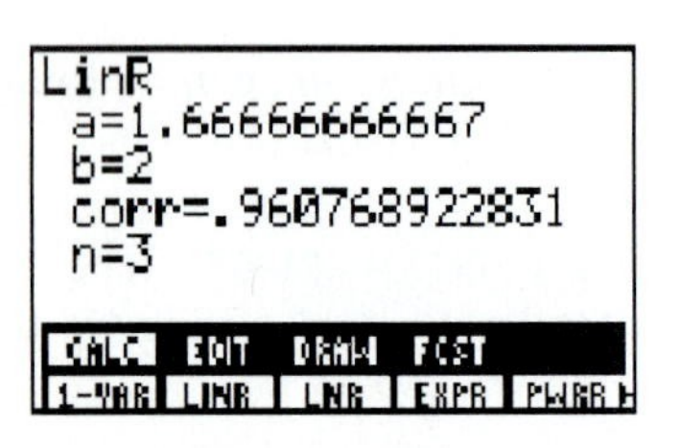

FIGURE 12

1. Press STAT F2 ENTER ENTER to obtain a table used for entering the data.

2. Press F5 to clear all previous data.

3. Enter the data for the three points by pressing **1** ENTER **4** ENTER **2** ENTER **5** ENTER **3** ENTER **8**.

4. Press STAT F1 ENTER ENTER F2. See Figure 12. The least-squares line has equation $y = bx + a$; that is, $y = 2x + \frac{5}{3}$. (*Note*: The number to the right of "`corr=`" is called the correlation coefficient. If the absolute value of the number is close to 1, then the straight line is a good fit to the points.)

5. If desired, the straight line (and the three points) can be graphed with the following steps:

   (a) First press GRAPH F1 and deselect all functions.

   (b) Press STAT F3 to invoke the statistical DRAW menu. (If any graphs appear, press F5 to clear them.)

   (c) Press F4 to draw the least-squares line, and press F2 to draw the three points. See Figure 13.

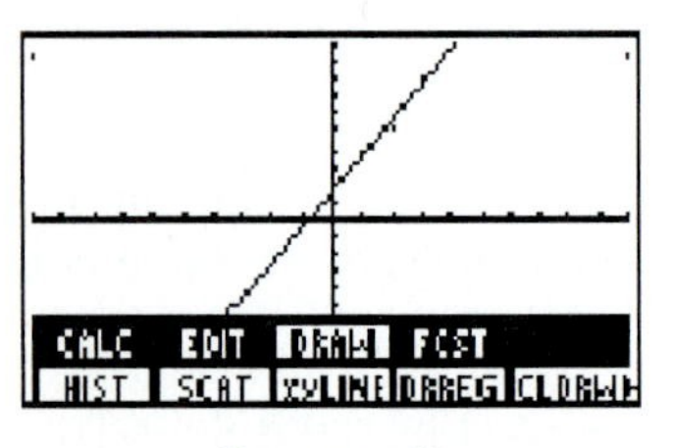

FIGURE 13

For a discussion of obtaining the least-squares line with Visual Calculus, see Appendix A, Part II.

VERIFICATION OF THE PROPERTIES OF SLOPE It is convenient to verify the properties of slope in the order 2, 3, 1. Verification of Slope Property 4 is outlined in Exercise 60.

Verification of Slope Property 2 Suppose that the equation of the line is $y = mx + b$. Then, since (x_2, y_2) is on the line, we have $y_2 = mx_2 + b$. Similarly, since (x_1, y_1) is on the line, we have $y_1 = mx_1 + b$. Subtracting, we see that

$$y_2 - y_1 = mx_2 - mx_1$$
$$y_2 - y_1 = m(x_2 - x_1),$$

so that

$$\frac{y_2 - y_1}{x_2 - x_1} = m.$$

This is the formula for the slope m stated in Slope Property 2.

Verification of Slope Property 3 The equation $y - y_1 = m(x - x_1)$ may be put into the form

$$y = mx + \underbrace{(y_1 - mx_1)}_{b}. \tag{3}$$

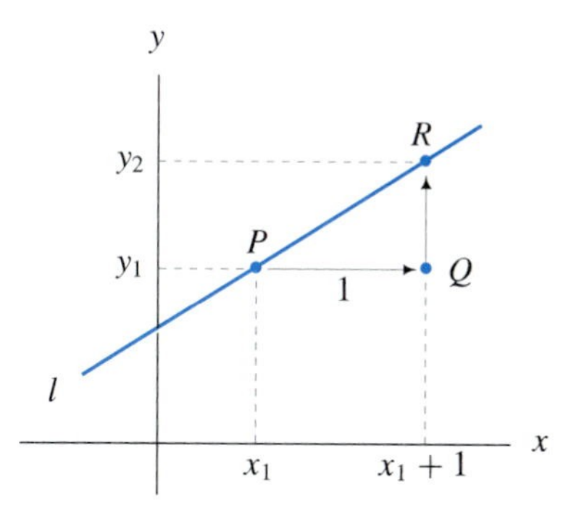

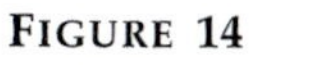
FIGURE 14

This is the equation of a line with slope m. Furthermore, the point (x_1, y_1) is on this line because equation (3) remains true when we plug in x_1 for x and y_1 for y. Thus the equation $y - y_1 = m(x - x_1)$ corresponds to the line of slope m passing through (x_1, y_1).

Verification of Slope Property 1 Let $P = (x_1, y_1)$ be on a line l and let y_2 be the y-coordinate of the point on l obtained by moving P one unit to the right and then moving vertically to return to the line. (See Figure 14.) By Slope Property 2, the slope m of this line l satisfies

$$m = \frac{y_2 - y_1}{(x_1 + 1) - x_1} = \frac{y_2 - y_1}{1} = y_2 - y_1.$$

Thus the difference of the y-coordinates of R and Q is m. This is Slope Property 1.

PRACTICE PROBLEMS 1.4

Find the slopes of the following lines.

1. The line whose equation is $x = 3y - 7$.
2. The line passing through the points $(2, 5)$ and $(2, 8)$.

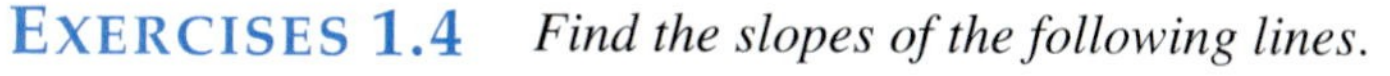

EXERCISES 1.4

Find the slopes of the following lines.

1. $y = 2 - 5x$
2. $y = -5x$
3. $y = 2$
4. $y = \frac{1}{3}(x + 2)$
5. $y = \dfrac{2x - 1}{7}$
6. $y = \frac{1}{4}$
7. $2x + 3y = 6$
8. $x - y = 2$

Find the equations of the following lines.

9. Slope is 3; y-intercept is $(0, -1)$.
10. Slope is $\frac{1}{2}$; y-intercept is $(0, 0)$.
11. Slope is 1; $(1, 2)$ on line.
12. Slope is $-\frac{1}{3}$; $(6, -2)$ on line.
13. Slope is -7; $(5, 0)$ on line.
14. Slope is $\frac{1}{2}$; $(2, -3)$ on line.
15. Slope is 0; $(7, 4)$ on line.
16. Slope is $-\frac{2}{3}$; $(0, 5)$ on line.
17. $(2, 1)$ and $(4, 2)$ on line.
18. $(5, -3)$ and $(-1, 3)$ on line.
19. $(0, 0)$ and $(1, -2)$ on line.
20. $(2, -1)$ and $(3, -1)$ on line.
21. Parallel to $y = -2x + 1$; $(\frac{1}{2}, 5)$ on line.
22. Parallel to $3x + y = 7$; $(-1, -1)$ on line.
23. Parallel to $3x - 6y = 1$; $(1, 0)$ on line.

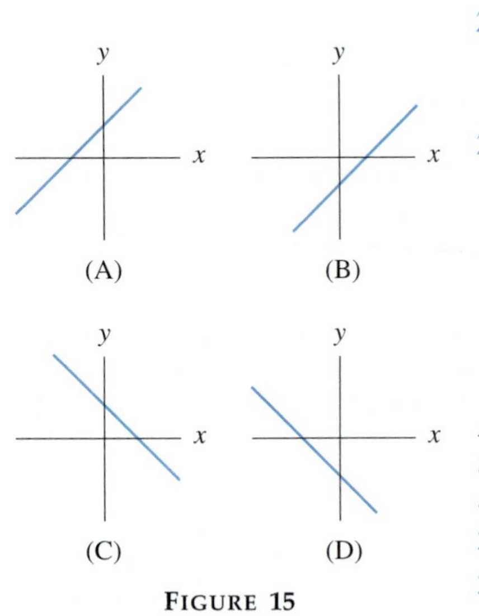

FIGURE 15

24. Parallel to $5x + 2y = -4$; $(0, 17)$ on line.

25. Each of the lines (A), (B), (C), and (D) in Figure 15 is the graph of one of the equations (a), (b), (c), and (d). Match each equation with its graph.
(a) $x + y = 1$ (b) $x - y = 1$ (c) $x + y = -1$ (d) $x - y = -1$

26. Table 1 gives some points on the line $y = mx + b$. Find m and b.

Table 1 POINTS ON A LINE

x	4.8	4.9	5	5.1	5.2
y	3.6	4.8	6	7.2	8.4

In Exercises 27–30, refer to a line of slope m. Suppose you begin at a point on the line and move h units in the x-direction. How many units must you move in the y-direction to return to the line?

27. $m = \frac{1}{2}, h = 4$

28. $m = 2, h = \frac{1}{4}$

29. $m = -3, h = .25$

30. $m = .2, h = 5$

In each of Exercises 31–34, we specify a line by giving the slope and one point on the line. We give the first coordinate of some points on the line. Without deriving the equation of the line, find the second coordinate of each of the points.

31. Slope is 2, $(1, 3)$ on line; $(2,\ )$; $(3,\ )$; $(0,\ )$.

32. Slope is -3, $(2, 2)$ on line; $(3,\ )$; $(4,\ )$; $(1,\ )$.

33. Slope is $-\frac{1}{4}$, $(-1, -1)$ on line; $(0,\ )$; $(1,\ )$; $(-2,\ )$.

34. Slope is $\frac{1}{3}$, $(-5, 2)$ on line; $(-4,\ )$; $(-3,\ )$; $(-2,\ )$.

For each pair of lines in the following figures, determine the one with the greater slope.

35. 36.

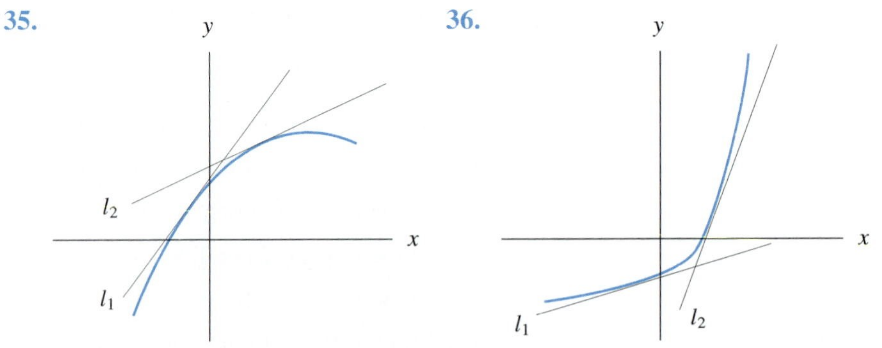

Find the equation and sketch the graph of the following lines.

37. With slope -2 and y-intercept $(0, -1)$.

38. With slope $\frac{1}{3}$ and y-intercept $(0, 1)$.

39. Through $(2, 0)$ with slope $\frac{4}{5}$.

40. Through $(-1, 3)$ with slope 0.

In Exercises 41–46, find an equation of a line with the given property. Each exercise has more than one correct answer.

41. y-intercept is $(0, 5)$.

42. x-intercept is $(9, 0)$.

43. Parallel to the line $4x + 5y = 6$.

44. Horizontal.

45. Slope is -2.

46. Vertical.

47. A salesperson's weekly pay depends on the volume of sales. If she sells x units of goods, then her pay is $y = 5x + 60$ dollars. Give an interpretation of the slope and the y-intercept of this straight line.

48. The demand equation for a monopolist is $y = -.02x + 7$, where x is the number of units produced and y is the price. That is, in order to sell x units of goods, the price must be $y = -.02x + 7$ dollars. Interpret the slope and y-intercept of this line.

49. For tax purposes, businesses are allowed to regard equipment as decreasing in value (or depreciating) each year. The amount of depreciation may be taken as an income tax deduction. Suppose that the value y of a piece of equipment x years after its purchase is given by $y = 500{,}000 - 50{,}000x$. Interpret the y-intercept and the slope of the graph.

50. In some cities the daily car rental charge is \$35 plus \$.30 for each mile the car is driven. Suppose a car is to be rented for three days. Express the total rental expense as a function of the number x of miles driven. (Assume that for each fraction of a mile driven, the same fraction of \$.30 is charged.)

51. In biochemistry, such as in the study of enzyme kinetics, one encounters a linear function of the form $f(x) = (K/V)x + 1/V$, where K and V are constants.

(a) If $f(x) = .2x + 50$, find K and V so that $f(x)$ may be written in the form $f(x) = (K/V)x + 1/V$.

(b) Find the x-intercept and y-intercept of the line $y = (K/V)x + 1/V$ (in terms of K and V).

52. The constants K and V in Exercise 51 are often determined from experimental data. Suppose that a line is drawn through data points and has x-intercept $(-500, 0)$ and y-intercept $(0, 60)$. Determine K and V so that the line is the graph of the function $f(x) = (K/V)x + 1/V$. [*Hint*: Use Exercise 51(b).]

53. A gas company will pay a property owner \$5000 for the right to drill on the land for natural gas and \$.10 for each thousand cubic feet of gas extracted from the land. Express the amount of money the landowner will receive as a function of the amount of gas extracted from the land.

54. In 1992, a patient paid \$300 per day for a semiprivate hospital room and \$1500 for an appendectomy operation. Express the total amount paid for an appendectomy as a function of the number of days of hospital confinement.

55. When a baseball thrown at 85 miles per hour is hit by a bat swung at x miles per hour, the ball travels $6x - 40$ feet.* (This formula assumes that $50 \le x \le$

---

* Robert K. Adair, *The Physics of Baseball* (New York: Harper & Row, 1990).

90 and that the bat is 35 inches long, weighs 32 ounces, and strikes a waist-high pitch so that the plane of the swing lies at 35° from the horizontal.) How fast must the bat be swung in order for the ball to travel 350 feet?

56. Temperatures of 32°F and 212°F correspond to temperatures of 0°C and 100°C. Suppose the linear equation $y = mx + b$ converts Fahrenheit temperatures to Celsius temperatures. Find m and b. What is the Celsius equivalent of 98.6°F?

57. Along one part of the cog railway on Mt. Washington in New Hampshire, the slope of the rails is .371. At this point, how much higher are the passengers in the first row of a rail car above those in the last row? The first and last rows of seats are about 38 feet apart.

58. The pressure p of water on a diver's body is a linear function of the diver's depth, x. At the water's surface, the pressure is 1 atmosphere. At a depth of 100 feet, the pressure is about 3.92 atmospheres.

(a) Find the linear function that relates p to x.

(b) Compute the pressure at a depth of 10 fathoms (60 feet).

59. (a) Draw the graph of any function $f(x)$ that passes through the point (3, 2).

(b) Choose a point to the right of $x = 3$ on the x-axis and label it $3 + h$.

(c) Draw the straight line through the points $(3, f(3))$ and $(3 + h, f(3 + h))$.

(d) What is the slope of this straight line (in terms of h)?

60. Prove Slope Property 4 of straight lines. (*Hint*: If $y = mx + b$ and $y = m'x + b'$ are two lines, then they have a point in common if and only if the equation $mx + b = m'x + b'$ has a solution x.)

61. Figure 16 shows the coordinates of the points on the least-squares line corresponding to the three points used in the discussion on finding best linear fits with graphing calculators. What is the least-squares error for this line?

62. Figure 17 shows the coordinates of the points on the least-squares line corresponding to the four points used in Exercise 63. What is the least-squares error for this line?

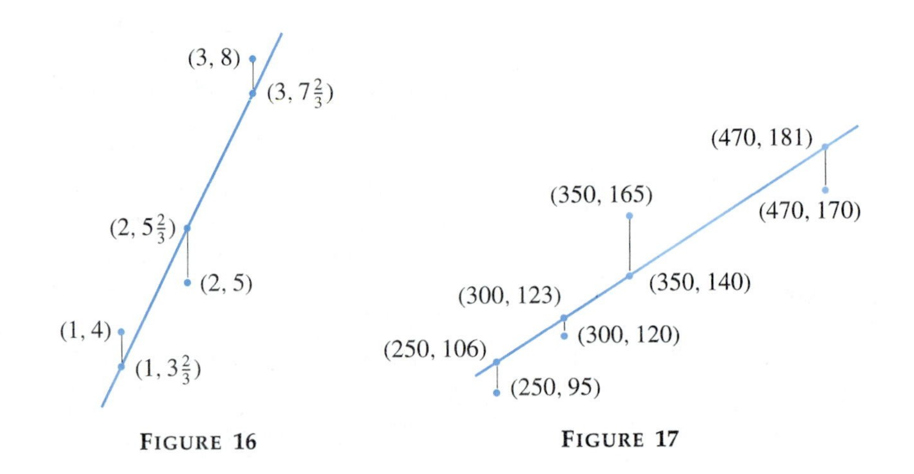

FIGURE 16 FIGURE 17

In the remaining exercises, use a graphing utility to obtain the least-squares lines.

63. The following table* gives the crude male death rate for lung cancer in 1950 and the per capita consumption of cigarettes in 1930 in various countries.

Table 2 CIGARETTE CONSUMPTION

Country	Cigarette Consumption (per capita)	Lung Cancer Deaths (per million males)
Norway	250	95
Sweden	300	120
Denmark	350	165
Australia	470	170

(a) Use the method of least squares to obtain the equation of the straight line that best fits these data. (See Figure 17.)

(b) In 1930 the per capita cigarette consumption in Finland was 1100. Use the equation of the straight line found in part (a) to estimate the male lung cancer death rate in Finland in 1950.

64. An ecologist wished to know whether certain species of aquatic insects have their ecological range limited by temperature. He collected the following data relating the average daily temperature at different portions of a creek with the elevation of that portion of the creek (above sea level).**

Table 3

Elevation (kilometers)	Average Temperature (degrees Celsius)
2.7	11.2
2.8	10
3.0	8.5
3.5	7.5

(a) Find the straight line that provides the best least-squares fit to these data.

(b) Use the linear function to estimate the average daily temperature for this creek at altitude 3.2 kilometers.

65. The accompanying table shows the 1988 price of a gallon of fuel and the consumption of motor fuel for several countries.

(a) Find the straight line that provides the best least-squares fit to these data.

* These data were obtained from *Smoking and Health*, Report of the Advisory Committee to the Surgeon General of the Public Health Service, U.S. Department of Health, Education, and Welfare, Washington, D.C., Public Health Service Publication No. 1103, p. 176.

** The authors express their thanks to Dr. J. David Allen, Department of Zoology, University of Maryland, for providing this data.

Table 4

Country	Price per Gallon in U.S. Dollars	Tons of Motor Fuel per 1000 Persons
United States	\$1.00	1400
England	\$2.20	620
Sweden	\$2.80	700
Franve	\$3.10	580
Italy	\$3.85	420

(b) In 1988, the price of gas in Holland was \$3.00 per gallon. Use the straight line of part (a) to estimate the amount of motor fuel per 1000 people in Holland.

SOLUTIONS TO PRACTICE PROBLEMS 1.4

1. We solve for y in terms of x:

$$y = \frac{1}{3}x + \frac{7}{3}.$$

The slope of the line is the coefficient of x, that is, $\frac{1}{3}$.

2. The line passing through these two points is a vertical line; therefore, its slope is undefined.

1.5 OTHER IMPORTANT FUNCTIONS

In addition to the linear functions studied in Section 1.4, other families of functions will appear throughout the text. They are introduced here, along with some examples of why they are important for applications.

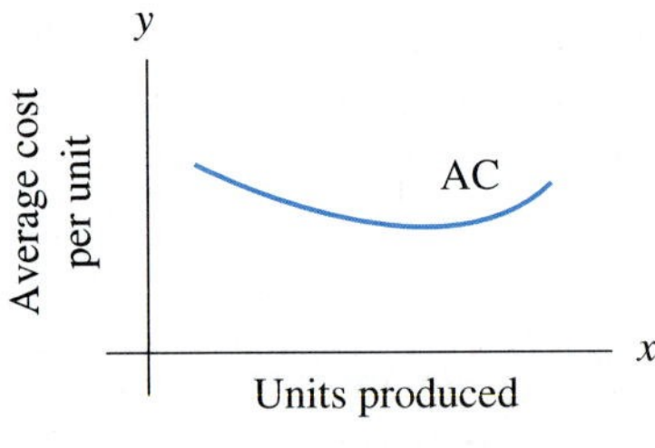

FIGURE 1 Average cost curve

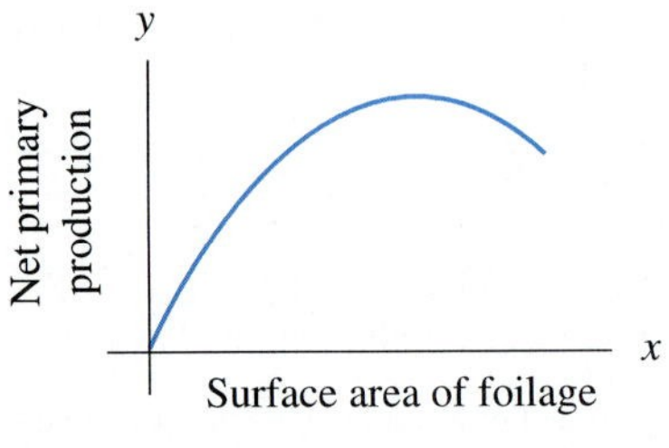

FIGURE 2 Production of nutrients

QUADRATIC FUNCTIONS Economists utilize average cost curves that relate the average unit costs of manufacturing a commodity to the number of units to be produced. (See Figure 1.) Ecologists use curves that relate the net primary production of nutrients in a plant to the surface area of the foliage. (See Figure 2.) Each of the curves is bowl-shaped, opening either up or down. The simplest functions whose graphs resemble these curves are the quadratic functions.

A *quadratic function* is a function of the form

$$f(x) = ax^2 + bx + c,$$

where a, b, and c are constants and $a \neq 0$. The domain of such a function consists of all numbers. The graph of a quadratic function is called a *parabola*. Two typical parabolas are drawn in Figures 3 and 4.

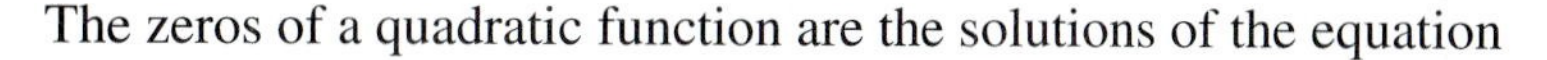

The zeros of a quadratic function are the solutions of the equation

$$ax^2 + bx + c = 0.$$

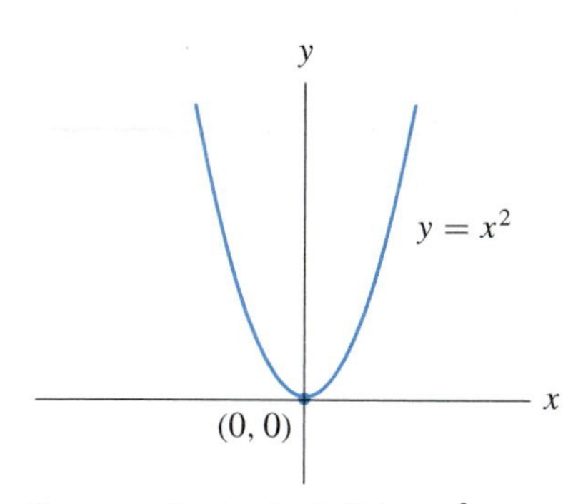

FIGURE 3 Graph of $f(x) = x^2$

A graphing utility will provide good estimates of the zeros, and this is often sufficient in an applied problem. When an exact description of the zeros is needed, there are two possibilities—factoring and the quadratic formula. Both methods are reviewed in Appendix B at the end of the text. Exercises 7–18 in this section will help you determine whether you need to consult the appendix.

POLYNOMIAL AND RATIONAL FUNCTIONS A *polynomial function* of degree n is one of the form

$$f(x) = a_n x^n + a_{n-1} x^{n-1} + \cdots + a_0,$$

where n is a nonnegative integer and $a_0, a_1, \ldots, a_n$ are constants and $a_n \neq 0$. Some examples of polynomial functions are

$$f(x) = 5x^3 - 3x^2 - 2x + 4 \quad \text{and} \quad g(x) = x^4 - x + 1.$$

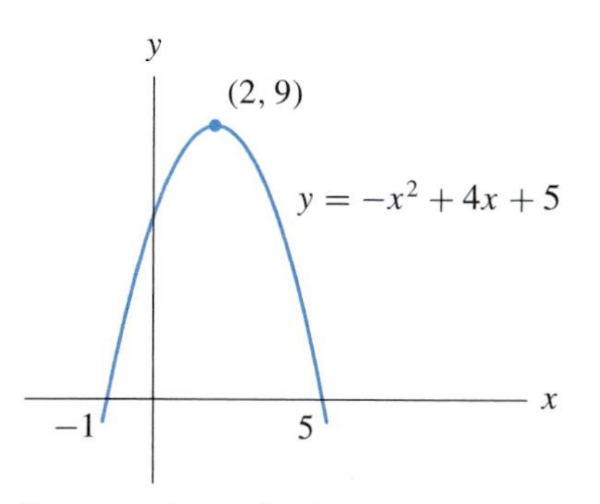

FIGURE 4 Graph of $f(x) = -x^2 + 4x + 5$

Of course, linear and quadratic functions are special cases of polynomial functions. The domain of a polynomial function consists of all numbers.

A function expressed as the quotient of two polynomials is called a *rational function*. Some examples are

$$h(x) = \frac{x^2 + 1}{x} \quad \text{and} \quad k(x) = \frac{x + 3}{x^2 - 4}.$$

The domain of a rational function excludes all values of x for which the denominator is zero. For example, the domain of $h(x)$ excludes $x = 0$, whereas the domain of $k(x)$ excludes $x = 2$ and $x = -2$. As we shall see, both polynomials and rational functions arise in applications of calculus.

Rational functions are used in environmental studies as *cost-benefit* models. The cost of removing a pollutant from the atmosphere is estimated as a function of the percentage of the pollutant removed. The higher the percentage removed, the greater the "benefit" to the people who breathe that air. The issues here are complex, of course, and the definition of "cost" is debatable. The cost to remove a small percentage of pollutant may be fairly low. But the removal cost of the final 5% of the pollutant, for example, may be terribly expensive.

EXAMPLE 1 Suppose a cost-benefit function is given by

$$f(x) = \frac{50x}{105 - x}, \qquad 0 \le x \le 100,$$

where x is the percentage of some pollutant to be removed and $f(x)$ is the associated cost (in millions of dollars). See Figure 5. Find the costs to remove 70%, 95%, and 100% of the pollutant.

Solution The cost to remove 70% is

$$f(70) = \frac{50(70)}{105 - 70} = 100 \text{ (million dollars)}.$$

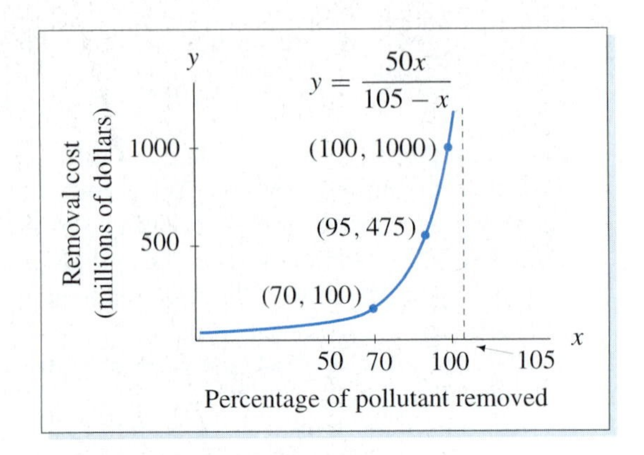

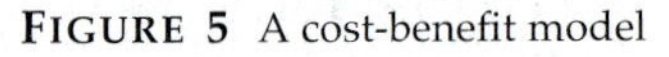
FIGURE 5 A cost-benefit model

Similar calculations show that

$$f(95) = 475 \quad \text{and} \quad f(100) = 1000.$$

Observe that the cost to remove the last 5% of the pollutant is $f(100) - f(95) = 1000 - 475 = 525$ million dollars. This is more than five times the cost to remove the first 70% of the pollutant. ■

THE ABSOLUTE VALUE FUNCTION The absolute value function of a number x is denoted by $|x|$ and is defined by

$$|x| = \begin{cases} x & \text{if } x \text{ is positive or zero} \\ -x & \text{if } x \text{ is negative.} \end{cases}$$

For example, $|5| = 5$, $|0| = 0$, and $|-3| = -(-3) = 3$.

The function defined for all numbers x by

$$f(x) = |x|$$

is called the *absolute value function*. Its graph coincides with the graph of the equation $y = x$ for $x \geq 0$ and with the graph of the equation $y = -x$ for $x < 0$. (See Figure 6.)

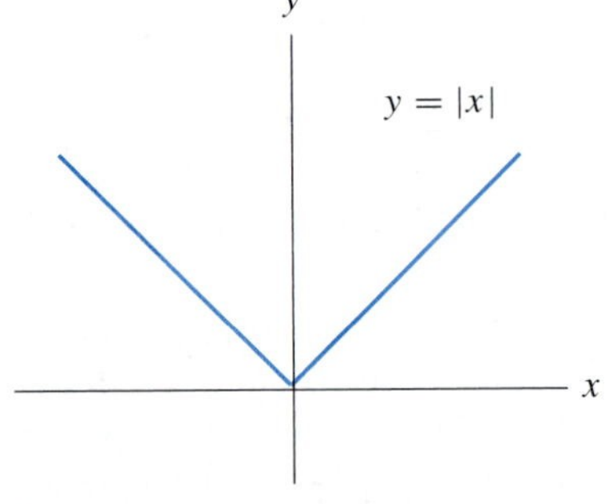

FIGURE 6 Graph of the absolute value function

EXPONENTIAL FUNCTIONS A function of the form

$$f(x) = b^x$$

is called an *exponential function*, because the variable x is in the exponent. The number b is called the base of the exponential function. If f is the exponential function with base 2, that is, if $f(x) = 2^x$, then

$$f(0) = 2^0 = 1, \quad f(1) = 2^1 = 2, \quad f(4) = 2^4 = 2 \cdot 2 \cdot 2 \cdot 2 = 16,$$

and

$$f(-1) = 2^{-1} = \frac{1}{2}, \quad f\left(\frac{1}{2}\right) = 2^{1/2} = \sqrt{2}, \quad f\left(\frac{3}{5}\right) = \left(2^{1/5}\right)^3.$$

In general, if b is positive and x is any rational number, say $x = m/n$ (with m and n integers), b^x is defined as $(b^{1/n})^m$. For irrational values of x (such as $\sqrt{2}$ or π), it is possible to define b^x by first approximating x with rational numbers and then applying a limiting process. See Exercises 39 and 40. We omit the details and simply assume henceforth that b^x can be defined for all numbers x in such a way that the usual laws of exponents remain valid.

For reference, here are the laws of exponents:

(i) $b^x \cdot b^y = b^{x+y}$

(ii) $b^{-x} = \dfrac{1}{b^x}$

(iii) $\dfrac{b^x}{b^y} = b^x \cdot b^{-y} = b^{x-y}$

(iv) $(b^y)^x = b^{xy}$

(v) $a^x b^x = (ab)^x$

(vi) $\dfrac{a^x}{b^x} = \left(\dfrac{a}{b}\right)^x$

Property (iv) may be used to change the appearance of an exponential function. For instance, the function $f(x) = 8^x$ may also be written as $f(x) = (2^3)^x = 2^{3x}$, and $g(x) = \left(\frac{1}{9}\right)^x$ may be written as $g(x) = (1/3^2)^x = (3^{-2})^x = 3^{-2x}$.

EXAMPLE 2 Use properties of exponents to write the following functions in the form 2^{kx} for a suitable constant k.

(a) $4^{5x/2}$ (b) $(2^{4x} \cdot 2^{-x})^{1/2}$ (c) $8^{x/3} \cdot 16^{3x/4}$ (d) $\dfrac{10^x}{5^x}$

Solution (a) First express the base 4 as a power of 2, and then use Property (iv):

$$4^{5x/2} = (2^2)^{5x/2} = 2^{2(5x/2)} = 2^{5x}.$$

(b) First use Property (i) to simplify the quantity inside the parentheses, and then use Property (iv):

$$(2^{4x} \cdot 2^{-x})^{1/2} = \left(2^{4x-x}\right)^{1/2} = \left(2^{3x}\right)^{1/2} = 2^{(3/2)x}.$$

(c) First express the bases 8 and 16 as powers of 2, and then use (iv) and (i):

$$8^{x/3} \cdot 16^{3x/4} = \left(2^3\right)^{x/3} \cdot \left(2^4\right)^{3x/4} = 2^x \cdot 2^{3x} = 2^{4x}.$$

(d) Use (v) to change the numerator 10^x, and then cancel the common term 5^x:

$$\frac{10^x}{5^x} = \frac{(2 \cdot 5)^x}{5^x} = \frac{2^x \cdot 5^x}{5^x} = 2^x.$$

An alternative method is to use Property (vi):

$$\frac{10^x}{5^x} = \left(\frac{10}{5}\right)^x = 2^x.$$ ■

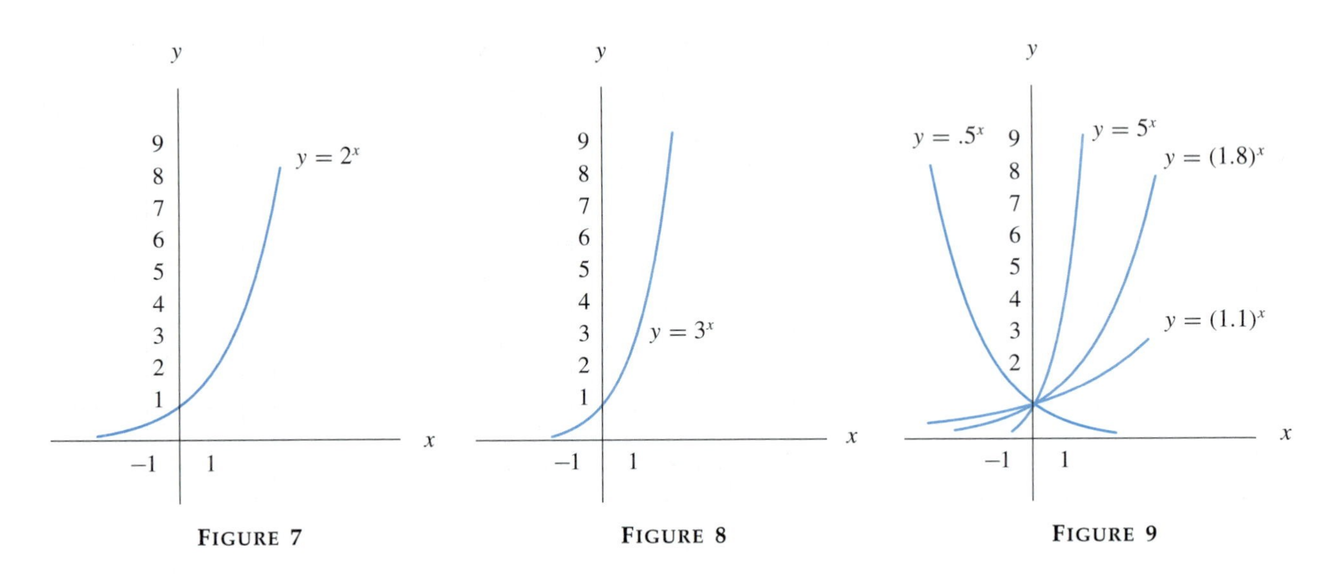

FIGURE 7 FIGURE 8 FIGURE 9

GRAPHS OF EXPONENTIAL FUNCTIONS For a positive base b, the graph of $f(x) = b^x$ is easy to produce on a graphing utility. Figures 7–9 show several sample curves. Notice that because the graph of 2^x is always rising, the function 2^x can never assume the same y-value twice. That is, the only way 2^r can equal 2^s is to have $r = s$.

When $b > 1$, the graph of b^x has the same basic shape as the graph of 2^x. The graph increases from left to right, although when b is close to 1 the graph rises rather slowly at first. Similarly, when $0 < b < 1$, the graph of b^x resembles the graph of $(.5)^x$ in Figure 9. The graph decreases from left to right. Thus, for any $b > 0$, the equation $b^r = b^s$ implies that $r = s$. This fact is useful when solving certain equations involving exponents.

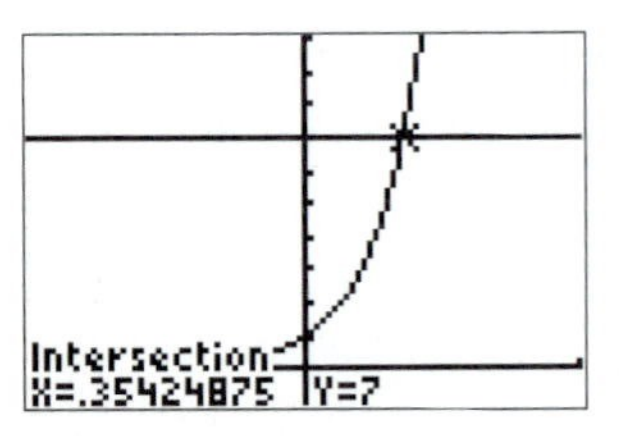

FIGURE 10 Graphs of $Y_1 = 3^{5x}$ and $Y_2 = 7$; $[-1, 1]$ *by* $[-1, 10]$

EXAMPLE 3 Let $f(x) = 3^{5x}$.

(a) Determine all x for which $f(x) = 27$.

(b) Determine all x for which $f(x) = 7$.

Solution (a) Since $27 = 3^3$, we must determine all x for which $3^{5x} = 3^3$. Equating exponents, we have $5x = 3$ and so $x = 3/5$.

(b) This equation cannot be solved with the method of part (a) since 7 cannot easily be written in the form 3^r. An approximate answer can be obtained with a graphing utility by finding the intersection of the two functions `Y1 =3^(5X)` and $Y_2 = 7$. From Figure 10, $x \approx .354$. ■

EXAMPLE 4 Marketing studies have demonstrated that if advertising and other promotion of a particular product are stopped and if other market conditions remain fairly constant, then the number of sales in the tth month following cessation of promotional effort can be approximated by a function of the form

$$S(t) = S_0 2^{-t/T},$$

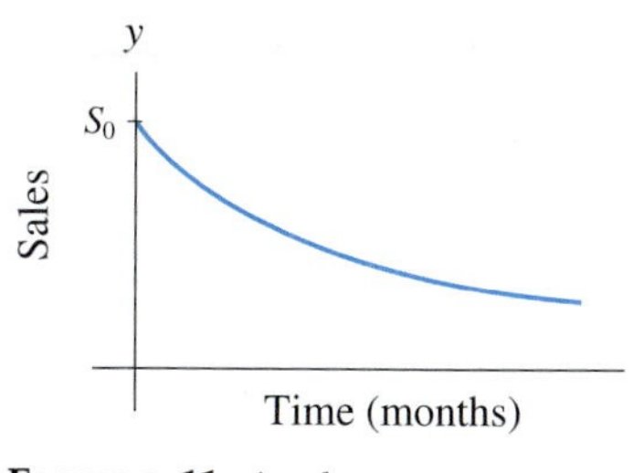

FIGURE 11 A sales curve

where S_0 and T are constants that depend on the type of product, the number of years of prior advertising, and other characteristics of the market. See Figure 11. Compute $S(t)$ for simple values of t to discover the market significance of the constants S_0 and T.

Solution When the advertising stops, and $t = 0$,

$$S(0) = S_0 2^0 = S_0 \cdot 1 = S_0.$$

Therefore, S_0 is the number of sales in the last month during which advertising occurred. When $t = T$,

$$S(T) = S_0 2^{-T/T} = \frac{1}{2} S_0,$$

which shows that T is the time (in months) at which the monthly sales are one-half what they were when the advertising stopped. ■

PRACTICE PROBLEMS 1.5

1. Let $f(x) = x^2 - 13x + 36$. Factor the quadratic polynomial and thereby find the zeros of f. Use a graphing utility to check the factorization.
2. Can a function such as $f(x) = 5^{3x}$ be written in the form $f(x) = b^x$ (for all x)? If so, what is b?

EXERCISES 1.5

Each of the functions in Exercises 1–6 has the form $ax^2 + bx + c$. Identify a, b, and c.

1. $y = 3x^2 - 4x$
2. $y = \dfrac{x^2 - 6x + 2}{3}$
3. $y = 3x - 2x^2 + 1$
4. $y = 3 - 2x + 4x^2$
5. $y = \dfrac{x^2}{2} + \dfrac{x}{6} + 8$
6. $y = 5 - \frac{1}{4}x^2$

Use a graphing utility to solve the equations in Exercises 7–12. Confirm your answers by the quadratic formula, converting any square roots you find to decimal form.

7. $5x^2 - 4x - 1 = 0$
8. $x^2 - 4x + 5 = 0$
9. $-2x^2 + 3x - 4 = 0$
10. $x^2 - \sqrt{2}\,x - \frac{5}{4} = 0$
11. $\frac{3}{2}x^2 - 6x + 5 = 0$
12. $9x^2 - 12x + 4 = 0$

Factor the polynomials in Exercises 13–18. Check your work with a graphing utility by evaluating the function at the zeros obtained from the factorization.

13. $x^2 + 8x + 15$
14. $x^2 - 10x + 16$
15. $x^2 - 13$
16. $3x - x^2$
17. $30 - 4x - 2x^2$
18. $5x^2 - 1$
19. Let $f(x)$ be the cost-benefit function from Example 1. If 70% of the pollutant has been removed, what is the added cost to remove another 5%? How does this compare with the cost to remove the final 5% of the pollutant? (See Example 1.)

20. Suppose that the cost (in millions of dollars) to remove x percent of a certain pollutant is given by the cost-benefit function

$$f(x) = \frac{20x}{102 - x} \quad \text{for } 0 \le x \le 100.$$

(a) Find the cost to remove 85% of the pollutant.
(b) Find the cost to remove the final 5% of the pollutant.

Evaluate each of the functions in Exercises 21–24 at the given value of x.

21. $f(x) = x^{100}$, $x = -1$
22. $f(x) = x^5$, $x = \frac{1}{2}$
23. $f(x) = |x|$, $x = 10^{-2}$
24. $f(x) = |x|$, $x = -2.5$

Write each function in Exercises 25–38 in the form 2^{kx} or 3^{kx}, for a suitable constant k.

25. $4^x, (\sqrt{3})^x, \left(\frac{1}{9}\right)^x$
26. $27^x, \left(\sqrt[3]{2}\right)^x, \left(\frac{1}{8}\right)^x$
27. $8^{2x/3}, 9^{3x/2}, 16^{-3x/4}$
28. $9^{-x/2}, 8^{4x/3}, 27^{-2x/3}$
29. $\left(\frac{1}{4}\right)^{2x}, \left(\frac{1}{8}\right)^{-3x}, \left(\frac{1}{81}\right)^{x/2}$
30. $\left(\frac{1}{9}\right)^{2x}, \left(\frac{1}{27}\right)^{x/3}, \left(\frac{1}{16}\right)^{-x/2}$
31. $2^{3x} \cdot 2^{-5x/2}, 3^{2x} \cdot \left(\frac{1}{3}\right)^{2x/3}$
32. $2^{5x/4} \cdot \left(\frac{1}{2}\right)^x, 3^{-2x} \cdot 3^{5x/2}$
33. $\left(2^{-3x} \cdot 2^{-2x}\right)^{2/5}, \left(9^{1/2} \cdot 9^4\right)^{x/9}$
34. $\left(3^{-x} \cdot 3^{x/5}\right)^5, \left(16^{1/4} \cdot 16^{-3/4}\right)^{3x}$
35. $\frac{3^{4x}}{3^{2x}}, \frac{2^{5x+1}}{2 \cdot 2^{-x}}, \frac{9^{-x}}{27^{-x/3}}$
36. $\frac{2^x}{6^x}, \frac{3^{-5x}}{3^{-2x}}, \frac{16^x}{8^{-x}}$
37. $6^x \cdot 3^{-x}, \frac{15^x}{5^x}, \frac{12^x}{2^{2x}}$
38. $7^{-x} \cdot 14^x, \frac{2^x}{6^x}, \frac{3^{2x}}{18^x}$

39. Use a graphing utility to fill in the following table of values of 2^x for values of x that approach $\sqrt{2} = 1.41421335\ldots$. Then use the graphing utility to compute $2^{\sqrt{2}}$ directly (which is probably accurate to more than 10 decimal places). To how many decimal places does the final value in the table agree with the number you found for $2^{\sqrt{2}}$?

x	1	1.4	1.41	1.414	1.4142	1.41421	1.414213
2^x							

40. Use a graphing utility to fill in the following table of values of 2^x, for values of x that approach $\pi = 3.141592654\ldots$. To how many decimal places does the final entry in the table agree with the value of 2^π computed by your graphing utility?

x	3	3.1	3.14	3.142	3.1416	3.141591	3.141593
2^x							

Solve the following equations for x.

41. $5^{2x} = 5^2$
42. $10^{-x} = 10^2$
43. $(2.5)^{2x+1} = (2.5)^5$
44. $(3.2)^{x-3} = (3.2)^5$

45. $10^{1-x} = 100$

46. $2^{4-x} = 8$

47. $3(2.7)^{5x} = 8.1$

48. $4(2.7)^{2x-1} = 10.8$

49. $(2^{x+1} \cdot 2^{-3})^2 = 2$

50. $(3^{2x} \cdot 3^2)^4 = 3$

51. $2^{3x} = 4 \cdot 2^{5x}$

52. $3^{5x} \cdot 3^x - 3 = 0$

53. $(1 + x)2^{-x} - 5 \cdot 2^{-x} = 0$

54. $(2 - 3x)5^x + 4 \cdot 5^x = 0$

The expressions in Exercises 55–58 may be factored as shown. Find the missing factors.

55. $2^{3+h} = 2^3(\quad)$

56. $3^{2+h} = 9(\quad)$

57. $2^{x+h} - 2^x = 2^x(\qquad)$

58. $3^{x+h} + 3^x = 3^x(\qquad)$

SOLUTIONS TO PRACTICE PROBLEMS 1.5

1. $f(x) = x^2 - 13x + 36 = (x - 4)(x - 9)$. The zeros of f are 4 and 9 because they satisfy the equation $(x - 4)(x - 9) = 0$. To check with a graphing utility that $(x - 4)(x - 9)$ really is a factorization of $f(x)$, check whether the graph of $y = x^2 - 13x + 36$ passes through $(4, 0)$ and $(9, 0)$. Or, use the graphing utility to evaluate $f(4)$ and $f(9)$.
2. If $5^{3x} = b^x$ for all x, then when $x = 1$, $5^{3(1)} = b^1$, which says that $b = 5^3 = 125$. This value of b certainly works, because $5^{3x} = (5^3)^x = 125^x$.

1.6 OPERATIONS ON FUNCTIONS

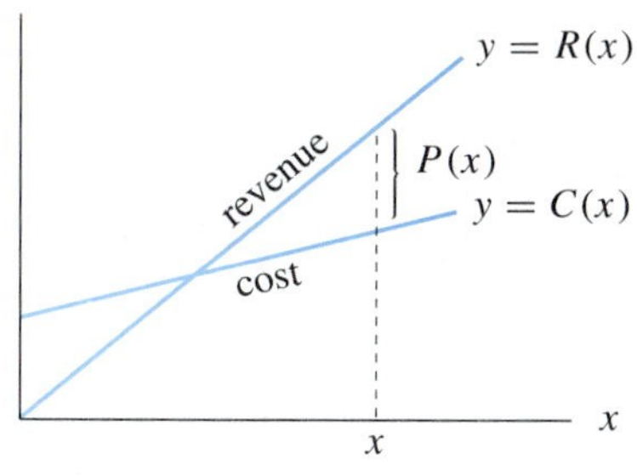

FIGURE 1 Profit equals revenue minus cost

Many functions encountered later in the text can be viewed as combinations of other functions. For example, let $P(x)$ represent the profit a company makes on the sale of x units of some commodity. If $R(x)$ denotes the revenue received from the sale of x units, and if $C(x)$ is the cost of producing x units, then

$$P(x) = R(x) - C(x)$$
$$[\text{profit}] = [\text{revenue}] - [\text{cost}].$$

Writing the profit function in this way makes it possible to predict the behavior of $P(x)$ from properties of $R(x)$ and $C(x)$. See Figure 1. For each x, the values of $R(x)$ and $C(x)$ are the y-coordinates of points on their respective graphs, and $P(x)$ is the difference of these two y-coordinates.

The first four examples here review the algebraic techniques needed to combine functions by addition, subtraction, multiplication, and division.

EXAMPLE 1 Let $f(x) = 3x + 4$ and $g(x) = 2x - 6$. Find

$$f(x) + g(x), \quad f(x) - g(x), \quad \frac{f(x)}{g(x)}, \quad \text{and} \quad f(x)g(x).$$

Solution For $f(x) + g(x)$ and $f(x) - g(x)$, we add or subtract corresponding terms:

$$f(x) + g(x) = (3x + 4) + (2x - 6) = 3x + 4 + 2x - 6 = 5x - 2$$
$$f(x) - g(x) = (3x + 4) - (2x - 6) = 3x + 4 - 2x + 6 = x + 10.$$

To compute $\frac{f(x)}{g(x)}$ and $f(x)g(x)$, we first substitute the formulas for $f(x)$ and $g(x)$:

$$\frac{f(x)}{g(x)} = \frac{3x+4}{2x-6}, \qquad f(x)g(x) = (3x+4)(2x-6).$$

The expression for $f(x)/g(x)$ is already in simplest form. To simplify the expression for $f(x)g(x)$, we carry out the multiplication indicated in $(3x+4)(2x-6)$. We must be careful to multiply each term of $3x+4$ by each term of $2x-6$. A common order for multiplying such expressions is: (1) the First terms, (2) the Outer terms, (3) the Inner terms, and (4) the Last terms. (This procedure may be remembered by the word FOIL.)

$$\begin{aligned} f(x)g(x) = (3x+4)(2x-6) &= 6x^2 - 18x + 8x - 24 \\ &= 6x^2 - 10x - 24. \end{aligned}$$

EXAMPLE 2 Let $g(x) = \frac{2}{x}$ and $h(x) = \frac{3}{x-1}$. Express $g(x) + h(x)$ as a rational function.

Solution First we have

$$g(x) + h(x) = \frac{2}{x} + \frac{3}{x-1}, \qquad x \neq 0, 1.$$

The restriction $x \neq 0, 1$ comes from the fact that $g(x)$ is defined only for $x \neq 0$ and $h(x)$ is defined only for $x \neq 1$. (A rational function is not defined for values of the variable for which the denominator is 0.) In order for us to add two fractions, their denominators must be the same. A common denominator for

$$\frac{2}{x} \quad \text{and} \quad \frac{3}{x-1}$$

is $x(x-1)$. If we multiply

$$\frac{2}{x} \quad \text{by} \quad \frac{x-1}{x-1},$$

we obtain an equivalent expression whose denominator is $x(x-1)$. Similarly, if we multiply

$$\frac{3}{x-1} \quad \text{by} \quad \frac{x}{x},$$

we obtain an equivalent expression whose denominator is $x(x-1)$. Thus

$$\begin{aligned}\frac{2}{x}+\frac{3}{x-1}&=\frac{2}{x}\cdot\frac{x-1}{x-1}+\frac{3}{x-1}\cdot\frac{x}{x}\\&=\frac{2(x-1)}{x(x-1)}+\frac{3x}{x(x-1)}\\&=\frac{2(x-1)+3x}{x(x-1)}\\&=\frac{5x-2}{x(x-1)}.\end{aligned}$$

So $g(x)+h(x)=\dfrac{5x-2}{x(x-1)}$. ■

EXAMPLE 3 Find $f(t)g(t)$, where

$$f(t)=\frac{t}{t-1}\quad\text{and}\quad g(t)=\frac{t+2}{t+1}.$$

Solution To multiply rational functions, multiply numerator by numerator and denominator by denominator:

$$f(t)g(t)=\frac{t}{t-1}\cdot\frac{t+2}{t+1}=\frac{t(t+2)}{(t-1)(t+1)}.$$

An alternative way of expressing $f(t)g(t)$ is obtained by carrying out the indicated multiplications:

$$f(t)g(t)=\frac{t^2+2t}{t^2+t-t-1}=\frac{t^2+2t}{t^2-1}.$$

The choice of which expression to use for $f(t)g(t)$ depends on the particular application. ■

EXAMPLE 4 Find $f(x)/g(x)$, where

$$f(x)=\frac{x}{x-3}\quad\text{and}\quad g(x)=\frac{x+1}{x-5}.$$

Solution The function is defined only for $x\neq 3$, and $g(x)$ is defined only for $x\neq 5$. The quotient $f(x)/g(x)$ is therefore not defined for $x=3, 5$. Moreover, the quotient is not defined for values of x for which $g(x)$ is equal to 0, that is, $x=-1$. Thus, the quotient is defined for $x\neq 3, 5, -1$. To divide $f(x)$ by $g(x)$, we multiply $f(x)$ by the reciprocal of $g(x)$:

$$\begin{aligned}\frac{f(x)}{g(x)}&=\frac{x}{x-3}\cdot\frac{x-5}{x+1}=\frac{x(x-5)}{(x-3)(x+1)}\\&=\frac{x^2-5x}{x^2-2x-3},\quad x\neq 3, -1, 5.\end{aligned}$$

■

VERTICAL TRANSLATION OF GRAPHS The sum of two functions is of particular interest when one of the functions is a constant function. For example, if $g(x) = 2$ for all x, and if

$$h(x) = f(x) + g(x) = f(x) + 2,$$

then the graph of h is obtained from that of f by adding 2 to the y-coordinates of the points on the graph of f. This action of moving the graph 2 units upward is called a vertical translation of the graph. Similarly, the graph of $h(x) = f(x) - 2$ is obtained by a vertical translation of the graph of f downward by 2 units. See Figure 2. The vertical translation is easy to see in Figure 2 because the graph of f is rather flat. The graphs in Figure 3 are also obtained by vertical translations of ± 2 units, even though they appear to approach each other. For instance, the points A, B, and C in Figure 3 have the same x-coordinates and are 2 units apart.

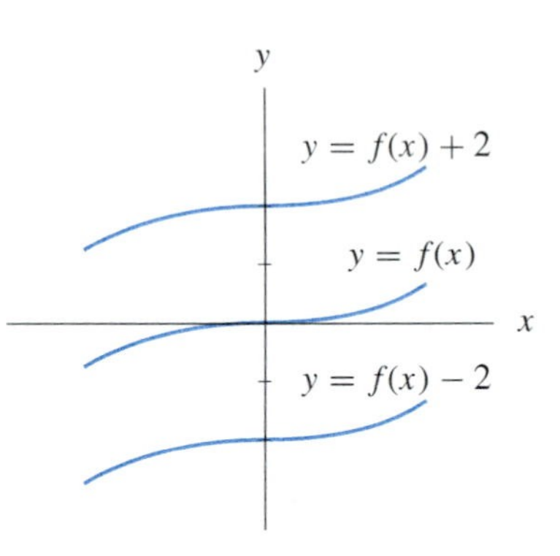

FIGURE 2

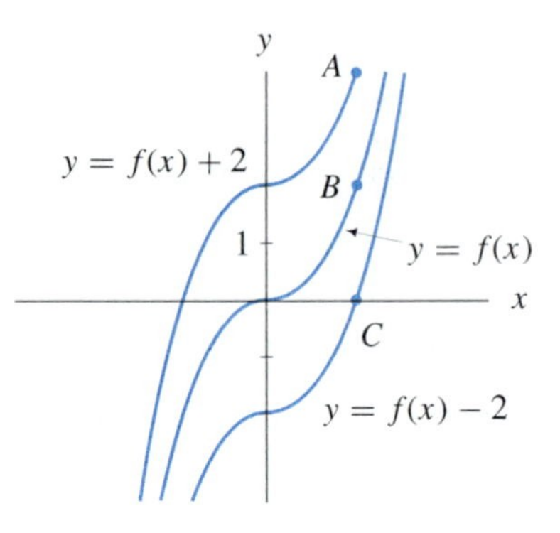

FIGURE 3

COMPOSITION OF FUNCTIONS Another important way of combining two functions $f(x)$ and $g(x)$ is to substitute the function $g(x)$ for every occurrence of the variable x in $f(x)$. The resulting function is called the *composition* (or *composite*) of $f(x)$ and $g(x)$ and is denoted by $f(g(x))$.

EXAMPLE 5 Let $f(x) = x^2 + 3x + 1$ and $g(x) = x - 5$. What is $f(g(x))$?

Solution We substitute $g(x)$ in place of each x in $f(x)$:

$$\begin{aligned} f(g(x)) &= [g(x)]^2 + 3g(x) + 1 \\ &= (x-5)^2 + 3(x-5) + 1 \\ &= (x^2 - 10x + 25) + (3x - 15) + 1 \\ &= x^2 - 7x + 11. \end{aligned}$$

■

Later in the text we shall need to study expressions of the form $f(x+h)$, where $f(x)$ is a given function and h represents some number. The meaning of $f(x+h)$ is that $x+h$ is to be substituted for each occurrence of x in the formula for $f(x)$. In fact, $f(x+h)$ is just a special case of $f(g(x))$, where $g(x) = x + h$.

EXAMPLE 6 If $f(x) = x^3$, find $f(x+h) - f(x)$.

Solution

$$\begin{aligned} f(x+h) &= (x+h)^3 = x^3 + 3x^2h + 3xh^2 + h^3 \\ f(x+h) - f(x) &= (x^3 + 3x^2h + 3xh^2 + h^3) - x^3 \\ &= 3x^2h + 3xh^2 + h^3. \end{aligned}$$

■

EXAMPLE 7 In a certain lake, the bass feed primarily on minnows, and the minnows feed on plankton. Suppose that the size of the bass population is a function $f(n)$ of the number n of minnows in the lake, and the number of minnows is a function $g(x)$ of the amount x of plankton in the lake. Express the size of the bass population as a function of the amount of plankton, if $f(n) = 50 + \sqrt{n/150}$ and $g(x) = 4x + 3$.

Solution We have $n = g(x)$. Substituting $g(x)$ for n in $f(n)$, we find the size of the bass population is given by

$$f(g(x)) = 50 + \sqrt{\frac{g(x)}{150}} = 50 + \sqrt{\frac{4x+3}{150}}.$$

```
Y1(4)+Y2(4)
                28
Y1(4)/Y2(4)
               -29
Y1(Y2(4))
                -1
```

FORMING COMBINATIONS OF FUNCTIONS WITH THE TI-82 OR TI-83 Suppose the function $f(x)$ has been assigned to Y_1 and $g(x)$ assigned to Y_2. Then Y_3 or any subsequent function can be set equal to either $\text{Y}_1 + \text{Y}_2$, $\text{Y}_1 - \text{Y}_2$, Y_1Y_2, Y_1/Y_2, or $\text{Y}_1(\text{Y}_2)$ to obtain $f(x) + g(x)$, $f(x) - g(x)$, $f(x)g(x)$, $f(x)/g(x)$, or $f(g(x))$.

Values of combinations of functions can be evaluated directly from the Home screen. For example, if Y_1 has been assigned $x^2 + 3x + 1$ and Y_2 has been assigned $x - 5$, then the screen on the left can be obtained.

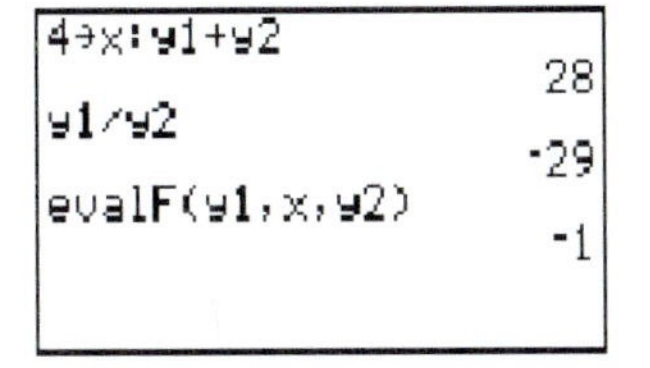

FORMING COMBINATIONS OF FUNCTIONS WITH THE TI-85 Suppose the function $f(x)$ has been assigned to `y1` and $g(x)$ has been assigned to `y2`. Then `y3` or any subsequent functions can be set equal to either `y1+y2`, `y1-y2`, `y1y2`, `y1/y2`, or `evalF(y1,x,y2)` to obtain $f(x) + g(x)$, $f(x) - g(x)$, $f(x)g(x)$, $f(x)/g(x)$, or $f(g(x))$. To display `evalF`, press [2nd] [CALC] [F1]. In general, `evalF(function1,variable,function2)` evaluates `function2`, assigns that value to `variable`, and then uses the value of `variable` to evaluate `function1`.

Values of combinations of functions can be evaluated directly from the Home screen. For example, if `y1` has been assigned $x^2 + 3x + 1$ and `y2` has been assigned $x - 5$, then the screen on the left can be obtained. [*Note*: The first line of the screen cannot be replaced by `y1(4)+y2(4)`.]

For a discussion of forming combinations of functions with Visual Calculus, see Appendix A, Part III.

PRACTICE PROBLEMS 1.6

1. Let $f(x) = x^5$, $g(x) = x^3 - 4x^2 + x - 8$.
   (a) Find $f(g(x))$. (b) Find $g(f(x))$.
2. Let $f(x) = x^2$. Calculate $\dfrac{f(1+h) - f(1)}{h}$ and simplify.

EXERCISES 1.6

Let $f(x) = x^2 + 1$, $g(x) = 9x$, and $h(x) = 5 - 2x^2$. Calculate the following functions.

1. $f(x) + g(x)$
2. $f(x) - h(x)$
3. $f(x)g(x)$
4. $g(x)h(x)$
5. $\dfrac{f(t)}{g(t)}$
6. $\dfrac{g(t)}{h(t)}$

In Exercises 7–12, express $f(x) + g(x)$ as a rational function. Carry out all multiplications.

7. $f(x) = \dfrac{2}{x-3}$, $g(x) = \dfrac{1}{x+2}$
8. $f(x) = \dfrac{3}{x-6}$, $g(x) = \dfrac{-2}{x-2}$

9. $f(x) = \dfrac{x}{x-8}, g(x) = \dfrac{-x}{x-4}$

10. $f(x) = \dfrac{-x}{x+3}, g(x) = \dfrac{x}{x+5}$

11. $f(x) = \dfrac{x+5}{x-10}, g(x) = \dfrac{x}{x+10}$

12. $f(x) = \dfrac{x+6}{x-6}, g(x) = \dfrac{x-6}{x+6}$

Let $f(x) = \dfrac{x}{x-2}$, $g(x) = \dfrac{5-x}{5+x}$, *and* $h(x) = \dfrac{x+1}{3x-1}$. *Express the following as rational functions.*

13. $f(x) - g(x)$

14. $f(t) - h(t)$

15. $f(x)g(x)$

16. $g(x)h(x)$

17. $\dfrac{f(x)}{g(x)}$

18. $\dfrac{h(s)}{f(s)}$

19. $f(x+1)g(x+1)$

20. $f(x+2) + g(x+2)$

21. $\dfrac{g(x+5)}{f(x+5)}$

22. $f\left(\dfrac{1}{t}\right)$

23. $g\left(\dfrac{1}{u}\right)$

24. $h\left(\dfrac{1}{x^2}\right)$

Let $f(x) = x^6$, $g(x) = \dfrac{x}{1-x}$, *and* $h(x) = x^3 - 5x^2 + 1$. *Calculate the following functions.*

25. $f(g(x))$

26. $h(f(t))$

27. $h(g(x))$

28. $g(f(x))$

29. $g(h(t))$

30. $f(h(x))$

31. If $f(x) = x^2$, find $f(x+h) - f(x)$ and simplify.

32. If $f(x) = 1/x$, find $f(x+h) - f(x)$ and simplify.

33. If $g(t) = 4t - t^2$, find $\dfrac{g(t+h) - g(t)}{h}$ and simplify.

34. If $g(t) = t^3 + 5$, find $\dfrac{g(t+h) - g(t)}{h}$ and simplify

35. After t hours of operation, an assembly line has assembled $A(t) = 20t - \frac{1}{2}t^2$ power lawn mowers, $0 \le t \le 10$. Suppose that the factory's cost of manufacturing x units is $C(x)$ dollars, where $C(x) = 3000 + 80x$.

(a) Express the factory's cost as a (composite) function of the number of hours of operation of the assembly line.

(b) What is the cost of the first 2 hours of operation?

36. During the first $\frac{1}{2}$ hour, the employees of a machine shop prepare the work area for the day's work. After that, they turn out 10 precision machine parts per hour, so that the output after t hours is $f(t)$ machine parts, where $f(t) = 10\left(t - \frac{1}{2}\right) = 10t - 5$, $\frac{1}{2} \le t \le 8$. The total cost of producing x machine parts is $C(x)$ dollars, where $C(x) = .1x^2 + 25x + 200$.

(a) Express the total cost as a (composite) function of t.

(b) What is the cost of the first 4 hours of operation?

37. Table 1 shows a conversion table for men's hat sizes for three countries. The function $g(x) = 8x + 1$ converts from British sizes to French sizes, and the function $f(x) = \frac{1}{8}x$ converts from French sizes to U.S. sizes. Determine the function $h(x) = f(g(x))$ and give its interpretation.

Table 1 CONVERSION TABLE FOR MEN'S HAT SIZES

Britain	$6\frac{1}{2}$	$6\frac{5}{8}$	$6\frac{3}{4}$	$6\frac{7}{8}$	7	$7\frac{1}{8}$	$7\frac{1}{4}$	$7\frac{3}{8}$
France	53	54	55	56	57	58	59	60
United States	$6\frac{5}{8}$	$6\frac{3}{4}$	$6\frac{7}{8}$	7	$7\frac{1}{8}$	$7\frac{1}{4}$	$7\frac{3}{8}$	$7\frac{1}{2}$

In the remaining exercises, use a graphing utility to obtain the graphs of the functions.

38. What happens to the graph of a function $f(x)$ when x is replaced by $x - .5$? That is, how is the graph of $f(x - .5)$ related to the graph of $f(x)$? What about the graph of $f(x + .5)$? Experiment with the function $f(x) = 4 + 8x^2 - x^4$, using the window $[-4, 4]$ *by* $[-25, 25]$. To measure how the graph moves, estimate the coordinates of corresponding points on the two graphs. Make a conjecture about how the graph of any function moves when x is replaced by $x - a$, where a is a positive number. What if a is negative?

39. What happens to the graph of a function $f(x)$ when every x in its formula is replaced by $-x$? That is, how is the graph of $f(-x)$ related to the graph of $f(x)$? Make a conjecture based on experiments with the functions:

(a) $f(x) = 6x - 3$; $[-3, 3]$ *by* $[-20, 20]$;

(b) $g(x) = x^4(x^2 - 1)^3(x - 2)$; $[-3, 3]$ *by* $[-40, 40]$.

40. How is the graph of $-f(-x)$ related to the graph of $f(x)$? Make a conjecture based on experiments with the functions in Exercise 39.

41. A function $f(x)$ is said to be an *even function* if $f(-x) = f(x)$ for every x in the domain of f; $f(x)$ is an *odd function* if $f(-x) = -f(x)$ for every x. Determine which of the following functions are even and which are odd. In each case, use the window $[-3, 3]$ *by* $[-10, 10]$. [To analyze a function $f(x)$, graph $f(x)$, $f(-x)$, and $-f(x)$ on the same coordinate system.]

(a) x^2 (b) x^3 (c) $x^3 - 6x$

(d) $4 + 8x^2 - x^4$ (e) $4x + \dfrac{1}{4x}$ (f) $\dfrac{1}{x^2 - 4}$

42. Referring to the examples in Exercise 41, explain what *even* and *odd* mean graphically.

43. Let $f(x) = x^2$, $g(x) = \sqrt{x}$, and $h(x) = f(g(x))$. Graph the three functions together in the window $[-4, 4]$ *by* $[-1, 7]$ and determine the domain and formula for $h(x)$.

44. Redo Exercise 43, where $h(x) = g(f(x))$.

45. Let $f(x) = x^2$ and $g(x) = x - 4$. Find a number a such that $f(g(a)) = g(f(a))$ and a number b such that $f(g(b)) \neq g(f(b))$.

SOLUTIONS TO PRACTICE PROBLEMS 1.6

1. (a) $f(g(x)) = [g(x)]^5 = (x^3 - 4x^2 + x - 8)^5$.

   (b) $g(f(x)) = [f(x)]^3 - 4[f(x)]^2 + f(x) - 8$

$$= (x^5)^3 - 4(x^5)^2 + x^5 - 8 = x^{15} - 4x^{10} + x^5 - 8.$$

2. $$\frac{f(1+h) - f(1)}{h} = \frac{(1+h)^2 - 1^2}{h} = \frac{1 + 2h + h^2 - 1}{h} = \frac{2h + h^2}{h} = 2 + h.$$

REVIEW OF THE FUNDAMENTAL CONCEPTS OF CHAPTER 1

1. Explain how to solve $f(x) = b$ geometrically from the graph of $y = f(x)$.
2. Explain how to find $f(a)$ geometrically from the graph of $y = f(x)$.
3. Explain the use of the vertical line test.
4. What is a zero of a function?
5. Define the slope of a line and give a physical description.
6. Suppose you know the slope and the coordinates of a point on a line. How could you draw the graph of the line without first finding its equation?
7. What is the point-slope form of the equation of a line?
8. Describe how to find an equation for a line when you know the coordinates of two points on the graph of the line.
9. What can you say about the slopes of parallel lines?
10. What is the least-squares line approximation to a set of data points?
11. Give two methods for finding the zeros of a quadratic function.
12. Define and give an example of each of the following types of functions.
    (a) exponential function
    (b) polynomial function
    (c) rational function
13. State as many laws of exponents as you can recall.
14. Without using a graphing utility, make a rough sketch of the graphs of the following functions:
    (a) $f(x) = \frac{1}{x}$ $(x > 0)$
    (b) $f(x) = x^2$
    (c) $f(x) = \sqrt{x}$
    (d) $f(x) = x^3$
    (e) $f(x) = 2^x$
15. What is function composition? Give an example.

Chapter 2
THE DERIVATIVE

2.1 THE SLOPE OF A CURVE AT A POINT

In order to extend the concept of slope from straight lines to more general curves, we must first discuss the notion of the tangent line to a curve at a point.

We have a clear idea of what is meant by the tangent line to a circle at a point P. It is the straight line that touches the circle at just the one point P. Let us focus on the region near P, designated by the dashed rectangle shown in Figure 1. The enlarged portion of the circle looks almost straight, and the straight line that it resembles is the tangent line. Further enlargements would make the circle near P look even straighter and have an even closer resemblance to the tangent line. In this sense, the tangent line to the circle at the point P is the straight line through P that best approximates the circle near P. In particular, the tangent line at P reflects the steepness of the circle at P. Thus it seems reasonable to define the *slope* of the circle at P to be the slope of the tangent line at P.

Similar reasoning leads us to a suitable definition of slope for an arbitrary curve at a point P. Consider the three graphs shown in Figure 2. In each case

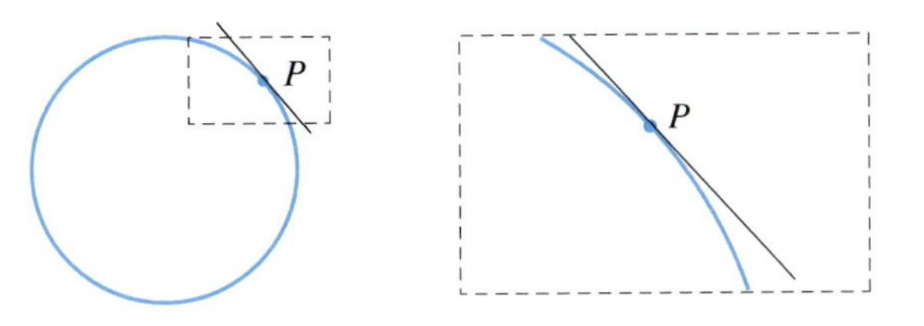

FIGURE 1 Enlarged portion of a circle

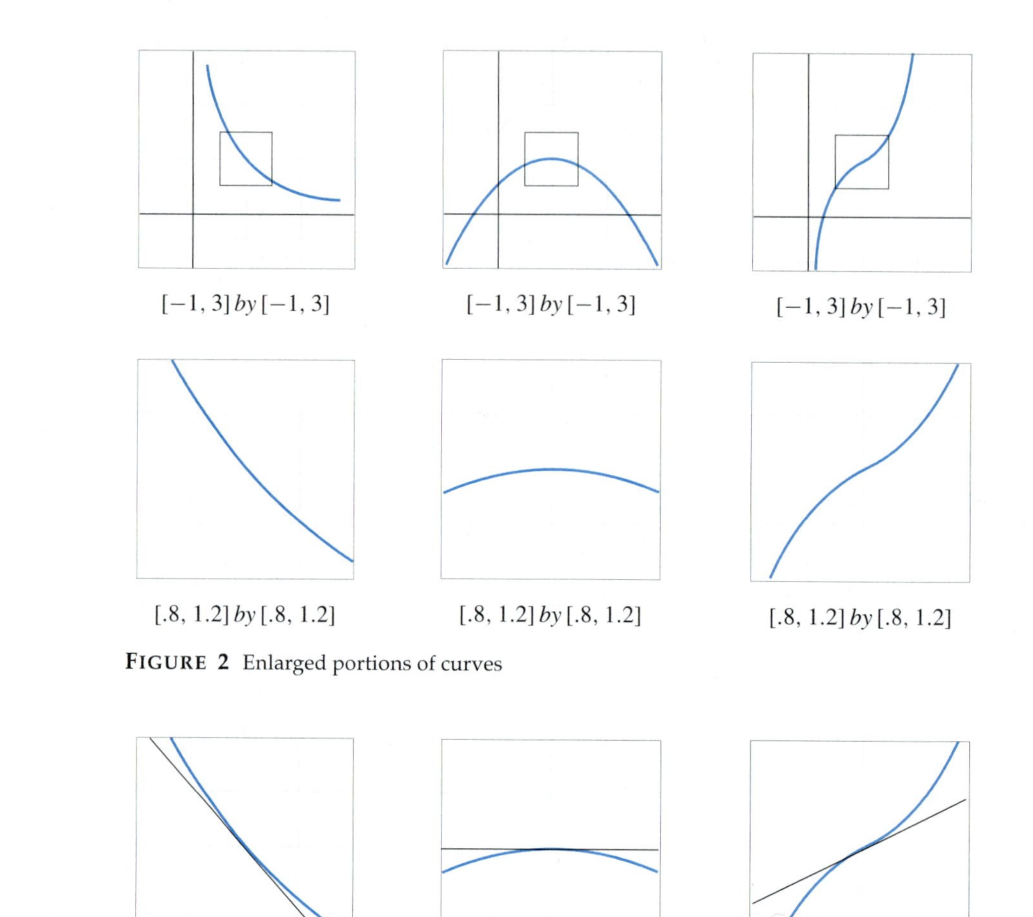

FIGURE 2 Enlarged portions of curves

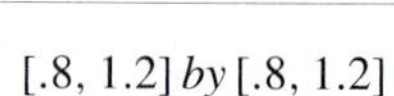

[.8, 1.2] by [.8, 1.2]   [.8, 1.2] by [.8, 1.2]   [.8, 1.2] by [.8, 1.2]

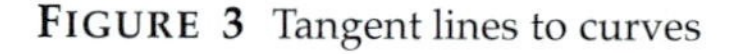

FIGURE 3 Tangent lines to curves

we have enlarged a small square around the point $P = (1, 1)$. Notice that the portion of each curve lying in the enlarged square looks almost straight. If we further magnify the curves near P, they would appear even straighter. Indeed, if we were to apply higher and higher magnification, the portion of each curve near P would approach a certain straight line more and more exactly. This straight line is called the *tangent line to the curve at* P. (See Figure 3.) It is the straight line through P that best approximates the curve near P. We define the *slope of a curve at the point* P to be the slope of the tangent line to the curve at P.

The portion of the curve near P can be, at least within an approximation, replaced by the tangent line at P. Therefore, the slope of the curve at P—that is, the slope of the tangent line at P—measures the rate of increase or decrease of the curve as it passes through P.

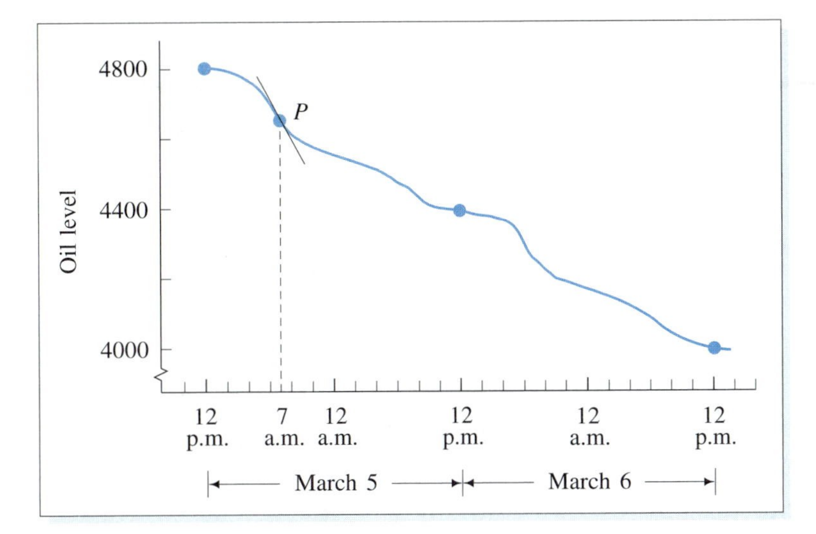

FIGURE 4 Oil level in a storage tank

EXAMPLE 1 An apartment complex has a storage tank to hold its heating oil. The tank was filled on January 1, but no more deliveries of oil will be made until some time in March. Let $f(t)$ be the number of gallons of fuel oil in the tank t days after January 1. The graph of $y = f(t)$ for a typical 2-day period appears in Figure 4. What is the physical significance of the slope of the graph at the point P?

Solution The curve near P is closely approximated by its tangent line. So think of the curve as replaced by its tangent line near P. Then the slope at P is just the rate of decrease of the oil level at 7 A.M. on March 5. ■

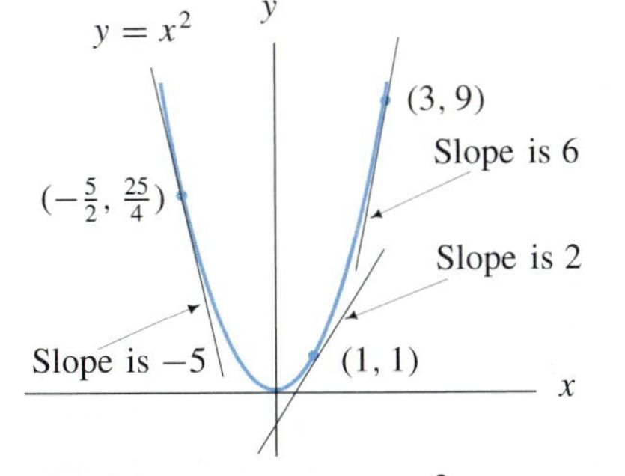

FIGURE 5 Graph of $y = x^2$

Notice that during the entire day of March 5, the graph in Figure 4 seems to be the steepest at 7 A.M. That is, the oil level is falling the fastest at that time. This corresponds to the fact that most people awake around 7 A.M., turn up their thermostats, take showers, and so on. Example 1 provides a typical illustration of the manner in which slopes can be interpreted as rates of change. We shall return to this idea in the next section.

If the point P is on the graph of $y = f(x)$ and the first coordinate of P is a, then the coordinates of P are $(a, f(a))$. The value of $f(a)$ gives the height of the point above (or below) the x-axis. The slope of the graph at P is denoted $f'(a)$. That is, $f'(a)$ is the slope of the tangent line at the point on the graph of $y = f(x)$ whose first coordinate is a. Figure 5 shows the slopes of the graph of $y = f(x)$ at several points, where $f(x) = x^2$. Referring to the graph, $f'(3) = 6$, $f'(1) = 2$, and $f'(-\frac{5}{2}) = -5$. *Note*: $f'(a)$ is also known as the *derivative of* $f(x)$ *at* $x = a$ and measures the steepness of the graph of $f(x)$ at $x = a$.

Definition: The number $f'(a)$, called *the derivative of* $f(x)$ *at* $x = a$, is the slope of the graph of $y = f(x)$ at the point where $x = a$.

Graphing utilities not only calculate values of functions, but also calculate values of derivatives and draw tangent lines at specified points. TI graphing calculators use the symbol dy/dx for the derivative. (This notation is discussed in Section 3.2.) Visual Calculus uses $f'(x)$ for the derivative.

SLOPES AND TANGENT LINES WITH A TI-82 OR TI-83

Suppose the function $f(x)$ has been assigned to Y1, Y1 is the only selected function, and an appropriate window has been specified. The following steps draw the tangent line through a point on the graph of $y = f(x)$ and give the slope of the tangent line:

1. Press [GRAPH] to display the graph of the function.
2. Press [2nd] [DRAW] **5** to select Tangent.
3. Move the cursor to any point on the graph. (On a TI-83, you also can just type in a value of x.)
4. Press [ENTER] to draw the tangent line through the point and to display the slope of the curve at that point. On a TI-82, the slope is displayed at the bottom of the screen as $dy/dx = slope$. See Figure 6(a). On a TI-83, the x-coordinate of the point and the equation of the tangent line are displayed. See Figure 6(b).
5. Repeat steps 2–4 to draw additional tangent lines. *Note 1*: To remove all tangent lines, press [2nd] [DRAW] **1**. *Note 2*: If Step 2 is replaced by [2nd] [CALC] **6**, then when the Enter key is pressed, $dy/dx = slope$ will be displayed at the bottom of the screen, but the tangent line will not be drawn.

For a discussion of obtaining slopes and tangent lines with Visual Calculus, see Appendix A, Part IV.

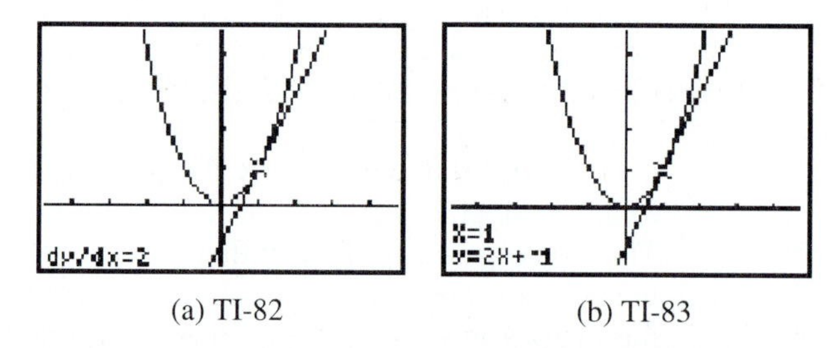

(a) TI-82 (b) TI-83

FIGURE 6 Tangent line to the graph of $y = x^2$ at $(1, 1)$

SLOPES AND TANGENT LINES WITH A TI-85

Suppose the function $f(x)$ has been assigned to y1, y1 is the only selected function, and an appropriate window has been specified. The following steps draw the tangent line through a point on the graph of $y = f(x)$ and give the slope of the tangent line:

1. Press [GRAPH] [F5] to display the graph of the function.
2. Press [MORE] [F1] [MORE] [MORE] [F3] to select TANLN from the MATH menu.
3. Move the cursor to any point on the graph.

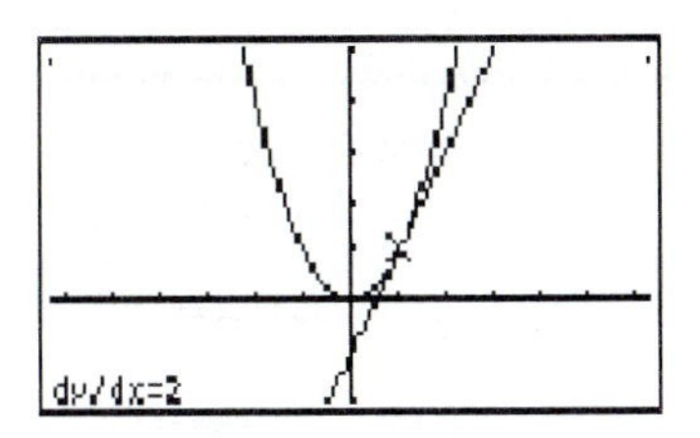

Tangent line to the graph of $y = x^2$ at $(1, 1)$ with a TI-85

4. Press [ENTER] to draw the tangent line through the point and to display the slope of the curve at that point. The slope is displayed at the bottom of the screen as $dy/dx = slope$. See the screen on the left.
5. To draw another tangent line, press [GRAPH] and then repeat steps 2–4.

   (*Note 1*: To remove all tangent lines, press [GRAPH] [MORE] [F2] [MORE] [F5]. *Note 2*: If step 2 is replaced by [MORE] [F1] [F4], then when the Enter key is pressed, $dy/dx = slope$ will be displayed at the bottom of the screen, but the tangent line will not be drawn.)

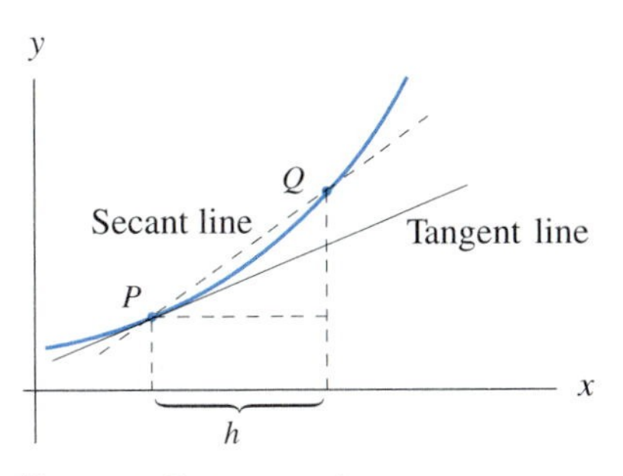

FIGURE 7 A secant line approximation to a tangent line

THE SECANT-LINE CALCULATION OF THE SLOPE AT A POINT Suppose $f(x)$ is a function such as $x^2 + 3x + 1$. A calculator or computer can determine a value such as $f(2)$ with the elementary arithmetic operations squaring, multiplication, and addition. But how is $f'(2)$ determined? We present one method here, the method used by the TI-82 and TI-83, and present another method later.

A fundamental idea for calculating the slope of the tangent line at a point P is to approximate the tangent line passing through P very closely by *secant lines*. A secant line at P is a straight line passing through P and a nearby point Q on the curve. (See Figure 7.) By taking Q very close to P, we can make the slope of the secant line approximately the slope of the tangent line with any desired degree of accuracy.

EXAMPLE 2 Consider the graph of $y = x^2$. Find the slopes of the secant lines through $P = (.5, .25)$ and the following choices of Q.

(a) $Q = (1, 1)$ (b) $Q = (.6, .36)$ (c) $Q = (.51, (.51)^2)$

Solution (a) See Figure 8.

$$[\text{slope of secant line}] = \frac{1 - .25}{1 - .5} = \frac{.75}{.5} = 1.5.$$

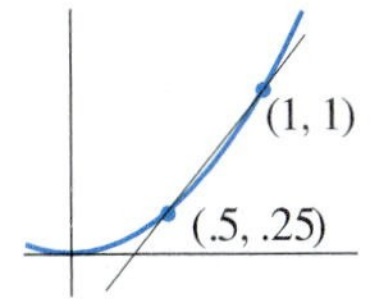

FIGURE 8

(b) See Figure 9 for a closer look at the curve near $(.5, .25)$.

$$[\text{slope of secant line}] = \frac{.36 - .25}{.6 - .5} = \frac{.11}{.1} = 1.1.$$

(c) $$[\text{slope of secant line}] = \frac{(.51)^2 - .25}{.51 - .5} = \frac{.2601 - .25}{.01} = 1.01.$$

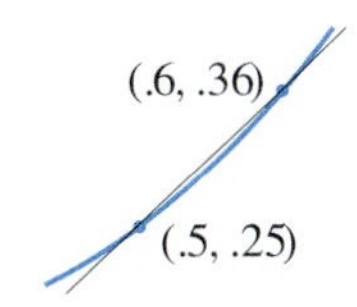

FIGURE 9

We can do much better than the calculations in Example 2! The next step is the key idea of this section. Imagine a point extremely close to $(.5, .25)$, whose x-coordinate is $.5 + h$, where h represents an extremely small number. For the point to be on the curve $y = x^2$, the y-coordinate must be $(.5 + h)^2$. We can calculate the slope of the secant line through $(.5, .25)$ and $(.5 + h, (.5 + h)^2)$ without specifying a particular value for h:

$$[\text{slope of secant line}] = \frac{(.5 + h)^2 - .25}{(.5 + h) - .5} = \frac{(.25 + h + h^2) - .25}{h}$$

$$= \frac{h + h^2}{h} = 1 + h.$$

As h gets smaller and smaller, this slope gets closer and closer to 1. So the slope of the tangent line must be exactly 1, because we can make the slope of the secant line as close to 1 as desired, simply by taking h sufficiently small.

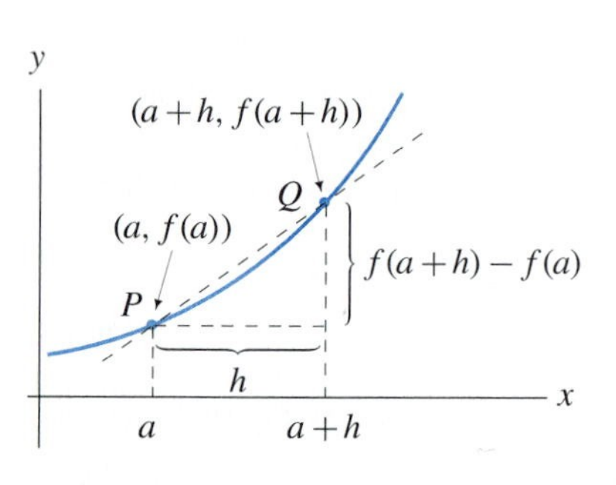

FIGURE 10 Computing the slope of a secant line

Here is the general idea. Consider a point $P = (a, f(a))$ on the graph of $y = f(x)$. Suppose a nearby point Q on the graph is h horizontal units away from P. Then Q has x-coordinate $a + h$ and y-coordinate $f(a + h)$. The slope of the secant line shown in Figure 10 is simply

$$[\text{slope of secant line}] = \frac{f(a+h) - f(a)}{(a+h) - a} = \frac{f(a+h) - f(a)}{h}.$$

In order to move Q closer to P along the curve, we let h get smaller and smaller, that is, we let h approach zero. Then the secant line approaches the tangent line, and so

$$[\text{slope of secant line}] \text{ approaches } [\text{slope of tangent line}],$$

that is,

$$\frac{f(a+h) - f(a)}{h} \quad \text{approaches } f'(a).$$

Since we can make the secant line as close to the tangent line as we wish by taking h sufficiently small, the quantity $[f(a+h) - f(a)]/h$ can be made to approximate $f'(a)$ to any desired degree of accuracy. Thus we arrive at the following method to compute the slope $f'(a)$:

> To calculate $f'(a)$, the slope of the graph of $y = f(x)$ at $x = a$:
>
> 1. First calculate $\dfrac{f(a+h) - f(a)}{h}$ for $h \neq 0$.
> 2. Then let h approach zero.
> 3. The quantity $\dfrac{f(a+h) - f(a)}{h}$ will approach $f'(a)$.

A graphing calculator can calculate

$$\frac{f(a+h) - f(a)}{h},$$

but cannot determine what happens as h approaches zero. Therefore, the calculator approximates $f'(a)$ by evaluating the quotient for a very small value of h and then rounding the answer. For the function $f(x) = x^2$, as often happens, the rounded value is exact. For some other functions, the rounded value will be a good approximation.

DERIVATIVES WITH THE TI-82 OR TI-83

The following steps calculate an approximation of the slope of $f(x)$ for any value of x. That is, they calculate $f'(a)$ for any a in the domain of $f(x)$.

1. Press [2nd] [QUIT] to invoke the Home screen.

2. Press [MATH] **8** to select `nDeriv(`.

3. Type in `f(X),X,a)`, where `f(X)` is the expression for the function and `a` is a number in the domain of $f(x)$.

4. Press [ENTER] to display the slope of the function $f(x)$ at the point $(a, f(a))$. [For instance, the instruction `nDeriv(X²,X,1)` produces the value `2`.]

Note: In step 3, the expression `f(X)` can be replaced by one of the functions in the list of functions. For instance, if X^2 has been assigned to Y_3, then the instruction `nDeriv(Y₃,X,1)` produces the value `2`.

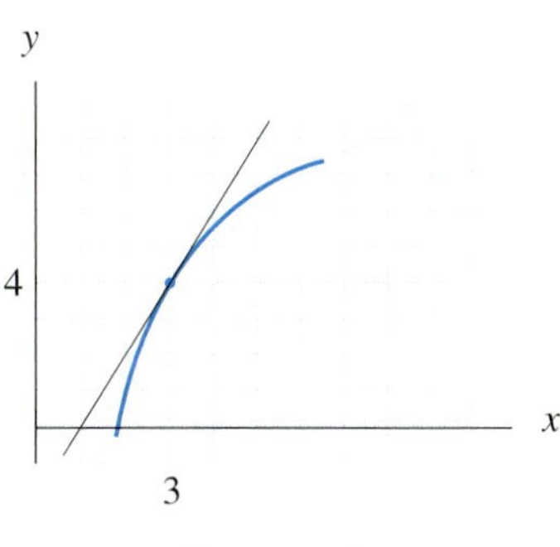

FIGURE 11

DERIVATIVES WITH A TI-85

The following steps calculate the slope of $f(x)$ for any value of x. That is, they calculate $f'(a)$ for any a in the domain of $f(x)$.

1. Press [2nd] [QUIT] to invoke the Home screen.

2. Press [2nd] [CALC] [F3] to select `der1`.

3. Type in `f(x),x,a)`, where `f(x)` is the expression for the function and `a` is a number in the domain of $f(x)$.

4. Press [ENTER] to display the slope of the function $f(x)$ at the point $(a, f(a))$. [For instance, the instruction `der1(x²,x,1)` produces the value `2`.]

Note: In step 3, the expression `f(x)` can be replaced by one of the functions in the list of functions. For instance, if x^2 has been assigned to `y3`, then the instruction `der1(y3,x,1)` produces the value `2`.

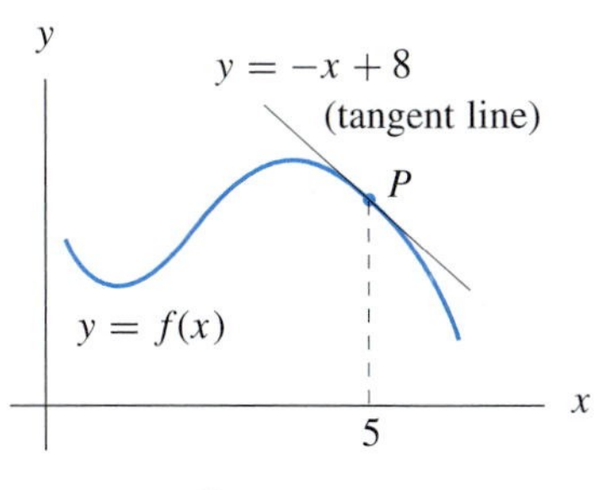

FIGURE 12

PRACTICE PROBLEMS 2.1

1. Refer to Figure 11.
   (a) What is the slope of the curve at $(3, 4)$?
   (b) What is the equation of the tangent line at the point where $x = 3$?
2. What is the equation of the tangent line to the graph of $y = \frac{1}{2}x + 1$ at the point $(4, 3)$?
3. Consider the curve $y = f(x)$ in Figure 12.
   (a) Find $f(5)$.
   (b) Find $f'(5)$.

EXERCISES 2.1

Trace the curves in Exercises 1–6 onto another piece of paper and sketch the tangent line in each case at the designated point P.

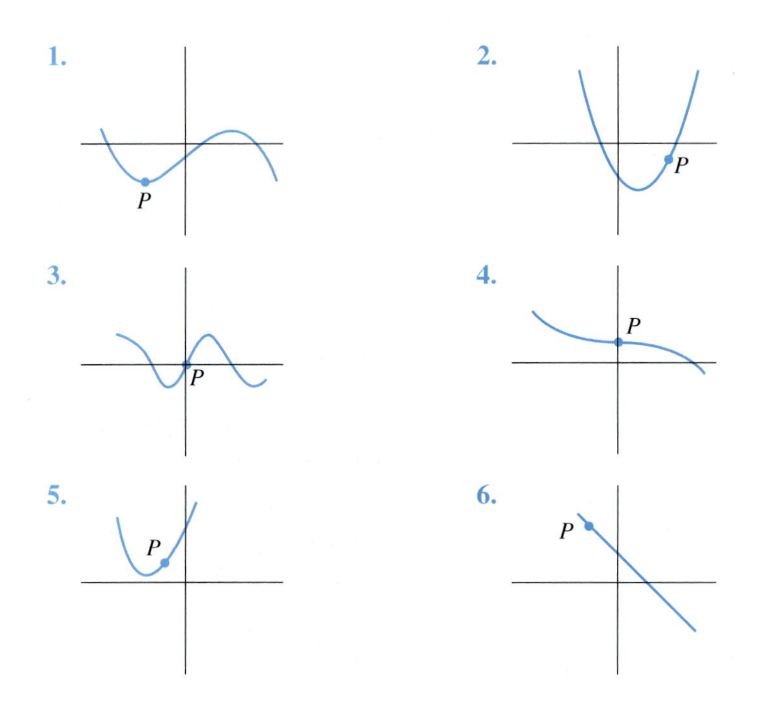

Estimate the slope of each of the following curves at the designated point P.

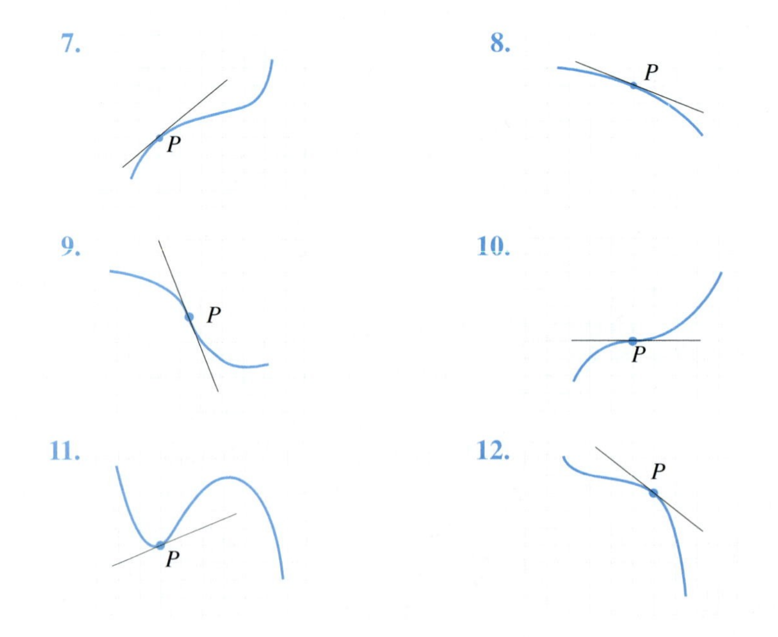

Exercises 13–18 refer to the points in Figure 13. Assign one of the following descriptors to each point: large positive slope, small positive slope, zero slope, small negative slope, large negative slope.

13. A 14. B 15. C

16. D 17. E 18. F

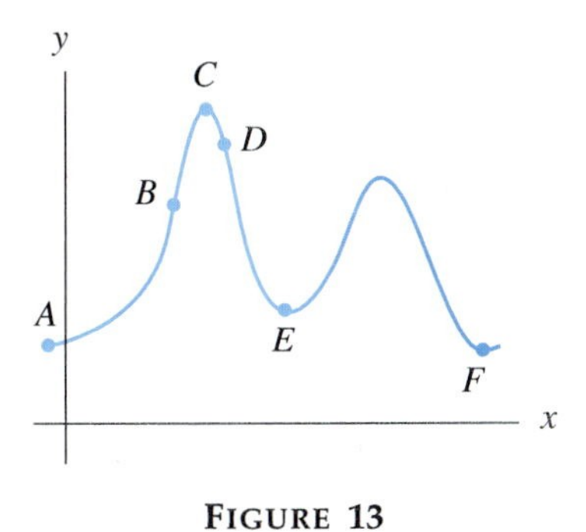

FIGURE 13

In Exercises 19–22, use a graphing utility to find the slope of the graph of the function at the given point and to draw the graph of the tangent line at that point. [Note: With a graphing calculator, select the ZDecimal (or ZDECM) window from the ZOOM menu so that your cursor will land on integer values of x. With Visual Calculus, this can be achieved by having the length of the x-interval be a power of 2.]

19. $f(x) = x^2 - 2x + 1$, $(2, 1)$

20. $f(x) = 2^x$, $(1, 2)$

21. $f(x) = \sqrt{x}$, $(4, 2)$

22. $f(x) = .5^x$, $(-1, 2)$

In Exercises 23–26, use a graphing utility to find the slope of the graph of the function at the point where $x = 2$. Then determine the equation of the tangent line at the point.

23. $f(x) = (x - 1)^3$

24. $f(x) = x - x^2$

25. $f(x) = (.25x - 1)^2$

26. $f(x) = .125x^4$

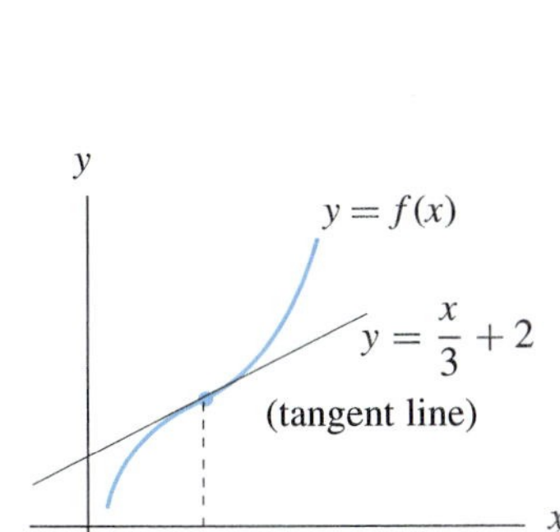

FIGURE 14

In Exercises 27–30, a function and the equation of its tangent line at the point $(1, 2)$ are given. Graph the function and its tangent line with the window

$$[.6, 1.4] \text{ by } [1.6, 2.4]$$

and then repeatedly enlarge with successive windows

$$[.8, 1.2] \text{ by } [1.8, 2.2], \quad [.9, 1.1] \text{ by } [1.9, 2.1], \quad [.95, 1.05] \text{ by } [1.95, 2.05], \ldots$$

until the two graphs are indistinguishable on your screen. Which window first produced the desired result?

27. $f(x) = \dfrac{4x}{1 + x^2}$, $y = 2$

28. $f(x) = 1 + \sqrt{x}$, $y = .5x + 1.5$

29. $f(x) = 1 + \dfrac{1}{x}$, $y = 3 - x$

30. $f(x) = 3 - x^3$, $y = -3x + 5$

31. Consider the curve $y = f(x)$ in Figure 14. Find $f(6)$ and $f'(6)$.

32. Consider the curve $y = f(x)$ in Figure 15. Find $f(1)$ and $f'(1)$.

33. The tangent line to the graph of $f(x)$ at the point where $x = -1$ has equation $y = 3x + 4$. Find $f(-1)$ and $f'(-1)$.

34. The tangent line to the graph of $f(x)$ at the point $(3, 4)$ passes through the point $(5, 10)$. Find $f(3)$ and $f'(3)$.

35. Find $f(4)$ and $f'(4)$, where $f(x) = 3x + 2$.

36. Find $f(4)$ and $f'(4)$, where $f(x) = -2$.

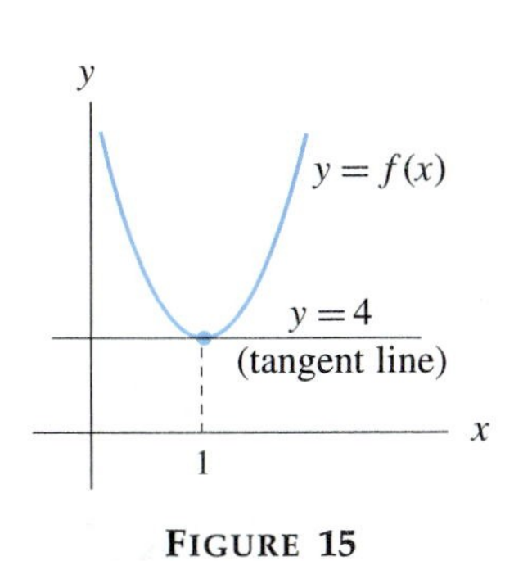

FIGURE 15

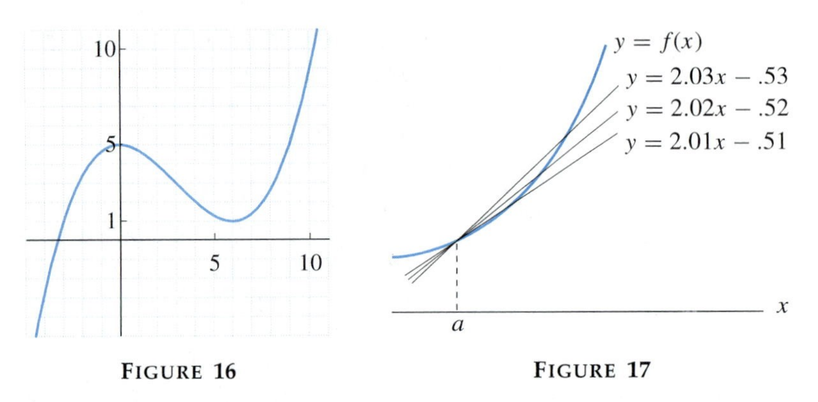

FIGURE 16

FIGURE 17

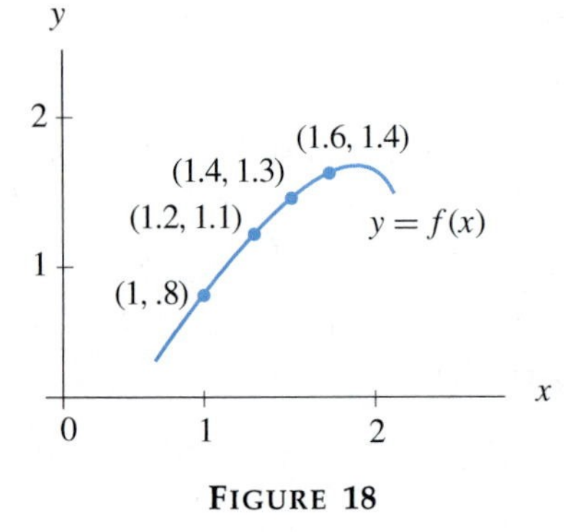

FIGURE 18

In Exercises 37–46, refer to the graph in Figure 16 and fill in the square with one of the relations $<$, $>$, *or* $=$.

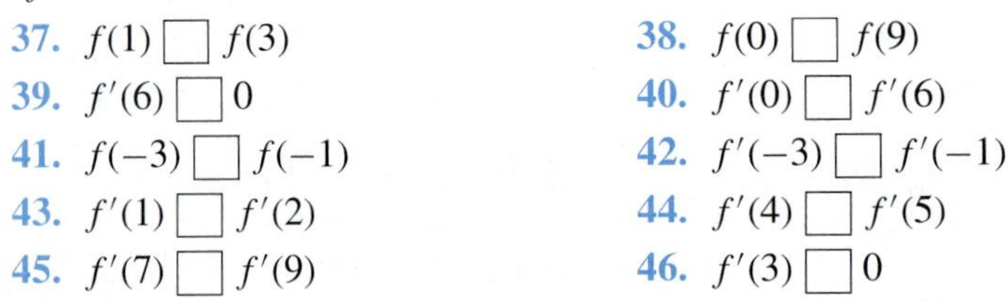

37. $f(1) \square f(3)$

38. $f(0) \square f(9)$

39. $f'(6) \square 0$

40. $f'(0) \square f'(6)$

41. $f(-3) \square f(-1)$

42. $f'(-3) \square f'(-1)$

43. $f'(1) \square f'(2)$

44. $f'(4) \square f'(5)$

45. $f'(7) \square f'(9)$

46. $f'(3) \square 0$

In Exercises 47–50, repeatedly zoom in on the graph of the function in the vicinity of the point $(3, 4)$ *until the curve looks nearly straight. Then read the coordinates of a point of the curve near the point* $(3, 4)$ *and use the two sets of coordinates to estimate the slope of the tangent line at* $x = 3$.

47. $f(x) = 2 + \sqrt{x+1}$

48. $f(x) = (x-1)^3 + x + 1$

49. $f(x) = \dfrac{40}{x^2+1}$

50. $f(x) = 2^{(x-1)}$

51. Consider the curve $y = f(x)$ in Figure 17. Find a and $f(a)$. Estimate $f'(a)$.

52. Consider the curve $y = f(x)$ in Figure 18. Estimate $f'(1)$.

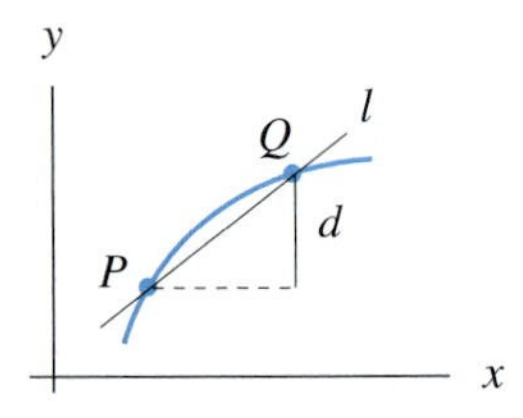

FIGURE 19

53. Let l be the line through the points P and Q in Figure 19.

(a) Suppose $P = (2, 4)$ and $Q = (5, 13)$. Find the slope of the line l and the length of the line segment d.

(b) As the point Q moves toward P, does the slope of the line l increase or decrease?

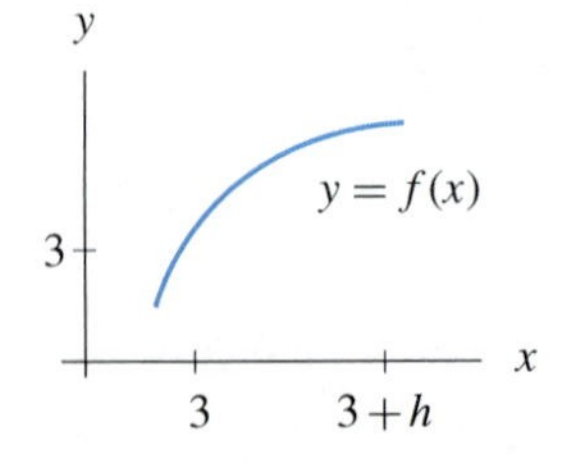

FIGURE 20

54. In Figure 20, h represents a positive number, and $3 + h$ is the number h units to the right of 3. Draw line segments on the graph having the following lengths.

(a) $f(3)$ (b) $f(3+h)$ (c) $f(3+h) - f(3)$ (d) h

(e) Draw a line of slope $\dfrac{f(3+h) - f(3)}{h}$.

55. Consider the function $f(x) = 5x^2 - 3$.

(a) Let $P = (1, 2)$ and $Q = (1.1, 3.05)$. Find the slope of the secant line through the points P and Q.

(b) Let $P = (1, 2)$ and $Q = (1+h, f(1+h))$, where $h = .01$. Find the slope of the secant line through the points P and Q.

(c) As h gets smaller and smaller, what value is approached by the slope of the secant line through the points $P = (1, 2)$ and $Q = (1+h, f(1+h))$?

SOLUTIONS TO PRACTICE PROBLEMS 2.1

1. (a) The slope of the curve at the point $(3, 4)$ is, by definition, the slope of the tangent line at $(3, 4)$. Note that the point $(4, 6)$ is also on the line. Therefore, the slope is

$$\frac{6-4}{4-3} = \frac{2}{1} = 2.$$

(b) Use the point-slope formula. The equation of the line passing through the point $(3, 4)$ and having slope 2 is

$$y - 4 = 2(x - 3)$$

or

$$y = 2x - 2.$$

2. The tangent line at $(4, 3)$ is, by definition, the line that best approximates the curve at $(4, 3)$. Since the "curve" in this case is itself a line, the curve and its tangent line at $(4, 3)$ (and at every other point) must be the same. Therefore, the equation is $y = \frac{1}{2}x + 1$.

3. (a) The number $f(5)$ is the y-coordinate of the point P. Since the tangent line passes through P, the coordinates of P satisfy the equation $y = -x + 8$. Since its x-coordinate is 5, its y-coordinate is $-5 + 8 = 3$. Therefore, $f(5) = 3$.

(b) The number $f'(5)$ is the slope of the tangent line at P, which is readily seen to be -1.

2.2 THE DERIVATIVE AS A RATE OF CHANGE

An important interpretation of the slope of a function at a point is as a rate of change. In this section, we examine this interpretation and discuss some of the applications in which it proves useful. The first step is to understand what is meant by "average rate of change" of a function $f(x)$. As we saw in the last section, the difference quotient

$$\frac{f(a+h) - f(a)}{h}$$

approaches $f'(a)$ as h approaches 0. We call this quotient (or ratio) *the average rate of change of $f(x)$ with respect to x over the interval from a to $a+h$*. Geometrically, this quotient is the slope of the secant line in Figure 1.

From the secant-line calculation of the derivative, we know that as h approaches 0, the slope of the secant line approaches $f'(a)$. Thus the average rate of change over the interval from a to $a + h$ approaches $f'(a)$ as h approaches 0. For this reason, we may interpret $f'(a)$ as the (instantaneous) rate of change of $y = f(x)$ exactly at the point where $x = a$.

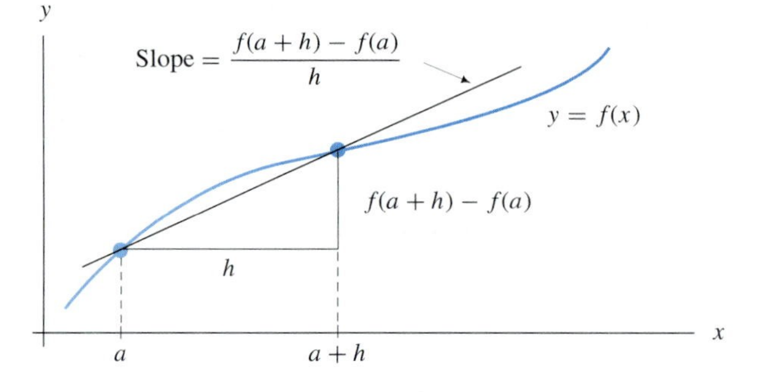

FIGURE 1 Average rate of change is the slope of the secant line

The derivative $f'(a)$ measures the rate of change of $f(x)$ at $x = a$.

In applications where a quantity is a function of time, t is a more convenient variable name than x. For instance, a function such as

$$h(t) = -16t^2 + 90t$$

gives the height (in feet) after t seconds of a ball thrown straight up into the air with an initial speed of 90 feet per second. The graph of $y = h(t)$ would be shown on a ty-coordinate system. (*Note*: With a graphing utility, use x instead of t to draw the graph or calculate the slope of the function at a point. That is, enter the function as $-16x^2 + 90x$.)

As a physical illustration of rate of change, consider the weight of a bear cub as a function of time, say $W = f(t)$. Suppose the bear's weight follows the growth pattern in Figure 2, and examine the time interval from $t = 3$ to $t = 7$ months. Then

$$h = 7 - 3 = 4 \text{ months.}$$

According to the graph, the weight changed during this time period from 9 to 33 pounds, so

$$f(7) - f(3) = 33 - 9 = 24 \text{ pounds.}$$

A growth of 24 pounds in 4 months amounts to an average gain of 6 pounds per month. That is, with $a = 3$ and $h = 4$,

$$\frac{f(a+h) - f(a)}{h} = \frac{f(3+4) - f(3)}{4} = \frac{f(7) - f(3)}{4} = \frac{24}{4} = 6.$$

This quotient is the slope of the secant line in Figure 2 through the points $(3, 9)$ and $(7, 33)$.

To determine the (instantaneous) rate of change of weight at 3 months, consider the bear cub's weight over a small time interval from $t = 3$ to $t = 3 + h$. As the length of this interval approaches zero, the average rate of weight change, $(f(3+h) - f(3))/h$, approaches $f'(3)$, the derivative of the weight function. Thus it is appropriate to call $f'(3)$ the rate of change of weight at $t = 3$.

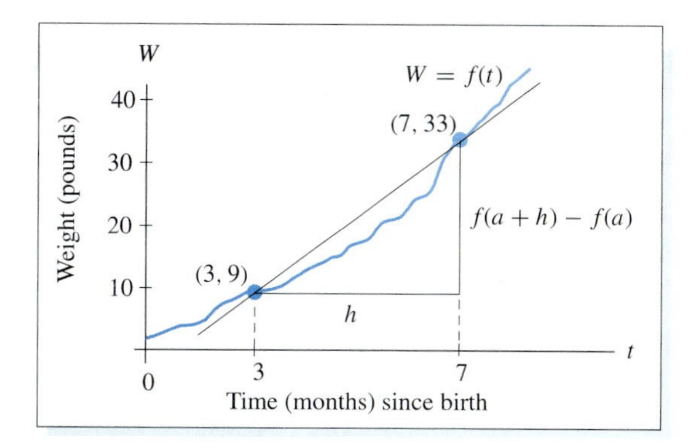

FIGURE 2 Weight gain of a bear

EXAMPLE 1 Suppose that $f(x) = x^2$.

(a) Calculate the average rate of change of $f(x)$ over the intervals 1 to 2, 1 to 1.1, and 1 to 1.01.

(b) Determine the (instantaneous) rate of change of $f(x)$ when $x = 1$.

Solution (a) The intervals are of the form 1 to $1 + h$ for $h = 1, .1$, and $.01$. The average rate of change is given by the ratio

$$\frac{f(1+h) - f(1)}{h} = \frac{(1+h)^2 - 1^2}{h}.$$

For the three given values of h, this expression has the following respective values:

$$h = 1: \quad \frac{(1+h)^2 - 1^2}{h} = \frac{2^2 - 1^2}{1} = \frac{4-1}{1} = 3;$$

$$h = .1: \quad \frac{(1+h)^2 - 1^2}{h} = \frac{(1.1)^2 - 1^2}{.1} = \frac{1.21 - 1}{.1} = \frac{.21}{.1} = 2.1;$$

$$h = .01: \quad \frac{(1+h)^2 - 1^2}{h} = \frac{(1.01)^2 - 1^2}{.01} = \frac{1.0201 - 1}{.01} = \frac{.0201}{.01} = 2.01.$$

Thus, the average rate of change of $f(x)$ over the interval 1 to 2 is 3 units per unit change in x. The average rate of change of $f(x)$ over the interval 1 to 1.1 is 2.1 units per unit change in x. The average rate of change of $f(x)$ over the interval 1 to 1.01 is 2.01 units per unit change in x.

(b) The instantaneous rate of change of $f(x)$ at $x = 1$ is equal to $f'(1)$. From Figure 5 of Section 2.1 we have $f'(1) = 2$. (This result also can be obtained with a graphing utility.) That is, the instantaneous rate of change is 2 units per unit change in x. Note how the average rates of change approach the instantaneous rate of changes as the intervals beginning at $x = 1$ shrink. ■

Convention: From now on, unless we explicitly use the word "average" when we refer to the "rate of change" of a function, we mean the "instantaneous" rate of change.

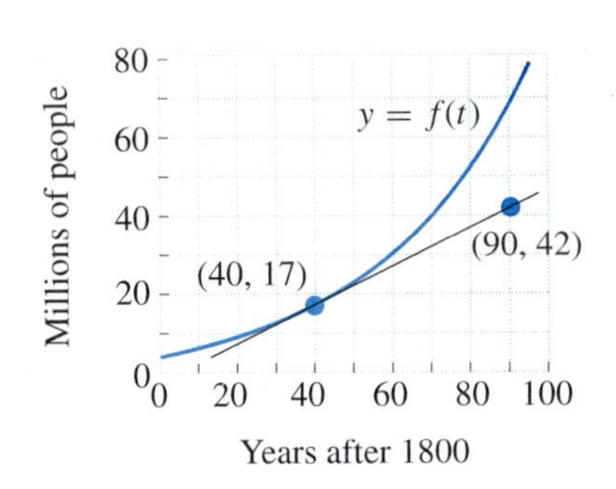

FIGURE 3 United States population from 1800 to 1900

EXAMPLE 2 The function $f(t)$ in Figure 3 gives the population of the United States (in millions) t years after 1800. The figure also shows the tangent line through the point (40, 17).

(a) What was the United States population in 1840?

(b) In what year was the population 40,000,000?

(c) How fast was the population growing in 1840?

(d) Was the population growing faster in 1810 or 1880?

Solution (a) Since 1840 is 40 years after 1800, the population in 1840 is $f(40)$. Since the point with coordinates (40, 17) is on the graph, $f(40) = 17$. That is, the population in 1840 was 17 million people.

(b) We must find the value of t for which $f(t) = 40$. The horizontal line $y = 40$ crosses the graph at the point (70, 40). Therefore, the population was 40 million people in the year 1870.

(c) The rate of growth of $f(t)$ at $t = 40$ is $f'(40)$; that is, the slope of the tangent line at $t = 40$. Since (40, 17) and (90, 42) are two points on the tangent line, the slope of the tangent line is

$$\frac{42 - 17}{90 - 40} = \frac{25}{50} = .5.$$

Therefore, in 1840 the population was growing at the rate of .5 million or 500,000 people per year.

(d) The graph is clearly steeper in 1880 than in 1810. Therefore, the population was growing faster in 1880 than in 1810. ■

EXAMPLE 3 Let $f(x)$ be the number of gallons of gas used by a car after it has been driven x miles.

(a) If the car gets 25 miles per gallon, what is $f(50)$?

(b) Is $f'(50)$ positive or negative?

(c) Would you expect $f'(50)$ to be greater for a subcompact car or a limousine?

Solution (a) 2. Two gallons of gas would be used after the car was driven 50 miles.

(b) $f'(2)$ is the rate at which gas is being used at a certain time. Since cars use more and more gas as they are driven, $f'(2)$ must be positive.

(c) Limousine, since gas guzzlers use gas at a greater rate. ■

FIGURE 4 A flask

EXAMPLE 4 Consider the flask in Figure 4. Suppose water is poured into the flask at a steady rate until it reaches the neck of the flask. Which of the graphs in Figure 5 most accurately describes $h(t)$, the height of water in the flask after t seconds?

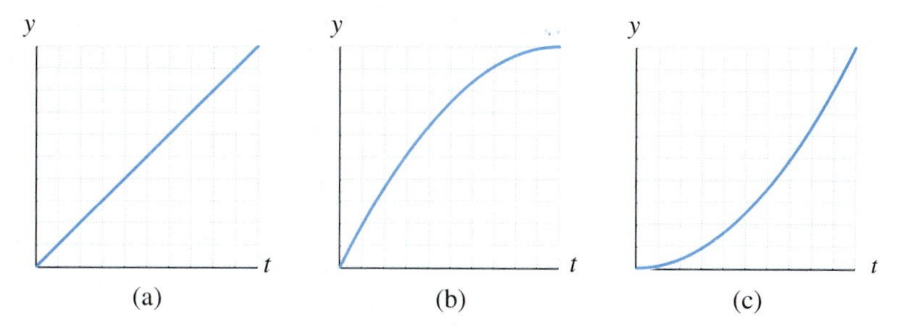

FIGURE 5 Water level as a function of time

Solution Due to the shape of the flask, the water level rises slowly at first and then faster and faster. Thus the function $h(t)$ increases at a slow rate at first and then increases at greater and greater rates as time progresses. That is, the slope must get greater and greater as time progresses. Therefore, Figure 5(c) most accurately describes $h(t)$. ■

APPROXIMATING THE CHANGE IN A FUNCTION Consider the function $f(x)$ for x near a. As we have just seen, we have the approximation

$$\frac{f(a+h)-f(a)}{h} \approx f'(a).$$

Multiplying both sides of this approximation by h, we have

$$f(a+h)-f(a) \approx f'(a)\cdot h. \qquad (1)$$

If x changes from a to $a+h$, then the change in the function is approximately $f'(a)$ times the change h in the value of x.

Figure 6 contains a geometric interpretation of (1). Given the small change h in x, the quantity $f'(a)\cdot h$ gives the corresponding change in y along the tangent line at $(a, f(a))$. In contrast, the quantity $f(a+h)-f(a)$ gives the change in y along the curve $y = f(x)$.

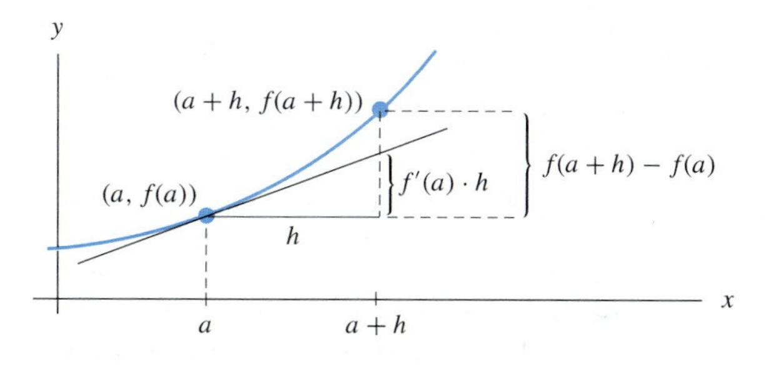

FIGURE 6 Change in y along the tangent line and along the graph of $y = f(x)$

EXAMPLE 5 Let the production function $p(x)$ give the number of units of goods produced when employing x units of labor. Suppose 5000 units of labor are currently employed, $p(5000) = 300$, and $p'(5000) = 2$.

(a) Interpret $p(5000) = 300$.

(b) Interpret $p'(5000) = 2$.

(c) Estimate the number of additional units of goods produced when x is increased from 5000 to $5000\frac{1}{2}$ units of labor.

(d) Estimate the change in the number of units of goods produced when x is decreased from 5000 to 4999 units of labor.

Solution (a) When 5000 units of labor are employed, 300 units of goods will be produced.

(b) If 5000 units of labor are currently employed and we consider adding more labor, productivity will increase at approximately the rate of 2 units of goods for each additional unit of labor.

(c) Here $h = \frac{1}{2}$. By (1), the change in $p(x)$ will be approximately

$$p'(5000) \cdot \frac{1}{2} = 2 \cdot \frac{1}{2} = 1.$$

About one additional unit will be produced.

(d) Here $h = -1$, since the amount of labor is reduced. The change in $p(x)$ will be approximately

$$p'(5000) \cdot (-1) = 2 \cdot (-1) = -2.$$

About two fewer units of goods will be produced. ■

THE MARGINAL CONCEPT IN ECONOMICS In economics, derivatives are often described by the adjective "marginal." For instance, if $C(x)$ is a cost function (the cost of producing x units of a commodity), then the value of the derivative $C'(a)$ is called the *marginal cost* at the production level of a units. Since the marginal cost is just a derivative, its value gives the rate at which costs are increasing with respect to the level of production, assuming that production is at level a. If we apply (1) to the cost function $C(x)$ and take h to be 1 unit, we have

$$C(a + 1) - C(a) \approx C'(a) \cdot 1 = C'(a). \tag{2}$$

The quantity $C(a + 1) - C(a)$ is the amount the cost rises when the production level is increased from a units to $a + 1$ units. See Figure 7. Economists interpret (2) by saying that the marginal cost is approximately the cost of producing one additional unit.

The derivative $P'(a)$ associated with a profit function $P(x)$ is called the *marginal profit* at the production level of a units. The value of a derivative of a revenue function is called a marginal revenue, and so on.

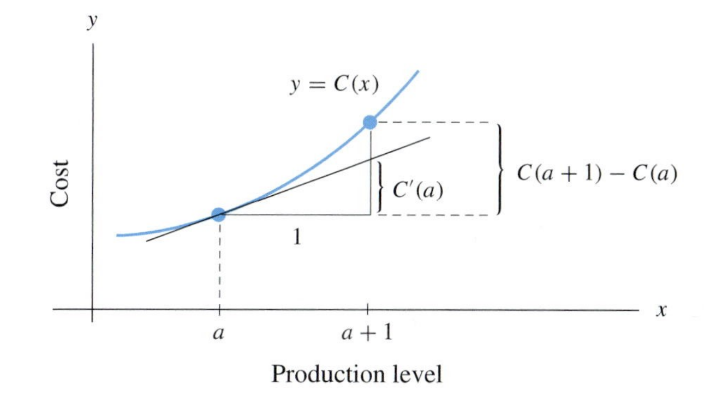

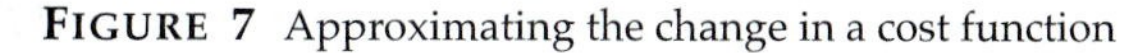

FIGURE 7 Approximating the change in a cost function

EXAMPLE 6 Suppose that the (daily) cost function is $C(x) = .005x^3 - 3x$ and production is proceeding at 1000 units per day.

(a) What is the extra cost of increasing production from 1000 to 1001 units per day?

(b) What is the marginal cost when $x = 1000$?

Solution (a) The change in cost when production is raised from 1000 to 1001 units per day is $C(1001) - C(1000)$, which equals (rounded to the nearest dollar)

$$\left[.005(1001)^3 - 3(1001)\right] - \left[.005(1000)^3 - 3(1000)\right]$$
$$= 5{,}012{,}012 - 4{,}997{,}000 = \$15{,}012.$$

```
Y1(1001)-Y1(1000
)
          15012.005
nDeriv(Y1,X,1000
)
              14997
```

(b) The marginal cost at production level 1000 is $C'(1000)$. A graphing utility gives the value 14,997. [*Note*: With a TI graphing calculator, set

```
Y1 = .005x^3-3x
```

and then evaluate `nDeriv(Y1,X,1000)` or `der1(y1,x,1000)`. With Visual Calculus, use the routine "1:Compute Values of a function of one variable."] Notice that 14,997 is close to the actual cost in (a) of increasing production by one unit. ■

VELOCITY An everyday illustration of rate of change is given by the velocity of a moving object. Suppose that we are driving a car along a straight road and at each time t we let $s(t)$ be our position on the road, measured from some convenient reference point. See Figure 8, where distances are positive to the right of the car. For the moment we shall assume that we are proceeding in only the positive direction along the road.

At any instant, the car's speedometer tells us how fast we are moving—that is, how fast our position $s(t)$ is changing. To show how the speedometer reading is related to our calculus concepts of a derivative, let us examine what is happening

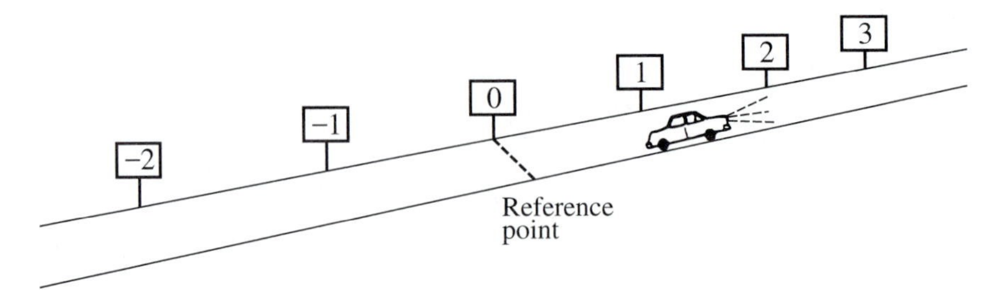

FIGURE 8 Position of a car traveling on a straight road

at a specific time, say $t = 1$. Consider a short time interval of duration h from $t = 1$ to $t = 1 + h$. Our car will move from position $s(1)$ to position $s(1 + h)$, a distance of $s(1 + h) - s(1)$. Thus the *average velocity* from $t = 1$ to $t = 1 + h$ is

$$\frac{[\text{distance traveled}]}{[\text{time elapsed}]} = \frac{s(1+h) - s(1)}{h}. \tag{3}$$

If the car is traveling at a steady speed during this time period, then the speedometer reading will equal the average velocity in (3).

From our discussion in Section 2.1 the ratio in (3) approaches the derivative $s'(1)$ as h approaches zero. For this reason we call $s'(1)$ the (instantaneous) velocity at $t = 1$. This number will agree with the speedometer reading at $t = 1$ because when h is very small, the car's speed will be nearly steady over the time interval from $t = 1$ to $t = 1+h$, and so the average velocity over this time interval will be nearly the same as the speedometer reading at $t = 1$.

The reasoning used for $t = 1$ holds for an arbitrary t as well. Thus the following definition makes sense.

If $s(t)$ denotes the position function of an object moving in a straight line, then the velocity $v(t)$ of the object at time t is given by

$$v(t) = s'(t).$$

In our discussion we assumed that the car moved in the positive direction. If the car moves in the opposite direction, the ratio (3) and the limiting value $s'(1)$ will be negative. So we interpret negative velocity as movement in the negative direction along the road.

EXAMPLE 7 When a ball is thrown straight up into the air, its position may be measured as the vertical distance from the ground rather than the distance from some reference point. Regard "up" as the positive direction, and let $s(t)$ be the height of the ball in feet after t seconds. Suppose that $s(t) = -16t^2 + 96t + 112$. Figure 9 contains a graph of the function $s(t)$ and a table giving various values of the function and its derivative. (*Note*: The graph does *not* show the path of the ball. The ball is rising *straight* up and then falling *straight* down.)

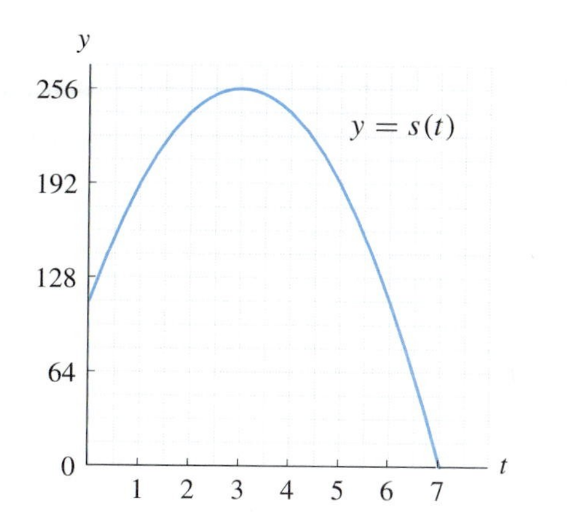

t	$s(t)$	$s'(t)$
0	112	96
0.5	156	80
1	192	64
1.5	220	48
2	240	32
2.5	252	16
3	256	0
3.5	252	−16
4	240	−32
4.5	220	−48
5	192	−64
5.5	156	−80
6	112	−96
6.5	60	−112
7	0	−128

FIGURE 9 Height of a ball as a function of time

(a) What is the height of the ball after 2 seconds?

(b) What is the velocity of the ball after 2 seconds?

(c) How high does the ball go and when does the ball reach its greatest height?

(d) When is the ball falling at the speed of 80 feet per second?

Solution (a) The answer can be read from the graph, read from the table, or calculated with the formula. From the graph, locate 2 on the t-axis and go up to the graph. The point on the graph corresponding to $t = 2$ is $(2, 240)$. Therefore, after 2 seconds the height of the ball is 240 feet. To answer the same question with the table, locate 2 in the t column and move to the right to read off $s(2)$ from the $s(t)$ column. To calculate $s(2)$ with the formula, substitute 2 for t:

$$s(2) = -16(2)^2 + 96(2) + 112 = 240.$$

(b) The velocity of the ball after 2 seconds is the slope of the tangent line at the point $(2, 240)$; that is, $s'(2)$. This value can be obtained from the table. After 2 seconds the ball is traveling at a velocity of 32 feet per second.

(c) The highest point on the graph of $s(t)$ is $(3, 256)$. Therefore, the greatest height, 256 feet, is attained after 3 seconds. (Notice that the velocity is 0 exactly at time $t = 3$.)

(d) The ball rises for the first 3 seconds, stops for an instant, and then falls. That is, values of $s'(t)$ are positive for $t < 3$ and negative for $t > 3$. The ball is falling at the speed of 80 feet per second when the velocity is -80. From the table, this velocity occurs when $t = 5.5$. ■

UNITS FOR RATE OF CHANGE Table 1 shows the units appearing in several examples from this section. In general,

$$[\text{units of measure for } f'(x)] = [\text{units of measure for } f(x)] \text{ per } [\text{unit of measure for } x].$$

Table 1 UNITS OF MEASURE

Example	Units for $f(x)$ or $f(t)$	Unit for x or t	Units for $f'(x)$ or $f'(t)$
Bear cub	pounds	month	pounds per month
Ball in air	feet	second	feet per second
U.S. population	millions of people	year	millions of people per year
Gasoline used	gallons	mile	gallons per mile

PRACTICE PROBLEMS 2.2

Let $f(t)$ be the temperature (in degrees Celsius) of a liquid at time t (in hours). The rate of temperature change at time a has the value $f'(a)$. Listed next are typical questions about $f(t)$ and its slope at various points. Match each question with the proper method of solution.

Questions:

1. What is the temperature of the liquid after 6 hours?
2. When is the temperature rising at the rate of 6 degrees per hour?
3. By how many degrees did the temperature rise during the first 6 hours?
4. When is the liquid's temperature only 6 degrees?
5. How fast is the temperature of the liquid changing after 6 hours?
6. What is the average rate of increase in the temperature during the first 6 hours?

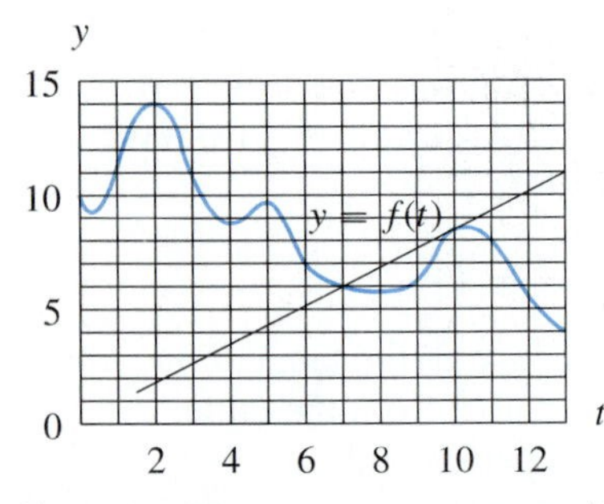

FIGURE 10 Interest rates on 3-month treasury bills (1979–1992)

Methods of Solution:

(a) Compute $f(6)$.

(b) Set $f(t) = 6$ and solve for t.

(c) Compute $[f(6) - f(0)]/6$.

(d) Compute $f'(6)$.

(e) Find a value of a for which $f'(a) = 6$.

(f) Compute $f(6) - f(0)$.

EXERCISES 2.2

1. Refer to Figure 10, where $f(t)$ is the interest rate (as a percent) on a 3-month treasury bill t years after January 1, 1979.
   (a) What was the interest rate on January 1, 1981?
   (b) How fast were interest rates rising on January 1, 1989?
   (c) Were interest rates falling faster on January 1, 1982, or January 1, 1985?
   (d) Were interest rates rising faster on January 1, 1980, or January 1, 1989?
2. Refer to Figure 11, where $f(t)$ is the average size of a farm (in thousands of acres) t years after January 1, 1900.
   (a) What was the average size of a farm on January 1, 1920?

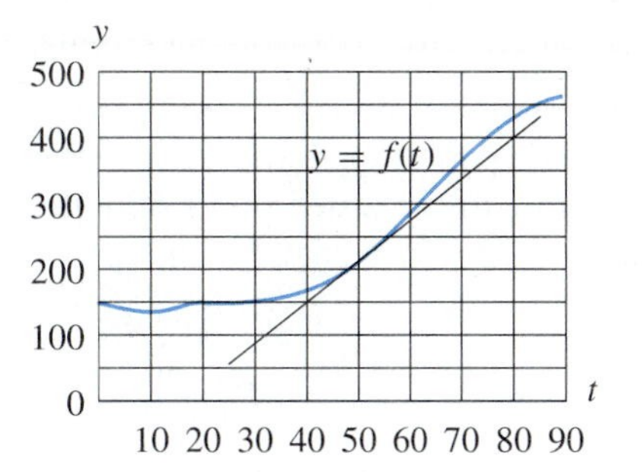

FIGURE 11 Average size of farms from 1900 to 1990

(b) How fast was the average farm size rising on January 1, 1950?

(c) Was the average farm size rising faster on January 1, 1960, or January 1, 1980?

(d) What was the average rate of change of $f(t)$ during the years from 1920 to 1960?

In Exercises 3–6, explain your reasoning in complete sentences.

3. Let $f(t)$ be the number of hours of daylight t days after the beginning of the year. Is $f'(25)$ positive or negative? Is $f'(250)$ positive or negative?

4. Let $f(x)$ be the average cost of an x carat diamond. Is $f'(1)$ positive or negative?

5. Let $f(x)$ be the number of people who purchase a certain car when the price is x dollars. Is $f'(12{,}000)$ positive or negative?

6. Let $f(t)$ be the temperature on a typical sunny day, t hours after the start of the day. (For instance, $t = 12$ corresponds to noon.) Is $f'(8)$ positive or negative? Is $f'(20)$ positive or negative?

7. Suppose a house was purchased on January 1, 1990. Let $f(t)$ be the value (in dollars) of the house after t years. Interpret $f(6) = 185{,}000$ and $f'(6) = 3000$. Estimate the value of the house on July 1, 1997.

8. Let $f(t)$ be the purchasing power of the dollar (in terms of 1990 dollars) t years after January 1, 1990. Interpret $f(6) = .79$ and $f'(6) = -.04$. Estimate the value of the dollar on September 1, 1996.

9. Let $f(t)$ be the number of farms (in millions) in the United States t years after January 1, 1900. Interpret $f(53) = 5.4$ and $f'(53) = -.1$. Estimate the number of farms on January 1, 1954.

10. A flu epidemic has struck a college campus. Let $f(t)$ be the total number of students who have contracted the flu after t days. Interpret $f(12) = 1500$ and $f'(12) = 100$. Estimate $f(13)$.

11. Let $f(t)$ be the temperature of a cup of coffee, t minutes after it has been poured. Interpret $f(4) = 120$ and $f'(4) = -5$. Estimate the temperature of the coffee after 4 minutes and 6 seconds, that is, after 4.1 minutes.

12. Suppose 5 mg of a drug is injected into the bloodstream. Let $f(t)$ be the amount present in the bloodstream after t hours. Interpret $f(3) = 2$ and $f'(3) = -.5$. Estimate the number of milligrams of the drug in the bloodstream after $3\frac{1}{2}$ hours.

13. Let $C(x)$ be the cost of manufacturing x bicycles per day in a certain factory. Interpret $C(50) = 5000$ and $C'(50) = 45$. Estimate the cost of manufacturing 52 bicycles per day.

14. Let $P(x)$ be the profit from selling x units of goods. Interpret $P(1000) = 23{,}500$ and $P'(1000) = 200$. Estimate the profit from selling 1003 units of goods.

15. Let $R(x)$ be the revenue from selling x units of goods. Interpret $R(200) = 30{,}000$ and $R'(200) = 120$. Estimate the revenue from selling 199 units of goods.

16. Let $f(t)$ be the number of Japanese yen one dollar could buy t years after January 1, 1990. Interpret $f(4) = 110$ and $f'(4) = -10$. Estimate the number of yen one dollar could buy on January 1, 1995.

17. Water is heating in a kettle. During a 5-minute period the temperature rises from 20 °C to 60 °C. Let $f(t)$ be the temperature of the water after t minutes. Discuss the three graphs of $y = f(t)$ in Figure 12 with respect to changes in the rate of heating as time progresses.

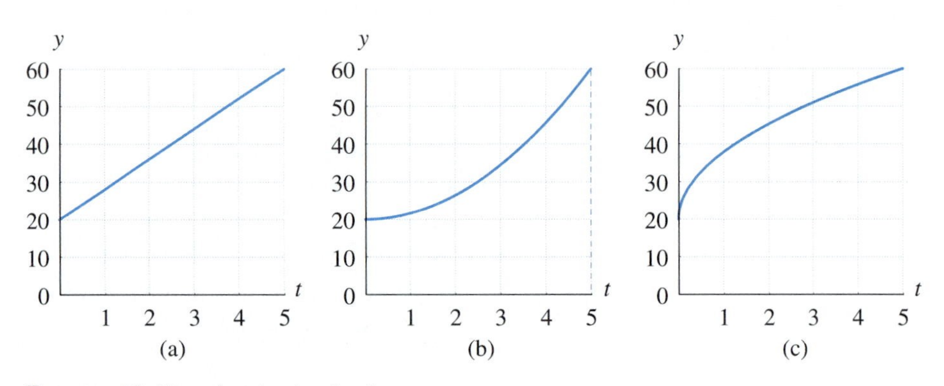

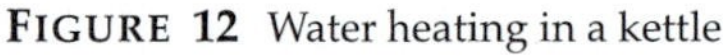
FIGURE 12 Water heating in a kettle

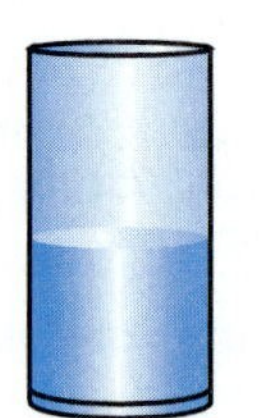
FIGURE 13 Cylindrical glass

FIGURE 14 Hemispherical glass

18. When a kettle of water is heated, the energy required to raise the temperature one degree is proportional to the current temperature. For instance, more energy is required to raise the temperature from 200 degrees to 201 degrees than from 100 degrees to 101 degrees. Which of the graphs in Figure 12 best represents the temperature of water in a boiling kettle after t minutes?

19. Consider the cylindrical glass in Figure 13. Suppose water is poured into the glass at a steady rate. Which of the graphs in Figure 5 from Example 4 most accurately describes $h(t)$, the height of water in the glass after t seconds?

20. Consider the hemispherical glass in Figure 14. Suppose water is poured into the glass at a steady rate. Which of the graphs in Figure 5 from Example 4 most accurately describes $h(t)$, the height of water in the glass after t seconds?

Exercises 21–26 refer to the ball in Example 7.

21. How fast is the ball rising after 2.5 seconds?

22. When is the ball rising at the speed of 48 feet per second?

23. When does the ball hit the ground?

24. At what time during the descent does the ball have the same height from which it was thrown?

25. When is the ball falling at the speed of 16 feet per second?

26. What is the velocity of the ball after 6.5 seconds?

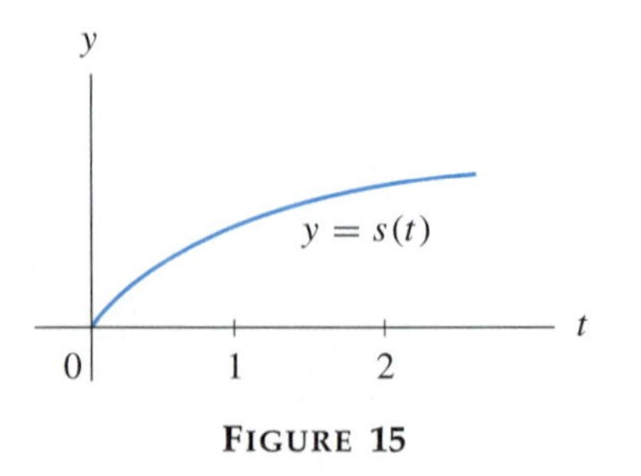

FIGURE 15

27. The function $s(t)$ in Figure 15 gives the distance traveled by a car after t hours. Is the car going faster at time $t = 1$ or $t = 2$?

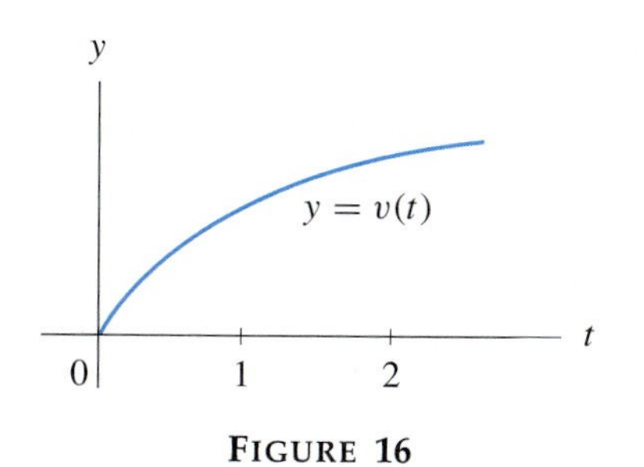

FIGURE 16

28. The function $v(t)$ in Figure 16 gives the velocity of a car after t hours. Is the car going faster at time $t = 1$ or $t = 2$?

Exercises 29–32 refer to Figure 17, where $s(t)$ is the number of feet traveled by a person after t seconds of walking along a straight path.

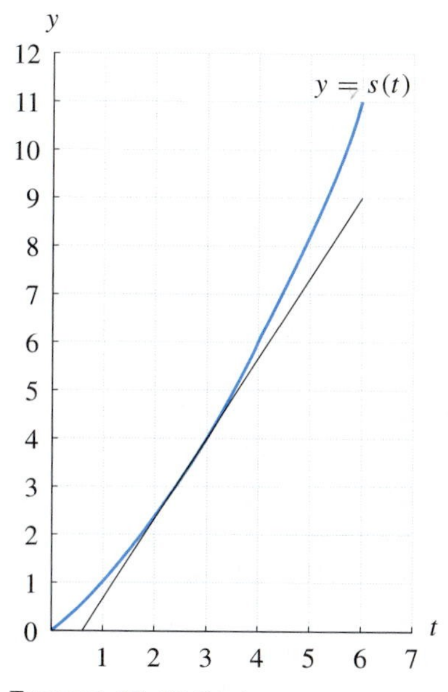

FIGURE 17 Walker's progress

29. How far has the person traveled after 6 seconds?

30. What is the person's average velocity from time $t = 1$ to $t = 4$?

31. What is the person's velocity at time $t = 3$?

32. Without evaluating velocities, determine whether the person is traveling faster at time $t = 5$ or at time $t = 6$.

33. Table 2 gives a car's trip-meter reading (in miles) at 1 hour into a trip and at several nearby times. What is the average speed during the time interval from 1 to 1.05 hours? Estimate the speed at time 1 hour into the trip.

Table 2 TRIP-METER READINGS AT SEVERAL TIMES

Time	.96	.97	.98	.99	1	1.01	1.02	1.03	1.04	1.05
Trip meter	43.2	43.7	44.2	44.6	45	45.4	45.8	46.3	46.8	47.4

34. A car is traveling from New York to Boston and is halfway between the two cities. Let $s(t)$ be the distance from New York during the next minute. Match each behavior with the corresponding graph of $s(t)$ in Figure 18.

(a) The car travels at a steady speed.
(b) The car is stopped.
(c) The car is backing up.
(d) The car is accelerating.
(e) The car is decelerating.

In Exercises 35–38, decide whether the statement is true or false. [*Refer to approximation (1).*]

35. If $x = a$ and $f'(a)$ is large and positive, then a small increase in x causes a large increase in the function value.

36. If $x = a$ and $f'(a)$ is small and positive, then a small increase in x causes a very small increase in the function value.

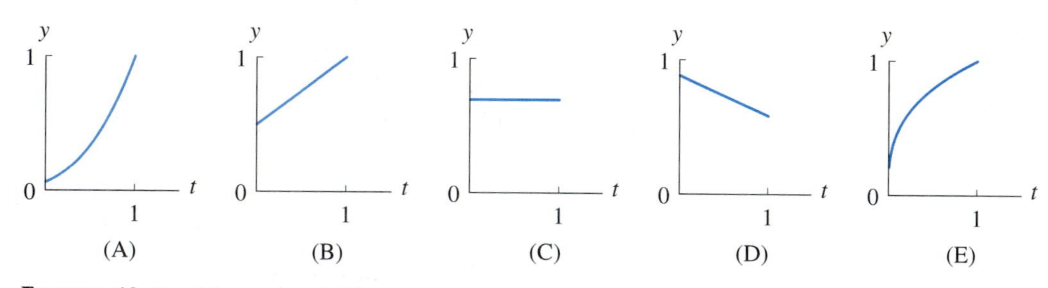

FIGURE 18 Possible graphs of $s(t)$

37. If $x = a$ and $f'(a)$ is small and negative, then a small increase in x causes a very small decrease in the function value.

38. If $x = a$ and $f'(a)$ is large and negative, then a small increase in x causes a large decrease in the function value.

39. Let $P(x)$ be the profit from producing (and selling) x units of goods. Match each question with the proper solution.

Questions:

(A) What is the profit from producing 1000 units of goods?
(B) For what level of production will the marginal profit be 1000 dollars?
(C) What is the marginal profit from producing 1000 units of goods?
(D) For what level of production will the profit be 1000 dollars?

Solutions:

(a) Determine $P'(1000)$.
(b) Find a value of a for which $P'(a) = 1000$.
(c) Set $P(x) = 1000$ and solve for x.
(d) Compute $P(1000)$.

40. Let $s(t)$ be the height (in feet) after t seconds of a ball thrown straight up into the air. Match each question with the proper solution.

Questions:

(A) What will be the velocity of the ball after 3 seconds?
(B) When will the velocity be 3 feet per second?
(C) What is the average velocity during the first 3 seconds?
(D) When will the ball be 3 feet above the ground?
(E) When will the ball hit the ground?
(F) How high will the ball be after 3 seconds?
(G) Assume the ball reached its greatest height after 4 seconds. How far did the ball travel during the first 3 seconds?

Solutions:

(a) Set $s(t) = 0$ and solve for t.
(b) Determine $s'(3)$.
(c) Compute $s(3)$.
(d) Set $s(t) = 3$ and solve for t.
(e) Find a value of a for which $s'(a) = 3$.
(f) Compute $[s(3) - s(0)]/3$.
(g) Compute $s(3) - s(0)$.

In the remaining exercises, use a graphing utility to determine any derivatives that are required.

41. Suppose the weight of a cancerous tumor after t weeks is $W(t) = .1t^2$ grams. By what amount does the weight of the tumor increase during the fifth week, that is, from time $t = 4$ to $t = 5$? What is the rate of growth of the tumor at time $t = 4$?

42. After an advertising campaign, the sales of a product often increase and then decrease. Suppose that t days after the end of the advertising, the daily sales are $f(t) = -3t^2 + 32t + 100$ units. By what amount do sales change during the fourth day; that is, from time $t = 3$ to $t = 4$? At what rate are the sales changing when $t = 2$?

43. An analysis of the daily output of a factory assembly line shows that about $60t + t^2 - \frac{1}{12}t^3$ units are produced after t hours of work, $0 \le t \le 8$. What is the instantaneous rate of production (in units per hour) when $t = 2$?

44. Liquid is pouring into a large vat. After t hours, there are $5t - \sqrt{t}$ gallons in the vat. At what instantaneous rate is the liquid flowing into the vat (in gallons per hour) when $t = 4$?

SOLUTIONS TO PRACTICE PROBLEMS 2.2

1. Method (a). The question involves $f(t)$, the temperature at time t. Since the time is given, compute $f(6)$.
2. Method (e). The question involves the rate of change of temperature, which is given by a slope. The interrogative "when" indicates that the time is unknown. Find that time, call it a, for which $f'(a) = 6$.
3. Method (f). The question asks for the change in the value of the function from time 0 to time 6, $f(6) - f(0)$.
4. Method (b). The question involves $f(t)$, and the time is unknown. Set $f(t) = 6$ and solve for t.
5. Method (d). The question involves $f'(t)$, and the time is given. Compute $f'(6)$.
6. Method (c). The question asks for the average rate of change of the function during the time interval 0 to 6, $[f(6) - f(0)]/6$.

2.3 THE DERIVATIVE FUNCTION

Figure 1(a) shows the graph of $f(x) = x^2$ and its tangent lines at $x = -\frac{5}{2}$, $x = 1$, and $x = 3$. From the figure, we see that $f'(-\frac{5}{2}) = -5$, $f'(1) = 2$, and $f'(3) = 6$. Some other slopes are $f'(-1) = -2$, $f'(0) = 0$, and $f'(2) = 4$. The slope of the curve at the point $(x, f(x))$ depends on the value of x; that is, it is a function of x. It turns out that the slope of the curve at the point $(x, f(x))$ is $2x$; that is, $f'(x) = 2x$. We say that the function $f'(x) = 2x$ is the *derivative* of the function $f(x) = x^2$.

If $f(x) = x^2$, then the derivative is the function $2x$. That is,

$$f'(x) = 2x.$$

In general, for any function $f(x)$, the slope of its graph at the point $(x, f(x))$ depends on x and is thus a function of x. This function is denoted $f'(x)$ and is

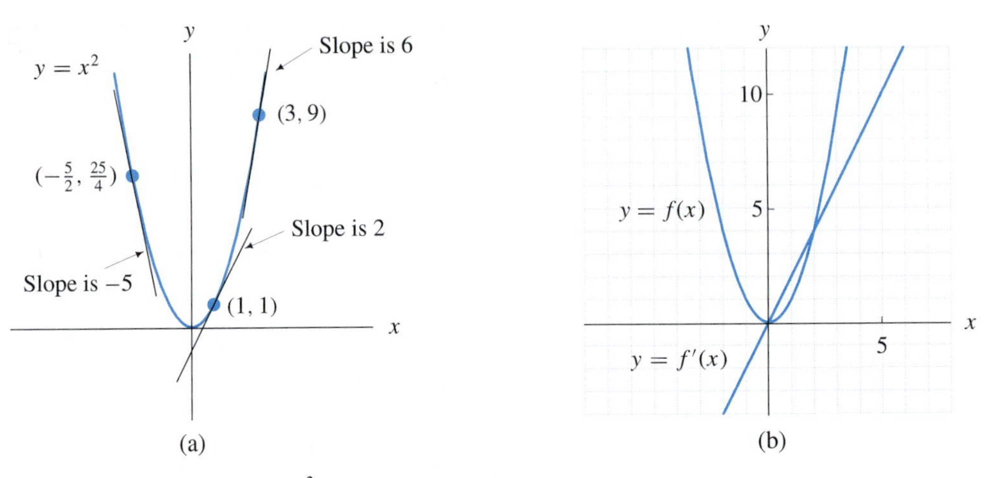

FIGURE 1 Graph of $f(x) = x^2$, its slopes, and derivative

called the derivative of the function $f(x)$. Graphing utilities can draw the graph of the derivative of any function. Figure 1(b) contains the graph of $f(x)$ and $f'(x)$ where $f(x) = x^2$, as produced by Visual Calculus.

As with the function $f(x) = x^2$, whenever $f(x)$ is given by a formula, so is $f'(x)$. An algebraic process called *differentiation* is required to obtain formulas for $f'(x)$. We develop techniques for differentiation in Chapter 3.

For a discussion of graphing $f'(x)$ and $f''(x)$ with Visual Calculus, see Appendix A, Part V.

GRAPHING $f^{0}(x)$ WITH A TI CALCULATOR Suppose the function $f(x)$ has been assigned to $\mathtt{Y}_1$. With a TI-82 or TI-83, set $\mathtt{Y}_2$ = `nDeriv(Y1,X,X)`. With a TI-85, set `y2 = der1(y1,x,x)`. $\mathtt{Y}_2$ will be the derivative of $f(x)$, that is, $f'(x)$. You can now select one or both of the functions to graph. To obtain the graphs in Figure 1, use $\mathtt{Y}_1 = \mathtt{X}^2$ and a window of $[-8, 8]$ *by* $[-4, 12]$.

Often an appropriate window for $f(x)$ will be either too small to accommodate $f'(x)$ or too large to provide a clear picture of $f'(x)$. In such cases, the graphs of $f(x)$ and $f'(x)$ should be obtained separately with different windows.

Figure 2 shows the graph of $f(x) = \frac{1}{30}x^3 - .4x^2 + 1.8x + 1$ with the slopes indicated at several points. Figure 3 shows the graph of $f'(x)$. Notice that in Figure 2, $f(x)$ has slope .6 at the point where $x = 2$. Therefore, the point (2, .6) is on the graph of $f'(x)$.

EXAMPLE 1 Refer to the graphs in Figures 2 and 3.

(a) What is the slope of $f(x)$ when $x = 1$?

(b) At what point(s) on the graph of $f(x)$ is the slope equal to .3?

Solution (a) The slope of $f(x)$ at $x = 1$ is $f'(1)$. Use Figure 3 to determine $f'(1)$. Counting over one unit on the x-axis, we must go up 1.1 units to reach the graph of $f'(x)$. Therefore, $f'(1) = 1.1$.

(b) The slope of $f(x)$ is .3 at those values of x for which $f'(x) = .3$. Therefore, we must find those values of x for which $f'(x) = .3$. From the graph in Figure

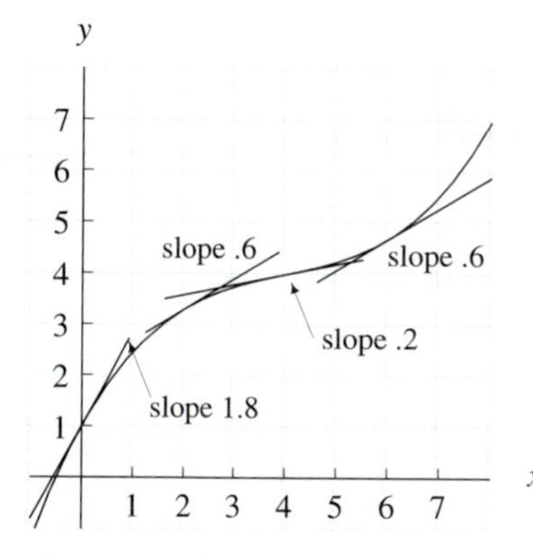

FIGURE 2 Graph of $f(x)$

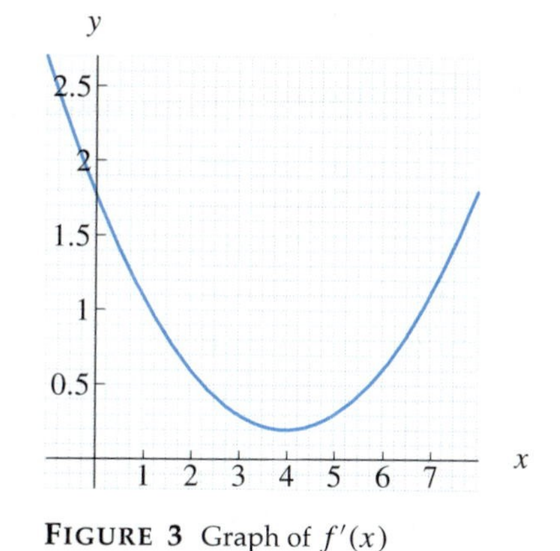

FIGURE 3 Graph of $f'(x)$

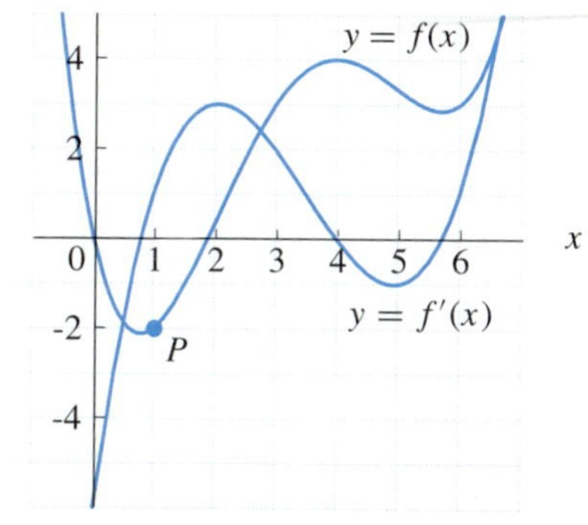

FIGURE 4 A function and its derivative

3, the horizontal line $y = .3$ crosses the graph of $f'(x)$ in two places, the points $(3, .3)$ and $(5, .3)$. Therefore, the slope of $f(x)$ equals .3 when $x = 3$ and $x = 5$; that is, at the points $(3, f(3))$ and $(5, f(5))$. Since $f(3) = 3.7$ and $f(5) = 4\frac{1}{6}$, the two points are $(3, 3.7)$ and $(5, 4\frac{1}{6})$. ■

EXAMPLE 2 Consider the function $f(x)$ in Figure 4.

(a) Find the equation of the tangent line at the point P.

(b) Clearly, the graph of $y = f(x)$ has a horizontal tangent line when $x = 4$. How could this have been determined from the graph of $y = f'(x)$?

Solution (a) The coordinates of the point P are $(1, -2)$. The slope of the graph of $y = f(x)$ at P is $f'(1)$. Looking at the graph of $f'(x)$, we see that the value of $f'(1)$ is 1. Therefore, the equation of the tangent line is

$$y - (-2) = 1(x - 1)$$
$$y + 2 = x - 1$$
$$y = x - 3.$$

(b) Since a horizontal line has slope 0, the graph of $f(x)$ has a horizontal tangent line at the point $(a, f(a))$ if $f'(a) = 0$. The condition $f'(a) = 0$ means that $f'(x)$ has a zero at $x = a$. The graph of $f'(x)$ reveals that $f'(x)$ crosses the x-axis at $x = 4$. ■

THE SECOND DERIVATIVE

Since $f'(x)$ is a full-fledged function, we can also talk about *its* slope at a point. Figure 5 is identical to Figure 4 except that a tangent line has been drawn to $f'(x)$ at the point $(1, 1)$ and this line has been labeled with its slope, $\frac{25}{6}$. That is, the slope of $f'(x)$ at $x = 1$ is $\frac{25}{6}$. Using the prime notation, we write $f''(1) = \frac{25}{6}$. The term $f''(1)$ is read *f prime prime of 1* or *f double-prime of 1* and is called the *second derivative of* $f(x)$ *at* $x = 1$. In general, for any number a in the domain of $f(x)$, $f''(x)$ is the slope of the graph of $f'(x)$ at $x = a$. The function $f''(x)$ is called the second derivative of $f(x)$.

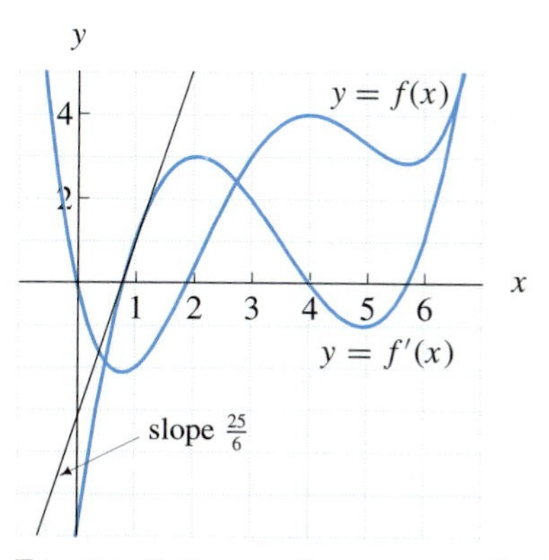

FIGURE 5 Tangent line drawn to $f'(x)$

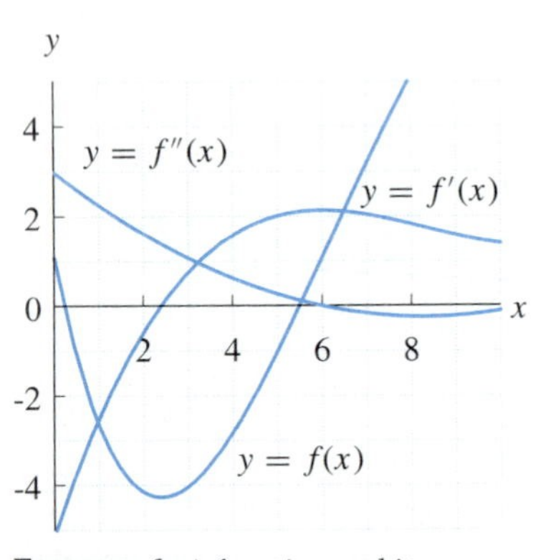

FIGURE 6 A function and its derivatives

EXAMPLE 3 Consider the function $f(x)$ in Figure 6.

(a) Find $f(5)$, $f'(5)$, and $f''(5)$.
(b) Find the slope of $f'(x)$ at the point where $x = 3$.
(c) Find a value of x for which $f'(x)$ has a horizontal tangent line.

Solution (a) The values of $f(5)$, $f'(5)$, and $f''(5)$ are found by looking at the vertical line $x = 5$ and observing where it crosses the graphs of $f(x)$, $f'(x)$, and $f''(x)$. The values are $f(5) = -1$, $f'(5) = 2$, and $f''(5) = .25$.

(b) By definition, the slope of $f'(x)$ at the point where $x = 3$ is $f''(3)$ which is 1.

(c) Looking at the graph of $y = f'(x)$, we see that it has a horizontal tangent line somewhere near $x = 6$. But where? The question is easier to answer by looking at the graph of $y = f''(x)$, which is zero wherever $f'(x)$ has a horizontal tangent. $f''(x)$ is zero at $x = 6$. Therefore, $f'(x)$ has a horizontal tangent line at $x = 6$. ■

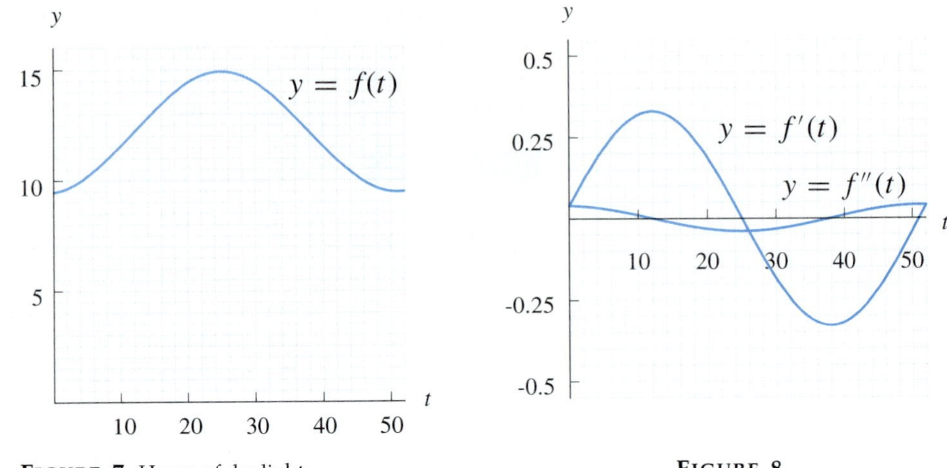

FIGURE 7 Hours of daylight

FIGURE 8

EXAMPLE 4 The number of hours of daylight t weeks after the beginning of the year in Washington, DC, is given by the function $f(t)$ in Figure 7. The graphs of $f'(t)$ and $f''(t)$ are given in Figure 8.

(a) How many hours of daylight are there after 40 weeks?
(b) After 32 weeks, how fast is the number of hours of daylight decreasing?
(c) When are there 14 hours of daylight per day?
(d) When is the number of hours of daylight increasing at the rate of 15 minutes per week?
(e) Approximately when does the graph of $f(t)$ have a horizontal tangent line?
(f) Approximately when does the graph of $f'(t)$ have a horizontal tangent line?

Solution (a) $f(40)$, which is 11.5.

(b) Since $f'(32) = -.25$, after 32 weeks the number of hours of daylight is changing at the rate of $-.25$ hours per week. That is, after 32 weeks the number of hours of daylight is decreasing at the rate of 15 minutes per week.

(c) The number of hours of daylight is given by the function $f(t)$ in Figure 7. Therefore, we must solve $f(t) = 14$. The horizontal line $y = 14$ crosses the graph of $f(t)$ twice, at $t = 18$ and at $t = 32$. Therefore, there are 14 hours of daylight per day after 18 weeks and after 32 weeks.

(d) The rate of change in the number of daylight hours is given by the function $f'(t)$ in Figure 8. Therefore, we must solve $f'(t) = .25$. (Time is measured in hours, and 15 minutes is .25 hours.) The horizontal line $y = .25$ crosses the graph of $f'(t)$ twice, at $t = 6$ and at $t = 18$. Therefore, the number of hours of daylight is changing at the rate of 15 minutes per week after 7 weeks and after 19 weeks.

(e) The function $f(t)$ has a horizontal tangent line where its first derivative is zero, that is, where $f'(t)$ crosses the t-axis. As Figure 8 shows, this happens twice, at $t = 25$ and at $t = 51$. (*Note*: $t = 25$ corresponds to June 21 and $t = 51$ corresponds to December 21. These dates are the summer and winter solstices, the longest and shortest days of the year.)

(f) The graph of $f'(t)$ has a horizontal tangent line when its derivative, $f''(t)$, is zero. The graph of $f''(t)$ crosses the t-axis when $t = 12$ and $t = 38$. (*Note*: $t = 12$ corresponds to March 21 and $t = 23$ corresponds to September 23. These dates are the spring and autumnal equinoxes. At equinoxes, there are 12 hours of daylight and 12 hours of night.) ■

The graph of $f''(x)$ can be easily obtained with a graphing utility.

GRAPHING $f''(x)$ WITH A TI CALCULATOR Suppose the function $f(x)$ has been assigned to Y_1. With a TI-82 or TI-83, set Y_2 = `nDeriv(Y1,X,X)` and set Y_3 = `nDeriv(Y2,X,X)`. Y_3 will be the derivative of $f'(x)$, that is, $f''(x)$. You can now select one or more of the three functions to graph. With a TI-85, set `y2 = der2(y1,x,x)`.

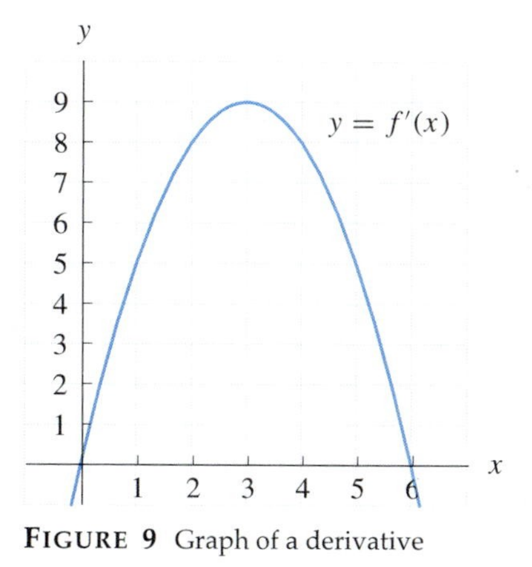

FIGURE 9 Graph of a derivative

THE SECOND DERIVATIVE AND MOTION

Let $s(t)$ be the position of an object at time t. For instance, $s(t)$ might be the height of a ball or the distance traveled by a car at time t. Then $s'(t)$ is the velocity of the object at time t. The rate of change of velocity at time t, $s''(t)$, is called the *acceleration at time t*. The velocity function is often denoted $v(t)$ and the acceleration function is denoted $a(t)$. Therefore, $v(t) = s'(t)$ and $a(t) = v'(t) = s''(t)$.

PRACTICE PROBLEMS 2.3

Figure 9 contains the graph of $f'(x)$.

1. What is the slope of $f(x)$ when $x = 3$?
2. For what values of x does the graph of $y = f(x)$ have slope 5?
3. Is the graph of $y = f(x)$ steeper at $x = 1$ or $x = 2$?

EXERCISES 2.3

Exercises 1–8 refer to the graphs of $f(x)$ and its derivative in Figure 10.

1. What is the slope of $f(x)$ when $x = 0$?

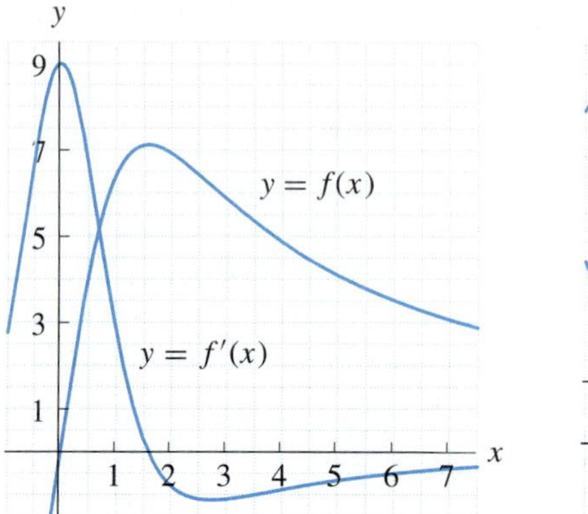

FIGURE 10 A function and its derivative

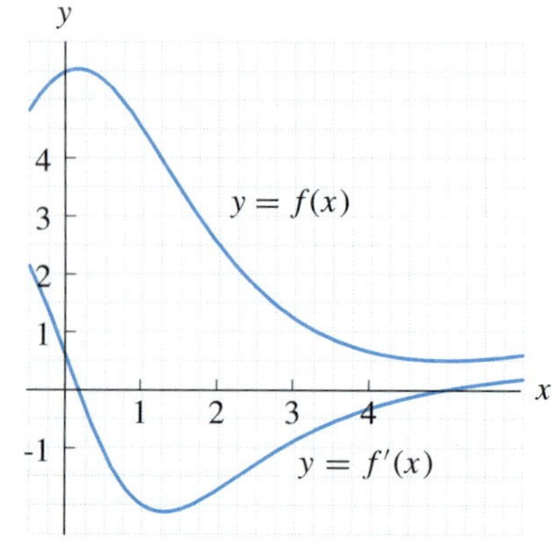

FIGURE 11 A function and its derivative

2. What is the equation of the tangent line to $f(x)$ at $x = 6$?

3. Solve $f'(x) = -1$ for x.

4. What is the rate of change of $f(x)$ with respect to x when $x = 0$?

5. By how much does $f(x)$ change as x increases from 6 to $6+h$, where $h = 1$?

6. Find a value of x for which the rate of change of $f(x)$ with respect to x is -1.

7. Is the graph of $y = f(x)$ steeper at $x = .5$ or $x = 1$?

8. What is the value of $f(x)$ when $x = \frac{1}{2}$?

Exercises 9–16 refer to the graphs of $f(x)$ and its derivative in Figure 11.

9. Find the coordinates of a point on the graph of $f(x)$ where the tangent line is horizontal.

10. Find the equation of the tangent line to $f(x)$ at $x = 1$.

11. What is the value of $f(x)$ when $x = 0$?

12. What is the rate of change of $f(x)$ with respect to x when $x = 0$?

13. Find a number a for which the slope of $y = f(x)$ at the point $(a, f(a))$ is -1.5.

14. By how much does $f(x)$ change as x increases from 2 to $2+h$, where $h = \frac{1}{2}$?

15. Looking only at the graph of $f(x)$, determine whether $f'(3)$ is greater or less than $f'(4)$. Confirm your answer by looking at the graph of $f'(x)$.

16. Looking only at the graph of $f(x)$, determine whether $f'(0)$ is greater or less than $f'(1)$. Confirm your answer by looking at the graph of $f'(x)$.

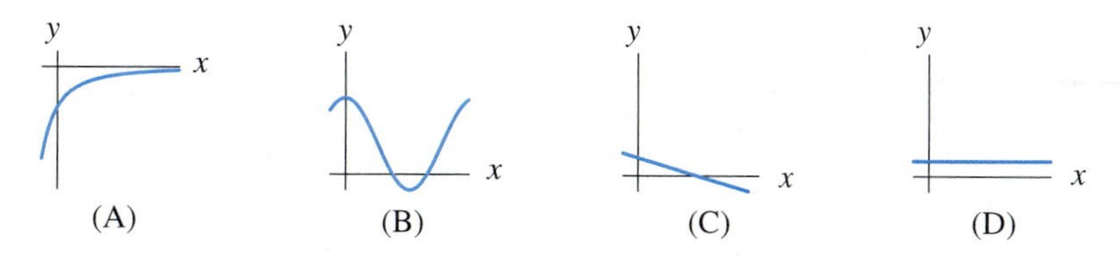

FIGURE 12 Derivatives of the functions from Exercises 17–20

In Exercises 17–20, find the graph in Figure 12 that is the derivative of $f(x)$.

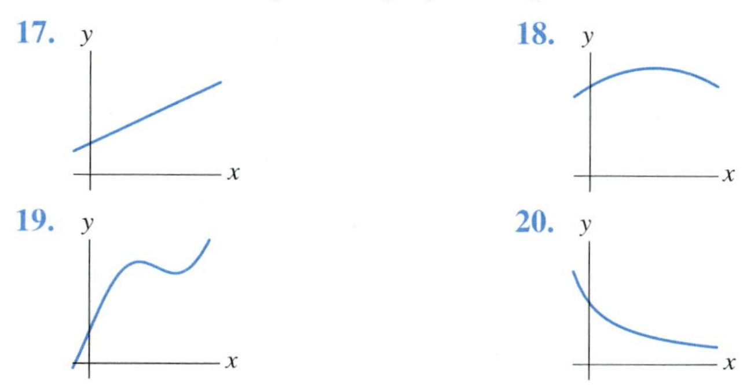

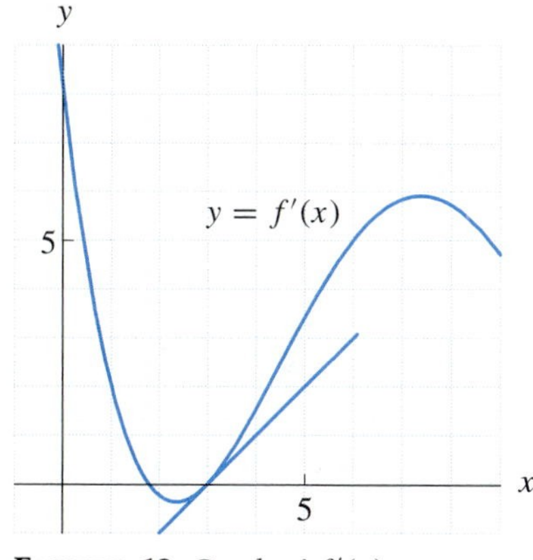

FIGURE 13 Graph of $f'(x)$

Exercises 21–28 refer to the graph of $f'(x)$ in Figure 13. This graph is used to answer questions about $f(x)$ and $f''(x)$.

21. What is the slope of $f(x)$ when $x = 6$?

22. Suppose $f(1) = 3$. What is the equation of the tangent line to $f(x)$ at $x = 1$?

23. Find a value of x for which the graph of $y = f(x)$ has slope 2.

24. Is the graph of $y = f(x)$ steeper at $x = 4$ or $x = 5$?

25. Estimate $f(1.5) - f(1)$.

26. By approximately how much does $f(x)$ increase as x increases from 6 to 6.25?

27. What is $f''(3)$?

28. Is $f''(9)$ positive or negative?

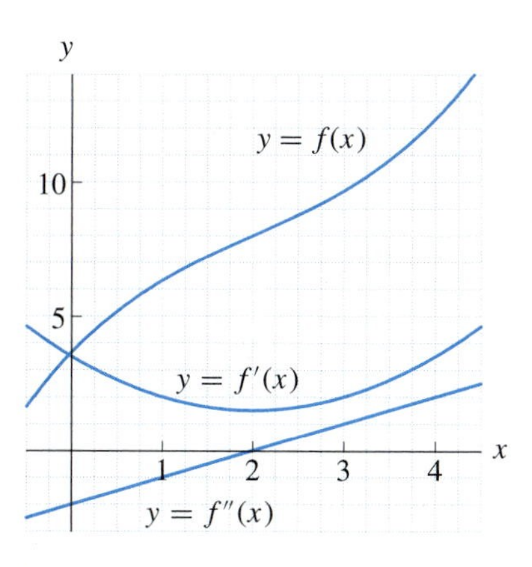

FIGURE 14 Graphs of $f(x)$ and its derivatives

Exercises 29–34 refer to the graphs of $f(x)$ and its first and second derivatives as shown in Figure 14.

29. Find the slope of the tangent line to $f'(x)$ at $x = 4$.

30. For what value of x does $f'(x)$ have a horizontal tangent line?

31. What is the equation of the tangent line to $f'(x)$ at $x = 1$?

32. Find a value of x for which the rate of change of $f'(x)$ with respect to x is -2.

33. Find two values of x at which $f(x)$ has slope 3.

34. Find the coordinates of a point on the graph of $f'(x)$ at which the tangent line has slope 1.

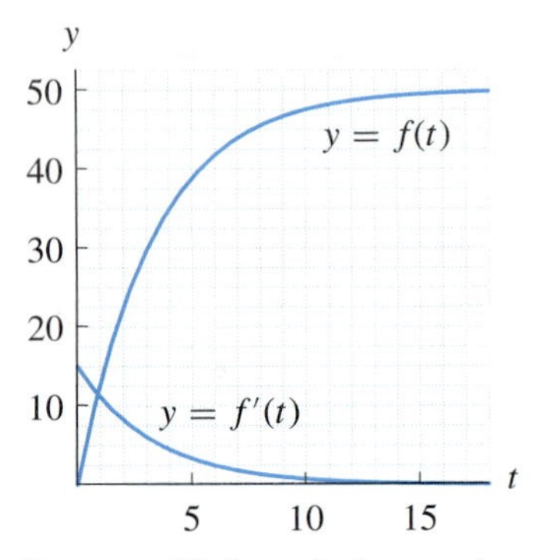

FIGURE 15 Spread of a news item by mass media

Exercises 35–38 refer to a news item broadcast by mass media to a potential audience of 50,000 people. The number of people (in thousands) who will have heard the news after t days is given by the function $f(t)$ in Figure 15.

35. How many people will have heard the news after 10 days?

36. At what rate is the news spreading initially?

37. When will 22,500 people have heard the news?

38. Approximately when will the news be spreading at the rate of 2500 people per day?

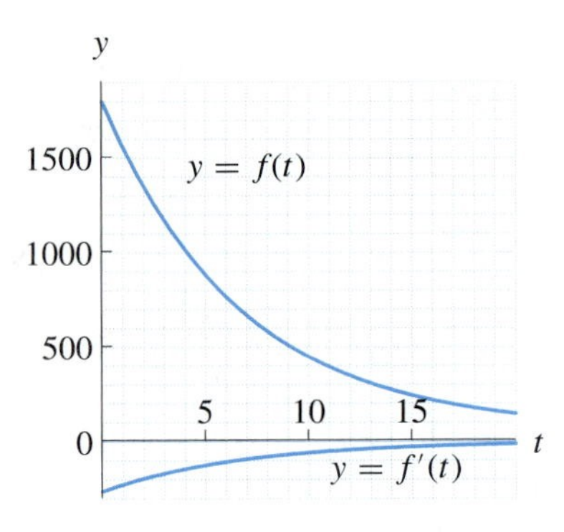

FIGURE 16 Cooling steel

Exercises 39–42 refer to a rod of molten steel with a temperature of 1800 °F that is immersed in a large vat of water at temperature 60 °F. The temperature of the rod after t seconds is given by the function $f(t)$ in Figure 16.

39. What is the temperature of the rod after 11 seconds?

40. At what rate is the temperature of the rod changing after 6 seconds?

41. Approximately when is the temperature of the rod 200 degrees?

42. Approximately when is the rod cooling at the rate of 200 degrees per second?

In Exercises 43–48, the cost function for a manufacturer is given by the function $C(x)$ in Figure 17(a). The graphs of $C'(x)$ and $C''(x)$ are given in Figure 17(b).

43. What is the cost of manufacturing 14 units of goods?

44. What is the marginal cost when 22 units of goods are manufactured?

45. At what level of production is the cost $1300?

46. At what level of production is the marginal cost $17?

47. At what level of production does the marginal cost function have a horizontal tangent line?

48. Approximately what is the slope of the marginal cost function when 48 units of goods are manufactured?

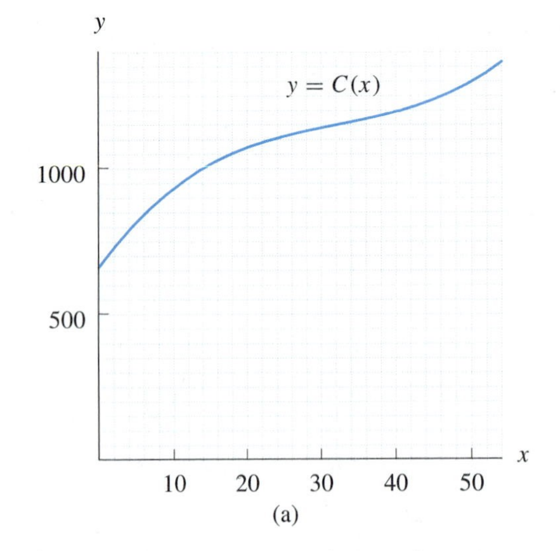

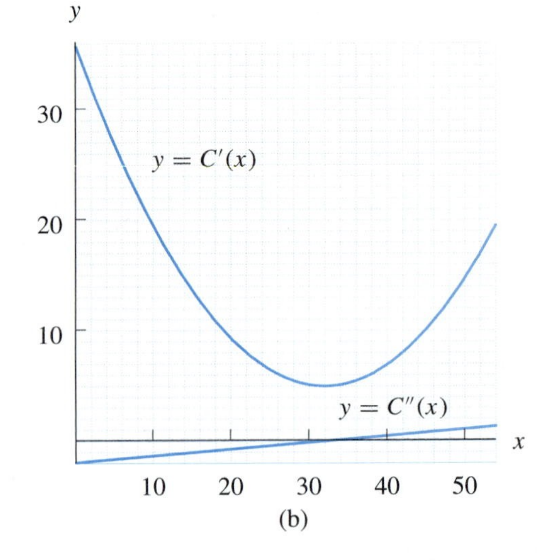

FIGURE 17 A cost curve and its two derivatives

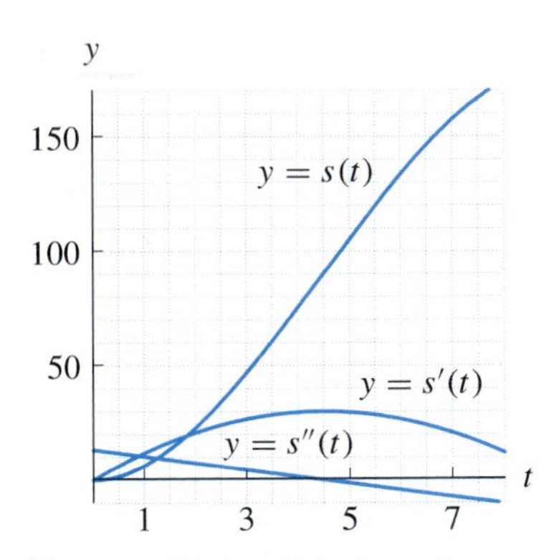

FIGURE 18 A vehicle in motion

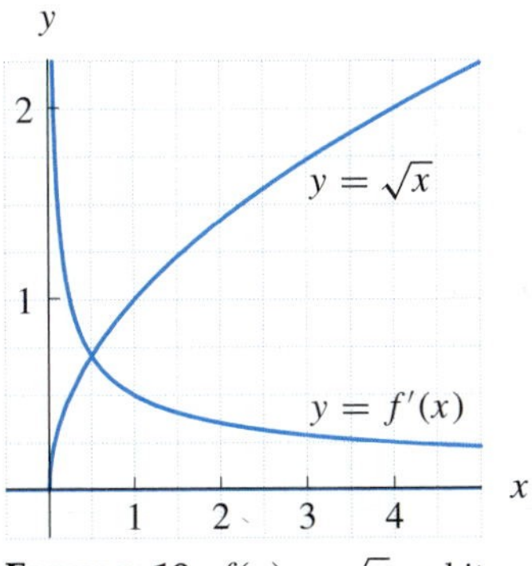

FIGURE 19 $f(x) = \sqrt{x}$ and its derivative

Exercises 49–54 concern an eight-second test run in which a vehicle accelerates for four seconds and then decelerates. The function $s(t)$ in Figure 18 gives the number of feet traveled after t seconds.

49. How far has the vehicle traveled after 3.5 seconds?

50. What is the velocity after 2 seconds?

51. What is the acceleration after 1 second?

52. When will the vehicle have traveled 120 feet?

53. When, during the second half of the test run, will the vehicle be traveling at the rate of 20 feet per second?

54. What is the greatest velocity and at what time is this greatest velocity reached? How far has the vehicle traveled at this time?

55. Figure 19 shows the graphs of $f(x) = \sqrt{x}$ and its derivative.

(a) Find $f'(1)$ and $f'(4)$.

(b) The formula for $f'(x)$ has the form $\dfrac{1}{k\sqrt{x}}$ for some constant k. Determine the formula for $f'(x)$.

(c) Use the formula for $f'(x)$ to find the slope of $f(x)$ at $x = 16$.

(d) Use the formula for $f'(x)$ to find the equation of the tangent line to the graph of $y = f(x)$ at $x = 9$.

56. (a) Let $f(x) = mx + b$. What is the formula for $f'(x)$?

(b) If $f(x) = x^2$, what is the formula for $f''(x)$?

In the remaining exercises, use a graphing utility to obtain the graphs of the functions.

57. Use the graph of $f(x) = x\sqrt{9 - x}$ and its derivative to answer the following questions. An appropriate window is $[-7, 9]$ *by* $[-11, 11]$.

(a) What is the slope of $f(x)$ when $x = 0$?

(b) Find an approximate solution to $f(x) = 10$.

(c) For what value of x does the graph of $f(x)$ have a horizontal tangent line?

(d) Find an approximate value of x for which the graph of $y = f(x)$ has slope -3.

(e) Is the graph of $y = f(x)$ steeper at $x = 2$ or $x = 3$?

(f) Is the slope of $y = f(x)$ greater at $x = 3$ or $x = 5$?

58. Use the graph of $f(x) = 1 + 2x + 18/x$ and its first derivative to answer the following questions. An appropriate window is $[0, 4]$ *by* $[-20, 40]$.

(a) What is $f'(2)$?

(b) For what value of x does the graph of $f(x)$ have a horizontal tangent line?

(c) What is the rate of change of $f(x)$ with respect to x when $x = 1$?

(d) What is the equation of the tangent line to $f(x)$ when $x = 1$?

59. Use the graph of the second derivative of $f(x) = (9 - x^2)/(3 + x^2)$ to answer the following questions. An appropriate window is $[-4, 4]$ *by* $[-4, 1]$.

(a) What is $f''(3)$?

(b) For what values of x does the graph of $f'(x)$ have a horizontal tangent line?

(c) For what value of x does the graph of $f'(x)$ have slope -2.35?

(d) Suppose the slope of $f(x)$ at the point $(3, 0)$ is $-.5$. What is the equation of the tangent line to the graph of $f'(x)$ when $x = 3$?

60. Use the graph of the second derivative of $f(x) = x^4 - 4x^3 + 10$ to answer the following questions. An appropriate window is $[-3, 5]$ *by* $[-30, 100]$.

(a) Solve $f''(x) = 36$.

(b) For what values of x does the graph of $f'(x)$ have a horizontal tangent line?

(c) For what values of x does the graph of $f'(x)$ have slope -9.

(d) Is $f'(x)$ steeper at $x = 2$ or $x = 3$?

61. A ball dropped from the roof of a building has height $s(t) = 1024 - 16t^2$ feet after t seconds. Sketch the graphs of $s(t)$ and $s'(t)$ in the window $[0, 8]$ *by* $[-300, 1100]$.

(a) How high is the ball after 2 seconds?

(b) When is the height 124 feet?

(c) What is the velocity after 6 seconds?

(d) When is the ball falling at a speed of 72 feet per second?

(e) What is the acceleration after 2 seconds?

62. The number of bacteria (in thousands) in a culture after t days is given by

$$f(t) = 175 + 15t(.9)^t.$$

Sketch the graph of $f(t)$ in the window $[0, 64]$ *by* $[0, 230]$ and sketch the graphs of $f'(t)$ and $f''(t)$ in the window $[0, 64]$ *by* $[-2.1, .5]$.

(a) How large is the culture after 28 days?

(b) How fast is the culture changing after 35 days?

(c) When are there 194,000 bacteria?

(d) When is the size of the culture decreasing at the rate of 1000 bacteria per day?

(e) When does $f'(t)$ have a horizontal tangent line? (*Note*: At that time, the size of the culture is decreasing at the fastest rate.)

(f) When does $f(t)$ have a horizontal tangent line? (*Note*: At that time, the culture attains its greatest size.)

63. The length of a certain weed after t weeks is

$$f(t) = \frac{30}{1 + 25(.6)^t} \text{ centimeters.}$$

Sketch the graphs of $f(t)$ and $f'(t)$ in the window $[0, 16]$ *by* $[0, 30]$, and sketch the graph of $f''(t)$ in the window $[0, 16]$ *by* $[-1, 1]$.

(a) What is the length of the weed after 1 week?

(b) How fast is the weed growing after 10 weeks?

(c) Approximately when is the weed 10 centimeters long?

(d) Approximately when is the weed growing at the rate of 2 centimeters per day?

(e) At what time does $f'(t)$ have a horizontal tangent line? (*Note*: At this time, the weed is growing at the greatest rate.)

64. The revenue from manufacturing and selling x units of a certain commodity is

$$R(x) = \frac{1000x}{4 + x}$$

thousand dollars during a year. Sketch the graphs of $R(x)$ and $R'(x)$ in the window $[0, 16]$ *by* $[-50, 800]$.

(a) How much money is earned from the sale of 4 units of goods?

(b) For what level of sales is the revenue \$750,000?

(c) What is the marginal revenue when 1 unit of goods is sold?

(d) For what level of sales is the marginal revenue \$40,000?

65. Draw the graphs of $f(x) = x^3$ and its first derivative. Use the window $[-8, 8]$ *by* $[-8, 8]$.

(a) Find $f'(0)$, $f'(1)$, $f'(2)$, and $f'(3)$. (*Note*: They are whole numbers.)

(b) Guess the formula for $f'(x)$.

(c) Use the formula for $f'(x)$ to find the slope of $f(x)$ at $x = 5$.

(d) Use the formula for $f'(x)$ to find the equation of the tangent line to the graph of $y = f(x)$ at $x = 4$.

66. Draw the graphs of

$$f(x) = \frac{1}{x}$$

and its first derivative. Use the window $[-1, 4]$ *by* $[-4, 4]$.

(a) Find $f'(1)$. (*Note*: It is a whole number.)

(b) Observe that $f'(4)$ is $-1/16$.

(c) The value of $f'(3)$ is $-1/9$. Guess the formula for $f'(x)$.

(d) Use the formula for $f'(x)$ to find the equation of the tangent line to the graph of $y = f(x)$ at $x = 1/2$.

SOLUTIONS TO PRACTICE PROBLEMS 2.3

1. The slope of $f(x)$ at $x = 3$ is $f'(3)$. This value can be read from the graph. Since $(3, 9)$ is on the graph of $f'(x)$, $f'(3)$ is 9.

2. The graph of $y = f(x)$ has slope 5 at any value of x for which $f'(x) = 5$. The horizontal line $y = 5$ crosses the graph of $f'(x)$ twice, when $x = 1$ and $x = 5$.

3. $f'(1) = 5$ and $f'(2) = 8$. Therefore, the function $f(x)$ has greater slope when $x = 2$.

2.4 DESCRIBING GRAPHS OF FUNCTIONS

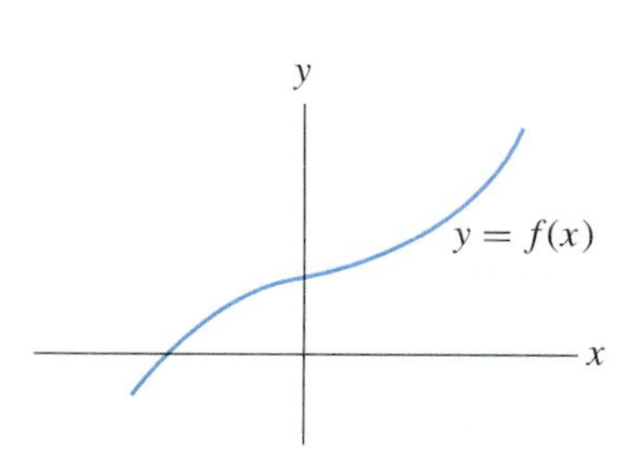

FIGURE 1 An increasing function

Let's examine the graph of a typical function, such as the one in Figure 1, and introduce some terminology to describe its behavior. First observe that the graph is either rising or falling, depending on whether we look at it from left to right or from right to left. To avoid confusion, we shall always follow the accepted practice of reading a graph from left to right.

Let's now examine the behavior of a function $f(x)$ in an interval throughout which it is defined. We say that a function $f(x)$ is *increasing in the interval* if the graph continuously rises as x goes from left to right through the interval. That is, whenever x_1 and x_2 are in the interval with $x_1 < x_2$, then we have $f(x_1) < f(x_2)$. We say that $f(x)$ is increasing at $x = c$ provided that $f(x)$ is increasing in some open interval on the x-axis that contains the point c.

We say that a function $f(x)$ is *decreasing in an interval* provided that the graph continuously falls as x goes from left to right through the interval. That is, whenever x_1 and x_2 are in the interval with $x_1 < x_2$, then we have $f(x_1) > f(x_2)$. We say that $f(x)$ is decreasing at $x = c$ provided that $f(x)$ is decreasing in some open interval that contains the point c. Figure 2 shows graphs that are increasing and decreasing at $x = c$. Observe in Figure 2(d) that when $f(c)$ is negative and $f(x)$ is decreasing, the values of $f(x)$ become *more* negative. When $f(c)$ is negative and $f(x)$ is increasing, as in Figure 2(e), the values of $f(x)$ become *less* negative.

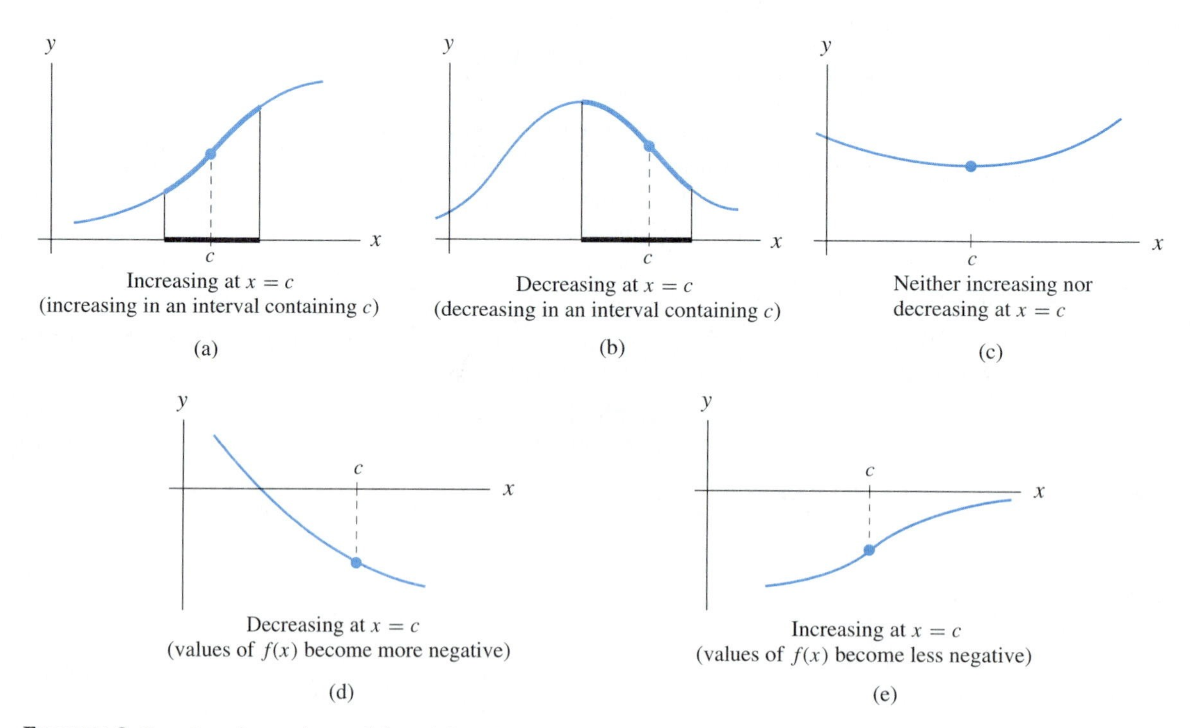

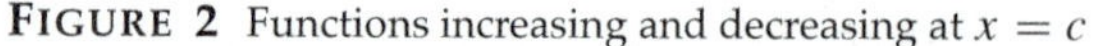
FIGURE 2 Functions increasing and decreasing at $x = c$

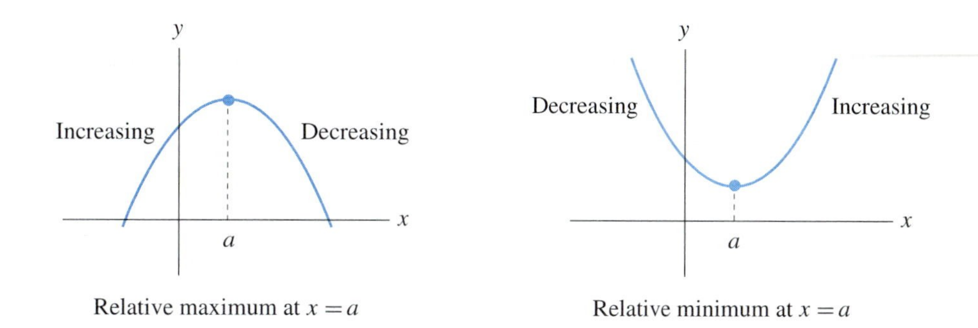

FIGURE 3 Relative extreme points

EXTREME POINTS A *relative extreme point* of a function is a point at which its graph changes from increasing to decreasing, or vice versa. We distinguish the two possibilities in an obvious way. A *relative maximum point* is a point at which the graph changes from increasing to decreasing; a *relative minimum point* is a point at which the graph changes from decreasing to increasing. (See Figure 3.) The adjective "relative" in these definitions indicates that a point is maximal or minimal only relative to nearby points on the graph.

The *maximum value* (or *absolute maximum value*) of a function is the largest value that the function assumes on its domain. The *minimum value* (or *absolute minimum value*) of a function is the smallest value that the function assumes on its domain. Functions may or may not have maximum or minimum values. (See Figure 4.) However, it can be shown that a function with no breaks whose domain is an interval of the form $a \leq x \leq b$ has both a maximum and a minimum value.

Maximum values and minimum values of functions usually occur at relative maximum points and relative minimum points, as in Figure 4(a). However, they can occur at endpoints of the domain, as in Figure 4(b). If so, we say that the function has an *endpoint extreme* (or *endpoint extremum*).

Relative maximum points and endpoint maximum points are higher than any nearby points. The maximum value of a function is the y-coordinate of the highest point on its graph. (The highest point is called the *absolute maximum point*.) Similar considerations apply to minima.

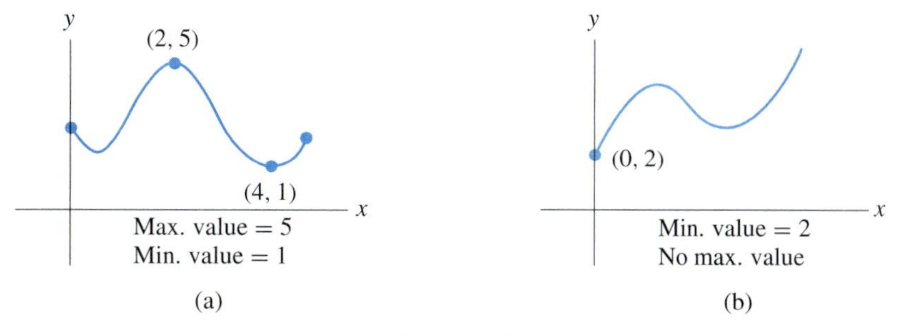

FIGURE 4

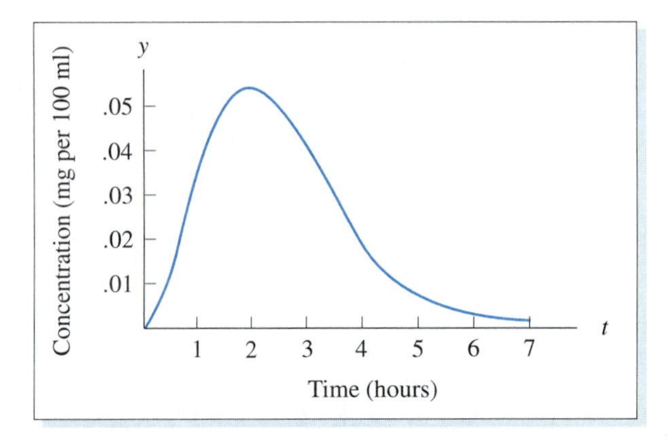

FIGURE 5 A drug time-concentration curve

EXAMPLE 1 When a drug is injected intramuscularly (into a muscle), the concentration of the drug in the veins has the time-concentration curve shown in Figure 5. Describe this graph, using the terms introduced above.

Solution Initially (when $t = 0$), there is no drug in the veins. When the drug is injected into the muscle, it begins to diffuse into the bloodstream. The concentration in the veins increases until it reaches its maximum value at $t = 2$. After this time the concentration begins to decrease, as the body's metabolic processes remove the drug from the blood. Eventually the drug concentration decreases to a level so small that, for all practical purposes, it is zero. ■

CHANGING SLOPE An important but subtle feature of a graph is the way the graph's slope *changes* (as we look from left to right). The graphs in Figure 6 are both increasing, but there is a fundamental difference in the way they are increasing. Graph I, which describes the U.S. national debt per person, is steeper for 1970 than for 1960. That is, the *slope* of graph I *increases* as we move from left to right. A newspaper description of graph I might read,

> U.S. public debt per capita rose at an increasing rate during the decade from 1960 to 1970.

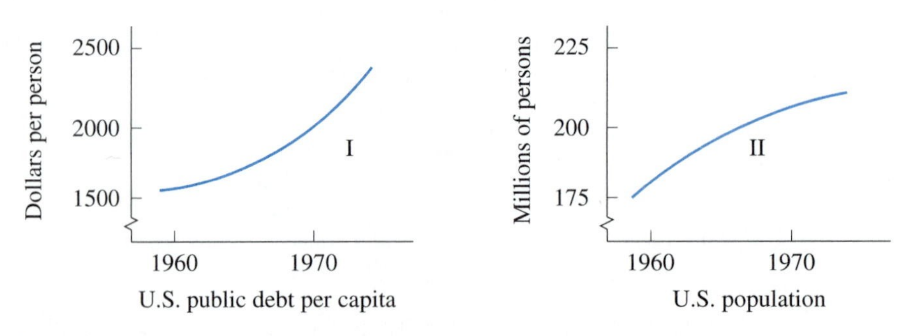

FIGURE 6 Increasing and decreasing slopes

In contrast, the *slope* of graph II *decreases* as we move from left to right. Although the U.S. population is rising each year, the rate of increase declines throughout the decade from 1960 to 1970. That is, the slope becomes less positive. The media might say,

During the 1960s, U.S. population rose at a decreasing rate.

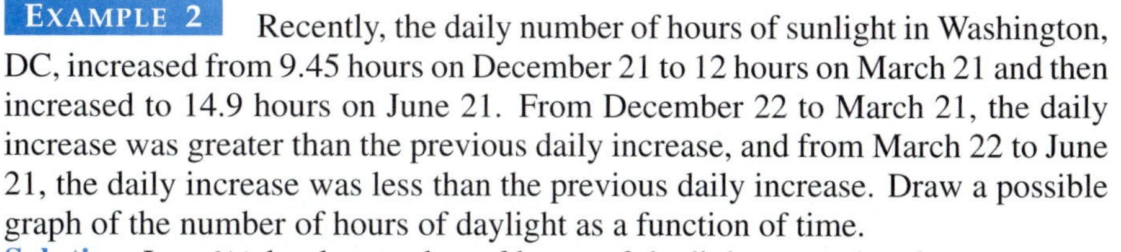

EXAMPLE 2 Recently, the daily number of hours of sunlight in Washington, DC, increased from 9.45 hours on December 21 to 12 hours on March 21 and then increased to 14.9 hours on June 21. From December 22 to March 21, the daily increase was greater than the previous daily increase, and from March 22 to June 21, the daily increase was less than the previous daily increase. Draw a possible graph of the number of hours of daylight as a function of time.

Solution Let $f(t)$ be the number of hours of daylight t months after December 21. See Figure 7. The first part of the graph, December 21 to March 21, is increasing at an increasing rate. The second part of the graph, March 21 to June 21, is increasing at a decreasing rate. ■

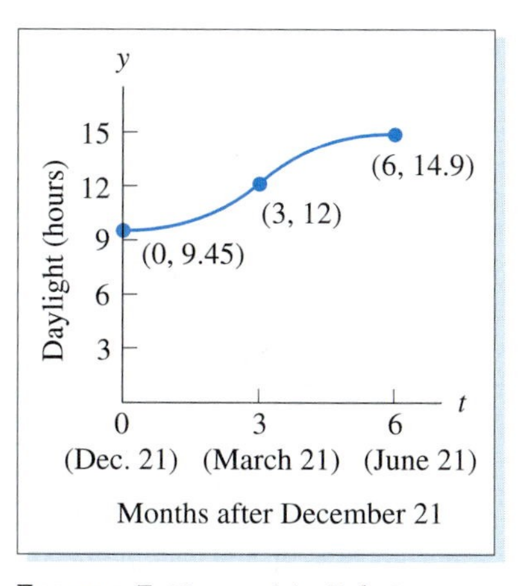

FIGURE 7 Hours of daylight in Washington, DC

Warning Recall that when a negative quantity decreases, it becomes more negative. (Think about the temperature outside when it is below zero and the temperature is falling.) So if the slope of a graph is negative and the slope is decreasing, then the slope is becoming more negative, as in Figure 8. This technical use of the term "decreasing" runs counter to our intuition, because in popular discourse, "decrease" often means to become smaller in size.

It is true that the curve in Figure 9 is becoming "less steep" in a nontechnical sense (since steepness, if it were defined, would probably refer to the magnitude of the slope). However, the slope of the curve in Figure 9 is increasing because it is becoming less negative. The popular press would probably describe the curve in Figure 9 as decreasing at a decreasing rate, since the rate of fall tends to taper off. Since this terminology is potentially confusing, we shall not use it.

CONCAVITY The U.S. debt and population graphs in Figure 6 may also be described in geometric terms: Graph I opens up and lies above its tangent line at each point, whereas graph II opens down and lies below its tangent line at each point (Figure 10).

We say that a function $f(x)$ is *concave up* at $x = a$ if there is an open interval on the x-axis containing a throughout which the graph of $f(x)$ lies above its

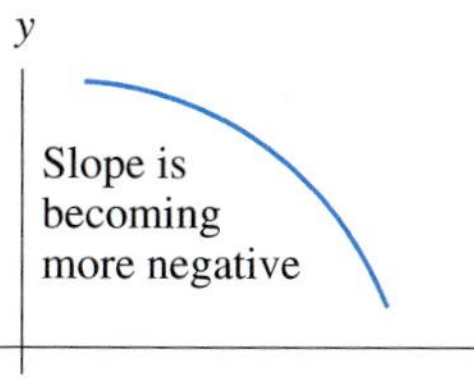

FIGURE 8 Slope is decreasing

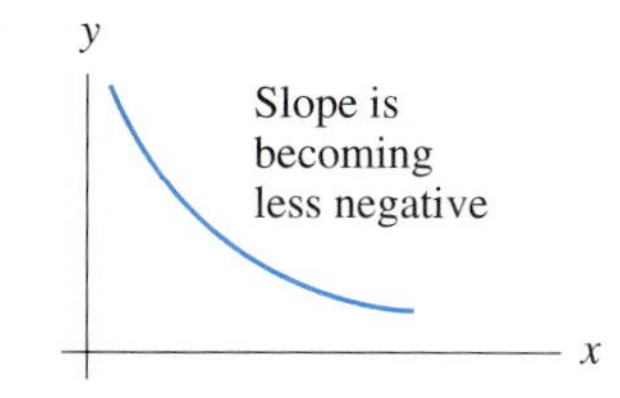

FIGURE 9 Slope is increasing

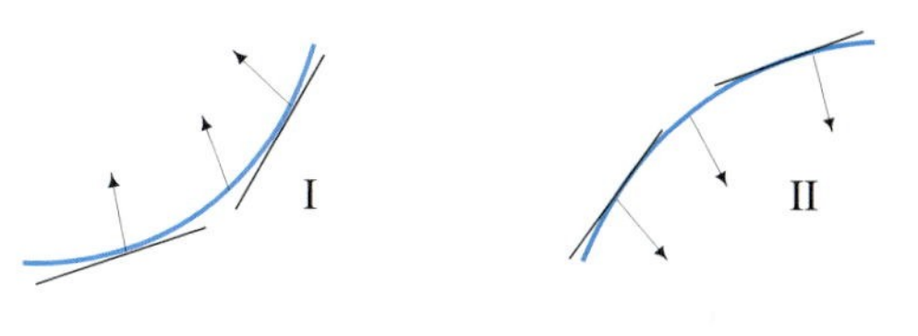

FIGURE 10 Relationship between concavity and tangent lines

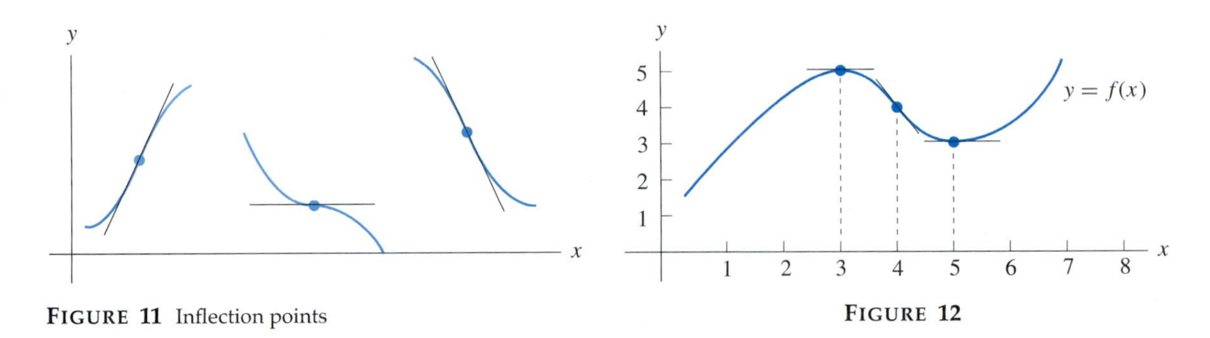

FIGURE 11 Inflection points

FIGURE 12

tangent line. Equivalently, $f(x)$ is concave up at $x = a$ if the slope of the graph increases as we move from left to right through $(a, f(a))$. Graph I is an example of a function that is concave up at each point.

Similarly, we say that a function $f(x)$ is *concave down* at $x = a$ if there is an open interval on the x-axis containing a throughout which the graph of $f(x)$ lies below its tangent line. Equivalently, $f(x)$ is concave down at $x = a$ if the slope of the graph decreases as we move from left to right through $(a, f(a))$. Graph II is concave down at each point.

An *inflection point* is a point on the graph of a function at which the function is continuous and at which the graph changes from concave up to concave down, or vice versa. At such a point, the graph crosses its tangent line (Figure 11). (The continuity condition means that the graph cannot break at an inflection point.)

EXAMPLE 3 Use the terms defined earlier to describe the graph shown in Figure 12.

Solution (a) For $x < 3$, $f(x)$ is increasing and concave down.

(b) Relative maximum point at $x = 3$.

(c) For $3 < x < 4$, $f(x)$ is decreasing and concave down.

(d) Inflection point at $x = 4$.

(e) For $4 < x < 5$, $f(x)$ is decreasing and concave up.

(f) Relative minimum point at $x = 5$.

(g) For $x > 5$, $f(x)$ is increasing and concave up. ■

INTERCEPTS, UNDEFINED POINTS, AND ASYMPTOTES A point at which a graph crosses the y-axis is called a y-intercept, and a point at which it crosses the x-axis is called an x-intercept. The x-coordinate of an x-intercept is sometimes called a "zero" of the function, since the function has the value zero there. (See Figure 13.)

Some functions are not defined for all values of x. For instance, $f(x) = 1/x$ is not defined for $x = 0$, and $f(x) = \sqrt{x}$ is not defined for $x < 0$. (See Figure 14.) Many functions that arise in applications are defined only for $x \geq 0$. A properly drawn graph should leave no doubt as to the values of x for which the function is defined.

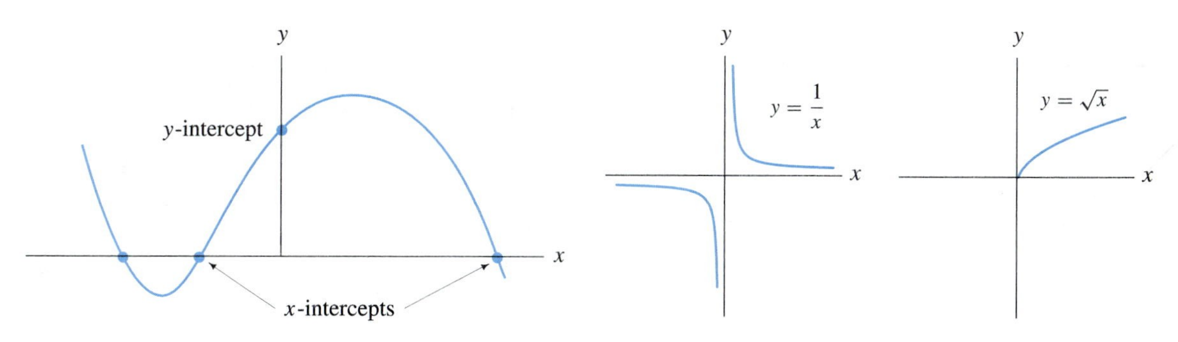

FIGURE 13 Intercepts of a graph

FIGURE 14 Graphs with undefined points

Graphs in applied problems sometimes straighten out and approach some straight line as x gets large (Figure 15). Such a straight line is called an *asymptote* of the graph. The most common asymptotes are horizontal as in (a) and (b) of Figure 15. In Example 1 the t-axis is an asymptote of the drug-concentration curve.

The horizontal asymptotes of a graph can be determined by observing the behavior of $f(x)$ as x gets large. For instance, if there is a number, call it L, such that the value of $f(x)$ gets arbitrarily close to L as x gets large and positive, then the horizontal line $y = L$ is an asymptote of $f(x)$. Similar results apply as x gets large and negative. For instance, the graph of

$$y = 1 + \frac{2}{1 + 2^{-x}}$$

in Figure 16 has both $y = 3$ and $y = 1$ as horizontal asymptotes.

Occasionally, a graph will approach a vertical line as x approaches some fixed value, as in Figure 17. Such a line is a *vertical asymptote*. Most often, we expect a vertical asymptote at a value x that would result in division by zero in the

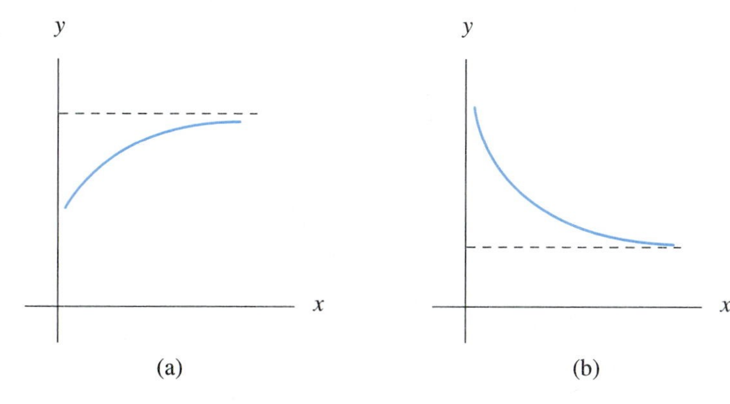

FIGURE 15 Graphs that approach asymptotes as x gets large

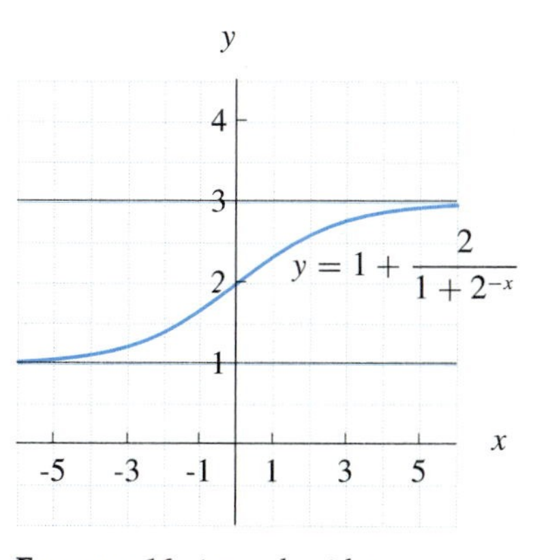

FIGURE 16 A graph with two horizontal asymptotes

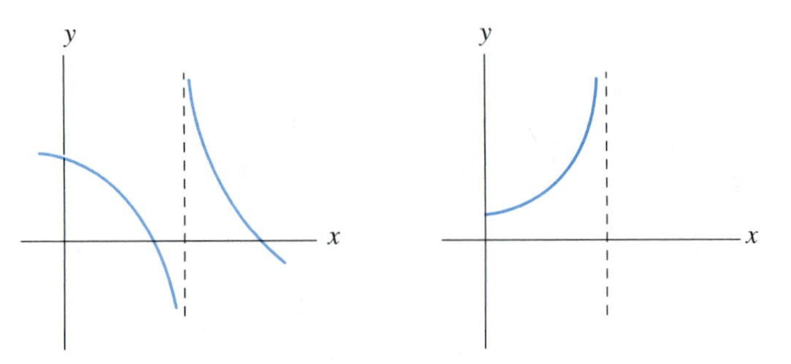

FIGURE 17 Examples of vertical asymptotes

definition of $f(x)$. For example, $f(x) = 1/(x - 3)$ has the line $x = 3$ as a vertical asymptote.

We now have six categories for describing the graph of a function:

1. Intervals in which the function is increasing (resp., decreasing), relative maximum points, relative minimum points
2. Maximum value, minimum value
3. Intervals in which the function is concave up (resp., concave down), inflection points
4. x-intercept, y-intercept
5. Undefined points
6. Asymptotes

For us, the first three categories will be the most important. However, the last three categories should not be forgotten.

PRACTICE PROBLEMS 2.4

1. Does the slope of the curve in Figure 18 increase or decrease as x increases?
2. At what value of x is the slope of the curve in Figure 19 minimized?

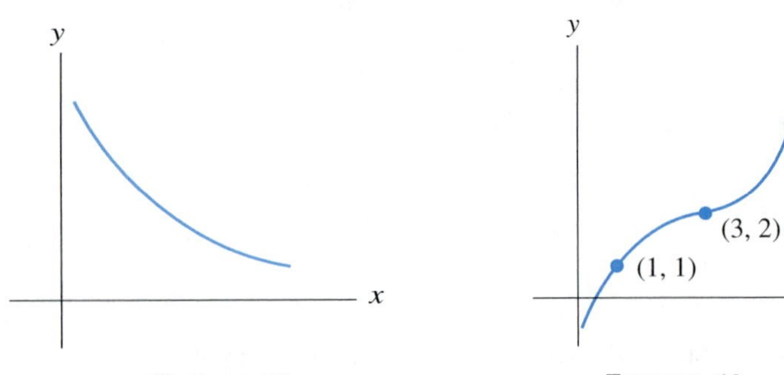

FIGURE 18

FIGURE 19

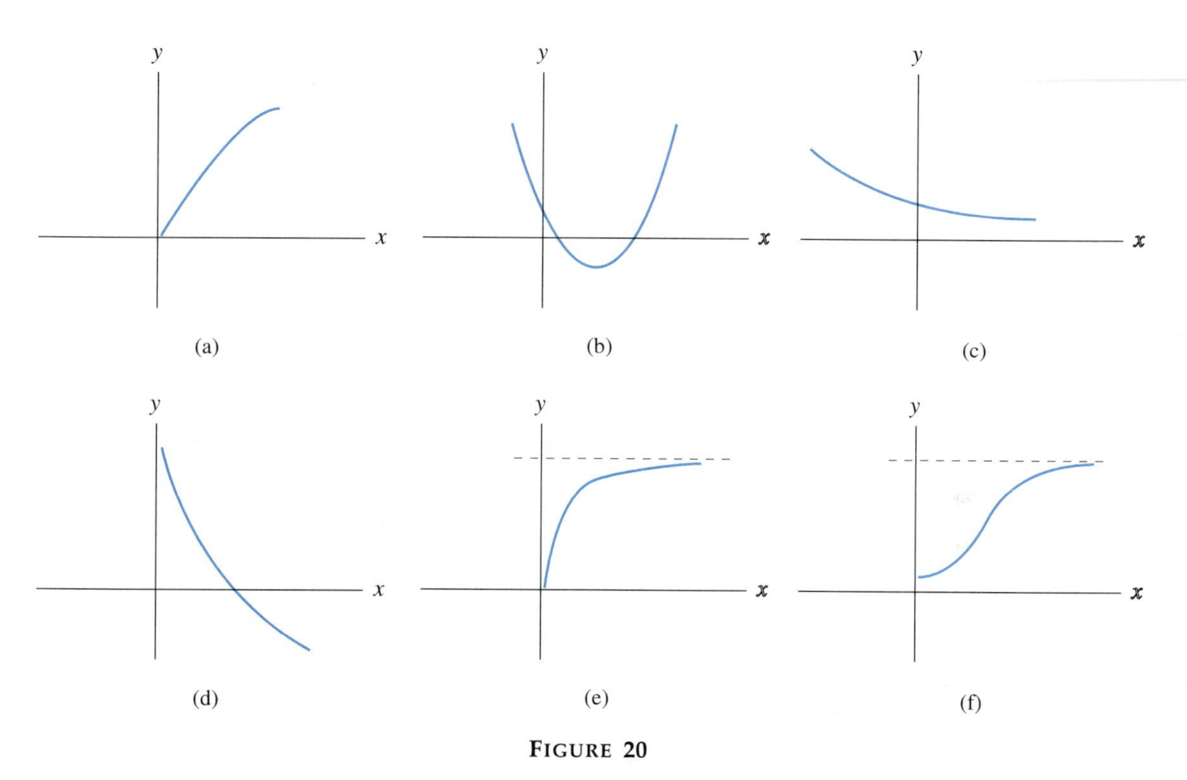

FIGURE 20

EXERCISES 2.4

Exercises 1–4 refer to graphs (a)–(f) in Figure 20.

1. Which functions are increasing for all x?
2. Which functions are decreasing for all x?
3. Which functions have the property that the slope always increases as x increases?
4. Which functions have the property that the slope always decreases as x increases?

Describe each of the following graphs. Your description should include each of the six categories mentioned earlier.

5. 6.

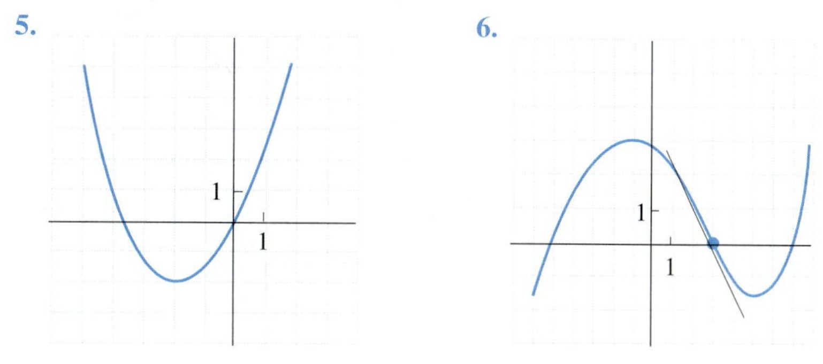

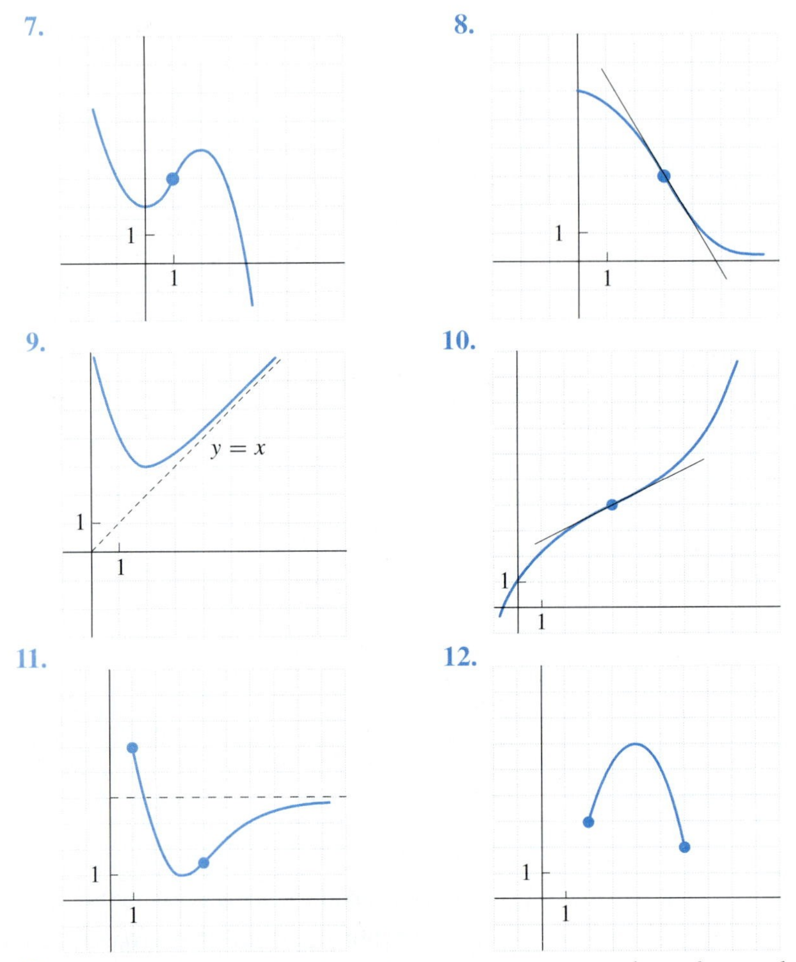

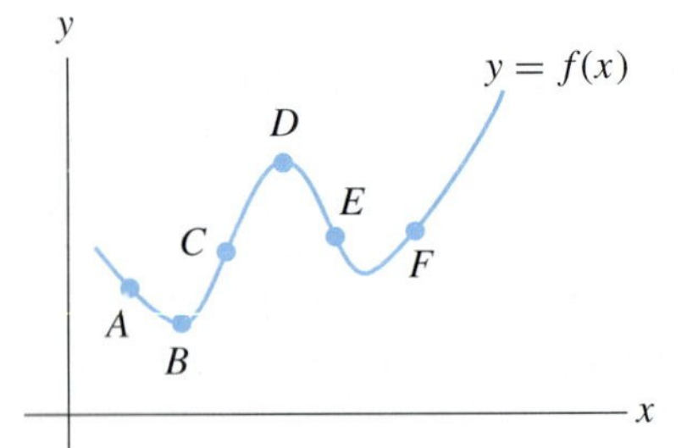

FIGURE 21

13. Describe the way the *slope* changes as you move along the graph (from left to right) in Exercise 5.

14. Describe the way the *slope* changes on the graph in Exercise 6.

15. Describe the way the *slope* changes on the graph in Exercise 8.

16. Describe the way the *slope* changes on the graph in Exercise 10.

Exercises 17 and 18 refer to the graph in Figure 21.

17. (a) At which labeled points is the function increasing?
 (b) At which labeled points is the graph concave up?
 (c) Which labeled point has the most positive slope?

18. (a) At which labeled points is the function decreasing?
 (b) At which labeled points is the graph concave down?
 (c) Which labeled point has the most negative slope (that is, negative and with the greatest magnitude)?

In Exercises 19–22, draw the graph of a function $y = f(x)$ with the stated properties.

19. Both the function and the slope increase as x increases.

20. The function increases and the slope decreases as x increases.

21. The function decreases and the slope increases as x increases. (*Note*: The slope is negative but becomes less negative.)

22. Both the function and the slope decrease as x increases. (*Note*: The slope is negative and becomes more negative.)

23. The annual world consumption of oil rises each year. Furthermore, the amount of the annual *increase* in oil consumption is also rising each year. Sketch a graph that could represent the annual world consumption of oil.

24. In certain professions the average annual income has been rising at an increasing rate. Let $f(T)$ denote the average annual income at year T for persons in one of these professions and sketch a graph that could represent $f(T)$.

25. At noon a child's temperature is 101° and is rising at an increasing rate. At 1 P.M. the child is given medicine. After 2 P.M. the temperature is still increasing but at a decreasing rate. The temperature reaches a peak of 103° at 3 P.M. and decreases to 100° by 5 P.M. Draw a possible graph of the function $T(t)$, the child's temperature at time t.

26. The number of parking tickets given out each year in the District of Columbia increased from 114,000 in 1950 to 1,500,000 in 1990. Also, each year's increase was greater than the increase for the previous year. Draw a possible graph of the annual number of parking tickets as a function of time. [*Note*: If $f(t)$ is the yearly rate t years after 1950, then $f(0) = 114$ and $f(40) = 1500$ thousand tickets.]

27. Let $C(x)$ denote the total cost of manufacturing x units of some product. Then $C(x)$ is an increasing function for all x. For small values of x, the rate of increase of $C(x)$ decreases. (This is because of the savings that are possible with "mass production.") Eventually, however, for large values of x, the cost $C(x)$ increases at an increasing rate. (This happens when production facilities are strained and become less efficient.) Sketch a graph that could represent $C(x)$.

28. One method of determining the level of blood flow through the brain requires the person to inhale air containing a fixed concentration of N_2O, nitrous oxide. During the first minute, the concentration of N_2O in the jugular vein grows at an increasing rate to a level of .25%. Thereafter it grows at a decreasing rate and reaches a concentration of about 4% after 10 minutes. Draw a possible graph of the concentration of N_2O in the vein as a function of time.

29. Suppose that some organic waste products are dumped into a lake at time $t = 0$, and suppose that the oxygen content of the lake at time t is given by the graph in Figure 22. Describe the graph in physical terms. Indicate the significance of the inflection point at $t = b$.

30. Figure 23 gives the U.S. electrical energy production in quadrillion kilowatt-hours from 1910 ($t = 10$) to 1990 ($t = 90$). In what year was the level of production growing at the greatest rate?

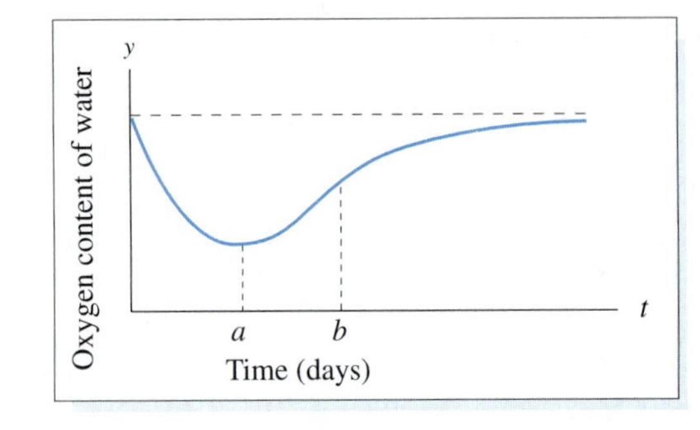

FIGURE 22 A lake's recovery from pollution

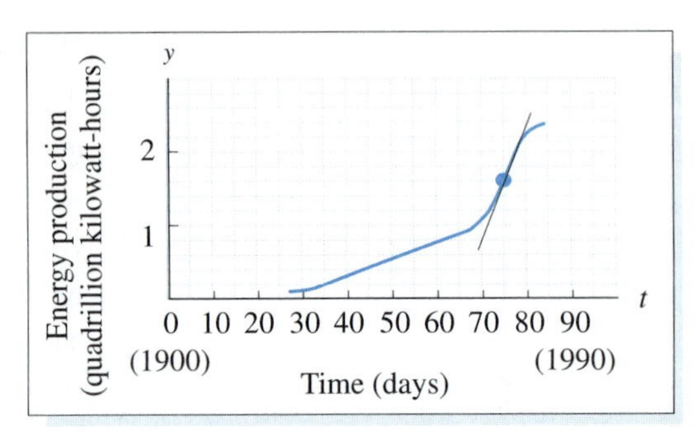

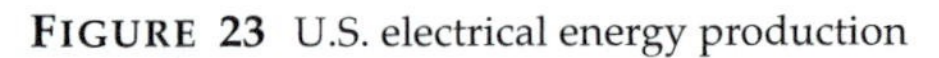

FIGURE 23 U.S. electrical energy production

31. Figure 24 gives the number of U.S. farms in millions from 1920 ($t = 20$) to 1990 ($t = 90$). In what year was the number of farms decreasing most rapidly?

32. Figure 25 shows the graph of the consumer price index for the years 1976 ($t = 0$) through 1990 ($t = 14$). This index measures how much a basket of commodities that cost \$100 in the beginning of 1976 would cost at any given time. In what year was the rate of increase of the index greatest? The least?

In Exercises 33 and 34, describe the graph using the terms introduced in this section. In particular, point out when it is increasing and/or decreasing at the greatest rates and identify those rates.

33. In any given locality, tap water temperature varies during the year. In Dallas, Texas, the tap water temperature (in degrees Fahrenheit) t days after the beginning of a year is given approximately by the graph of the function* in Figure 26.

34. Figure 27 shows the percentage of first-year college students intending to major in business** from 1970 to 1992. Time is measured in years and $t = 0$ corresponds to 1970.

---

* See D. Rapp, *Solar Energy* (Englewood Cliffs, N.J.: Prentice-Hall, Inc., 1981), p. 171.

** *Statistical Abstracts of the United States*, 1994, p. 180, Table 285.

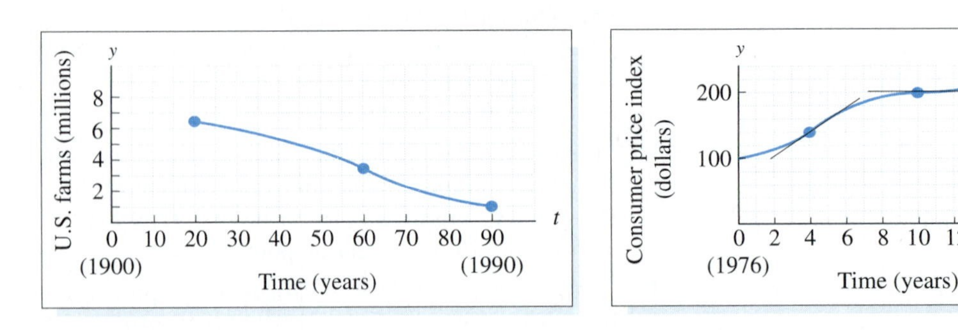

FIGURE 24 Number of U.S. farms

FIGURE 25 Consumer price index

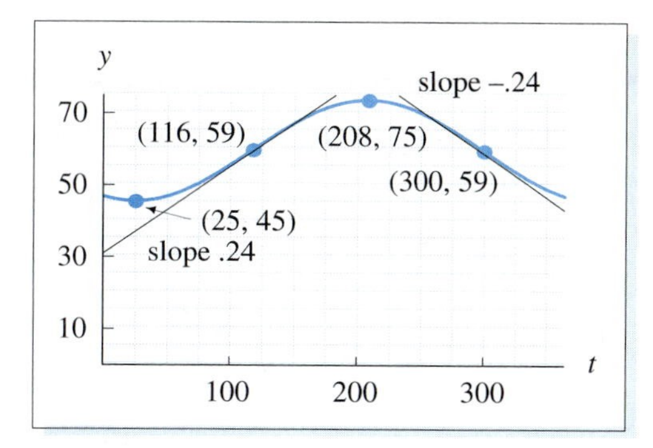

FIGURE 26 Tap water temperature in Dallas, Texas

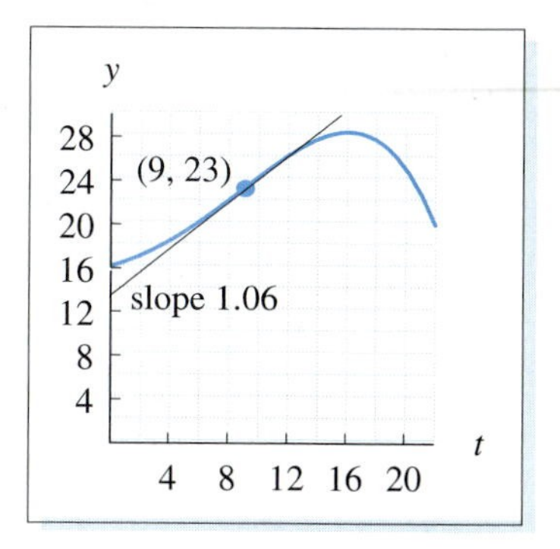

FIGURE 27 Percentage of first-year college students intending to major in business

Exercises 35–39 refer to Figure 28. The population (in millions) of the United States (excluding Alaska and Hawaii) t years after 1800 is given by the function $f(t)$ in Figure 28(a). The graphs of $f'(t)$ and $f''(t)$ are shown in Figures 28(b) and (c).

35. What was the population of the United States in 1820?
36. Approximately when was the population 90 million?
37. How fast was the population growing in 1870?
38. When was the population growing at the rate of 200,000 people per year?
39. In what year did the graph of $f(t)$ have an inflection point? [*Note*: In that year, the graph of $f'(t)$ had a horizontal tangent line and the population was growing at the greatest rate.] How large was the population at that time, and how fast was it growing?

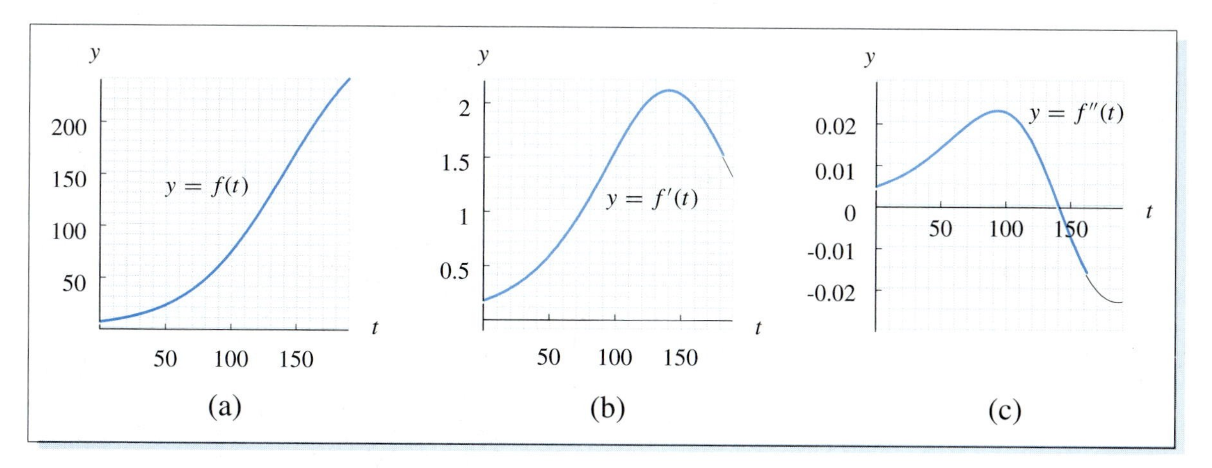

FIGURE 28 Population of the United States from 1800 to 1994

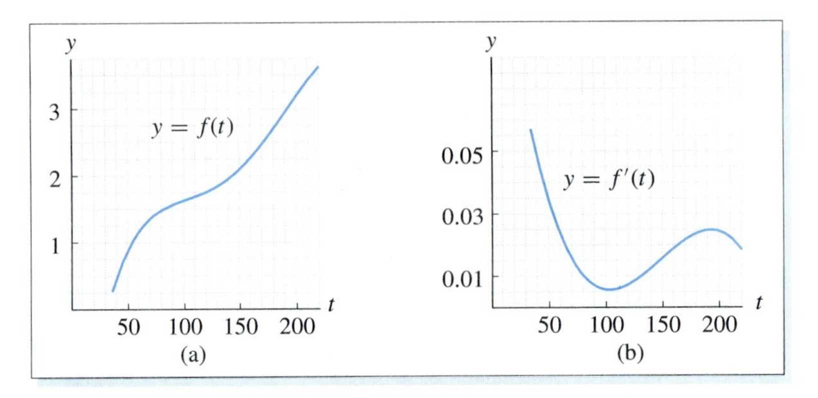

FIGURE 29 The growth of a tropical grass

Exercises 40-46 refer to Figure 29. The height (in meters) of the tropical bunch-grass elephantmillet, t days after mowing, is given by the graph of $f(t)$ in Figure 29(a). The graph of $f'(t)$ is given in Figure 29(b).*

40. How tall was the grass after 160 days?
41. When was the grass 1.75 meters high?
42. How fast was the grass growing after 80 days?
43. When was the grass growing at the rate of .035 meters per day?
44. During the first 200 days, when was the grass growing slowest?
45. During the period from the 100th to the 220th day, approximately when was the grass growing fastest?
46. During the period from the 50th to the 150th day, approximately when was the grass growing fastest?
47. Let $s(t)$ be the distance (in feet) traveled by a parachutist after t seconds from the time of opening the chute, and suppose that $s(t)$ has the line $y = -15t$ as an asymptote. What does this imply about the velocity of the parachutist? (*Note*: Distance traveled downward is given a negative value.)
48. Let $P(t)$ be the population of a bacteria culture after t days and suppose that $P(t)$ has the line $y = 25{,}000{,}000$ as an asymptote. What does this imply about the size of the population?

In Exercises 49–52, sketch the graph of a function having the given properties.

49. Defined for $0 \leq x \leq 10$; relative maximum point at $x = 3$; absolute maximum value at $x = 10$.
50. Relative maximum points at $x = 1$ and $x = 5$; relative minimum point at $x = 3$; inflection points at $x = 2$ and $x = 4$.
51. Defined and increasing for all $x \geq 0$; inflection point at $x = 5$; asymptotic to the line $y = (3/4)x + 5$.
52. Defined for $x \geq 0$; absolute minimum value at $x = 0$; relative maximum point at $x = 4$; asymptotic to the line $y = (x/2) + 1$.

* Woodard and Prine, "Crop Quality & Utilization," *Crop Science*, **33**(1993), 818–824.

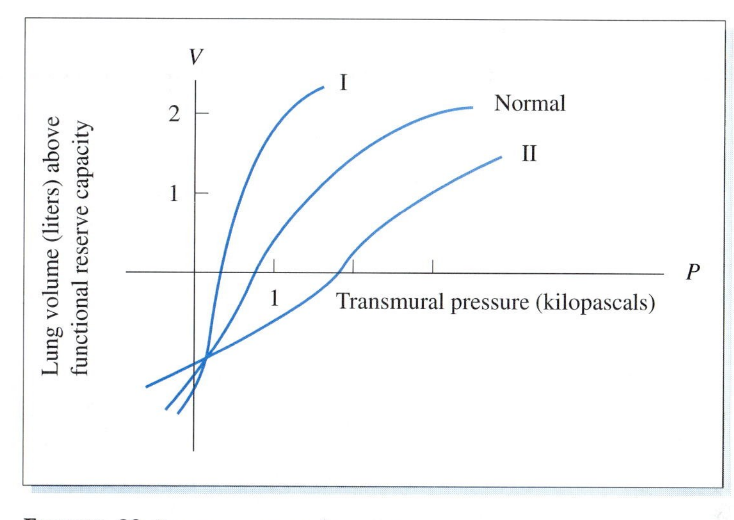

FIGURE 30 Lung pressure-volume curves

53. Consider a smooth curve with no undefined points.
  (a) If it has two relative maximum points, must it have a relative minimum point?
  (b) If it has two relative extreme points, must it have an inflection point?

54. Suppose the function $f(x)$ has a relative minimum at $x = a$ and a relative maximum at $x = b$. Must $f(a)$ be less than $f(b)$?

The difference between the pressure inside the lungs and the pressure surrounding the lungs is called the transmural pressure (*or transmural pressure gradient*). *Figure 30 shows for three different persons how the volume of the lungs is related to the transmural pressure, based on static measurement taken while there is no air flowing through the mouth. (The functional reserve capacity mentioned on the vertical axis is the volume of air in the lungs at the end of a normal expiration.) The rate of change in lung volume with respect to transmural pressure is called the* lung compliance.

55. If the lungs are less flexible than normal, an increase in pressure will cause a smaller change in lung volume than in a normal lung. In this case, is the compliance relatively high or low?

56. Most lung diseases cause a decrease in lung compliance. However, the compliance of a person with emphysema is higher than normal. Which curve (I or II) in the figure could correspond to a person with emphysema?

In the remaining exercises, use a graphing utility to obtain the graphs of the functions. Graph the function in the given window and answer the following questions. (Many of your answers will be approximate.)

(a) *What is the (absolute) maximum value of $f(x)$ over the given interval? For what value of x does it occur?*

(b) *What is the (absolute) minimum value of $f(x)$ over the given interval? For what value of x does it occur?*

57. $f(x) = x + x^2 - x^3$, $[-2, 2]$ *by* $[-10, 10]$

58. $f(x) = \frac{1}{3}x^3 - x^2 - 3x + 2$, $[-3, 5]$ *by* $[-10, 10]$

59. $f(x) = 2x^3(x-2)^2$, $[-.5, 2.5]$ *by* $[-2, 8]$

60. $f(x) = (x^4 - 6x^3)/32$, $[-2, 7]$ *by* $[-5, 12]$

61. $f(x) = (x-1)^2(x-3)^2$, $[0, 4]$ *by* $[0, 10]$

62. $f(x) = \dfrac{3x^2}{x^2+1}$, $[-4, 4]$ *by* $[-.5, 3.5]$

SOLUTIONS TO PRACTICE PROBLEMS 2.4

1. The curve is concave up, so the slope increases. Even though the curve itself is decreasing, the slope becomes less negative as we move from left to right.

2. At $x = 3$. We have drawn in tangent lines at various points (Figure 31). Note that as we move from left to right, the slopes decrease steadily until the point $(3, 2)$, at which time they start to increase. This is consistent with the fact that the graph is concave down (hence, slopes are decreasing) to the left of $(3, 2)$ and concave up (hence, slopes are increasing) to the right of $(3, 2)$. Extreme values of slopes always occur at inflection points.

FIGURE 31

2.5 THE FIRST AND SECOND DERIVATIVE RULES

We shall now show how properties of the graph of a function $f(x)$ are determined by properties of the derivatives, $f'(x)$ and $f''(x)$. These relationships will provide the key to the curve-sketching and optimization problems discussed in the rest of the chapter.

We begin with a discussion of the first derivative of a function $f(x)$. Suppose that for some value of x, say $x = a$, the derivative $f'(a)$ is positive. Then the tangent line at $(a, f(a))$ has positive slope and is therefore a rising line (moving from left to right, of course). Since the graph of $f(x)$ near $(a, f(a))$ resembles its tangent line, the function must be increasing at $x = a$. Similarly, when $f'(a) < 0$, the function is decreasing at $x = a$. (See Figure 1.)

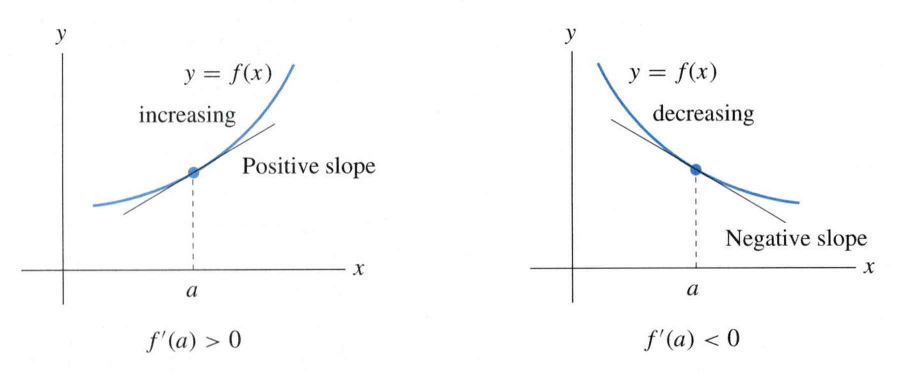

FIGURE 1 Illustration of first derivative rule

Thus, we have the following useful result.

> *First Derivative Rule* If $f'(a) > 0$, then $f(x)$ is increasing at $x = a$. If $f'(a) < 0$, then $f(x)$ is decreasing at $x = a$.

When $f'(a) = 0$, the first derivative rule is not decisive. In this case the function $f(x)$ might be increasing or decreasing or have a relative extreme point at $x = a$.

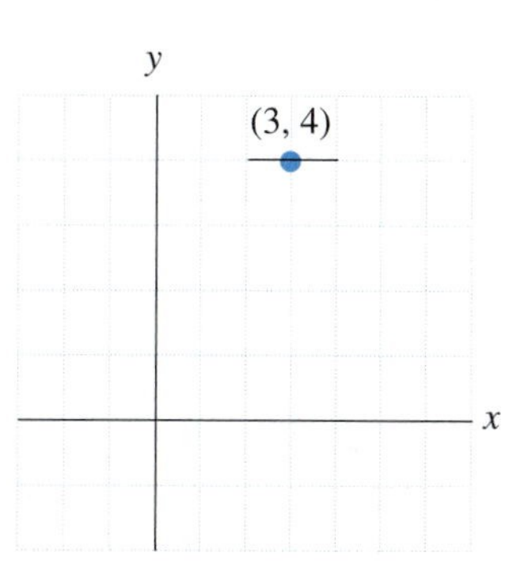

FIGURE 2

EXAMPLE 1 Sketch the graph of a function $f(x)$ that has all the following properties:

(a) $f(3) = 4$.

(b) $f'(x) > 0$ for $x < 3$, $f'(3) = 0$, and $f'(x) < 0$ for $x > 3$.

Solution The only specific point on the graph is $(3, 4)$ [property (a)]. We plot this point and then use the fact that $f'(3) = 0$ to sketch the tangent line at $x = 3$ (Figure 2).

From property (b) and the first derivative rule, we know that $f(x)$ must be increasing for x less than 3 and decreasing for x greater than 3. A graph with these properties might look like the curve in Figure 3. ■

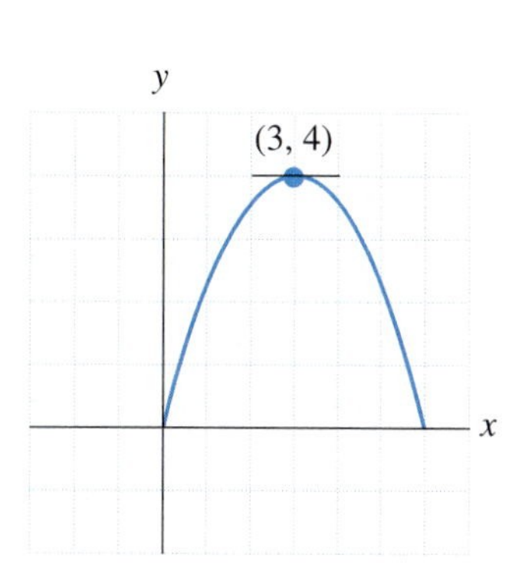

FIGURE 3

The second derivative of a function $f(x)$ gives useful information about the concavity of the graph of $f(x)$. Suppose that $f''(a)$ is negative. Then since $f''(x)$ is the derivative of $f'(x)$, we conclude that $f'(x)$ has a negative derivative at $x = a$. In this case, $f'(x)$ must be a decreasing function at $x = a$; that is, the slope of the graph of $f(x)$ is decreasing as we move from left to right on the graph near $(a, f(a))$. (See Figure 4.) This means that the graph of $f(x)$ is concave down at $x = a$. A similar analysis shows that if $f''(a)$ is positive, then $f(x)$ is concave up at $x = a$. Thus we have the following rule.

> *Second Derivative Rule* If $f''(a) > 0$, then $f(x)$ is concave up at $x = a$. If $f''(a) < 0$, then $f(x)$ is concave down at $x = a$.

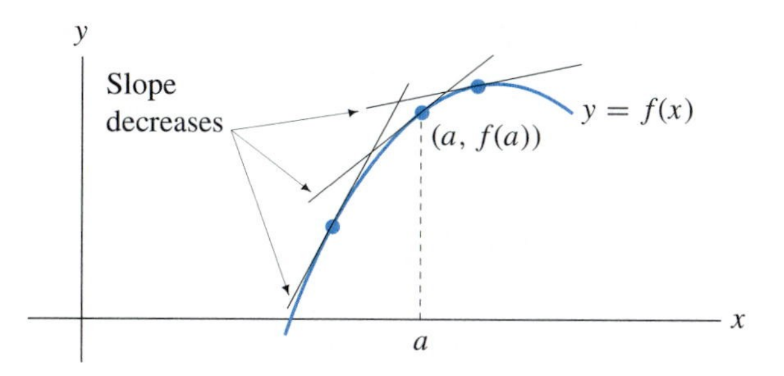

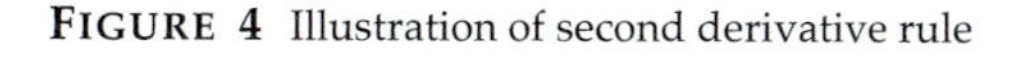

FIGURE 4 Illustration of second derivative rule

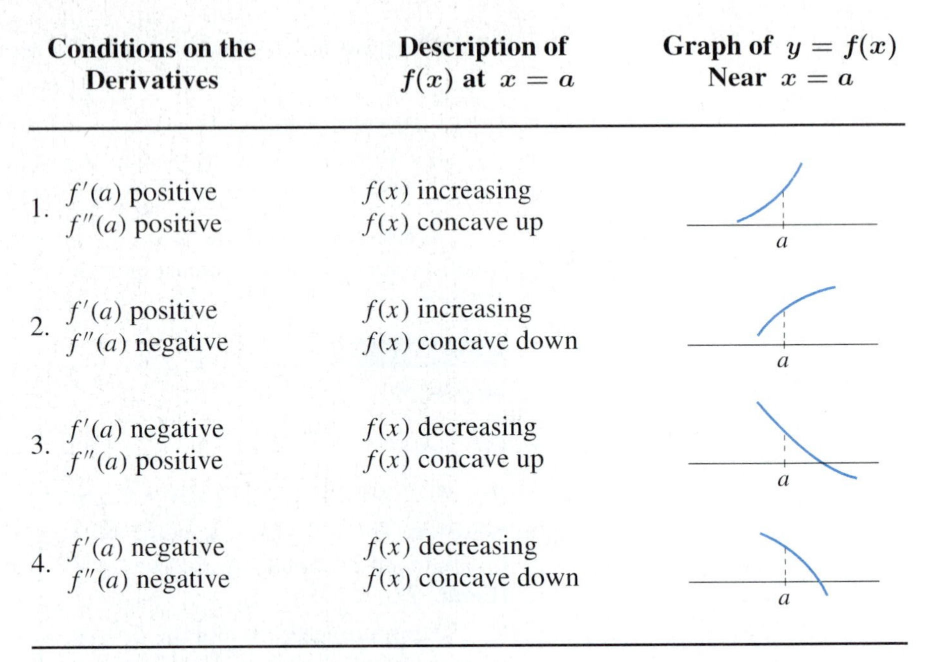

	Conditions on the Derivatives	Description of $f(x)$ at $x = a$	Graph of $y = f(x)$ Near $x = a$
1.	$f'(a)$ positive $f''(a)$ positive	$f(x)$ increasing $f(x)$ concave up	a
2.	$f'(a)$ positive $f''(a)$ negative	$f(x)$ increasing $f(x)$ concave down	a
3.	$f'(a)$ negative $f''(a)$ positive	$f(x)$ decreasing $f(x)$ concave up	a
4.	$f'(a)$ negative $f''(a)$ negative	$f(x)$ decreasing $f(x)$ concave down	a

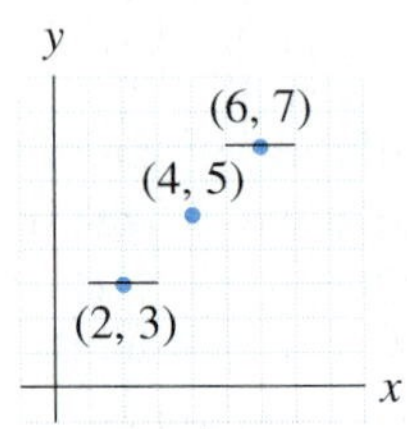

FIGURE 5

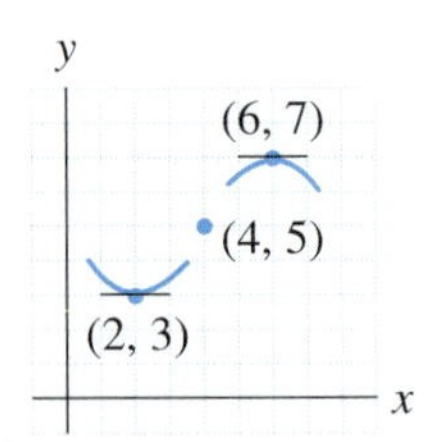

FIGURE 6

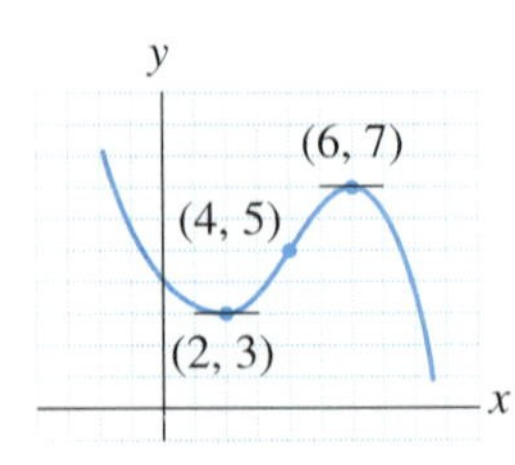

FIGURE 7

When $f''(a) = 0$, the second derivative rule gives no information. In this case, the function might be concave up, concave down, or neither at $x = a$.

The above chart shows how a graph may combine the properties of increasing, decreasing, concave up, and concave down.

EXAMPLE 2 Sketch the graph of a function $f(x)$ with all the following properties:

(a) (2, 3), (4, 5), and (6, 7) are on the graph.

(b) $f'(6) = 0$ and $f'(2) = 0$.

(c) $f''(x) > 0$ for $x < 4$, $f''(4) = 0$, and $f''(x) < 0$ for $x > 4$.

Solution First we plot the three points from property (a) and then sketch two tangent lines, using the information from property (b). (See Figure 5.) From property (c) and the second derivative rule, we know that $f(x)$ is concave up for $x < 4$. In particular, $f(x)$ is concave up at (2, 3). Also, $f(x)$ is concave down for $x > 4$, in particular, at (6, 7). Note that $f(x)$ must have an inflection point at $x = 4$ because the concavity changes there. We now sketch small portions of the curve near (2, 3) and (6, 7). (See Figure 6.) We can now complete the sketch (Figure 7), taking care to make the curve concave up for $x < 4$ and concave down for $x > 4$. ■

EXAMPLE 3 Consider the graph in Figure 8 of $f'(x)$, the derivative of the function $f(x)$.

(a) Explain why $f(x)$ must have a relative maximum at $x = 11$.

(b) Explain why $f(x)$ must have an inflection point at $x = 4$.

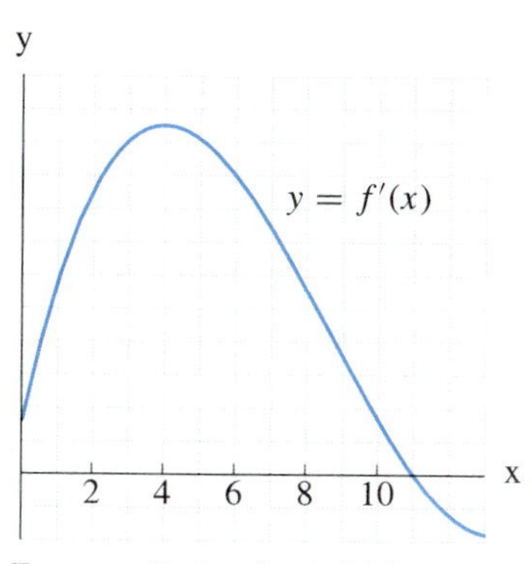

FIGURE 8 Graph of $f'(x)$

Solution (a) Since $f'(11) = 0$, $f(x)$ has a horizontal tangent line at $x = 11$. Consider the behavior of $f(x)$ for $10 \leq x \leq 12$. For x to the left of 11, $f'(x)$ is positive, and therefore the function $f(x)$ is increasing. For x to the right of 11, $f'(x)$ is negative, and therefore $f(x)$ is decreasing. That is, at $x = 11$ the function $f(x)$ changes from increasing to decreasing. This is the definition of a relative maximum point. (*Note*: In general, functions have relative extreme points where the graphs of their derivatives cross the x-axis.)

(b) $f'(x)$ has a relative maximum point at $x = 4$. That is, as we move from left to right through $x = 4$, $f'(x)$ increases until we reach $x = 4$ and then decreases. That is, the slope of $f(x)$ increases until we reach $x = 4$, and then decreases. That is, $f(x)$ is concave up until we reach $x = 4$, and then is concave down. (Increasing slope implies concave up, decreasing slope implies concave down.) Since concavity changes at $x = 4$, we have an inflection point there. (*Note*: In general, functions have inflection points where the graphs of their first derivatives have relative extreme points.) ■

PRACTICE PROBLEMS 2.5

1. Make a good sketch of the function $f(x)$ near the point where $x = 2$, given that $f(2) = 5$, $f'(2) = 1$, and $f''(2) = -3$.
2. The graph of $f(x) = x^3$ is shown in Figure 9.
   (a) Is the function increasing at $x = 0$?
   (b) Compute $f'(0)$.
   (c) Reconcile your answers to parts (a) and (b) with the first derivative rule.
3. The graph of $y = f'(x)$ is shown in Figure 10. Explain why $f(x)$ must have a relative minimum point at $x = 3$.

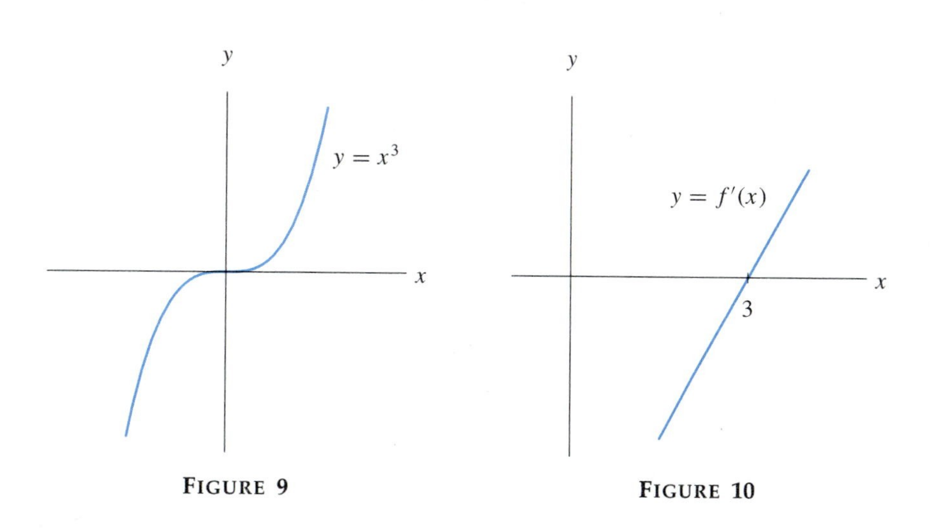

FIGURE 9

FIGURE 10

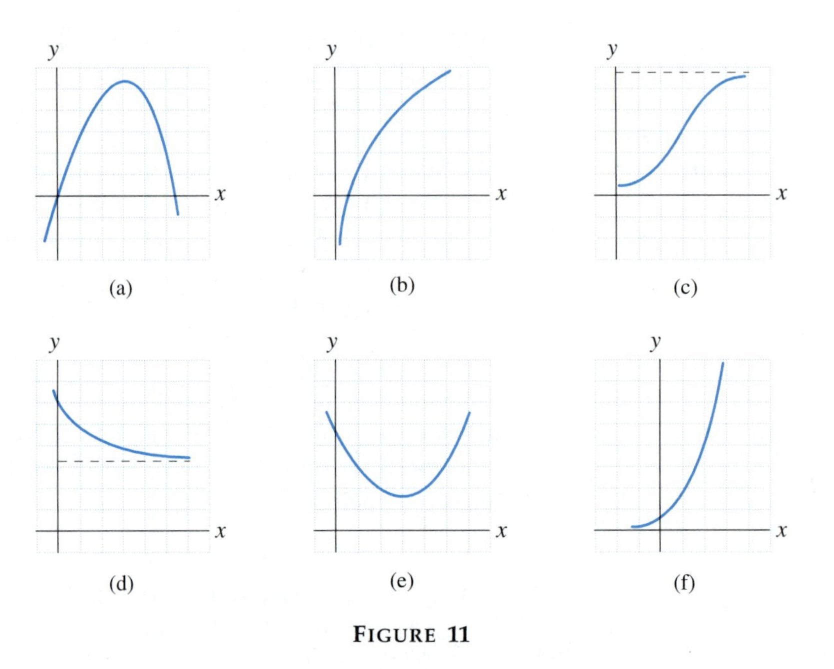

FIGURE 11

EXERCISES 2.5

Exercises 1–4 refer to the functions whose graphs are shown in Figure 11.

1. Which functions have a positive first derivative for all x?
2. Which functions have a negative first derivative for all x?
3. Which functions have a positive second derivative for all x?
4. Which functions have a negative second derivative for all x?
5. Which one of the graphs in Figure 12 could represent a function $f(x)$ for which $f(a) > 0$, $f'(a) = 0$, and $f''(a) < 0$?
6. Which one of the graphs in Figure 12 could represent a function $f(x)$ for which $f(a) = 0$, $f'(a) < 0$, and $f''(a) > 0$?

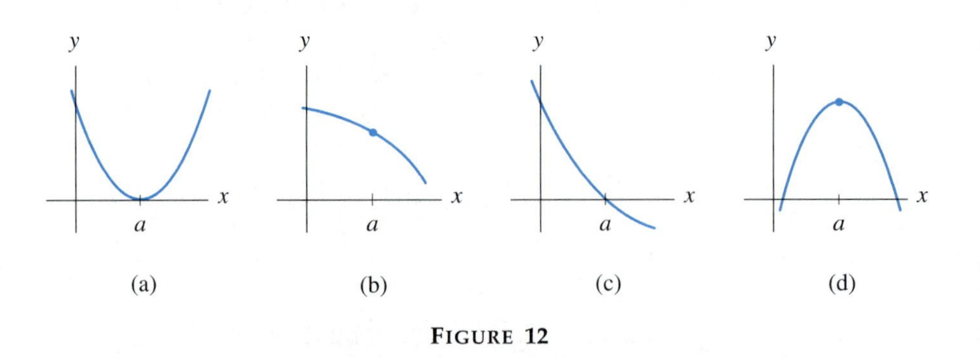

FIGURE 12

In Exercises 7–12, sketch the graph of a function that has the properties described.

7. $f(2) = 1$; $f'(2) = 0$; concave up for all x.
8. $f(-1) = 0$; $f'(x) < 0$ for $x < -1$; $f'(-1) = 0$; and $f'(x) > 0$ for $x > -1$.
9. $f(3) = 5$; $f'(x) > 0$ for $x < 3$; $f'(3) = 0$; and $f'(x) > 0$ for $x > 3$.
10. $(-2, -1)$ and $(2, 5)$ are on the graph; $f'(-2) = 0$ and $f'(2) = 0$; $f''(x) > 0$ for $x < 0$; $f''(0) = 0$, $f''(x) < 0$ for $x > 0$.
11. $(0, 6)$, $(2, 3)$, and $(4, 0)$ are on the graph; $f'(0) = 0$ and $f'(4) = 0$; $f''(x) < 0$ for $x < 2$; $f''(2) = 0$, $f''(x) > 0$ for $x > 2$.
12. $f(x)$ defined only for $x \geq 0$; $(0, 0)$ and $(5, 6)$ are on the graph; $f'(x) > 0$ for $x \geq 0$; $f''(x) < 0$ for $x < 5$; $f''(5) = 0$, $f''(x) > 0$ for $x > 5$.

In Exercises 13–18, use the given information to make a good sketch of the function $f(x)$ near $x = 3$.

13. $f(3) = 4$, $f'(3) = -\frac{1}{2}$, $f''(3) = 5$
14. $f(3) = -2$, $f'(3) = 0$, $f''(3) = 1$
15. $f(3) = 1$, $f'(3) = 0$, inflection point at $x = 3$, $f'(x) > 0$ for $x > 3$
16. $f(3) = 4$, $f'(3) = -\frac{3}{2}$, $f''(3) = -2$
17. $f(3) = -2$, $f'(3) = 2$, $f''(3) = 3$
18. $f(3) = 3$, $f'(3) = 1$, inflection point at $x = 3$, $f''(x) < 0$ for $x > 3$
19. Refer to the graph in Figure 13. Fill in each entry of the grid with POS, NEG, or 0.

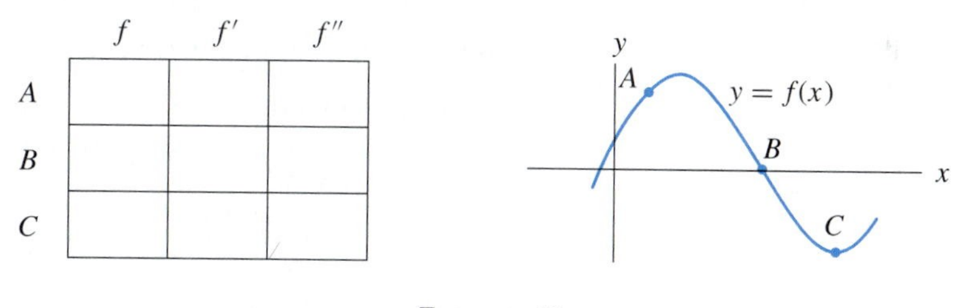

FIGURE 13

20. The first and second derivatives of the function $f(x)$ have the values given in Table 1.

    (a) Find the x-coordinates of all relative extreme points.
    (b) Find the x-coordinates of all inflection points.

Table 1 VALUES OF THE FIRST TWO DERIVATIVES OF A FUNCTION

x	$0 \leq x < 2$	2	$2 < x < 3$	3	$3 < x < 4$	4	$4 < x \leq 6$
$f'(x)$	Positive	0	Negative	Negative	Negative	0	Negative
$f''(x)$	Negative	Negative	Negative	0	Positive	0	Negative

FIGURE 14

Exercises 21–32 refer to Figure 14, which contains the graph of $f'(x)$, the derivative of the function $f(x)$.

21. Explain why $f(x)$ must be increasing at $x = 6$.
22. Explain why $f(x)$ must be decreasing at $x = 4$.
23. Explain why $f(x)$ has a relative maximum at $x = 3$.
24. Explain why $f(x)$ has a relative minimum at $x = 5$.
25. Explain why $f(x)$ must be concave up at $x = 0$.
26. Explain why $f(x)$ must be concave down at $x = 2$.
27. Explain why $f(x)$ has an inflection point at $x = 1$.
28. Explain why $f(x)$ has an inflection point at $x = 4$.
29. If $f(6) = 3$, what is the equation of the tangent line to the graph of $y = f(x)$ at $x = 6$?
30. If $f(6) = 8$, what is an approximate value of $f(6.5)$?
31. If $f(0) = 3$, what is an approximate value of $f(.25)$?
32. If $f(0) = 3$, what is the equation of the tangent line to the graph of $y = f(x)$ at $x = 0$?
33. Suppose melting snow causes a river to overflow its banks and $h(t)$ is the number of inches of water on Main Street t hours after the melting begins.
    (a) If $h'(100) = \frac{1}{3}$, by approximately how much will the water level change during the next half-hour?
    (b) Which of the following two conditions is the best news?
        (i) $h(100) = 3, h'(100) = 2, h''(100) = -5$
        (ii) $h(100) = 3, h'(100) = -2, h''(100) = 5$
34. Suppose $T(t)$ is the temperature on a hot summer day at time t hours.
    (a) If $T'(10) = 4$, by approximately how much will the temperature rise from 10:00 to 10:45?
    (b) Which of the following two conditions is the best news?
        (i) $T(10) = 95, T'(10) = 4, T''(10) = -3$
        (ii) $T(10) = 95, T'(10) = -4, T''(10) = 3$

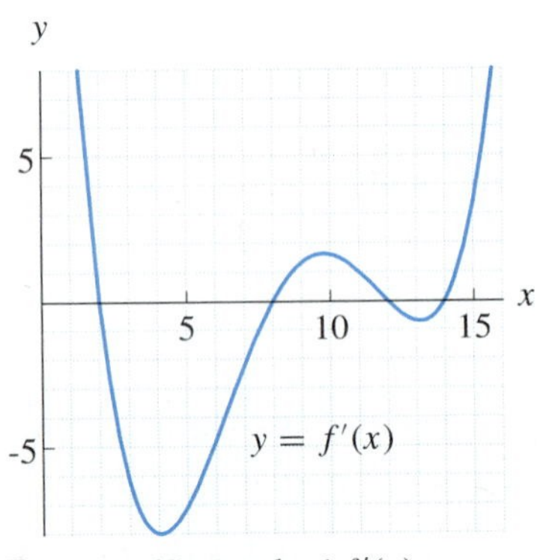

FIGURE 15 Graph of $f'(x)$

Exercises 35–46 refer to Figure 15, which contains the graph of $f'(x)$, the derivative of the function $f(x)$.

35. How many relative extreme points does $f(x)$ have?
36. How many relative minimum points does $f(x)$ have?
37. How many inflection points does $f(x)$ have?
38. How many relative maximum points does $f(x)$ have?
39. Is $f''(8)$ positive or negative?
40. Is $f''(12)$ positive or negative?
41. Explain why the tangent line to $f(x)$ at $x = 2$ lies above the graph of $f(x)$.
42. Explain why the tangent line to $f(x)$ at $x = 14$ lies below the graph of $f(x)$.

43. Suppose $f(4) = 1$. What is the equation of the tangent line to the graph of $f(x)$ at the point $(4, 1)$?

44. Suppose $f(11) = 5$. Make a good sketch of the function $f(x)$ near the point where $x = 11$.

45. Where does $f(x)$ have its greatest value on the interval $6 \leq x \leq 7$?

46. Where does $f(x)$ have its greatest value on the interval $9 \leq x \leq 10$?

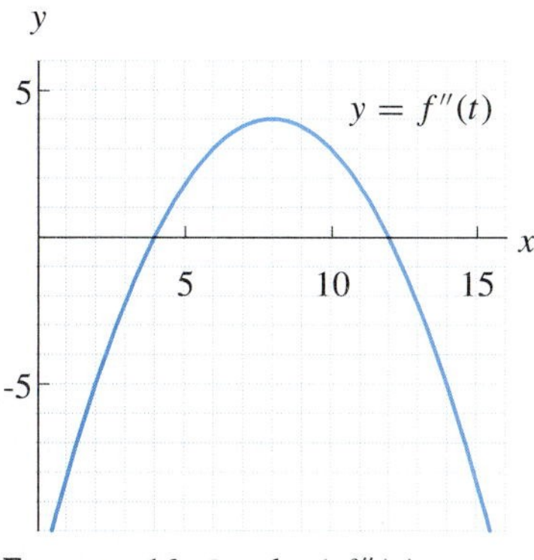

FIGURE 16 Graph of $f''(x)$

Exercises 47–52 refer to Figure 16, which contains the graph of $f''(x)$, the second derivative of the function $f(x)$.

47. Explain why $f(x)$ must be concave up at $x = 8$.

48. Explain why $f(x)$ must be concave down at $x = 3$.

49. Explain why $f(x)$ must have an inflection point at $x = 12$.

50. Suppose $f(6) = 5$ and $f'(6) = \frac{1}{2}$. Make a good sketch of the function $f(x)$ near the point where $x = 6$.

51. Suppose $f'(8) = 0$. What conclusion can you make about the graph of $f(x)$ at $x = 8$?

52. Suppose $f'(14) = 0$. What conclusion can you make about the graph of $f(x)$ at $x = 14$?

In Exercises 53–56, determine which function is the derivative of the other.

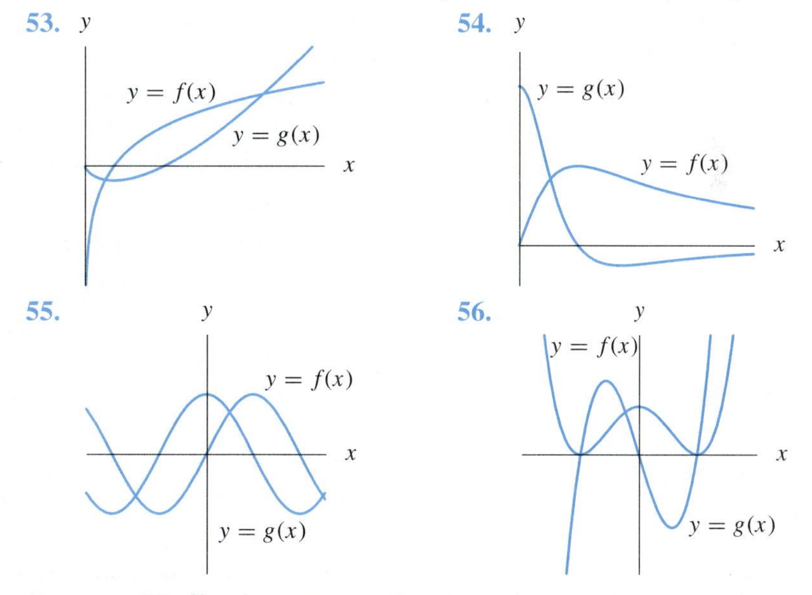

Exercises 57–62 refer to Figure 17. After a drug is taken orally, the amount of the drug in the bloodstream after t hours is given by the function $f(t)$ in Figure 17.

57. How many units of the drug are in the bloodstream after 7 hours?

58. At what rate is the level of the drug in the bloodstream increasing after 1 hour?

59. Approximately when (while the level is decreasing) is the level of the drug in the bloodstream 20 units?

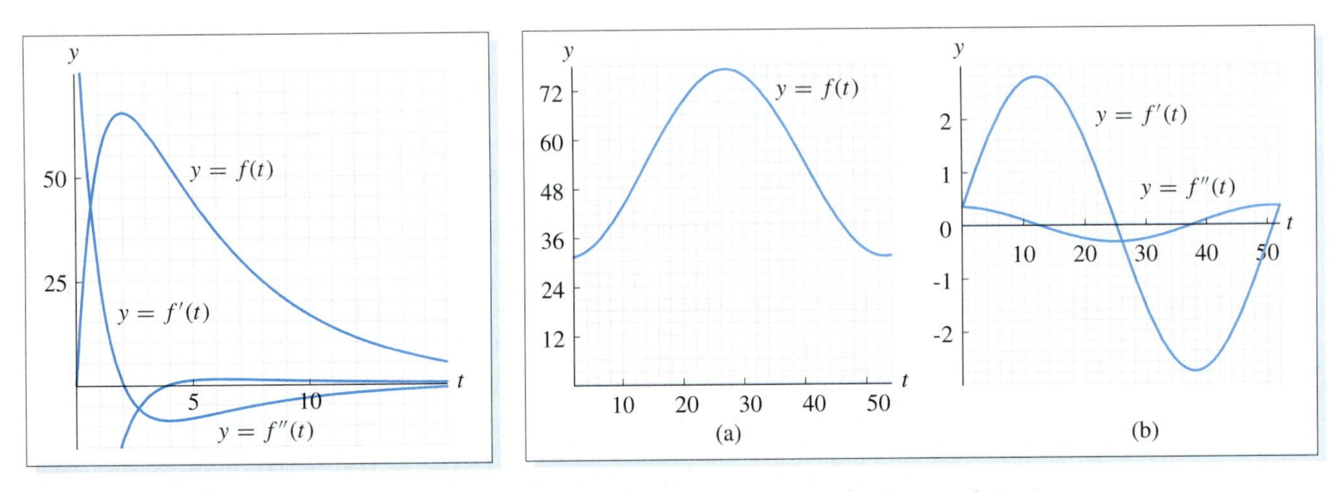

FIGURE 17 Level of the drug in the bloodstream

FIGURE 18 A temperature curve and its two derivatives

60. Approximately when is the level of the drug in the bloodstream decreasing at the rate of 5 units per hour?

61. What is the greatest level of the drug in the bloodstream and when is this level reached?

62. When is the level of the drug in the bloodstream increasing fastest?

Exercises 63–68 refer to Figure 18. The average weekly temperature in Washington, DC, t weeks after the beginning of the year is given by the function $f(t)$ in Figure 18(a). The first and second derivatives of $f(t)$ are shown in Figure 18(b).

63. What is the average weekly temperature after 18 weeks?

64. After 20 weeks, how fast is the temperature changing?

65. When is the average weekly temperature 39 degrees?

66. When is the average weekly temperature falling at the rate of 1 degree per day?

67. When is the average weekly temperature greatest? Least?

68. When is the average weekly temperature increasing fastest? Decreasing fastest?

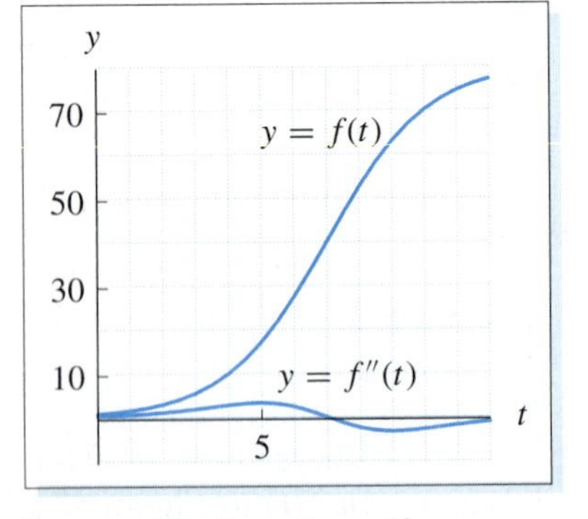

FIGURE 19 Percentage of homes having a TV and a VCR

Exercises 69–71 refer to Figure 19. The function $f(t)$ in Figure 19 gives the percentage of homes (with TV sets) that also had a VCR in year t, where $t = 0$ corresponds to 1980.

69. In what year did 10% of the homes have a VCR?

70. What percentage of the homes had a VCR in 1990?

71. When was the percentage of homes with VCRs growing at the greatest rate? Approximately what percentage of homes had VCRs at that time?

In the remaining exercises, use a graphing utility to obtain the graphs of the functions. In Exercises 72–75, let $f(x)$ be the given function, but display only the graph of $f'(x)$. [With a calculator, set `Y1 = f(x)`, set `Y2 = nDeriv(Y1,X,X)` or `y2=der1(y1,x,x)`, select only Y_2, *and press* GRAPH. *With Visual Calculus, graph the function with the* "Analyze a Function and Its Derivatives" *routine, and then press* 0 1 *as soon as the function appears.*] *Use the graph of $f'(x)$ to determine the important features, such as relative extreme points and inflection points, of the graph of $f(x)$. Then check your conclusions by displaying the graph of $f(x)$.*

72. $f(x) = 4x^3 - 6x^2 - 9x + 5$; $[-2, 2]$ *by* $[-25, 25]$
73. $f(x) = -x^3 + 9x^2 - 15x$; $[-2, 6]$ *by* $[-15, 30]$
74. $f(x) = 3x^5 - 20x^3 - 120x$; $[-4, 4]$ *by* $[-325, 325]$
75. $f(x) = 2x^2(.5)^x$; $[-1, 7]$ *by* $[-1, 2.5]$

In Exercises 76–79, graph the given function and use the graph of $f(x)$ to determine the important features, such as x-intercepts and relative extreme points, of the graph of $f'(x)$. Then check your conclusions by displaying the graph of $f'(x)$.

76. $f(x) = 2x^3 - 8x + 1$; $[-4, 4]$ *by* $[-35, 35]$
77. $f(x) = x^3 - 9x^2 + 15x + 5$; $[-1, 7]$ *by* $[-25, 25]$
78. $f(x) = x^4 - x^2$; $[-2, 2]$ *by* $[-1, 1]$
79. $f(x) = 3x^5 - 5x^3$; $[-2, 2]$ *by* $[-4, 4]$
80. Find the minimum value of the slope of $f(x) = x^3 - 9x^2 + 20x + 5$. Use the window $[-2, 6]$ *by* $[-20, 20]$.
81. Find the maximum value of the slope of $f(x) = 6x^2 - x^3$. Use the window $[-2, 6]$ *by* $[-20, 35]$.
82. Find the maximum value of the slope of

$$f(x) = \frac{1}{.25 + (.5)^x}.$$

Use the window $[0, 8]$ *by* $[-1, 4]$.

83. Find the minimum value of the slope of

$$f(x) = \frac{20}{3 + x^2}.$$

Use the window $[-4, 4]$ *by* $[-5, 7]$.

84. A rumor is spreading through a city. The number of people who have heard the rumor after t days is given by $f(t) = 100{,}000/(1 + 9.134(.8)^t)$. Sketch the graph of $f(t)$ in the window $[0, 32]$ *by* $[0, 100000]$ and sketch the graphs of $f'(t)$ and $f''(t)$ in the window $[0, 32]$ *by* $[-500, 6000]$.
   (a) Approximately how many people have heard the rumor after 14 days?
   (b) How fast is the rumor spreading after 8 days?
   (c) When will 61,437 people have heard the rumor?
   (d) When is the rumor spreading at the rate of 1320 people per day?
   (e) When is the rumor spreading fastest?

85. A cancerous tumor* has volume $f(t) = 1.825^3(1-1.6(.657)^t)^3$ milliliters after t weeks, with $t > 1$. Sketch the graphs of $f(t)$, $f'(t)$, $f''(t)$ in the window $[1, 17]$ *by* $[-.5, 6]$.

(a) Approximately how large is the tumor after 14 weeks?
(b) Approximately when will the tumor have a volume of 4 milliliters?
(c) Approximately how fast is the tumor growing after 5 weeks?
(d) Approximately when is the tumor growing at the rate of .5 milliliters per week?
(e) Approximately when is the tumor growing at the fastest rate?

86. The number of farms in the United States t years after 1925 is

$$f(t) = 1.5 + \frac{4.7}{1 + \left(\dfrac{t}{40.91}\right)^4}$$

million. Sketch the graph of $f(t)$ in the window $[0, 64]$ *by* $[0, 6.4]$ and sketch the graphs of $f'(t)$ and $f''(t)$ in the window $[0, 64]$ *by* $[-.13, .005]$.

(a) Approximately how many farms were there in 1963?
(b) At what rate was the number of farms declining in 1947?
(c) In what year were there about 6 million farms?
(d) During which years was the number of farms declining at the rate of 80,000 farms per year?
(e) During which year was the number of farms declining fastest?

SOLUTIONS TO PRACTICE PROBLEMS 2.5

1. Since $f(2) = 5$, the point $(2, 5)$ is on the graph [Figure 20(a)]. Since $f'(2) = 1$, the tangent line at the point $(2, 5)$ has slope 1. Draw in the tangent line [Figure 20(b)]. Near the point $(2, 5)$ the graph looks approximately like the tangent line. Since $f''(2) = -3$, a negative number, the graph is concave down at the point $(2, 5)$. Now we are ready to sketch the graph [Figure 20(c)].

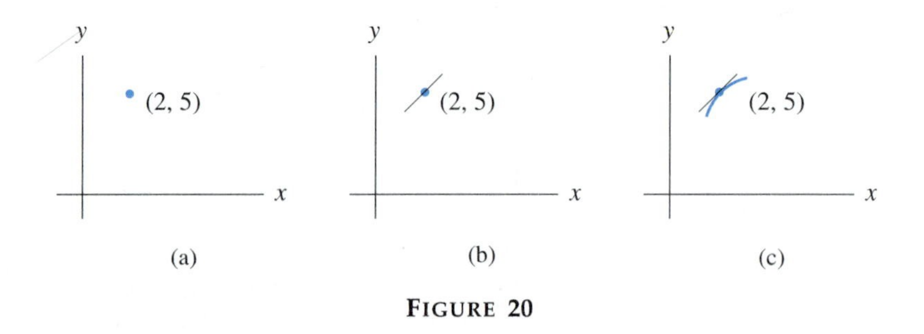

FIGURE 20

* Baker, Goddard, Clark, and Whimster, "Proportion of Necrosis in Transplanted Murine Adenocarcinomas and Its Relationship to Tumor Growth," *Growth, Development & Aging*, **54**(1990), 85–93.

2. (a) Yes. The graph is steadily increasing as we pass through the point (0, 0).
   (b) Since $f'(x) = 3x^2$, $f'(0) = 3 \cdot 0^2 = 0$.
   (c) There is no contradiction here. The first derivative rule says that if the derivative is positive, the function is increasing. However, it does not say that this is the only condition under which a function is increasing. As we have just seen, sometimes we can have the first derivative zero and the function still increasing.
3. Since $f'(x)$, the derivative of $f(x)$, is negative to the left of $x = 3$ and positive to the right of $x = 3$, $f(x)$ is decreasing to the left of $x = 3$ and increasing to the right of $x = 3$. Therefore, by the definition of a relative minimum point, $f(x)$ has a relative minimum point at $x = 3$.

2.6 LIMITS AND THE DERIVATIVE

The notion of a limit is one of the fundamental ideas of calculus. Indeed, any "theoretical" development of calculus rests on an extensive use of limits. Even in this book, where we have adopted an intuitive viewpoint, limit arguments are used occasionally (although in an informal way). In this section we give a brief introduction to limits and their role in calculus. As we shall see, the limit concept will allow us to define the notion of a derivative independently of our geometric reasoning.

Actually, we have already considered a limit in our discussion of the derivative, although we did not use the term "limit." Using the geometric reasoning of Section 2.1, we have the following procedure for calculating the derivative of a function $f(x)$ at $x = a$. First calculate the *difference quotient*

$$\frac{f(a+h) - f(a)}{h},$$

where h is a nonzero number. Next, allow h to assume both positive and negative numbers arbitrarily close to zero but different from zero. In symbols, we write $h \to 0$ and say h approaches zero. The values of the difference quotient then approach the value of the derivative $f'(a)$. We say that the number $f'(a)$ is the *limit* of the difference quotient as h approaches zero, and in symbols we write

$$f'(a) = \lim_{h \to 0} \frac{f(a+h) - f(a)}{h}. \tag{1}$$

As a numerical example of formula (1), consider the case of the derivative of $f(x) = x^2$ at $x = 2$. The difference quotient in this case has the form

$$\frac{f(2+h) - f(2)}{h} = \frac{(2+h)^2 - 2^2}{h}.$$

Table 1 gives some typical values of this difference quotient for progressively smaller values of h, both positive and negative. It is clear that the values of the

Table 1

h	$\dfrac{(2+h)-2^2}{h}$	h	$\dfrac{(2+h)-2^2}{h}$
1	$\dfrac{(2+1)^2-2^2}{1}=5$	-1	$\dfrac{(2+(-1))^2-2^2}{-1}=3$
.1	$\dfrac{(2+.1)^2-2^2}{.1}=4.10$	$-.1$	$\dfrac{(2+(-.1))^2-2^2}{-.1}=3.9$
.01	$\dfrac{(2+.01)^2-2^2}{.01}=4.01$	$-.01$	$\dfrac{(2+(-.01))^2-2^2}{-.01}=3.99$
.001	$\dfrac{(2+.001)^2-2^2}{.001}=4.001$	$-.001$	$\dfrac{(2+(-.001))^2-2^2}{-.001}=3.999$
.0001	$\dfrac{(2+.0001)^2-2^2}{.0001}=4.0001$	$-.0001$	$\dfrac{(2+(-.0001))^2-2^2}{-.0001}=3.9999$

difference quotient are approaching 4 as $h \to 0$. In other words, 4 *is the limit* of the difference quotient as $h \to 0$. Thus

$$\lim_{h\to 0} \frac{(2+h)^2-2^2}{h} = 4.$$

Since the values of the difference quotient approach the derivative $f'(2)$, we conclude that $f'(2) = 4$.

Our discussion of the behavior of the derivative has been based on an intuitive geometric concept of the tangent line. However, the limit on the right in (1) may be considered independently of its geometric interpretation. In fact, we may use (1) to *define* $f'(a)$. We say that f is *differentiable* at $x = a$ if

$$\frac{f(a+h)-f(a)}{h}$$

approaches some number as $h \to 0$, and we denote this limiting number by $f'(a)$. If the difference quotient

$$\frac{f(a+h)-f(a)}{h}$$

does not approach any specific number as $h \to 0$, we say that f is *nondifferentiable* at $x = a$. Essentially all the functions in this text are differentiable at all points in their domain. A few exceptions are described in Section 2.7.

To better understand the limit concept used to define the derivative, it will be helpful to look at limits in a more general setting. If we let

$$g(h) = \frac{(2+h)^2-2^2}{h},$$

then Table 1 gives values of $g(h)$ as $h \to 0$. These values obviously approach 4. We may express this by writing

$$\lim_{h \to 0} g(h) = 4.$$

The preceding discussion suggests the following definition: Let $g(x)$ be a function, a, a number. We say that the number L *is the limit of* $g(x)$ *as* x *approaches* a provided that, as x gets arbitrarily close (but not equal) to a, the values of $g(x)$ approach L. In this case we write

$$\lim_{x \to a} g(x) = L.$$

If, as x approaches a, the values of $g(x)$ do *not* approach a specific number, then we say that the limit of $g(x)$ as x approaches a *does not exist*. Let us give some further examples of limits.

EXAMPLE 1 Determine $\lim_{x \to 2}(3x - 5)$.

Solution Let us make a table of values of X approaching 2 and the corresponding values of $Y_1 = 3X{-}5$:

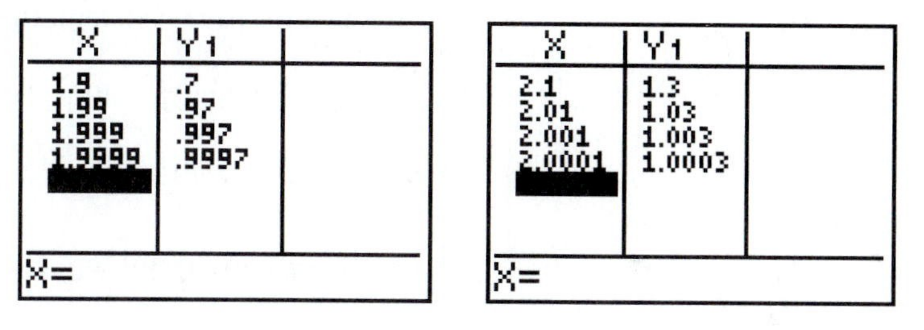

X	Y1
1.9	.7
1.99	.97
1.999	.997
1.9999	.9997

X=

X	Y1
2.1	1.3
2.01	1.03
2.001	1.003
2.0001	1.0003

X=

As x approaches 2, we see that $3x - 5$ approaches 1. In terms of our notation,

$$\lim_{x \to 2}(3x - 5) = 1.$$

EXAMPLE 2 For each of the functions in Figure 1, determine if $\lim_{x \to 2} g(x)$ exists. (The circles drawn on the graphs are meant to represent breaks in the graph, indicating that the functions under consideration are not defined at $x = 2$.)

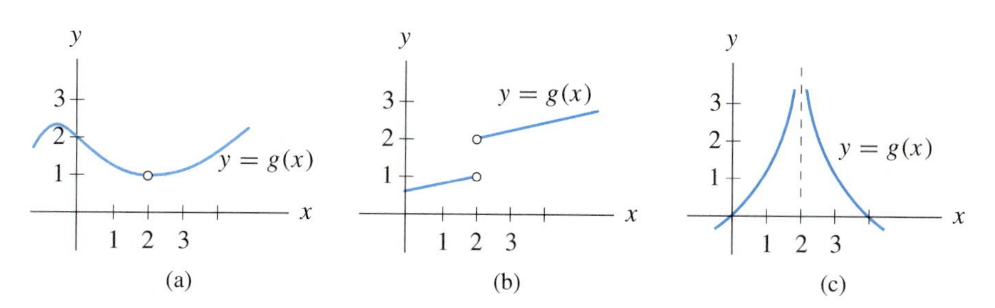

FIGURE 1 Functions with an undefined point

Solution (a) $\lim_{x\to 2} g(x) = 1$. We can see that as x gets closer and closer to 2, the values of $g(x)$ get closer and closer to 1. This is true for values of x to both the right and the left of 2.

(b) $\lim_{x\to 2} g(x)$ does not exist. As x approaches 2 from the right, $g(x)$ approaches 2. However, as x approaches 2 from the left, $g(x)$ approaches 1. In order for a limit to exist, the function must approach the *same* value from each direction.

(c) $\lim_{x\to 2} g(x)$ does not exist. As x approaches 2, the values of $g(x)$ become larger and larger and do not approach a fixed number. ■

The following limit theorems, which we cite without proof, allow us to reduce the computation of limits for combinations of functions to computations of limits involving the constituent functions.

Limit Theorems Suppose that $\lim_{x\to a} f(x)$ and $\lim_{x\to a} g(x)$ both exist. Then we have the following results:

I. If k is a constant, then $\lim_{x\to a} k \cdot f(x) = k \cdot \lim_{x\to a} f(x)$.

II. If r is a positive constant, then $\lim_{x\to a} [f(x)]^r = \left[\lim_{x\to a} f(x)\right]^r$.

III. $\lim_{x\to a} [f(x) + g(x)] = \lim_{x\to a} f(x) + \lim_{x\to a} g(x)$.

IV. $\lim_{x\to a} [f(x) - g(x)] = \lim_{x\to a} f(x) - \lim_{x\to a} g(x)$.

V. $\lim_{x\to a} [f(x) \cdot g(x)] = \left[\lim_{x\to a} f(x)\right] \cdot \left[\lim_{x\to a} g(x)\right]$.

VI. If $\lim_{x\to a} g(x) \neq 0$, then $\lim_{x\to a} \dfrac{f(x)}{g(x)} = \dfrac{\lim_{x\to a} f(x)}{\lim_{x\to a} g(x)}$.

EXAMPLE 3 Use the limit theorems to compute the following limits.

(a) $\lim_{x\to 2} x^3$ (b) $\lim_{x\to 2} 5x^3$ (c) $\lim_{x\to 2}(5x^3 - 15)$

(d) $\lim_{x\to 2} \sqrt{5x^3 - 15}$ (e) $\lim_{x\to 2} \left(\sqrt{5x^3 - 15}/x^5\right)$

Solution (a) Since $\lim_{x\to 2} x = 2$, we have by Limit Theorem II that

$$\lim_{x\to 2} x^3 = \left(\lim_{x\to 2} x\right)^3 = 2^3 = 8.$$

(b)
$$\begin{aligned} \lim_{x\to 2} 5x^3 &= 5 \lim_{x\to 2} x^3 && \text{(Limit Theorem I with } k = 5\text{)} \\ &= 5 \cdot 8 && \text{[by part (a)]} \\ &= 40. \end{aligned}$$

(c) $\lim_{x\to 2}(5x^3 - 15) = \lim_{x\to 2} 5x^3 - \lim_{x\to 2} 15$ (Limit Theorem IV).

Note that

$$\lim_{x\to 2} 15 = 15.$$

This is because the constant function $g(x) = 15$ always has the value 15, and so its limit as x approaches *any* number is 15. By part (b), $\lim_{x\to 2} 5x^3 = 40$. Thus

$$\lim_{x\to 2}(5x^3 - 15) = 40 - 15 = 25.$$

(d)
$$\begin{aligned} \lim_{x\to 2} \sqrt{5x^3 - 15} &= \lim_{x\to 2}(5x^3 - 15)^{1/2} \\ &= \left[\lim_{x\to 2}(5x^3 - 15)\right]^{1/2} && \text{[Limit Theorem II with } r = \tfrac{1}{2},\ f(x) = 5x^3 - 15] \\ &= 25^{1/2} && \text{[by part (c)]} \\ &= 5. \end{aligned}$$

(e) The limit of the denominator is $\lim_{x\to 2} x^5$, which is $2^5 = 32$, a nonzero number. So by Limit Theorem VI, we have

$$\lim_{x\to 2} \frac{\sqrt{5x^3 - 15}}{x^5} = \frac{\lim_{x\to 2} \sqrt{5x^3 - 15}}{\lim_{x\to 2} x^5} = \frac{5}{32} \quad \text{[by part (d)].}$$

The following facts, which may be deduced by repeated applications of the various limit theorems, are extremely handy in evaluating limits.

> *Limit of a Polynomial* Let $p(x)$ be a polynomial function, a any number. Then
>
> $$\lim_{x\to a} p(x) = p(a).$$

> *Limit of a Rational Function* Let $r(x) = p(x)/q(x)$ be a rational function, where $p(x)$ and $q(x)$ are polynomials. Let a be a number such that $q(a) \neq 0$. Then
>
> $$\lim_{x\to a} r(x) = r(a).$$

In other words, to determine a limit for a polynomial or a rational function, simply evaluate the function at $x = a$, provided, of course, that the function is defined at $x = a$. For instance, we can rework the solution to Example 3(c) as follows:

$$\lim_{x\to 2}(5x^3 - 15) = 5(2)^3 - 15 = 25.$$

Many situations require algebraic simplifications before the limit theorems can be applied.

EXAMPLE 4 Compute the following limits.

(a) $\lim_{x\to 3} \frac{x^2-9}{x-3}$ (b) $\lim_{x\to 0} \frac{\sqrt{x+4}-2}{x}$

Solution (a) The function

$$\frac{x^2-9}{x-3}$$

is not defined when $x = 3$, since

$$\frac{3^2-9}{3-3} = \frac{0}{0},$$

which is undefined. That causes no difficulty, since the limit as x approaches 3 depends only on the values of x *near* 3 and excludes consideration of the value at $x = 3$ itself. To evaluate the limit, note that $x^2 - 9 = (x-3)(x+3)$. So for $x \neq 3$,

$$\frac{x^2-9}{x-3} = \frac{(x-3)(x+3)}{x-3} = x+3.$$

As x approaches 3, $x + 3$ approaches 6. Therefore,

$$\lim_{x\to 3} \frac{x^2-9}{x-3} = 6.$$

(b) Since the denominator approaches zero when taking the limit, we may not apply Limit Theorem VI directly. However, if we first apply an algebraic trick, the limit may be evaluated. Multiply numerator and denominator by $\sqrt{x+4}+2$.

$$\begin{aligned}\frac{\sqrt{x+4}-2}{x} \cdot \frac{\sqrt{x+4}+2}{\sqrt{x+4}+2} &= \frac{(x+4)-4}{x(\sqrt{x+4}+2)}\\ &= \frac{x}{x(\sqrt{x+4}+2)}\\ &= \frac{1}{\sqrt{x+4}+2}.\end{aligned}$$

Thus

$$\begin{aligned}\lim_{x\to 0} \frac{\sqrt{x+4}-2}{x} &= \lim_{x\to 0} \frac{1}{\sqrt{x+4}+2}\\ &= \frac{\lim_{x\to 0} 1}{\lim_{x\to 0}(\sqrt{x+4}+2)} \qquad \text{(Limit Theorem VI)}\\ &= \frac{1}{4}.\end{aligned}$$

■

The basic differentiation rules are obtained from the limit definitions of the derivative. The next example (and Exercises 29–40) illustrates this process. There are three main steps.

Using Limits to Calculate a Derivative

1. Write the difference quotient $\dfrac{f(a+h)-f(a)}{h}$.

2. Simplify the difference quotient.

3. Find the limit as $h \to 0$.

EXAMPLE 5 Use limits to compute the derivative $f'(5)$ for the following functions.

(a) $f(x) = 15 - x^2$ (b) $f(x) = \dfrac{1}{2x-3}$

Solution In each case, we must calculate

$$\lim_{h\to 0} \frac{f(5+h)-f(5)}{h}.$$

(a)

$$\frac{f(5+h)-f(5)}{h} = \frac{[15-(5+h)^2]-(15-5^2)}{h} \quad \text{(step 1)}$$

$$= \frac{[15-(25+10h+h^2)]-(15-25)}{h} \quad \text{(step 2)}$$

$$= \frac{-10h-h^2}{h} = -10-h.$$

Therefore,

$$f'(5) = \lim_{h\to 0}(-10-h) = -10. \quad \text{(step 3)}$$

(b)

$$\frac{f(5+h)-f(5)}{h} = \frac{\dfrac{1}{2(5+h)-3} - \dfrac{1}{2(5)-3}}{h} \quad \text{(step 1)}$$

$$= \frac{\dfrac{1}{7+2h} - \dfrac{1}{7}}{h} = \frac{\dfrac{7-(7+2h)}{(7+2h)7}}{h} \quad \text{(step 2)}$$

$$= \frac{-2h}{(7+2h)7\cdot h} = \frac{-2}{(7+2h)7} = \frac{-2}{49+14h}.$$

$$f'(5) = \lim_{h\to 0} \frac{-2}{49+14h} = -\frac{2}{49}. \quad \text{(step 3)}$$

■

Remark When computing the limit in Example 5, we considered only values of h near zero (and not $h = 0$ itself). Therefore, we were freely able to divide both numerator and denominator by h.

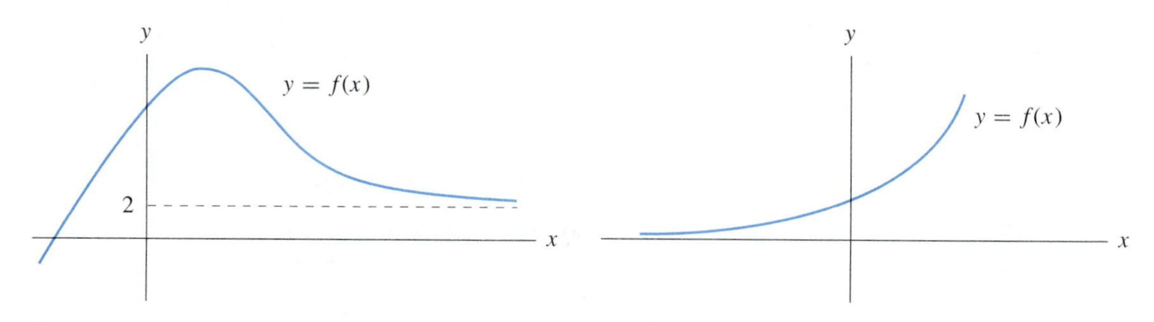

FIGURE 2 Function with a limit as x approaches infinity

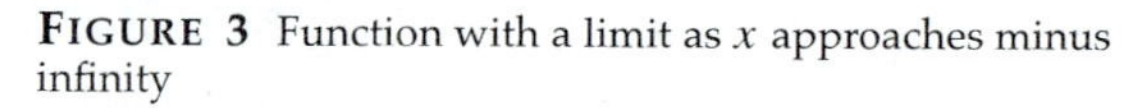
FIGURE 3 Function with a limit as x approaches minus infinity

INFINITY AND LIMITS Consider the function $f(x)$ whose graph is sketched in Figure 2. As x grows large, the value of $f(x)$ approaches 2. In this circumstance, we say that 2 is the *limit of $f(x)$ as x approaches infinity*. Infinity is denoted by the symbol ∞.

The preceding limit statement is expressed in the following notation:

$$\lim_{x\to\infty} f(x) = 2.$$

In a similar vein, consider the function whose graph is sketched in Figure 3. As x grows large in the negative direction, the value of $f(x)$ approaches 0. In this circumstance, we say that 0 is the *limit of $f(x)$ as x approaches minus infinity*. In symbols,

$$\lim_{x\to-\infty} f(x) = 0.$$

EXAMPLE 6 Calculate the following limits.

(a) $\displaystyle\lim_{x\to\infty} \frac{1}{x^2+1}$ (b) $\displaystyle\lim_{x\to\infty} \frac{x+1}{x-1}$

Solution (a) As x increases without bound, so does x^2+1. Therefore, $1/(x^2+1)$ approaches zero as x approaches ∞.

(b) Both $x+1$ and $x-1$ increase without bound as x does. To determine the limit of their quotient, we employ an algebraic trick. Divide both numerator and denominator by x to obtain

$$\lim_{x\to\infty} \frac{x+1}{x-1} = \lim_{x\to\infty} \frac{1+\dfrac{1}{x}}{1-\dfrac{1}{x}}.$$

As x increases without bound, $1/x$ approaches zero, so that both $1+(1/x)$ and $1-(1/x)$ approach 1. Thus the desired limit is $1/1 = 1$. ■

PRACTICE PROBLEMS 2.6

Determine which of the following limits exist. Compute the limits that exist.

1. $\lim_{x\to 6} \frac{x^2 - 4x - 12}{x - 6}$

2. $\lim_{x\to 6} \frac{4x + 12}{x - 6}$

EXERCISES 2.6

For each of the following functions $g(x)$, determine whether or not $\lim_{x\to 3} g(x)$ exists. If so, give the limit.

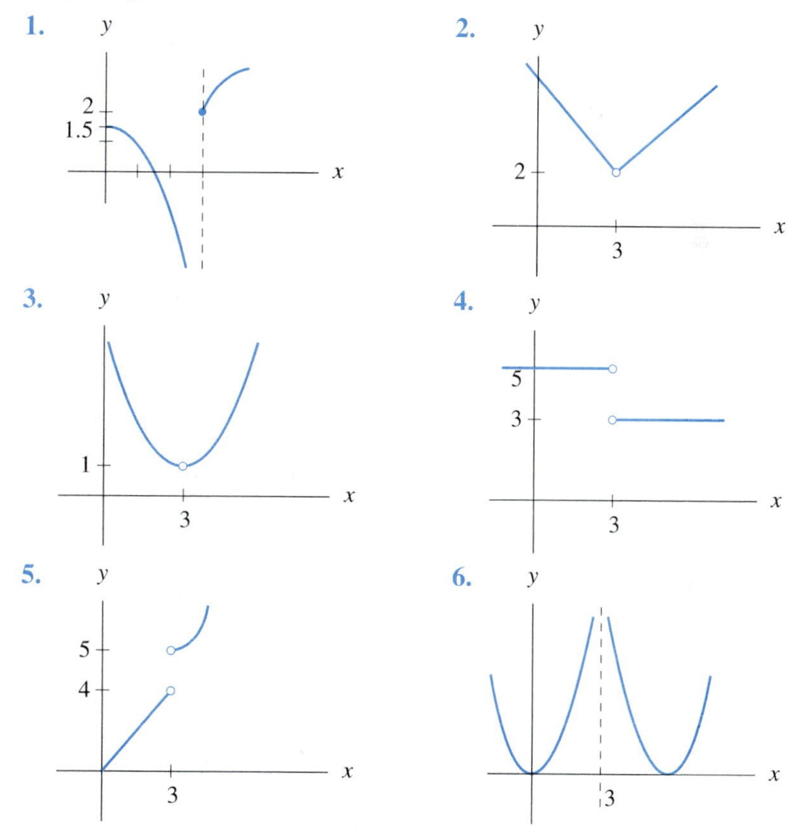

Determine which of the following limits exist. Compute the limits that exist.

7. $\lim_{x\to 1}(1 - 6x)$

8. $\lim_{x\to 2} \frac{x}{x - 2}$

9. $\lim_{x\to 3} \sqrt{x^2 + 16}$

10. $\lim_{x\to 4}(x^3 - 7)$

11. $\lim_{x\to 5} \frac{x^2 + 1}{5 - x}$

12. $\lim_{x\to 6}\left(\sqrt{6x} + 3x - \frac{1}{x}\right)(x^2 - 4)$

13. $\lim_{x\to 7}(x + \sqrt{x - 6})(x^2 - 2x + 1)$

14. $\lim_{x\to 8} \frac{\sqrt{5x - 4} - 1}{3x^2 + 2}$

15. $\lim_{x\to 9} \frac{\sqrt{x^2 - 5x - 36}}{8 - 3x}$

16. $\lim_{x\to 10}(2x^2 - 15x - 50)^{20}$

17. $\lim_{x\to 0} \frac{x^2+3x}{x}$

18. $\lim_{x\to 1} \frac{x^2-1}{x-1}$

19. $\lim_{x\to 2} \frac{-2x^2+4x}{x-2}$

20. $\lim_{x\to 3} \frac{x^2-x-6}{x-3}$

21. $\lim_{x\to 4} \frac{x^2-16}{4-x}$

22. $\lim_{x\to 5} \frac{2x-10}{x^2-25}$

23. $\lim_{x\to 6} \frac{x^2-6x}{x^2-5x-6}$

24. $\lim_{x\to 7} \frac{x^3-2x^2+3x}{x^2}$

25. $\lim_{x\to 8} \frac{x^2+64}{x-8}$

26. $\lim_{x\to 9} \frac{1}{(x-9)^2}$

27. $\lim_{x\to 0} \frac{-2}{\sqrt{x+16}+7}$

28. $\lim_{x\to 0} \frac{4x}{x(x^2+3x+5)}$

Use limits to compute the following derivatives.

29. $f'(3)$ where $f(x) = x^2+1$

30. $f'(2)$ where $f(x) = x^3$

31. $f'(0)$ where $f(x) = x^3+3x+1$

32. $f'(0)$ where $f(x) = x^2+2x+2$

33. $f'(3)$ where $f(x) = \frac{1}{2x+5}$

34. $f'(4)$ where $f(x) = \sqrt{2x-1}$

35. $f'(2)$ where $f(x) = \sqrt{5-x}$

36. $f'(3)$ where $f(x) = \frac{1}{7-2x}$

37. $f'(0)$ where $f(x) = \sqrt{1-x^2}$

38. $f'(2)$ where $f(x) = (5x-4)^2$

39. $f'(0)$ where $f(x) = (x+1)^3$

40. $f'(0)$ where $f(x) = \sqrt{x^2+x+1}$

Each limit in Exercises 41–46 is a definition of $f'(a)$. Determine the function $f(x)$ and the value of a.

41. $\lim_{h\to 0} \frac{(9+h)^{1/2}-3}{h}$

42. $\lim_{h\to 0} \frac{(2+h)^3-8}{h}$

43. $\lim_{h\to 0} \frac{\frac{1}{10+h}-.1}{h}$

44. $\lim_{h\to 0} \frac{(64+h)^{1/3}-4}{h}$

45. $\lim_{h\to 0} \frac{(3(1+h)^2+4)-7}{h}$

46. $\lim_{h\to 0} \frac{(1+h)^{-1/2}-1}{h}$

Compute the following limits.

47. $\lim_{x\to\infty} \frac{1}{x^2}$

48. $\lim_{x\to -\infty} \frac{1}{x^2}$

49. $\lim_{x\to\infty} \frac{1}{x-8}$

50. $\lim_{x\to\infty} \frac{1}{3x+5}$

51. $\lim_{x\to\infty} \frac{2x+1}{x+2}$

52. $\lim_{x\to\infty} \frac{x^2+x}{x^2-1}$

In the remaining exercises, use a graphing utility to obtain the graphs of the functions. In Exercises 53–58, graph the function and examine values of $f(x)$ near $x = a$ in order to make a good guess of the value of $\lim_{x\to a} f(x)$. Start with the window shown and then use smaller windows as necessary.

53. $\lim_{x\to 3} \dfrac{x^2 - x - 6}{x - 3}$; [1, 5] *by* [0, 7]

54. $\lim_{x\to 5} \dfrac{x^2 - 5x}{x - 5}$; [3, 7] *by* [0, 7]

55. $\lim_{x\to 0} (x^2)^x$; [−1, 1] *by* [0, 3]

56. $\lim_{x\to 4} \dfrac{\sqrt{5x - 4} - 4}{x - 4}$; [3, 5] *by* [0, 1]

57. $\lim_{x\to 1} \dfrac{x - 1}{\sqrt{x} - 1}$; [0, 2] *by* [0, 3]

58. $\lim_{x\to 1} \dfrac{2^x - 2}{x - 1}$; [0, 2] *by* [0, 2]

In Exercises 59–62, graph the function and examine values of $f(x)$ as x gets large in order to make a good guess of the value of $\lim_{x\to\infty} f(x)$. Start with the window shown and then use other windows as necessary.

59. $\lim_{x\to\infty} \dfrac{3x^2 + 100}{7x + x^2}$; [0, 50] *by* [0, 5]

60. $\lim_{x\to\infty} \dfrac{3x^3 - 10}{100x + 2x^3}$; [0, 50] *by* [0, 2]

61. $\lim_{x\to\infty} \dfrac{x^3}{2^x}$; [0, 15] *by* [0, 5]

62. $\lim_{x\to\infty} \dfrac{\sqrt{3 + 4x^2}}{x}$; [0, 10] *by* [0, 10]

SOLUTIONS TO PRACTICE PROBLEMS 2.6

1. The function under consideration is a rational function. Since the denominator has the value 0 at $x = 6$, we cannot immediately determine the limit by just evaluating the function at $x = 6$. Also,

$$\lim_{x\to 6} (x - 6) = 0.$$

Since the function in the denominator has limit 0, we cannot apply Limit Theorem VI. However, since the definition of limit considers only values of x different from 6, the quotient can be simplified by factoring and canceling:

$$\frac{x^2 - 4x - 12}{x - 6} = \frac{(x + 2)(x - 6)}{(x - 6)} = x + 2 \quad \text{for } x \neq 6.$$

Now $\lim_{x\to 6} (x + 2) = 8$. Therefore, $\lim_{x\to 6} \dfrac{x^2 - 4x - 12}{x - 6} = 8$.

2. No limit exists. It is easily seen that

$$\lim_{x\to 6} (4x + 12) = 36 \quad \text{and} \quad \lim_{x\to 6} (x - 6) = 0.$$

As x approaches 6, the denominator gets very small and the numerator approaches 36. For example, if $x = 6.00001$, then the numerator is 36.00004 and the denominator is .00001. The quotient is 3,600,004. As x approaches 6 even more closely, the quotient gets arbitrarily large and cannot possibly approach a limit.

2.7 DIFFERENTIABILITY AND CONTINUITY

In the preceding section we defined differentiability of $f(x)$ at $x = a$ in terms of a limit. If this limit does not exist, then we say that $f(x)$ is *nondifferentiable* at

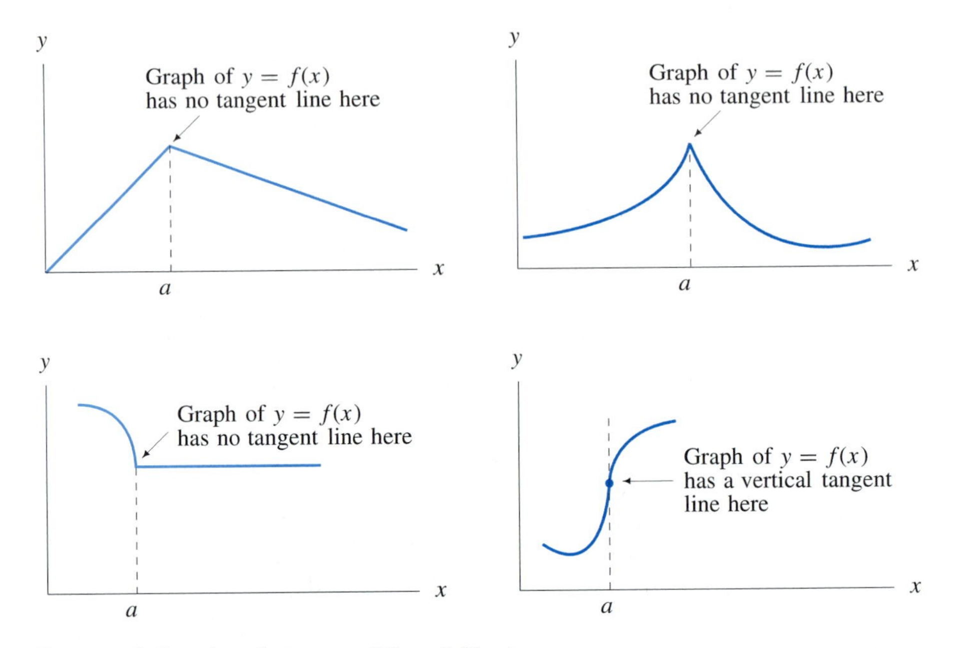

FIGURE 1 Functions that are nondifferentiable at $x = a$

$x = a$. Geometrically, the nondifferentiability of $f(x)$ at $x = a$ can manifest itself in several ways. First of all, the graph of $f(x)$ could have no tangent line at $x = a$. Second, the graph could have a vertical tangent line at $x = a$. (Recall that slope is not defined for vertical lines.) Some of the various geometric possibilities are illustrated in Figure 1.

The following example illustrates how nondifferentiable functions can arise in practice.

EXAMPLE 1 A railroad company charges \$10 per mile to haul a boxcar up to 200 miles and \$8 per mile for each mile exceeding 200. In addition, the railroad charges a \$1000 handling charge per boxcar. Graph the cost of sending a boxcar x miles.

Solution If x is at most 200 miles, then the cost $C(x)$ is given by $C(x) = 1000 + 10x$ dollars. The cost for 200 miles is $C(200) = 1000 + 2000 = 3000$ dollars. If x exceeds 200 miles, then the total cost will be

$$C(x) = \underbrace{3000}_{\substack{\text{cost of first} \\ \text{200 miles}}} + \underbrace{8(x - 200)}_{\substack{\text{cost of miles in} \\ \text{excess of 200}}} = 1400 + 8x.$$

Thus

$$C(x) = \begin{cases} 1000 + 10x, & 0 < x \leq 2000 \\ 1400 + 8x, & x > 200. \end{cases}$$

The graph of $C(x)$ is sketched in Figure 2. Note that $C(x)$ is not differentiable at $x = 200$. ■

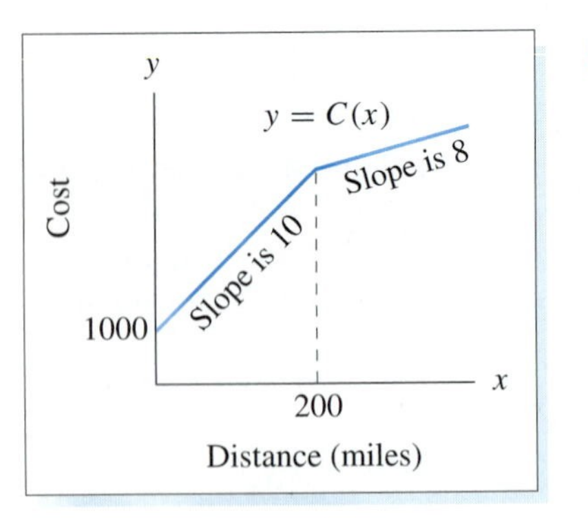

FIGURE 2 Cost of hauling a boxcar

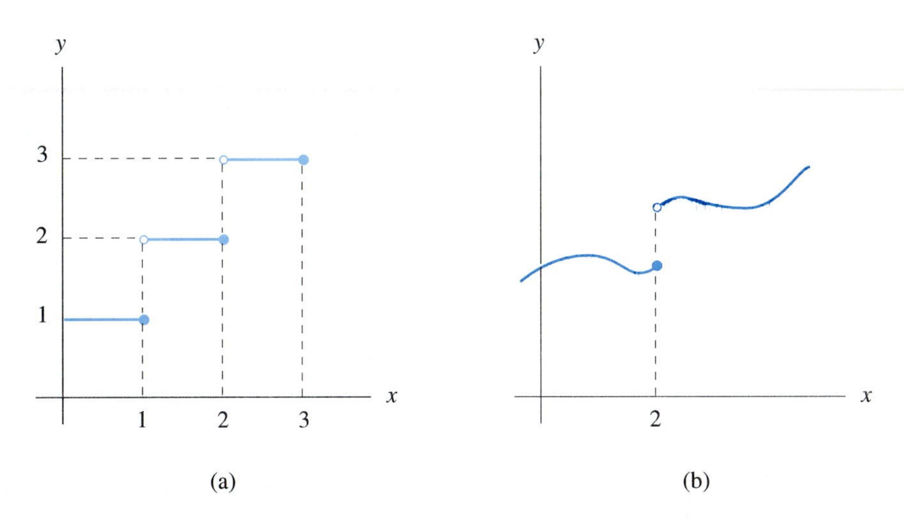

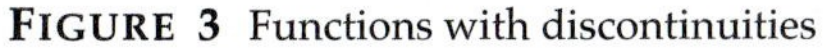
FIGURE 3 Functions with discontinuities

Closely related to the concept of differentiability is that of continuity. We say that a function $f(x)$ is *continuous* at $x = a$ provided that, roughly speaking, its graph has no breaks (or gaps) as it passes through the point $(a, f(a))$. That is, $f(x)$ is continuous at $x = a$ provided that we can draw the graph through $(a, f(a))$ without lifting our pencil from the paper. The functions whose graphs are drawn in Figures 1 and 2 are continuous for all values of x. By contrast, however, the function whose graph is drawn in Figure 3(a) is not continuous (we say it is *discontinuous*) at $x = 1$ and $x = 2$, since the graph has breaks there. Similarly, the function whose graph is drawn in Figure 3(b) is discontinuous at $x = 2$.

Discontinuous functions can occur in applications, as the following example shows.

EXAMPLE 2 Suppose that a manufacturing plant is capable of producing 15,000 units in one shift of 8 hours. For each shift worked, there is a fixed cost of \$2000 (for light, heat, etc.). Suppose that the variable cost (the cost of labor and raw materials) is \$2 per unit. Graph the cost $C(x)$ of manufacturing x units.

Solution If $x \leq 15{,}000$, a single shift will suffice, so that

$$C(x) = 2000 + 2x, \quad 0 \leq x \leq 15{,}000.$$

If x is between 15,000 and 30,000, one extra shift will be required, and

$$C(x) = 4000 + 2x, \quad 15{,}000 < x \leq 30{,}000.$$

If x is between 30,000 and 45,000, the plant will need to work three shifts, and

$$C(x) = 6000 + 2x, \quad 30{,}000 < x \leq 45{,}000.$$

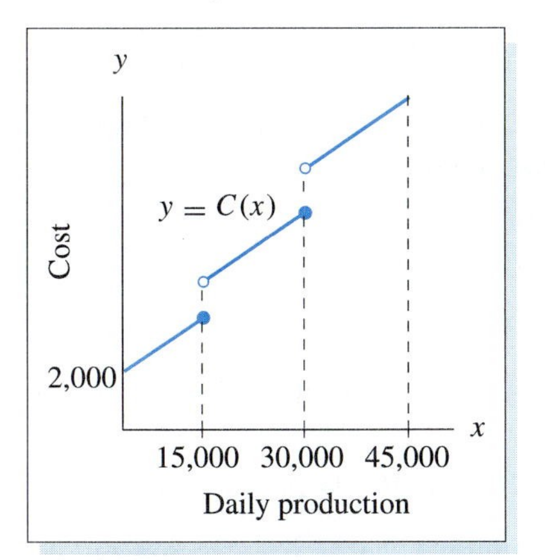

FIGURE 4 Cost function of a manufacturing plant

The graph of $C(x)$ for $0 \leq x \leq 45{,}000$ is drawn in Figure 4. Note that the graph has breaks at two points. ■

The relationship between differentiability and continuity is this:

Theorem 1 If $f(x)$ is differentiable at $x = a$, then $f(x)$ is continuous at $x = a$.

Note, however, that the converse statement is definitely false: A function may be continuous at $x = a$ but still not be differentiable there. The functions whose graphs are drawn in Figure 1 provide examples of this phenomenon.

Just as with differentiability, the notion of continuity can be phrased in terms of limits. In order for $f(x)$ to be continuous at $x = a$, the values of $f(x)$ for all x near a must be close to $f(a)$ (otherwise, the graph would have a break at $x = a$). In fact, the closer x is to a, the closer $f(x)$ must be to $f(a)$ (again, in order to avoid a break in the graph). In terms of limits, we must therefore have

$$\lim_{x \to a} f(x) = f(a).$$

Conversely, an intuitive argument shows that if the limit relation above holds, then the graph of $y = f(x)$ has no breaks at $x = a$.

> *Limit Definition of Continuity* A function $f(x)$ is continuous at $x = a$ provided the following limit relation holds:
>
> $$\lim_{x \to a} f(x) = f(a). \tag{1}$$

In order for (1) to hold, three conditions must be fulfilled:

1. $f(x)$ must be defined at $x = a$.
2. $\lim_{x \to a} f(x)$ must exist.
3. The limit $\lim_{x \to a} f(x)$ must have the value $f(a)$.

A function will fail to be continuous at $x = a$ when any one of these conditions fails to hold. The various possibilities are illustrated in the next example.

EXAMPLE 3 Determine whether the functions whose graphs are drawn in Figure 5 are continuous at $x = 3$. Use the limit definition.

Solution (a) Here $\lim_{x \to 3} f(x) = 2$. However, $f(3) = 4$. So

$$\lim_{x \to 3} f(x) \neq f(3)$$

and $f(x)$ is not continuous at $x = 3$. (Geometrically, this is clear. The graph has a break at $x = 3$.)

(b) $\lim_{x \to 3} g(x)$ does not exist, so $g(x)$ is not continuous at $x = 3$.

(c) $\lim_{x \to 3} h(x)$ does not exist, so $h(x)$ is not continuous at $x = 3$.

(d) $f(x)$ is not defined at $x = 3$, so $f(x)$ is not continuous at $x = 3$. ■

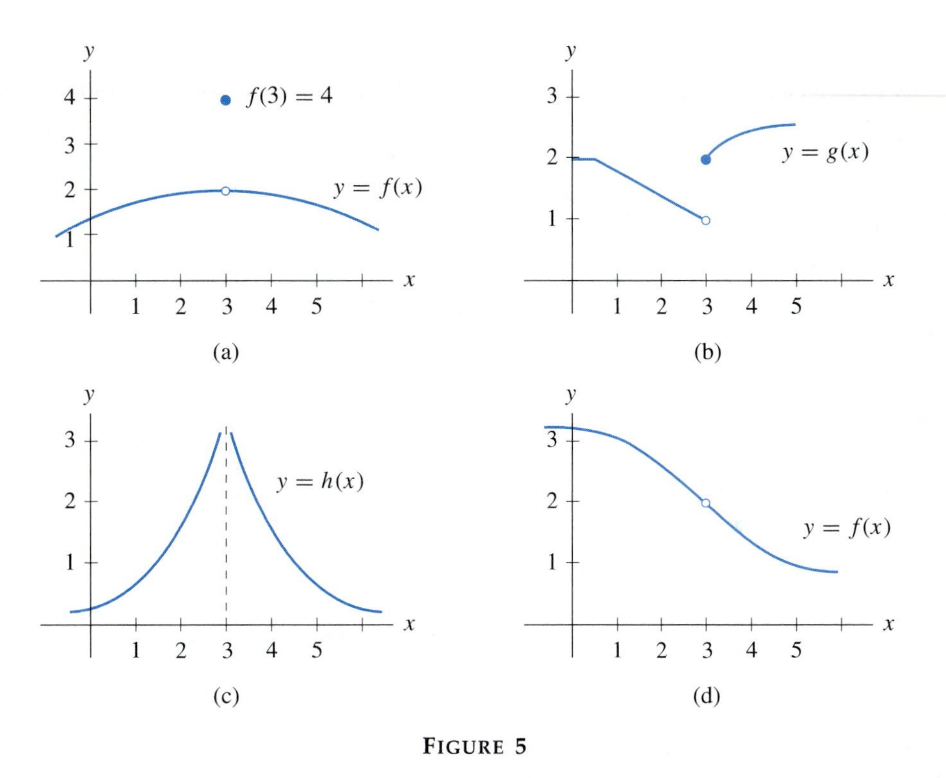

FIGURE 5

Using our result on the limit of a polynomial function (Section 2.6), we see that

$$p(x) = a_0 + a_1 x + \cdots + a_n x^n, \quad a_0, \ldots, a_n \text{ constants},$$

is continuous for all x. Similarly, a rational function

$$\frac{p(x)}{q(x)}, \quad p(x), q(x) \text{ polynomials},$$

is continuous for all x for which $q(x) \neq 0$.

PRACTICE PROBLEMS 2.7

Let

$$f(x) = \begin{cases} \dfrac{x^2 - x - 6}{x - 3} & \text{for } x \neq 3 \\ 4 & \text{for } x = 3. \end{cases}$$

1. Is $f(x)$ continuous at $x = 3$?
2. Is $f(x)$ differentiable at $x = 3$?

EXERCISES 2.7

Is the function whose graph is drawn in Figure 6 continuous at the following values of x?

1. $x = 0$
2. $x = -3$
3. $x = 3$
4. $x = .001$
5. $x = -2$
6. $x = 2$

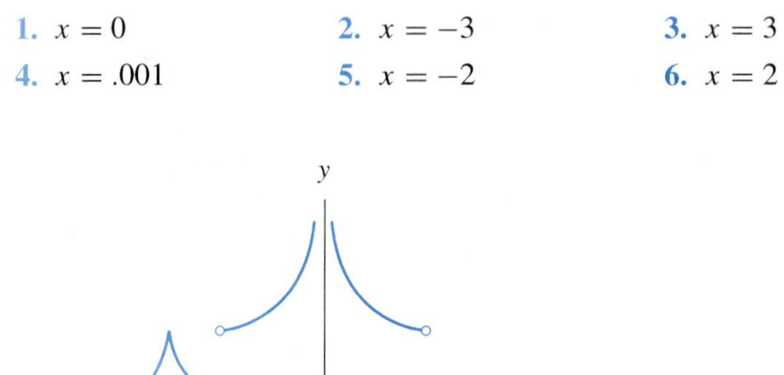
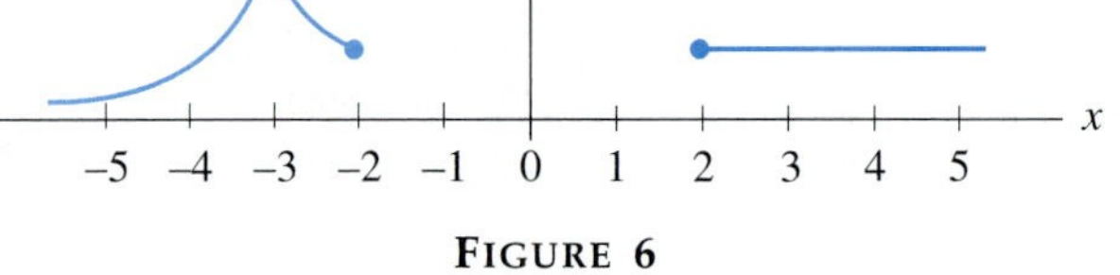

FIGURE 6

Is the function whose graph is drawn in Figure 6 differentiable at the following values of x?

7. $x = 0$
8. $x = -3$
9. $x = 3$
10. $x = .001$
11. $x = -2$
12. $x = 2$

Determine whether each of the following functions is continuous and/or differentiable at $x = 1$.

13. $f(x) = x^2$

14. $f(x) = \dfrac{1}{x}$

15. $f(x) = \begin{cases} x + 2 & \text{for } -1 \le x \le 1 \\ 3x & \text{for } 1 < x < 5 \end{cases}$

16. $f(x) = \begin{cases} x & \text{for } 1 \le x \le 2 \\ x^3 & \text{for } 0 \le x < 1 \end{cases}$

17. $f(x) = \begin{cases} 2x - 1 & \text{for } 0 \le x \le 1 \\ 1 & \text{for } 1 < x \end{cases}$

18. $f(x) = \begin{cases} x & \text{for } x \ne 1 \\ 2 & \text{for } x = 1 \end{cases}$

19. $f(x) = \begin{cases} \dfrac{1}{x - 1} & \text{for } x \ne 1 \\ 0 & \text{for } x = 1 \end{cases}$

20. $f(x) = \begin{cases} x - 1 & \text{for } 0 \le x < 1 \\ 1 & \text{for } x = 1 \\ 2x - 2 & \text{for } x > 1 \end{cases}$

The functions in Exercises 21–26 are defined for all x except for one value of x. If possible, define $f(x)$ at the exceptional point in a way that makes $f(x)$ continuous for all x.

21. $f(x) = \dfrac{x^2 - 7x + 10}{x - 5}, x \neq 5$

22. $f(x) = \dfrac{x^2 + x - 12}{x + 4}, x \neq -4$

23. $f(x) = \dfrac{x^3 - 5x^2 + 4}{x^2}, x \neq 0$

24. $f(x) = \dfrac{x^2 + 25}{x - 5}, x \neq 5$

25. $f(x) = \dfrac{(6 + x)^2 - 36}{x}, x \neq 0$

26. $f(x) = \dfrac{\sqrt{9 + x} - \sqrt{9}}{x}, x \neq 0$

SOLUTIONS TO PRACTICE PROBLEMS 2.7

1. The function $f(x)$ is defined at $x = 3$, namely, $f(3) = 4$. When computing $\lim_{x \to 3} f(x)$, we exclude consideration of $x = 3$; therefore, we can simplify the expression for $f(x)$ as follows:

$$f(x) = \frac{x^2 - x - 6}{x - 3} = \frac{(x - 3)(x + 2)}{x - 3} = x + 2.$$

Clearly,

$$\lim_{x \to 3} f(x) = \lim_{x \to 3}(x + 2) = 5.$$

Since $\lim_{x \to 3} f(x) = 5 \neq 4 = f(3)$, $f(x)$ is not continuous at $x = 3$.

2. There is no need to compute any limits in order to answer this question. By Theorem 1, since $f(x)$ is not continuous at $x = 3$, it cannot possibly be differentiable there.

REVIEW OF THE FUNDAMENTAL CONCEPTS OF CHAPTER 2

1. Give a physical description of what is meant by the slope of $f(x)$ at the point $(2, f(2))$.
2. What does $f'(2)$ represent?
3. Explain why $f'(2)$ is the limit of slopes of secant lines through the point $(2, f(2))$.
4. Explain how to calculate $f'(2)$ as the limit of slopes of secant lines through the point $(2, f(2))$.
5. What is meant by the average rate of change of a function over an interval?
6. How is an (instantaneous) rate of change related to average rates of change?
7. What expression involving a derivative gives an approximation to $f(a+h) - f(a)$?
8. Describe marginal cost in your own words.
9. How do you determine the proper units for a rate of change? Give an example.
10. Explain the relationship between derivatives and velocity and acceleration.

11. State as many terms used to describe graphs of functions as you can recall.
12. What is the difference between having a relative maximum at $x = 2$ and having an absolute maximum at $x = 2$?
13. What is an asymptote? Give an example.
14. State the first derivative rule. Second derivative rule.
15. In your own words, explain the meaning of $\lim_{x \to 2} f(x) = 3$. Give an example of such a function $f(x)$.
16. Give the limit definition of $f'(2)$, that is, the slope of $f(x)$ at the point $(2, f(2))$.
17. State as many limit theorems as you can recall.
18. In your own words, explain the meaning of $\lim_{x \to \infty} f(x) = 3$. Give an example of such a function $f(x)$.
19. In your own words, explain the meaning of "$f(x)$ is continuous at $x = 2$." Give an example of such a function $f(x)$.

Chapter 3

Algebraic Differentiation and Its Applications

Calculus techniques can be applied to a wide variety of problems in real life. We consider many examples in this chapter. In each case we construct a function as a "mathematical model" of some problem and then analyze the function and its derivative in order to gain information about the original problem. Our principal method for analyzing a function will be to sketch its graph and determine its extreme points and inflection points. For this reason we devote the first part of the chapter to differentiating functions and to curve sketching.

3.1 Some Rules for Differentiation

When a function $f(x)$ is given by a formula, the derivative function $f'(x)$ can also be given by a formula. The process of obtaining the formula for $f'(x)$ is called *differentiation.*

For constant and linear functions, the formulas are obvious. As Figure 1 shows, the graph of a constant function is a horizontal line and so the slope at any point is 0. As Figure 2 shows, the graph of the linear function $f(x) = mx + b$ is a straight line of slope m and so the slope is m at every point. Thus we have

The derivative of a constant function $f(x) = b$ is zero; that is,

$$f'(x) = 0.$$

The derivative of a linear function $f(x) = mx + b$ is m; that is,

$$f'(x) = m.$$

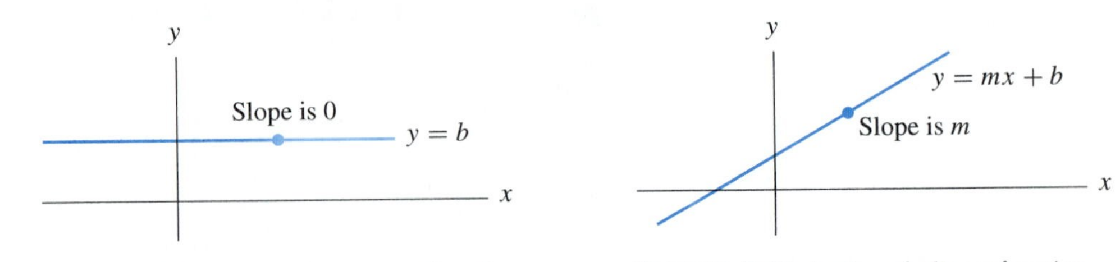

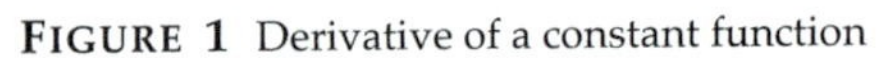
FIGURE 1 Derivative of a constant function

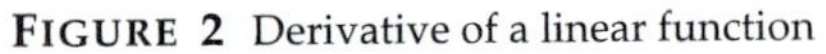
FIGURE 2 Derivative of a linear function

Next, consider the function $f(x) = x^2$. As indicated in Section 2.1 (and proved at the end of this section), the slope of the graph of $y = x^2$ at the point (x, y) is equal to $2x$.

> If $f(x) = x^2$, then the derivative is the function $2x$. That is,
> $$f'(x) = 2x.$$

In Exercise 65 of Section 2.3, we discovered that the slope of the graph of $y = x^3$ at the point (x, y) is $3x^2$. This can be restated in terms of derivatives as follows:

> If $f(x) = x^3$, then the derivative is $3x^2$. That is,
> $$f'(x) = 3x^2.$$

We should, at this stage at least, keep the geometric meaning of these formulas clearly in mind. Figure 3 shows the graphs of x^2 and x^3 together with the interpretations of their derivative functions.

One of the reasons calculus is so useful is that it provides general techniques that can be easily used to determine derivatives. One such general rule, which contains the formulas above as special cases, is the so-called power rule.

> *Power Rule* Let r be any number and let $f(x) = x^r$. Then $f'(x) = rx^{r-1}$.

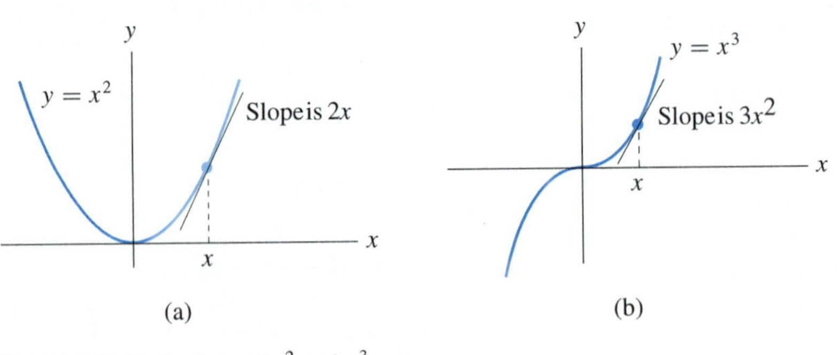

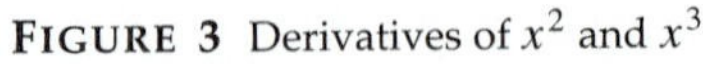
FIGURE 3 Derivatives of x^2 and x^3

Indeed, if $r = 2$, then $f(x) = x^2$ and $f'(x) = 2x^{2-1} = 2x$. If $r = 3$, then $f(x) = x^3$ and $f'(x) = 3x^{3-1} = 3x^2$. We shall prove the power rule in Section 4.6. Until then, we shall use it to calculate derivatives.

EXAMPLE 1 Let $f(x) = \sqrt{x}$. What is $f'(x)$?

Solution Recall that $\sqrt{x} = x^{1/2}$. We may apply the power rule with $r = \frac{1}{2}$:

$$f(x) = x^{1/2};$$

$$f'(x) = \frac{1}{2}x^{1/2-1} = \frac{1}{2}x^{-1/2}$$

$$= \frac{1}{2} \cdot \frac{1}{x^{1/2}} = \frac{1}{2\sqrt{x}}.$$

Another important special case of the power rule occurs for $r = -1$, corresponding to $f(x) = x^{-1}$. In this case, $f'(x) = (-1)x^{-1-1} = -x^{-2}$. However, since $x^{-1} = 1/x$ and $x^{-2} = 1/x^2$, the power rule for $r = -1$ may also be written as follows:*

$$\text{If } f(x) = \frac{1}{x}, \text{ then } f'(x) = -\frac{1}{x^2} \quad (x \neq 0). \tag{1}$$

EXAMPLE 2 Find the slope of the curve $y = 1/x$ at $(2, \frac{1}{2})$.

Solution Set $f(x) = 1/x$. The point $(2, 1/2)$ corresponds to $x = 2$, so in order to find the slope at this point, we compute $f'(2)$. From formula (1) we find that

$$f'(2) = -\frac{1}{2^2} = -\frac{1}{4}.$$

Thus the slope of $y = 1/x$ at the point $(2, \frac{1}{2})$ is $-\frac{1}{4}$. (See Figure 4.)

Two additional rules of differentiation greatly extend the number of functions we can differentiate.

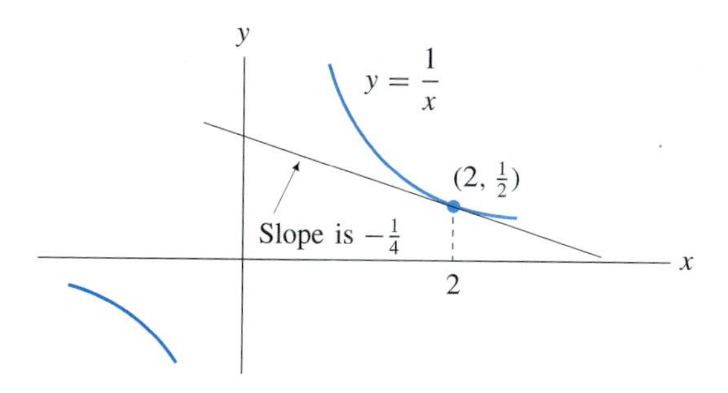

FIGURE 4 Derivative of $1/x$

* The formula gives $f'(x)$ for $x \neq 0$. The derivative of $f(x)$ is not defined at $x = 0$ since $f(x)$ itself is not defined there.

1. Constant-Multiple Rule

$$[k \cdot f(x)]' = k \cdot f'(x), \quad k \text{ a constant.}$$

2. Sum Rule

$$[f(x) + g(x)]' = f'(x) + g'(x).$$

We shall discuss these rules and then prove them.

THE CONSTANT-MULTIPLE RULE Starting with a function $f(x)$, we can multiply it by a constant number k in order to obtain a new function $k \cdot f(x)$. For instance, if $f(x) = x^2 - 4x + 1$ and $k = 2$, then

$$2f(x) = 2(x^2 - 4x + 1) = 2x^2 - 8x + 2.$$

The constant-multiple rule says that the derivative of the new function $k \cdot f(x)$ is just k times the derivative of the original function.* In other words, when faced with the differentiation of a constant times a function, simply carry along the constant and differentiate the function.

EXAMPLE 3 Calculate.

(a) $(2x^5)'$ (b) $\left(\frac{x^3}{4}\right)'$ (c) $\left(-\frac{3}{x}\right)'$

Solution (a) With $k = 2$ and $f(x) = x^5$, we have

$$(2x^5)' = 2 \cdot (x^5)' = 2(5x^4) = 10x^4.$$

(b) Write $\frac{x^3}{4}$ in the form $\frac{1}{4} \cdot (x^3)$. Then

$$\left(\frac{x^3}{4}\right)' = \frac{1}{4}(x^3)' = \frac{1}{4}(3x^2) = \frac{3}{4}x^2.$$

(c) Write $-\frac{3}{x}$ in the form $-3 \cdot \frac{1}{x}$. Then

$$\left(-\frac{3}{x}\right)' = (-3) \cdot \left(\frac{1}{x}\right)' = (-3) \cdot \frac{-1}{x^2} = \frac{3}{x^2}.$$

EXAMPLE 4 Let $f(x) = 5\sqrt{x}$. Calculate $f''(x)$.

Solution Begin with $f(x) = 5x^{1/2}$. Differentiate once to get $f'(x)$ and then differentiate again to get $f''(x)$.

$$f'(x) = 5 \cdot \left(\sqrt{x}\right)' = 5\left(x^{1/2}\right)' = \frac{5}{2}x^{-1/2}.$$

$$f''(x) = \frac{5}{2}\left(x^{-1/2}\right)' = \frac{5}{2}\left(-\frac{1}{2}x^{-3/2}\right) = -\frac{5}{4}x^{-3/2}.$$

---

* More precisely, the constant-multiple rule asserts that if $f(x)$ is differentiable at $x = a$, then so is the function $k \cdot f(x)$, and the derivative of $k \cdot f(x)$ at $x = a$ may be computed using the given formula.

This answer may also be written as $-\dfrac{5}{4x^{3/2}}$. ■

THE SUM RULE To differentiate a sum of functions, differentiate each function individually and add the derivatives together.* Another way of saying this is "the derivative of a sum of functions is the sum of the derivatives."

EXAMPLE 5 Find each of the following.

(a) $(x^3 + 5x)'$ (b) $\left(x^4 - \dfrac{3}{x^2}\right)'$ (c) $(2x^7 - x^5 + 8)'$

Solution (a) Let $f(x) = x^3$ and $g(x) = 5x$. Then

$$(x^3 + 5x)' = (x^3)' + (5x)' = 3x^2 + 5.$$

(b) The sum rule applies to differences as well as sums (see Exercise 44). Indeed, by the sum rule,

$$\begin{aligned}\left(x^4 - \frac{3}{x^2}\right)' &= \left(x^4\right)' + \left(-\frac{3}{x^2}\right)' && \text{(sum rule)}\\ &= \left(x^4\right)' - 3\left(x^{-2}\right)' && \text{(constant-multiple rule)}\\ &= 4x^3 - 3\left(-2x^{-3}\right)\\ &= 4x^3 + 6x^{-3}.\end{aligned}$$

After some practice, one usually omits most or all of the intermediate steps and simply writes

$$\left(x^4 - \frac{3}{x^2}\right)' = 4x^3 + 6x^{-3}.$$

(c) We apply the sum rule repeatedly and use the fact that the derivative of a constant function is 0:

$$\begin{aligned}(2x^7 - x^5 + 8)' &= (2x^7)' - (x^5)' + (8)'\\ &= 2(7x^6) - 5x^4 + 0\\ &= 14x^6 - 5x^4.\end{aligned}$$

■

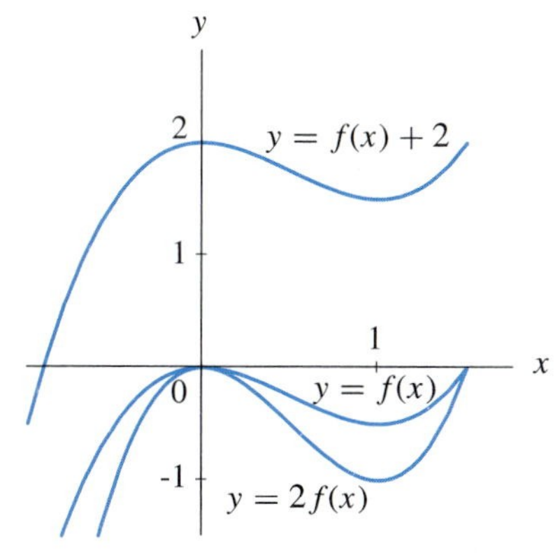

FIGURE 5 Two effects of a constant on the graph of $f(x)$

Remark The differentiation of a function plus a constant is different from the differentiation of a constant times a function. Figure 5 shows the graphs of $f(x)$, $f(x) + 2$, and $2 \cdot f(x)$, where $f(x) = x^3 - \frac{3}{2}x^2$. For each x, the graphs of $f(x)$ and $f(x) + 2$ have the same slope. In contrast, for each x, the slope of the graph of $2 \cdot f(x)$ is twice the slope of the graph of $f(x)$. Upon differentiation, an added constant disappears, whereas a constant that multiplies a function is carried along.

---

* More precisely, the sum rule asserts that if both $f(x)$ and $g(x)$ are differentiable at $x = a$, then so is $f(x) + g(x)$, and the derivative (at $x = a$) of the sum is then the sum of the derivatives (at $x = a$).

PROOFS OF THE CONSTANT-MULTIPLE AND SUM RULES

Let us verify both rules when x has the value a. Recall that if $f(x)$ is differentiable at $x = a$, then its derivative is the limit

$$\lim_{h \to 0} \frac{f(a+h) - f(a)}{h}.$$

CONSTANT-MULTIPLE RULE We assume that $f(x)$ is differentiable at $x = a$. We must prove that $k \cdot f(x)$ is differentiable at $x = a$ and that its derivative there is $k \cdot f'(a)$. This amounts to showing that the limit

$$\lim_{h \to 0} \frac{k \cdot f(a+h) - k \cdot f(a)}{h}$$

exists and has the value $k \cdot f'(a)$. However,

$$\begin{aligned}
\lim_{h \to 0} \frac{k \cdot f(a+h) - k \cdot f(a)}{h} &= \lim_{h \to 0} k \left[\frac{f(a+h) - f(a)}{h} \right] \\
&= k \cdot \lim_{h \to 0} \frac{f(a+h) - f(a)}{h} \qquad \text{(by Limit Theorem I)} \\
&= k \cdot f'(a) \quad [\text{since } f(x) \text{ is differentiable at } x = a],
\end{aligned}$$

which is what we desired to show.

SUM RULE We assume that both $f(x)$ and $g(x)$ are differentiable at $x = a$. We must prove that $f(x) + g(x)$ is differentiable at $x = a$ and that its derivative is $f'(a) + g'(a)$. That is, we must show that the limit

$$\lim_{h \to 0} \frac{[f(a+h) + g(a+h)] - [f(a) + g(a)]}{h}$$

exists and equals $f'(a) + g'(a)$. Using Limit Theorem III and the fact that $f(x)$ and $g(x)$ are differentiable at $x = a$, we have

$$\begin{aligned}
&\lim_{h \to 0} \frac{[f(a+h) + g(a+h)] - [f(a) + g(a)]}{h} \\
&\qquad = \lim_{h \to 0} \left[\frac{f(a+h) - f(a)}{h} + \frac{g(a+h) - g(a)}{h} \right] \\
&\qquad = \lim_{h \to 0} \frac{f(a+h) - f(a)}{h} + \lim_{h \to 0} \frac{g(a+h) - g(a)}{h} \\
&\qquad = f'(a) + g'(a).
\end{aligned}$$

PRACTICE PROBLEMS 3.1

1. Find the derivative $(x)'$.
2. Let $f(x) = 1/x^4$.
   (a) Find its derivative.
   (b) Find $f'(2)$.

EXERCISES 3.1

In Exercises 1–4, rewrite each function below as a sum, where each term in the sum is a constant times a power of x.

1. $f(x) = \frac{2}{x} + \sqrt{x}$
2. $f(x) = \frac{1}{\sqrt{x}} + \frac{3}{x^2}$
3. $f(x) = (\sqrt{x})^3 + \frac{5}{x^{-3}}$
4. $f(x) = \frac{1}{x^{2/3}} + 7\sqrt[3]{x}$

In Exercises 5–22, find $f'(x)$.

5. $f(x) = x^4$
6. $f(x) = x^{99}$
7. $f(x) = x^{3/2}$
8. $f(x) = x^{5/3}$
9. $f(x) = \frac{1}{x^2}$
10. $f(x) = x^{-5}$
11. $f(x) = 3x^{-6}$
12. $f(x) = \frac{1}{2}\sqrt{x}$
13. $f(x) = 1 - \frac{x}{5}$
14. $f(x) = \frac{1}{3}$
15. $f(x) = 0$
16. $f(x) = x^2 + 3x - 2$
17. $f(x) = .25x^4 - 3x^2 + \pi$
18. $f(x) = .01x^{50} + \frac{1}{3}x^6 - x$
19. $f(x) = \frac{x}{3} + \frac{3}{x}$
20. $f(x) = \frac{2x^2}{5} + \frac{4}{\sqrt{x}}$
21. $f(x) = \frac{3}{2\sqrt{x}} + 2\sqrt{x}$
22. $f(x) = \frac{5}{3x} + \frac{1}{x^{-2/3}}$

In Exercises 23–30, find $f''(x)$.

23. $f(x) = 2x^3 + 3x - 1$
24. $f(x) = 2 - x + 7x^2$
25. $f(x) = x - \sqrt{x}$
26. $f(x) = \frac{1}{x} + 5x^2$
27. $f(x) = \frac{2x - 1}{5}$
28. $f(x) = \pi x^2 + 2\pi x$
29. $f(x) = 3x^{2/3}$
30. $f(x) = \frac{-5}{x^{1/5}}$

In Exercises 31 and 32, find the slope of the graph of $f(x)$ at the designated point.

31. $f(x) = x + \frac{1}{x}$; $(.5, 2.5)$
32. $f(x) = 2\sqrt{x} + x^2$; $(1, 3)$

In Exercises 33 and 34, find the equation of the tangent line to the curve at the point where $x = 1$. (You can check your answer by graphing the function and the line with a graphing utility and confirming that the line is tangent to the curve at the proper point.)

33. $y = 4x^{3/2} - \frac{3}{2}x^4$
34. $y = \frac{4}{x^3} + \frac{1}{\sqrt{x}}$

In Exercises 35 and 36, find the x-coordinates of all points on the curve where the tangent line is horizontal.

35. $y = \frac{1}{3}x^3 - \frac{1}{2}x^2 - 6x + 10$
36. $y = 3x + \frac{12}{x}$
37. A market finds that if it prices a certain product so as to sell x units each week, then the revenue received will be approximately $2x - .001x^2$ dollars.

(a) Find the marginal revenue at a sales level of 600 units.
(b) At what sales level will the marginal revenue be \$1.10 per unit?

38. Suppose the revenue from producing (and selling) x units of a product is given by $R(x) = .01x^2 - 3x$ dollars.
(a) Find the marginal revenue at a production level of 1800 units.
(b) Find the production level where the revenue is \$1800.

39. A manufacturer estimates that the hourly cost of producing x units of a product on an assembly line is $.1x^3 - 6x^2 + 136x + 200$ dollars.
(a) Compute $C(21) - C(20)$, the extra cost of raising the production from 20 to 21 units.
(b) Find the marginal cost when the production level is 20 units.

40. Suppose that the profit from producing x units of a product is given by $P(x) = .003x^3 + .01x$ dollars.
(a) Compute the additional profit gained from increasing sales from 100 to 101 units.
(b) Find the marginal profit at a production level of 100 units.

41. The profit from manufacturing and selling x units of goods is $f(x) = -.01x^2 + 50x - 10{,}000$ dollars. Find and interpret $f(2000)$ and $f'(2000)$.

42. Let $f(x) = x^4 - 12x^2 + 100$. Is $f(x)$ increasing or decreasing at $x = 2$? Is $f(x)$ concave up or concave down at $x = 2$?

43. Let $f(x) = 6\sqrt{x} - x$. Is $f(x)$ increasing or decreasing at $x = 4$? Is $f(x)$ concave up or concave down at $x = 4$?

44. Using the sum rule and the constant-multiple rule, show that for any functions $f(x)$ and $g(x)$,
$$[f(x) - g(x)]' = f'(x) - g'(x).$$

45. Suppose you found the equation of the tangent line to the graph of $y = f(x)$ at the point (2, 4), graphed both the tangent line and the curve with a graphing utility, and obtained the display in Figure 6. What went wrong?

46. Repeat Exercise 45 for the display in Figure 7.

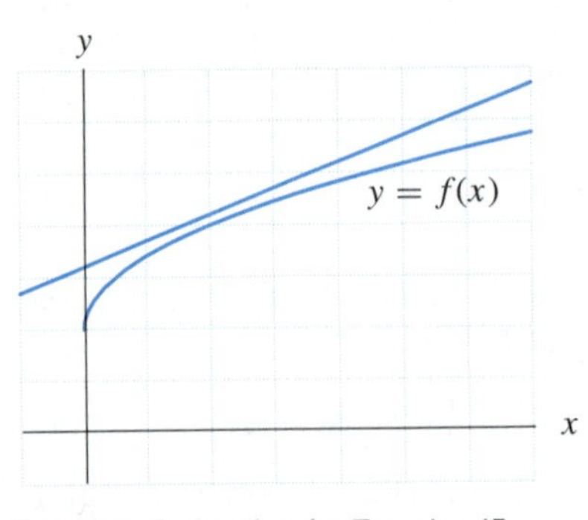

FIGURE 6 Display for Exercise 45

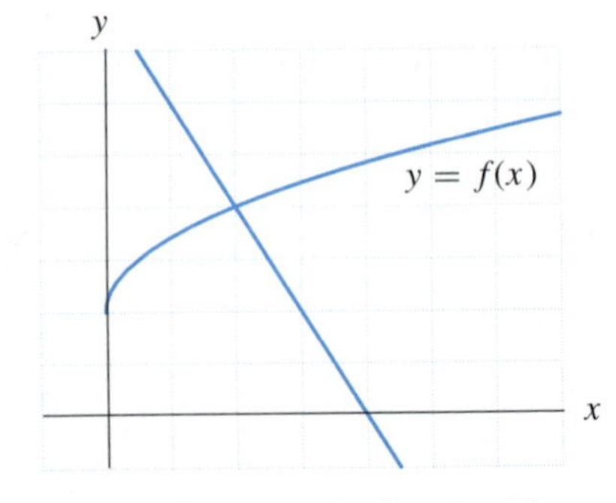

FIGURE 7 Display for Exercise 46

Solutions to Practice Problems 3.1

1. The problem asks for the derivative of the function $y = x$, a straight line of slope 1. Therefore, $(x)' = 1$. The result can also be obtained from the power rule with $r = 1$. If $f(x) = x^1$, then $f'(x) = 1 \cdot x^{1-1} = x^0 = 1$. See Figure 8.

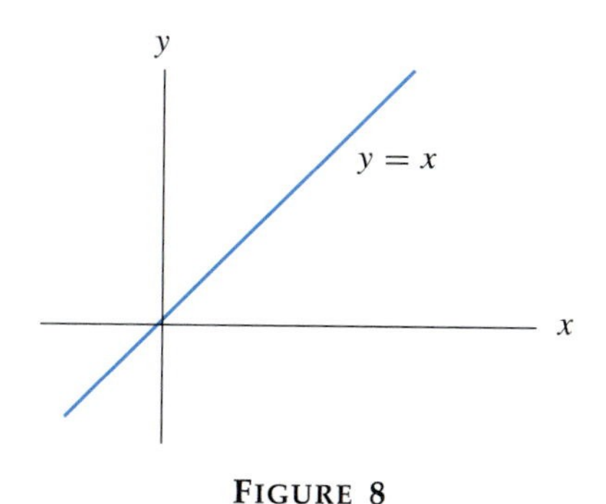

FIGURE 8

2. (a) The function $1/x^4$ can be written as the power function x^{-4}. Here $r = -4$. Therefore,

$$f'(x) = (-4)x^{(-4)-1} = -4x^{-5} = \frac{-4}{x^5}.$$

   (b) $f'(2) = -4/2^5 = -4/32 = -\frac{1}{8}$.

3.2 More About Derivatives

Alternate Notation for the Derivative The operation of forming a derivative $f'(x)$ from a function $f(x)$ is also indicated by the symbol $\frac{d}{dx}$ (read "the derivative with respect to x"). Thus

$$\frac{d}{dx} f(x) = f'(x).$$

For example,

$$\frac{d}{dx}\left(x^6\right) = 6x^5, \quad \frac{d}{dx}\left(x^{5/3}\right) = \frac{5}{3}x^{2/3}, \quad \frac{d}{dx}\left(\frac{1}{x}\right) = -\frac{1}{x^2}.$$

When working with an equation of the form $y = f(x)$, we often write $\frac{dy}{dx}$ as a symbol for the derivative $f'(x)$. For example, if $y = x^6$, we may write

$$\frac{dy}{dx} = 6x^5.$$

The corresponding notation for the second derivative is $\frac{d^2}{dx^2}$. This symbolic notation arises from differentiating $\frac{d}{dx} f(x)$; that is,

$$f'(x) = \frac{d}{dx} f(x),$$
$$f''(x) = \frac{d}{dx}\left[\frac{d}{dx} f(x)\right].$$

If we evaluate the derivative $f'(x)$ at a specific value of x, say $x = a$, we get a number $f'(a)$ that gives the slope of the curve $y = f(x)$ at the point $(a, f(a))$. Another way of writing $f'(a)$ is

$$\left.\frac{dy}{dx}\right|_{x=a}.$$

If we have a second derivative $f''(x)$, then its value when $x = a$ is written

$$f''(a) \quad \text{or} \quad \left.\frac{d^2y}{dx^2}\right|_{x=a}.$$

EXAMPLE 1 If $y = x^4 - 5x^3 + 7$, find $\left.\frac{d^2y}{dx^2}\right|_{x=3}$.

Solution

$$\frac{dy}{dx} = \frac{d}{dx}(x^4 - 5x^3 + 7) = 4x^3 - 15x^2,$$

$$\frac{d^2y}{dx^2} = \frac{d}{dx}(4x^3 - 15x^2) = 12x^2 - 30x,$$

$$\left.\frac{d^2y}{dx^2}\right|_{x=3} = 12(3)^2 - 30(3) = 108 - 90 = 18.$$

■

VARIABLES OTHER THAN x AND y In the differentiation rules from the previous section the letter x was used to denote the variable. Any letter could have been used. For instance, with t as the variable, the derivative of t^3 is $3t^2$. Also, with t as the variable, the notation $\frac{d}{dx}$ becomes $\frac{d}{dt}$. For instance,

$$\frac{d}{dt}(t^3) = 3t^2 \quad \text{and} \quad \frac{d^2}{dt^2}(t^3) = 6t.$$

Letters other than y can be used to represent functions. In applied situations, suggestive letters are usually chosen. For instance, we might use P to represent population or A to represent area.

```
nDeriv(X^3-2X²+3
X,X,-2)
          23.000001
nDeriv(nDeriv(X^
3-2X²+3X,X,X),X,
-2)
                -16
```

Solution to Example 2 with a TI-82 or TI-83

```
der1(x^3-2 x²+3 x,x,-
2)
                 23
der2(x^3-2x²+3x,x,-2)
                -16
evalF nDer der1 der2 fnInt
```

Solution to Example 2 with a TI-85

EXAMPLE 2 If $s = t^3 - 2t^2 + 3t$, find

$$\left.\frac{ds}{dt}\right|_{t=-2} \quad \text{and} \quad \left.\frac{d^2s}{dt^2}\right|_{t=-2}.$$

Solution

$$\frac{ds}{dt} = \frac{d}{dt}(t^3 - 2t^2 + 3t) = 3t^2 - 4t + 3.$$

$$\left.\frac{ds}{dt}\right|_{t=-2} = 3(-2)^2 - 4(-2) + 3 = 12 + 8 + 3 = 23.$$

To find the value of the second derivative at $t = -2$, we must first differentiate $\frac{ds}{dt}$.

$$\frac{d^2s}{dt^2} = \frac{d}{dt}(3t^2 - 4t + 3) = 6t - 4.$$

$$\left.\frac{d^2s}{dt^2}\right|_{t=-2} = 6(-2) - 4 = -12 - 4 = -16.$$

■

THE GENERAL POWER RULE Frequently, we will encounter expressions of the form $[g(x)]^r$ — for instance, $(x^3+5)^2$, where $g(x) = x^3+5$ and $r = 2$. The general power rule says that, to differentiate $[g(x)]^r$, we must first treat $g(x)$ as if it were simply an x, form $r[g(x)]^{r-1}$, and then multiply it by the "correction factor" $g'(x)$. That is,

General Power Rule* Let r be a real number. Then

$$\frac{d}{dx}[g(x)]^r = r \cdot [g(x)]^{r-1} \cdot g'(x).$$

Thus,

$$\begin{aligned}\frac{d}{dx}(x^3+5)^2 &= 2(x^3+5)^1 \cdot \frac{d}{dx}(x^3+5)\\ &= 2(x^3+5)\cdot(3x^2)\\ &= 6x^2(x^3+5).\end{aligned}$$

In this special case it is easy to verify that the general power rule gives the correct answer. We first expand $(x^3+5)^2$ and then differentiate:

$$(x^3+5)^2 = (x^3+5)(x^3+5) = x^6+10x^3+25.$$

From the constant-multiple rule and the sum rule, we have

$$\begin{aligned}\frac{d}{dx}(x^3+5)^2 &= \frac{d}{dx}(x^6+10x^3+25)\\ &= 6x^5+30x^2+0\\ &= 6x^2(x^3+5).\end{aligned}$$

The two methods give the same answer.

Note that if we set $g(x) = x$ in the general power rule, we recover the power rule. So the general power rule contains the power rule as a special case.

EXAMPLE 3 Differentiate $\sqrt{1-x^2}$.

Solution

$$\begin{aligned}\frac{d}{dx}\left(\sqrt{1-x^2}\right) &= \frac{d}{dx}\left[(1-x^2)^{1/2}\right]\\ &= \frac{1}{2}(1-x^2)^{-1/2}\cdot\frac{d}{dx}(1-x^2)\\ &= \frac{1}{2}(1-x^2)^{-1/2}\cdot(-2x)\\ &= \frac{-x}{(1-x^2)^{1/2}} = \frac{-x}{\sqrt{1-x^2}}.\end{aligned}$$

■

* More precisely, the general power rule asserts that if $g(x)$ is differentiable at $x = a$ and if $[g(x)]^r$ and $[g(x)]^{r-1}$ are both defined at $x = a$, then $[g(x)]^r$ is also differentiable at $x = a$ and its derivative is given by the formula stated.

EXAMPLE 4 Differentiate $y = \dfrac{1}{x^3 + 4x}$.

Solution

$$
\begin{aligned}
y &= \frac{1}{x^3 + 4x} = (x^3 + 4x)^{-1}. \\
\frac{dy}{dx} &= (-1)(x^3 + 4x)^{-2} \cdot \frac{d}{dx}(x^3 + 4x) \\
&= \frac{-1}{(x^3 + 4x)^2}(3x^2 + 4) \\
&= -\frac{3x^2 + 4}{(x^3 + 4x)^2}.
\end{aligned}
$$

■

In Chapter 2 we analyzed applications graphically. In the next two examples we solve application problems algebraically.

Solution to Example 5 with a TI-82 or TI-83

EXAMPLE 5 A common clinical procedure for studying a person's calcium metabolism (the rate at which the body assimilates and uses calcium) is to inject some chemically "labeled" calcium into the bloodstream and then measure how fast this calcium is removed from the blood by the person's bodily processes. Suppose that t days after an injection of calcium, the amount A of the labeled calcium remaining in the blood is $A = t^{-3/2}$ for $t \geq .5$, where A is measured in suitable units.* See Figure 3. How fast (in units of calcium per day) is the body removing calcium from the blood when $t = 1$ day?

Solution The rate of change (per day) of calcium in the blood is given by the derivative

$$\frac{dA}{dt} = -\frac{3}{2}t^{-5/2}.$$

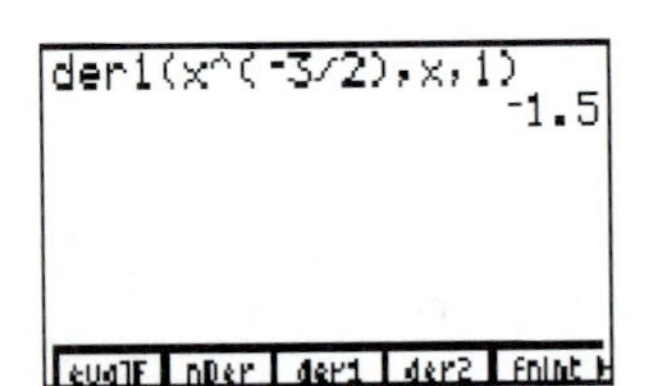

Solution to Example 5 with a TI-85

When $t = 1$, this rate equals

$$\left.\frac{dA}{dt}\right|_{t=1} = -\frac{3}{2}(1)^{-5/2} = -\frac{3}{2}.$$

The amount of calcium in the blood is changing at the rate of $-\frac{3}{2}$ units per day when $t = 1$. The negative sign indicates that the amount of calcium is decreasing rather than increasing. ■

EXAMPLE 6 Suppose a ball is thrown straight up into the air and its height after t seconds is $s(t) = -16t^2 + 128t + 5$ feet.

(a) What is the velocity after 2 seconds?

(b) What is the acceleration after 2 seconds?

(c) At what time is the velocity -32 feet per second? (The negative sign indicates that the ball's height is decreasing; that is, the ball is falling.)

(d) When is the ball at a height of 117 feet?

---

* For a discussion of this mathematical model, see J. Defares, I. Sneddon, and M. Wise, *An Introduction to the Mathematics of Medicine and Biology* (Chicago: Year Book Publishers, Inc., 1973), pp. 609–619.

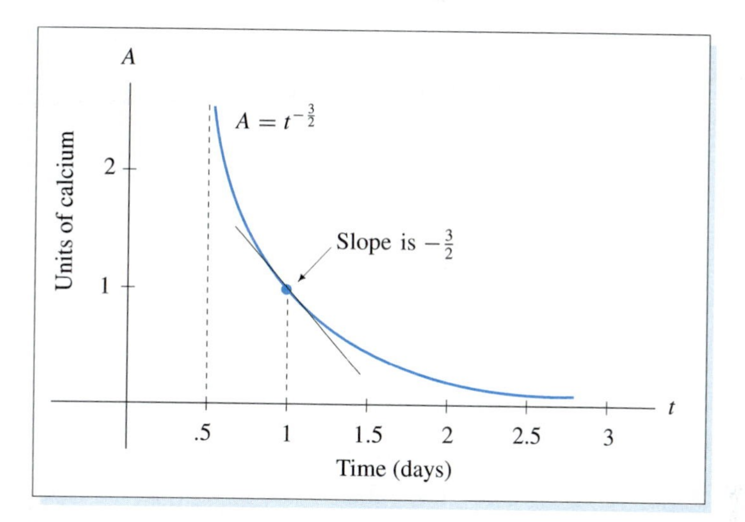

FIGURE 3 Labeled calcium in the blood

Solution (a) The velocity is the rate of change of the height function, so

$$v(t) = s'(t) = -32t + 128.$$

The velocity when $t = 2$ is $v(2) = -32(2) + 128 = 64$ feet per second.

(b) $a(t) = v'(t) = -32$. The acceleration is -32 feet per second squared for all t. This constant acceleration is due to the downward (and therefore negative) force of gravity.

(c) Since the velocity is given and the time is unknown, we set $v(t) = -32$ and solve for t:

$$\begin{aligned} -32t + 128 &= -32 \\ -32t &= -160 \\ t &= 5. \end{aligned}$$

The velocity is -32 feet per second when t is 5 seconds.

(d) The question here involves the height function, not the velocity. Since the height is given and the time is unknown, we set $s(t) = 117$ and solve for t:

$$\begin{aligned} -16t^2 + 128t + 5 &= 117 \\ -16(t^2 - 8t + 7) &= 0 \\ -16(t - 1)(t - 7) &= 0. \end{aligned}$$

The ball is at a height of 117 feet when $t = 1$ and when $t = 7$ seconds. ■

PRACTICE PROBLEMS 3.2

1. Differentiate $g(r) = 2\pi rh$.

2. Differentiate the function $y = \dfrac{x + (x^5 + 1)^{10}}{3}$.

EXERCISES 3.2

In Exercises 1–16, find the first derivative.

1. $f(x) = (3x^2 + x)^{10}$
2. $f(x) = (1 - 2x)^3$
3. $f(x) = (5 - x)^{3/2}$
4. $f(x) = (x^3 - 4x^2)^{.2}$
5. $y = \dfrac{1}{x^2 + x + 1}$
6. $y = 5\sqrt{3x^2 + x}$
7. $y = \sqrt{-2x + 1}$
8. $y = \dfrac{1}{(x^2 - 7)^5}$
9. $f(t) = (t^4 + 5t^2)^3$
10. $f(t) = \dfrac{1}{t^3 + 3t}$
11. $s = \sqrt{1 - t}$
12. $w = \dfrac{5}{3t^5 - 2t}$
13. $v = \frac{4}{3}\pi r^3$
14. $f(r) = \pi h r^2 + 2\pi r$
15. $f(s) = 2\pi h s^2$
16. $A = (3t + 4)^5$
17. Find $\dfrac{d}{dx}(3x^2 - \frac{1}{2}x + 1)$.
18. Find $\dfrac{d}{dx}(5x^3 + 2x)^4$.
19. Find $\dfrac{d}{dt}(4t - 3t^2)^3$.
20. Find $\dfrac{d}{dt}\sqrt{t^2 + 1}$.

Compute the following.

21. $\left.\dfrac{d}{dx}(2x^2 - 3)\right|_{x=5}$
22. $\left.\dfrac{d}{dt}(1 - 2t - 3t^2)\right|_{t=-1}$
23. $\left.\dfrac{d}{dz}(z^2 - 4)^3\right|_{z=1}$
24. $\left.\dfrac{d}{dT}\left(\dfrac{1}{3T + 1}\right)\right|_{T=2}$
25. $\left.\dfrac{d^2}{dx^2}(3x^3 - x^2 + 7x - 1)\right|_{x=2}$
26. $\dfrac{d}{dt}\left(\dfrac{dv}{dt}\right)$, where $v = 2t^{-3}$
27. $\dfrac{d}{dP}\left(\dfrac{dy}{dP}\right)$, where $y = \dfrac{k}{2P - 1}$
28. $\left.\dfrac{d^2V}{dr^2}\right|_{r=2}$, where $V = ar^3$
29. $f'(3)$ and $f''(3)$, where $f(x) = \sqrt{10 - 2x}$
30. $g'(2)$ and $g''(2)$, where $g(T) = (3T - 5)^{10}$
31. Suppose a company finds that the revenue R generated by spending x dollars on advertising is given by $R = 1000 + 80x - .02x^2$ dollars, for $0 \le x \le 2000$. Find and interpret $\left.\dfrac{dR}{dx}\right|_{x=1500}$.
32. A supermarket finds that its average daily volume of business V (in thousands of dollars) and the number of hours t the store is open for business each day are approximately related by the formula

$$V = 20\left(1 - \frac{100}{100 + t^2}\right), \quad 0 \le t \le 24.$$

Find and interpret $\left.\dfrac{dV}{dt}\right|_{t=10}$.

33. Write the constant-multiple rule using $\frac{d}{dx}$ notation.

34. Write the sum rule using $\frac{d}{dx}$ notation.

35. Suppose that the weight in grams of a cancerous tumor at time t is $W(t) = .1t^2$, where t is measured in weeks.

    (a) How large is the tumor after 3 weeks?
    (b) How fast is the tumor growing after 3 weeks?
    (c) When will the tumor weigh 3.2 grams?
    (d) When will the tumor be growing at the rate of 3.2 grams per week?

36. After an advertising campaign, the sales of a product often increase and then decrease. Suppose that t days after the end of the advertising, the daily sales are $-3t^2 + 32t + 100$ units.

    (a) What is the daily level of sales after 1 day?
    (b) At what rate are the daily level of sales increasing after 1 day?
    (c) When are the daily level of sales increasing at the rate of 14 units per day?
    (d) When will 152 units be sold per day?
    (e) Are the daily sales increasing or decreasing after 6 days?

37. A toy rocket fired straight up into the air has height $s(t) = 160t - 16t^2$ feet after t seconds.

    (a) What is the rocket's initial velocity (when $t = 0$)?
    (b) How high is the rocket after 3 seconds?
    (c) How fast is the rocket rising after 3 seconds?
    (d) When is the rocket descending at the rate of 96 feet per second?
    (e) At what time will the rocket hit the ground?
    (f) At what velocity will the rocket be traveling just as it smashes into the ground?
    (g) What is the acceleration when $t = 3$?

38. A helicopter is rising straight up in the air. Its distance from the ground t seconds after takeoff is $s(t)$ feet, where $s(t) = t^2 + t$.

    (a) How long will it take for the helicopter to rise 20 feet?
    (b) Find the velocity and the acceleration of the helicopter when it is 20 feet above the ground.
    (c) How high will the helicopter be when it is rising at the speed of 15 feet per second?

39. Let $f(t)$ be the amount of oxygen (in suitable units) in a lake t days after sewage is dumped into the lake, and suppose that $f(t)$ is given approximately by

$$f(t) = 1 - \frac{10}{t+10} + \frac{100}{(t+10)^2}.$$

    (a) What was the initial oxygen content of the lake?

(b) After 5 days, is the oxygen content of the lake increasing or decreasing? Is it increasing or decreasing after 12 days?

(c) What is the average daily decrease in oxygen content during the first 10 days after sewage is dumped into the lake?

40. A bicyclist passes a mailbox while traveling on a straight road. His distance (in feet) from the mailbox t seconds after passing it is given approximately by the function

$$s(t) = \frac{(2t-6)^3}{9} + 24; \quad 0 \le t \le 10.$$

(a) How fast is the cyclist going as he passes the mailbox? Is he accelerating or decelerating?

(b) How far from the mailbox is the cyclist after 7.5 seconds?

(c) When is he accelerating at the rate of 16 ft/sec^2?

(d) The cyclist encounters a stop sign during the 10 seconds after he passes the mailbox. How far is the stop sign from the mailbox?

(e) What is the cyclist's average speed during the 10 seconds after he passes the mailbox?

In the remaining exercises, use a graphing utility to obtain the graphs of the functions and answer the questions when appropriate.

41. The weekly cost for a manufacturer is given by the function $C(x) = .006x^3 - .7x^2 + 32x + 250$.

(a) Graph $C(x)$, $C'(x)$, and $C''(x)$ for $0 \le x \le 75$.

(b) What is the cost of manufacturing 10 units of goods per week?

(c) For what level of production will the cost be \$742?

(d) What is the marginal cost when 72 units of goods are manufactured?

(e) At what level of production is the marginal cost \$7?

(f) Determine whether the cost curve is concave up or concave down at $x = 50$ and interpret the result.

(g) At what level of production does the marginal cost function have a horizontal tangent line?

(h) Approximately what is the slope of the marginal cost function when 49 units of goods are manufactured?

42. When \$1000 is deposited into the bank at $x\%$ interest compounded annually for 10 years, the balance is

$$f(x) = 1000\left(1 + \frac{x}{100}\right)^{10} \quad \text{dollars.}$$

(a) Graph $f(x)$ and $f'(x)$.

(b) What will the balance be at 6% interest?

(c) For what interest rate will the balance be \$2600?

(d) If the interest rate is 8% and interest is increasing by 1%, by approximately how much will the balance increase?

SOLUTIONS TO PRACTICE PROBLEMS 3.2

1. The expression $2\pi rh$ contains two numbers, 2 and π, and two letters, r and h. The notation $g(r)$ tells us that the expression $2\pi rh$ is to be regarded as a function of r. Therefore, h—and hence $2\pi h$—is to be treated as a constant, and differentiation is done with respect to the variable r. That is,

$$g(r) = (2\pi h)r,$$
$$g'(r) = 2\pi h.$$

2. All three rules are required to differentiate this function:

$$\begin{aligned}
\frac{dy}{dx} &= \frac{d}{dx}\frac{1}{3}\cdot\left[x + (x^5+1)^{10}\right] \\
&= \frac{1}{3}\frac{d}{dx}\left[x + (x^5+1)^{10}\right] && \text{(constant-multiple rule)} \\
&= \frac{1}{3}\left[\frac{d}{dx}(x) + \frac{d}{dx}(x^5+1)^{10}\right] && \text{(sum rule)} \\
&= \frac{1}{3}\left[1 + 10(x^5+1)^9\cdot(5x^4)\right] && \text{(general power rule)} \\
&= \frac{1}{3}\left[1 + 50x^4(x^5+1)^9\right].
\end{aligned}$$

3.3 OBTAINING GRAPHS OF FUNCTIONS (INTRODUCTION)

There are two important reasons for obtaining good graphs of functions. First, a geometric "picture" of a function is often easier to comprehend than its abstract formula. Second, a good graph is needed for many applications of calculus.

A good graph of a function should reveal the important features of the function. In particular, the graph should show the relative extreme points and inflection points. When relevant, the graph also should show the intercepts and asymptotes. Since we have the assistance of a graphing utility, one important concern is to find a proper viewing window. This section focuses on graphs that reveal relative extreme and inflection points. The next section explores intercepts and asymptotes in greater detail.

EXAMPLE 1 Which of the following viewing windows is most appropriate for the function $f(x) = (x-2)^3 - 3.75(x-2)^2 + 4.5(x-2)$?

(a) $[-10, 10]$ *by* $[-10, 10]$
(b) $[-1, 5]$ *by* $[-75, 7]$
(c) $[-1, 5]$ *by* $[-.1, 2]$

Solution The three graphs are shown in Figure 1.

(a) This standard viewing window shows the general nature of the graph. Something interesting is happening at $x = 3$; but what is it?

(b) In (a), the graph is really only visible between $x = 1$ and $x = 5$. Figure 1(b) covers these values of x and a little more so that the y-axis will appear. Since

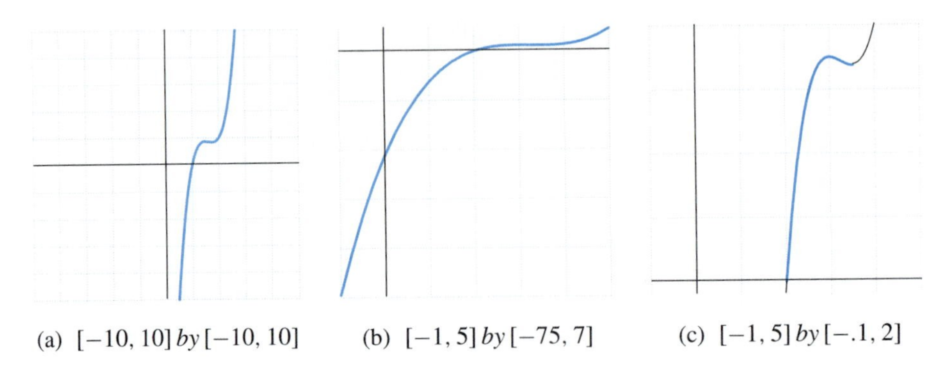

(a) $[-10, 10]$ *by* $[-10, 10]$ (b) $[-1, 5]$ *by* $[-75, 7]$ (c) $[-1, 5]$ *by* $[-.1, 2]$

FIGURE 1 Three views of the function in Example 1

$f(-1)$ is approximately -75 and $f(5)$ is 7, we have let the y-values range from -75 to 7. We can see many function values. However, we still don't know what is going on near $x = 3$. There might be an inflection point, or there might be more.

(c) Here the range of the y-values has been shortened to get a better picture of the situation near $x = 3$. This graph shows that the function has a relative maximum point, a relative minimum point, and an inflection point near $x = 3$. The exact locations of the extreme points and the inflection points can be obtained by looking at the graphs of $f'(x)$ and $f''(x)$ or by algebraically solving $f'(x) = 0$ and $f''(x) = 0$.

Each graph has its use. The first graph uses a standard window and gives a rough look at the function. The second graph shows the x- and y-intercepts. The third graph shows the relative extreme points and inflection point. Since many applied calculus problems require us to find extreme points or inflection points, graphs such as the third graph will often be most useful. ■

Our technique for obtaining good graphs involves an interplay between algebraic calculations and use of a graphing utility. We not only strive to display a good graph, but also seek to give the coordinates of all relative extreme points and inflection points. Although there is no rigid procedure that will always be used, the following general guidelines are helpful in obtaining a good graph of the function $f(x)$:

1. Algebraically determine the relative extreme points and inflection points.

   (a) Calculate $f'(x)$ and $f''(x)$.

   (b) Set $f'(x) = 0$ and solve for x to find all possible relative extreme points. Evaluate $f''(x)$ at each of these points. If $f''(x)$ is nonzero at one of these points, we will know the nature of the extreme point.

   (c) Set $f''(x) = 0$ and solve for x to find all possible inflection points.

2. With pencil and paper, plot the points found in step 1 and make a partial sketch of the graph near these points.

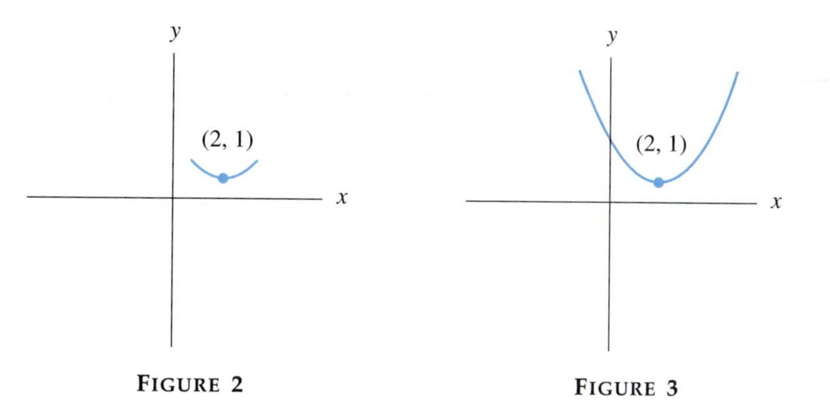

FIGURE 2

FIGURE 3

3. With a graphing utility, choose a window containing the x- and y-coordinates of the points found in step 1, both the x- and y-axes (if reasonable), and enough of the graph to see the general shape.

EXAMPLE 2 The graph of the quadratic function $f(x) = \frac{1}{4}x^2 - x + 2$ is a parabola and so has one relative extreme point. Find it and sketch the graph.

Solution The function and its first and second derivatives are

$$\begin{aligned} f(x) &= \frac{1}{4}x^2 - x + 2, \\ f'(x) &= \frac{1}{2}x - 1, \\ f''(x) &= \frac{1}{2}. \end{aligned}$$

Setting $f'(x) = 0$, we have $\frac{1}{2}x - 1 = 0$, so that $x = 2$. Thus, $f'(2) = 0$. Geometrically, this means that the graph of $f(x)$ will have a horizontal tangent line at the point where $x = 2$. To plot this point, we substitute the value 2 for x in the original expression for $f(x)$:

$$f(2) = \frac{1}{4}(2)^2 - (2) + 2 = 1.$$

Is (2, 1) a relative extreme point? In order to decide, we look at $f''(x)$. Since $f''(2) = \frac{1}{2}$, which is positive, the graph of $f(x)$ is concave up at $x = 2$. So a partial sketch of the graph near (2, 1) should look something like Figure 2.

We see that (2, 1) is a relative minimum point. In fact, it is the only relative extreme point, for there is no other place where the first derivative is zero. Since the graph has no other "turning points," it must be decreasing before it gets to (2, 1) and then increasing to the right of (2, 1). Note that since $f''(x)$ is positive (and equal to $\frac{1}{2}$) for all x, the graph is concave upward at each point and there are no inflection points. A completed rough sketch is given in Figure 3. From the rough sketch, a good domain is $-4 \leq x \leq 8$. Since $f(-4) = 10 = f(8)$ and we would like to see the x-axis if possible, a good range for the y-values is

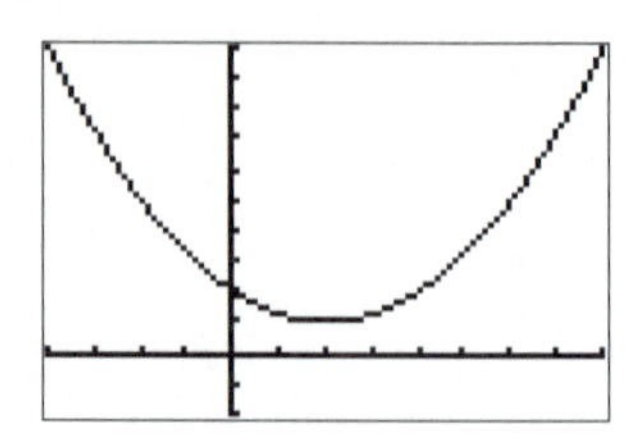

FIGURE 4 Window $[-4, 8]$ *by* $[-2, 10]$

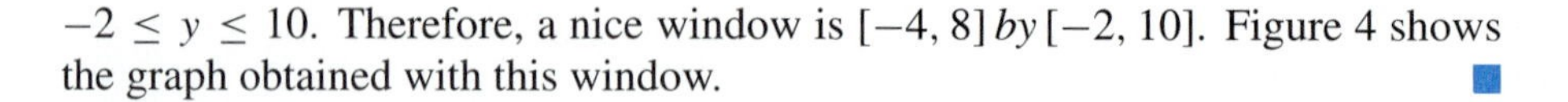

$-2 \leq y \leq 10$. Therefore, a nice window is $[-4, 8]$ *by* $[-2, 10]$. Figure 4 shows the graph obtained with this window. ■

EXAMPLE 3 Graph the function $y = x^3 - 33x^2 + 120x + 1500$.

Solution Let

$$f(x) = x^3 - 33x^2 + 120x + 1500.$$

Then

$$f'(x) = 3x^2 - 66x + 120,$$
$$f''(x) = 6x - 66.$$

To look for possible relative extreme points, set $f'(x) = 0$ and solve for x:

$$3x^2 - 66x + 120 = 0$$
$$3(x^2 - 22x + 40) = 0$$
$$3(x - 2)(x - 20) = 0$$
$$x = 2 \quad \text{or} \quad x = 20.$$

Substituting these values of x back into $f(x)$,we find that

$$f(2) = 2^3 - 33(2)^2 + 120(2) + 1500 = 1616,$$
$$f(20) = 20^3 - 33(20)^2 + 120(20) + 1500 = -1300.$$

Therefore, $(2, 1616)$ and $(20, -1300)$ are possible relative extreme points. (The values of $f(2)$ and $f(20)$ also can be obtained with a graphing utility.)

Substituting $x = 2$ and $x = 20$ into $f''(x)$, we obtain

$$f''(2) = 6(2) - 66 = -54,$$
$$f''(20) = 6(20) - 66 = 54.$$

By the second derivative test, $(2, 1616)$ is a relative maximum point and $(20, -1300)$ is a relative minimum point. Figure 5 shows a partial sketch of the graph near these points.

To look for possible inflection points, set $f''(x) = 0$ and solve for x:

$$6x - 66 = 0$$
$$6(x - 11) = 0$$
$$x = 11.$$

Substituting this value of x back into $f(x)$, we find that

$$f(11) = 11^3 - 33(11)^2 + 120(11) + 1500 = 158.$$

Therefore, $(11, 158)$ is a possible inflection point. See Figure 6.

Since $(2, 1616)$ and $(20, -1300)$ are the only turning points, the graph must be increasing before it gets to $(2, 1616)$, decreasing from $(2, 1616)$ to

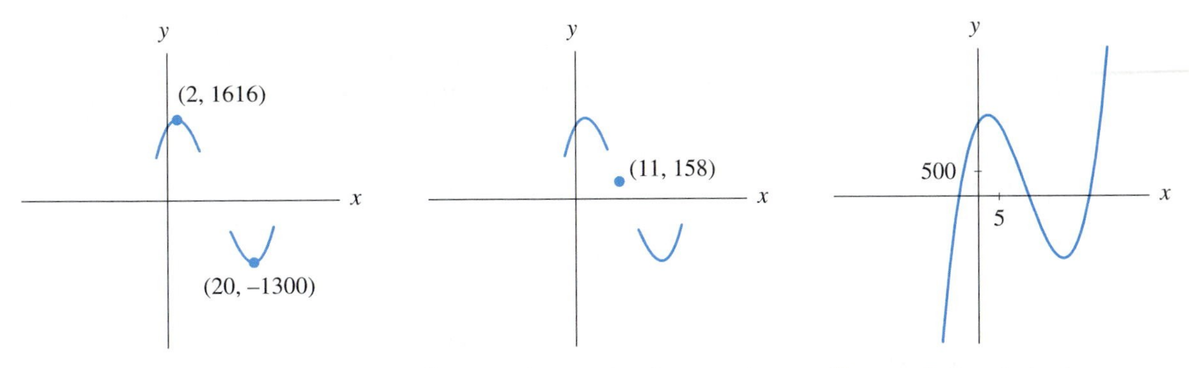

FIGURE 5 Relative extreme points   FIGURE 6 Possible inflection point   FIGURE 7 Rough sketch

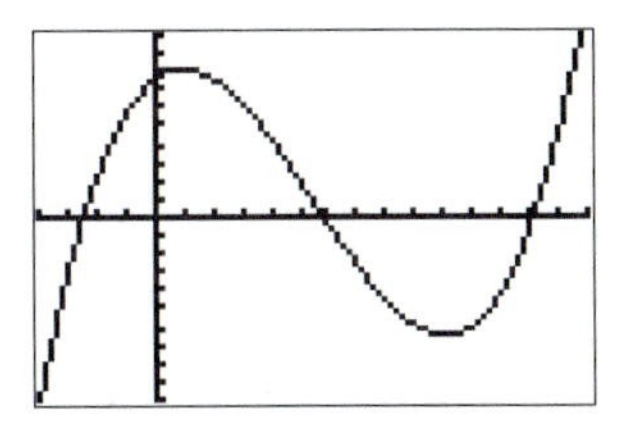

FIGURE 8 Window $[-8, 30]$ *by* $[-2000, 2000]$

$(20, -1300)$, and then increasing again to the right of $(20, -1300)$. Since the graph is concave down at $(2, 1616)$ and concave up at $(20, -1300)$, there must be an inflection point somewhere in between. Since the only possible candidate is $(11, 158)$, it must indeed be an inflection point. A rough sketch incorporating these properties appears in Figure 7. From the rough sketch, a nice window appears to be

$$[-8, 30]\ by\ [-2000, 2000].$$

Figure 8 shows the graph obtained with this window. ■

Note: The graphs of quadratic polynomials are always parabolas. They have the general shape of the curve in Figure 3, or that curve turned upside-down. The graphs of the two cubic polynomials shown in this section are similar. Each has a relative maximum point, a relative minimum point, and an inflection point in between them. There is one other possible shape for a cubic—a futuristic letter "S" with an inflection point and no relative extreme points. This type of cubic is considered in the next section.

PRACTICE PROBLEMS 3.3

1. Which of the curves in Figure 9 could possibly be the graph of a function of the form $f(x) = ax^2 + bx + c$, where $a \neq 0$?
2. Which of the curves in Figure 10 could possibly be the graph of a function of the form $f(x) = ax^3 + bx^2 + cx + d$, where $a \neq 0$?

EXERCISES 3.3

In Exercises 1–8, draw the graph in the designated window and use the trace feature of a graphing utility to estimate the coordinates of any relative extreme points and inflection points. Then determine the coordinates of these points algebraically.

1. $f(x) = x^2 + 2x + 1$; $[-3.5, 1.5]$ *by* $[-1, 4]$
2. $f(x) = -3x^2 - 2x + 6$; $[-2, 2]$ *by* $[-1, 7]$

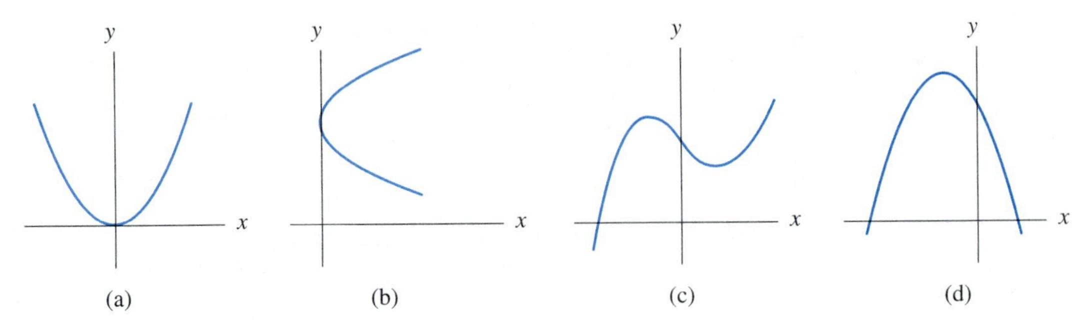

FIGURE 9

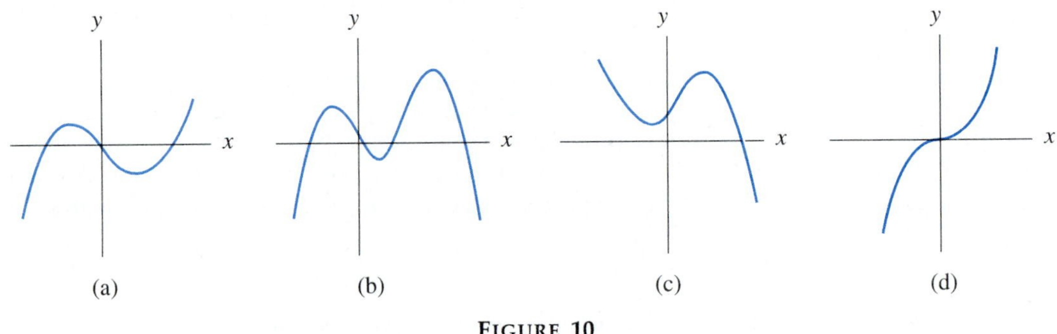

FIGURE 10

3. $f(x) = (3 - 2x)^2 - 4x$; $[0, 4]$ *by* $[-8, 2]$
4. $f(x) = 2x^3 - \frac{3}{2}x^2 - 1$; $[-2, 2]$ *by* $[-3, 1]$
5. $f(x) = (2x - 1)^3 + 1$; $[-.5, 1.5]$ *by* $[-.5, 2.5]$
6. $f(x) = (2x - 1)^3 + (1 - 2x)^2$; $[-.2, 1]$ *by* $[-.5, 1]$
7. $f(x) = (x^2 - 9)^2 + 3$; $[-5, 5]$ *by* $[-10, 100]$
8. $f(x) = \frac{3}{4}x^4 - 6x^2$; $[-4, 4]$ *by* $[-15, 2]$

In Exercises 9–24, use both algebraic calculations and a graphing utility to obtain a good graph of the function. Follow steps 1 through 3 given in this section. Then draw the graph shown on the screen, give the window used, and label all relative extreme points and inflection points.

9. $f(x) = 2x^2 - 4$
10. $f(x) = -.5x^2 + x - 4$
11. $f(x) = -x^2 - 8x - 10$
12. $f(x) = 3x^2 + 6x - 5$
13. $f(x) = x^2 - 16x + 260$
14. $f(x) = 3x^2 - 300x + 5$
15. $f(x) = 2x^3 + 9x^2 + 12x + 1$
16. $f(x) = x^3 - 3x + 3$
17. $f(x) = -x^3 + 30x^2 - 2000$
18. $f(x) = 2x^3 - 3x^2 - 36x + 20$
19. $f(x) = 12x - x^3$
20. $f(x) = 11 + 9x - 3x^2 - x^3$
21. $f(x) = x^3 + 3x^2 + 3x - 4$
22. $f(x) = 3 - 4x + 2x^2 - \frac{1}{3}x^3$
23. $f(x) = \frac{2}{3}x^3 + 6x^2 + 10x$
24. $f(x) = (6 - 2x)^3 + (6 - 2x)^2$

25. Draw the graph of $f(x) = \frac{1}{6}x^3 - x^2 + 3x + 3$ in the window $[-2, 6]$ *by* $[-10, 20]$. This shape is referred to as a futuristic "S." It has an inflection point at $x = 2$ but no relative extreme points. Enlarge the window a few times to convince yourself that there are no relative extreme points anywhere. What does this tell you about $f'(x)$?

26. Draw the graph of $f(x) = \frac{1}{6}x^3 - \frac{5}{2}x^2 + 13x - 20$ in the window $[0, 10]$ *by* $[-20, 30]$. This shape is referred to as a futuristic "S." Algebraically determine the coordinates of the inflection point. Zoom in and zoom out to convince yourself that there are no relative extreme points anywhere.

27. Draw the graph of $f(x) = 2x + \dfrac{18}{x} - 10$ in the window $[0, 16]$ *by* $[0, 16]$. In what ways is this graph like the graph of a parabola that opens upward? In what ways is it different?

28. Draw the graph of $f(x) = 3x + \dfrac{75}{x} - 25$ in the window $[0, 25]$ *by* $[0, 50]$. Use the trace feature of the graphing utility to estimate the coordinates of the relative minimum point. Then determine the coordinates algebraically. Convince yourself both graphically (with the graphing utility) and algebraically that this function has no inflection points.

SOLUTIONS TO PRACTICE PROBLEMS 3.3

1. Answer: (a) and (d). Curve (b) has the shape of a parabola, but it is not the graph of any function, since vertical lines cross it twice. Curve (c) has two relative extreme points, but the derivative of $f(x)$ is a linear function, which could not be zero for two different values of x.

2. Answer: (a), (c), (d). Curve (b) has three relative extreme points, but the derivative of $f(x)$ is a quadratic function, which could not be zero for three different values of x.

3.4 OBTAINING GRAPHS OF FUNCTIONS (CONCLUSION)

In Section 3.3 we discussed the main techniques for curve sketching. Here we add a few finishing touches and examine some slightly more complicated curves. In particular, we shall consider intercepts and vertical and horizontal asymptotes.

INTERCEPTS The relative extreme and inflection points are very important points on a curve. In addition, the x- and y-intercepts often have some intrinsic interest in an applied problem. The y-intercept is $(0, f(0))$. To find the x-intercepts on the graph of $f(x)$, we must find those values of x for which $f(x) = 0$. Since this can be a difficult (or impossible) algebraic problem, we shall find x-intercepts algebraically only when they are easy to find. Of course, a good graph on a calculator or computer will indicate where they lie and the "root," "zero," or "solve" routines can be used to approximate their values, as discussed in Section 1.2.

EXAMPLE 1 Obtain the graph of $y = \frac{1}{2}x^2 - 20x - 400$.

Solution Let

$$f(x) = \frac{1}{2}x^2 - 20x - 400.$$

Then

$$f'(x) = x - 20,$$
$$f''(x) = 1.$$

Since $f'(x) = 0$ only when $x = 20$ and since $f''(20)$ is positive, $f(x)$ must have a relative minimum at $x = 20$. The relative minimum point is $(20, f(20)) = (20, -600)$.

The y-intercept is $(0, f(0)) = (0, -400)$. To find the x-intercepts, we set $f(x) = 0$ and solve for x:

$$\frac{1}{2}x^2 - 20x - 400 = 0.$$

The expression for $f(x)$ is not easily factored, so we use the quadratic formula to solve the equation:

$$x = \frac{-(-20) \pm \sqrt{(-20)^2 - 4(\frac{1}{2})(-400)}}{2(\frac{1}{2})} = 20 \pm \sqrt{1200}.$$

The x-intercepts are $(20 - \sqrt{1200}, 0)$ and $(20 + \sqrt{1200}, 0)$. Since $\sqrt{1200} \approx 34.6$, the x-intercepts are approximately $(-14.6, 0)$ and $(54.6, 0)$. Figure 1 shows a sketch of the function as it might be drawn on a piece of paper. This sketch suggests $[-30, 70]$ *by* $[-600, 600]$ as a good window for a graphing utility. See Figure 2.

FIGURE 1 $y = \frac{1}{2}x^2 - 20x - 400$

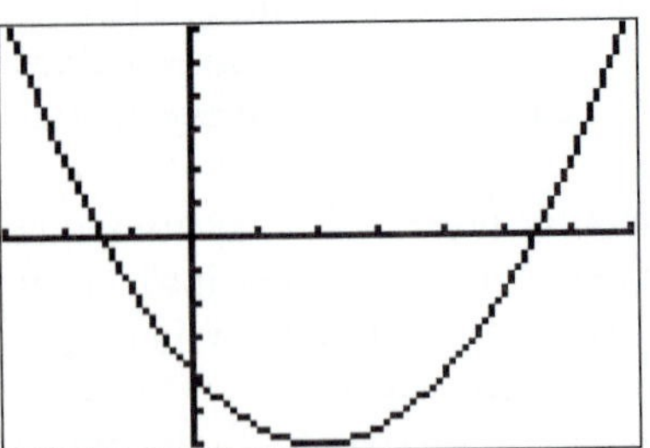

FIGURE 2 $[-30, 70]$ *by* $[-600, 600]$; Xscl = 10, Yscl = 100

EXAMPLE 2 Sketch the graph of $f(x) = \frac{1}{6}x^3 - \frac{3}{2}x^2 + 5x + 1$.

Solution

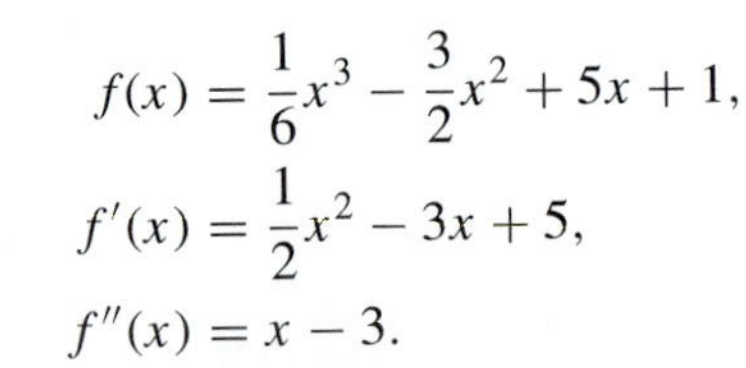

$$f(x) = \frac{1}{6}x^3 - \frac{3}{2}x^2 + 5x + 1,$$
$$f'(x) = \frac{1}{2}x^2 - 3x + 5,$$
$$f''(x) = x - 3.$$

Let us set $f'(x) = 0$ and try to solve for x:

$$\frac{1}{2}x^2 - 3x + 5 = 0. \quad (1)$$

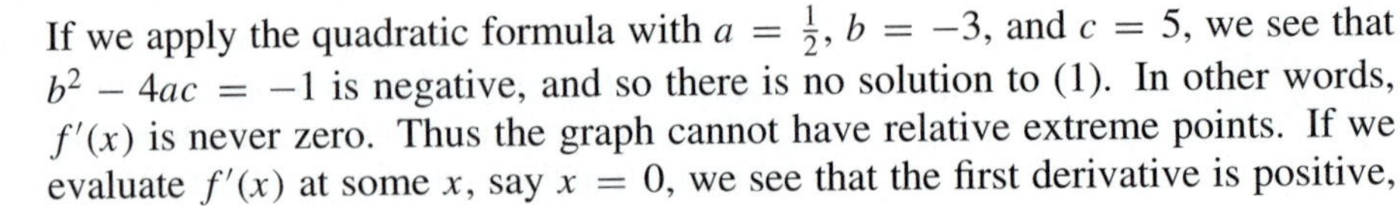

If we apply the quadratic formula with $a = \frac{1}{2}$, $b = -3$, and $c = 5$, we see that $b^2 - 4ac = -1$ is negative, and so there is no solution to (1). In other words, $f'(x)$ is never zero. Thus the graph cannot have relative extreme points. If we evaluate $f'(x)$ at some x, say $x = 0$, we see that the first derivative is positive,

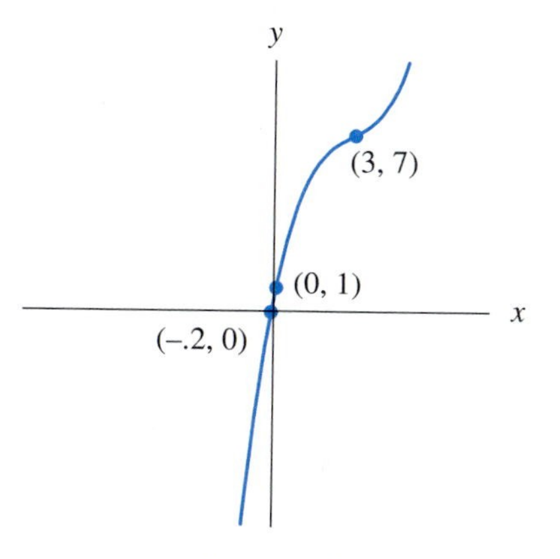

FIGURE 3

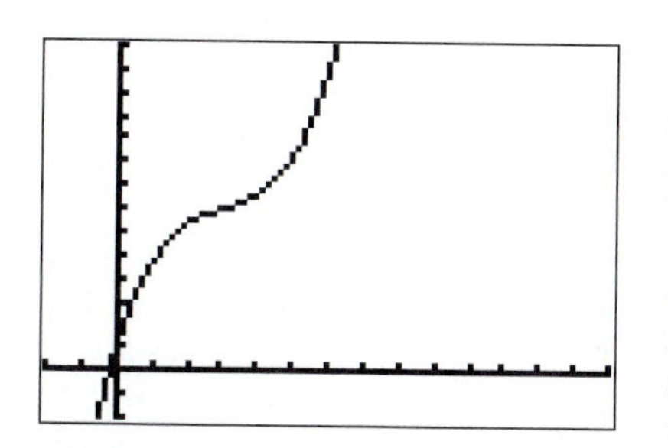
FIGURE 4
Window $[-2, 14]$ *by* $[-2, 14]$

and so $f(x)$ is increasing there. Since the graph of $f(x)$ is a smooth curve with no relative extreme points and no breaks, $f(x)$ must be increasing for all x. (If a function were increasing at $x = a$ and decreasing at $x = b$, then it would have a relative extreme point between a and b.)

Now let us check concavity.

	$f''(x) = x - 3$	Graph of $f(x)$
$x < 3$	Negative	Concave down
$x = 3$	Zero	Concavity reverses
$x > 3$	Positive	Concave up

The inflection point is $(3, f(3)) = (3, 7)$. The y-intercept is $(0, f(0)) = (0, 1)$.

Figure 3 shows a rough sketch using the information found so far. The x-intercept cannot easily be found algebraically. However, a graphing utility gives the x-intercept as approximately $(-.2, 0)$. Figure 4 shows an accurate graph displayed with the window $[-2, 14]$ *by* $[-2, 14]$. ■

ASYMPTOTES An asymptote is a straight line that a graph approaches. The graph of $y = 1/x$ $(x > 0)$ has the y-axis as a vertical asymptote and the x-axis as a horizontal asymptote. See Figure 5. In general, the graph of $y = f(x)$ has the vertical line $x = a$ as an asymptote if $f(x)$ approaches ∞ or $-\infty$ as x approaches a from the right or from the left. (Usually a will be a value at which the function is not defined.) The graph has $y = b$ as a horizontal asymptote if $\lim_{x\to\infty} f(x) = b$ or $\lim_{x\to-\infty} f(x) = b$. See Figure 6.

Graphs similar to the one in the next example arise in several applications later in this chapter.

EXAMPLE 3 Obtain the graph of $f(x) = x + \dfrac{1}{x}$, for $x > 0$.

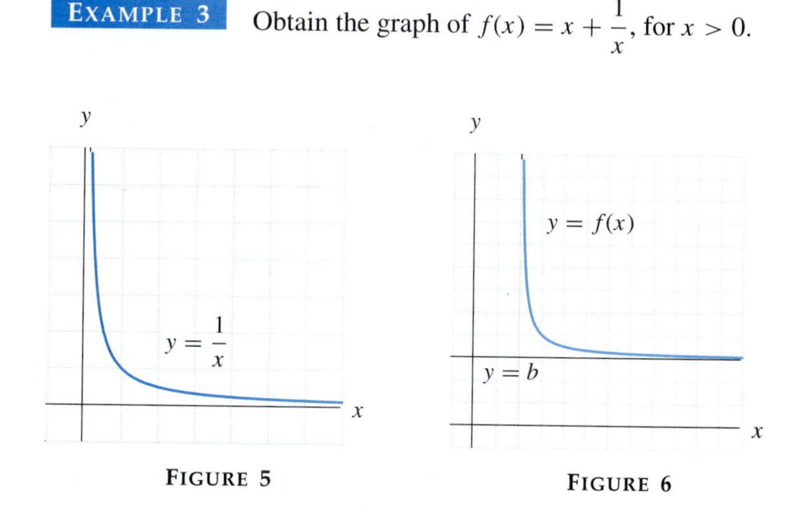

FIGURE 5

FIGURE 6

Solution

$$f(x) = x + \frac{1}{x},$$
$$f'(x) = 1 - \frac{1}{x^2},$$
$$f''(x) = \frac{2}{x^3}.$$

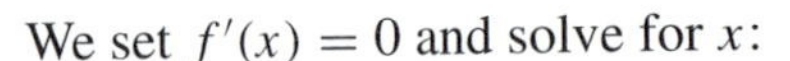

We set $f'(x) = 0$ and solve for x:

$$1 - \frac{1}{x^2} = 0$$
$$1 = \frac{1}{x^2}$$
$$x^2 = 1$$
$$x = 1.$$

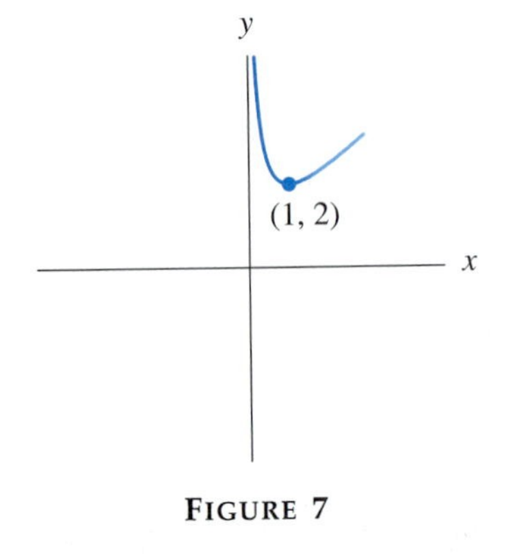

FIGURE 7

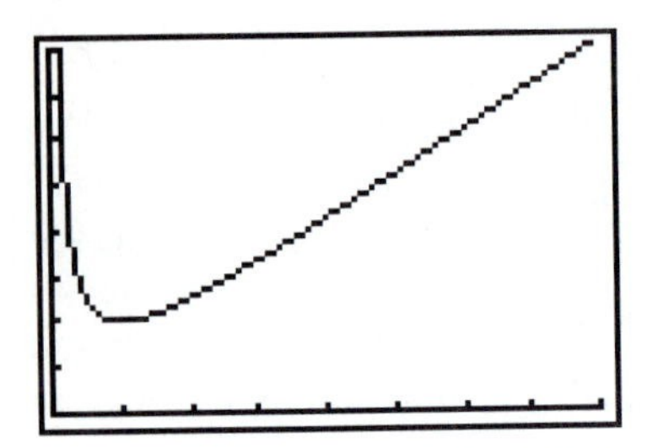

FIGURE 8 Window [0, 8] *by* [0, 8]

(We exclude the case $x = -1$ because we are only considering positive values of x.) The graph has a possible relative extreme point at $(1, f(1)) = (1, 2)$. Now, $f''(1) = 2 > 0$, and so the graph is concave up at $x = 1$ and $(1, 2)$ is a relative minimum point. In fact, $f''(x) = (2/x^3) > 0$ for all positive x, and therefore the graph is concave up at all points.

Before making a partial sketch of the graph, notice that as x approaches zero [a point at which $f(x)$ is not defined], the term $1/x$ in the formula for $f(x)$ becomes arbitrarily large. Thus $f(x)$ has the line $x = 0$ (that is, the y-axis) as a vertical asymptote. Draw the portion of the graph near $(1, 2)$ and extend the curve to the left so that it sidles up to the y-axis as x gets close to 0. See Figure 7. Since the first derivative is zero only at $x = 1$, there are no turning points to the right of $x = 1$. Therefore, the curve continues to rise as x gets large. Figure 8 shows an accurate graph displayed with the window $[0, 8]$ *by* $[0, 8]$. ■

Functions whose graphs have horizontal asymptotes arise in applications in Chapters 5 and 6. These functions are always nonnegative and their asymptotes are easily predicted. Special differentiation techniques will be required to determine the extreme and inflection points. For now, we will just be concerned with obtaining a good window.

EXAMPLE 4 Which of the following viewing windows is most appropriate for the function

$$f(x) = 2 - \frac{16x + 64}{x^3 + 64}, \quad x \geq 0?$$

(a) $[0, 6]$ *by* $[0, 2.5]$
(b) $[0, 100]$ *by* $[0, 2.5]$
(c) $[0, 25]$ *by* $[0, 2.5]$

Solution The three graphs are shown in Figure 9.
(a) This viewing window reveals a relative minimum point at $x = 2$ and an inflection point at about $x = 4$. The graph's only shortcoming is that it leaves us guessing about what happens to the right of $x = 6$. As we move to the right, does the graph have a relative maximum point, grow indefinitely, or level off?

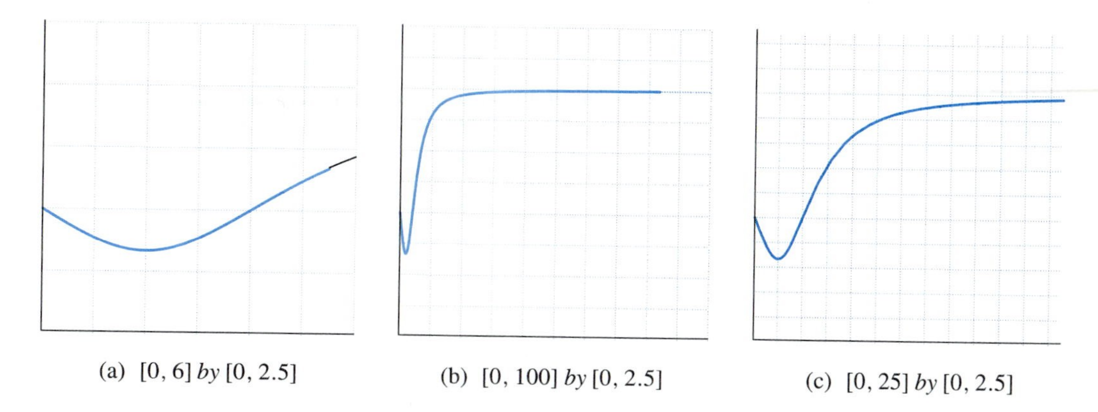

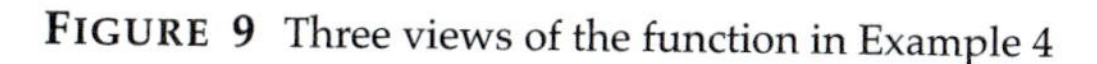
FIGURE 9 Three views of the function in Example 4

(b) This graph answers the question posed above. The graph levels off and approaches $y = 2$ as a horizontal asymptote. The shortcoming of this graph is that the range of values for x is so large that the interesting behavior at $x = 2$ and $x = 4$ is not clearly revealed.

(c) This viewing window, which is a compromise between (a) and (b), reveals all the important information about the graph of the function. ■

PRACTICE PROBLEMS 3.4

Consider the function $f(x) = 60x^2 - \pi x^3$.

1. Display the graph with a graphing utility in the window

$$[0, 20] \text{ by } [-500, 3500]$$

and estimate the coordinates of any relative extreme points, inflection points, and intercepts.

2. Algebraically determine the coordinates estimated in Problem 1.

EXERCISES 3.4

In Exercises 1–8, display the graph with a graphing utility in the designated window and estimate the coordinates of any relative extreme points, inflection points, and intercepts. Then determine the coordinates of these points algebraically.

1. $f(x) = x^2 - 3x + 1$; $[-2.5, 5.5]$ *by* $[-2, 12]$
2. $f(x) = x^2 - 5x + 5$; $[-1.5, 6.5]$ *by* $[-2, 12]$
3. $f(x) = 6 + 3x - x^2$; $[-4, 7]$ *by* $[-10, 10]$
4. $f(x) = 3x^3 - 2x^2 - x$; $[-2, 2]$ *by* $[-5, 5]$
5. $f(x) = 2x^3 - 7x^2 + 4x$; $[-2, 4]$ *by* $[-5, 5]$
6. $f(x) = \frac{1}{2}(3x - 2)^3$; $[-1, 3]$ *by* $[-6, 6]$
7. $f(x) = 42x^2 - \pi x^3$; $[-2, 14]$ *by* $[-100, 1200]$
8. $f(x) = (3 - x)^3 - (x - 3)^2$; $[0, 4]$ *by* $[-1, 1]$

In Exercises 9–24, use both algebraic calculations and a graphing utility to obtain a good graph of the function. Follow steps 1 through 3 given in the previous section. Then draw the graph appearing on the screen, give the window used, and label all relative extreme points, inflection points, intercepts, and vertical or horizontal asymptotes.

9. $f(x) = x^2 - 6x + 5$
10. $f(x) = 8 + 6x - 2x^2$
11. $f(x) = 8 - 2x^2$
12. $f(x) = 3x^2 - 2x$
13. $f(x) = (3x + 2)^2$
14. $f(x) = (6 - 3x)^2 + 1$
15. $f(x) = x^3 + 6x^2 + 9x$
16. $f(x) = 4x^3 - 64x^2 + 256x$
17. $f(x) = -.0006x^3 - .03x^2 + 2x + 20$
18. $f(x) = 120x^2 - 2x^3$
19. $f(x) = 220x - \frac{\pi}{2}x^2$
20. $f(x) = \dfrac{800{,}000}{x} + 5x$
21. $f(x) = 2500 + \dfrac{1{,}000{,}000}{x}$
22. $f(x) = -\left(\frac{\pi}{2} - 2\right)x^2 + 14x$
23. $f(x) = 2\pi\left(x^2 - \dfrac{16}{x^2}\right)$
24. $f(x) = 84\pi x^2 - 2\pi^2 x^3$
25. Graphs of cubic polynomials cross the x-axis at either one or three places. What does this imply about the possible number of relative extreme points of the graph of a polynomial of degree 4?
26. Graphs of quadratic polynomials cross the x-axis at either zero or two places. What does this imply about the possible number of inflection points of the graph of a polynomial of degree 4?
27. Graphs of polynomials of degree n have at most n x-intercepts. What does this imply about the possible number of relative extreme points of the graph of a polynomial of degree $n + 1$?
28. Graphs of polynomials of degree n have at most n x-intercepts. What does this imply about the possible number of inflection points of the graph of a polynomial of degree $n + 2$?
29. For the function $f(x) = x^4 - 8x^3 + 24x^2 - 32x + 17$, $f'(2) = 0$ and $f''(2) = 0$. Therefore, the second derivative test cannot be used to determine the nature of the graph of $f(x)$ at $x = 2$. Graph the function with a graphing utility in the window $[0, 4]$ *by* $[0, 4]$, and determine the nature of the graph at $x = 2$.
30. For the function $f(x) = x^5 - 10x^4 + 40x^3 - 80x^2 + 80x - 30$, $f'(2) = 0$ and $f''(2) = 0$. Therefore, the second derivative test cannot be used to determine the nature of the graph of $f(x)$ at $x = 2$. Graph the function with a graphing utility in the window $[0, 4]$ *by* $[0, 4]$, and determine the nature of the graph at $x = 2$.
31. When a glass of cold water is placed in a warm room, the temperature of the water after t minutes is

$$f(t) = \frac{70}{1 + .85^t} \quad \text{degrees.}$$

With a graphing utility, sketch the graph of the function in the window $[0, 30]$ *by* $[0, 100]$. What was the original temperature of the water? The graph has a horizontal asymptote. What is its equation, and what does it say about the temperature of the room?

32. A parachutist is in free fall. Her speed after t seconds is $f(t) = 176(1-(.5)^t)$ feet per second. With a graphing utility, sketch the graph of the function in the window $[0, 12]$ *by* $[0, 200]$. What is her initial speed? The graph has a horizontal asymptote. What is its equation and what does it say about the speed of the parachutist?

In Exercises 33–36, the graphs of the functions have the given lines as horizontal asymptotes. Determine a viewing window that clearly displays the main features of the graph.

33. $f(x) = 50 - \dfrac{80x+3}{x^3+5}$ $(x \geq 0)$; $y = 50$

34. $f(x) = \dfrac{100(x-1)}{x^5+25}$ $(x \geq 0)$; $y = 0$

35. $f(x) = \dfrac{300}{75(1+.95^x)}$; $y = 0$, $y = 4$

36. $f(x) = 2 - .96^x$; $y = 2$

37. If $f(x)$ has a horizontal asymptote, does $f'(x)$ also have a horizontal asymptote? If so, what is the asymptote? What about $f''(x)$?

38. If $f(x)$ has a vertical asymptote at $x = a$, what can you conclude about $f'(x)$ and $f''(x)$ at $x = a$?

39. A boy standing on the ledge of a building tosses a grapefruit straight up into the air. The height of the grapefruit after t seconds is

$$s(t) = 38 + 64t - 16t^2 \quad \text{feet.}$$

Sketch the graphs of $s(t)$, $s'(t)$, and $s''(t)$ in the window $[0, 5]$ *by* $[-110, 110]$.

(a) Find and interpret the y-intercepts of the graphs of $s(t)$, $s'(t)$, and $s''(t)$.
(b) How high is the grapefruit after 2 seconds?
(c) When is the grapefruit 86 feet high?
(d) How fast is the grapefruit rising after 1 second?
(e) When is the grapefruit falling at the rate of 64 feet per second?
(f) Find and interpret the t-intercepts of the graphs of $s(t)$ and $s'(t)$.
(g) How high does the grapefruit go?
(h) How fast is the grapefruit moving when it hits the ground?

SOLUTIONS TO PRACTICE PROBLEMS 3.4

1. Figure 10 shows the graph displayed in the indicated window with the estimates of the relevant points. Rough estimates can be obtained by tracing. However, better estimates are found by finding the roots of $f(x) = 0$, $f'(x) = 0$, and $f''(x) = 0$.

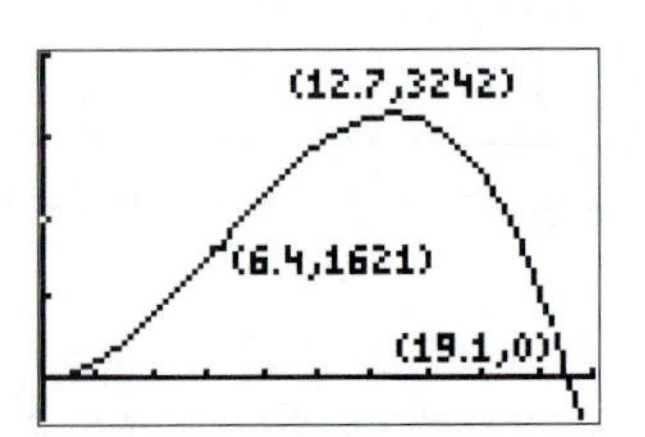

FIGURE 10 Graph of $f(x) = 60x^2 - \pi x^3$

2.
$$f(x) = 60x^2 - \pi x^3,$$
$$f'(x) = 120x - 3\pi x^2,$$
$$f''(x) = 120 - 6\pi x.$$

To find x-intercepts, set $f(x) = 0$:

$$60x^2 - \pi x^3 = 0$$
$$x^2(60 - \pi x) = 0$$
$$x = 0 \quad \text{or} \quad x = 60/\pi.$$

To find relative extreme points, set $f'(x) = 0$:

$$120x - 3\pi x^2 = 0$$
$$3x(40 - \pi x) = 0$$
$$x = 0 \quad \text{or} \quad x = 40/\pi.$$

$$f(0) = 0$$
$$f(40/\pi) = 60\left(\frac{40}{\pi}\right)^2 - \pi\left(\frac{40}{\pi}\right)^3$$
$$= \left(\frac{40}{\pi}\right)^2\left[60 - \pi\left(\frac{40}{\pi}\right)\right]$$
$$= \left(\frac{40}{\pi}\right)^2(20) = \frac{32{,}000}{\pi^2}.$$

The relative extreme points are $(0, 0)$ and $\left(\frac{40}{\pi}, \frac{32{,}000}{\pi^2}\right)$.

To find inflection points, set $f''(x) = 0$:

$$120 - 6\pi x = 0$$
$$6(20 - \pi x) = 0$$
$$x = 20/\pi.$$

$$f(20/\pi) = 60\left(\frac{20}{\pi}\right)^2 - \pi\left(\frac{20}{\pi}\right)^3$$
$$= \left(\frac{20}{\pi}\right)^2\left[60 - \pi\left(\frac{20}{\pi}\right)\right]$$
$$= \left(\frac{20}{\pi}\right)^2(40) = \frac{16{,}000}{\pi^2}.$$

The inflection point is $\left(\frac{20}{\pi}, \frac{16{,}000}{\pi^2}\right)$.

3.5 Optimization Problems

Among the most important applications of the derivative concept are "optimization" problems, in which some quantity must be maximized or minimized. Examples of such problems abound in many areas of life. An airline must decide how many flights to schedule between two cities in order to maximize its profits. A doctor wants to find the minimum amount of a drug that will produce a desired response in one of her patients. A manufacturer needs to determine how often to replace certain equipment in order to minimize maintenance and replacement costs.

Our purpose in this section is to illustrate how calculus can be used to solve optimization problems. In each example we will find or construct a function that provides a "mathematical model" for the problem. Then, by sketching the graph of this function, we will be able to determine the answer to the original optimization problem by locating the highest or lowest point on the graph. The y-coordinate of this point will be the maximum value or minimum value of the function.

The first two examples are quite simple because the functions to be studied are given explicitly.

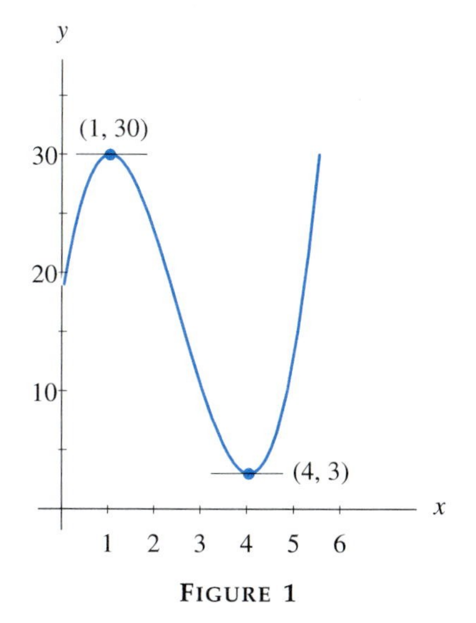

FIGURE 1

EXAMPLE 1 Find the minimum value of the function $f(x) = 2x^3 - 15x^2 + 24x + 19$ for $x \geq 0$.

Solution Using the curve-sketching techniques from Section 3.3, we obtain the graph in Figure 1. The lowest point on the graph is (4, 3). The minimum *value* of the function $f(x)$ is the y-coordinate of this point—namely, 3. ■

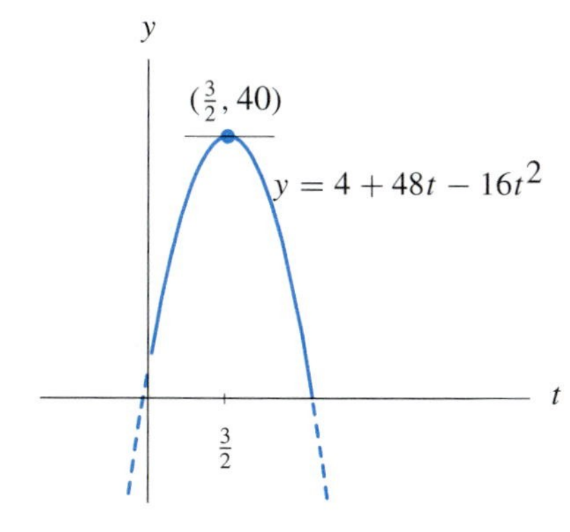

FIGURE 2

EXAMPLE 2 Suppose that a ball is thrown straight up into the air and its height after t seconds is $4 + 48t - 16t^2$ feet. Determine how long it will take for the ball to reach its maximum height and determine the maximum height.

Solution Consider the function $f(t) = 4 + 48t - 16t^2$. For each value of t, $f(t)$ is the height of the ball at time t. We want to find the value of t for which $f(t)$ is the greatest. Using the techniques of Section 3.3, we sketch the graph of $f(t)$. (See Figure 2.) Note that we may neglect the portions of the graph corresponding to points for which either $t < 0$ or $f(t) < 0$. [A negative value of $f(t)$ would correspond to the ball being underneath the ground.] We see that $f(t)$ is greatest when $t = \frac{3}{2}$. At this value of t, the ball attains a height of 40 feet. [Note that the curve in Figure 2 is the graph of $f(t)$, *not* a picture of the physical path of the ball.]

Answer The ball reaches its maximum height of 40 feet in 1.5 seconds. ■

EXAMPLE 3 A person wants to plant a rectangular garden along one side of a house, with a picket fence on the other three sides of the garden. Find the dimensions of the largest garden that can be enclosed using 40 feet of fencing.

Solution The first step is to make a simple diagram and assign letters to the quantities that may vary. Let us denote the dimensions of the rectangular garden by w and x (Figure 3). The phrase "largest garden" indicates that we must maximize

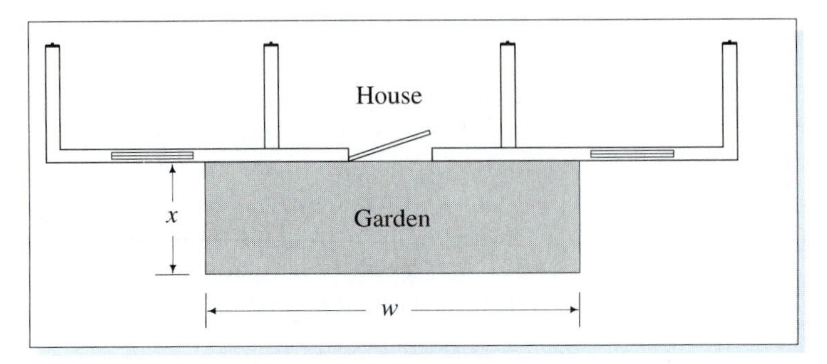

FIGURE 3

the area, A, of the garden. In terms of the variables w and x,

$$A = wx. \tag{1}$$

The fencing on three sides must total 40 running feet; that is,

$$2x + w = 40. \tag{2}$$

We now solve equation (2) for w in terms of x:

$$w = 40 - 2x. \tag{3}$$

Substituting this expression for w into equation (1), we have

$$A = (40 - 2x)x = 40x - 2x^2. \tag{4}$$

We now have a formula for the area A that depends on just one variable, and so we may graph A as a function of x. From the statement of the problem, the value of $2x$ can be at most 40, so that the domain of the function consists of x in the interval $0 < x < 20$.

Using curve-sketching techniques, we obtain the graph in Figure 4. We see from the graph that the area is maximized when $x = 10$. (The maximum area is 200 square feet, but this fact is not needed for the problem.) From equation (3) we find that when $x = 10$,

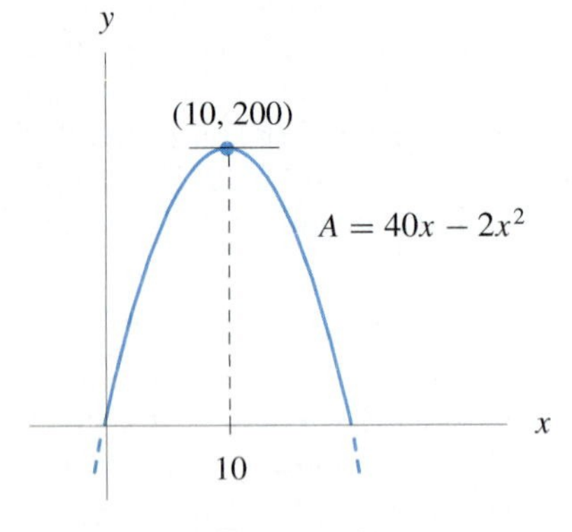

FIGURE 4

$$w = 40 - 2(10) = 20.$$

Answer $w = 20$ feet, $x = 10$ feet. ■

Equation (1) in Example 3 is called an *objective equation*. It expresses the quantity to be optimized (the area of the garden) in terms of the variables w and x. Equation (2) is called a *constraint equation* because it places a limit or constraint on the way x and w may vary.

EXAMPLE 4 The manager of a department store wants to build a 600-square-foot rectangular enclosure on the store's parking lot in order to display some equipment. Three sides of the enclosure will be built of redwood fencing, at a cost of

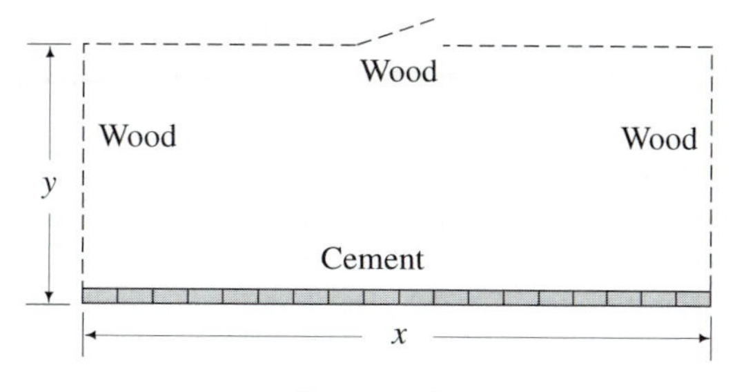

FIGURE 5

\$14 per running foot. The fourth side will be built of cement blocks, at a cost of \$28 per running foot. Find the dimensions of the enclosure that will minimize the total cost of the building materials.

Solution Let x be the length of the side built out of cement blocks and let y be the length of an adjacent side, as shown in Figure 5. The phrase "minimize the total cost" tells us that the objective equation should be a formula giving the total cost of the building materials.

$$\begin{aligned}
[\text{cost of redwood}] &= [\text{length of redwood fencing}] \times [\text{cost per foot}] \\
&= (x + 2y) \cdot 14 = 14x + 28y. \\
[\text{cost of cement blocks}] &= [\text{length of cement wall}] \times [\text{cost per foot}] \\
&= x \cdot 28.
\end{aligned}$$

If C denotes the total cost of the materials, then

$$\begin{aligned}
C &= (14x + 28y) + 28x \\
C &= 42x + 28y \qquad \text{(objective equation).} \qquad (5)
\end{aligned}$$

Since the area of the enclosure must be 600 square feet, the constraint equation is

$$xy = 600. \qquad (6)$$

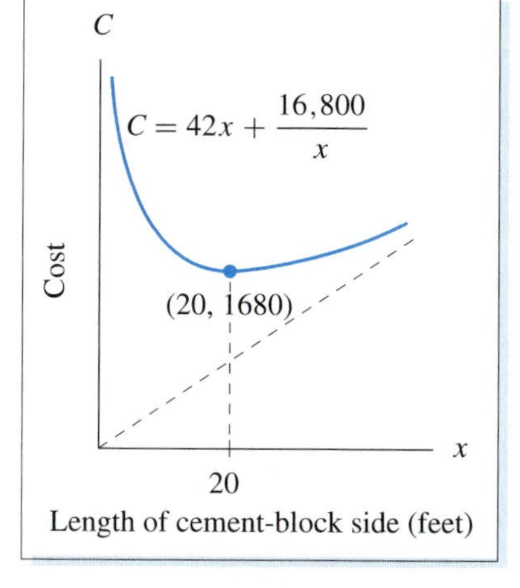

FIGURE 6

We simplify the objective equation by solving (6) for one of the variables, say y, and substituting into (5). Since $y = 600/x$,

$$C = 42x + 28\left(\frac{600}{x}\right) = 42x + \frac{16,800}{x}.$$

We now have C as a function of the single variable x. From the context, we must have $x > 0$, since a length must be positive. However, to any positive value for x, there is a corresponding value for C. So the domain of C consists of all $x > 0$. We may now sketch the graph of C (Figure 6). (A similar curve was sketched in Example 3 of Section 3.4.) The minimum total cost of \$1680 occurs where $x = 20$. From equation (6) we find that the corresponding value of y is $\frac{600}{20} = 30$.

Answer $x = 20$ feet, $y = 30$ feet. ■

EXAMPLE 5 U.S. parcel post regulations state that packages must have length plus girth of no more than 84 inches. Find the dimensions of the cylindrical package of greatest volume that is mailable by parcel post.

Solution Let l be the length of the package and let r be the radius of the circular end. (See Figure 7.) The phrase "greatest volume" tells us that the objective equation should express the volume of the package in terms of the dimensions l and r. Let V denote the volume. Then

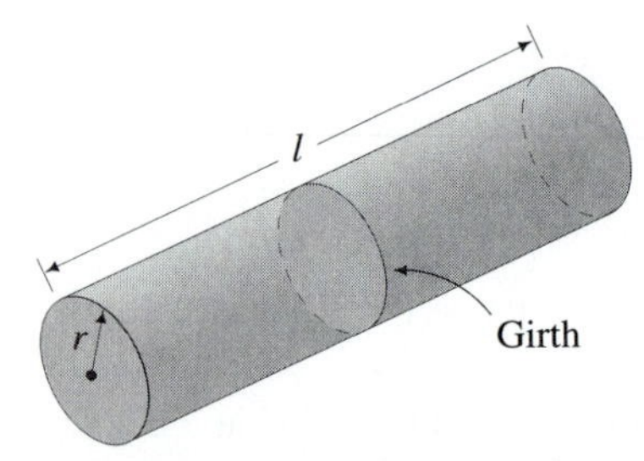

FIGURE 7 Cylindrical mailing package

$$V = [\text{area of base}] \cdot [\text{length}]$$

$$V = \pi r^2 l \qquad \text{(objective equation).} \tag{7}$$

The girth equals the circumference of the end—that is, $2\pi r$. Since we want the package to be as large as possible, we must use the entire 84 inches allowable:

$$\text{length} + \text{girth} = 84$$

$$l + 2\pi r = 84 \qquad \text{(constraint equation).} \tag{8}$$

We now solve equation (8) for one of the variables, say $l = 84 - 2\pi r$. Substituting this expression into (7), we obtain

$$V = \pi r^2(84 - 2\pi r) = 84\pi r^2 - 2\pi^2 r^3. \tag{9}$$

Let $f(r) = 84\pi r^2 - 2\pi^2 r^3$. Then, for each value of r, $f(r)$ is the volume of the parcel with end radius r that meets the postal regulations. We want to find that value of r for which $f(r)$ is as large as possible.

Using curve-sketching techniques, we obtain the graph of $f(r)$ in Figure 8. The domain excludes values of r that are negative and values of r for which the volume $f(r)$ is negative. Points corresponding to values of r not in the domain are shown with a dashed curve. We see that the volume is greatest when $r = 28/\pi \approx 8.9$. This value can also be found with a graphing utility by locating a zero of the derivative of $f(r)$. See Figure 9.

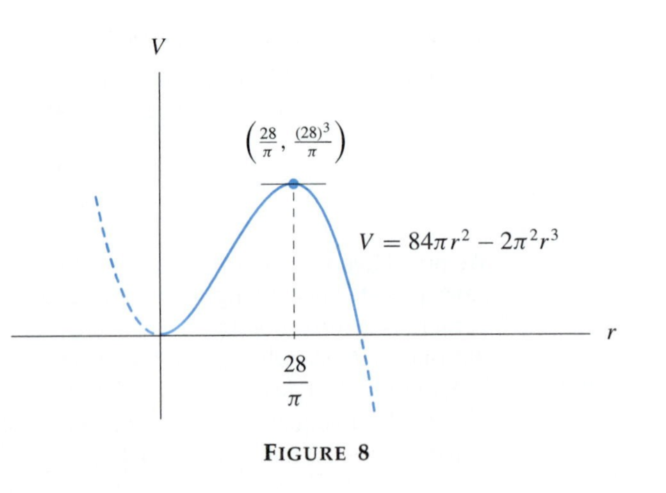

FIGURE 8

From (8) we find that the corresponding value of l is

$$l = 84 - 2\pi r = 84 - 2\pi\left(\frac{28}{\pi}\right)$$
$$= 84 - 56 = 28.$$

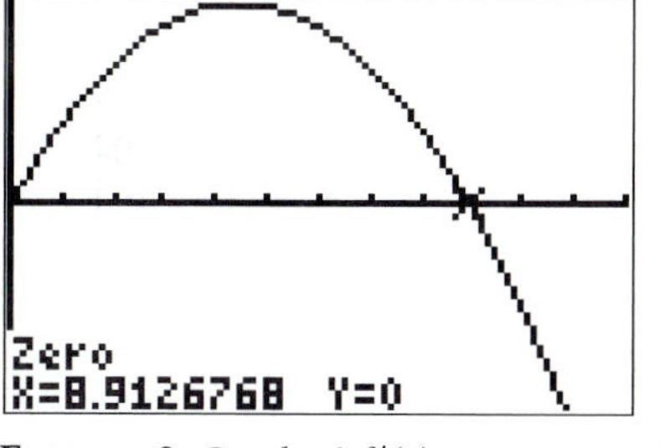

FIGURE 9 Graph of $f'(r)$

The girth when $r = 28/\pi$ is

$$2\pi r = 2\pi\left(\frac{28}{\pi}\right) = 56.$$

Answer $l = 28$ inches, $r = 28/\pi$ inches. ■

Suggestions for Solving an Optimization Problem

1. Draw a picture, if possible.
2. Decide what quantity Q is to be maximized or minimized.
3. Assign letters to other quantities that may vary.
4. Determine the "objective equation" that expresses Q as a function of the variables assigned in step 3.
5. Find the "constraint equation" that relates the variables to each other and to any constants that are given in the problem.
6. Use the constraint equation to simplify the objective equation in such a way that Q becomes a function of only one variable. Determine the domain of this function.
7. Sketch the graph of the function obtained in step 6 and use this graph to solve the optimization problem.

Optimization problems often involve geometric formulas. The most common formulas are shown in Figure 10.

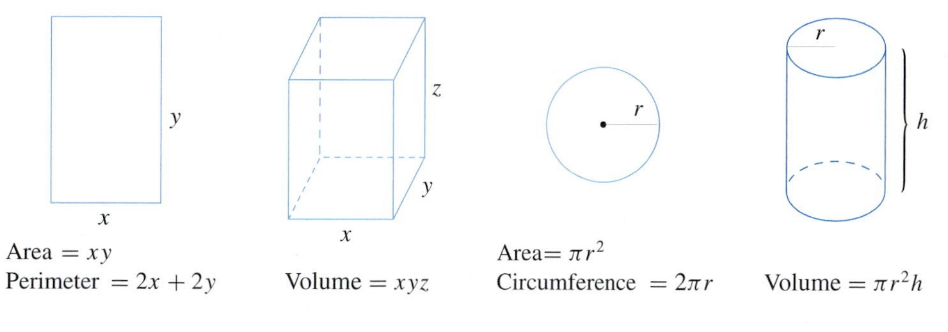

FIGURE 10

PRACTICE PROBLEMS 3.5

1. A canvas wind shelter for the beach has a back, two square sides, and a top (Figure 11). Suppose that 96 square feet of canvas are to be used. Find the dimensions of the shelter for which the space inside the shelter (i.e., the volume) will be maximized.

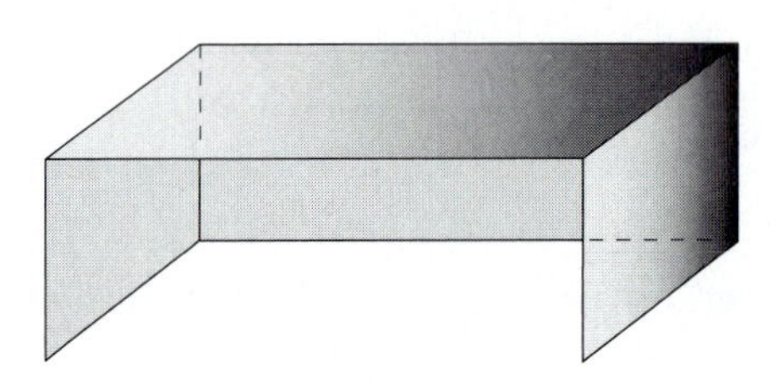

FIGURE 11

2. In Practice Problem 1, what are the objective equation and the constraint equation?

EXERCISES 3.5

1. For what x does the function $g(x) = 10 + 40x - x^2$ have its maximum value?
2. Find the maximum value of the function $f(x) = 12x - x^2$ and give the value of x where this maximum occurs.
3. Find the minimum value of $f(t) = t^3 - 6t^2 + 40$, $t \geq 0$, and give the value of t where this minimum occurs.
4. For what t does the function $f(t) = t^2 - 24t$ have its minimum value?
5. Three hundred and twenty dollars are available to fence in a rectangular garden. The fencing for the side of the garden facing the road costs \$6 per foot and the fencing for the other three sides costs \$2 per foot. [See Figure 12(a).] Consider the problem of finding the dimensions of the largest possible garden.
   (a) Determine the objective and constraint equations.
   (b) Express the quantity to be maximized as a function of x.
   (c) Find the optimal values of x and y.
6. Figure 12(b) shows an open rectangular box with a square base. Consider the problem of finding the values of x and h for which the volume is 32 cubic feet and the total surface area of the box is minimal. (The surface area is the sum of the areas of the five faces of the box.)
   (a) Determine the objective and constraint equations.
   (b) Express the quantity to be minimized as a function of x.
   (c) Find the optimal values of x and h.
7. Postal requirements specify that parcels must have length plus girth of at most 84 inches. Consider the problem of finding the dimensions of the square-ended rectangular package of greatest volume that is mailable.

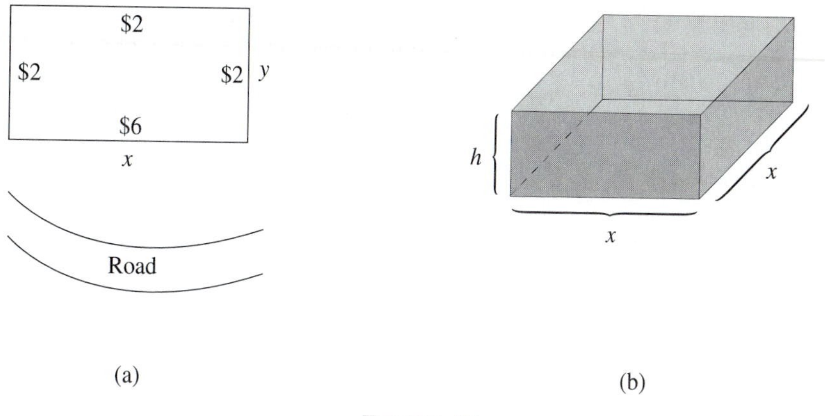

(a) (b)

FIGURE 12

(a) Draw a square-ended rectangular box. Label each edge of the square end with the letter x and label the remaining dimension of the box with the letter h.
(b) Express the length plus the girth in terms of x and h.
(c) Determine the objective and constraint equations.
(d) Express the quantity to be maximized as a function of x.
(e) Find the optimal values of x and h.

8. Consider the problem of finding the dimensions of the rectangular garden of area of 100 square meters for which the amount of fencing needed to surround the garden is as small as possible.
   (a) Draw a picture of a rectangle and select appropriate letters for the dimensions.
   (b) Determine the objective and constraint equations.
   (c) Find the optimal values for the dimensions.

9. A rectangular garden of area 75 square feet is to be surrounded on three sides by a brick wall costing \$10 per foot and on one side by a fence costing \$5 per foot. Find the dimensions of the garden such that the cost of materials is minimized.

10. A closed rectangular box with a square base and a volume of 12 cubic feet is to be constructed using two different types of materials. The top is made of a metal costing \$2 per square foot and the remainder of wood costing \$1 per square foot. Find the dimensions of the box for which the cost of materials is minimized.

11. Find the dimensions of the closed rectangular box with a square base and volume 8000 cubic centimeters that can be constructed with the least amount of material.

12. A canvas wind shelter for the beach has a back, two square sides, and a top. Find the dimensions for which the volume will be 250 cubic feet and that requires the least possible amount of canvas.

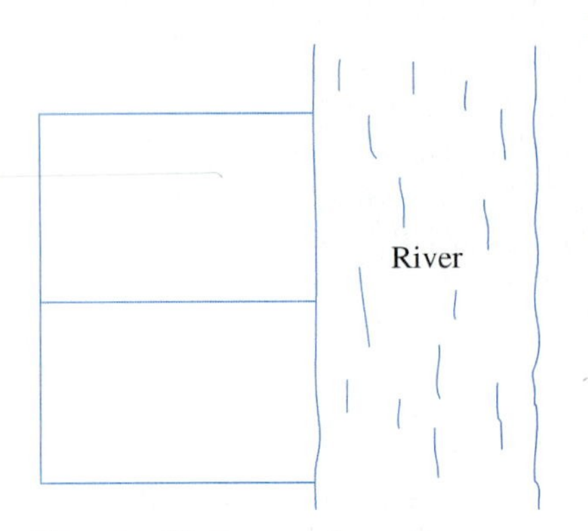

FIGURE 13 Rectangular pastures along a river

13. A farmer has \$1500 available to build an E-shaped fence along a straight river so as to create two identical rectangular pastures. (See Figure 13.) The materials for the side parallel to the river cost \$6 per foot and the materials for the three sections perpendicular to the river cost \$5 per foot. Find the dimensions for which the total area is as large as possible.

14. Find the dimensions of the rectangular garden of greatest area that can be fenced off (all four sides) with 300 meters of fencing.

15. Find two positive numbers, x and y, whose sum is 100 and whose product is as large as possible.

16. Find two positive numbers, x and y, whose product is 100 and whose sum is as small as possible.

17. Figure 14(a) shows a Norman window, which consists of a rectangle capped by a semicircular region. Find the value of x such that the perimeter of the window will be 14 feet and the area of the window will be as large as possible.

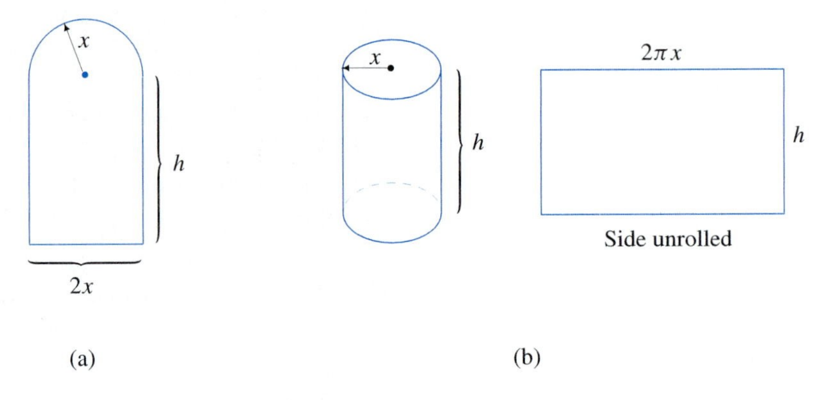

FIGURE 14

18. A large soup can is to be designed so that the can will hold 16π cubic inches (about 28 ounces) of soup. [See Figure 14(b).] Find the values of x and h for which the amount of metal needed is as small as possible.

19. In Example 3 one can solve the constraint equation (2) for x instead of w to get $x = 20-\frac{1}{2}w$. Substituting this for x in (1), one has $A = xw = (20-\frac{1}{2}w)w$. Sketch the graph of the equation $A = 20w-\frac{1}{2}w^2$ and show that the maximum occurs when $w = 20$ and $x = 10$.

20. A ship uses $5x^2$ dollars of fuel per hour when traveling at a speed of x miles per hour. The other expenses of operating the ship amount to \$2000 per hour. What speed minimizes the cost of a 500-mile trip? (*Hint*: Express cost in terms of speed and time. The constraint equation is *distance* = *speed* × *time*.)

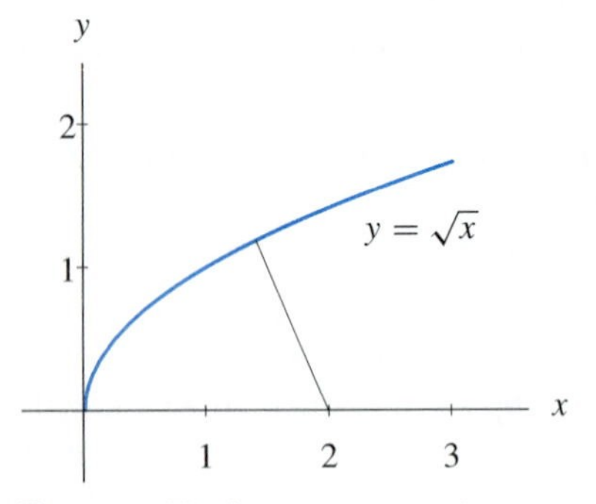

FIGURE 15 Shortest distance from a point to a curve

21. Find the point on the graph of $y = \sqrt{x}$ that is closest to the point $(2, 0)$. See Figure 15. [*Hint*: $\sqrt{(x-2)^2 + y^2}$ has its smallest value when $(x-2)^2 + y^2$ does. Therefore, just minimize the second expression.]

SOLUTIONS TO PRACTICE PROBLEMS 3.5

1. Since the sides of the wind shelter are square, we may let x represent the length of each side of the square. The remaining dimension of the wind shelter can be denoted by the letter h. (See Figure 16.) The volume of the shelter is x^2h, and this is to be maximized. Since we have learned to maximize only functions of a single variable, we must express h in terms of x. We must use the information that 96 feet of canvas are used—that is, $2x^2 + 2xh = 96$. (*Note*: The roof and the back each has area xh, and each end has area x^2.) We now solve this equation for h:

$$2x^2 + 2xh = 96$$
$$2xh = 96 - 2x^2$$
$$h = \frac{96}{2x} - \frac{2x^2}{2x} = \frac{48}{x} - x.$$

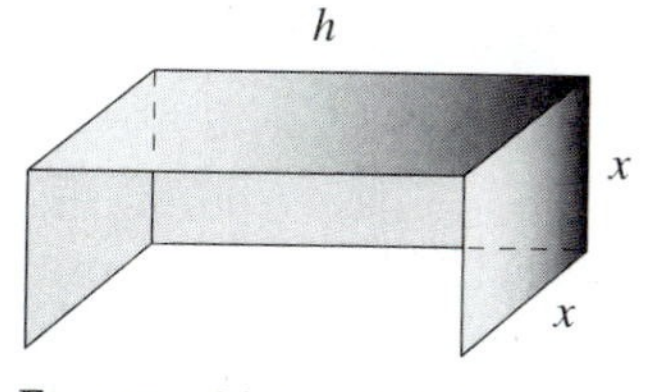

FIGURE 16 Wind shelter

The volume, V, is

$$x^2h = x^2\left(\frac{48}{x} - x\right) = 48x - x^3.$$

By sketching the graph of $V = 48x - x^3$, we see that V has a maximum value when $x = 4$. Then, $h = \frac{48}{4} - 4 = 12 - 4 = 8$. So each end of the shelter should be a 4-foot by 4-foot square and the shelter should be 8 feet long.

2. The objective equation is $V = x^2h$, since it expresses the volume (the quantity to be maximized) in terms of the variables. The constraint equation is $2x^2 + 2xh = 96$, for it relates the variables to each other; that is, it can be used to express one of the variables in terms of the other.

3.6 FURTHER OPTIMIZATION PROBLEMS

In this section we apply the optimization techniques developed in the preceding section to some practical situations.

INVENTORY CONTROL When a firm regularly orders and stores supplies for later use or resale, it must decide on the size of each order. If it orders enough supplies to last an entire year, the business will incur heavy *carrying costs*. Such costs include insurance, storage costs, and cost of capital that is tied up in inventory. To reduce these carrying costs, the firm could order smaller quantities of the supplies at frequent intervals. However, such a policy increases *ordering costs*. These might consist of minimum freight charges, the clerical costs of preparing the orders, and the costs of receiving and checking the orders when they arrive. Clearly, the firm must find an inventory ordering policy that lies between these two extremes.

The following example illustrates the use of calculus to minimize the firm's annual inventory cost, where

$$[\text{inventory cost}] = [\text{ordering cost}] + [\text{carrying cost}].$$

We assume that each order is the same size. The size of the order that minimizes the inventory cost is called the *economic order quantity*, commonly referred to in business as the EOQ.*

EXAMPLE 1 A supermarket manager wants to establish an optimal inventory policy for frozen orange juice. It is estimated that a total of 1200 cases will be sold at a steady rate during the next year. The manager plans to place several orders of the same size spaced equally throughout the year. Use the following data to determine the economic order quantity, that is, the order size that minimizes the total ordering and carrying costs.

1. The ordering cost for each delivery is $75.
2. It costs $8 to carry one case of orange juice in inventory for one year. (Carrying costs should be computed on the average inventory during the order-reorder period.)

Solution Let x be the order quantity and r the number of orders placed during the year. The number of cases of orange juice in inventory declines steadily from x cases (each time a new order is filled) to 0 cases at the end of each order-reorder period. Figure 1 shows that the average number of cases in storage during the year is $x/2$. Since the carrying cost for one case is $8 per year, the cost for $x/2$ cases is $8 \cdot (x/2)$ dollars. Now

$$\begin{aligned}[\text{inventory cost}] &= [\text{ordering cost}] + [\text{carrying cost}] \\ &= 75r + 8 \cdot \frac{x}{2} \\ &= 75r + 4x.\end{aligned}$$

If C denotes the inventory cost, then the objective equation is

$$C = 75r + 4x.$$

Since there are r orders of x cases each, the total number of cases ordered during the year is $r \cdot x$. Therefore, the constraint equation is

$$r \cdot x = 1200.$$

The constraint equation says that $r = 1200/x$. Substitution into the objective equation yields

$$C = \frac{90{,}000}{x} + 4x.$$

Figure 2 is the graph of C as a function of x, for $x > 0$. The total cost is at a minimum when $x = 150$. Therefore, the optimum inventory policy is to order 150 cases at a time and to place $1200/150 = 8$ orders during the year. ■

* See James C. Van Horne, *Financial Management and Policy*, 6th ed. (Englewood Cliffs, N.J.: Prentice-Hall, Inc., 1983), pp. 416–420.

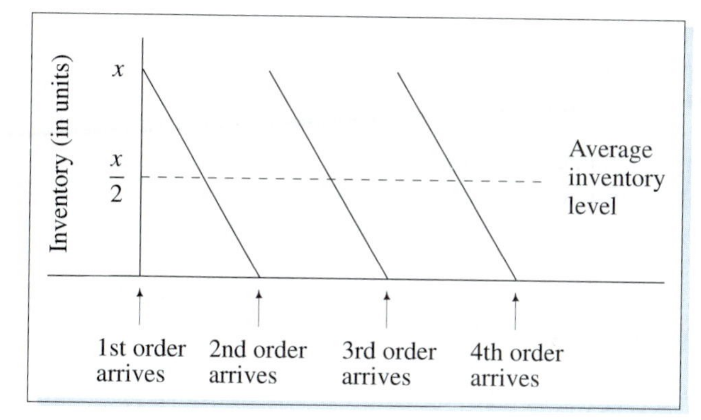

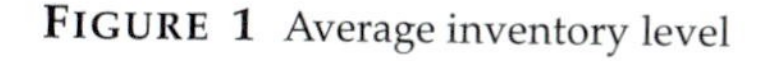

FIGURE 1 Average inventory level

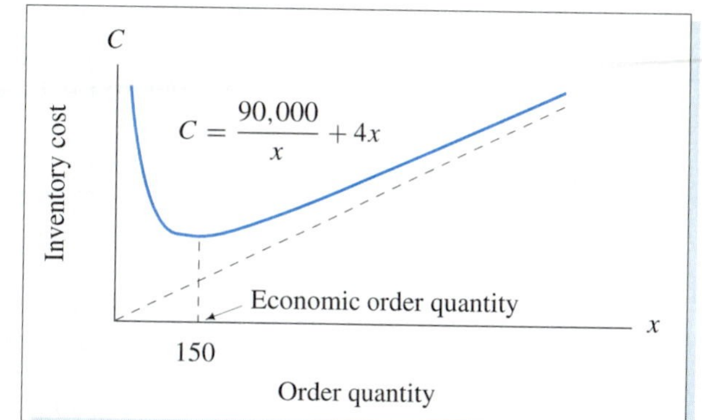

FIGURE 2 Cost function for inventory problem

EXAMPLE 2 What should the inventory policy of Example 1 be if sales of frozen orange juice increase fourfold (i.e., 4800 cases are sold each year), but all other conditions are the same?

Solution The only change in our previous solution is in the constraint equation, which now becomes

$$r \cdot x = 4800.$$

The objective equation is, as before,

$$C = 75r + 4x.$$

Since $r = 4800/x$,

$$C = 75 \cdot \frac{4800}{x} + 4x = \frac{360{,}000}{x} + 4x.$$

Now

$$C' = -\frac{360{,}000}{x^2} + 4.$$

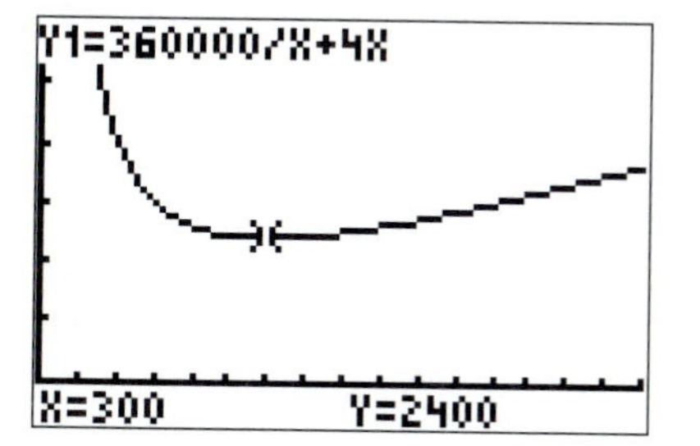

Setting $C' = 0$ yields

$$\frac{360{,}000}{x^2} = 4$$
$$90{,}000 = x^2$$
$$x = 300.$$

Therefore, the economic order quantity is 300 cases. ■

Notice that although the sales increased by a factor of 4, the economic order quantity increased by only a factor of 2 ($= \sqrt{4}$). In general, a store's inventory of an item should be proportional to the square root of the expected sales. (See Exercise 5 for a derivation of this result.) Many stores tend to keep their average inventories at a fixed percentage of sales. For example, each order may contain enough goods to last for 4 or 5 weeks. This policy is likely to create excessive inventories of high-volume items and uncomfortably low inventories of slower-moving items.

Manufacturers have an inventory-control problem similar to that of retailers. They have the carrying costs of storing finished products and the startup costs of setting up each production run. The size of the production run that minimizes the sum of these two costs is called the *economic lot size*. See Exercises 2 and 3.

CONTRACTION OF THE TRACHEA DURING COUGHING The next example develops a mathematical model for a physiological process.

EXAMPLE 3 When a person coughs, the trachea (windpipe) contracts. (See Figure 3.) Let

$$\begin{aligned} r_0 &= \text{normal radius of the trachea,} \\ r &= \text{radius during a cough,} \\ P &= \text{increase in air pressure in the trachea during a cough,} \\ v &= \text{velocity of air through trachea during a cough.} \end{aligned}$$

Use the following principles of fluid flow to determine how much the trachea should contract (that is, determine r) in order to create the greatest air velocity—that is, the most effective condition for clearing the lungs and the trachea.

1. $r_0 - r = aP$, for some positive constant a. (Experiment has shown that, during coughing, the decrease in the radius of the trachea is nearly proportional to the increase in the air pressure.)
2. $v = b \cdot P \cdot \pi r^2$, for some positive constant b. (The theory of fluid flow requires that the velocity of the air forced through the trachea be proportional to the product of the increase in the air pressure and the area of a cross section of the trachea.)

Solution In this problem the constraint equation (**1**) and the objective equation (**2**) are given directly. Solving equation (**1**) for P and substituting this result into equation (**2**), we have

$$v = b\left(\frac{r_0 - r}{a}\right)\pi r^2 = k(r_0 - r)r^2,$$

where $k = b\pi/a$. Now v is expressed as a function of the single variable r. To find the radius r at which the velocity v is a maximum, we first compute the derivatives:

$$\begin{aligned} v &= k(r_0 r^2 - r^3), \\ \frac{dv}{dr} &= k(2r_0 r - 3r^2) = kr(2r_0 - 3r), \\ \frac{d^2v}{dr^2} &= k(2r_0 - 6r). \end{aligned}$$

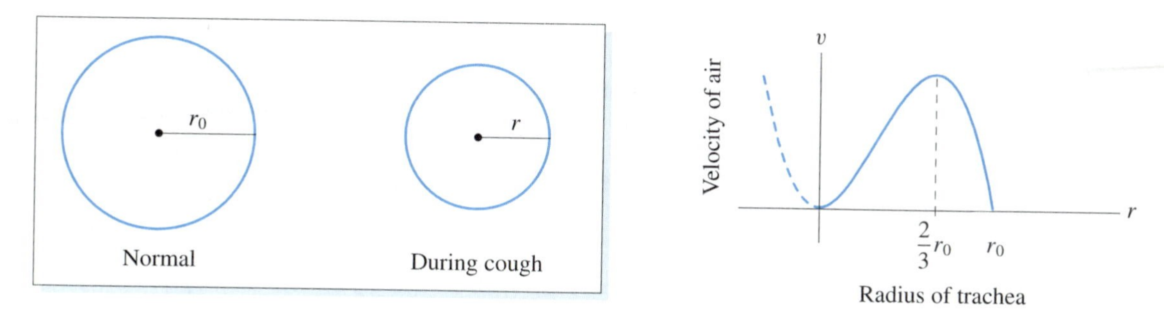

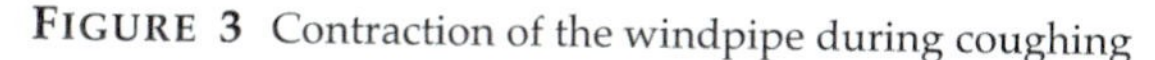
FIGURE 3 Contraction of the windpipe during coughing

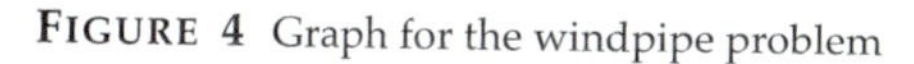
FIGURE 4 Graph for the windpipe problem

We see that $\frac{dv}{dr} = 0$ when $r = 0$ or when $2r_0 - 3r = 0$; that is, when $r = \frac{2}{3}r_0$. It is easy to see that $\frac{d^2v}{dr^2}$ is positive at $r = 0$ and is negative at $r = \frac{2}{3}r_0$. The graph of v as a function of r is drawn in Figure 4. The air velocity is maximized at $r = \frac{2}{3}r_0$. ■

When solving optimization problems, we look for the maximum or minimum point on a graph. From our discussion of curve sketching, we have seen that this point occurs either at a relative extreme point or at an endpoint of the domain of definition. In all our optimization problems so far, the maximum or minimum points were at relative extreme points. In the next example, the optimum point is an endpoint.

EXAMPLE 4 A rancher has 204 meters of fencing from which to build two corrals: one square and the other rectangular with length that is twice the width. Find the dimensions that result in the greatest combined area.

Solution Let x be the width of the rectangular corral and h be the length of each side of the square corral. (See Figure 5.) Let A be the combined area. Then

$$A = [\text{area of square}] + [\text{area of rectangle}] = h^2 + 2x^2.$$

The constraint equation is

$$204 = [\text{perimeter of square}] + [\text{perimeter of rectangle}] = 4h + 6x.$$

FIGURE 5 Two corrals

Since the perimeter of the rectangle cannot exceed 204, we must have $0 \le 6x \le 204$, or $0 \le x \le 34$. Solving the constraint equation for h and substituting into the objective equation leads to the function graphed in Figure 6. The graph reveals that the area is minimized when $x = 18$. However, the problem asks for the *maximum* possible area. From Figure 6 we see that this occurs at the endpoint where $x = 0$. Therefore, the rancher should build only the square corral, with $h = 204/4 = 51$ meters. In this example, the objective function has an endpoint extremum; namely, the maximum value occurs at the endpoint $x = 0$. ■

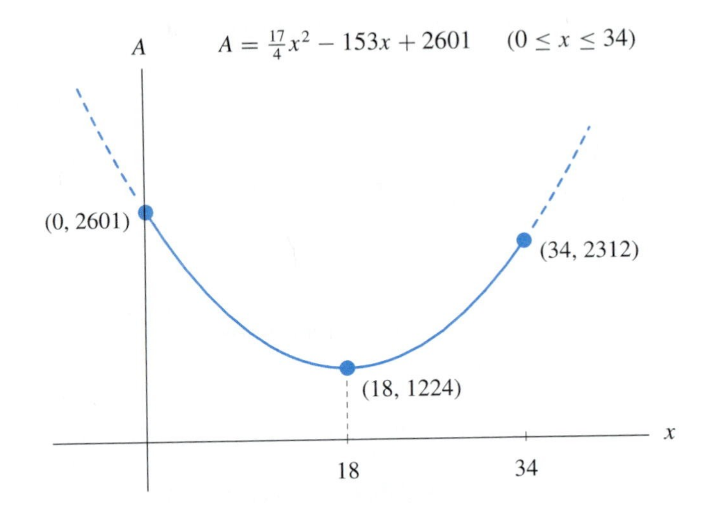

FIGURE 6 Combined area of the corrals

For certain types of optimization problems, the answer must be a whole number. Such is the case when we determine the optimum number of people for a certain task.

EXAMPLE 5 A political campaign wants to hire temporary workers to address and stuff 21,000 envelopes. Each worker can process 100 envelopes per hour and is paid \$6 per hour. In addition, a temp agency is paid \$10 per person to provide the workers and a supervisor is paid \$15 per hour. How many people should be hired in order to minimize the total cost?

Solution Let x be the number of people, h be the number of hours worked, and C be the total cost. Then,

$$\begin{aligned} C &= [\text{salary of workers}] + [\text{fee to temp agency}] + [\text{salary of supervisor}] \\ &= 6hx + 10x + 15h \qquad \text{(objective equation)}. \end{aligned}$$

Note: [salary of workers] is

$$[\text{hourly wage}] \cdot [\text{number of hours}] \cdot [\text{number of workers}] = 6hx.$$

The constraint equation is

$$\begin{aligned} [\text{envelopes processed per hour}] \cdot [\text{number of hours}] &= [\text{number of envelopes}] \\ 100x \cdot h &= 21{,}000. \end{aligned}$$

We now solve the constraint equation for one of the variables, say

$$h = \frac{210}{x}.$$

Substituting this expression into the objective equation yields

$$C = 6\left(\frac{210}{x}\right)x + 10x + 15\left(\frac{210}{x}\right) = 1260 + 10x + \frac{3150}{x}.$$

We now have C as a function of the single variable x and so can sketch the graph of C (Figure 7). A graphing utility gives (17.748, 1614.96) as the relative extreme point. Algebraically, the x-coordinate of the point is $\sqrt{315}$. Since 17.748 people cannot be employed, the correct answer must be either 17 or 18. When $x = 17$, $C = \$1615.29$. When $x = 18$, $C = \$1615$. Therefore, the least cost is incurred when 18 workers are hired. ■

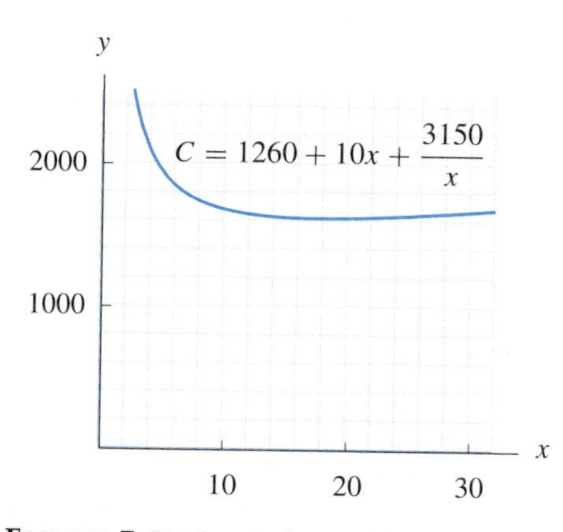

FIGURE 7 Total cost of processing envelopes

PRACTICE PROBLEMS 3.6

1. In the inventory problem of Example 1, suppose that the sales of frozen orange juice increase ninefold; that is, 10,800 cases are sold each year. What is the new economic order quantity?
2. In the envelope processing problem of Example 5, suppose that the hourly wage for workers is increased to \$7. What is the new optimum number of workers to be hired?

EXERCISES 3.6

1. A California distributor of sporting equipment expects to sell 10,000 cases of tennis balls during the coming year at a steady rate. Yearly carrying costs (to be computed on the average number of cases in stock during the year) are \$10 per case, and the cost of placing an order with the manufacturer is \$80.
   (a) Find the inventory cost incurred if the distributor orders 500 cases at a time during the year.
   (b) Determine the economic order quantity; that is, the order quantity that minimizes the inventory cost.
2. The Great American Tire Co. expects to sell 600,000 tires of a particular size and grade during the next year. Sales tend to be roughly the same from month to month. Setting up each production run costs the company \$15,000. Carrying costs, based on the average number of tires in storage, amount to \$5 per year for one tire.

(a) Determine the costs incurred if there are 10 production runs during the year.

(b) Find the economic lot size (i.e., the production run size that minimizes the overall cost of producing the tires).

3. Foggy Optics, Inc. makes laboratory microscopes. Setting up each production run costs \$2500. Insurance costs, based on the average number of microscopes in the warehouse, amount to \$20 per microscope per year. Storage costs, based on the maximum number of microscopes in the warehouse, amount to \$15 per microscope per year. Suppose that the company expects to sell 1600 microscopes at a fairly uniform rate throughout the year. Determine the number of production runs that will minimize the company's overall expenses.

4. A bookstore is attempting to determine the economic order quantity for a popular book. The store sells 8000 copies of this book a year. The store figures that it costs \$40 to process each new order for books. The carrying cost (due primarily to interest payments) is \$2 per book, to be figured on the maximum inventory during an order-reorder period. How many times a year should orders be placed?

5. A store manager wants to establish an optimal inventory policy for an item. Sales are expected to be at a steady rate and should total Q items sold during the year. Each time an order is placed, a cost of h dollars is incurred. Carrying costs for the year will be s dollars per item, to be figured on the average number of items in storage during the year. Show that the total inventory cost is minimized when each order calls for $\sqrt{2hQ/s}$ items.

6. Refer to the inventory problem of Example 1. Suppose that the distributor offers a discount of \$1 per case for orders of 600 or more cases. Should the manager change the quantity ordered?

7. Starting with a 100-foot-long stone wall, a farmer would like to construct a rectangular enclosure by adding 400 feet of fencing as shown in Figure 8(a). Find the values of x and w that result in the greatest possible area.

8. Rework Exercise 7 for the case where only 200 feet of fencing is added to the stone wall.

9. A rectangular corral of 54 square meters is to be fenced off and then divided by a fence into two sections, as shown in Figure 8(b). Find the dimensions of the corral so that the amount of fencing required is minimized.

10. Referring to Exercise 9, suppose that the cost of the fencing for the boundary is \$5 per meter and the dividing fence costs \$2 per meter. Find the dimensions of the corral that minimizes the cost of the fencing.

11. Design an open rectangular box with square ends, having a volume of 36 cubic inches, that minimizes the amount of material required for construction.

12. A storage shed is to be built in the shape of a box with a square base. It is to have a volume of 150 cubic feet. The concrete for the base costs \$4 per square foot, the material for the roof costs \$2 per square foot, and the material for the

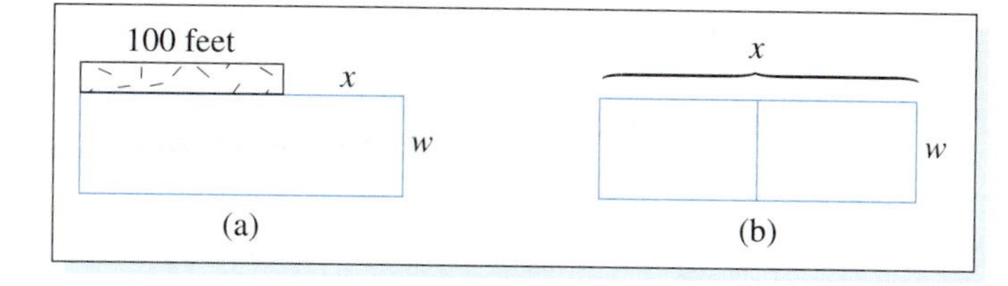

FIGURE 8 Rectangular enclosures

sides costs $2.50 per square foot. Find the dimensions of the most economical shed.

13. A supermarket is to be designed as a rectangular building with a floor area of 12,000 square feet. The front of the building will be mostly glass and will cost $70 per running foot for materials. The other three walls will be constructed of brick and cement block, at a cost of $50 per running foot. Ignore all other costs (labor, cost of foundation and roof, etc.) and find the dimensions of the base of the building that will minimize the cost of the materials for the four walls of the building.

14. A certain airline requires that rectangular packages carried on an airplane by passengers be such that the sum of the three dimensions is at most 120 centimeters. Find the dimensions of the square-ended rectangular package of greatest volume that meets this requirement.

15. An athletic field [Figure 9(a)] consists of a rectangular region with a semicircular region at each end. The perimeter will be used for a 440-yard track. Find the value of x for which the area of the rectangular region is as large as possible.

16. An open rectangular box is to be constructed by cutting square corners out of a 16 by 16-inch piece of cardboard and folding up the flaps. [See Figure 9(b).] Find the value of x for which the volume of the box will be as large as possible.

17. A closed rectangular box is to be constructed with a base that is twice as long as it is wide. Suppose that the total surface area must be 27 square feet. Find the dimensions of the box that will maximize the volume.

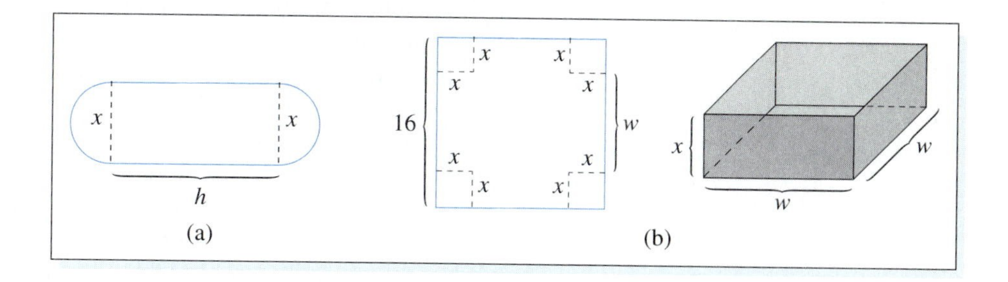

FIGURE 9

18. Consider a parabolic arch whose shape may be represented by the graph of $y = 9 - x^2$, where the base of the arch lies on the x-axis from $x = -3$ to $x = 3$. Find the dimensions of the rectangular window of maximum area that can be constructed inside the arch.

19. Advertising for a certain product is terminated, and t weeks later the weekly sales are $f(t)$ cases, where $f(t) = 100(t + 8)^{-1} - 4000(t + 8)^{-2}$. At what time is the weekly sales amount falling the fastest?

20. An open rectangular box of volume 400 cubic inches has a square base and a partition down the middle. See Figure 10. Find the dimensions of the box for which the amount of material needed to construct the box is as small as possible.

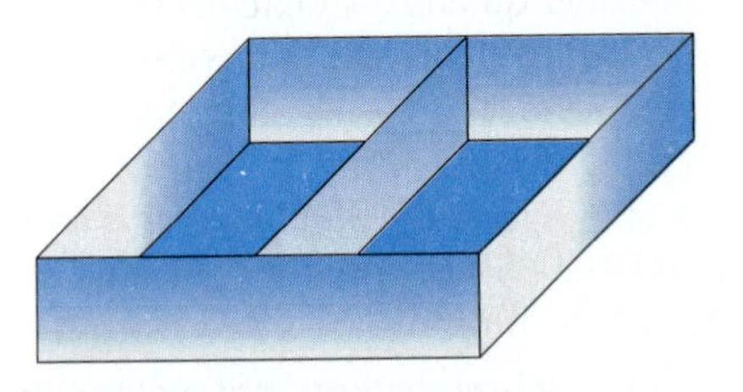

FIGURE 10 Open rectangular box with dividing partition

21. Suppose $f(x)$ is defined on the interval $0 \leq x \leq 5$ and $f'(x)$ is negative for all x. For what value of x will $f(x)$ have its greatest value?

22. The proprietor of a small business has one day to fill an order for 50 units of a particular product. For a price of \$40 dollars each, he can rent as many as 10 machines, each of which produces 5 units of the product per hour. He would have to hire one person to operate the machines at a salary of \$6 per hour (for up to 8 hours). How many machines should he rent so as to minimize the cost of filling the order?

23. A woman operates a small business making and selling hand-crafted bracelets. It costs her $100+15x$ dollars to produce x bracelets each week. She determines that if she charges d dollars per bracelet, she will be able to sell $x = 45 - .5d$ bracelets per week. How many bracelets should she make (and sell) so as to maximize her weekly profit? (*Hint*: Profit = Revenue − Cost, and Revenue = Quantity · Price.)

24. The Smiths are vacationing and have spent the afternoon on a scenic island near their cottage on the beach. To get back to the cottage, they must take a boat to some point on the shore, and then take a shuttle bus the rest of the way to the cottage (see Figure 11). The boat travels at 9 miles per hour, and the shuttle bus travels at 15 miles per hour. (*Note*: time = distance/rate.)
    (a) Suppose it has been a long day, and the Smiths are eager to get back to the cottage. Find x (miles) so as to minimize traveling time.
    (b) Now suppose the Smiths had a wonderful day, and they don't want it to end. Find x so as to *maximize* the time it takes them to get back to the cottage.

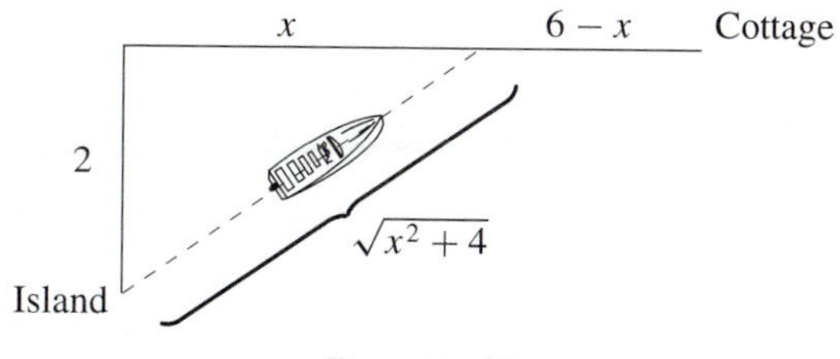

FIGURE 11

SOLUTIONS TO PRACTICE PROBLEMS 3.6

1. This problem can be solved in the same manner that Example 2 was solved. However, the comment made at the end of Example 2 indicates that the economic order quantity should increase by a factor of 3, since $3 = \sqrt{9}$. Therefore, the economic order quantity is $3 \cdot 150 = 450$ cases.

2. The salary of workers changes from $6hx$ to $7hx$. After the substitution,

$$C = 7\left(\frac{210}{x}\right)x + 10x + 15\left(\frac{210}{x}\right) = 1470 + 10x + \frac{3150}{x}.$$

This new expression for C differs from the original expression only in the constant term and so has the same solution, $x = 18$ workers.

3.7 APPLICATIONS OF CALCULUS TO BUSINESS AND ECONOMICS

In recent years economic decision-making has become more and more mathematically oriented. Faced with huge masses of statistical data, depending on hundreds or even thousands of different variables, business analysts and economists have increasingly turned to mathematical methods to help them describe what is happening, predict the effects of various policy alternatives, and choose reasonable courses of action from the myriad of possibilities. Among the mathematical methods employed is calculus. In this section we illustrate just a few of the many applications of calculus to business and economics. All our applications will center around what economists call the *theory of the firm*. In other words, we study the activity of a business (or possibly a whole industry) and restrict our analysis to a time period during which background conditions (such as supplies of raw materials, wage rates, taxes) are fairly constant. We then show how calculus can help the management of such a firm make vital production decisions.

Management, whether or not it knows calculus, utilizes many functions of the sort we have been considering. Examples of such functions are

$C(x)$ = cost of producing x units of the product,

$R(x)$ = revenue generated by selling x units of the product,

$P(x) = R(x) - C(x)$ = the profit (or loss) generated by producing and selling x units of the product.

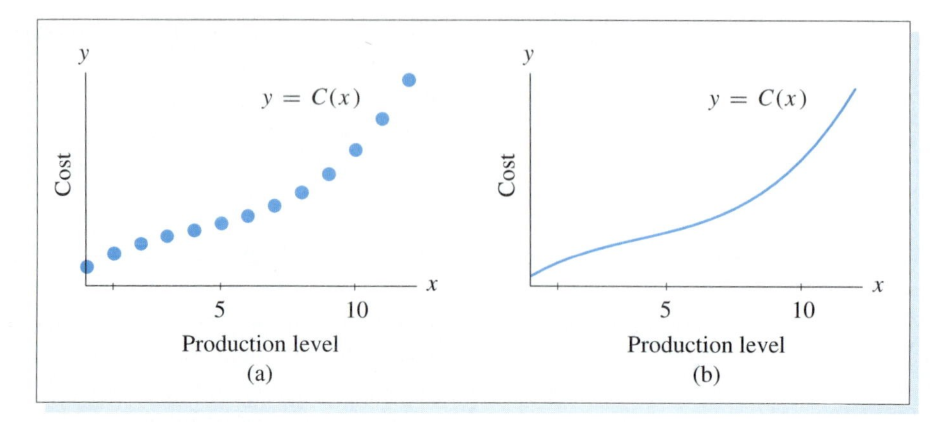

FIGURE 1 A cost function

Note that the functions $C(x)$, $R(x)$, and $P(x)$ often are defined only for nonnegative integers—that is, for $x = 0, 1, 2, 3, \ldots$. The reason is that it does not make sense to speak about the cost of producing -1 cars or the revenue generated by selling 3.62 refrigerators. Thus each of the functions may give rise to a set of discrete points on a graph, as in Figure 1(a). In studying these functions, however, economists usually draw a smooth curve through the points and assume that $C(x)$ is actually defined for all positive x. Of course, we must often interpret answers to problems in light of the fact that x is, in most cases, a nonnegative integer.

COST FUNCTIONS If we assume that a cost function $C(x)$ has a smooth graph as in Figure 1(b), we can use the tools of calculus to study it. A typical cost function is analyzed in Example 1.

EXAMPLE 1 Suppose that the cost function for a manufacturer is given by $C(x) = (10^{-6})x^3 - .003x^2 + 5x + 1000$ dollars.

(a) Describe the behavior of the marginal cost.
(b) Sketch the graph of $C(x)$.

Solution The first two derivatives of $C(x)$ are given by

$$C'(x) = (3 \cdot 10^{-6})x^2 - .006x + 5,$$
$$C''(x) = (6 \cdot 10^{-6})x - .006.$$

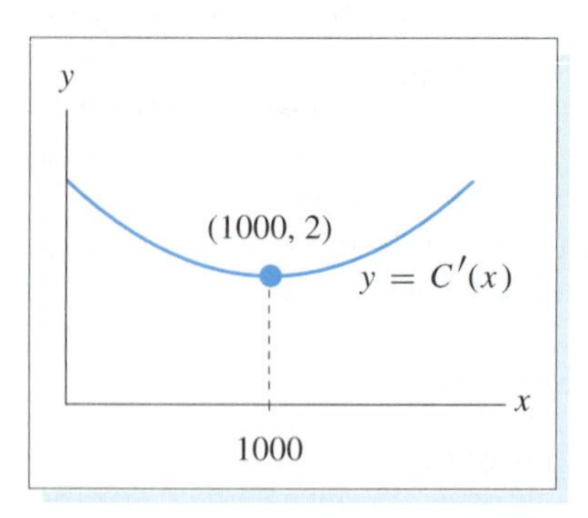

FIGURE 2 A marginal cost function

Let us sketch the marginal cost $C'(x)$ first. From the behavior of $C'(x)$, we will be able to graph $C(x)$. The marginal cost function $y = (3 \cdot 10^{-6})x^2 - .006x + 5$ has as its graph a parabola that opens upward. Since $y' = C''(x) = .000006(x - 1000)$, we see that the parabola has a horizontal tangent at $x = 1000$. So the minimum value of $C'(x)$ occurs at $x = 1000$. The corresponding y-coordinate is

$$(3 \cdot 10^{-6})(1000)^2 - .006 \cdot (1000) + 5 = 3 - 6 + 5 = 2.$$

The graph of $y = C'(x)$ is shown in Figure 2. Consequently, at first the marginal cost decreases. It reaches a minimum of 2 at production level 1000 and increases thereafter. This answers part (a). Let us now graph $C(x)$. From the graph of $C'(x)$ shown in Figure 2 , we see that $C'(x)$ is never zero, so the graph of $C(x)$ has no relative extreme points. Since $C'(x)$ is always positive, $C(x)$ is always increasing (as any cost curve should be). Moreover, since $C'(x)$ decreases for x less than 1000 and increases for x greater than 1000, we see that $C(x)$ is concave down for x less than 1000, is concave up for x greater than 1000, and has an inflection point at $x = 1000$. Also, $C(0) = 1000$, so the y-intercept of $C(x)$ is $(0, 1000)$. The graph of $C(x)$ is drawn in Figure 3. Note that the inflection point of $C(x)$ occurs at the value of x for which marginal cost is a minimum. ■

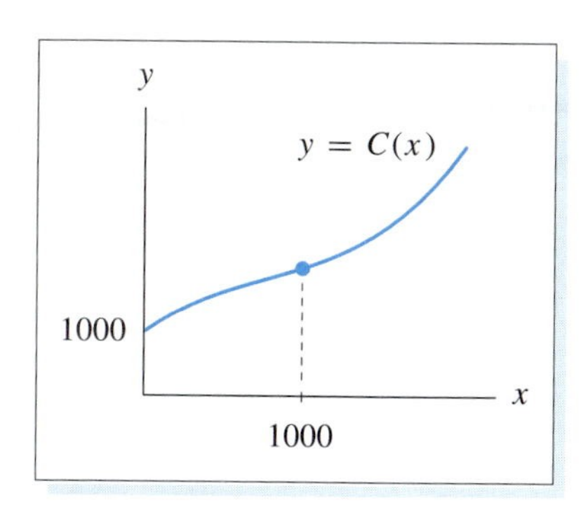

FIGURE 3 A cost function

Actually, most marginal cost functions have the same general shape as the marginal cost curve of Example 1. For when x is small, production of additional units is subject to economies of production, which lowers unit costs. Thus, for x small, marginal cost decreases. However, increased production eventually leads to overtime, use of less efficient, older plants, and competition for scarce raw materials. As a result, the cost of additional units will increase for very large x. So we see that $C'(x)$ initially decreases and then increases.

REVENUE FUNCTIONS In general, a business is concerned not only with its costs, but also with its revenues. Recall that if $R(x)$ is the revenue received from the sale of x units of some commodity, then the derivative $R'(x)$ is called the *marginal revenue*. Economists use this to measure the rate of increase in revenue per unit increase in sales.

If x units of a product are sold at a price p per unit, then the total revenue $R(x)$ is given by

$$R(x) = x \cdot p.$$

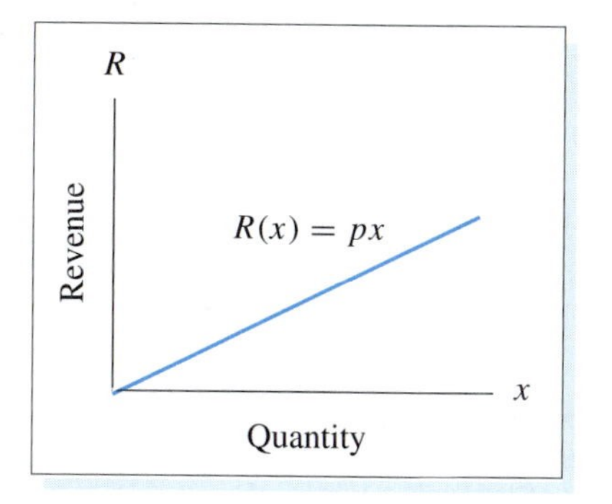

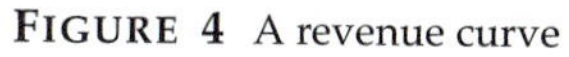

FIGURE 4 A revenue curve

If a firm is small and is in competition with many other companies, its sales have little effect on the market price. Then since the price is constant as far as the one firm is concerned, the marginal revenue $R'(x)$ equals the price p [that is, $R'(x)$ is the amount that the firm receives from the sale of one additional unit]. In this case, the revenue function will have a graph as in Figure 4.

An interesting problem arises when a single firm is the only supplier of a certain product or service—that is, when the firm has a monopoly. Consumers will buy large amounts of the commodity if the price per unit is low and less if the price is raised. For each quantity x, let $f(x)$ be the highest price per unit that can be set in order to sell x units to consumers. Since selling greater quantities requires a lowering of the price, $f(x)$ will be a decreasing function. Figure 5 shows a typical "demand curve" that relates the quantity demanded, x, to the price $p = f(x)$.

The *demand equation* $p = f(x)$ determines the total revenue function. If the firm wants to sell x units, the highest price it can set is $f(x)$ dollars per unit, and so the total revenue from the sale of x units is

$$R(x) = x \cdot p = x \cdot f(x). \tag{1}$$

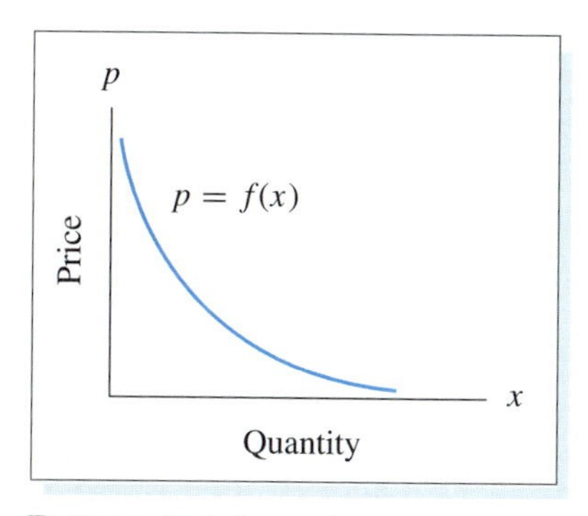

FIGURE 5 A demand curve

The concept of a demand curve applies to an entire industry (with many producers) as well as to a single monopolistic firm. In this case, many producers offer the same product for sale. If x denotes the total output of the industry, then $f(x)$

is the market price per unit of output and $x \cdot f(x)$ is the total revenue earned from the sale of the x units.

EXAMPLE 2 The demand equation for a certain product is $p = 6 - \frac{1}{2}x$. Find the level of production that results in maximum revenue.

Solution In this case, the revenue function $R(x)$ is

$$R(x) = x \cdot p = x\left(6 - \frac{1}{2}x\right) = 6x - \frac{1}{2}x^2.$$

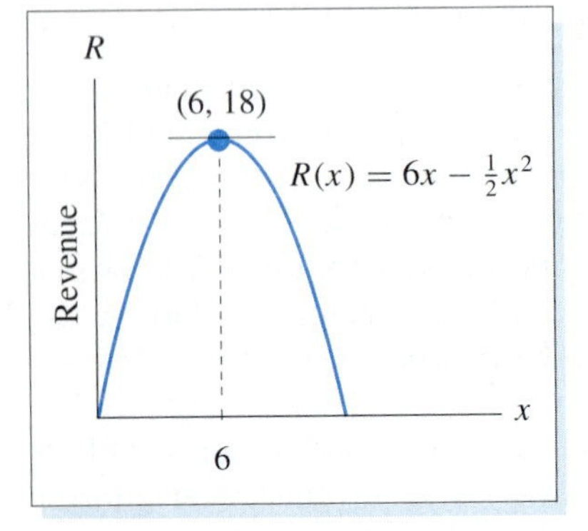

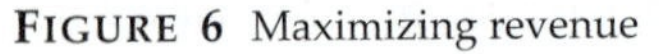
FIGURE 6 Maximizing revenue

The marginal revenue is given by

$$R'(x) = 6 - x.$$

The graph of $R(x)$ is a parabola that opens downward (Figure 6). It has a horizontal tangent precisely at those x for which $R'(x) = 0$—that is, for those x at which marginal revenue is 0. The only such x is $x = 6$. The corresponding value of revenue is

$$R(6) = 6 \cdot 6 - \frac{1}{2}(6)^2 = 18.$$

Thus the rate of production resulting in maximum revenue is $x = 6$, which results in total revenue of 18. ■

EXAMPLE 3 The WMA Bus Lines offers sightseeing tours of Washington, DC. One of the tours, priced at \$7 per person, had an average demand of about 1000 customers per week. When the price was lowered to \$6, the weekly demand jumped to about 1200 customers. Assuming that the demand equation is linear, find the tour price that should be charged per person in order to maximize the total revenue each week.

Solution First we must find the demand equation. Let x be the number of customers per week and let p be the price of a tour ticket. Then $(x, p) = (1000, 7)$ and $(x, p) = (1200, 6)$ are on the demand curve (Figure 7). Using the point-slope formula for the line through these two points, we have

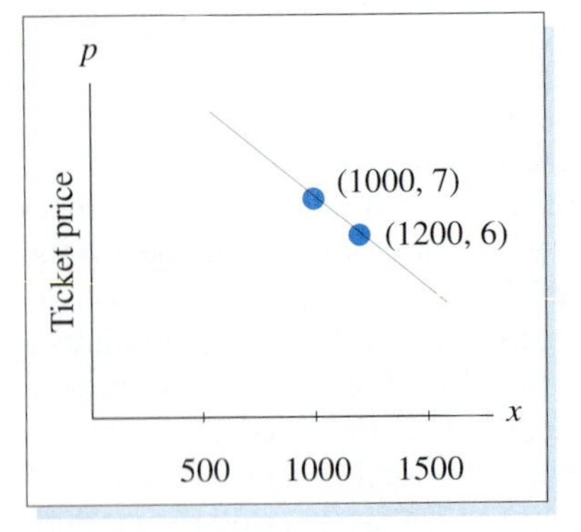

FIGURE 7 A demand curve

$$p - 7 = \frac{7-6}{1000-1200} \cdot (x - 1000) = -\frac{1}{200}(x - 1000) = -\frac{1}{200}x + 5,$$

so

$$p = 12 - \frac{1}{200}x. \tag{2}$$

From equation (1) we obtain the revenue function

$$R(x) = x \cdot p = x\left(12 - \frac{1}{200}x\right) = 12x - \frac{1}{200}x^2.$$

The marginal revenue is

$$R'(x) = 12 - \frac{1}{100}x = -\frac{1}{100}(x - 1200).$$

Using $R(x)$ and $R'(x)$, we can sketch the graph of $R(x)$ (Figure 8). The maximum revenue occurs when the marginal revenue is zero—that is, when $x = 1200$. The price corresponding to this number of customers is found from the demand equation (2),

$$p = 12 - \frac{1}{200}(1200) = 6.$$

Thus the price of \$6 is most likely to bring the greatest revenue per week. ■

PROFIT FUNCTIONS Once we know the cost function $C(x)$ and the revenue function $R(x)$, we can compute the profit function $P(x)$ from

$$P(x) = R(x) - C(x).$$

EXAMPLE 4 Suppose that the demand equation for a monopolist is $p = 100 - .01x$ and the cost function is $C(x) = 50x + 10{,}000$. Find the value of x that maximizes the profit, and determine the corresponding price and total profit for this level of production. (See Figure 9.)

Solution The total revenue function is

$$R(x) = x \cdot p = x(100 - .01x) = 100x - .01x^2.$$

Hence the profit function is

$$\begin{aligned} P(x) &= R(x) - C(x) \\ &= 100x - .01x^2 - (50x + 10{,}000) \\ &= -.01x^2 + 50x - 10{,}000. \end{aligned}$$

The graph of this function is a parabola that opens downward. (See Figure 10.) Its highest point will be where the curve has zero slope—that is, where the marginal profit $P'(x)$ is zero. Now

$$P'(x) = -.02x + 50 = -.02(x - 2500).$$

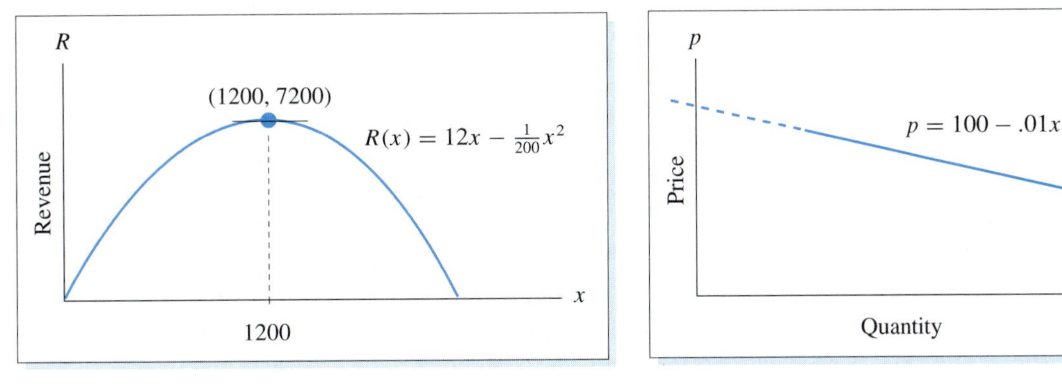

FIGURE 8 Maximizing revenue

FIGURE 9 A demand curve

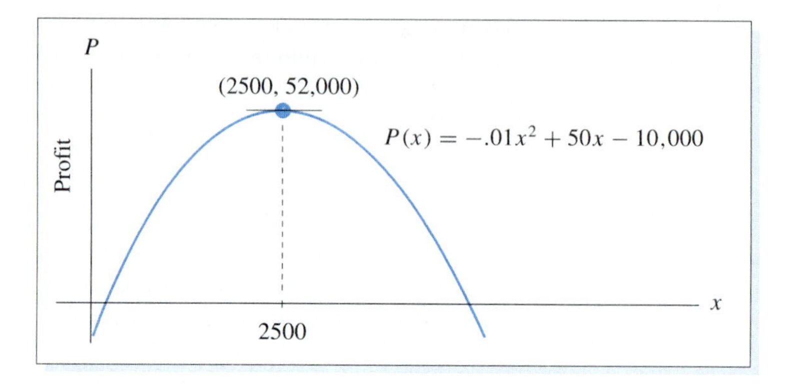

FIGURE 10 Maximizing profit

So $P'(x) = 0$ when $x = 2500$. The profit for this level of production is

$$P(2500) = -.01(2500)^2 + 50(2500) - 10{,}000 = 52{,}500.$$

Finally, we return to the demand equation to find the highest price that can be charged per unit to sell all 2500 units:

$$p = 100 - .01(2500) = 100 - 25 = 75.$$

Answer Produce 2500 units and sell them at \$75 per unit. The profit will be \$52,500. ■

EXAMPLE 5 Rework Example 4 under the condition that the government imposes an excise tax of \$10 per unit.

Solution For each unit sold, the manufacturer will have to pay \$10 to the government. In other words, $10x$ dollars are added to the cost of producing and selling x units. The cost function is now

$$C(x) = (50x + 10{,}000) + 10x = 60x + 10{,}000.$$

The demand equation is unchanged by this tax, so the revenue function is still

$$R(x) = 100x - .01x^2.$$

Proceeding as before, we have

$$\begin{aligned} P(x) &= R(x) - C(x) \\ &= 100x - .01x^2 - (60x + 10{,}000) \\ &= -.01x^2 + 40x - 10{,}000, \\ P'(x) &= -.02x + 40 = -.02(x - 2000). \end{aligned}$$

The graph of $P(x)$ is still a parabola that opens downward, and the highest point is where $P'(x) = 0$—that is, where $x = 2000$. (See Figure 11.) The corresponding profit is

$$P(2000) = -.01(2000)^2 + 40(2000) - 10{,}000 = 30{,}000.$$

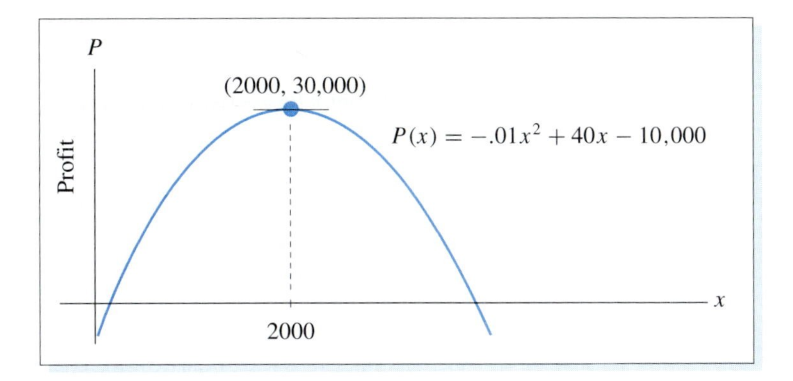

FIGURE 11 Profit after an excise tax

From the demand equation, $p = 100 - .01x$, we find the price that corresponds to $x = 2000$:

$$p = 100 - .01(2000) = 80.$$

Answer Produce 2000 units and sell them at \$80 per unit. The profit will be \$30,000. ■

Notice in Example 5 that the optimal price is raised from \$75 to \$80. If the monopolist wishes to maximize profits, he or she should pass only half the \$10 tax on to the consumer. The monopolist cannot avoid the fact that profits will be substantially lowered by the imposition of the tax. This is one reason why industries lobby against taxation.

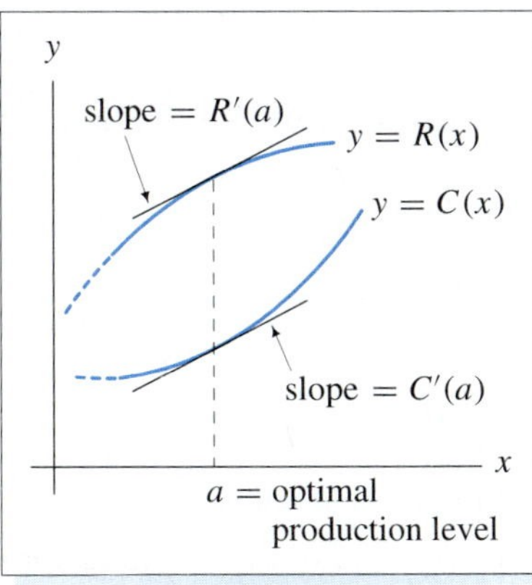

FIGURE 12

SETTING PRODUCTION LEVELS Suppose that a firm has cost function $C(x)$ and revenue function $R(x)$. In a free-enterprise economy the firm will set production x in such a way as to maximize the profit function

$$P(x) = R(x) - C(x).$$

We have seen that if $P(x)$ has a maximum at $x = a$, then $P'(a) = 0$. In other words, since $P'(x) = R'(x) - C'(x)$,

$$R'(a) - C'(a) = 0,$$
$$R'(a) = C'(a).$$

Thus profit is maximized at a production level for which marginal revenue equals marginal cost. (See Figure 12.)

PRACTICE PROBLEMS 3.7

1. Rework Example 4 by finding the production level at which marginal revenue equals marginal cost.
2. Rework Example 4 under the condition that the fixed cost is increased from \$10,000 to \$15,000.

EXERCISES 3.7

1. Given the cost function $C(x) = x^3 - 6x^2 + 13x + 15$, find the minimum marginal cost.
2. Suppose that a total cost function is $C(x) = .0001x^3 - .06x^2 + 12x + 100$. Is the marginal cost increasing, decreasing, or not changing at $x = 100$? Find the minimum marginal cost.
3. The revenue function for a one-product firm is

$$R(x) = 200 - \frac{1600}{x+8} - x.$$

Find the value of x that results in maximum revenue.

4. The revenue function for a particular product is $R(x) = x(4 - .0001x)$. Find the largest possible revenue.
5. A one-product firm estimates that its daily total cost function (in suitable units) is $C(x) = x^3 - 6x^2 + 13x + 15$ and its total revenue function is $R(x) = 28x$. Find the value of x that maximizes the daily profit.
6. A small tie shop sells ties for \$3.50 each. The daily cost function is estimated to be $C(x)$ dollars, where x is the number of ties sold on a typical day and $C(x) = .0006x^3 - .03x^2 + 2x + 20$. Find the value of x that will maximize the store's daily profit.
7. The demand equation for a certain commodity is $p = \frac{1}{12}x^2 - 10x + 300$, $0 \leq x \leq 60$. Find the value of x and the corresponding price p that maximize the revenue.
8. The demand equation for a product is $p = 2 - .001x$. Find the value of x and the corresponding price p that maximize the revenue.
9. Some years ago it was estimated that the demand for steel approximately satisfied the equation $p = 256 - 50x$, and the total cost of producing x units of steel was $C(x) = 182 + 56x$. (The quantity x was measured in millions of tons and the price and total cost were measured in millions of dollars.) Determine the level of production and the corresponding price that maximize the profits.
10. Consider a rectangle in the xy-plane, with corners at $(0, 0)$, $(a, 0)$, $(0, b)$, and (a, b). Suppose that (a, b) lies on the graph of the equation $y = 30 - x$. Find a and b such that the area of the rectangle is maximized. What economic interpretation can be given to your answer if the equation $y = 30 - x$ represents a demand curve and y is the price corresponding to the demand x?
11. Until recently hamburgers at the city sports arena cost \$2 each. The food concessionaire sold an average of 10,000 hamburgers on a game night. When the price was raised to \$2.40, hamburger sales dropped off to an average of 8000 per night.
    (a) Assuming a linear demand curve, find the price of a hamburger that will maximize the nightly hamburger revenue.
    (b) Suppose that the concessionaire has fixed costs of \$1000 per night and the variable cost is \$.60 per hamburger. Find the price of a hamburger that will maximize the nightly hamburger profit.

12. When the average ticket price for a concert at the opera house was \$50, attendance averaged 4000 people per performance. When the ticket price was raised to \$52, attendance declined to an average of 3800 people per performance. What should the average ticket price be in order to maximize the revenue for the opera house? (Assume a linear demand curve.)

13. The monthly demand equation for an electric utility company is estimated to be

$$p = 60 - (10^{-5})x,$$

where p is measured in dollars and x is measured in thousands of kilowatt-hours. The utility has fixed costs of \$7,000,000 per month and variable costs of \$30 per 1000 kilowatt-hours of electricity generated, so that the cost function is

$$C(x) = 7 \cdot 10^6 + 30x.$$

(a) Find the value of x and the corresponding price for 1000 kilowatt-hours that maximize the utility's profit.

(b) Suppose that rising fuel costs increase the utility's variable costs from \$30 to \$40 so that its new cost function is

$$C_1(x) = 7 \cdot 10^6 + 40x.$$

Should the utility pass all this increase of \$10 per 1000 kilowatt-hours on to consumers? Explain your answer.

14. The demand equation for a monopolist is $p = 200 - 3x$, and the cost function is $C(x) = 75 + 80x - x^2$, $0 \le x \le 40$.

(a) Determine the value of x and the corresponding price that maximize the profit.

(b) Suppose that the government imposes a tax on the monopolist of \$4 per unit quantity produced. Determine the new price that maximizes the profit.

(c) Suppose that the government imposes a tax of T dollars per unit quantity produced so that the new cost function is

$$C(x) = 75 + (80 + T)x - x^2, \quad 0 \le x \le 40.$$

Determine the new value of x that maximizes the monopolist's profit as a function of T. Assuming that the monopolist cuts back production to this level, express the tax revenues received by the government as a function of T. Finally, determine the value of T that will maximize the tax revenue received by the government.

SOLUTIONS TO PRACTICE PROBLEMS 3.7

1. The revenue function is $R(x) = 100x - .01x^2$, so the marginal revenue function is $R'(x) = 100 - .02x$. The cost function is $C(x) = 50x + 10{,}000$, so the marginal cost function is $C'(x) = 50$. Let us now equate the two marginal functions and solve for x.

$$\begin{aligned} R'(x) &= C'(x) \\ 100 - .02x &= 50 \\ -.02x &= -50 \\ x &= \frac{-50}{-.02} = \frac{5000}{2} = 2500. \end{aligned}$$

   Of course, we obtain the same level of production as before.

2. If the fixed cost is increased from \$10,000 to \$15,000, the new cost function will be $C(x) = 50x + 15{,}000$, but the marginal cost function will still be $C'(x) = 50$. Therefore, the solution will be the same: 2500 units should be produced and sold at \$75 per unit. (Increases in fixed costs should not necessarily be passed on to the consumer if the objective is to maximize profit.)

REVIEW OF THE FUNDAMENTAL CONCEPTS OF CHAPTER 3

1. State the power rule, constant multiple rule, and sum rule.
2. Give two different notations for the first derivative of $f(x)$ at $x = 2$. Second derivative.
3. State the general power rule and give an example.
4. Outline a method for locating the relative extreme points of a function.
5. Outline an algebraic procedure for obtaining the graph of a function.
6. What is an objective equation?
7. What is a constraint equation?
8. Outline the procedure for solving an optimization problem.
9. How are the cost, revenue, and profit functions related?

Chapter 4

Exponential and Logarithmic Functions

When an investment grows steadily at 15% per year, the rate of growth of the investment at any time is proportional to the value of the investment at that time. When a bacteria culture grows in a laboratory dish, the rate of growth of the culture at any moment is proportional to the total number of bacteria in the dish at that moment. These situations are examples of what is called *exponential growth.* A pile of radioactive uranium ^{235}U decays at a rate that at each moment is proportional to the amount of ^{235}U present. This decay of uranium (and of radioactive elements in general) is called *exponential decay*. Both exponential growth and exponential decay can be described and studied in terms of exponential functions and the natural logarithm function. The properties of these functions are investigated in this chapter. Subsequently, we shall explore a wide range of applications in fields such as business, biology, archeology, public health, and psychology.

4.1 The Exponential Function e^x

Let us begin by examining the graphs of the exponential functions shown in Figure 1. They all pass through (0, 1) but with different slopes there. Notice that the graph of 5^x is quite steep at $x = 0$, while the graph of $(1.1)^x$ is nearly horizontal at $x = 0$. It turns out that at $x = 0$, the graph of 2^x has a slope of approximately .69, while the graph of 3^x has a slope of approximately 1.1.

Evidently, there is a particular value of the base b, between 2 and 3, where the graph of b^x has slope *exactly* 1 at $x = 0$. We denote this special value of b by the letter e, and we call

$$f(x) = e^x$$

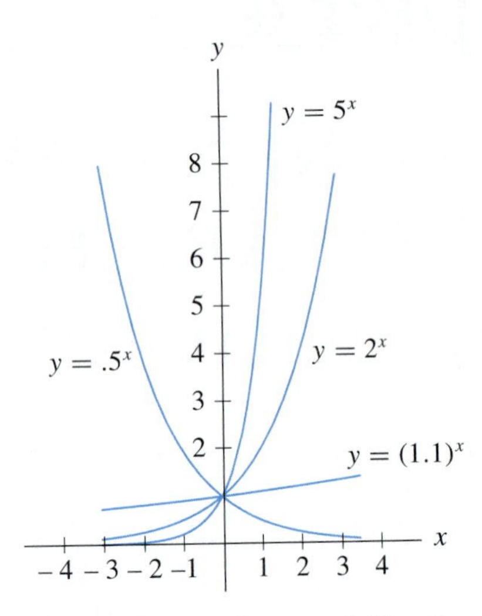

FIGURE 1 Several exponential functions

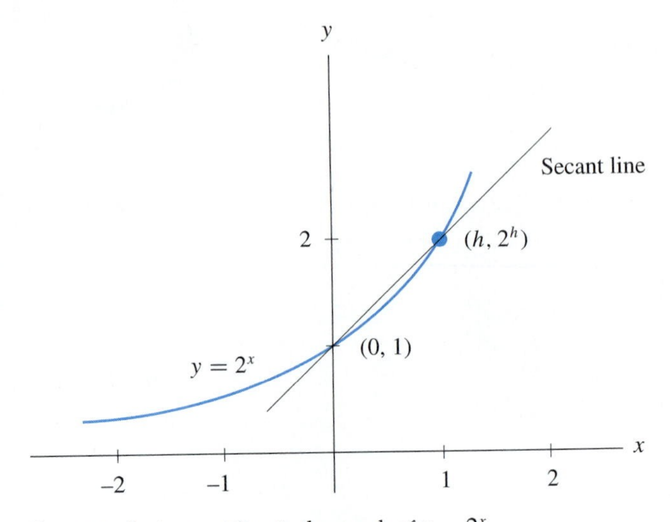

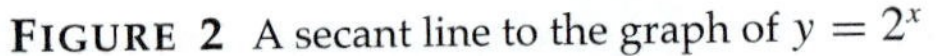

FIGURE 2 A secant line to the graph of $y = 2^x$

the exponential function. The number e is an important constant of nature that has been calculated to thousands of decimal places. To 10 significant digits, we have $e = 2.718281828$. For our purposes, it is usually sufficient to think of e as "approximately 2.7."

Our goal in this section is to find a formula for the derivative of e^x. It turns out that the calculations for e^x and 2^x are very similar. Since many people are more comfortable working with 2^x rather than e^x, we shall first analyze the graph of 2^x. Then we shall draw the appropriate conclusions about the graph of e^x.

Before computing the slope of $y = 2^x$ at an arbitrary x, let us consider the special case $x = 0$. Denote the slope at $x = 0$ by m. We shall use the secant-line approximation of the derivative to approximate m. We proceed by constructing a secant line in Figure 2. The slope of the secant line through $(0, 1)$ and $(h, 2^h)$ is

$$\frac{2^h - 1}{h}.$$

As h approaches zero, the slope of the secant line approaches the slope of $y = 2^x$ at $x = 0$. That is,

$$m = \lim_{h \to 0} \frac{2^h - 1}{h}. \tag{1}$$

We can estimate the value of m by taking h smaller and smaller. When $h = .1$, we have $2^h \approx 1.072$ (from a calculator), and

$$\frac{2^h - 1}{h} \approx \frac{.072}{.1} = .72.$$

When $h = .01$, we have $2^h \approx 1.00696$, and

$$\frac{2^h - 1}{h} \approx \frac{.00696}{.01} \approx .696.$$

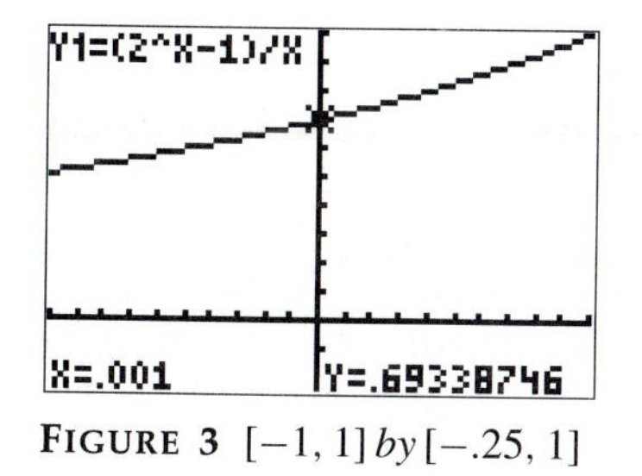

FIGURE 3 $[-1, 1]$ by $[-.25, 1]$

When $h = .001$, $2^h \approx 1.0006934$, so that

$$\frac{2^h - 1}{h} \approx \frac{.0006934}{.001} \approx .693. \qquad \text{(See Figure 3.)}$$

Thus it is reasonable to conclude that $m \approx .69$.* Since m equals the slope of $y = 2^x$ at $x = 0$, we have

$$m = \frac{d}{dx}(2^x)\bigg|_{x=0} \approx .69. \tag{2}$$

Now that we have estimated the slope of $y = 2^x$ at $x = 0$, let us compute the slope for an arbitrary value of x. We construct a secant line through $(x, 2^x)$ and a nearby point $(x + h, 2^{x+h})$ on the graph. The slope of the secant line is

$$\frac{2^{x+h} - 2^x}{h}. \tag{3}$$

By law of exponents (i), we have $2^{x+h} - 2^x = 2^x(2^h - 1)$ so that, by (1), we see that

$$\lim_{h \to 0} \frac{2^{x+h} - 2^x}{h} = \lim_{h \to 0} 2^x \frac{2^h - 1}{h} = 2^x \lim_{h \to 0} \frac{2^h - 1}{h} = m2^x. \tag{4}$$

However, the slope of the secant (3) approaches the derivative of 2^x as h approaches zero. Consequently, we have

$$\frac{d}{dx}(2^x) = m2^x, \quad \text{where } m = \frac{d}{dx}(2^x)\bigg|_{x=0}. \tag{5}$$

EXAMPLE 1 Calculate (a) $\frac{d}{dx}(2^x)\Big|_{x=3}$ and (b) $\frac{d}{dx}(2^x)\Big|_{x=-1}$.

Solution (a) $\frac{d}{dx}(2^x)\Big|_{x=3} = m \cdot 2^3 = 8m \approx 8(.69) = 5.52.$

(b) $\frac{d}{dx}(2^x)\Big|_{x=-1} = m \cdot 2^{-1} = .5m \approx .5(.69) = .345.$ ■

The calculations just carried out for $y = 2^x$ can be carried out for $y = b^x$, where b is any positive number. Equation (5) will read exactly the same except that 2 will be replaced by b. Thus we have the following formula for the derivative of the function $f(x) = b^x$:

* To 10 decimal places, m is .6931471806.

$$\frac{d}{dx}(b^x) = mb^x, \qquad \text{where } m = \frac{d}{dx}(b^x)\bigg|_{x=0}. \tag{6}$$

Our calculations showed that if $b = 2$, then $m \approx .69$. If $b = 3$, then it turns out that $m \approx 1.1$. (See Exercise 1.) Obviously, the derivative formula in (6) is simplest when $m = 1$, that is, when the graph of b^x has slope 1 at $x = 0$. As we said earlier, this special value of b is denoted by the letter e. Thus the number e has the property that

$$m = \frac{d}{dx}(e^x)\bigg|_{x=0} = 1, \tag{7}$$

and so from (6)

$$\frac{d}{dx}(e^x) = 1 \cdot e^x = e^x. \tag{8}$$

The graphical interpretation of (7) is that the curve $y = e^x$ has slope 1 at $x = 0$. The graphical interpretation of (8) is that the slope of the curve $y = e^x$ at an arbitrary value of x is exactly equal to the value of the function e^x at that point. (See Figure 4.)

The function e^x is the same type of function as 2^x and 3^x except that taking derivatives of e^x is much easier. For this reason, functions based on e^x are used in almost all applications that require an exponential-type function to describe physical phenomena.

USING e^x WITH A TI CALCULATOR

For a discussion of using e^x with Visual Calculus, see Appendix A, Part VI.

The function e^x appears above the LN button. To graph the function, set Y_1 equal to e^x with the following keystrokes:

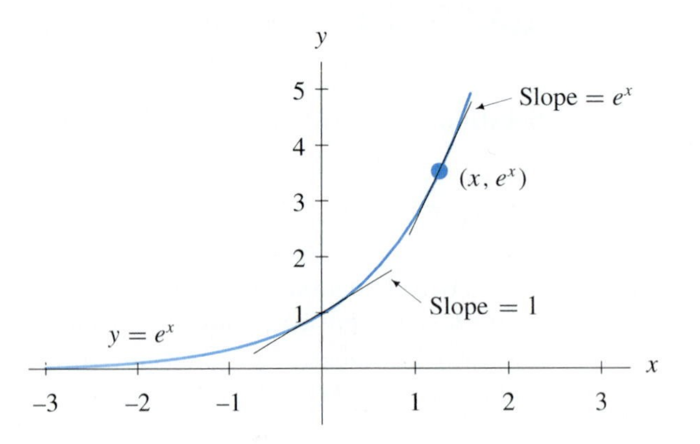

FIGURE 4 Fundamental properties of e^x

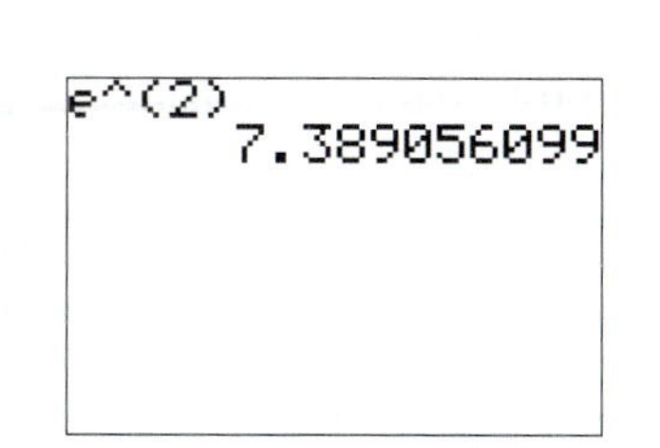

FIGURE 5

(TI-82) [2nd] [e^x] [X, T, θ]

(TI-83) [2nd] [e^x] [X, T, θ, n] [)]

(TI-85) [2nd] [e^x] [x-VAR]

The screen will show `Y1=e^X`, `Y1=e^(X)`, or `y1=e^x`. A function such as e^{3x} is entered as `Y1=e^3X` or as `Y1=e^(3X)`.

Values of e^x can be calculated by pushing a few buttons. To evaluate the function at $x = a$, press [2nd] [e^x], type in the number a (followed by a parenthesis when using a TI-83), and then press [ENTER]. See Figure 5.

THE FUNCTIONS e^{kx} Exponential functions of the form e^{kx} occur in many applications. Figure 6 shows the graphs of several functions of this type when k is a positive number. These curves $y = e^{kx}$, k positive, have several properties in common:

1. (0, 1) is on the graph.

2. The graph lies strictly above the x-axis (e^{kx} is never zero).

3. The x-axis is an asymptote as x becomes large negatively.

4. The graph is always increasing and concave up.

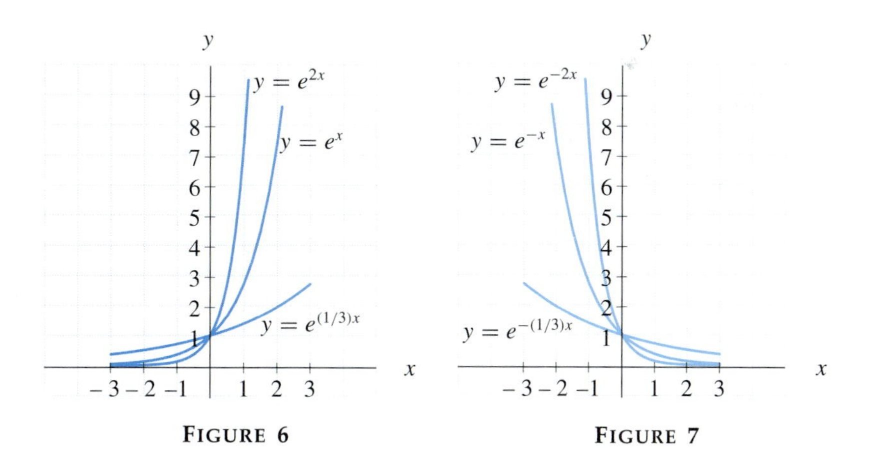

FIGURE 6 FIGURE 7

When k is negative, the graph of $y = e^{kx}$ is decreasing. (See Figure 7.) Note the following properties of the curves $y = e^{kx}$, k negative:

1. (0, 1) is on the graph.

2. The graph lies strictly above the x-axis.

3. The x-axis is an asymptote as x becomes large positively.

4. The graph is always decreasing and concave up.

THE FUNCTIONS b^x If b is a positive number, then the function b^x may be written in the form e^{kx} for some k. For example, take $b = 2$. From Figure 4 it is clear that there is some value of x such that $e^x = 2$. Call this value k, so that $e^k = 2$. Then

$$2^x = (e^k)^x = e^{kx}$$

for all x. In general, if b is any positive number, there is a value of x, say $x = k$, such that $e^k = b$. In this case, $b^x = (e^k)^x = e^{kx}$. Thus all the curves $y = b^x$ discussed earlier can be written in the form $y = e^{kx}$. This enables us to focus on exponential functions with base e instead of studying 2^x, 3^x, and so on.

Exponential functions of the form $f(x) = e^{kx}$ can be differentiated with the general power rule:

$$\begin{aligned} f(x) &= e^{kx} = (e^x)^k, \\ f'(x) &= k(e^x)^{k-1}(e^x)' && \text{[general power rule]} \\ &= k(e^x)^{k-1}(e^x) && [\textit{Note}\text{: } (e^x) = (e^x)'.] \\ &= k(e^x)^{k-1+1} && [\text{since } (e^x)^r \cdot (e^x)^s = (e^x)^{r+s}] \\ &= k(e^x)^k \\ &= ke^{kx}. \end{aligned}$$

That is,

$$\boxed{\frac{d}{dx}(e^{kx}) = ke^{kx}.} \tag{9}$$

EXAMPLE 2 Differentiate the following exponential functions.

(a) e^{-2x} (b) $3e^{5x}$ (c) $3e^{kx}$, where k is a constant
(d) Ce^{kx}, where C and k are constants

Solution (a) Here $k = -2$ and $\frac{d}{dx}(e^{-2x}) = (-2)e^{-2x}$.

(b) $\frac{d}{dx}(3e^{5x}) = 3\frac{d}{dx}(e^{5x}) = 3 \cdot 5e^{5x} = 15e^{5x}$.

(c)
$$\begin{aligned} \frac{d}{dx}(3e^{kx}) &= 3\frac{d}{dx}(e^{kx}) \\ &= 3 \cdot ke^{kx} && \text{[by (9)]} \\ &= 3ke^{kx}. \end{aligned}$$

(d) $\frac{d}{dx}(Ce^{kx}) = C\frac{d}{dx}(e^{kx}) = Cke^{kx}$. ■

The result of part (d) may be summarized in an extremely useful fashion as follows: Suppose that we let $y = Ce^{kx}$. By part (d) we have

$$y' = Cke^{kx} = k \cdot (Ce^{kx}) = ky.$$

In other words, the derivative of the function Ce^{kx} is k times the function itself. Let us record this fact.

> Let C, k be any constants and let $y = Ce^{kx}$. Then y satisfies the equation
> $$y' = ky.$$

The equation $y' = ky$ involves an unknown function y and its derivative y'. Any equation involving one or more derivatives of an unknown function is called a *differential equation.*

PRACTICE PROBLEMS 4.1

1. Write the function $e^{2x} \cdot e^{-5x}$ in the form e^{kx}.
2. Differentiate $f(t) = 20e^{-\frac{t}{4}}$.

EXERCISES 4.1

x	3^x
0	1.00000
.001	1.00110
.010	1.01105
.100	1.11612

1. Use the table at the left to show that
$$\left.\frac{d}{dx}(3^x)\right|_{x=0} \approx 1.1.$$
That is, calculate the slope
$$\frac{3^h - 1}{h}$$
of the secant line passing through the point $(0, 1)$ and $(h, 3^h)$. Take $h = .1, .01$, and $.001$.

x	$(2.7)^x$
0	1.00000
.001	1.00099
.010	1.00998
.100	1.10443

2. Use the table at the left to show that
$$\left.\frac{d}{dx}((2.7)^x)\right|_{x=0} \approx .99.$$
That is, calculate
$$\frac{(2.7)^h - 1}{h}$$
for $h = .1, .01$, and $.001$.

x	e^x
0	1.00000
.001	1.00100
.005	1.00501
.010	1.01005

3. Consider the secant line on the graph of e^x passing through $(0, 1)$ and (h, e^h). Its slope is
$$\frac{e^h - 1}{h}.$$
Use the table at the left to compute this quantity for $h = .01, .005$, and $.001$.
4. Use the fact that $e^{2+x} = e^2 \cdot e^x$ to find
$$\frac{d}{dx}(e^{2+x}).$$
[Remember that e^2 is just a constant—approximately $(2.7)^2$.]

In Exercises 5–10, use properties of exponents to write the functions in the form e^{kx}.

5. $(e^x)^2$ **6.** $\dfrac{e^{3x}}{e^x}$ **7.** $\dfrac{1}{e^{-2x}}$

8. $e^{2x} \cdot e^{3x}$ **9.** $\dfrac{e^{-5x}}{e^{3x}}$ **10.** $\dfrac{1}{e^{.2x}}$

11. Simplify $e^{1-x} \cdot e^{2x}$.

12. Simplify $e^3 \cdot e^{x+1}$.

Differentiate the following functions.

13. e^{-x} **14.** e^{10x} **15.** $5e^x$

16. $20e^{.5x}$ **17.** $5e^{-.2x}$ **18.** $\dfrac{1}{e^x}$

19. $\dfrac{1}{e^{3x}}$ **20.** $\dfrac{3}{e^{-2x}}$ **21.** $e^{-\frac{x}{3}}$

22. $\dfrac{e^x - e^{-x}}{2}$ **23.** $(e^x - e^{-x})^3$ **24.** $(x^2 + e^{2x})^3$

25. $(1 + 5e^x)^4$ **26.** $\dfrac{e^x + e^{-x}}{e^x}$ **27.** $e^t \cdot e^{3t}$

28. $\dfrac{100}{1 + 5e^{-.2t}}$ **29.** $\dfrac{25}{1 + 4e^{-.5t}}$ **30.** $2(1 - e^{-.3t})$

31. The value of a computer t years after purchase is $v(t) = 2000e^{-.35t}$ dollars. How much is the computer worth after 4 years? At what rate (in dollars per year) is the computer depreciating after 4 years?

32. A painting purchased in 1994 for \$100,000 is estimated to be worth $G(t) = 100{,}000e^{.2t}$ dollars after t years. How much will the painting be worth in 1999? At what rate (in dollars per year) will the painting be appreciating in 1999?

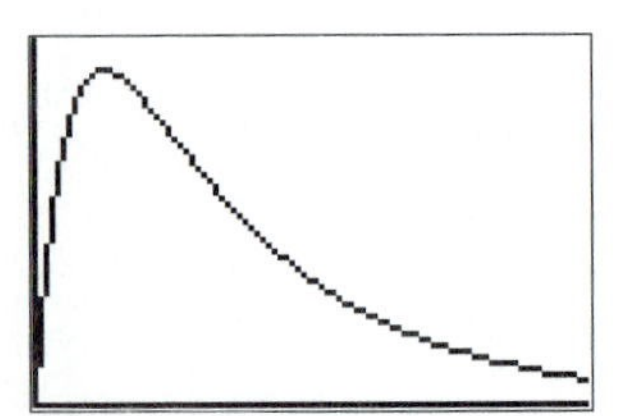

FIGURE 8 [0, 16] *by* [0, 70]
Level of drug

33. When a drug is taken orally, the amount of the drug in the bloodstream after t hours is $f(t) = 120(e^{-.2t} - e^{-t})$ units. See Figure 8. How many units of the drug are in the bloodstream after 6 hours? At what rate is the level of the drug in the bloodstream decreasing after 6 hours?

34. The spread of information by radio follows the model $f(t) = 4000(1 - e^{-.25t})$, where $f(t)$ is the number of people who have heard the information after t days. How many people will have heard the news after 3 days? At what rate (in people per day) will the news be spreading after 3 days?

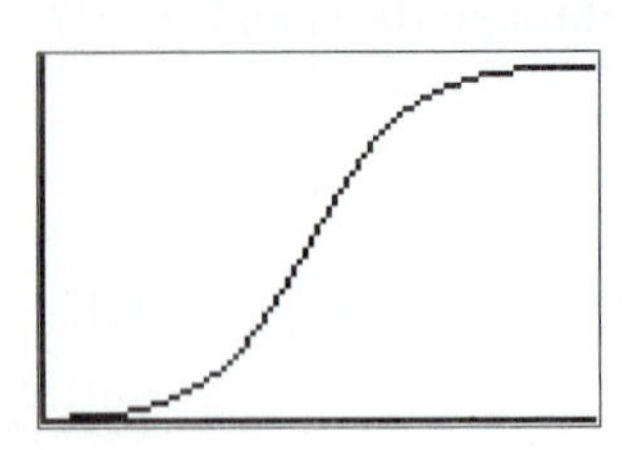

FIGURE 9
[0, 21] *by* [0, 4100]
Information spread by word of mouth

35. The spread of information by word of mouth follows the model

$$f(t) = \frac{4000}{1 + 200e^{-.52t}},$$

where $f(t)$ is the number of people who have heard the information after t days. See Figure 9. What are $f(3)$ and $f'(3)$, and what do they tell about the situation?

36. When a rod of molten steel with a temperature of 2800 °F is placed in a large vat of water at temperature 70 °F, the temperature of the rod after t seconds is $f(t) = 70(1 + 39e^{-.2t})$ degrees. What are $f(10)$ and $f'(10)$, and what do they tell about the situation?

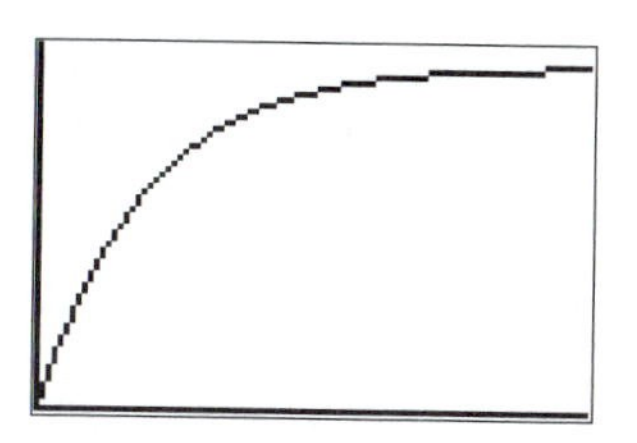

FIGURE 10
[0, 32] *by* [0, 210]
Speed of parachutist

37. The speed of a parachutist during free fall is $f(t) = 200(1 - e^{-.16t})$ feet per second. See Figure 10. How fast is the parachutist falling after 10 seconds? What is the parachutist's acceleration after 10 seconds?

38. The world population t years after 1995 is $6.5e^{.015t}$ billion people. What is the world population in 1998? At what rate (in billions of people per year) is the population growing in 1998?

In Exercises 39–42, give a differential equation satisfied by the function.

39. $y = 10e^{4x}$

40. $y = 5e^{-.2x}$

41. $y = e^{-.15x}$

42. $y = 2e^{x/5}$

43. As h approaches 0, what value is approached by the difference quotient $\dfrac{e^h - 1}{h}$? (*Hint*: $1 = e^0$.)

44. As h approaches 0, what value is approached by $\dfrac{e^{2h} - 1}{h}$? (*Hint*: $1 = e^0$.)

In the remaining exercises, use a graphing utility to obtain the graphs of the functions. In Exercises 45–48, find the equation of the tangent line to the graph of the given function at the given value of x. Then graph the function and the tangent line together to confirm your answer.

45. $f(x) = e^x$, $x = 0$.

46. $f(x) = e^{-2x}$, $x = 0$.

47. $f(x) = e^{-.6x}$, $x = 2$. (*Note*: Use .3 as the approximate value of $e^{-1.2}$.)

48. $f(x) = e^{.7x}$ at $x = 1$. (*Note*: Use 2 as the approximate value of $e^{.7}$.)

49. Simultaneously graph e^{2x} and its derivative in the window $[-4, 4]$ *by* $[0, 8]$. (*Note*: With a TI calculator, use `nDeriv` or `der1` to obtain the derivative. With Visual Calculus, use the "Analyze a Function and Its Derivatives" routine and press **1** to obtain the derivative.) Trace along the graphs and observe that at each value of x the value of the derivative is twice the value of the function.

50. Demonstrate that e^x grows faster than any power function. For instance, graph $\dfrac{x^n}{e^x}$ for $n = 3$ with the window $[0, 16]$ *by* $[0, 25]$ and observe that the function approaches zero as x gets large. Repeat for $n = 4$ and $n = 5$.

51. Display the graph of $\dfrac{10^x - 1}{x}$ and use it to estimate

$$\frac{d}{dx}(10^x)\bigg|_{x=0}.$$

Then use the graphing utility to find the value of the derivative.

52. Investigate the effect of constants on the graph of e^x. Graph the following functions for x between -2 and 4. Use appropriate windows.

(i) e^x (ii) $e^{x/2}$ (iii) e^{x^2} (iv) e^{2x} (v) $2e^x$ (vi) $(e^x)^2$

(a) Which of the graphs are exactly the same? Explain why.
(b) Arrange the functions in an order such that their graphs grow faster as $x \to \infty$.

53. (a) Graph $y = e^x$.
(b) Zoom in on the region near $x = 0$ until the curve appears as a straight line, and estimate the slope of the line. This number is an estimate of $\frac{d}{dx}e^x$ at $x = 0$. Compare your answer with the actual slope, 1.
(c) Repeat parts (a) and (b) for $y = 2^x$. Observe that the slope at $x = 0$ is not 1.

54. The graph of e^{x-2} may be obtained by translating the graph of e^x two units to the right. Find a constant C such that the graph of Ce^x is the same as the graph of e^{x-2}. Verify your result by graphing both functions.

SOLUTIONS TO PRACTICE PROBLEMS 4.1

1. By the additive property of exponents, $e^{2x} \cdot e^{-5x} = e^{2x+(-5x)} = e^{-3x}$. Here $k = -3$.

2. $f(t) = 20e^{-\frac{t}{4}} = 20e^{(-\frac{1}{4})t}$ has the form $f(t) = Ce^{kt}$, with $k = -\frac{1}{4}$. Therefore, $f'(t) = -\frac{1}{4} \cdot f(t) = -\frac{1}{4}\left(20e^{-\frac{t}{4}}\right) = -5e^{-\frac{t}{4}}$.

4.2 THE PRODUCT AND QUOTIENT RULES

We can now differentiate functions of the form Ce^{kx}. The product and quotient rules allow us to differentiate functions such as x^2e^{3x} and $\frac{x}{e^x}$. These rules are also useful when differentiating products and quotients of functions discussed earlier in the text.

We observed in our discussion of the sum rule for derivatives that the derivative of the sum of two differentiable functions is the sum of the derivatives. Unfortunately, however, the derivative of the product $f(x)g(x)$ is *not* the product of the derivatives. Rather, the derivative of a product is determined from the following rule:

Product Rule

$$\frac{d}{dx}[f(x)g(x)] = f(x)g'(x) + g(x)f'(x).$$

The derivative of the product of two functions is the first function times the derivative of the second plus the second function times the derivative of the first. At the end of this section we show why this statement is true.

EXAMPLE 1 Show that the product rule works for the case $f(x) = x^2$, $g(x) = x^3$.

Solution Since $x^2 \cdot x^3 = x^5$, we know that

$$\frac{d}{dx}[x^2 \cdot x^3] = \frac{d}{dx}[x^5] = 5x^4.$$

On the other hand, using the product rule,

$$\begin{aligned}\frac{d}{dx}(x^2 \cdot x^3) &= x^2\frac{d}{dx}(x^3) + x^3\frac{d}{dx}(x^2)\\ &= x^2(3x^2) + x^3(2x)\\ &= 3x^4 + 2x^4 = 5x^4.\end{aligned}$$

Thus the product rule gives the correct answer. ■

EXAMPLE 2 Differentiate the product x^2e^{3x}.

Solution Let $f(x) = x^2$ and $g(x) = e^{3x}$. Then

$$\begin{aligned}\frac{d}{dx}(x^2e^{3x}) &= x^2 \cdot \frac{d}{dx}(e^{3x}) + e^{3x} \cdot \frac{d}{dx}(x^2)\\ &= x^2 \cdot 3e^{3x} + e^{3x} \cdot 2x\\ &= (3x^2 + 2x)e^{3x}.\end{aligned}$$ ■

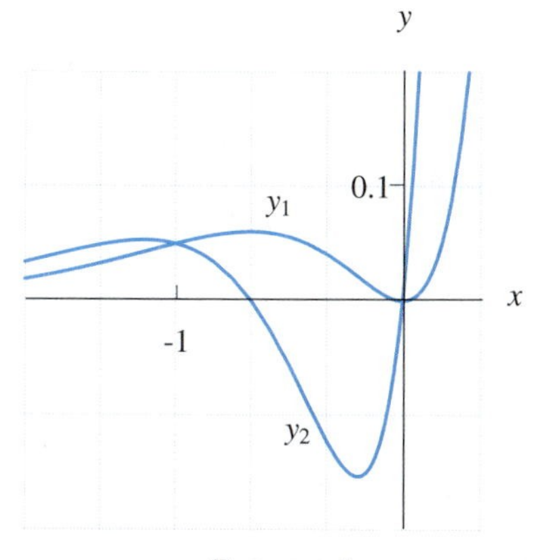

FIGURE 1

Note: A graphing utility can be used to determine whether the answer is reasonable. One method is to graph both the function and its alleged derivative to see if the graphs are consistent. Figure 1 shows the graphs of $y_1 = x^2e^{3x}$ and $y_2 = (3x^2 + 2x)e^{3x}$ with the window $[-1.67, .33]$ *by* $[-.2, .2]$. The curve y_1 has relative extreme points at $x = -\frac{2}{3}$ and $x = 0$ and the curve y_2 has zeros there, as is appropriate. The curve y_2 is positive when y_1 is increasing and is negative when y_1 is decreasing. The inflection points of y_1 appear to line up with the relative extreme points y_2. Therefore, y_2 is consistent with the derivative of y_1.

EXAMPLE 3 Apply the product rule to $y = g(x) \cdot g(x)$.

Solution $\frac{d}{dx}[g(x) \cdot g(x)] = g(x) \cdot g'(x) + g(x) \cdot g'(x) = 2g(x)g'(x).$

This answer is the same as that given by the general power rule:

$$\frac{d}{dx}[g(x) \cdot g(x)] = \frac{d}{dx}[g(x)]^2 = 2g(x)g'(x).$$ ■

THE QUOTIENT RULE Another useful formula for differentiating functions is the quotient rule.

Quotient Rule

$$\frac{d}{dx}\left[\frac{f(x)}{g(x)}\right] = \frac{g(x)f'(x) - f(x)g'(x)}{[g(x)]^2}.$$

One must be careful to remember the order of the terms in this formula because of the minus sign in the numerator. The numerator of the derivative is *the bottom function times the derivative of the top function* minus *the top function times the derivative of the bottom function*.

EXAMPLE 4 Differentiate $\dfrac{e^x}{1+x}$.

Solution Let $f(x) = e^x$ and $g(x) = 1 + x$. Then

$$\begin{aligned}\frac{d}{dx}\left(\frac{e^x}{1+x}\right) &= \frac{(1+x)\cdot\frac{d}{dx}(e^x) - e^x\cdot\frac{d}{dx}(1+x)}{(1+x)^2}\\ &= \frac{(1+x)\cdot e^x - e^x\cdot 1}{(1+x)^2}\\ &= \frac{e^x + xe^x - e^x}{(1+x)^2}\\ &= \frac{xe^x}{(1+x)^2}.\end{aligned}$$

EXAMPLE 5 Suppose that the total cost of manufacturing x units of a certain product is given by the function $C(x)$. Then the *average cost per unit*, AC, is defined by

$$\mathrm{AC} = \frac{C(x)}{x}.$$

Recall that *marginal cost*, MC, is defined by

$$\mathrm{MC} = C'(x).$$

Show that at the level of production where the average cost is at a minimum, the average cost equals the marginal cost.

Solution In practice, the marginal cost and average cost curves will have the general shapes shown in Figure 2. The minimum point on the average cost curve will

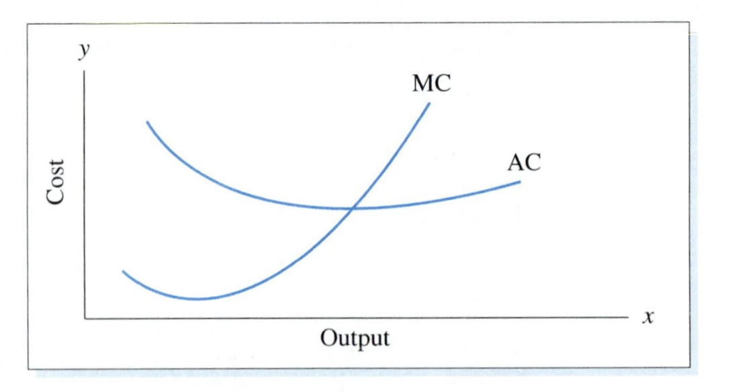

FIGURE 2 Marginal cost and average cost functions

occur when $\frac{d}{dx}(AC) = 0$. To compute the derivative, we need the quotient rule,

$$\frac{d}{dx}(AC) = \frac{d}{dx}\left[\frac{C(x)}{x}\right] = \frac{x \cdot C'(x) - C(x)}{x^2}.$$

Setting the derivative equal to zero and multiplying by x^2, we obtain

$$\begin{aligned} 0 &= x \cdot C'(x) - C(x) \\ C(x) &= x \cdot C'(x) \\ \frac{C(x)}{x} &= C'(x) \\ \text{AC} &= \text{MC}. \end{aligned}$$

Thus, when the output x is chosen so that the average cost is minimized, the average cost equals the marginal cost. ■

At the end of Section 4.1 we saw that if $y = Ce^{kx}$, then y satisfies the differential equation $y' = ky$. Very often an applied problem will involve a function $y = f(x)$ that satisfies the differential equation $y' = ky$. It can be shown that y must then necessarily be an exponential function of the form Ce^{kx}. That is, we have the following result:

> Suppose that $y = f(x)$ satisfies the differential equation
>
> $$y' = ky.$$
>
> Then y is an exponential function of the form
>
> $$Ce^{kx}, \quad C \text{ a constant}.$$

(1)

In Exercise 43 we use the product rule to verify this result.

EXAMPLE 6 Determine all functions $y = f(x)$ such that $y' = -.2y$.

Solution The equation $y' = -.2y$ has the form $y' = ky$ with $k = -.2$. Therefore, any solution of the equation has the form

$$y = Ce^{-.2x},$$

where C is a constant. ■

VERIFICATION OF THE PRODUCT AND QUOTIENT RULES

VERIFICATION OF THE PRODUCT RULE From our discussion of limits we compute the derivative of $f(x)g(x)$ at $x = a$ as the limit

$$\left.\frac{d}{dx}[f(x)g(x)]\right|_{x=a} = \lim_{h \to 0} \frac{f(a+h)g(a+h) - f(a)g(a)}{h}.$$

Let us add and subtract the quantity $f(a)g(a+h)$ in the numerator. After factoring and applying Limit Theorem III, we obtain

$$\lim_{h\to 0} \frac{[f(a+h)g(a+h) - f(a)g(a+h)] + [f(a)g(a+h) - f(a)g(a)]}{h}$$
$$= \lim_{h\to 0} g(a+h) \cdot \frac{f(a+h) - f(a)}{h} + \lim_{h\to 0} f(a) \cdot \frac{g(a+h) - g(a)}{h}.$$

This expression may be rewritten by Limit Theorems I and V as

$$\lim_{h\to 0} g(a+h) \cdot \lim_{h\to 0} \frac{f(a+h) - f(a)}{h} + f(a) \lim_{h\to 0} \frac{g(a+h) - g(a)}{h}.$$

Note, however, that since $g(x)$ is differentiable at $x = a$, it is continuous there, so that $\lim_{h\to 0} g(a+h) = g(a)$. Therefore, the expression above equals

$$g(a)f'(a) + f(a)g'(a).$$

VERIFICATION OF THE QUOTIENT RULE From the general power rule, we know that

$$\frac{d}{dx}\left[\frac{1}{g(x)}\right] = \frac{d}{dx}[g(x)]^{-1} = (-1)[g(x)]^{-2} \cdot g'(x).$$

We can now derive the quotient rule from the product rule:

$$\begin{aligned}
\frac{d}{dx}\left[\frac{f(x)}{g(x)}\right] &= \frac{d}{dx}\left[\frac{1}{g(x)} \cdot f(x)\right] \\
&= \frac{1}{g(x)} \cdot f'(x) + f(x) \cdot \frac{d}{dx}\left[\frac{1}{g(x)}\right] \\
&= \frac{g(x)f'(x)}{[g(x)]^2} + f(x) \cdot (-1)[g(x)]^{-2} \cdot g'(x) \\
&= \frac{g(x)f'(x) - f(x)g'(x)}{[g(x)]^2}.
\end{aligned}$$

PRACTICE PROBLEMS 4.2

1. For what values of x does the curve $y = x^2e^{3x}$ have a horizontal tangent line?
2. Differentiate $y = \dfrac{5}{x^4 - x^3 + 1}$.

EXERCISES 4.2

Differentiate the functions in Exercises 1–12.

1. xe^x
2. $\dfrac{e^x}{x}$
3. $\dfrac{e^{2x}}{1+x}$
4. $(1+x^2)e^x$
5. $\dfrac{1}{xe^x - 1}$
6. $xe^{\frac{x}{2}}$

7. $(2x^3 - 5x)(3x + 1)$
8. $\dfrac{x}{2x + 3}$
9. $\dfrac{x - 2}{x + 2}$
10. $(2x - 1)(x^2 - 3)$
11. $(3x - 4)e^{5x}$
12. $\dfrac{e^{2x}}{1 + e^{3x}}$
13. Find the equation of the tangent line to the curve $y = (1 + x)e^x$ at $x = 0$.
14. Find the equation of the tangent line to the curve $y = \dfrac{x}{1 + e^x}$ at $x = 0$.
15. Find the equation of the tangent line to the curve $y = \dfrac{x}{e^x}$ at $x = 1.1$. (*Note*: Use 3 as the approximate value of $e^{1.1}$.)
16. Find the equation of the tangent line to the curve $y = (2x + 3)e^{-.6x}$ at $x = 2$. (*Note*: Use .3 as the approximate value of $e^{-1.2}$.)
17. The number of bacteria in a culture after t days is given by $f(t) = 900(20 + te^{-.04t})$, for $0 \le t \le 100$. How large is the culture after 75 days? At what rate is the culture changing after 75 days?
18. The cost of producing x units of a certain commodity is $C(x) = 1000 + x^{.5}e^{(.3x+2)}$ dollars. Suppose 4 units of goods are produced. Determine the cost and the marginal cost.
19. The revenue from selling x units of goods is $R(x) = 100x\sqrt{5 - .1x}$ dollars. Determine the revenue and marginal revenue when 10 units of goods are produced.
20. The relationship between the area of the pupil of the eye and the intensity of light was analyzed by B. H. Crawford.* Crawford concluded that the area of the pupil is

$$f(x) = \frac{160x^{-.4} + 94.8}{4x^{-.4} + 15.8} \qquad (0 \le x \le 37)$$

square millimeters when x units of light are entering the eye per unit time. How large is the pupil when 20 units of light are entering the eye per unit time? When 20 units of light are entering the eye per unit time, what is the rate of change of pupil size with respect to a unit change in light intensity?

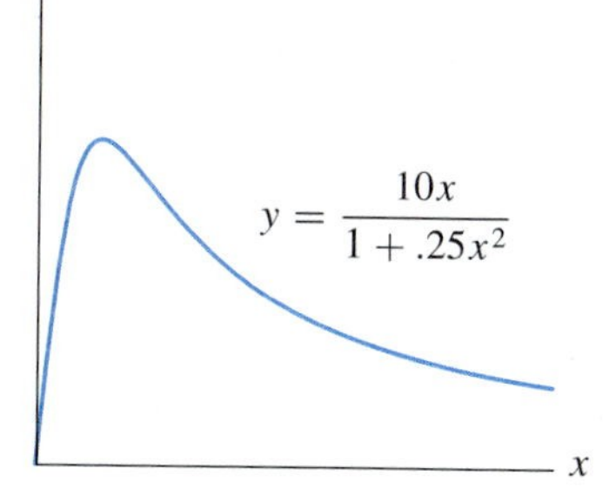

FIGURE 3

21. Figure 3 shows the graph of

$$y = \frac{10x}{1 + .25x^2}$$

for $x \ge 0$. Find the coordinates of the maximum point.
22. Figure 4 shows the graph of $y = xe^{-.2x}$ for $x \ge 0$. Find the coordinates of the maximum point and the inflection point.
23. Figure 5 shows the graph of $y = (5 - 2x)e^{2x}$ for $0 \le x \le 2.5$. Find the x-coordinates of the maximum point and the inflection point.

* "The Dependence of Pupil Size upon the External Light Stimulus Under Static and Variable Conditions," *Proc. Royal Society Series B*, **121** (1937), 376–395.

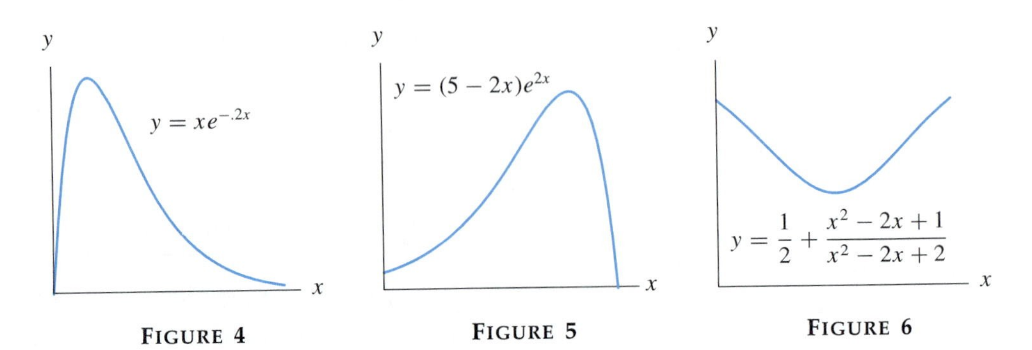

FIGURE 4 FIGURE 5 FIGURE 6

24. Figure 6 shows the graph of

$$y = \frac{1}{2} + \frac{x^2 - 2x + 1}{x^2 - 2x + 2}$$

for $0 \leq x \leq 2$. Find the coordinates of the minimum point.

In Exercises 25–28, determine all solutions of the differential equation.

25. $y' = -4y$

26. $y' = \frac{1}{3}y$

27. $y' = \frac{y}{2}$

28. $y' = -.016y$

29. A sugar refinery can produce x tons of sugar per week at a weekly cost of $.2x^2 + 5x + 2250$ dollars. Find the level of production for which the average cost is at a minimum and show that the average cost equals the marginal cost at that level of production.

30. A cigar manufacturer produces x cases of cigars per day at a daily cost of $50x(x+200)/(x+100)$ dollars. Show that his cost increases and his average cost decreases as the output x increases.

31. Let $R(x)$ be the revenue function from the sale of x units of a product. The *average revenue per unit* is defined by $\text{AR} = R(x)/x$. Show that at the level of production where the average revenue is maximized, the average revenue equals the marginal revenue.

32. Let $s(t)$ be the number of miles a car travels in t hours. Then the average velocity during the first t hours is $\overline{v}(t) = s(t)/t$ miles per hour. Suppose that the average velocity is maximized at time t_0. Show that at this time the average velocity $\overline{v}(t_0)$ equals the instantaneous velocity $s'(t_0)$. [*Hint*: Compute the derivative of $\overline{v}(t)$.] See Example 5.

33. The width of a rectangle is increasing at a rate of 3 inches per second and its length is increasing at the rate of 4 inches per second. At what rate is the area of the rectangle increasing when its width is 5 inches and its length is 6 inches? [*Hint*: Let $W(t)$ and $L(t)$ be the widths and lengths, respectively, at time t.]

34. A company plans to decrease the amount of sulfur dioxide escaping from its smokestacks. The estimated cost-benefit function is

$$f(x) = \frac{3x}{105 - x}, \quad 0 \leq x \leq 100,$$

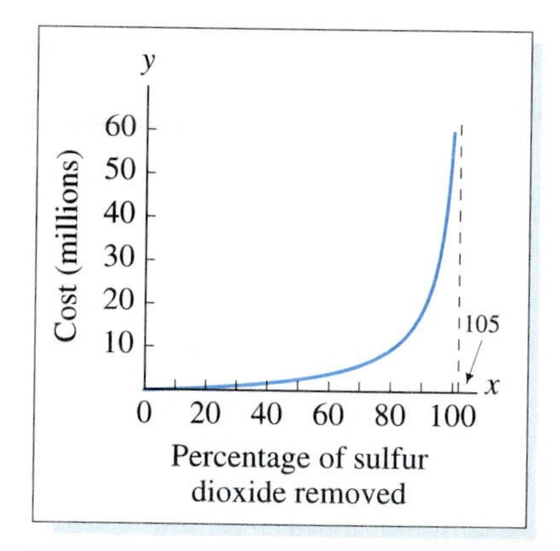

FIGURE 7 A cost-benefit function

where $f(x)$ is the cost in millions of dollars for eliminating x percent of the total sulfur dioxide. (See Figure 7.) Find the value of x at which the rate of increase of the cost-benefit function is 1.4 million dollars per unit. (Each unit is 1 percentage point increase in pollutant removed.)

35. Show that the derivative of $\frac{x^4}{x^2+1}$ is not $\frac{4x^3}{2x}$.

36. Show that the derivative of $(5x^3)(2x^4)$ is not $(15x^2)(8x^3)$.

37. Suppose that $f(x)$ is a function whose derivative is $f'(x) = \frac{1}{1+x^2}$. Find the derivative of $\frac{f(x)}{1+x^2}$.

38. Suppose that $f(x)$ and $g(x)$ are differentiable functions such that $f(1) = 2$, $f'(1) = 3$, $g(1) = 4$, and $g'(1) = 5$. Find $\left.\frac{d}{dx}[f(x)g(x)]\right|_{x=1}$.

39. Consider the functions of Exercise 38. Find $\left.\frac{d}{dx}\left[\frac{f(x)}{g(x)}\right]\right|_{x=1}$.

40. Suppose that $f(x)$ is a function whose derivative is $f'(x) = 1/x$. Find the derivative of $xf(x) - x$.

41. Let $f(x) = 1/x$ and $g(x) = x^3$.

(a) Show that the product rule yields the correct derivative of $(1/x)x^3 = x^2$.

(b) Compute the product $f'(x)g'(x)$ and note that it is *not* the derivative of $f(x)g(x)$.

42. The derivative of $(x^3 - 4x)/x$ is obviously $2x$ for $x \neq 0$, because $(x^3 - 4x)/x = x^2 - 4$ for $x \neq 0$. Verify that the quotient rule gives the same derivative.

43. Verify the result (1). [*Hint*: Let $g(x) = f(x)e^{-kx}$. Show that $g'(x) = 0$. You may assume that only a constant function has a zero derivative.]

In Exercises 44 and 45, use the fact that in 1991 the population of the United States was 252,160,000 people and growing at the rate of 3,750,000 people per year.

44. In 1991, the per capita consumption of gasoline in the United States was 52.3 gallons and growing at the rate of .2 gallons per year. At what rate was the total consumption of gasoline in the United States increasing in 1991?

45. In 1991, 11,801 million pints of ice cream were consumed in the United States, and the quantity consumed was growing at the rate of 212 million pints per year. At what rate was the per capita consumption of ice cream increasing in 1991?

In the remaining exercises, use a graphing utility to obtain the graphs of the functions.

46. Graph $f(x) = xe^x$ and $g(x) = x + x^2 + .5x^3$ together in the window $[-2, 2]$ *by* $[-2, 8]$. For what values of x are the two graphs indistinguishable? Show that at $x = 0$ the two functions have the same values for their first, second, and third derivatives.

In Exercises 47–50, differentiate $f(x)$ and then compare the graphs of $f(x)$ and $f'(x)$ in the given window to check that the expression for $f'(x)$ is reasonable.

47. $f(x) = xe^{-x}$, $[0, 4]$ *by* $[-.2, 1]$

48. $f(x) = x^3e^{-x}$, $[0, 8]$ *by* $[-1, 2]$

49. $f(x) = \dfrac{e^x - e^{-x}}{2}$, $[-2, 2]$ *by* $[-4, 4]$

50. $f(x) = (x + 1)e^{-x}$, $[-1, 3]$ *by* $[-1, 2]$

51. The *Laffer curve*, named after the American economist Arthur Laffer, gives the relationship between tax rates and government revenues. As tax rates increase from zero, revenues increase until a maximum is reached. But if tax rates are further increased, they discourage government spending and business investment, thereby reducing revenues. Suppose the Laffer curve for a country is

$$f(x) = \frac{10x(100 - x)}{50 + x}.$$

That is, if the tax rate is x percent, then the revenue from taxes is $f(x)$ million dollars.

(a) Graph $f(x)$ with the window $[0, 100]$ *by* $[0, 300]$.

(b) How much revenue is generated when the tax rate is 30%?

(c) For what tax rate(s) is the revenue 150 million dollars?

(d) If the tax rate is currently 50%, what is the rate of change of revenue with respect to change in the tax rate?

(e) For what tax rate is the rate of change of revenue with respect to change in the tax rate 6.5 million dollars per percent?

(f) Find the tax rate that results in the greatest revenue. Differentiate $f(x)$ and solve $f'(x) = 0$ to find the exact value of the tax rate.

SOLUTIONS TO PRACTICE PROBLEMS 4.2

1. Let $f(x) = x^2e^{3x}$. From Example 2, $f'(x) = (3x^2+2x)e^{3x}$. Set $f'(x) = 0$ and solve for x. Since e^{3x} is never zero, $f'(x)$ will be zero only if $3x^2 + 2x = 0$. Now $3x^2 + 2x = x(3x + 2) = 0$ if $x = 0$ or $x = -\frac{2}{3}$. Therefore, the curve has a horizontal tangent line when $x = 0$ and when $x = -\frac{2}{3}$.

2. At first glance, differentiation appears to require the quotient rule. However, if we rewrite the function as $y = 5(x^4 - x^3 + 1)^{-1}$, we can differentiate it with the general power rule:

$$y' = -5(x^4 - x^3 + 1)^{-2}(4x^3 - 3x^2) = \frac{-20x^3 + 15x^2}{(x^4 - x^3 + 1)^2}.$$

4.3 THE CHAIN RULE

Although we can now differentiate many types of functions, certain functions built from exponential functions still evade us. One such function is the well-known *normal* function from statistics: $f(x) = \frac{1}{\sqrt{2\pi}} e^{-\frac{1}{2}x^2}$. In this section we develop a powerful differentiation technique called the chain rule, which allows us to differentiate the normal function and many other functions that arise in applications. Actually, the familiar general power rule is a special case of the chain rule.

A useful way of combining functions $f(x)$ and $g(x)$ is to replace each occurrence of the variable x in $f(x)$ by the function $g(x)$. The resulting function is called the *composition* (or composite) of $f(x)$ and $g(x)$ and is denoted $f(g(x))$.*

EXAMPLE 1 Let $f(x) = e^{3x}$, $g(x) = x^2$. Find

(a) $f(g(x))$ (b) $g(f(x))$.

Solution (a) Replace the x in $f(x)$ by $g(x)$ to obtain

$$f(g(x)) = e^{3 \cdot g(x)} = e^{3x^2}.$$

(b) Replace the x in $g(x)$ by $f(x)$ to obtain

$$g(f(x)) = (f(x))^2 = (e^{3x})^2 = e^{6x}.$$

■

EXAMPLE 2 Write the following functions as composites of simpler functions.

(a) $h(x) = (x^5 + 9x + 3)^8$
(b) $k(x) = \sqrt{4x^2 + 1}$

Solution (a) $h(x) = f(g(x))$, where the outside function is the power function, $(\ldots)^8$, that is, $f(x) = x^8$. Inside this power function is $g(x) = x^5 + 9x + 3$.
(b) $k(x) = f(g(x))$, where the outside function is the square root function, $f(x) = \sqrt{x}$, and the inside function is $g(x) = 4x^2 + 1$. ■

A function of the form $[g(x)]^r$ is a composite $f(g(x))$, where the outside function is $f(x) = x^r$. We have already given a rule for differentiating this function—namely,

$$\frac{d}{dx}[g(x)]^r = r[g(x)]^{r-1}g'(x).$$

The *chain rule* has the same form, except that the outside function $f(x)$ can be *any* differentiable function.

> *The Chain Rule* To differentiate $f(g(x))$, first differentiate the outside function $f(x)$ and substitute $g(x)$ for x in the result. Then multiply by the derivative of the inside function $g(x)$. Symbolically,
>
> $$\frac{d}{dx}f(g(x)) = f'(g(x))g'(x).$$

EXAMPLE 3 Use the chain rule to compute the derivative of $f(g(x))$, where $f(x) = e^{3x}$ and $g(x) = x^2$.

* See Section 1.6 for additional information on function composition.

Solution

$$f'(x) = 3e^{3x}, \qquad g'(x) = 2x,$$
$$f'(g(x)) = 3e^{3x^2}.$$

Finally, by the chain rule,

$$\begin{aligned}\frac{d}{dx} f(g(x)) &= f'(g(x))g'(x)\\ &= 3e^{3x^2} \cdot 2x\\ &= 6xe^{3x^2}.\end{aligned}$$

Since $f(g(x))$ is e^{3x^2}, we have

$$\frac{d}{dx} e^{3x^2} = 6xe^{3x^2}.$$

■

The function differentiated in Example 3 also could have been written as $e^{g(x)}$, where $g(x) = 3x^2$. Let's apply the chain rule to the general case $f(g(x))$, where $f(x) = e^x$. By the chain rule,

$$\begin{aligned}\frac{d}{dx}(e^{g(x)}) &= f'(g(x))g'(x)\\ &= f(g(x))g'(x) \qquad \text{[since } f'(x) = f(x)\text{]}\\ &= e^{g(x)}g'(x).\end{aligned}$$

So we have the following result:

Chain Rule for Exponential Functions Let $g(x)$ be any differentiable function. Then

$$\frac{d}{dx}\left(e^{g(x)}\right) = e^{g(x)}g'(x). \tag{1}$$

EXAMPLE 4 Differentiate e^{3x^2} using the rule in (1).

Solution Here $g(x) = 3x^2$, $g'(x) = 6x$, so

$$\begin{aligned}\frac{d}{dx}(e^{3x^2}) &= e^{3x^2} \cdot 6x\\ &= 6xe^{3x^2}.\end{aligned}$$

■

EXAMPLE 5 Differentiate $e^{x-\frac{1}{x}}$.

Solution

$$\begin{aligned}\frac{d}{dx}\left(e^{x-\frac{1}{x}}\right) &= e^{x-\frac{1}{x}} \cdot \frac{d}{dx}\left(x - \frac{1}{x}\right)\\ &= e^{x-\frac{1}{x}}\left(1 + \frac{1}{x^2}\right).\end{aligned}$$

■

EXAMPLE 6 Use (1) to differentiate e^{kx}.

Solution $$\frac{d}{dx}(e^{kx}) = e^{kx}\cdot\frac{d}{dx}(kx) = e^{kx}\cdot k = ke^{kx}.$$ ■

VERIFICATION OF THE CHAIN RULE Suppose that $f(x)$ and $g(x)$ are differentiable, and let $x = a$ be a number in the domain of $f(g(x))$. Since every differentiable function is continuous, we have

$$\lim_{h\to 0} g(a+h) = g(a),$$

which implies that

$$\lim_{h\to 0}[g(a+h) - g(a)] = 0. \tag{2}$$

Now $g(a)$ is a number in the domain of f, and the limit definition of the derivative gives us

$$f'(g(a)) = \lim_{k\to 0}\frac{f(g(a)+k) - f(g(a))}{k}. \tag{3}$$

Let $k = g(a+h) - g(a)$. By equation (2), k approaches zero as h approaches zero. Also, $g(a+h) = g(a) + k$. Therefore, (3) may be rewritten in the form

$$f'(g(a)) = \lim_{h\to 0}\frac{f(g(a+h)) - f(g(a))}{g(a+h) - g(a)}. \tag{4}$$

[Strictly speaking, we must assume that the denominator in (4) is never zero. This assumption may be avoided by a somewhat different and more technical argument, which we omit.] Finally, we show that the function $f(g(x))$ has a derivative at $x = a$. We use the limit definition of the derivative, Limit Theorem V, and (4) above.

$$\begin{aligned}
\frac{d}{dx}[f(g(x))]\bigg|_{x=a} &= \lim_{h\to 0}\frac{f(g(a+h)) - f(g(a))}{h}\\
&= \lim_{h\to 0}\left[\frac{f(g(a+h)) - f(g(a))}{g(a+h) - g(a)}\cdot\frac{g(a+h) - g(a)}{h}\right]\\
&= \lim_{h\to 0}\frac{f(g(a+h)) - f(g(a))}{g(a+h) - g(a)}\cdot\lim_{h\to 0}\frac{g(a+h) - g(a)}{h}\\
&= f'(g(a))\cdot g'(a).
\end{aligned}$$

PRACTICE PROBLEMS 4.3

Consider the function $h(x) = (2x^3 - 5)^5 + e^{2x^3-5}$.

1. Write $h(x)$ as a composite function, $f(g(x))$.
2. Compute $f'(x)$ and $f'(g(x))$.
3. Use the chain rule to differentiate $h(x)$.

EXERCISES 4.3

Compute $f(g(x))$ and $g(f(x))$, where $f(x)$ and $g(x)$ are the following.

1. $f(x) = e^x$, $g(x) = 5x$

2. $f(x) = 3x - 1$, $g(x) = e^{-x}$

3. $f(x) = x\sqrt{x+1}$, $g(x) = e^{2x}$

4. $f(x) = \dfrac{1}{x}$, $g(x) = e^x$

Each function below may be viewed as a composite function $f(g(x))$. Find $f(x)$ and $g(x)$.

5. $(1 + e^x)^3$

6. $\dfrac{2}{x + e^x}$

7. $\dfrac{e^{3x} + 1}{e^{3x} - 1}$

8. $\sqrt{x^2 + 1}$

9. $(5e^x + 1)^{-1/2}$

10. e^{e^x}

Differentiate the functions in Exercises 11–22 using one or more of the differentiation rules discussed thus far.

11. e^{x^2+3x}

12. e^{x-2x^3}

13. xe^{x^2}

14. $\dfrac{e^{x^2}}{x}$

15. $5e^{3+\frac{2}{x}}$

16. $4e^{\sqrt{x}}$

17. $2e^{(x-1)/x}$

18. $5x + e^2$

19. $e^3 \cdot x^2$

20. $(e^3)^x$

21. e^{1+e^x}

22. $e^{2e^{3x}}$

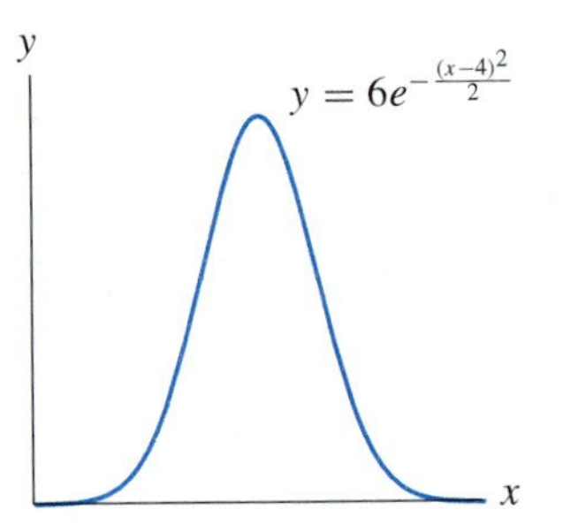

FIGURE 1

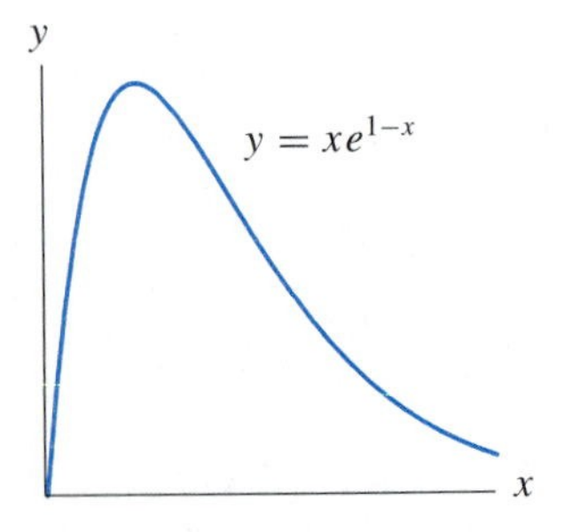

FIGURE 2

23. Figure 1 contains the graph of $y = 6e^{-\frac{(x-4)^2}{2}}$ for $0 \le x \le 8$. Find the coordinates of the maximum point and the inflection points.

24. Figure 2 contains the graph of $y = xe^{1-x}$ for $0 \le x \le 5$. Find the coordinates of the maximum point and the inflection point.

25. Find the equation of the tangent line to the curve $y = 5e^{x^2-3x}$ at the point $(3, 5)$.

26. Find the equation of the tangent line to the curve $y = 4e^{3x-\frac{12}{x}}$ at the point $(2, 4)$.

27. Let a and b be positive numbers. A curve whose equation is $y = e^{-ae^{-bx}}$ is called a *Gompertz growth curve*. These curves are used in biology to describe certain types of population growth. Compute the derivative of $y = e^{-2e^{-.01x}}$.

28. In an expression of the form $f(g(x))$, $f(x)$ is called the *outer* function and $g(x)$ is called the *inner* function. Verbalize the chain rule using the words "inner" and "outer."

In the remaining exercises, use a graphing utility to obtain the graphs of the functions. In Exercises 29–32, differentiate $f(x)$ and then compare the graphs of $f(x)$ and $f'(x)$ in the given window to check that the expression for $f'(x)$ is reasonable.

29. $f(x) = e^{\left(\frac{1}{1+x^2}\right)}$; $[-4, 4]$ *by* $[-2, 3]$

30. $f(x) = x^2e^{-x^2}$; $[0, 4]$ *by* $[-.5, .75]$

31. $f(x) = \sqrt{x}\, e^{-x^2}$; $[0, 2]$ *by* $[-.75, .75]$

32. $f(x) = \dfrac{e^{x^2}}{x^2}$; $[0, 2]$ *by* $[-15, 15]$

33. (Analysis of the growth of a tumor) A cancerous tumor has volume $f(t) = 1.825^3(1 - 1.6e^{-.4196t})^3$ milliliters after t weeks, with $t \geq 2$.*
   (a) Graph $f(t)$, $f'(t)$, $f''(t)$ for $2 \leq t \leq 15$.
   (b) Approximately how large is the tumor after 14 weeks?
   (c) Approximately when will the tumor have a volume of 4 milliliters?
   (d) Approximately how fast is the tumor growing after 5 weeks?
   (e) Approximately when is the tumor growing at the rate of .38 milliliters per week?
   (f) Approximately when is the tumor growing at the fastest rate?

SOLUTIONS TO PRACTICE PROBLEMS 4.3

1. Let $f(x) = x^5 + e^x$ and $g(x) = 2x^3 - 5$.
2. $f'(x) = 5x^4 + e^x$, $f'(g(x)) = 5(2x^3 - 5)^4 + e^{2x^3-5}$.
3. We have $g'(x) = 6x^2$. Then, from the chain rule and the result of Problem 2, we have

$$h'(x) = f'(g(x))g'(x) = [5(2x^3 - 5)^4 + e^{2x^3-5}](6x^2).$$

4.4 THE NATURAL LOGARITHM FUNCTION

As a preparation for the definition of the natural logarithm, we shall make a geometrical digression. In Figure 1 we have plotted several pairs of points. Observe how they are related to the line $y = x$.

The points $(5, 7)$ and $(7, 5)$, for example, are the same distance from the line $y = x$. If we were to plot the point $(5, 7)$ with wet ink and then fold the page along the line $y = x$, the ink would produce a second blot at the point $(7, 5)$. If we think of the line $y = x$ as a mirror, then $(7, 5)$ is the mirror image of $(5, 7)$. We say that $(7, 5)$ is the *reflection* of $(5, 7)$ through the line $y = x$. Similarly, $(5, 7)$ is the reflection of $(7, 5)$ through the line $y = x$.

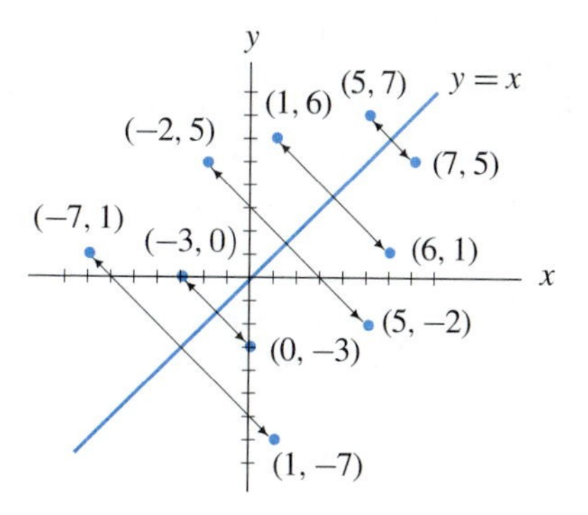

FIGURE 1 Reflections of points through the line $y = x$

Now let us consider all points lying on the graph of the exponential function $y = e^x$ [see Figure 2(a)]. If we reflect each such point through the line $y = x$, we obtain a new graph [see Figure 2(b)]. For each positive x, there is exactly one value of y such that (x, y) is on the new graph. We call this value of y the *natural logarithm of* x, denoted $\ln x$. Thus the reflection of the graph of $y = e^x$ through the line $y = x$ is the graph of the natural logarithm function $y = \ln x$.

We may deduce some properties of the natural logarithm function from an inspection of its graph:

* Baker, Goddard, Clark, and Whimster, "Proportion of Necrosis in Transplanted Murine Adenocarcinomas and Its Relationship to Tumor Growth," *Growth, Development, and Aging*, **54** (1990), 85–93.

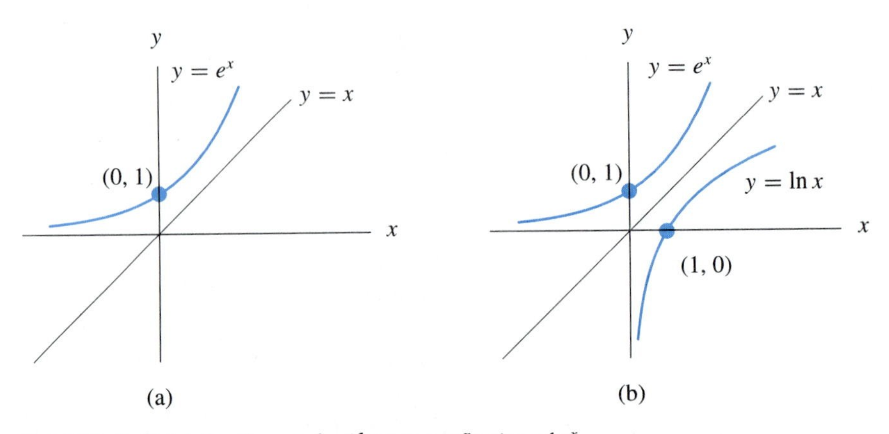

FIGURE 2 Obtaining the graph of $\ln x$ as a reflection of e^x

1. The point $(1, 0)$ is on the graph of $y = \ln x$ [because $(0, 1)$ is on the graph of $y = e^x$]. In other words,

$$\boxed{\ln 1 = 0.} \tag{1}$$

2. $\ln x$ is defined only for positive values of x.
3. $\ln x$ is negative for x between 0 and 1.
4. $\ln x$ is positive for x greater than 1.
5. $\ln x$ is an increasing, concave down function.

Let us study the relationship between the natural logarithm and exponential functions more closely. From the way in which the graph of $\ln x$ was obtained, we know that (a, b) is on the graph of $\ln x$ if and only if (b, a) is on the graph of e^x. However, a typical point on the graph of $\ln x$ is of the form $(a, \ln a)$, $a > 0$. So for any positive value of a, the point $(\ln a, a)$ is on the graph of e^x. That is,

$$e^{\ln a} = a.$$

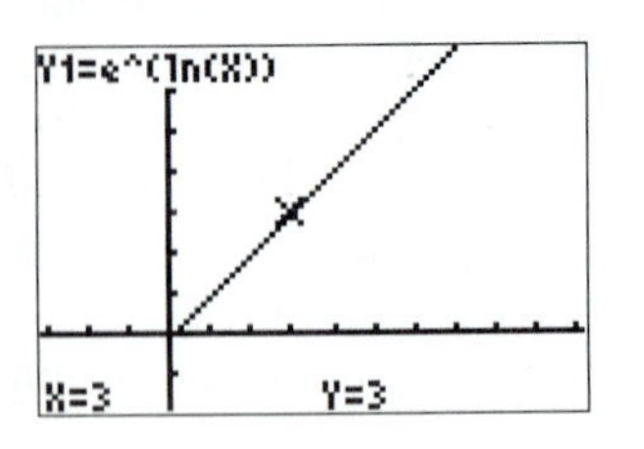

Since a was an arbitrary positive number, we have the following important relationship between the natural logarithm and exponential functions:

$$\boxed{e^{\ln x} = x, \qquad \text{for } x > 0.} \tag{2}$$

Equation (2) can be put into verbal form:

> For each positive number x, $\ln x$ is that exponent to which we must raise e in order to get x.

If b is any number, then e^b is positive and hence $\ln(e^b)$ makes sense. What is $\ln(e^b)$? Since (b, e^b) is on the graph of e^x, we know that (e^b, b) must be on the graph of $\ln x$. That is, $\ln(e^b) = b$. Thus we have shown that

$$\boxed{\ln(e^x) = x, \qquad \text{for any } x.} \tag{3}$$

The identities (2) and (3) express the fact that the natural logarithm function is the *inverse* of the exponential function (for $x > 0$). For instance, if we take a number x and compute e^x, then by (3), we can undo the effect of the exponentiation by taking the natural logarithm; that is, the logarithm of e^x equals the original number x. Similarly, if we take a positive number x and compute $\ln x$, then by (2), we can undo the effect of the logarithm by raising e to the $\ln x$ power; that is, $e^{\ln x}$ equals the original number x.

For a discussion of using $\ln x$ with Visual Calculus, see Appendix A, Part VII.

USING $\ln x$ WITH A TI CALCULATOR

The function $\ln x$ is invoked with the [LN] button. To graph the function, set Y_1 (or `y1`) equal to $\ln x$ with the following keystrokes:

(TI-82) [LN] [X, T, θ]

(TI-83) [LN] [X, T, θ, n] [)]

(TI-85) [LN] [x-VAR]

The screen will show either `Y1 = lnX`, `Y1 = ln(X)`, or `y1=lnx`.

Values of $\ln x$ can be calculated with the push of a button. To evaluate the function, press [LN], type in the number (followed by a parenthesis when using a TI-83), and then press [ENTER]. See Figure 3.

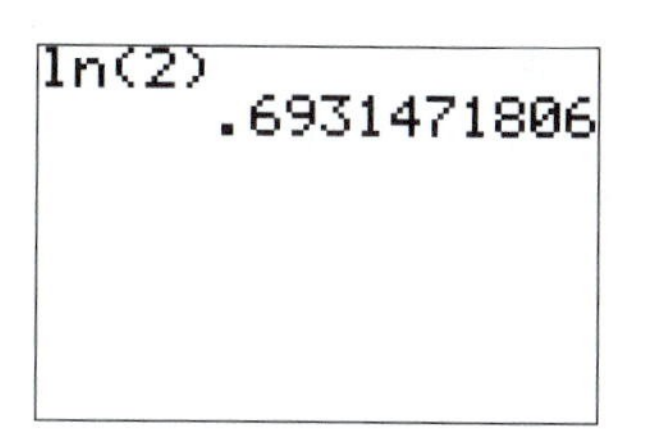

FIGURE 3

EXAMPLE 1 Solve the equation $5e^{x-3} = 4$ for x.

Solution First divide each side by 5:

$$e^{x-3} = .8.$$

Taking the logarithm of each side and using (3), we have

$$\begin{aligned} \ln(e^{x-3}) &= \ln .8 \\ x - 3 &= \ln .8 \\ x &= 3 + \ln .8. \end{aligned}$$

[If desired, the numerical value of x can be obtained with a graphing utility, namely, $x = 3 - .22314 = 2.77686$ (to five decimal places).] ■

EXAMPLE 2 Solve the equation $2\ln x + 7 = 0$ for x.

Solution

$$\begin{aligned} 2\ln x &= -7 \\ \ln x &= -3.5 \\ e^{\ln x} &= e^{-3.5} \\ x &= e^{-3.5} \qquad \text{[by (2)].} \end{aligned}$$

■

OTHER EXPONENTIAL AND LOGARITHMIC FUNCTIONS In our discussion of the exponential function, we mentioned that all exponential functions of the form b^x, where b is a fixed positive number, can be expressed in terms of *the* exponential function e^x. Now we can be quite explicit. For since $b = e^{\ln b}$, we see that

$$b^x = (e^{\ln b})^x = e^{(\ln b)x}.$$

Hence we have shown that

$$\boxed{b^x = e^{kx}, \qquad \text{where } k = \ln b.}$$

The natural logarithm function is sometimes called the *logarithm to the base* e, for it is the inverse of the exponential function e^x. If we reflect the graph of the function $y = 2^x$ through the line $y = x$, we obtain the graph of a function called the *logarithm to the base* 2, denoted by $\log_2 x$. Similarly, if we reflect the graph of $y = 10^x$ through the line $y = x$, we obtain the graph of a function called the *logarithm to the base* 10, denoted by $\log_{10} x$. (See Figure 4.)

Logarithms to the base 10 are sometimes called *common* logarithms. Common logarithms are usually introduced into algebra courses for the purpose of simplifying certain arithmetic calculations. However, with the advent of computers and calculators, the need for common logarithms has diminished. It can be shown that

$$\log_{10} x = \frac{1}{\ln 10} \cdot \ln x,$$

so that $\log_{10} x$ is simply a constant multiple of $\ln x$. However, we shall not need this fact. Common logarithms are calculated on calculators with the LOG key and in Visual Calculus with the function log.

The natural logarithm function is used in calculus because differentiation and integration formulas are simpler than for $\log_{10} x$ or $\log_2 x$, and so on. (Recall

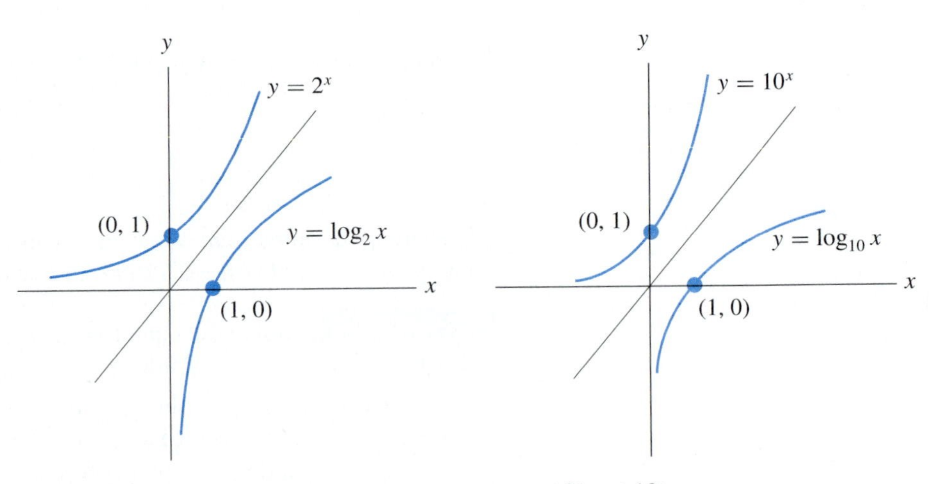

FIGURE 4 Graphs of $\log_2 x$ and $\log_{10} x$ as reflections of 2^x and 10^x

that we prefer the function e^x over the functions 10^x and 2^x for the same reason.) Also, $\ln x$ arises "naturally" in the process of solving certain differential equations that describe various growth processes.

PRACTICE PROBLEMS 4.4

1. Find $\ln e$.
2. Solve $e^{-3x} = 2$ using the natural logarithm function.

EXERCISES 4.4

1. Find $\ln(1/e)$.
2. Find $\ln(\sqrt{e})$.
3. If $e^{-x} = 1.7$, write x in terms of the natural logarithm.
4. If $e^x = 3.5$, write x in terms of the natural logarithm.
5. If $\ln x = 2.2$, write x using the exponential function.
6. If $\ln x = -5.7$, write x using the exponential function.

Simplify the following expressions.

7. $\ln e^2$
8. $e^{\ln 1.37}$
9. $e^{e^{\ln 1}}$
10. $\ln(e^{.73 \ln e})$
11. $e^{5 \ln 1}$
12. $\ln(\ln e)$

Solve the following equations for x.

13. $e^{2x} = 5$
14. $e^{3x-1} = 4$
15. $\ln(4 - x) = \frac{1}{2}$
16. $\ln 3x = 2$
17. $\ln x^2 = 6$
18. $e^{x^2} = 7$
19. $6e^{-.00012x} = 3$
20. $2 - \ln x = 0$
21. $\ln 5x = \ln 3$
22. $\ln(x^2 - 3) = 0$
23. $\ln(\ln 2x) = 0$
24. $3 \ln x = 8$
25. $2e^{x/3} - 9 = 0$
26. $4 - 3e^{x+6} = 0$
27. $300e^{.2x} = 1800$
28. $750e^{-.4x} = 375$
29. $e^{5x} \cdot e^{\ln 5} = 2$
30. $e^{x^2-5x+6} = 1$
31. $4e^x \cdot e^{-2x} = 6$
32. $(e^x)^2 \cdot e^{2-3x} = 4$

In Exercises 33–36, find the coordinates of each relative extreme point of the given function and determine if the point is a relative maximum point or a relative minimum point.

33. $f(x) = e^{-x} + 3x$
34. $f(x) = 5x - 2e^x$
35. $f(x) = \frac{1}{3}e^{2x} - x + \frac{1}{2}\ln\frac{3}{2}$
36. $f(x) = 5 - \frac{1}{2}x - e^{-3x}$
37. When a drug or vitamin is administered intramuscularly (into a muscle), the concentration in the blood at time t after the injection can be approximated by a function of the form $f(t) = c(e^{-k_1 t} - e^{-k_2 t})$. The graph of $f(t) = 5(e^{-.01t} - e^{-.51t})$, for $t \geq 0$, has the general shape shown in Figure 14 of Section 1.1. Find the value of t at which this function reaches its maximum value.
38. Under certain geographic conditions, the wind velocity v at a height x centimeters above the ground is given by $v = K \ln(x/x_0)$, where K is a positive constant (depending on the air density, average wind velocity, etc.), and x_0 is a roughness parameter (depending on the roughness of the vegetation on

the ground).* Suppose that $x_0 = .7$ centimeters (a value that applies to lawn grass 3 centimeters high) and $K = 300$ centimeters per second.

(a) At what height above the ground is the wind velocity zero?

(b) At what height is the wind velocity 1200 centimeters per second?

39. Find k such that $2^x = e^{kx}$ for all x.

40. Find k such that $3^x = e^{kx}$ for all x.

In the remaining exercises, use a graphing utility to obtain the graphs of the functions.

41. Graph $y = e^x$ and $y = 5$ together and determine the x-coordinate of their point of intersection. Express this number in terms of a logarithm.

42. Use the technique of Exercise 41 to estimate $\log_{10} 6$.

43. Graph $y = \ln x$ and $y = 2$ together and determine the x-coordinate of their point of intersection. Express this number in terms of a power of e.

44. Use the technique of Exercise 43 to estimate $e^{1.5}$.

45. The national health expenditures (in billions of dollars) from 1960 to 1994 are approximately given by $f(t) = 27e^{.1t}$, where year 0 corresponds to 1960. [*Note*: The answers to parts (b)–(e) can be obtained either algebraically or from the graphs. Try both methods.]

(a) Graph $f(t)$ and $f'(t)$.

(b) How much money was spent in 1976?

(c) How fast were expenditures rising in 1980?

(d) When did expenditures reach 375 billion dollars?

(e) When were expenditures rising at the rate of 6 billion dollars per year?

46. The atmospheric pressure at an altitude of x kilometers is $f(x) = 1035e^{-.12x}$ grams per square centimeter. [*Note*: The answers to parts (b)–(e) can be obtained either algebraically or from the graphs. Try both methods.]

(a) Graph $f(x)$ and $f'(x)$.

(b) What is the pressure at an altitude of 4 km?

(c) At what altitude is the pressure 568 grams per square centimeter?

(d) At an altitude of 24 km, at what rate is the atmospheric pressure changing with respect to change in altitude?

(e) At what altitude is -10 grams per square cm per km the rate of change in atmospheric pressure with respect to change in altitude?

47. The speed of a parachutist during free fall after x seconds is $f(x) = 200(1 - e^{-.16t})$ feet per second. [*Note*: The answers to parts (b)–(e) can be obtained either algebraically or from the graphs. Try both methods.]

(a) Graph $f(x)$ and $f'(x)$.

(b) How fast is the parachutist falling after 25 seconds?

(c) What is the parachutist's acceleration after 10 seconds?

---

* G. Cox, B. Collier, A. Johnson, and P. Miller, *Dynamic Ecology* (Englewood Cliffs, N. J.: Prentice-Hall, Inc., 1973), pp. 113–115.

(d) After approximately how many seconds is the parachutist falling at the speed of 56 feet per second?

(e) Approximately when is the parachutist accelerating at a rate of 4 feet per second?

(f) What is the terminal speed; that is, the greatest speed that can be approached?

48. $\ln x$ grows slower than any function of the form $\sqrt[n]{x}$. The following steps demonstrate this fact for $n = 3$ and $n = 4$.

(a) Graph $\ln x$ and $\sqrt[3]{x}$ together in the window [0, 200] *by* [0, 6]. From what value of x on is $\sqrt[3]{x}$ greater than $\ln x$?

(b) Show that $\dfrac{\ln x}{\sqrt[3]{x}}$ approaches zero as x gets large.

(c) Repeat (a) for $\ln x$ and $\sqrt[4]{x}$ in the window [0, 10000] *by* [0, 10].

(d) Show that $\dfrac{\ln x}{\sqrt[4]{x}}$ approaches zero as x gets large.

SOLUTIONS TO PRACTICE PROBLEMS 4.4

1. Answer: 1. The number $\ln e$ is that exponent to which e must be raised in order to obtain e.

2. Take the logarithm of each side and use (3) to simplify the left side:

$$\begin{aligned} \ln e^{-3x} &= \ln 2 \\ -3x &= \ln 2 \\ x &= -\frac{\ln 2}{3}. \end{aligned}$$

4.5 THE DERIVATIVE OF $\ln x$

Let us now compute the derivative of $\ln x$ for $x > 0$. Since $e^{\ln x} = x$, we have

$$\frac{d}{dx}(e^{\ln x}) = \frac{d}{dx}(x) = 1. \tag{1}$$

On the other hand, if we differentiate $e^{\ln x}$ by the chain rule, we find that

$$\frac{d}{dx}(e^{\ln x}) = e^{\ln x} \cdot \frac{d}{dx}(\ln x) = x \cdot \frac{d}{dx}(\ln x), \tag{2}$$

where the last equality used the fact that $e^{\ln x} = x$.

By combining equations (1) and (2) we obtain

$$x \cdot \frac{d}{dx}(\ln x) = 1.$$

In other words,

$$\frac{d}{dx}(\ln x) = \frac{1}{x}, \qquad x > 0. \tag{3}$$

By combining this differentiation formula with the chain rule, product rule, and quotient rule, we may differentiate many functions involving $\ln x$.

EXAMPLE 1 Differentiate.

(a) $(\ln x)^5$ (b) $x \ln x$ (c) $\ln(x^3 + 5x^2 + 8)$

Solution (a) By the general power rule,

$$\begin{aligned}\frac{d}{dx}(\ln x)^5 &= 5(\ln x)^4 \cdot \frac{d}{dx}(\ln x)\\ &= 5(\ln x)^4 \cdot \frac{1}{x}\\ &= \frac{5(\ln x)^4}{x}.\end{aligned}$$

(b) By the product rule,

$$\begin{aligned}\frac{d}{dx}(x \ln x) &= x \cdot \frac{d}{dx}(\ln x) + (\ln x) \cdot 1\\ &= x \cdot \frac{1}{x} + \ln x\\ &= 1 + \ln x.\end{aligned}$$

(c) By the chain rule,

$$\begin{aligned}\frac{d}{dx}\ln(x^3 + 5x^2 + 8) &= \frac{1}{x^3 + 5x^2 + 8} \cdot \frac{d}{dx}(x^3 + 5x^2 + 8)\\ &= \frac{3x^2 + 10x}{x^3 + 5x^2 + 8}.\end{aligned}$$

Let $g(x)$ be any differentiable function. For any value of x for which $g(x)$ is positive, the function $\ln(g(x))$ is defined. For such a value of x, the derivative is given by the chain rule as

$$\frac{d}{dx}[\ln g(x)] = \frac{1}{g(x)} \cdot \frac{d}{dx}g(x) = \frac{g'(x)}{g(x)}. \tag{4}$$

Example 1(c) illustrates a special case of this formula.

EXAMPLE 2 The function $f(x) = (\ln x)/x$ has a relative extreme point for some $x > 0$. Find the point, and determine whether it is a relative maximum or a relative minimum point.

Solution By the quotient rule,

$$f'(x) = \frac{x \cdot \frac{1}{x} - \ln x \cdot 1}{x^2} = \frac{1 - \ln x}{x^2},$$

$$f''(x) = \frac{x^2\left(-\frac{1}{x}\right) - (1 - \ln x)(2x)}{x^4} = \frac{2\ln x - 3}{x^3}.$$

If we set $f'(x) = 0$, then

$$\begin{aligned} 1 - \ln x &= 0 \\ \ln x &= 1 \\ e^{\ln x} &= e^1 = e \\ x &= e. \end{aligned}$$

Therefore, the only possible relative extreme point is at $x = e$. When $x = e$, $f(e) = (\ln e)/e = 1/e$. Furthermore,

$$f''(e) = \frac{2\ln e - 3}{e^3} = -\frac{1}{e^3} < 0,$$

which implies that the graph of $f(x)$ is concave down at $x = e$. Therefore, $(e, 1/e)$ is a relative maximum point of the graph of $f(x)$. ■

The next example introduces a function that will be needed later when we study integration.

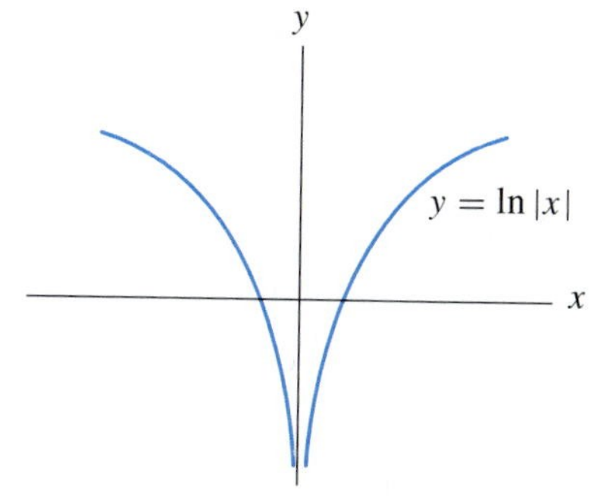

FIGURE 1 Graph of $\ln|x|$

EXAMPLE 3 The function $\ln|x|$ is defined for all nonzero values of x. Its graph is sketched in Figure 1. Compute the derivative of $\ln|x|$.

Solution If x is positive, then $|x| = x$, so

$$\frac{d}{dx}\ln|x| = \frac{d}{dx}\ln x = \frac{1}{x}.$$

If x is negative, then $|x| = -x$; and, by (4),

$$\begin{aligned} \frac{d}{dx}\ln|x| &= \frac{d}{dx}\ln(-x) \\ &= \frac{1}{-x}\cdot\frac{d}{dx}(-x) \\ &= \frac{1}{-x}\cdot(-1) = \frac{1}{x}. \end{aligned}$$ ■

Therefore, we have established the following useful fact:

$$\frac{d}{dx}\ln|x| = \frac{1}{x}, \qquad x \neq 0.$$

PRACTICE PROBLEMS 4.5

Differentiate.

1. $f(x) = \dfrac{1}{\ln(x^4+5)}$

2. $f(x) = \ln(\ln x)$

EXERCISES 4.5

Differentiate the following functions.

1. $\ln 2x$
2. $\ln x^2$
3. $\ln(x+5)$
4. $x^2 \ln x$
5. $\dfrac{1}{x}\ln(x+1)$
6. $\sqrt{\ln x}$
7. $e^{\ln x + x}$
8. $\ln\left(\dfrac{x}{x-3}\right)$
9. $4 + \ln\left(\dfrac{x}{2}\right)$
10. $\ln\sqrt{x}$
11. $(\ln x)^2 + \ln x$
12. $\ln(x^3+2x+1)$
13. $\ln(kx)$, k constant
14. $\dfrac{x}{\ln x}$
15. $\dfrac{x}{(\ln x)^2}$
16. $(\ln x)e^{-x}$
17. $e^{2x}\ln x$
18. $(\ln x + 1)^3$
19. $\ln(e^{5x}+1)$
20. $\ln\left(e^{e^x}\right)$

Find.

21. $\dfrac{d}{dt}(t^2 \ln 4)$
22. $\dfrac{d^2}{dx^2}\ln(1+x^2)$
23. $\dfrac{d^2}{dt^2}(\ln t)^3$
24. Find the slope of the graph of $y = \ln|x|$ at $x = 3$ and $x = -3$.
25. Write the equation of the tangent line to the graph of $y = \ln(x^2+e)$ at $x = 0$.
26. The function $f(x) = (\ln x + 1)/x$ has a relative extreme point for $x > 0$. Find the coordinates of the point. Is it a relative maximum point?
27. The function $f(x) = (\ln x)/\sqrt{x}$ has a relative extreme point for $x > 0$. Find the coordinates of the point. Is it a relative maximum point?
28. The function $f(x) = x/(\ln x + x)$ has a relative extreme point for $x > 1$. Find the coordinates of the point. Is it a relative minimum point?
29. Suppose that the total revenue function for a manufacturer is $R(x) = 300\ln(x+1)$, so that the sale of x units of a product brings in about $R(x)$ dollars. Suppose also that the total cost of producing x units is $C(x)$ dollars, where $C(x) = 2x$. Find the value of x at which the profit function $R(x)-C(x)$ will be maximized. Show that the profit function has a relative maximum and not a relative minimum point at this value of x.
30. Evaluate $\lim\limits_{h\to 0}\dfrac{\ln(7+h)-\ln 7}{h}$.

31. Find the maximum area of a rectangle in the first quadrant with one corner at the origin, two sides on the coordinate axes, and one corner on the graph of $y = -\ln x$.

In the remaining exercises, use a graphing utility to obtain the graphs of the functions and answer the questions when appropriate. In Exercises 32–35, differentiate $f(x)$ and then compare the graphs of $f(x)$ and $f'(x)$ in the given window to check that the expression for $f'(x)$ is reasonable.

32. $f(x) = \dfrac{x}{\ln x}$; $[0, 3]$ *by* $[-20, 20]$

33. $f(x) = x - \ln x$; $[0, 4]$ *by* $[-4, 4]$

34. $f(x) = e^{.5x} - \ln(1 + x^2)$; $[0, 2]$ *by* $[-.5, 1.2]$

35. $f(x) = x^2 \ln x$; $[0, 1]$ *by* $[-.5, 2]$

In Exercises 36 and 37, sketch the graphs of the functions and identify all relative extreme points and inflection points.

36. $f(x) = \ln x + \dfrac{1}{x} - \dfrac{1}{2}$

37. $f(x) = 1 + \ln(x^2 - 6x + 10)$

38. (Analysis of the effectiveness of an insect repellent) Human hands covered with cotton fabrics that had been impregnated with the insect repellent DEPA were inserted for five minutes into a test chamber containing 200 female mosquitoes.* The function $f(x) = 26.48 - 14.09 \ln x$ gives the number of mosquito bites received when the concentration was x percent. [*Note*: The answers to parts (b)–(e) can be obtained either algebraically or from the graphs. You might consider trying both methods.]
   (a) Graph $f(x)$ and $f'(x)$ for $0 \le x \le 100$.
   (b) How many bites were received when the concentration was 3.25%?
   (c) What concentration resulted in 15 bites?
   (d) At what rate is the number of bites changing with respect to concentration of DEPA when $x = 2.75$?
   (e) For what concentration does the rate of change of bites with respect to concentration equal -10 bites per percentage increase in concentration?

39. (Spread of information by mass media) The spread of a piece of information in a town follows the model $f(t) = P(1 - e^{-kt})$, where $f(t)$ is the number of people who have heard the information after t days, P is the town population, and k is a constant. Use the values $P = 40{,}000$ and $k = .25$. [*Note*: The answers to parts (b)–(e) can be obtained either algebraically or from the graphs. You might consider trying both methods.]
   (a) Graph $f(t)$ and $f'(t)$ for $0 \le t \le 32$.
   (b) What is $f(3)$, and what does it tell about the situation?

---

* Rao, K. M., Prakash, S., Kumar, S., Suryanarayana, M. V. S., Bhagwat, M. M., Gharia, M. M., and Bhavsar, R. B., "N-diethylphenylacetamide in Treated Fabrics, as a Repellent Against Aedes aegypti and Culex quinquefasclatus (Deptera: Culiciae)," *Journal of Medical Entomology*, **28** (January 1991), 1.

(c) What is $f'(3)$, and what does it tell about the situation?
(d) When will 80% of the town's population have heard the information?
(e) When will the information be spreading at the greatest rate?

40. (Revenue function) Suppose that the demand equation for a certain commodity is $p = \dfrac{45}{\ln x}$, $3 \le x \le 19$. Let $R(x)$ be the revenue function. [*Note*: The answer to parts (b) and (d) can be obtained either algebraically or from the graphs. You might consider trying both methods.]
(a) Graph $R(x)$ and $R'(x)$ in the window $[3, 19]$ *by* $[-30, 300]$.
(b) How much money is received from the sale of 13 units of goods?
(c) For what sales level will \$270 be received?
(d) What is the marginal revenue when 4 units of goods are sold?
(e) For what sales level(s) will the marginal revenue be \$11 per unit?
(f) For what sales level will the marginal revenue be greatest?

41. (Cost function) Suppose that the cost function for a certain commodity is

$$C(x) = 2 + \frac{10\ln(x+1)}{4 - .3x}$$

thousand dollars, $0 \le x \le 8$.
(a) Graph $C(x)$ and $C'(x)$ in the window $[0, 8]$ *by* $[0, 16]$.
(b) What is the cost of producing 3.5 units of goods?
(c) How many units of goods can be produced for \$13,000?
(d) What is the marginal cost when 6.5 units of goods are produced?
(e) For what level of production will the marginal cost be \$1500?

SOLUTIONS TO PRACTICE PROBLEMS 4.5

1. Here $f(x) = [\ln(x^4+5)]^{-1}$. By the general power rule,

$$f'(x) = -[\ln(x^4+5)]^{-2} \cdot \frac{d}{dx}\ln(x^4+5)$$
$$= -[\ln(x^4+5)]^{-2} \cdot \frac{4x^3}{x^4+5}.$$

2. $f'(x) = \dfrac{d}{dx}\ln(\ln x) = \dfrac{1}{\ln x} \cdot \dfrac{d}{dx}\ln x = \dfrac{1}{\ln x} \cdot \dfrac{1}{x} = \dfrac{1}{x \ln x}.$

4.6 PROPERTIES OF THE NATURAL LOGARITHM FUNCTION

The natural logarithm function $\ln x$ possesses many of the familiar properties of logarithms to the base 10 (or common logarithms) that are encountered in algebra.

Let x and y be positive numbers, b any number.

LI $\ln(xy) = \ln x + \ln y.$

LII $\ln\left(\frac{1}{x}\right) = -\ln x.$

LIII $\ln\left(\frac{x}{y}\right) = \ln x - \ln y.$

LIV $\ln(x^b) = b \ln x.$

VERIFICATION OF LI By equation (2) of Section 4.4, we have $e^{\ln(xy)} = xy$, $e^{\ln x} = x$, and $e^{\ln y} = y$. Therefore,

$$e^{\ln(xy)} = xy = e^{\ln x} \cdot e^{\ln y} = e^{\ln x + \ln y}.$$

By equating exponents, we get LI.

VERIFICATION OF LII Since $e^{\ln(1/x)} = 1/x$, we have

$$e^{\ln(1/x)} = \frac{1}{x} = \frac{1}{e^{\ln x}} = e^{-\ln x}.$$

By equating exponents, we get LII.

VERIFICATION OF LIII By LI and LII we have

$$\ln\left(\frac{x}{y}\right) = \ln\left(x \cdot \frac{1}{y}\right) = \ln x + \ln\left(\frac{1}{y}\right) = \ln x - \ln y.$$

VERIFICATION OF LIV Since $e^{\ln(x^b)} = x^b$, we have

$$e^{\ln(x^b)} = x^b = (e^{\ln x})^b = e^{b \ln x}.$$

Equating exponents, we get LIV.

These properties of the natural logarithm should be learned thoroughly. You will find them useful in many calculations involving $\ln x$ and the exponential function.

EXAMPLE 1 Simplify $\ln 5 + 2 \ln 3$.

Solution

$$\begin{aligned} \ln 5 + 2\ln 3 &= \ln 5 + \ln 3^2 && \text{(LIV)} \\ &= \ln 5 + \ln 9 \\ &= \ln 45. && \text{(LI)} \end{aligned}$$

■

EXAMPLE 2 Simplify $\frac{1}{2}\ln(4t) - \ln(t^2+1)$.

Solution

$$\begin{aligned}\frac{1}{2}\ln(4t) - \ln(t^2+1) &= \ln[(4t)^{1/2}] - \ln(t^2+1) && \text{(LIV)}\\ &= \ln(2\sqrt{t}) - \ln(t^2+1)\\ &= \ln\left(\frac{2\sqrt{t}}{t^2+1}\right). && \text{(LIII)}\end{aligned}$$

EXAMPLE 3 Simplify $\ln x + \ln 3 + \ln y - \ln 5$.

Solution Use (LI) twice and (LIII) once.

$$\begin{aligned}(\ln x + \ln 3) + \ln y - \ln 5 &= \ln 3x + \ln y - \ln 5\\ &= \ln 3xy - \ln 5\\ &= \ln\left(\frac{3xy}{5}\right).\end{aligned}$$

EXAMPLE 4 Differentiate $f(x) = \ln[x(x+1)(x+2)]$.

Solution First rewrite $f(x)$, using (LI):

$$f(x) = \ln[x(x+1)(x+2)] = \ln x + \ln(x+1) + \ln(x+2).$$

Then $f'(x)$ is easily calculated:

$$f'(x) = \frac{1}{x} + \frac{1}{x+1} + \frac{1}{x+2}.$$

Let us now use properties of the natural logarithm function to finally establish the power rule. Let $f(x) = x^r$. Then

$$\ln f(x) = \ln x^r = r \ln x.$$

Differentiation of this equation yields

$$\begin{aligned}\frac{f'(x)}{f(x)} &= r \cdot \frac{1}{x}\\ f'(x) &= r \cdot \frac{1}{x} \cdot f(x) = r \cdot \frac{1}{x} \cdot x^r = rx^{r-1}.\end{aligned}$$

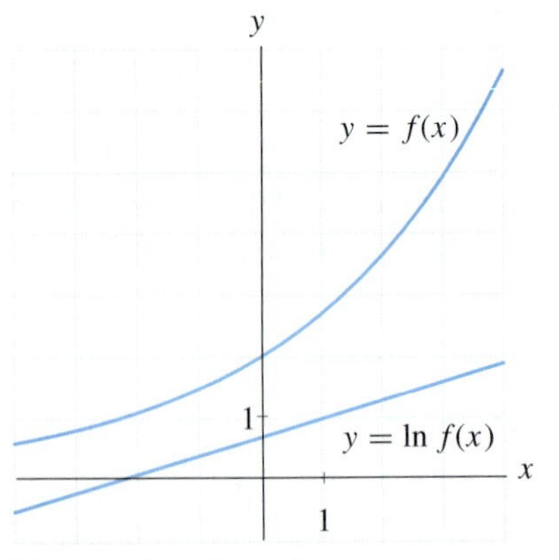

FIGURE 1 A logarithmic transformation

LINEARIZING EXPONENTIAL CURVES The logarithm function transforms exponential curves to straight lines. Figure 1 contains the graphs of $y = f(x)$ and $y = \ln f(x)$, where $f(x) = 2e^{.3x}$. [These graphs are easily obtained with a TI calculator or Visual Calculus. With a TI calculator, set `Y1 = 2e^(.3X)` and `Y2 =`

`ln(Y1)` and graph the two functions in the window $[-4, 4]$ *by* $[-1, 7]$. With Visual Calculus, use the "Graphs of Functions" routine, set `f(x)=2e^(3x)` and `g(x)=ln f(x)`.]

In general, suppose $f(x)$ is a function of exponential type. Then,

$$\begin{aligned} f(x) &= Ce^{kx} \\ \ln f(x) &= \ln(Ce^{kx}) \\ &= \ln C + \ln e^{kx} && \text{(LI)} \\ &= \ln C + kx && (\ln e^u = u) \\ &= \underbrace{k}_{m}\, x + \underbrace{\ln C}_{b}. \end{aligned}$$

That is, $\ln f(x)$ has the form $mx + b$ and is thus a straight line of slope k and y-intercept $(0, \ln C)$. Conversely, if a function $f(x)$ has the property that $\ln f(x)$ is a linear function, then $f(x)$ is a function of exponential type. To summarize,

> The function $f(x)$ has the form Ce^{kx} if and only if the graph of $\ln f(x)$ is a straight line of slope k and y-intercept $(0, \ln C)$. (1)

EXAMPLE 5 Find the function, $f(x)$, of exponential type whose graph contains the points (2, 15) and (6, 2).

x	$f(x)$	$\ln f(x)$
2	15	2.708
6	2	.693

Solution The table at the left shows the values of $f(x)$ and $\ln f(x)$ at the two values of x. Since the graph of $\ln f(x)$ is a straight line, its slope is

$$m = \frac{2.708 - .693}{2 - 6} = \frac{2.015}{-4} \approx -.5.$$

By the point-slope formula, the equation of the line is

$$\begin{aligned} y - 2.708 &= -.5(x - 2) \\ y &= -.5x + 3.708. \end{aligned}$$

By (1),

$$\begin{aligned} \ln C &= 3.708 \\ C &= e^{3.708} = 40.8. \end{aligned}$$

Therefore, $f(x) = 40.8e^{-.5x}$. ■

PRACTICE PROBLEM 4.6

1. Differentiate $f(x) = \ln\left[\dfrac{e^x\sqrt{x}}{(x+1)^6}\right]$.

EXERCISES 4.6

Simplify the following expressions.

1. $\ln 5 + \ln x$

2. $\ln x^5 - \ln x^3$

3. $\frac{1}{2}\ln 9$

4. $3\ln\frac{1}{2} + \ln 16$

5. $\ln 4 + \ln 6 - \ln 12$

6. $\ln 2 - \ln x + \ln 3$

7. $e^{2\ln x}$

8. $\frac{3}{2}\ln 4 - 5\ln 2$

9. $5\ln x - \frac{1}{2}\ln y + 3\ln z$

10. $e^{\ln x^2 + 3\ln y}$

11. $\ln x - \ln x^2 + \ln x^4$

12. $\frac{1}{2}\ln xy + \frac{3}{2}\ln\frac{x}{y}$

13. Which is larger, $2\ln 5$ or $3\ln 3$?

14. Which is larger, $\frac{1}{2}\ln 16$ or $\frac{1}{3}\ln 27$?

15. Which of the following is the same as $4 \cdot \ln 2x$?

(a) $\ln 8x$ (b) $8 \cdot \ln x$ (c) $\ln 8 + \ln x$ (d) $\ln 16x^4$

16. Which of the following is the same as $\ln(9x) - \ln(3x)$?

(a) $\ln 6x$ (b) $\ln(9x)/\ln(3x)$ (c) $6 \cdot \ln x$ (d) $\ln 3$

17. Which of the following is the same as $\dfrac{\ln 8x^2}{\ln 2x}$?

(a) $\ln 4x$ (b) $4x$ (c) $\ln 8x^2 - \ln 2x$ (d) none of these

18. Which of the following is the same as $\ln 9x^2$?

(a) $2 \cdot \ln 9x$ (b) $3x \cdot \ln x$ (c) $2 \cdot \ln 3x$ (d) none of these

Differentiate.

19. $\ln[(x+5)(2x-1)(4-x)]$

20. $\ln[x^3(x+1)^4]$

21. $\ln\left[\dfrac{(x+1)(3x-2)}{x+2}\right]$

22. $\ln\left[\dfrac{x^2}{(3-x)^3}\right]$

23. $\ln\left[\dfrac{\sqrt{x}}{x^2+1}\right]$

24. $\ln[e^{x^2}(x^4+x^2+1)]$

25. There are substantial empirical data to show that if x and y measure the sizes of two organs of a particular animal, then x and y are related by an *allometric equation* of the form

$$\ln y - k\ln x = \ln c,$$

where k and c are positive constants that depend only on the type of parts or organs that are measured and are constant among animals belonging to the same species.* Solve this equation for y in terms of x, k, and c.

26. In the study of epidemics, one finds the equation

$$\ln(1-y) - \ln y = C - rt,$$

where y is the fraction of the population that has a specific disease at time t. Solve the equation for y in terms of t and the constants C and r.

---

* E. Batschelet, *Introduction to Mathematics for Life Scientists* (New York: Springer-Verlag, 1971), pp. 305–307.

In Exercises 27 and 28, find the function of exponential type whose graph contains the given points.

27. (1, 6), (3, 14)

28. (−2, 8), (3, 1)

In Exercises 29 and 30, determine whether the three points lie on the graph of a function of exponential type.

29. (2, 1.2), (4, 4.8), (5, 9.6)

30. (2, 2.5), (3, 1.5), (4, 1)

31. Suppose the function $y = f(x)$ satisfies a differential equation of the form $y' = ky$ and the points (0, 4) and (3, 5) are on the graph of $f(x)$. Find $f(x)$.

32. Suppose the function $y = f(x)$ satisfies a differential equation of the form $y' = ky$ and the points (1, 5) and (4, 3) are on the graph of $f(x)$. Find k.

In the remaining exercises, use a graphing utility to obtain the graphs of the functions and answer the questions when appropriate.

33. Graph $\ln(ex)$ and $\ln x$ together in the window $[0, 10]$ *by* $[-4, 4]$. What relationship do you discern between them? Use a property of logarithms to show why the relationship holds.

34. Graph $\ln\left(\frac{x}{e}\right)$ and $\ln x$ together in the window $[0, 10]$ *by* $[-3, 3]$. What relationship do you discern between them? Use a property of logarithms to show why the relationship holds.

35. Graph $y = 2^x$ and $y = e^{x \cdot \ln 2}$ together in the window $[-4, 4]$ *by* $[-1, 16]$. They should be identical. Explain why.

36. (a) Graph $y = \ln x$.
    (b) Zoom in on the curve near $x = 2$ until the curve appears as a straight line, and estimate the slope of the line. This number is an estimate of $\frac{d}{dx} \ln x$ at $x = 2$. Compare your answer with the actual slope, $\frac{1}{2}$.
    (c) Repeat (b) at $x = 3$.

37. (Velocity of a rocket) The velocity of a rocket, t seconds after blastoff, is

$$v(t) = 3 \ln\left(\frac{1}{1 - .03t}\right)$$

feet per second for $0 \le t \le 32$.
    (a) Graph $v(t)$ and $v'(t)$.
    (b) What is the velocity after 10 seconds?
    (c) Approximately when is the rocket traveling at the rate of 2 feet per second?
    (d) How fast is the rocket accelerating after 10 seconds?
    (e) Approximately when is the rocket accelerating at the rate of .9 feet per second?

38. (Spread of information by word of mouth) The spread of a piece of information in a town by word of mouth follows the model

$$f(t) = \frac{P}{1 + Be^{-kt}},$$

where $f(t)$ is the number of people who have heard the information after t days, P is the town population, and B and k are constants. Use the values $P = 40{,}000$, $B = 200$, and $k = .52$. [*Note*: The answers to parts (b)–(e) can be obtained either algebraically or from the graphs. You might consider trying both methods.]

(a) Graph $f(t)$ and $f'(t)$ for $0 \le t \le 100$.
(b) What is $f(3)$, and what does it tell about the situation?
(c) What is $f'(3)$, and what does it tell about the situation?
(d) When will 80% of the town's population have heard the information?
(e) When will the information be spreading at the greatest rate?

SOLUTIONS TO PRACTICE PROBLEM 4.6

1. Use the properties of the natural logarithm to express $f(x)$ as a sum of simple functions before differentiating.

$$\begin{aligned} f(x) &= \ln\left[\frac{e^x\sqrt{x}}{(x+1)^6}\right] \\ &= \ln e^x + \ln\sqrt{x} - \ln(x+1)^6 \\ &= x + \frac{1}{2}\ln x - 6\ln(x+1). \\ f'(x) &= 1 + \frac{1}{2x} - \frac{6}{x+1}. \end{aligned}$$

REVIEW OF THE FUNDAMENTAL CONCEPTS OF CHAPTER 4

1. What is e?
2. State as many features of the graph of $y = e^{kx}$ (when k positive) as you can recall. When k negative.
3. Give the differential equation satisfied by $y = Ce^{kt}$.
4. State the product rule and quotient rule.
5. State the chain rule. Give an example.
6. What is a logarithm?
7. State as many features of the graph of $y = \ln x$ as you can recall.
8. State the two key equations giving the relationships between e^x and $\ln x$. (*Hint*: The right side of each equation is just x.)
9. What is the difference between a natural logarithm and a common logarithm.
10. Give the formula that converts a function of the form b^x to an exponential function with base e.
11. State the differentiation formula for each of the following functions:
    (a) $f(x) = e^{kx}$
    (b) $f(x) = e^{g(x)}$
    (c) $f(x) = \ln g(x)$
12. State as many properties of the natural logarithm function as you can recall.
13. Explain in detail how ln converts functions of exponential type into linear functions.

Chapter 5

APPLICATIONS OF THE EXPONENTIAL AND LOGARITHMIC FUNCTIONS

Earlier, we introduced the exponential functions e^x and the natural logarithm function $\ln x$ and studied their most important properties. From the way we introduced these functions, it is by no means clear that they have any substantial connection with the physical world. However, as this chapter will demonstrate, the exponential and natural logarithm functions intrude into the study of many physical problems, often in a very curious and unexpected way.

Here the most significant fact that we require is that the exponential function is uniquely characterized by its differential equation. In other words, we will constantly make use of the following fact, which was stated previously.

> The function* $y = Ce^{kt}$ satisfies the differential equation
>
> $$y' = ky.$$
>
> Conversely, if $y = f(t)$ satisfies the differential equation above, then $y = Ce^{kt}$ for some constant C.

If $f(t) = Ce^{kt}$, then, by setting $t = 0$, we have

$$f(0) = Ce^0 = C.$$

Therefore, C is the value of $f(t)$ at $t = 0$.

* Note that we use the variable t instead of x. The reason is that, in most applications, the variable of our exponential function is time. The variable t will be used throughout this chapter.

5.1 EXPONENTIAL GROWTH AND DECAY

In biology, chemistry, and economics it is often necessary to study the behavior of a quantity that is increasing as time passes. If, at every instant, the rate of increase of the quantity is proportional to the quantity at that instant, then we say that the quantity is *growing exponentially* or is *exhibiting exponential growth*. A simple example of exponential growth is exhibited by the growth of bacteria in a culture. Under ideal laboratory conditions a bacteria culture grows at a rate proportional to the number of bacteria present. It does so because the growth of the culture is accounted for by the division of the bacteria. The more bacteria there are at a given instant, the greater the possibilities for division and hence the more rapid is the rate of growth.

Let us study the growth of a bacteria culture as a typical example of exponential growth. Suppose that $P(t)$ denotes the number of bacteria in a certain culture at time t. The rate of growth of the culture at time t is $P'(t)$. We assume this rate of growth is proportional to the size of the culture at time t, so that

$$P'(t) = kP(t), \tag{1}$$

where k is a positive constant of proportionality. If we let $y = P(t)$, then (1) can be written as

$$y' = ky.$$

Therefore, from our discussion at the beginning of this chapter, we see that

$$y = P(t) = P_0 e^{kt}, \tag{2}$$

where $P_0 = P(0)$ is the number of bacteria in the culture at time $t = 0$. The number k is called the *growth constant*.

EXAMPLE 1 Suppose that a certain bacteria culture grows at a rate proportional to its size. At time $t = 0$, approximately 20,000 bacteria are present. In 5 hours there are 400,000 bacteria. Determine a function that expresses the size of the culture as a function of time, measured in hours.

Solution Let $P(t)$ be the number of bacteria present at time t. By assumption, $P(t)$ satisfies a differential equation of the form $y' = ky$, so $P(t)$ has the form

$$P(t) = P_0 e^{kt},$$

where the constants P_0 and k must be determined. The value of P_0 and k can be obtained from the data that give the population size at two different times. We are told that

$$P(0) = 20{,}000, \quad P(5) = 400{,}000. \tag{3}$$

The first condition immediately implies that $P_0 = 20{,}000$, so

$$P(t) = 20{,}000e^{kt}.$$

Using the second condition in (3), we have

$$\begin{aligned} 20{,}000e^{k\cdot 5} &= P(5) = 400{,}000 \\ e^{5k} &= 20 \\ 5k &= \ln 20 \\ k &= \frac{\ln 20}{5} \approx .60. \end{aligned} \tag{4}$$

So we may take

$$P(t) = 20{,}000e^{.6t}.$$

This function is a mathematical model of the growth of the bacteria culture. (See Figure 1.) ■

EXAMPLE 2 Suppose that a colony of fruit flies is growing according to the exponential law $P(t) = P_0e^{kt}$, and suppose that the size of the colony doubles in 12 days. Determine the growth constant k.

Solution We do not know the initial size of the population at $t = 0$. However, we are told that $P(12) = 2P(0)$; that is,

$$\begin{aligned} P_0e^{k\cdot 12} &= 2P_0 \\ e^{12k} &= 2 \\ 12k &= \ln 2 \\ k &= \frac{1}{12}\ln 2 \approx .058. \end{aligned}$$

■

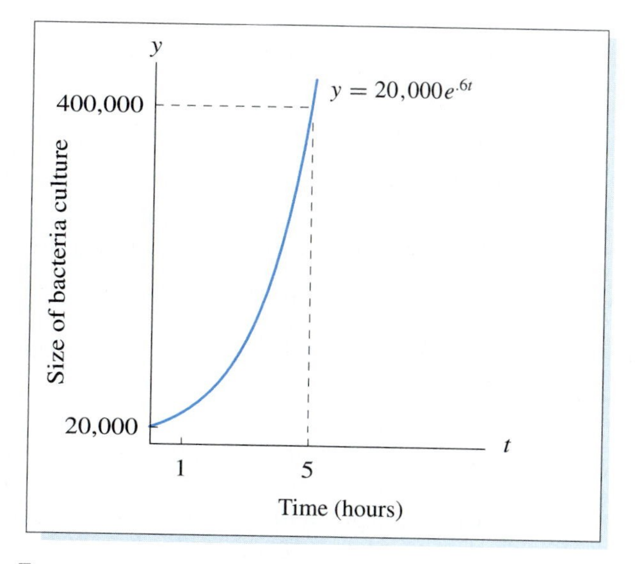

FIGURE 1 A model for bacteria growth

Notice that the initial size P_0 of the population was not given in Example 2. We were able to determine the growth constant because we were told the amount of time required for the colony to double in size. Thus the growth constant does not depend on the initial size of the population. This property is characteristic of exponential growth.

EXAMPLE 3 Suppose that the initial size of the colony in Example 2 was 300.

(a) At what time will the colony contain 1800 fruit flies?

(b) How fast will the colony be growing at that time?

Solution (a) From Example 2 we have $P(t) = P_0e^{.058t}$. Since $P(0) = 300$, we conclude that

$$P(t) = 300e^{.058t}.$$

Now that we have an explicit formula for the size of the colony, we can set $P(t) = 1800$ and solve for t:

$$\begin{aligned} 300e^{.058t} &= 1800 \\ e^{.058t} &= 6 \\ .058t &= \ln 6 \\ t &= \frac{\ln 6}{.058} \approx 31 \text{ days.} \end{aligned}$$

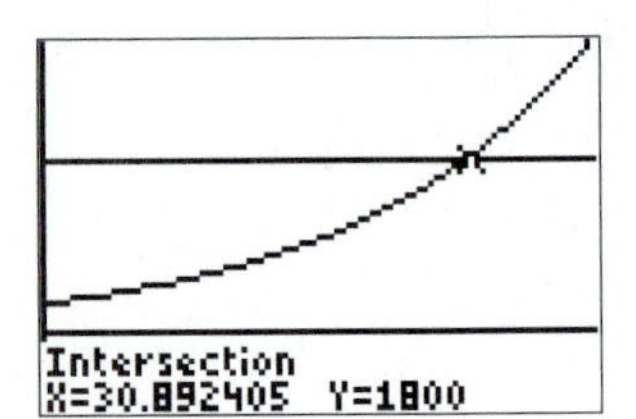

FIGURE 2 Y1=1800
Y2=300e^.058X

(b) $P(t)$ has the formula $P(t) = 300e^{.058t}$ and satisfies the differential equation $y' = .058y$. That is, $P'(t) = .058P(t)$.

$$P'(31) = .058P(31) = .058(1800) = 104.4.$$

When the population is 1800, the colony will be growing at the rate of 104.4 fruit flies per day. ■

Note 1: Part (a) of Example 3 also can be answered with a graphing utility. See Figure 2. However, the algebraic solution is preferable since it is straightforward and gives the exact solution.

Note 2: Part (b) of Example 3 can be answered without first answering part (a).

Note 3: The *growth constant*, .058, is a measure of how fast the population is growing. However, it is different from the *rate of growth*. The rate of growth (for instance, 104.4 after 31 days) is given by the derivative of the function and is always changing.

Population Size	Day
300	0
600	12
1,200	24
2,400	36
4,800	48
9,600	60
19,200	72
⋮	⋮

The table to the left shows the growth of the colony in Examples 2 and 3. Recall from Example 2 that the size of the colony doubles every 12 days. Notice that 1800 is exactly halfway between 1200 ($t = 24$) and 2400 ($t = 36$). It is incorrect to guess that the population will equal 1800 when t is halfway between $t = 24$ and $t = 36$—that is, when $t = 30$. We saw in Example 3 that it takes approximately 31 days for the size of the colony to reach 1800 fruit flies.

EXAMPLE 4 Table 1 gives the U.S. population* for several years. This data lies approximately on the graph of an exponential curve. Find the exponential growth equation that best fits this data.

Table 1 U.S. POPULATION

Year	1988	1989	1990	1991	1992
Pop. (in millions)	245	247	250	253	255

Solution Let $f(t)$ be the U.S. population (in millions), t years after 1988. Since the graph of $f(t)$ is nearly exponential, the graph of $\ln f(t)$ is nearly linear. Table 2 gives the values of $\ln f(t)$. A graphing utility gives the slope and y-intercept of the regression (that is, least-squares) line for the data in Table 2 as $m = .0104$, $b = 5.5002$.

Table 2

Years after 1988	0	1	2	3	4
$\ln f(t)$	5.501	5.509	5.521	5.533	5.541

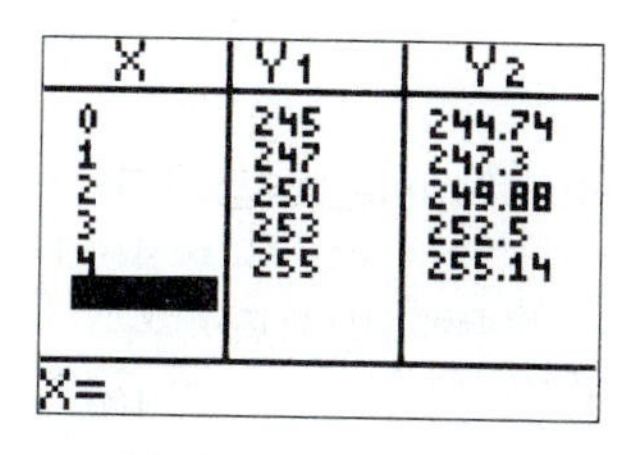

FIGURE 3
Y_1 = actual data
Y_2=244.74e^(.0104X)

Therefore,

$$\ln f(t) \approx 5.5002 + .0104t$$

$$e^{\ln f(t)} \approx e^{5.5002+.0104t} = e^{5.5002}e^{.0104t}$$

$$f(t) \approx 244.74e^{.0104t}.$$

The function $f(t) = 244.74e^{.0104t}$ best describes the growth of the U.S. population. Figure 3 shows that the population given by $f(t)$ is close to the actual population given in Table 1. ■

EXPONENTIAL DECAY An example of negative exponential growth, or *exponential decay*, is given by the disintegration of a radioactive element such as uranium 235. It is known that, at any instant, the rate at which a radioactive substance is decaying is proportional to the amount of the substance that has not yet disintegrated. If $P(t)$ is the quantity present at time t, then $P'(t)$ is the rate of decay. Of course, $P'(t)$ must be negative, since $P(t)$ is decreasing. Thus we may write $P'(t) = kP(t)$ for some negative constant k. To emphasize the fact that the constant is negative, k is often replaced by $-\lambda$, where λ is a positive constant.** Then $P(t)$ satisfies the differential equation

$$P'(t) = -\lambda P(t). \tag{5}$$

The general solution of (5) has the form

$$P(t) = P_0e^{-\lambda t}$$

* *Statistical Abstract of the United States*, 1994, p. 8.
** λ is the Greek lowercase letter lambda.

for some positive number P_0. We call such a function an *exponential decay function*. The constant λ is called the *decay constant*.

EXAMPLE 5 The decay constant for strontium 90 is $\lambda = .0244$, where the time is measured in years. How long will it take for a quantity P_0 of strontium 90 to decay to one-half its original size?

Solution We have

$$P(t) = P_0 e^{-.0244t}.$$

Next, set $P(t)$ equal to $\frac{1}{2}P_0$ and solve for t:

$$\begin{aligned} P_0 e^{-.0244t} &= \frac{1}{2}P_0 \\ e^{-.0244t} &= \frac{1}{2} = .5 \\ -.0244t &= \ln .5 \\ t &= \frac{\ln .5}{-.0244} \approx 28 \text{ years.} \end{aligned}$$

The *half-life* of a radioactive element is the length of time required for a given quantity of that element to decay to one-half its original size. Thus strontium 90 has a half-life of about 28 years. (See Figure 4.) Notice from Example 5 that the half-life does not depend on the initial amount P_0.

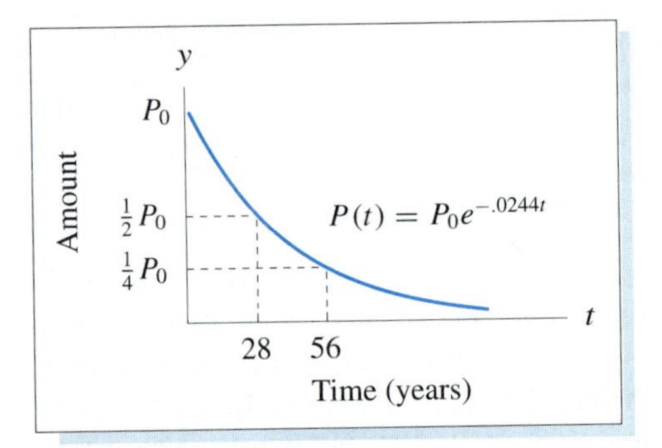

FIGURE 4 Half-life of radioactive strontium 90

EXAMPLE 6 Radioactive carbon 14 has a half-life of about 5730 years. Find its decay constant.

Solution If P_0 denotes the initial amount of carbon 14, then the amount after t years will be

$$P(t) = P_0 e^{-\lambda t}.$$

After 5730 years, $P(t)$ will equal $\frac{1}{2}P_0$. That is,

$$P_0 e^{-\lambda(5730)} = P(5730) = \frac{1}{2}P_0 = .5P_0.$$

Solving for λ gives

$$\begin{aligned} e^{-5730\lambda} &= .5 \\ -5730\lambda &= \ln .5 \\ \lambda &= \frac{\ln .5}{-5730} \approx .00012. \end{aligned}$$

■

One of the problems connected with aboveground nuclear explosions is the radioactive debris that falls on plants and grass, thereby contaminating the food supply of animals. Strontium 90 is one of the most dangerous components of "fallout" because it has a relatively long half-life and because it is chemically similar to calcium and is absorbed into the bone structure of animals (and humans) who eat contaminated food. Iodine 131 is also produced by nuclear explosions, but it presents less of a hazard because it has a half-life of only 8 days.

EXAMPLE 7 If dairy cows eat hay containing too much iodine 131, their milk will be unfit to drink. Suppose that some hay contains 10 times the maximum allowable level of iodine 131. For how many days should the hay be stored before it is fed to dairy cows?

Solution Let P_0 be the amount of iodine 131 present in the hay. Then the amount at time t is $P(t) = P_0 e^{-\lambda t}$ (t in days). The half-life of iodine 131 is 8 days, so

$$\begin{aligned} P_0 e^{-8\lambda} &= .5 P_0 \\ e^{-8\lambda} &= .5 \\ -8\lambda &= \ln .5 \\ \lambda &= \frac{\ln .5}{-8} \approx .087, \end{aligned}$$

and

$$P(t) = P_0 e^{-.087t}.$$

Now that we have the formula for $P(t)$, we want to find t such that $P(t) = \frac{1}{10} P_0$. We have

$$P_0 e^{-.087t} = .1 P_0,$$

so

$$\begin{aligned} e^{-.087t} &= .1 \\ -.087t &= \ln .1 \\ t &= \frac{\ln .1}{-.087} \approx 26 \text{ days}. \end{aligned}$$

■

RADIOCARBON DATING Knowledge about radioactive decay is valuable to social scientists who want to estimate the age of objects belonging to ancient civilizations. Several different substances are useful for radioactive-dating techniques; the most common is radiocarbon, ^{14}C. Carbon 14 is produced in the upper atmosphere when cosmic rays react with atmospheric nitrogen. Because the ^{14}C eventually decays, the concentration of ^{14}C cannot rise above certain levels. An equilibrium is reached when ^{14}C is produced at the same rate as it decays. Scientists usually assume that the total amount of ^{14}C in the biosphere has remained constant over the past 50,000 years. Consequently, it is assumed that the *ratio* of ^{14}C to ordinary nonradioactive carbon 12, ^{12}C, has been constant during this same period. (The ratio is about one part ^{14}C to 10^{12} parts of ^{12}C.) Both ^{14}C and ^{12}C are in the atmosphere in the form of carbon dioxide. All living vegetation and most forms of animal life contain ^{14}C and ^{12}C in the same proportion as the atmosphere. The reason is that plants absorb carbon dioxide through photosynthesis. The ^{14}C and ^{12}C in plants are distributed through the various food chains to almost all animal life.

When an organism dies, it stops replacing its carbon, and therefore the amount of ^{14}C begins to decrease through radioactive decay. (The ^{12}C in the dead organism remains constant.) At a later date, the ratio of ^{14}C to ^{12}C can be measured in order to determine when the organism died. (See Figure 5.)

EXAMPLE 8 A parchment fragment was discovered that had about 80% of the ^{14}C level found today in living matter. Estimate the age of the parchment.

Solution We assume that the original ^{14}C level in the parchment was the same as the level in living organisms today. Consequently, about eight-tenths of the original ^{14}C remains. From Example 6 we obtain the formula for the amount of ^{14}C present t years after the parchment was made from an animal skin:

$$P(t) = P_0 e^{-.00012t},$$

where P_0 = initial amount. We want to find t such that $P(t) = .8P_0$:

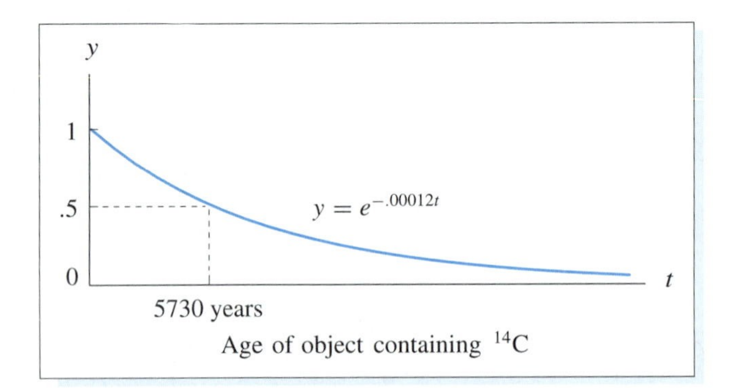

FIGURE 5 ^{14}C to ^{12}C ratio compared to the ratio in living plants

$$P_0e^{-.00012t} = .8P_0$$
$$e^{-.00012t} = .8$$
$$-.00012t = \ln .8$$
$$t = \frac{\ln .8}{-.00012} \approx 1860 \text{ years old.}$$

A SALES DECAY CURVE Marketing studies* have demonstrated that if advertising and other promotion of a particular product are stopped and if other market conditions remain fairly constant, then, at any time t, the sales of that product will be declining at a rate proportional to the amount of current sales at t. (See Figure 6.) If S_0 is the number of sales in the last month during which advertising occurred and if $S(t)$ is the number of sales in the tth month following the cessation of promotional effort, then a good mathematical model for $S(t)$ is

$$S(t) = S_0e^{-\lambda t},$$

where λ is a positive number called the *sales decay constant.* The value of λ depends on many factors, such as the type of product, the number of years of prior advertising, the number of competing products, and other characteristics of the market.

THE TIME CONSTANT Consider an exponential decay function $y = Ce^{-\lambda t}$. In Figure 7, we have drawn the tangent line to the decay curve when $t = 0$. The slope there is the initial rate of decay. If the decay process were to continue at this rate, the decay curve would follow the tangent line and y would be zero at

* M. Vidale and H. Wolfe, "An Operations-Research Study of Sales Response to Advertising," *Operations Research*, **5** (1957), 370–381. Reprinted in F. Bass et al., *Mathematical Models and Methods in Marketing* (Homewood, Ill.: Richard D. Irwin, Inc., 1961).

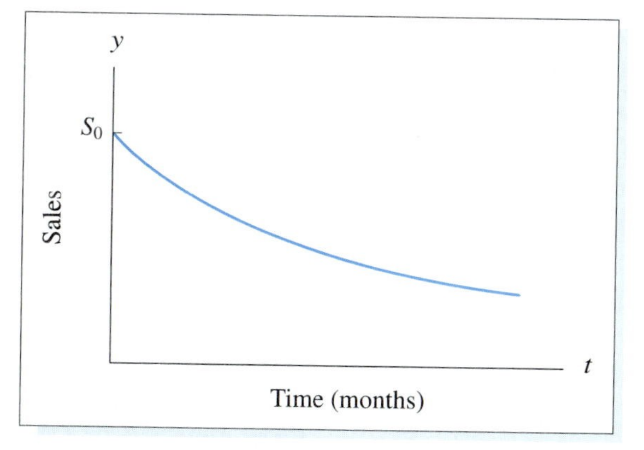

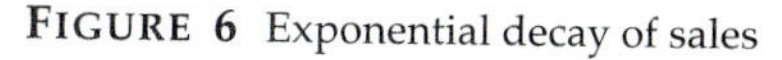

FIGURE 6 Exponential decay of sales

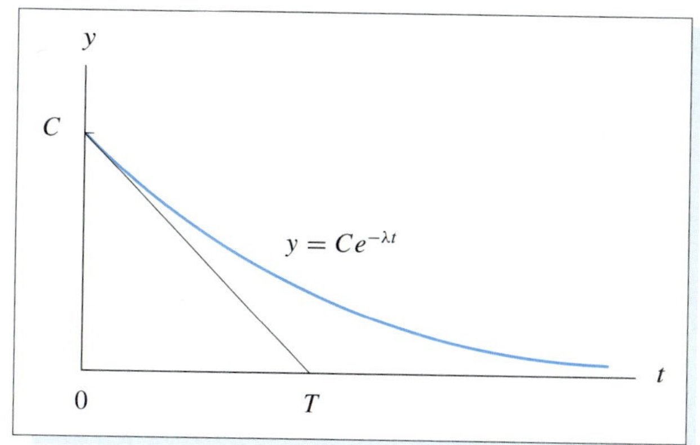

FIGURE 7 The time constant T in exponential decay: $T = 1/\lambda$

some time T. This time is called the *time constant* of the decay curve. It can be shown (see Exercise 37) that $T = 1/\lambda$ for the curve $y = Ce^{-\lambda t}$. Thus $\lambda = 1/T$, and the decay curve can be written in the form

$$y = Ce^{-t/T}.$$

If one has experimental data that tend to lie along an exponential decay curve, then the numerical constants for that curve may be obtained from Figure 7. First, sketch the curve and estimate the y-intercept, C. Then sketch an approximate tangent line and from this estimate the time constant, T. This procedure is sometimes used in biology and medicine.

PRACTICE PROBLEMS 5.1

1. (a) Solve the differential equation $P'(t) = -.6P(t)$, $P(0) = 50$.
   (b) Solve the differential equation $P'(t) = kP(t)$, $P(0) = 4000$, where k is some constant.
   (c) Find the value of k in part (b) for which $P(2) = 100P(0)$.
2. Under ideal conditions a colony of *E. coli* bacteria can grow by a factor of 100 every 2 hours. If initially 4000 bacteria are present, how long will it take before there are 1,000,000 bacteria?

EXERCISES 5.1

1. Let $P(t)$ be the population (in millions) of a certain city t years after 1990, and suppose that $P(t)$ satisfies the differential equation

$$P'(t) = .02P(t), \quad P(0) = 3.$$

   (a) Find the formula for $P(t)$.
   (b) What was the initial population, that is, the population in 1990?
   (c) What was the population in 1995?
   (d) How fast is the population growing when it reaches 4 million people?
   (e) When will the population reach 4 million people?
   (f) What is the growth constant?
   (g) How fast is the population growing initially?
   (h) When is the population growing at the rate of 70,000 people per year?
   (i) After how many years will the initial population double?
2. Approximately 10,000 bacteria are placed in a culture. Let $P(t)$ be the number of bacteria present in the culture after t hours, and suppose that $P(t)$ satisfies the differential equation

$$P'(t) = .55P(t).$$

   (a) Find the formula for $P(t)$.
   (b) What is $P(0)$?
   (c) How many bacteria are there after 5 hours?
   (d) How fast is the number of bacteria growing when it reaches 100,000?
   (e) When will the number of bacteria reach 100,000?
   (f) What is the growth constant?

(g) How fast is the number of bacteria growing initially?

(h) When is the number of bacteria growing at the rate of 34,000 bacteria per hour?

(i) After how many hours will the initial number of bacteria triple?

3. Determine the growth constant of a population that is growing at a rate proportional to its size, where the population doubles in size every 40 days.

4. Suppose a population is growing exponentially with growth constant .04. In how many years will the current population double?

5. Suppose a population is growing exponentially with growth constant .05. In how many years will the current population triple?

6. Determine the growth constant of a population that is growing at a rate proportional to its size, where the population triples in size every 10 years.

7. The rate of growth of a certain cell culture is proportional to its size. In 10 hours a population of 1 million cells grew to 9 million. How large will the cell culture be after 15 hours?

8. The world's population was 5.51 billion on January 1, 1993, and 5.69 billion on January 1, 1995. Assume that at any time the population grows at a rate proportional to the population at that time. In what year will the world's population reach 7 billion?

9. A certain bacteria culture grows at a rate proportional to its size, and it doubles every half-hour. Suppose that the culture contains 3 million bacteria at time $t = 0$ (with time in hours).

(a) At what time will there be 600 million bacteria present?

(b) At what time will the culture be growing at the rate of 600 million bacteria per hour?

10. Mexico City is expected to become the most heavily populated city in the world by the end of this century. At the beginning of 1990, 20.2 million people lived in the metropolitan area of Mexico City, and the population was growing exponentially. The 1995 population was 23 million. (Part of the growth is due to immigration.) If this trend continues, how large will the population be in the year 2000?

11. The population of a city t years after 1990 satisfies the differential equation $y' = .02y$. What is the growth constant? How fast will the population be growing when the population reaches 3 million people? At what level of population will the population be growing at the rate of 100,000 people per year?

12. A colony of bacteria is growing exponentially with growth constant .4, with time measured in hours. Determine the size of the colony when it is growing at the rate of 200,000 bacteria per hour. Determine the rate at which the colony will be growing when its size is 1,000,000.

13. Let $M(t)$ be the median price of a home in the United States, where $t = 0$ corresponds to 1986. During the second part of the 1980s, $M(t)$ satisfied the differential equation $M'(t) = .05M(t)$, $M(0) = 80{,}000$. Assuming $M(t)$ continues to satisfy this equation in the 1990s, estimate the median price of a home in the year 2000.

14. The consumer price index (CPI) gives a measure of the prices of commodities commonly purchased by consumers. For example, an increase of 10% in the CPI corresponds to a 10% average increase in the prices of consumer goods. Let $f(t)$ be the CPI at time t, where time is measured in years since January 1, 1980. Suppose that $f(t)$ satisfies the differential equation $f'(t) = .12f(t)$, $f(0) = 100$. (This means that at each time t, the CPI is rising at an annual rate of 12%, and the index is set equal to 100 on January 1, 1980.) How many years will it take for the CPI to double?

15. A common infection of the urinary tract in humans is caused by the bacterium *E. coli*. The infection is generally noticed when the bacteria colony reaches a population of about 10^8. The colony doubles in size about every 20 minutes. When a full bladder is emptied, about 90% of the bacteria are eliminated. Suppose that at the beginning of a certain time period, a person's bladder and urinary tract contain 10^8 *E. coli* bacteria. During an interval of T minutes, the person drinks enough liquid to fill the bladder. Find the value of T such that if the bladder is emptied after T minutes, about 10^8 bacteria will still remain. (*Note*: The average bladder holds about 1 liter of urine. It is seldom possible to eliminate an *E. coli* infection by diuresis without drugs—such as by drinking large amounts of water.)

16. In 1950 the world's population required 1×10^9 hectares* of arable land for food growth and in 1980 2×10^9 hectares were required. Current population trends indicate that if $A(t)$ denotes the amount of land needed t years after 1950, then $A'(t) = kA(t)$ for some constant k.

    (a) Derive a formula for $A(t)$.

    (b) The total amount of arable land on the earth's surface is estimated at 3.2×10^9 hectares. In what year will the earth exhaust its supply of land for growing food? [Data based on the Club of Rome's report, *The Limits to Growth* by D. H. Meadows, D. L. Meadows, J. Randers, and W. Behrens III (New York: Universe Books, 1972).]

17. Table 3 gives the per capita national health expenditures** for several years. This data lies approximately on the graph of an exponential curve. Find the differential equation (with initial value) that describes this data. If this trend continues, what will be the per capita expenditures in the year 2000?

Table 3 U.S. PER CAPITA PUBLIC HEALTH EXPENDITURES

Year	1987	1988	1989	1990	1991
Dollars	1962	2146	2352	2601	2868

* A hectare equals 2.471 acres.

** *Statistical Abstract of the United States*, 1994, p. 107.

18. Table 4 gives the average tuition of private colleges* for several years. This data lies approximately on the graph of an exponential curve. Find the exponential growth equation that best fits this data. If this trend continues, when will the average tuition reach \$15,000?

Table 4 AVERAGE TUITION OF PRIVATE COLLEGES

Year	1989	1990	1991
Dollars	7461	8174	8772

19. The population (in millions) of a state t years after 1970 is given by the graph of the exponential function $y = P(t)$ with growth constant .025 in Figure 8. [*Hint*: In parts (c) and (d), use the differential equation satisfied by $P(t)$.]

(a) What was the population in 1974?

(b) When was the population 10 million?

(c) How fast was the population growing in 1974?

(d) When was the population growing at the rate of 275,000 people per year?

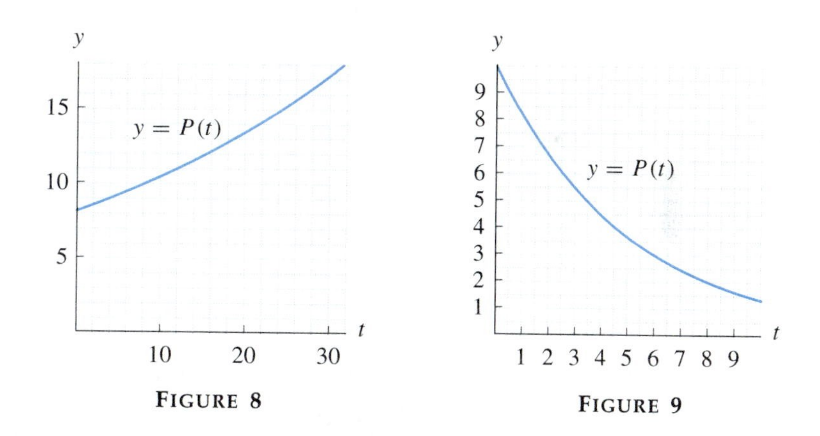

FIGURE 8 FIGURE 9

20. A sample of radioactive material decays over time (measured in hours) with decay constant .2. The graph of the exponential function $y = P(t)$ in Figure 9 gives the number of grams remaining after t hours. [*Hint*: In parts (c) and (d), use the differential equation satisfied by $P(t)$.]

(a) How much was remaining after 4 hours?

(b) What is the half-life of the radioactive material?

(c) How fast was the sample decaying after 6 hours?

(d) When was the population decaying at the rate of .4 grams per hour?

* *Statistical Abstract of the United States*, 1994, p. 8.

21. A sample of 8 grams of radioactive material is placed in a vault. Let $P(t)$ be the amount remaining after t years, and suppose that $P(t)$ satisfies the differential equation

$$P'(t) = -.05P(t).$$

(a) Find the formula for $P(t)$.
(b) What is $P(0)$?
(c) How much of the material will remain after 10 years?
(d) How fast is the sample disintegrating when just one gram remains?
(e) When will just one gram remain?
(f) What is the decay rate, that is, the decay constant?
(g) How fast is the sample disintegrating initially?
(h) When is the sample disintegrating at the rate of .14 grams per year?
(i) What is the half-life of the radioactive material?
(j) After how many years will just 2 grams remain?

22. Radium 226 is used in cancer radiotherapy, as a neutron source for some research purposes, and as a constituent of luminescent paints. Let $P(t)$ be the number of grams of radium 226 in a sample remaining after t years, and suppose that $P(t)$ satisfies the differential equation

$$P'(t) = -.00043P(t), \quad P(0) = 12.$$

(a) Find the formula for $P(t)$.
(b) What was the initial amount?
(c) Approximately how much of the radium will remain after 943 years?
(d) Approximately when will just one gram remain?
(e) How fast is the sample disintegrating when just one gram remains?
(f) What is the decay constant?
(g) How fast is the radium disintegrating initially?
(h) When is the sample disintegrating at the rate of .004 grams per year?
(i) What is the half-life of radium?
(j) After how many years will just 3 grams remain?

23. The decay constant for the radioactive element cesium 137 is .023 when time is measured in years. Find the half-life of cesium 137.

24. Radioactive cobalt 60 has a half-life of 5.3 years. Find the decay constant of cobalt 60.

25. A radioactive element has a half-life of 20 days. Find its decay constant. How much time is required for 75% of a sample of the element to disintegrate?

26. Five grams of a certain radioactive material decays to 3 grams in 1 year. After how many years will just 1 gram remain?

27. An island in the Pacific Ocean is contaminated by fallout from nuclear explosion. If the strontium 90 is 100 times the level that scientists believe is "safe," how many years will it take for the island once again to be "safe" for human habitation? The half-life of strontium 90 is 28 years.

28. By 1974 the United States had an estimated 80 million gallons of radioactive products from nuclear power plants and other nuclear reactors. These waste products were stored in various sorts of containers (made of such materials as stainless steel and cement), and the containers were buried in the ground and the ocean. Scientists feel that the waste products must be prevented from contaminating the rest of the earth until more than 99.99% of the radioactivity is gone (i.e., until the level is less than .0001 times the original level). If a storage cylinder contains waste products whose half-life is 1500 years, how many years must the container survive without leaking? (*Note*: Some of the containers are already leaking.)

29. Let $f(t)$ be the value of the dollar t years after January 1, 1990, where the value is in terms of the purchasing power on January 1, 1990, and $f(0) = 1.00$. Suppose that the rate of decrease of $f(t)$ at any time t is proportional to $f(t)$. Give the formula for $f(t)$ if by January 1, 1992, the dollar had lost 15% of its purchasing power.

30. A sample of radioactive material satisfies the differential equation $y' = -.05y$, where time is measured in days. What is the decay constant? How fast will the sample be disintegrating when the sample size is 2 grams? For what sample size will the sample size be decreasing at the rate of 3 grams per day?

31. A sample of radioactive material has decay constant .25, where time is measured in hours. How fast will the sample be disintegrating when the sample size is 12 grams? For what sample size will the sample size be decreasing at the rate of 2 grams per hour?

32. In 1947, a cave with beautiful prehistoric wall paintings was discovered in Lascaux, France. Some charcoal found in the cave contained 20% of the ^{14}C expected in living trees. How old are the Lascaux cave paintings? (Recall that the decay constant for ^{14}C is .00012.)

33. According to legend, in the fifth century King Arthur and his knights sat at a huge round table. A round table alleged to have belonged to King Arthur was found at Winchester Castle in England. In 1976 carbon dating revealed the amount of radiocarbon in the table to be 91% of the radiocarbon present in living wood. Could the table possibly have belonged to King Arthur? Why?

34. A 4500-year-old wooden chest was found in the tomb of the twenty-fifth century B.C. Chaldean king Meskalumdug of Ur. What percentage of the original ^{14}C would you expect to find in the wooden chest? (Recall that the decay constant for ^{14}C is .00012.)

35. Sandals woven from strands of tree bark were found recently in Fort Rock Creek in Oregon. The bark contained 34% of the level of ^{14}C found in living bark. Approximately how old are the sandals?

36. Many scientists believe there have been four ice ages in the past 1 million years. Before the technique of carbon dating was known, geologists erroneously believed that the retreat of the Fourth Ice Age began about 25,000 years ago. In 1950, logs from ancient spruce trees were found under glacial

debris near Two Creeks, Wisconsin. Geologists determined that these trees had been crushed by the advance of ice during the Fourth Ice Age. Wood from the spruce trees contained 27% of the level of ^{14}C found in living trees. Approximately how long ago did the Fourth Ice Age actually occur?

37. Let T be the time constant of the curve $y = Ce^{-\lambda t}$ as defined in Figure 7. Show that $T = 1/\lambda$. (*Hint*: Express the slope of the tangent line in Figure 7 in terms of C and T. Then set this slope equal to the slope of the curve $y = Ce^{-\lambda t}$ at $t = 0$.)

38. Suppose that a person is given an injection of 300 milligrams of penicillin at time 0, and let $f(t)$ be the amount (in milligrams) of the penicillin present in the person's bloodstream t hours after the injection. Then the amount of penicillin present decays exponentially, and a typical formula for $f(t)$ is $f(t) = 300e^{-(2/3)t}$.
    (a) What is the initial rate of decay of the penicillin?
    (b) What is the time constant for this decay curve $y = f(t)$? (See the discussion accompanying Figure 7.)

39. (Jeopardy-style TI questions) The population of a certain country is growing exponentially. The population in t years is $4e^{.02t}$ million. Set Y_1 = `4e^(.02X)` and set Y_2 = `.08e^(.02X)`, the derivative of Y_1. Match each of the following answers with its corresponding question.

    Answers:
    (a) Store the number 0 in the variable X and then display the value of Y_1.
    (b) Graph Y_1 and the equation $Y_3 = 2$. Then use "intersect" or "ISECT" to find the x-coordinate of their intersection point.
    (c) Store the number 2 in the variable X and then display the value of Y_2.
    (d) Graph Y_1 and the equation $Y_3 = 8$. Then use "intersect" or "ISECT" to find the x-coordinate of their intersection point.
    (e) Store the number 2 in the variable X and then display the value of Y_1.
    (f) Graph Y_2 and the equation $Y_3 = 2$. Then use "intersect" or "ISECT" to find the x-coordinate of their intersection point.

    Questions:
    (A) How fast will the population be growing in 2 years?
    (B) How long will it take for the initial population to double?
    (C) What will the size of the population be in 2 years?
    (D) What is the initial size of the population?
    (E) When will the size of the population be 2 million?
    (F) When will the population be growing at the rate of 2 million people per year?

40. (Jeopardy-style Visual Calculus questions) The population of a certain country is growing exponentially. The population in t years is $4e^{.02t}$ million. In the "Analyze a Function and Its Derivatives" routine, set `f(x)=4exp(.02x)` and set $0 \le x \le 32$. Graph the function and press [1] to display the derivative. Match each of the following answers with its corresponding question.

Answers:

(a) Use the left-arrow key to move the crosshairs all the way to the left. Then read the value of $f(x)$.

(b) Move the crosshairs until the value of $f(x)$ is as close to 2 as possible. Then read the value of x.

(c) Move the crosshairs until the value of x is 2. Then read the value of $f'(x)$.

(d) Move the crosshairs until the value of $f(x)$ is as close to 8 as possible. Then read the value of x.

(e) Move the crosshairs until the value of x is 2. Then read the value of $f(x)$.

(f) Move the crosshairs until the value of $f'(x)$ is as close to 2 as possible. Then read the value of x.

Questions:

(A) How fast will the population be growing in 2 years?

(B) How long will it take for the initial population to double?

(C) What will be the size of the population in 2 years?

(D) What is the initial size of the population?

(E) When will the size of the population be 2 million?

(F) When will the population be growing at the rate of 2 million people per year?

In the remaining exercises, use a graphing utility to obtain the graphs of the functions and answer the questions posed.

41. Redo Exercise 1, but use a graphing utility to answer the questions when possible.

42. Redo Exercise 2, but use a graphing utility to answer the questions when possible.

SOLUTIONS TO PRACTICE PROBLEMS 5.1

1. (a) Answer: $P(t) = 50e^{-.6t}$. Differential equations of the type $y' = ky$ have as their solution $P(t) = Ce^{kt}$, where C is $P(0)$.

(b) Answer: $P(t) = 4000e^{kt}$. This problem is like the previous one except that the constant is not specified. Additional information is needed if one wants to determine a specific value of k.

(c) Answer: $P(t) = 4000e^{2.3t}$. From the solution to part (b) we know that $P(t) = 4000e^{kt}$. We are given that $P(2) = 100P(0) = 100(4000) = 400{,}000$. So

$$\begin{aligned} P(2) &= 4000e^{k(2)} = 400{,}000 \\ e^{2k} &= 100 \\ 2k &= \ln 100 \\ k &= \frac{\ln 100}{2} \approx 2.3. \end{aligned}$$

2. Let $P(t)$ be the number of bacteria present after t hours. We must first find an expression for $P(t)$ and then determine the value of t for which $P(t) = 1{,}000{,}000$. From the discussion at the beginning of the section we know that $P'(t) = k \cdot P(t)$. Also, we are given that $P(2)$ (the population after 2 hours) is $100P(0)$ (100 times the initial population). From Problem 1(c) we have an expression for $P(t)$:

$$P(t) = 4000e^{2.3t}.$$

Now we must solve $P(t) = 1{,}000{,}000$ for t:

$$\begin{aligned} 4000e^{2.3t} &= 1{,}000{,}000 \\ e^{2.3t} &= 250 \\ 2.3t &= \ln 250 \\ t &= \frac{\ln 250}{2.3} \approx 2.4. \end{aligned}$$

Therefore, after approximately 2.4 hours there will be 1,000,000 bacteria.

5.2 COMPOUND INTEREST

When money is deposited in a savings account, interest is paid at stated intervals. If this interest is added to the account and thereafter earns interest itself, then the interest is called *compound interest.* The original amount deposited is called the *principal amount.* The principal amount plus the compound interest is called the *compound amount.* The interval between interest payments is referred to as the *interest period.* In formulas for compound interest, the interest rate is expressed as a decimal rather than a percent. Thus 6% is written as .06.

If \$1000 is deposited at 6% annual interest, compounded annually, the compound amount at the end of the first year will be

$$A_1 = \underset{\text{principal}}{1000} + \underset{\text{interest}}{1000(.06)} = 1000(1 + .06).$$

At the end of the second year the compound amount will be

$$\begin{aligned} A_2 &= \underset{\text{compound amount}}{A_1} + \underset{\text{interest}}{A_1(.06)} = A_1(1 + .06) \\ &= [1000(1 + .06)](1 + .06) = 1000(1 + .06)^2. \end{aligned}$$

At the end of 3 years

$$\begin{aligned} A_3 &= A_2 + A_2(.06) = A_2(1 + .06) \\ &= [1000(1 + .06)^2](1 + .06) = 1000(1 + .06)^3. \end{aligned}$$

After n years the compound amount will be

$$A = 1000(1 + .06)^n.$$

In this example the interest period was 1 year. The important point to note, however, is that at the end of each interest period the compound amount grew by a factor of $(1+.06)$. In general, if the interest rate is i instead of .06, the compound amount will grow by a factor of $(1+i)$ at the end of each interest period.

Suppose that a principal amount P is invested at a compound interest rate i per interest period, for a total of n interest periods. Then the compound amount A at the end of the nth period will be

$$A = P(1+i)^n. \tag{1}$$

EXAMPLE 1 Suppose that \$5000 is invested at 8% per year, with interest compounded annually. What is the compound amount after 3 years?

Solution Substituting $P = 5000$, $i = .08$, and $n = 3$ into formula (1), we have

$$\begin{aligned} A &= 5000(1 + .08)^3 = 5000(1.08)^3 \\ &= 5000(1.259712) = 6298.56 \text{ dollars.} \end{aligned}$$

It is common practice to state the interest rate as a percent per year ("per annum"), even though each interest period is often shorter than 1 year. If the annual rate is r and if interest is paid and compounded m times per year, then the interest rate i for each period is given by

$$[\text{rate per period}]\ i = \frac{r}{m} = \frac{[\text{annual interest rate}]}{[\text{periods per year}]}.$$

Many banks pay interest quarterly. If the stated annual rate is 5%, then $i = .05/4 = .0125$.

If interest is compounded for t years, with m interest periods each year, there will be a total of mt interest periods. If in formula (1) we replace n by mt and replace i by r/m, we obtain the following formula for the compound amount:

$$A = P\left(1 + \frac{r}{m}\right)^{mt}, \tag{2}$$

where

P = principal amount,

r = interest rate per annum,

m = number of interest periods per year,

t = number of years.

EXAMPLE 2 Suppose that \$1000 is deposited in a savings account that pays 6% per annum, compounded quarterly. If no additional deposits or withdrawals are made, how much will be in the account at the end of 1 year?

Solution We use (2) with $P = 1000$, $r = .06$, $m = 4$, and $t = 1$.

$$A = 1000\left(1 + \frac{.06}{4}\right)^4 = 1000(1.015)^4$$

$$\approx 1000(1.06136355) \approx 1061.36 \text{ dollars.}$$

Note that the \$1000 in Example 2 earned a total of \$61.36 in (compound) interest. This is 6.136% of \$1000. Savings institutions sometimes advertise this rate as the *effective* annual interest rate. That is, the savings institutions mean that *if* they paid interest only once a year, they would have to pay a rate of 6.136% in order to produce the same earnings as their 6% rate compounded quarterly. The stated annual rate of 6% is often called the *nominal rate*.

The effective annual rate can be increased by compounding the interest more often. Some savings institutions compound interest monthly or even daily.

EXAMPLE 3 Suppose that the interest in Example 2 were compounded monthly. How much would be in the account at the end of 1 year? What about the case when 6% annual interest is compounded daily? What is the effective annual rate in each case?

Solution For monthly compounding, $m = 12$. From (2) we have

$$A = 1000\left(1 + \frac{.06}{12}\right)^{12} = 1000(1.005)^{12}$$

$$\approx 1000(1.06167781) \approx 1061.68 \text{ dollars.}$$

The effective rate in this case is 6.168%.

A "bank year" usually consists of 360 days (in order to simplify calculations). So, for daily compounding, we take $m = 360$. Then

$$A = 1000\left(1 + \frac{.06}{360}\right)^{360} \approx 1000(1.00016667)^{360}$$

$$\approx 1000(1.06183133) \approx 1061.83 \text{ dollars.}$$

With daily compounding, the effective rate is 6.183%.

What would happen if the interest in Example 3 were compounded more often than once a day? Would the total interest be much more than \$61.83 if the interest were compounded every hour? Every minute? To answer these questions, we shall connect the notion of compound interest to the exponential function. Recall that the compound amount A is given by

$$A = P\left(1 + \frac{r}{m}\right)^{mt} = P\left(1 + \frac{r}{m}\right)^{(m/r)\cdot rt}.$$

If we set $h = r/m$, then $1/h = m/r$, and

$$A = P(1+h)^{(1/h)\cdot rt}.$$

As the frequency of compounding is increased, m gets large and $h = r/m$ approaches 0. To determine what happens to the compound amount, we must therefore examine the limit

$$\lim_{h \to 0} P(1+h)^{(1/h)\cdot rt}.$$

The following remarkable fact is proved in the appendix at the end of this section:

$$\lim_{h \to 0}(1+h)^{1/h} = e.$$

Using this fact together with two limit theorems, we have

$$\lim_{h \to 0} P(1+h)^{(1/h)rt} = P\left[\lim_{h \to 0}(1+h)^{1/h}\right]^{rt} = Pe^{rt}.$$

These calculations show that the compound amount calculated from the formula

$$P\left(1+\frac{r}{m}\right)^{mt}$$

gets closer to Pe^{rt} as the number m of interest periods per year is increased. When the formula

$$A = Pe^{rt} \tag{3}$$

is used to calculate the compound amount, we say that the interest is *compounded continuously*.

We can now answer the question posed following Example 3. Suppose that \$1000 is deposited for 1 year in an account paying 6% per annum. Then $P = 1000$, $r = .06$, and $t = 1$. If the interest is compounded continuously, the compound amount at the end of 1 year will be

$$1000e^{.06} \approx 1061.84 \text{ dollars.}$$

Recall from Example 3 that daily compounding of the 6% interest produced \$1061.83. Consequently, more frequent compounding (such as every hour or every second) will produce at most 1 cent more.

In recent years, when banks wanted to offer the maximum effective rate of interest permitted by law, many banks and savings institutions advertised savings accounts paying interest compounded continuously. However, as we have seen, the effect of compounding continuously is practically the same as compounding daily, unless the principal P is quite large.

When interest is compounded continuously, the compound amount $A(t)$ is an exponential function of the number of years t that interest is earned, $A(t) = Pe^{rt}$. Hence $A(t)$ satisfies the differential equation

$$\frac{dA}{dt} = rA.$$

The rate of growth of the compound amount is proportional to the amount of money present. Since the growth comes from the interest, we conclude that under continuous compounding, interest is earned continuously at a rate of growth proportional to the amount of money present.

The formula $A = Pe^{rt}$ contains four variables. (Remember that the letter e here represents a specific constant, $e = 2.718\ldots$.) In a typical problem, we are given values for three of these variables and must solve for the remaining variable.

EXAMPLE 4 One thousand dollars is invested at 5% interest compounded continuously.

(a) Give the formula for $A(t)$, the compound amount after t years.

(b) How much will be in the account after 6 years?

(c) After 6 years, at what rate will $A(t)$ be increasing?

(d) How long is required for the initial investment to double?

Solution (a) $P = 1000$ and $r = .05$. By (3), $A(t) = 1000e^{.05t}$.

(b) $A(6) = 1000e^{.05(6)} = 1000e^{.3} = \$1349.86$.

(c) Rate of growth is different from interest rate. Interest rate is fixed at 5% and does not change with time. On the other hand, the rate of growth $A'(t)$ is always changing. Since $A(t) = 1000e^{.05t}$, $A'(t) = 1000 \cdot .05e^{.05t} = 50e^{.05t}$.

$$A'(6) = 50e^{.05(6)} = 50e^{.3} = \$67.49.$$

After 6 years the investment is growing at the rate of \$67.49 per year.

There is an easier way to answer part (c). Since $A(t)$ satisfies the differential equation $A'(t) = rA(t)$,

$$A'(6) = .05A(6) = .05 \cdot 1349.86 = \$67.49.$$

(d) We must find t such that $A(t) = \$2000$. So we set $1000e^{.05t} = 2000$ and solve for t algebraically. (See Figure 1 for a graphical solution.)

$$\begin{aligned} 1000e^{.05t} &= 2000 \\ e^{.05t} &= 2 \\ \ln e^{.05t} &= \ln 2 \\ .05t &= \ln 2 \\ t &= \frac{\ln 2}{.05} \approx 13.86 \text{ years.} \end{aligned}$$

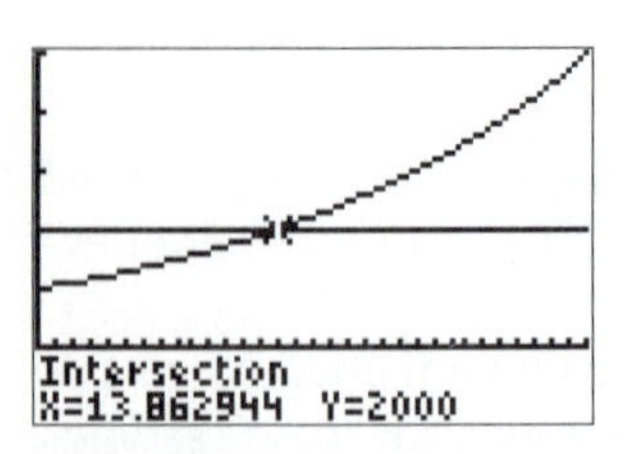

FIGURE 1 Y$_1$=2000
Y$_2$=1000e^(.05X)

Remark The calculations in Example 4 would be essentially unchanged after the first step if the initial amount of the investment were changed from $1000 to any arbitrary amount P. When this investment doubles, the compound amount will be $2P$. So one sets $2P = Pe^{.05t}$ and solves for t as we did above, to conclude that at 5% interest compounded continuously any amount doubles in about 13.86 years.

If P dollars are invested today, the formula $A = Pe^{rt}$ gives the value of this investment after t years (assuming continuously compounded interest). We say that P is the *present value* of the amount A to be received in t years. If we solve for P in terms of A, we obtain

$$P = Ae^{-rt}. \tag{4}$$

The concept of the present value of money is an important theoretical tool in business and economics. Problems involving depreciation of equipment, for example, may be analyzed by calculus techniques when the present value of money is computed from (4) using continuously compounded interest.

EXAMPLE 5 Find the present value of $5000 to be received in 2 years if money can be invested at 12% compounded continuously.

Solution Use (4) with $A = 5000$, $r = .12$, and $t = 2$:

$$\begin{aligned} P &= 5000e^{-(.12)(2)} = 5000e^{-.24} \\ &\approx 5000(.78663) = 3933.15 \text{ dollars.} \end{aligned}$$

APPENDIX A LIMIT FORMULA FOR e

For $h \neq 0$, we have

$$\ln(1+h)^{1/h} = (1/h)\ln(1+h).$$

Taking the exponential of both sides, we find that

$$(1+h)^{1/h} = e^{(1/h)\ln(1+h)}.$$

Since the exponential function is continuous,

$$\lim_{h\to 0}(1+h)^{1/h} = e^{\left[\lim_{h\to 0}(1/h)\ln(1+h)\right]}. \tag{5}$$

To examine the limit inside the exponential function, we note that $\ln 1 = 0$, and hence

$$\lim_{h\to 0}\left(\frac{1}{h}\right)\ln(1+h) = \lim_{h\to 0}\frac{\ln(1+h)-\ln 1}{h}.$$

The limit on the right is a difference quotient of the type used to compute a derivative. In fact,

$$\lim_{h\to 0}\frac{\ln(1+h)-\ln 1}{h} = \left.\frac{d}{dx}\ln x\right|_{x=1} = \left.\frac{1}{x}\right|_{x=1} = 1.$$

Thus the limit inside the exponential function in (5) is 1. That is,

$$\lim_{h\to 0}(1+h)^{1/h} = e^{[1]} = e.$$

PRACTICE PROBLEMS 5.2

1. One thousand dollars is to be invested in a bank for 4 years. Would 8% interest compounded semiannually be better than $7\frac{3}{4}$% interest compounded continuously?
2. A building was bought for \$150,000 and sold 10 years later for \$400,000. What interest rate (compounded continuously) was earned on the investment?

EXERCISES 5.2

1. Suppose \$1000 is deposited in a savings account at 10% interest compounded annually. What is the compound amount after 2 years?
2. Five thousand dollars is deposited in a savings account at 6% interest compounded monthly. Give the formula that describes the compound amount after 4 years.
3. Ten thousand dollars is invested at 8% interest compounded quarterly. Give the formula that describes the value of the investment after 3 years.
4. What is the effective annual rate of interest of a savings account paying 8% interest compounded semiannually?
5. One thousand dollars is invested at 7% interest compounded continuously. Compute the value of the investment at the end of 6 years.
6. A condominium was purchased in 1989 for \$100,000. If it appreciated at 12% compounded continuously, how much was it worth in 1996?
7. Five hundred dollars is deposited in a savings account paying 7% interest compounded daily. *Estimate* the balance in the account at the end of 3 years.
8. Ten thousand dollars is deposited into a money market fund paying 8% interest compounded continuously. How much interest will be earned during the first half-year if this rate of 8% does not change?
9. Pablo Picasso's *Angel Fernandez de Soto* was acquired in 1946 for a post-war splurge of \$22,220. The painting was sold in 1995 for \$29.1 million, the fifth highest price ever paid for a Picasso painting. What rate of interest compounded continuously did this investment earn?
10. One thousand dollars is deposited in a savings account at 6% interest compounded continuously. How many years are required for the balance in the account to reach \$2500?
11. Ten thousand dollars is to be invested in a highly speculative venture for 1 year. Would you rather receive 40% interest compounded semiannually or 39% interest compounded continuously?
12. A lot purchased in 1976 for \$5000 was appraised at \$60,000 in 1995. If the lot continues to appreciate at the same rate, when will it be worth \$100,000?

13. Ten thousand dollars is invested at 6.5% interest compounded continuously. When will the investment be worth \$41,787?

14. How many years are required for an investment to double in value if it is appreciating at the rate of 4% compounded continuously?

15. A farm purchased in 1985 for \$1,000,000 was valued at \$3,000,000 in 1995. If the farm continues to appreciate at the same rate (with continuous compounding), when will it be worth \$10,000,000?

16. Find the present value of \$1000 payable at the end of 3 years if money may be invested at 8% with interest compounded continuously.

17. Find the present value of \$2000 to be received in 10 years if money may be invested at 8% with interest compounded continuously.

18. A parcel of land bought in 1990 for \$10,000 was worth \$16,000 in 1995. If the land continues to appreciate at this rate, in what year will it be worth \$45,000?

19. One hundred dollars is deposited in a savings account at 7% interest compounded continuously. What is the effective annual rate of return?

20. In a certain town, property values tripled from 1980 to 1995. If this trend continues, when will property values be at five times their 1980 level? (Use an exponential model for the property value at time t.)

21. How much money must you invest now at 4.5% interest compounded continuously in order to have \$10,000 at the end of 5 years?

22. Investment A is currently worth \$70,200 and is growing at the rate of 13% per year compounded continuously. Investment B is currently worth \$60,000 and is growing at the rate of 14% per year compounded continuously. After how many years will the two investments have the same value?

23. Suppose that the present value of \$1000 to be received in 2 years is \$559.90. What rate of interest, compounded continuously, was used to compute this present value?

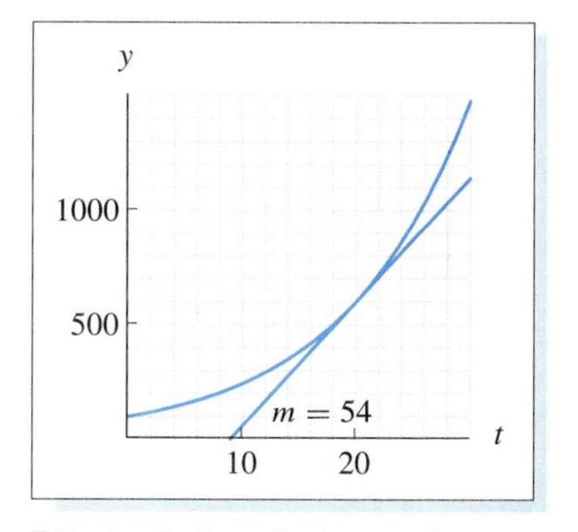

FIGURE 2 Growth of money in a savings account

24. Two thousand dollars is deposited in a savings account at 5% interest compounded continuously. Let $f(t)$ be the compound amount after t years. Find and interpret $f'(2)$.

25. A few years after money is deposited into the bank, the compound amount is \$1000 and is growing at the rate of \$60 per year. What interest rate (compounded continuously) is the money earning?

26. The current balance in a savings account is \$1230 and the interest rate is 4.5% compounded continuously. At what rate is the compound amount currently growing?

27. The curve in Figure 2 shows the growth of money in a savings account with interest compounded continuously.
    (a) What is the balance after 20 years?
    (b) At what rate is the money growing after 20 years?
    (c) Use the answers to (a) and (b) to determine the interest rate.

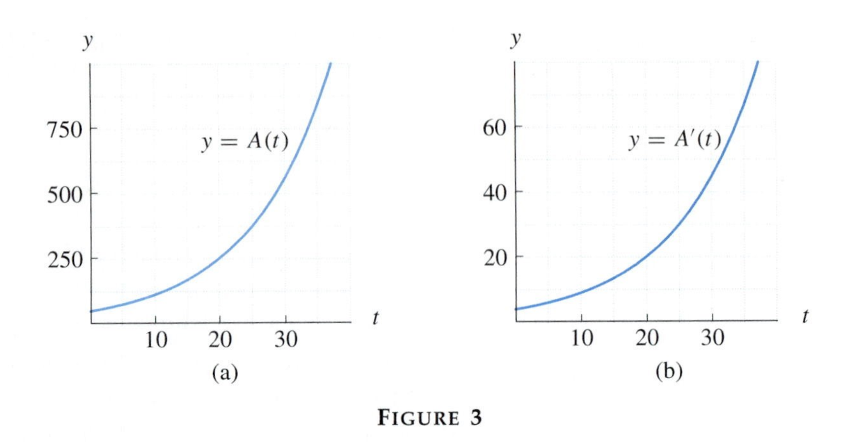

FIGURE 3

28. The function $f(t)$ in Figure 3(a) gives the balance in a savings account after t years with interest compounded continuously. Figure 3(b) shows the derivative of $f(t)$.
    (a) What is the balance after 25 years?
    (b) How fast is the balance increasing after 25 years?
    (c) Use the answers to (a) and (b) to determine the interest rate.
    (d) When is the balance \$250?
    (e) When is the balance increasing at the rate of \$20 per year?
    (f) Why do the graphs of $A(t)$ and $A'(t)$ look the same?

29. A small amount of money is deposited in a savings account with interest compounded continuously. Let $A(t)$ be the balance in the account after t years. Match each of the following answers with its corresponding question.

    Answers:
    (a) Pe^{rt}
    (b) $A(3)$
    (c) $A(0)$
    (d) $A'(3)$
    (e) Solve $A'(t) = 3$ for t.
    (f) Solve $A(t) = 3$ for t.
    (g) $y' = ky$
    (h) Solve $A(t) = 3A(0)$ for t.

    Questions:
    (A) How fast will the balance be growing in 3 years?
    (B) Give the general form of the function $A(t)$.
    (C) How long will it take for the initial deposit to triple?
    (D) Find the balance after 3 years.
    (E) When will the balance be 3 dollars?
    (F) When will the balance be growing at the rate of 3 dollars per year?
    (G) What was the principal amount?
    (H) Give a differential equation satisfied by $A(t)$.

In the remaining exercises, use a graphing utility to obtain the graphs of the function.

30. Convince yourself that

$$\lim_{h\to 0}(1+h)^{1/h} = e$$

by graphing $f(x) = (1+x)^{1/x}$ near $x = 0$ and examining values of $f(x)$ for x near 0. For instance, use a window such as $[-.01, .01]$ *by* $[0, 4]$.

31. Convince yourself that daily compounding is nearly the same as continuous compounding by graphing $100(1 + (.05/360))^{360x}$ together with $100e^{.05x}$ in the window $[0, 64]$ *by* $[0, 2500]$. The two graphs should appear the same on the screen. Approximately how far apart are they when $x = 32$? When $x = 64$?

SOLUTIONS TO PRACTICE PROBLEMS 5.2

1. Let us compute the balance after 4 years for each type of interest.

8% compounded semiannually: Use formula (2). Here $P = 1000$, $r = .08$, $m = 2$ ("semiannually" means that there are two interest periods per year), and $t = 4$. Therefore,

$$A = 1000\left(1 + \frac{.08}{2}\right)^{2\cdot 4} = 1000(1.04)^8 = 1368.57.$$

$7\frac{3}{4}$*% compounded continuously*: Use the formula $A = Pe^{rt}$, where $P = 1000$, $r = .0775$, and $t = 4$. Then

$$A = 1000e^{(.0775)\cdot 4} = 1000e^{.31} = 1363.43.$$

Therefore, 8% compounded semiannually is better.

2. If the \$150,000 had been compounded continuously for 10 years at interest rate r, the balance would be $150{,}000e^{r\cdot 10}$. The question asks: For what value of r will the balance be 400,000? We need just solve an equation for r.

$$\begin{aligned} 150{,}000e^{r\cdot 10} &= 400{,}000 \\ e^{r\cdot 10} &\approx 2.67 \\ r \cdot 10 &= \ln 2.67 \\ r &= \frac{\ln 2.67}{10} \approx .098. \end{aligned}$$

Therefore, the investment earned 9.8% interest per year.

5.3 APPLICATIONS OF THE NATURAL LOGARITHM FUNCTION TO ECONOMICS

In this section we consider two applications of the natural logarithm to the field of economics. Our first application is concerned with relative rates of change and the second with elasticity of demand.

RELATIVE RATES OF CHANGE The *logarithmic derivative* of a function $f(t)$ is defined by the equation

$$\frac{d}{dt} \ln f(t) = \frac{f'(t)}{f(t)}. \tag{1}$$

The quantity on either side of equation (1) is often called the *relative rate of change of* $f(t)$ *per unit change of* t. Indeed, this quantity compares the rate of change of $f(t)$ [namely, $f'(t)$] with $f(t)$ itself. The *percentage rate of change* is the relative rate of change of $f(t)$ expressed as a percentage.

A simple example will illustrate these concepts. Suppose that $f(t)$ denotes the average price per pound of sirloin steak at time t and $g(t)$ denotes the average price of a new car (of a given make and model) at time t, where $f(t)$ and $g(t)$ are given in dollars and time is measured in years. Then the ordinary derivatives $f'(t)$ and $g'(t)$ may be interpreted as the rate of change of the price of a pound of sirloin steak and of a new car, respectively, where both are measured in dollars per year. Suppose that, at a given time t_0, we have $f(t_0) = \$5.25$ and $g(t_0) = \$12{,}000$. Moreover, suppose that $f'(t_0) = \$.75$ and $g'(t_0) = \$1500$. Then at time t_0 the price per pound of steak is increasing at a rate of \$.75 per year, while the price of a new car is increasing at a rate of \$1500 per year. Which price is increasing more quickly? It is not meaningful to say that the car price is increasing faster simply because \$1500 is larger than \$.75. We must take into account the vast difference between the actual cost of a car and the cost of steak. The usual basis of comparison of price increases is the percentage rate of increase. In other words, at $t = t_0$, the price of sirloin steak is increasing at the percentage rate

$$\frac{f'(t_0)}{f(t_0)} = \frac{.75}{5.25} \approx .143 = 14.3\%$$

per year, but at the same time the price of a new car is increasing at the percentage rate

$$\frac{g'(t_0)}{g(t_0)} = \frac{1500}{12{,}000} \approx .125 = 12.5\%$$

per year. Thus the price of sirloin steak is increasing at a faster percentage rate than the price of a new car.

Economists often use relative rates of change (or percentage rates of change) when discussing the growth of various economic quantities, such as national income or national debt, because such rates of change can be meaningfully compared.

EXAMPLE 1 Suppose that a certain school of economists modeled the Gross National Product of the United States at time t (measured in years from January 1, 1990) by the formula

$$f(t) = 3.4 + .04t + .13e^{-t},$$

where the Gross National Product is measured in trillions of dollars. What was the predicted percentage rate of growth (or decline) of the economy at $t = 0$ and $t = 1$?

Solution Since

$$f'(t) = .04 - .13e^{-t},$$

we see that

$$\frac{f'(0)}{f(0)} = \frac{.04 - .13}{3.4 + .13} = -\frac{.09}{3.53} \approx -2.6\%,$$

$$\frac{f'(1)}{f(1)} = \frac{.04 - .13e^{-1}}{3.4 + .04 + .13e^{-1}} = -\frac{.00782}{3.4878} \approx -.2\%.$$

So on January 1, 1990, the economy is predicted to contract at a relative rate of 2.6% per year; on January 1, 1991, the economy is predicted to be still contracting but only at a relative rate of .2% per year. ■

EXAMPLE 2 Suppose that the value in dollars of a certain business investment at time t may be approximated empirically by the function

$$f(t) = 750{,}000e^{.6\sqrt{t}}.$$

Use a logarithmic derivative to describe how fast the value of the investment is increasing when $t = 5$ years.

Solution We have

$$\begin{aligned}\frac{f'(t)}{f(t)} &= \frac{d}{dt}\ln f(t) = \frac{d}{dt}\left(\ln 750{,}000 + \ln e^{.6\sqrt{t}}\right)\\ &= \frac{d}{dt}\left(\ln 750{,}000 + .6\sqrt{t}\right)\\ &= (.6)\left(\frac{1}{2}\right)t^{-1/2} = \frac{.3}{\sqrt{t}}.\end{aligned}$$

When $t = 5$,

$$\frac{f'(5)}{f(5)} = \frac{.3}{\sqrt{5}} \approx .1342 = 13.4\%.$$

Thus, when $t = 5$ years, the value of the investment is increasing at the relative rate of 13.4% per year. ■

In certain mathematical models, it is assumed that for a limited period of time, the percentage rate of change of a particular function is constant. The following example shows that such a function must be an exponential function.

EXAMPLE 3 Suppose that the function $f(t)$ has a constant relative rate of change k. Show that $f(t) = Ce^{kt}$ for some constant C.

Solution We are given that

$$\frac{d}{dt}\ln f(t) = k.$$

That is,

$$\frac{f'(t)}{f(t)} = k.$$

Hence $f'(t) = kf(t)$. But this is just the differential equation satisfied by exponential functions. Therefore, we must have $f(t) = Ce^{kt}$ for some constant C. ■

ELASTICITY OF DEMAND In Section 3.7 we considered demand equations for monopolists and for entire industries. Recall that a demand equation expresses, for each quantity q to be produced, the market price that will generate a demand of exactly q. For instance, the demand equation

$$p = 150 - .01x \tag{2}$$

says that in order to sell x units, the price must be set at $150 - .01x$ dollars. To be specific: In order to sell 6000 units, the price must be set at $150 - .01(6000) = \$90$ per unit.

Equation (2) may be solved for x in terms of p to yield

$$x = 100(150 - p). \tag{3}$$

This last equation gives quantity in terms of price. If we let the letter q represent quantity, equation (3) becomes

$$q = 100(150 - p). \tag{3$'$}$$

This equation is of the form $q = f(p)$, where in this case $f(p)$ is the function $f(p) = 100(150 - p)$. In what follows it will be convenient always to write our demand functions so that the quantity q is expressed as a function $f(p)$ of the price p.

Usually, raising the price of a commodity lowers demand. Therefore, the typical demand function $q = f(p)$ is decreasing and has a negative slope everywhere.

A demand function $q = f(p)$ relates the quantity demanded to the price. Therefore, the derivative $f'(p)$ compares the change in quantity demanded with the change in price. By way of contrast, the concept of elasticity is designed to compare the *relative* rate of change of the quantity demanded with the *relative* rate of change of price.

Let us be more explicit. Consider a particular demand function $q = f(p)$ and a particular price p. Then at this price, the ratio of the relative rates of change of the quantity demanded and the price is given by

$$\frac{\text{[relative rate of change of quantity]}}{\text{[relative rate of change of price]}} = \frac{\dfrac{d}{dp}\ln f(p)}{\dfrac{d}{dp}\ln p} = \frac{f'(p)/f(p)}{1/p} = \frac{pf'(p)}{f(p)}.$$

Since $f'(p)$ is always negative for a typical demand function, the quantity $pf'(p)/f(p)$ will be negative for all values of p. For convenience, economists prefer to work with positive numbers, and therefore the *elasticity of demand* is taken to be this quantity multiplied by -1.

The elasticity of demand $E(p)$ at price p for the demand function $q = f(p)$ is defined to be

$$E(p) = \frac{-pf'(p)}{f(p)}.$$

EXAMPLE 4 Suppose that the demand function for a certain metal is $q = 100 - 2p$, where p is the price per pound and q is the quantity demanded (in millions of pounds).

(a) What quantity can be sold at \$30 per pound?

(b) Determine the function $E(p)$.

(c) Determine and interpret the elasticity of demand at $p = 30$.

(d) Determine and interpret the elasticity of demand at $p = 20$.

Solution (a) In this case, $q = f(p)$, where $f(p) = 100 - 2p$. When $p = 30$, we have $q = f(30) = 100 - 2(30) = 40$. Therefore, 40 million pounds of the metal can be sold. We also say that the *demand* is 40 million pounds.

(b)
$$E(p) = \frac{-pf'(p)}{f(p)} = \frac{-p(-2)}{100 - 2p} = \frac{2p}{100 - 2p}.$$

(c) The elasticity of demand at price $p = 30$ is $E(30)$.

$$E(30) = \frac{2(30)}{100 - 2(30)} = \frac{60}{40} = \frac{3}{2}.$$

When the price is set at \$30 per pound, a small increase in price will result in a relative rate of decrease in quantity demanded of about $\frac{3}{2}$ times the relative rate of increase in price. For example, if the price is increased from \$30 by 1%, then the quantity demanded will decrease by about 1.5%.

(d) When $p = 20$, we have

$$E(20) = \frac{2(20)}{100 - 2(20)} = \frac{40}{60} = \frac{2}{3}.$$

When the price is set at \$20 per pound, a small increase in price will result in a relative rate of decrease in quantity demanded of only $\frac{2}{3}$ of the relative rate of increase of price. For example, if the price is increased from \$20 by 1%, the quantity demanded will decrease by $\frac{2}{3}$ of 1%. ■

Economists say that demand is *elastic* at price p_0 if $E(p_0) > 1$ and *inelastic* at price p_0 if $E(p_0) < 1$. In Example 4, the demand for the metal is elastic at \$30 per pound and inelastic at \$20 per pound.

The significance of the concept of elasticity may perhaps best be appreciated by studying how revenue, $R(p)$, responds to changes in price. Recall that

$$[\text{revenue}] = [\text{quantity}] \cdot [\text{price per unit}],$$

that is,

$$R(p) = f(p) \cdot p.$$

If we differentiate $R(p)$ using the product rule, we find that

$$\begin{aligned} R'(p) &= \frac{d}{dp}[f(p) \cdot p] = f(p) \cdot 1 + p \cdot f'(p) \\ &= f(p)\left[1 + \frac{pf'(p)}{f(p)}\right] \\ &= f(p)[1 - E(p)]. \end{aligned} \tag{4}$$

Now suppose that demand is elastic at some price p_0. Then $E(p_0) > 1$ and $1 - E(p_0)$ is negative. Since $f(p)$ is always positive, we see from (4) that $R'(p_0)$ is negative. Therefore, by the first derivative rule, $R(p)$ is decreasing at p_0. So an increase in price will result in a decrease in revenue, and a decrease in price will result in an increase in revenue. On the other hand, if demand is inelastic at p_0, then $1 - E(p_0)$ will be positive and hence $R'(p_0)$ will be positive. In this case an increase in price will result in an increase in revenue. We may summarize this as follows:

> The change in revenue is in the opposite direction of the change in price when demand is elastic and in the same direction when demand is inelastic.

Figure 1 shows the elasticity and revenue curves for the demand function of Example 4. Notice that demand is inelastic ($E(p) < 1$) to the left of $p = 25$ and the revenue function is increasing to the left of $p = 25$.

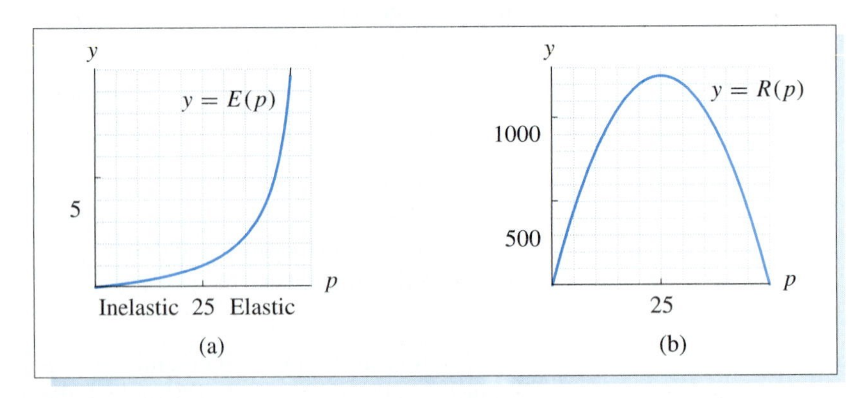

FIGURE 1 Elasticity and revenue functions for the demand function $q = 100 - 2p$

PRACTICE PROBLEMS 5.3

The current toll for the use of a certain toll road is \$2.50. A study conducted by the state highway department determined that with a toll of p dollars, q cars will use the road each day, where $q = 60{,}000e^{-.5p}$.

1. Compute the elasticity of demand at $p = 2.5$.
2. Is demand elastic or inelastic at $p = 2.5$?
3. If the state increases the toll slightly, will the revenue increase or decrease?

EXERCISES 5.3

Determine the percentage rate of change of the functions at the points indicated.

1. $f(t) = t^2$ at $t = 10$ and $t = 50$
2. $f(t) = t^{10}$ at $t = 10$ and $t = 50$
3. $f(x) = e^{.3x}$ at $x = 10$ and $x = 20$
4. $f(x) = e^{-.05x}$ at $x = 1$ and $x = 10$
5. $f(t) = e^{.3t^2}$ at $t = 1$ and $t = 5$
6. $G(s) = e^{-.05s^2}$ at $s = 1$ and $s = 10$
7. $f(p) = 1/(p + 2)$ at $p = 2$ and $p = 8$
8. $g(p) = 5/(2p + 3)$ at $p = 1$ and $p = 11$
9. Suppose that the annual sales S (in dollars) of a company may be approximated empirically by the formula

$$S = 50{,}000\sqrt{e^{\sqrt{t}}},$$

where t is the number of years beyond some fixed reference date. Use a logarithmic derivative to determine the percentage rate of growth of sales at $t = 4$.

10. Suppose that the price of wheat per bushel at time t (in months) is approximated by

$$f(t) = 4 + .001t + .01e^{-t}.$$

What is the percentage rate of change of $f(t)$ at $t = 0$? $t = 1$? $t = 2$?

11. Suppose an investment grows at a continuous 12% rate per year. In how many years will the value of the investment double?
12. Suppose that the value of a piece of property is growing at a continuous r% rate per year and that the value doubles in 3 years. Find r.

For each demand function, find $E(p)$ and determine if demand is elastic or inelastic (or neither) at the indicated price.

13. $q = 700 - 5p$, $p = 80$
14. $q = 600e^{-.2p}$, $p = 10$
15. $q = 400(116 - p^2)$, $p = 6$
16. $q = (77/p^2) + 3$, $p = 1$
17. $q = p^2e^{-(p+3)}$, $p = 4$
18. $q = 700/(p + 5)$, $p = 15$
19. Currently, 1800 people ride a certain commuter train each day and pay \$4 for a ticket. The number of people q willing to ride the train at price p is $q = 600(5 - \sqrt{p})$. The railroad would like to increase its revenue.
    (a) Is demand elastic or inelastic at $p = 4$?
    (b) Should the price of a ticket be raised or lowered?
20. A company can sell $q = 9000/(p + 60) - 50$ radios at a price of p dollars per radio. The current price is \$30.

(a) Is demand elastic or inelastic at $p = 30$?

(b) If the price is lowered slightly, will revenue increase or decrease?

21. A movie theater has a seating capacity of 3000 people. The number of people attending a show at price p dollars per ticket is $q = (18{,}000/p) - 1500$. Currently, the price is $6 per ticket.

    (a) Is demand elastic or inelastic at $p = 6$?

    (b) If the price is lowered, will revenue increase or decrease?

22. A subway charges 65 cents per person and has 10,000 riders each day. The demand function for the subway is $q = 2000\sqrt{90 - p}$.

    (a) Is demand elastic or inelastic at $p = 65$?

    (b) Should the price of a ride be raised or lowered in order to increase the amount of money taken in by the subway?

23. A country that is the major supplier of a certain commodity wishes to improve its balance-of-trade position by lowering the price of the commodity. The demand function is $q = 1000/p^2$.

    (a) Compute $E(p)$.

    (b) Will the country succeed in raising its revenue?

24. Show that any demand function of the form $q = a/p^m$ has constant elasticity m.

A cost function $C(x)$ gives the total cost of producing x units of a product. The elasticity of cost at quantity x *is defined to be*

$$E_c(x) = \frac{\frac{d}{dx}\ln C(x)}{\frac{d}{dx}\ln x}.$$

25. Show that $E_c(x) = x \cdot C'(x)/C(x)$.

26. Show that $E_c(x)$ is equal to the marginal cost divided by the average cost.

27. Let $C(x) = (1/10)x^2 + 5x + 300$. Show that $E_c(50) < 1$. (Hence when producing 50 units, a small relative increase in production results in an even smaller relative increase in total cost. Also, the average cost of producing 50 units is greater than the marginal cost at $x = 50$.)

28. Let $C(x) = 1000e^{.02x}$. Determine and simplify the formula for $E_c(x)$. Show that $E_c(60) > 1$, and interpret this result.

In the remaining exercises, use a graphing utility to obtain the graphs of the functions.

29. Sketch the revenue curve for the demand function $q = 500e^{-.5p}$ with the window $[0, 8]$ *by* $[0, 400]$, and determine the values of p for which demand is elastic.

30. Sketch the revenue curve for the demand function $q = \sqrt{64 - p^2}$ with the window $[0, 8]$ *by* $[0, 32]$, and determine the values of p for which demand is inelastic.

SOLUTIONS TO PRACTICE PROBLEMS 5.3

1. The demand function is $f(p) = 60{,}000e^{-.5p}$.

$$f'(p) = -30{,}000e^{-.5p},$$

$$E(p) = \frac{-pf'(p)}{f(p)} = \frac{-p(-30{,}000)e^{-.5p}}{60{,}000e^{-.5p}} = \frac{p}{2},$$

$$E(2.5) = \frac{2.5}{2} = 1.25.$$

2. The demand is elastic, because $E(2.5) > 1$.
3. Since demand is elastic at \$2.50, a slight change in price causes revenue to change in the *opposite* direction. Hence revenue will decrease.

REVIEW OF THE FUNDAMENTAL CONCEPTS OF CHAPTER 5

1. What differential equation is key to solving exponential growth and decay problems? State a result about the solution to this differential equation.
2. What is the significance of a growth constant? Decay constant?
3. Explain how radiocarbon dating works.
4. State the formula for each of the following quantities.
   (a) compound amount of P dollars after n interest periods at an interest rate i per period
   (b) compound amount of P dollars in t years at interest rate r compounded continuously
   (c) present value of A dollars in t years at interest rate r compounded continuously
5. What is the difference between a relative rate of change and a percentage rate of change?
6. Define the elasticity of demand, $E(p)$, for a demand function. How is $E(p)$ used?

Chapter 6

ANTIDIFFERENTIATION AND DIFFERENTIAL EQUATIONS

A *differential equation* is an equation expressing a relationship between an unknown function $y = f(x)$ and one or more of its derivatives. Examples of such equations are

$$y' = 6x + 3,$$
$$y' = 6y,$$
$$y' = .05y(1000 - y).$$

As we shall see, many physical processes can be described by differential equations. In this chapter we explore some topics in differential equations and use our resulting knowledge to study problems from many different fields, including business, genetics, and ecology. Several types of differential equations can be solved by antidifferentiation, the reverse of the differentiation process.

6.1 ANTIDIFFERENTIATION

We have developed several techniques for calculating the derivative $F'(x)$ of a function $F(x)$. In many applications, however, it is necessary to proceed in reverse. We are given the derivative $F'(x)$ and must determine the function $F(x)$. The process of determining $F(x)$ from $F'(x)$ is called *antidifferentiation*. The next example gives a typical application involving antidifferentiation.

EXAMPLE 1 During the early 1970s, the annual worldwide rate of oil consumption grew exponentially with a growth constant of about .07. At the beginning of 1970, the rate was about 16.1 billion barrels of oil per year. Let $R(t)$ denote the rate of oil consumption at time t, where t is the number of years since the beginning of 1970. Then a reasonable model for $R(t)$ is given by

$$R(t) = 16.1e^{.07t}. \tag{1}$$

Use this formula for $R(t)$ to determine the total amount of oil that would have been consumed from 1970 to 1980 had this rate of consumption continued throughout the decade.

Solution Let $T(t)$ be the total amount of oil consumed from time 0 (1970) until time t. We wish to calculate $T(10)$, the amount of oil consumed from 1970 to 1980. We do this by first determining a formula for $T(t)$. Since $T(t)$ is the total oil consumed, the derivative $T'(t)$ is the *rate* of oil consumption, namely, $R(t)$. Thus, although we do not yet have a formula for $T(t)$, we do know that

$$T'(t) = 16.1e^{.07t}.$$

Thus the problem of determining a formula for $T(t)$ has been reduced to a problem of antidifferentiation: Find a function whose derivative is $R(t)$. We shall solve this particular problem after developing some techniques for solving antidifferentiation problems in general. ■

Suppose that $f(x)$ is a given function and $F(x)$ is a function having $f(x)$ as its derivative—that is, $F'(x) = f(x)$. We call $F(x)$ an *antiderivative* of $f(x)$.

EXAMPLE 2 Find an antiderivative of $f(x) = x^2$.

Solution One such function is $F(x) = \frac{1}{3}x^3$, since

$$F'(x) = \frac{1}{3} \cdot 3x^2 = x^2.$$

Another antiderivative is $F(x) = \frac{1}{3}x^3 + 2$, since

$$\frac{d}{dx}\left(\frac{1}{3}x^3 + 2\right) = \frac{1}{3} \cdot 3x^2 + 0 = x^2.$$

In fact, if C is any constant, the function $F(x) = \frac{1}{3}x^3 + C$ is also an antiderivative of x^2, since

$$\frac{d}{dx}\left(\frac{1}{3}x^3 + C\right) = \frac{1}{3} \cdot 3x^2 + 0 = x^2.$$

(The derivative of a constant function is zero.) ■

EXAMPLE 3 Find an antiderivative of the function $f(x) = 2x - (1/x^2)$.

Solution Since

$$\frac{d}{dx}(x^2) = 2x \quad \text{and} \quad \frac{d}{dx}\left(\frac{1}{x}\right) = -\frac{1}{x^2},$$

we see that one antiderivative of $f(x)$ is given by

$$F(x) = x^2 + \frac{1}{x}.$$

However, any function of the form $x^2 + (1/x) + C$, C a constant, will do, since

$$\frac{d}{dx}\left(x^2 + \frac{1}{x} + C\right) = 2x - \frac{1}{x^2} + 0 = 2x - \frac{1}{x^2}.$$ ■

Using the same reasoning as in Examples 2 and 3, we see that if $F(x)$ is an antiderivative of $f(x)$, then so is $F(x) + C$, where C is any constant. Thus if we know one antiderivative $F(x)$ of a function $f(x)$, we can write down an infinite number of antiderivatives of $f(x)$ by adding all possible constants C to $F(x)$. It turns out that in this way we obtain *all* antiderivatives of $f(x)$. That is, we have the following fundamental result.

Theorem I If $F_1(x)$ and $F_2(x)$ are two antiderivatives of the same function $f(x)$, then $F_1(x)$ and $F_2(x)$ differ by a constant. In other words, there is a constant C such that

$$F_2(x) = F_1(x) + C.$$

Geometrically, the graph of any antiderivative $F_2(x)$ is obtained by shifting the graph of $F_1(x)$ vertically. See Figure 1.

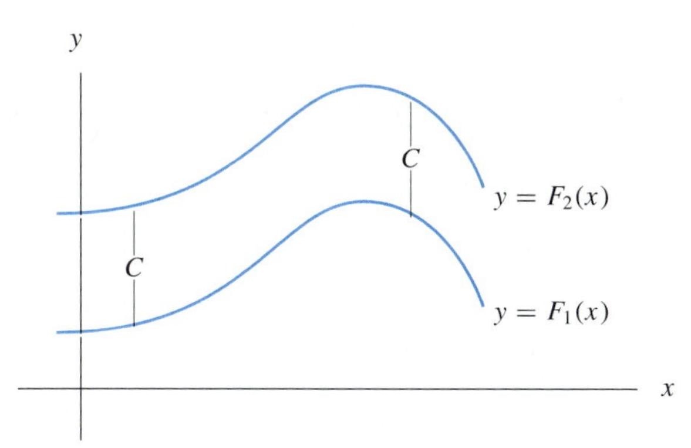

FIGURE 1 Two antiderivatives of the same function

Our verification of Theorem I is based on the following fact, which is important in its own right.

Theorem II If $F'(x) = 0$ for all x, then $F(x) = C$ for some constant C.

It is easy to see why Theorem II is reasonable. (A formal proof of the theorem requires an important theoretical result called the mean value theorem.) If

$F'(x) = 0$ for all x, then the curve $y = F(x)$ has slope equal to zero at every point. Thus the tangent line to $y = F(x)$ at any point is horizontal, which implies that the graph of $y = F(x)$ is a horizontal line. (Try to draw the graph of a function with a horizontal tangent everywhere. There is no choice but to keep your pencil moving on a constant, horizontal line!) If the horizontal line is $y = C$, then $F(x) = C$ for all x.

VERIFICATION OF THEOREM I If $F_1(x)$ and $F_2(x)$ are two antiderivatives of $f(x)$, then the function $F(x) = F_2(x) - F_1(x)$ has the derivative

$$F'(x) = F_2'(x) - F_1'(x) = f(x) - f(x) = 0.$$

So, by Theorem II, we know that $F(x) = C$ for some constant C. In other words, $F_2(x) - F_1(x) = C$, so that

$$F_2(x) = F_1(x) + C,$$

which is Theorem I.

Using Theorem I, we can find *all* antiderivatives of a given function once we know one antiderivative. For instance, since one antiderivative of x^2 is $\frac{1}{3}x^3$ (Example 2), all antiderivatives of x^2 have the form $\frac{1}{3}x^3+C$, where C is a constant.

Suppose that $f(x)$ is a function whose antiderivatives are $F(x) + C$. The standard way to express this fact is to write

$$\int f(x)\,dx = F(x) + C.$$

The symbol $\int$ is called an *integral sign*. The entire notation $\int f(x)\,dx$ is called an *indefinite integral* and stands for antidifferentiation of the function $f(x)$. We always record the variable of interest by prefacing it by the letter d. For example, if the variable of interest is t rather than x, then we write $\int f(t)\,dt$ for the antiderivative of $f(t)$.

EXAMPLE 4 Determine

(a) $\int x^r\,dx$, r a constant $\neq -1$. (b) $\int e^{kx}\,dx$, k a constant $\neq 0$.

Solution (a) By the constant-multiple and power rules,

$$\frac{d}{dx}\left(\frac{1}{r+1}x^{r+1}\right) = \frac{1}{r+1}\cdot\frac{d}{dx}x^{r+1} = \frac{1}{r+1}\cdot(r+1)x^r = x^r.$$

Thus $x^{r+1}/(r+1)$ is an antiderivative of x^r. Letting C represent any constant, we have

$$\int x^r\,dx = \frac{1}{r+1}x^{r+1} + C, \quad r \neq -1. \tag{2}$$

(b) An antiderivative of e^{kx} is e^{kx}/k, since

$$\frac{d}{dx}\left(\frac{1}{k}e^{kx}\right) = \frac{1}{k}\cdot\frac{d}{dx}e^{kx} = \frac{1}{k}(ke^{kx}) = e^{kx}.$$

Hence

$$\int e^{kx}\,dx = \frac{1}{k}e^{kx} + C, \quad k \neq 0. \tag{3}$$

Formula (2) does not give an antiderivative of x^{-1} because $1/(r+1)$ is undefined for $r = -1$. However, we know that for $x \neq 0$, the derivative of $\ln|x|$ is $1/x$. Hence $\ln|x|$ is an antiderivative of $1/x$, and we have

$$\int \frac{1}{x}\,dx = \ln|x| + C, \quad x \neq 0. \tag{4}$$

Formulas (2), (3), and (4) were each obtained by "reversing" a familiar differentiation rule. In a similar fashion, one may use the sum rule and constant-multiple rule for derivatives to obtain rules for antiderivatives:

$$\int [f(x) + g(x)]\,dx = \int f(x)\,dx + \int g(x)\,dx, \tag{5}$$

$$\int kf(x)\,dx = k\int f(x)\,dx, \quad k \text{ a constant.} \tag{6}$$

In words, (5) says that a sum of functions may be antidifferentiated term by term, and (6) says that a constant multiple may be moved through the integral sign. ■

EXAMPLE 5 Compute

$$\int \left(t^{-3} + 7e^{5t} + \frac{4}{t}\right) dt.$$

Solution Using the preceding rules, we have

$$\begin{aligned}
\int \left(t^{-3} + 7e^{5t} + \frac{4}{t}\right) dt &= \int t^{-3}\,dt + \int 7e^{5t}\,dt + \int \frac{4}{t}\,dt \\
&= \int t^{-3}\,dt + 7\int e^{5t}\,dt + 4\int \frac{1}{t}\,dt \\
&= \frac{1}{-2}t^{-2} + 7\left(\frac{1}{5}e^{5t}\right) + 4\ln|t| + C \\
&= \frac{-1}{2t^2} + \frac{7}{5}e^{5t} + 4\ln|t| + C.
\end{aligned}$$ ■

After some practice, most of the intermediate steps shown in the solution of Example 5 can be omitted.

A function has infinitely many different antiderivatives, corresponding to the various choices of the constant C. In applications, it is often necessary to satisfy an additional condition, which then determines a specific value of C.

EXAMPLE 6 Find the function $f(x)$ for which $f'(x) = x^2 - 2$ and $f(1) = \frac{4}{3}$.

Solution The unknown function $f(x)$ is an antiderivative of $x^2 - 2$. One antiderivative of $x^2 - 2$ is $\frac{1}{3}x^3 - 2x$. Therefore, by Theorem I,

$$f(x) = \frac{1}{3}x^3 - 2x + C, \qquad C \text{ a constant.}$$

Figure 2 shows the graphs of $f(x)$ for several choices of C. We want the function whose graph passes through $(1, \frac{4}{3})$. To find the value of C that makes $f(1) = \frac{4}{3}$, we set

$$\frac{4}{3} = f(1) = \frac{1}{3}(1)^3 - 2(1) + C = -\frac{5}{3} + C$$

and find $C = \frac{4}{3} + \frac{5}{3} = 3$. Therefore, $f(x) = \frac{1}{3}x^3 - 2x + 3$. ■

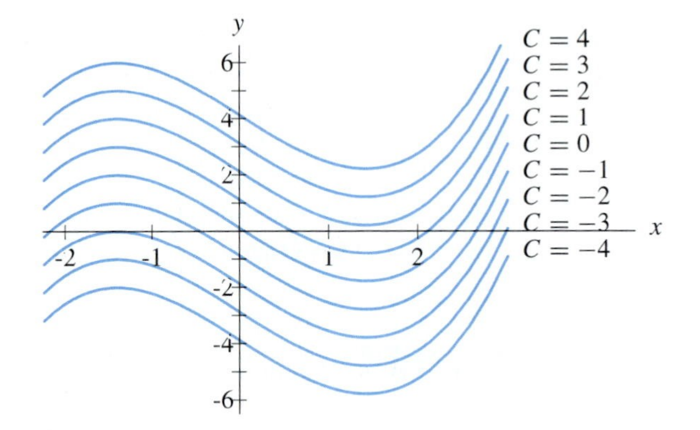

FIGURE 2 Several antiderivatives of $x^2 - 2$

PRACTICE PROBLEMS 6.1

1. Determine each of the following.

   (a) $\displaystyle\int t^{7/2}\,dt$ (b) $\displaystyle\int \left(\frac{x^3}{3} + \frac{3}{x^3} + \frac{3}{x}\right) dx$

2. Find the value of k that makes the antidifferentiation formula true:

$$\int (1 - 2x)^3\,dx = k(1 - 2x)^4 + C.$$

EXERCISES 6.1

Find all antiderivatives of each of the following functions.

1. $f(x) = x$
2. $f(x) = 9x^8$
3. $f(x) = e^{3x}$
4. $f(x) = e^{-3x}$
5. $f(x) = 3$
6. $f(x) = -4x$

In Exercises 7–22, find the value of k that makes the antidifferentiation formula true. (Note: You can check your answer without looking in the answer section. How?)

7. $\int x^{-5}\,dx = kx^{-4} + C$
8. $\int x^{1/3}\,dx = kx^{4/3} + C$
9. $\int \sqrt{x}\,dx = kx^{3/2} + C$
10. $\int \frac{6}{x^3}\,dx = \frac{k}{x^2} + C$
11. $\int \frac{10}{t^6}\,dt = kt^{-5} + C$
12. $\int \frac{3}{\sqrt{t}}\,dt = k\sqrt{t} + C$
13. $\int 5e^{-2t}\,dt = ke^{-2t} + C$
14. $\int 3e^{t/10}\,dt = ke^{t/10} + C$
15. $\int 2e^{4x-1}\,dx = ke^{4x-1} + C$
16. $\int \frac{4}{e^{3x+1}}\,dx = \frac{k}{e^{3x+1}} + C$
17. $\int (x-7)^{-2}\,dx = k(x-7)^{-1} + C$
18. $\int \sqrt{x+1}\,dx = k(x+1)^{3/2} + C$
19. $\int (x+4)^{-1}\,dx = k\ln|x+4| + C$
20. $\int \frac{5}{(x-8)^4}\,dx = \frac{k}{(x-8)^3} + C$
21. $\int (3x+2)^4\,dx = k(3x+2)^5 + C$
22. $\int (2x-1)^3\,dx = k(2x-1)^4 + C$

Determine the following.

23. $\int (x^2 - x - 1)\,dx$
24. $\int (x^3 + 6x^2 - x)\,dx$
25. $\int \left(\frac{2}{\sqrt{x}} - 3\sqrt{x}\right)dx$
26. $\int \left[\frac{\sqrt{t}}{4} - 4(t-3)^{-2}\right]dt$
27. $\int \left(4 - 5e^{-5t} + \frac{e^{2t}}{3}\right)dt$
28. $\int (e^2 + 3t^2 - 2e^{3t})\,dt$

Find all functions $f(t)$ with the following property.

29. $f'(t) = t^{3/2}$
30. $f'(t) = \frac{4}{6+t}$
31. $f'(t) = 0$
32. $f'(t) = t^2 - 5t - 7$

Find all functions $f(x)$ with the following properties.

33. $f'(x) = x,\ f(0) = 3$
34. $f'(x) = 8x^{1/3},\ f(1) = 4$

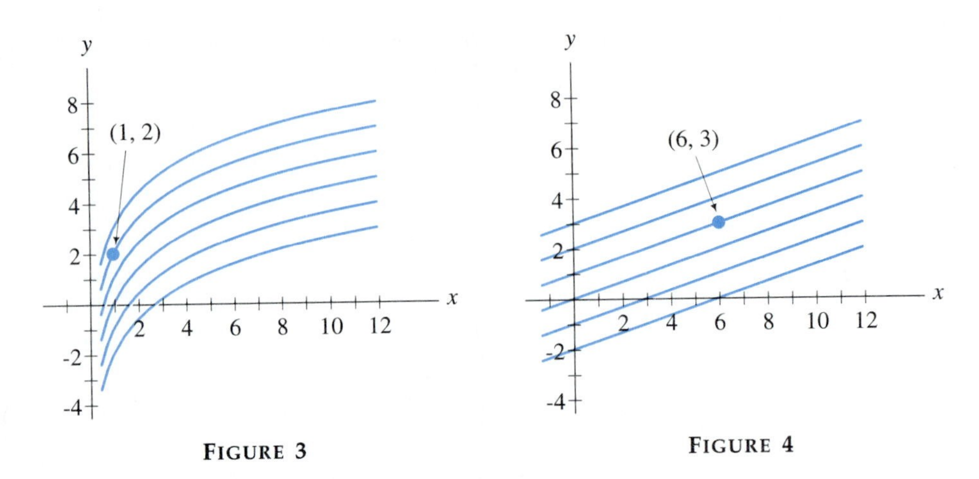

FIGURE 3 FIGURE 4

35. $f'(x) = \sqrt{x} + 1,\ f(4) = 0$

36. $f'(x) = x^2 + \sqrt{x},\ f(1) = 3$

37. Figure 3 shows the graphs of several functions $f(x)$ for which $f'(x) = 2/x$. Find the expression for the function $f(x)$ whose graph passes through $(1, 2)$

38. Figure 4 shows the graphs of several functions $f(x)$ for which $f'(x) = 1/3$. Find the expression for the function $f(x)$ whose graph passes through $(6, 3)$.

39. Which of the following is $\int \ln x\, dx$?

(a) $\dfrac{1}{x} + C$ (b) $x \cdot \ln x - x + C$ (c) $\dfrac{1}{2} \cdot (\ln x)^2 + C$

40. Which of the following is $\int x\sqrt{x+1}\, dx$?

(a) $\frac{2}{5}(x+1)^{5/2} - \frac{2}{3}(x+1)^{3/2} + C$ (b) $\frac{1}{2}x^2 \cdot \frac{2}{3}(x+1)^{3/2} + C$

41. Figure 5 contains the graph of a function $F(x)$. On the same coordinate system, draw the graph of the function $G(x)$ having the properties $G(0) = 0$ and $G'(x) = F'(x)$ for each x.

42. Figure 6 contains an antiderivative of the function $f(x)$. Draw the graph of another antiderivative of $f(x)$.

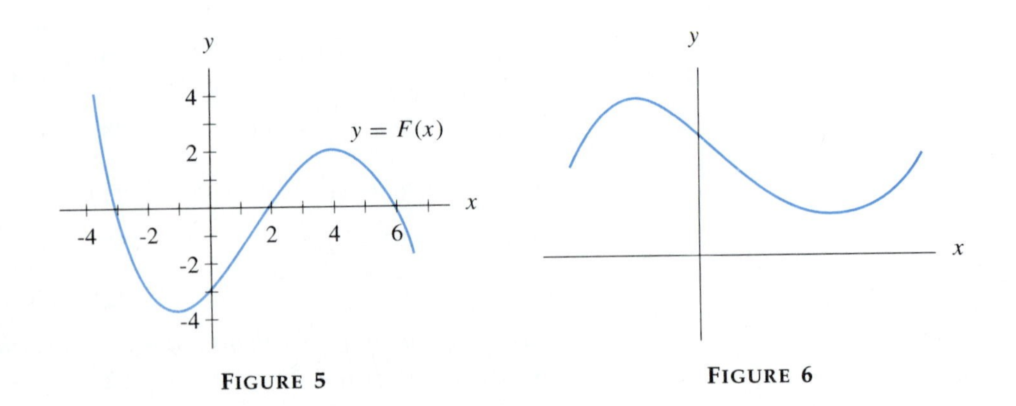

FIGURE 5 FIGURE 6

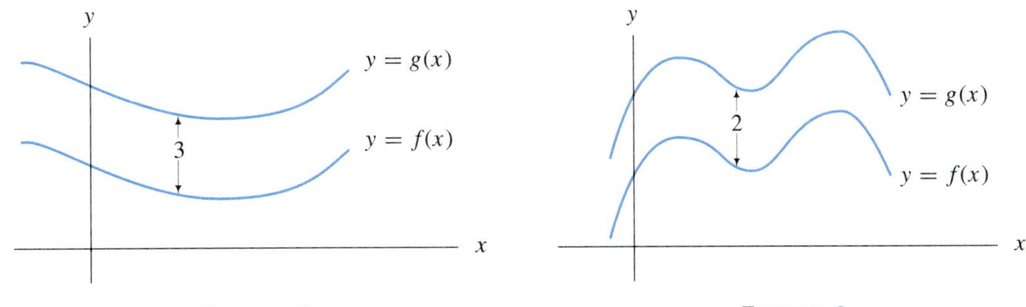

FIGURE 7

FIGURE 8

43. The function $g(x)$ in Figure 7 was obtained by shifting the graph of $f(x)$ up three units. If $f'(5) = \frac{1}{4}$, what is $g'(5)$?

44. The function $g(x)$ in Figure 8 was obtained by shifting the graph of $f(x)$ up two units. What is the derivative of $h(x) = g(x) - f(x)$?

In the remaining exercises, use a graphing utility to obtain the graphs of the functions. In Exercises 45–48, find an antiderivative of $f(x)$, call it $F(x)$, and then compare the graphs of $F(x)$ and $f(x)$ in the given window to check that the expression for $F(x)$ is reasonable. [That is, determine whether the two graphs are consistent. When $F(x)$ has a relative extreme point, $f(x)$ should be zero. When $F(x)$ is increasing, $f(x)$ should be positive. And so on.]

45. $f(x) = 2x - e^{-.02x}$; $[-10, 10]$ *by* $[-20, 100]$

46. $f(x) = e^{2x} + e^{-x} + \frac{1}{2}x^2$; $[-2.4, 1.7]$ *by* $[-10, 10]$

47. $f(x) = 8x^2 - 3e^{-x} + x^3$; $[-3, 2]$ *by* $[-4, 7]$

48. $f(x) = \frac{1}{10}(x^3 - 9x^2 + 5x) + 3$; $[-4, 11]$ *by* $[-14, 7]$

SOLUTIONS TO PRACTICE PROBLEMS 6.1

1. (a) $\int t^{7/2}\,dt = \frac{1}{\frac{9}{2}}t^{9/2} + C = \frac{2}{9}t^{9/2} + C.$

(b)
$$\int\left(\frac{x^3}{3} + \frac{3}{x^3} + \frac{3}{x}\right)dx = \int\left(\frac{1}{3}\cdot x^3 + 3x^{-3} + 3\cdot\frac{1}{x}\right)dx$$
$$= \frac{1}{3}\left(\frac{1}{4}x^4\right) + 3\left(-\frac{1}{2}x^{-2}\right) + 3\ln|x| + C$$
$$= \frac{1}{12}x^4 - \frac{3}{2}x^{-2} + 3\ln|x| + C.$$

2. Since we are told that the antiderivative has the general form $k(1-2x)^4$, all we have to do is determine the value of k for which
$$\frac{d}{dx}[k(1-2x)^4] = (1-2x)^3.$$
Differentiating, we obtain
$$\frac{d}{dx}[k(1-2x)^4] = 4k(1-2x)^3(-2) \quad \text{or} \quad -8k(1-2x)^3,$$
which is supposed to equal $(1-2x)^3$. Therefore, $-8k = 1$, so $k = -\frac{1}{8}$.

6.2 DIFFERENTIAL EQUATIONS

Example 6 of the previous section states:

Find the function $f(x)$ for which $f'(x) = x^2 - 2$ and $f(1) = \frac{4}{3}$.

Let y denote the unknown function $f(x)$. The condition $f(1) = \frac{4}{3}$, called an *initial condition*, can be written as $y(1) = \frac{4}{3}$. With this notation, the statement of Example 6 becomes

$$\text{Solve } y' = x^2 - 2,\ y(1) = \frac{4}{3}. \tag{1}$$

The equation $y' = x^2 - 2$ is a differential equation, and the pair $y' = x^2 - 2$, $y(1) = \frac{4}{3}$ is called a "differential equation with initial condition," an "initial-value problem," or just simply a "differential equation."

In general, the solution to a differential equation of the form $y' = g(x)$ is the infinite collection of antiderivatives of $g(x)$. The solution to the initial-value problem $y' = g(x)$, $y(a) = b$ is the specific antiderivative whose graph contains the point (a, b). The method for solving $y' = g(x)$, $y(a) = b$ is as follows.

1. Solve $y' = g(x)$. The solution, $y = \int g(x)\,dx$, is an expression involving the constant C.
2. Substitute a for x in the expression found in step 1, set it equal to b, and solve the equation for C.
3. Write down the solution from step 1 with the value of C found in step 2.

Having introduced the basics of antidifferentiation, let us now solve the oil-consumption problem posed in Example 1 of Section 6.1.

SOLUTION OF OIL-CONSUMPTION PROBLEM The rate of oil consumption at time t is $R(t) = 16.1e^{.07t}$ billion barrels per year. Moreover, we observed that the total consumption $T(t)$, from time 0 to time t, is an antiderivative of $R(t)$. Also, $T(0) = 0$. Letting $y = T(t)$, we must solve the differential equation

$$y' = 16.1e^{.07t}, \qquad y(0) = 0.$$

The equation $y' = 16.1e^{.07t}$ has solution $\int 16.1e^{.07t}\,dt$. Using (3) and (6) from Section 6.1,

$$y = \int 16.1e^{.07}\,dt = \frac{16.1}{.07}e^{.07t}\,dt + C = 230e^{.07t} + C,$$

where C is a constant. Since $y(0) = 0$, the constant C must satisfy

$$0 = y(0) = 230e^{.07(0)} + C = 230 + C$$
$$C = -230.$$

Therefore,

$$y = 230e^{.07t} - 230 = 230(e^{.07t} - 1).$$

That is,

$$T(t) = 230(e^{.07t} - 1).$$

The total amount of oil that would have been consumed from 1970 to 1980 is

$$T(10) = 230(e^{.07(10)} - 1) \approx 233 \text{ billion barrels.}$$

EXAMPLE 1 A rocket is fired vertically into the air. Its velocity at t seconds after lift-off is $v(t) = 6t + .5$ meters per second. Before launch, the top of the rocket is 8 meters above the launch pad. Find the height of the rocket (measured from the top of the rocket to the launch pad) at time t.

Solution If $s(t)$ denotes the height of the rocket at time t, then $s'(t)$ is the rate at which the height is changing. That is, $s'(t) = v(t)$. When $t = 0$, the rocket's height is 8 meters. That is, $s(0) = 8$. Letting $y = s(t)$, we must solve the differential equation

$$y' = 6t + .5, \quad y(0) = 8.$$

The equation $y' = 6t + .5$ has solution $\int (6t + .5)\,dt$. Thus,

$$y = \int (6t + .5)\,dt = 3t^2 + .5t + C,$$

where C is constant. Since $y(0) = 8$, the constant C must satisfy

$$8 = y(0) = 3(0)^2 + .5(0) + C = C.$$

Thus $C = 8$ and

$$y = 3t^2 + .5t + 8.$$

That is,

$$s(t) = 3t^2 + .5t + 8.$$

EXAMPLE 2 A company's marginal cost function is $.015x^2 - 2x + 80$ dollars, where x denotes the number of units produced in one day. The company has fixed costs of $1000 per day.

(a) Find the cost of producing x units per day.

(b) Suppose the current production level is $x = 30$. Determine the increase in cost if the production level is raised to $x = 60$ units.

Solution (a) Let $C(x)$ be the cost of producing x units in one day. The derivative $C'(x)$ is the marginal cost. In other words, $C(x)$ is an antiderivative of the marginal cost function. The $1000 fixed costs are the costs incurred when producing 0 units. That is, $C(0) = 1000$. Letting $y = C(x)$, we must solve the differential equation

$$y' = .015x^2 - 2x + 80, \quad y(0) = 1000.$$

The equation $y' = .015x^2 - 2x + 80$ has solution $\int .015x^2 - 2x + 80\,dx$. Thus,

$$y = \int .015x^2 - 2x + 80\,dx = .005x^3 - x^2 + 80x + C,$$

where C is a constant. Since $y(0) = 1000$, the constant C must satisfy

$$1000 = y(0) = .005(0)^3 - (0)^2 + 80(0) + C = C$$
$$C = 1000.$$

Therefore,

$$y = .005x^3 - x^2 + 80x + 1000.$$

That is,

$$C(x) = .005x^3 - x^2 + 80x + 1000.$$

(b) The cost when $x = 30$ is $C(30)$, and the cost when $x = 60$ is $C(60)$. So the *increase* in cost when production is raised from $x = 30$ to $x = 60$ is $C(60) - C(30)$. We compute

$$C(60) = .005(60)^3 - (60)^2 + 80(60) + 1000 = 3280,$$
$$C(30) = .005(30)^3 - (30)^2 + 80(30) + 1000 = 2635.$$

Thus, the increase in cost is \$3280 – \$2635 = \$645. ■

GRAPHING ANTIDERIVATIVES WITH A TI CALCULATOR

For a discussion of graphing antiderivatives with Visual Calculus, see Appendix A, Part VIII.

When graphing antiderivatives with a TI calculator use x as the variable. The following steps produce the graph of the solution to (1).

1. Set $\mathtt{Y_1} = \mathtt{X}^2 - 2$.
2. Set `Y₂=fnInt(Y₁,X,1,X)+(4/3)`. (*Note*: `fnInt` is found in the MATH menu of a TI-82 or TI-83, and in the CALC menu of a TI-85.)
3. Select $\mathtt{Y_2}$ and deselect all other functions.
4. Set the window to $[-2.25, 2.25]$ *by* $[-7, 7]$.
5. Graph the function $\mathtt{Y_2}$. (*Note*: Graphing requires about 40 seconds.)

In general, the following steps graph the solution to the differential equation $y' = g(x)$, $y(a) = b$.

1. Set $\mathtt{Y_1} = \mathtt{g(X)}$.
2. Set `Y₂=fnInt(Y₁,X,a,X)+b`.
3. Select $\mathtt{Y_2}$ and deselect all other functions.
4. Set an appropriate window.
5. Graph $\mathtt{Y_2}$.

All differential equations discussed in this section have been of the form $y' = g(x)$ and are solved by antidifferentiation. In Chapters 4 and 5 we considered differential equations of the form $y' = ky$, which have solutions of the form Ce^{kx}. The following example shows another type of differential equation used to describe a physical process. (The solution to this type of differential equation is graphed in Section 6.3 and given explicitly in Section 6.4.) The example should be studied carefully, for it contains the key to understanding many similar problems that will appear in the exercises.

Newton's Law of Cooling

EXAMPLE 3 Suppose that a red-hot rod is plunged into a bath of cool water. Let $f(t)$ be the temperature of the rod at time t, and suppose that the water is maintained at a constant temperature of 10 °C. According to Newton's law of cooling, the rate of change of $f(t)$ is proportional to the difference between the two temperatures 10° and $f(t)$. Find a differential equation that describes this physical law.

Solution The two key ideas are "rate of change" and "proportional." The rate of change of $f(t)$ is the derivative $f'(t)$. Since this is proportional to the difference $10 - f(t)$, there exists a constant k such that

$$f'(t) = k[10 - f(t)]. \tag{2}$$

The term "proportional" does not tell us whether k is positive or negative (or zero). We must decide this, if possible, from the context of the problem. In the present situation, the steel rod is hotter than the water, so $10 - f(t)$ is negative. Also, $f(t)$ will decrease as time passes, so $f'(t)$ should be negative. Thus, to make $f'(t)$ negative in (2), k must be a positive number. From (2) we see that $y = f(t)$ satisfies a differential equation of the form $y' = k(10 - y)$, k a positive constant. ■

A constant function that satisfies a differential equation is called a *constant solution* of the differential equation. Constant solutions occur in many of the applied problems considered later in the chapter.

EXAMPLE 4 Find a constant solution of $y' = 3y - 12$.

Solution Let $f(t) = c$ for all t. Then $f'(t)$ is zero for all t. If $f(t)$ satisfies the differential equation

$$f'(t) = 3 \cdot f(t) - 12,$$

then $0 = 3 \cdot c - 12$, and so $c = 4$. This is the only possible value for a constant solution. Substitution shows that the function $f(t) = 4$ is indeed a solution of the differential equation. ■

PRACTICE PROBLEMS 6.2

Solve the following differential equations.

1. $y' = e^{3x}$
2. $y' = 3y$

EXERCISES 6.2

Solve the following differential equations.

1. $y' = 2x^4 - 5 + \dfrac{3}{x^2}$
2. $y' = 3x - 1$
3. $y' = e^3$
4. $y' = 4$
5. $y' = e^{\frac{x}{4}} - \dfrac{3}{x}$
6. $y' = e^{-.2x} + \dfrac{1}{x}$
7. $y' = 3x^2 - 5x + 6$, $y(0) = 5$
8. $y' = 4 - x^5$, $y(0) = 3$
9. $y' = e^{\frac{1}{2}x}$, $y(0) = -2$
10. $y' = e^{3x}$, $y(0) = .6$
11. $y' = -\dfrac{4}{x}$, $y(1) = 8$
12. $y' = \dfrac{2}{x^2}$, $y(1) = 3$
13. $y' = \frac{1}{4}y$, $y(0) = 6$
14. $y' = 3y$, $y(0) = .54$
15. Figure 1 shows the graphs of several functions that satisfy the differential equation $y' = x^3$. Find the expression for the function whose graph passes through $(1, -3)$.
16. Figure 2 shows the graphs of several functions that satisfy the differential equation $y' = -2/x^2$. Find the expression for the function whose graph passes through $(2, -1)$.

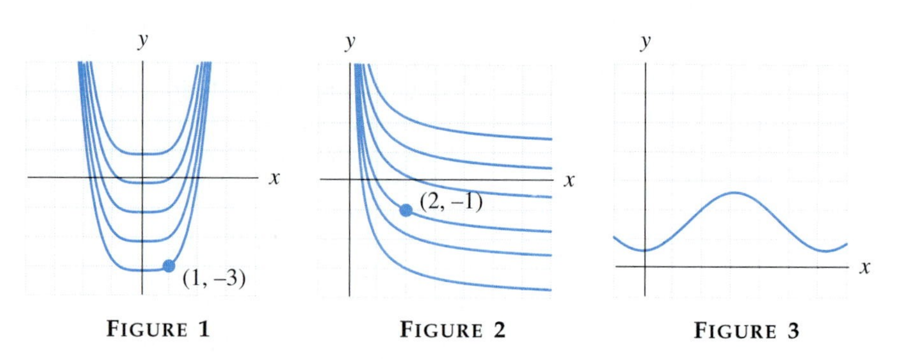

FIGURE 1 FIGURE 2 FIGURE 3

17. Which of the following is a solution to $y' = xe^x$?

    (a) $e^x - xe^x$ (b) $xe^x + e^x$ (c) $xe^x - e^x$

18. Which of the following is a solution to $y' = 9x^2 \ln x$?

    (a) $3x^3 \ln x + x^3$ (b) $3x^3 \ln x - x^3$ (c) $x^3 - 3x^3 \ln x$

19. Figure 3 contains the graph of a solution to a differential equation of the form $y' = g(x)$. On the same coordinate system, draw the graph of the solution to the differential equation satisfying the initial condition $y(0) = 2$.
20. The graph of $y = h(x)$ in Figure 4 was obtained by shifting the graph of $y = f(x)$ up three units. $f(x)$ is the solution to $y' = \ln x$, $y(2) = -1$. Find an initial-value problem having $h(x)$ as a solution.
21. The function $f(x) = \frac{1}{2}x^2 \ln x - \frac{1}{4}x^2 + 1$ satisfies the differential equation $y' = x \ln x$, $y(1) = .75$. Solve the differential equation $y' = x \ln x$, $y(1) = 2$.
22. The function $f(x) = x^2 e^x - 2xe^x + 2e^x + 1$ satisfies the differential equation $y' = x^2 e^x$, $y(1) = 3$. Solve the differential equation $y' = x^2 e^x$, $y(1) = 0$.

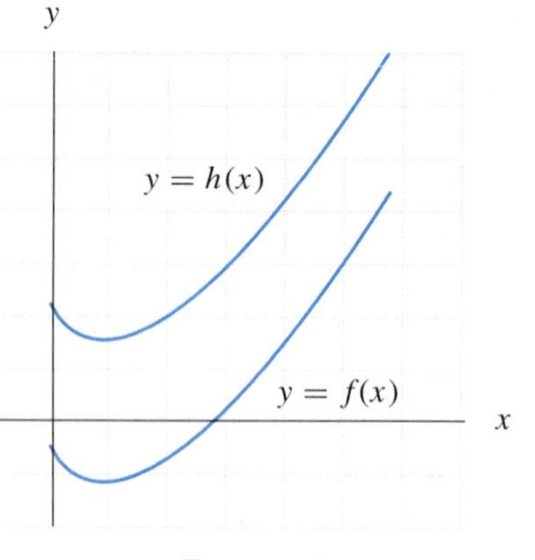

FIGURE 4

23. Let $f(t)$ denote the amount of capital invested by a certain business firm at time t. The rate of change of invested capital, $f'(t)$, is sometimes called the *rate of net investment*. Suppose that the management of the firm decides that the optimum level of investment should be C dollars and that, at any time, the rate of net investment should be proportional to the difference between C and the total capital invested. Construct a differential equation that describes this situation.

24. A cool object is placed in a room that is maintained at a constant temperature of 20 °C. The rate at which the temperature of the object rises is proportional to the difference between the room temperature and the temperature of the object. Let $y = f(t)$ be the temperature of the object at time t, and give a differential equation that describes the rate of change of $f(t)$.

25. Suppose the Consumer Products Safety Commission issues new regulations that affect the toy manufacturing industry. Every toy manufacturer will have to make certain changes in its manufacturing process. Let $f(t)$ be the fraction of manufacturers that have complied with the regulations within t months. Note that $0 \le f(t) \le 1$. Suppose that the rate at which new companies comply with the regulations is proportional to the fraction of companies who have not yet complied. Construct a differential equation satisfied by $f(t)$.

26. The Los Angeles Zoo plans to transport a California sea lion to the San Diego Zoo. The animal will be wrapped in a wet blanket during the trip. At any time t the blanket will lose water (owing to evaporation) at a rate proportional to the amount $f(t)$ of water in the blanket. Initially, the blanket will contain 2 gallons of seawater. Set up the differential equation satisfied by $f(t)$.

In the remaining exercises, use a graphing utility to obtain the graphs of the functions.

27. (Velocity of a rocket) The velocity of a rocket, t seconds after blastoff, is

$$v(t) = 30 \ln\left(\frac{1}{1 - .03t}\right)$$

feet per second for $0 \le t \le 32$. Consider the function $s(t)$, the height of the rocket after t seconds. Then $s(t)$ satisfies the differential equation $y' = v(t)$, $y(0) = 0$.

(a) Obtain the graph of the solution to this differential equation in the window $[0, 32]$ *by* $[0, 900]$.

(b) Approximately what is the height of the rocket after 17 seconds?

(c) Approximately when is the rocket 50 feet off the ground?

28. (SAT exams) Let $f(x)$ be the percentage of students who scored at most a grade of x on the SAT Math exam. Statisticians have determined that $f(x)$ satisfies the differential equation

$$y' = \frac{1}{100\sqrt{2\pi}} e^{-.5\left(\frac{x-500}{100}\right)^2}, \quad y(500) = .5.$$

(a) Obtain the graph of the solution to the differential equation in the window $[200, 800]$ *by* $[0, 1]$.

(b) Approximately how many students received a grade of 600 or less?

(c) What grade places a student in the top 25%?

29. Suppose that drilling an oil well has a fixed cost of \$10,000 and a marginal cost of $C'(x) = 10{,}000 + 8x \ln(x+1)$ dollars per foot, where x is the depth in feet. Then $C(x)$, the total cost of drilling x feet, satisfies the differential equation $y' = 10{,}000 + 8x \ln(x+1)$, $y(0) = 10{,}000$.

(a) Obtain the graph of the solution to this differential equation in the window $[0, 1000]$ *by* $[0, 36000000]$.

(b) Approximately how much does it cost to drill 636 feet?

(c) Approximately what depth can be drilled for \$10 million dollars?

30. A large company determines that if they produce x units of their product and if they sell their product at the highest price such that all x units will be sold, then their marginal revenue is $5000 - 2x$ dollars. [Recall that if $R(x)$ is the company's revenue at a production level of x units, then the marginal revenue, $R'(x)$, is the approximate increase in revenue when the company increases its level of production from x units to $x+1$ units. Since the company must lower the selling price to compensate for an increase in production, the marginal revenue is a decreasing function of x.] Find a formula for the revenue function $R(x)$ by solving an appropriate differential equation. Use the fact that the company's revenue is 0 at a production level of 0 units.

31. Find a constant solution of $y' = 6 - 2y$.

32. Find two constant solutions of $y' = 4y(y-7)$.

SOLUTIONS TO PRACTICE PROBLEMS 6.2

1. Since the differential equation has the form $y' = [\textit{expression in } x]$, it can be solved by antidifferentiation. Since $\int e^{3x}\,dx = \frac{1}{3}e^{3x} + C$, the solution is $y = \frac{1}{3}e^{3x} + C$.

2. This differential equation has the special form $y' = ky$ studied in Chapter 4. There its solution was found to be $y = Ce^{3x}$.

Note: In both cases there are infinitely many solutions, one for each choice of C. Had initial conditions been given, we would have obtained specific values for C. In Problem 1 the solutions are translates of each other. This is not so in Problem 2.

6.3 QUALITATIVE THEORY OF AUTONOMOUS DIFFERENTIAL EQUATIONS

In this section we present a technique for sketching solutions to differential equations of the form $y' = g(y)$ *without having to solve the differential equation.* This technique is valuable for three reasons. First, there are many differential equations for which explicit solutions cannot be written down. Second, even when an explicit solution is available, we still face the problem of determining its behavior. For example, does the solution increase or decrease? If it increases, does it approach an asymptote or does it grow arbitrarily large? Third, and probably most significant, in many applications the explicit formula for a solution is unnecessary; only a general knowledge of the behavior of the solution is needed. That is, a qualitative understanding of the solution is sufficient.

The theory introduced in this section is part of what is called the *qualitative theory of differential equations.* We shall limit our attention to differential equations of the form $y' = g(y)$. Such differential equations are called *autonomous.* The term "autonomous" here means "independent of time" and refers to the fact that the right-hand side of $y' = g(y)$ depends only on y and not on t. All the applications studied in the next section involve autonomous differential equations.

Throughout this section we consider the values of each solution $y = f(t)$ only for $t \geq 0$. To introduce the qualitative theory, let us examine the graphs of the various typical solutions of the differential equation $y' = g(y)$, where $g(y) = \frac{1}{2}(1-y)(4-y)$. The solution curves in Figure 1 illustrate the following properties.

■ **Property I** Corresponding to each zero of $g(y)$ there is a constant function that is a solution of the differential equation. Specifically, if $g(c) = 0$, the constant function $y = c$ is a solution. (The constant solutions in Figure 1 are $y = 1$ and $y = 4$.)

■ **Property II** The constant solutions divide the ty-plane into horizontal strips. Each nonconstant solution lies completely in one strip.

■ **Property III** Each nonconstant solution is either strictly increasing or decreasing.

■ **Property IV** Each nonconstant solution either is asymptotic to a constant solution or else increases or decreases without bound.

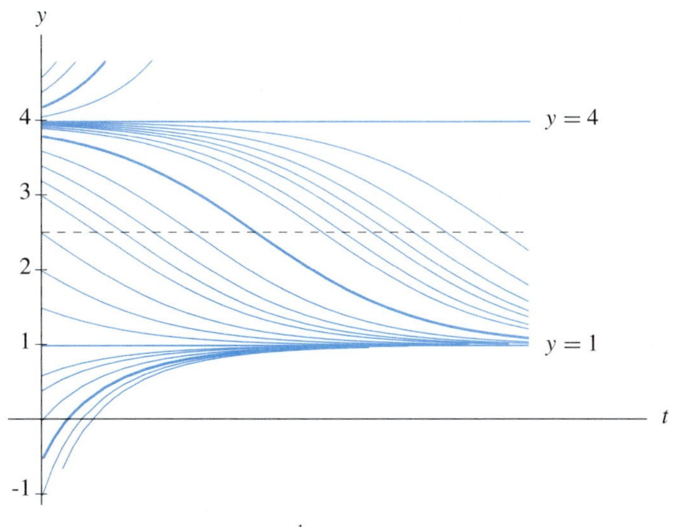

FIGURE 1 Solutions of $y' = \frac{1}{2}(1 - y)(4 - y)$

It can be shown that Properties I–IV are valid for the solutions of any autonomous differential equation $y' = g(y)$ provided that $g(y)$ is a "sufficiently well-behaved" function. We shall assume these properties in this chapter.

Using Properties I–IV, we can sketch the general shape of any solution curve by looking at the graph of the function $g(y)$ and the behavior of that graph near $y(0)$. The procedure for doing this is illustrated in the following example.

EXAMPLE 1 Sketch the solution to $y' = e^{-y} - 1$ that satisfies $y(0) = -2$.

Solution Here $g(y) = e^{-y} - 1$. On a yz-coordinate system we draw the graph of the function $z = g(y) = e^{-y} - 1$ [Figure 2(a)]. The function $g(y) = e^{-y} - 1$ has a zero when $y = 0$. Therefore, the differential equation $y' = e^{-y} - 1$ has the constant solution $y = 0$. We indicate this constant solution on a ty-coordinate system in Figure 2(b). To begin the sketch of the solution satisfying $y(0) = -2$, we locate this initial value of y on the (horizontal) y-axis in Figure 2(a) and on the (vertical) y-axis in Figure 2(b).

To determine whether the solution increases or decreases when it leaves the initial point $y(0)$ on the ty graph, we look at the yz graph and note that $z = g(y)$ is positive at $y = -2$ [Figure 3(a)]. Consequently, since $y' = g(y)$, the derivative of

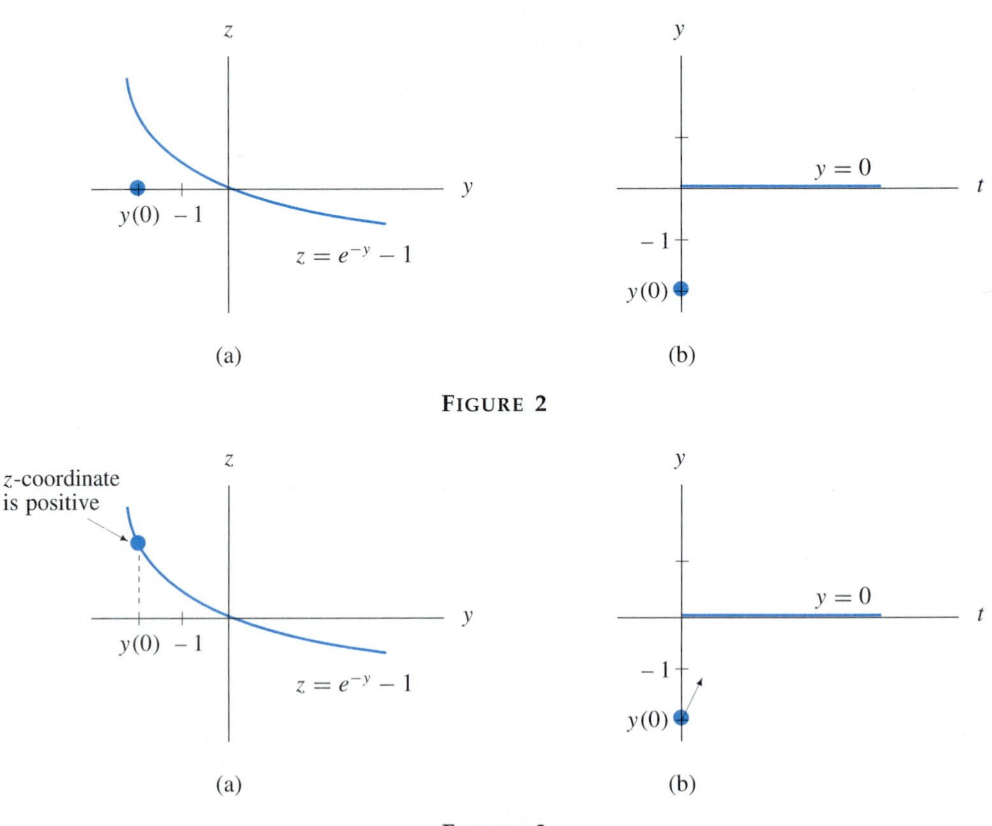

FIGURE 2

FIGURE 3

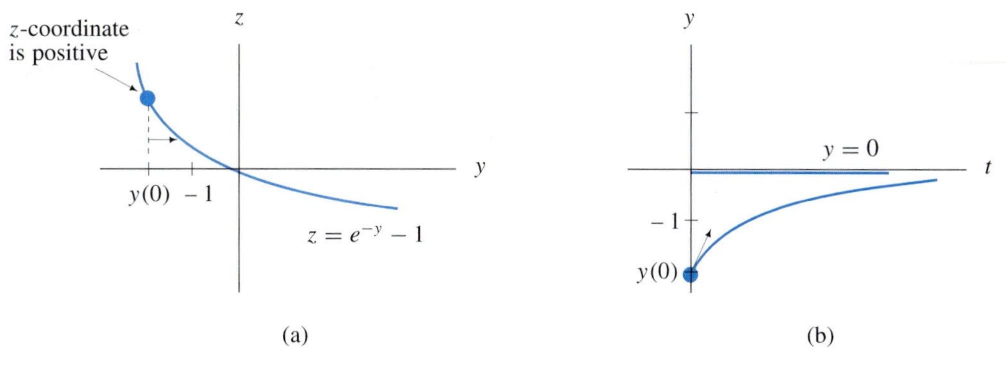

FIGURE 4

the solution is positive, which implies that the solution is increasing. We indicate this by an arrow at the initial point in Figure 3(b). Moreover, the solution y will increase asymptotically to the constant solution $y = 0$, by Properties III and IV of autonomous differential equations.

Next, we place an arrow in Figure 4(a) to remind us that y will move from $y = -2$ toward $y = 0$. As y moves to the right toward $y = 0$ in Figure 4(a), the z-coordinates of points on the graph of $g(y)$ becomes less positive; that is, $g(y)$ becomes less positive. Consequently, since $y' = g(y)$, the slope of the solution curve becomes less positive. Thus the solution curve is concave down, as we have shown in Figure 4(b). ■

An important point to remember when sketching solutions is that z-coordinates on the yz graph are values of $g(y)$, and since $y' = g(y)$, a z-coordinate gives the *slope* of the solution curve at the corresponding point on the ty graph.

EXAMPLE 2 Sketch the graphs of the solutions to $y' = y + 2$ satisfying

(a) $y(0) = 1$,

(b) $y(0) = -3$.

Solution Here $g(y) = y + 2$. The graph of $z = g(y)$ is a straight line of slope 1 and z-intercept 2. [See Figure 5(a).] This line crosses the y-axis only where $y = -2$. Thus the differential equation $y' = y + 2$ has one constant solution, $y = -2$. [See Figure 5(b).]

(a) We locate the initial value $y(0) = 1$ on the y-axes of both graphs in Figure 5. The corresponding z-coordinate on the yz graph is positive; therefore, the solution on the ty graph has positive slope and is increasing as it leaves the initial point. We indicate this by an arrow in Figure 5(b). Now, Property IV of autonomous differential equations implies that y will increase without bound from its initial value. As we let y increase from 1 in Figure 6(a), we see that the z-coordinates [i.e., values of $g(y)$] increase. Consequently, y' is increasing, so the graph of the solution must be concave up. We have sketched the solution in Figure 6(b).

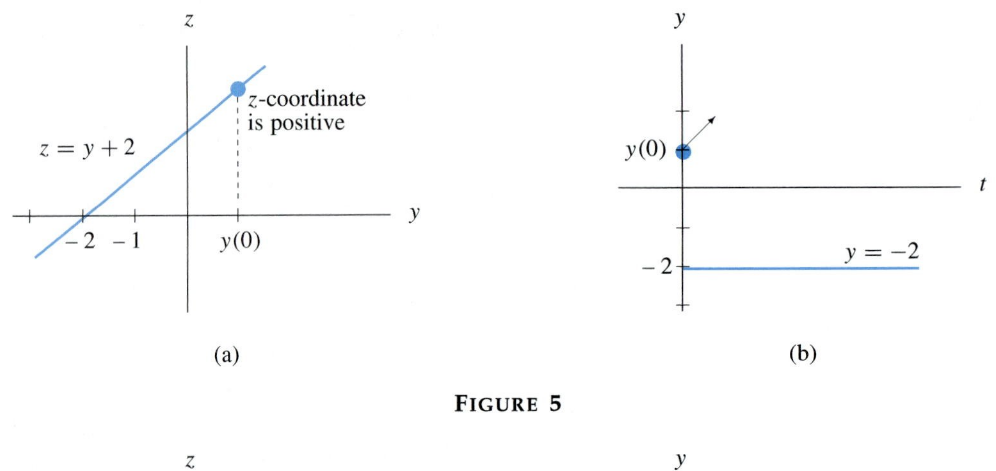

FIGURE 5

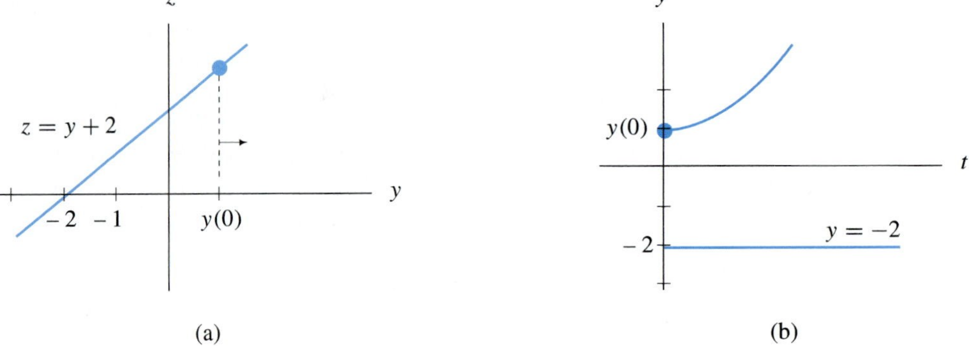

FIGURE 6

(b) Next we graph the solution for which $y(0) = -3$. From the graph of $z = y+2$, we see that z is negative when $y = -3$. This implies that the solution is decreasing as it leaves the initial point. (See Figure 7.) It follows that the values of y will continue to decrease without bound and become more and more negative. This means that on the yz graph y must move to the *left* [Figure 8(a)]. We now examine what happens to $g(y)$ as y moves to the left. (This is the opposite of the ordinary way to read a graph.) The z-coordinate becomes more negative; hence the slopes on the solution curve will become more negative. Thus the solution curve must be concave down, as in Figure 8(b). ■

From the preceding examples we can state a few rules for sketching a solution to $y' = g(y)$ with $y(0)$ given:

1. Sketch the graph of $z = g(y)$ on a yz-coordinate system. Find and label the zeros of $g(y)$.
2. For each zero c of $g(y)$, draw the constant solution $y = c$ on the ty-coordinate system.
3. Plot $y(0)$ on the y-axes of the two coordinate systems.

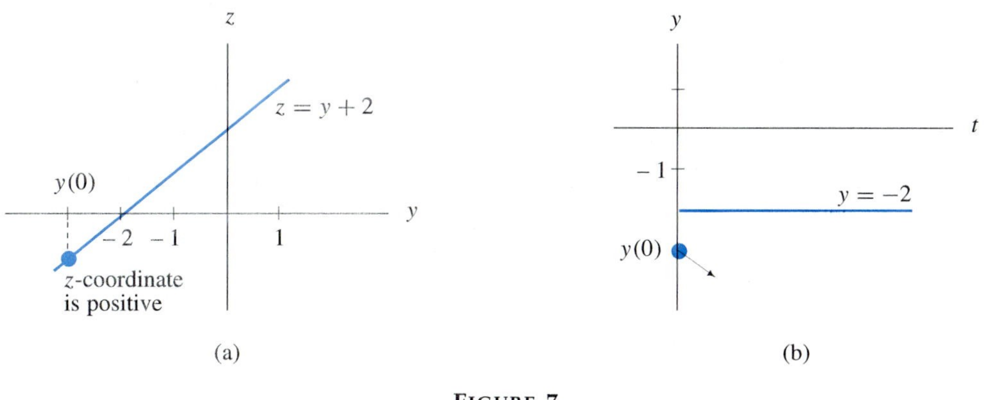

FIGURE 7

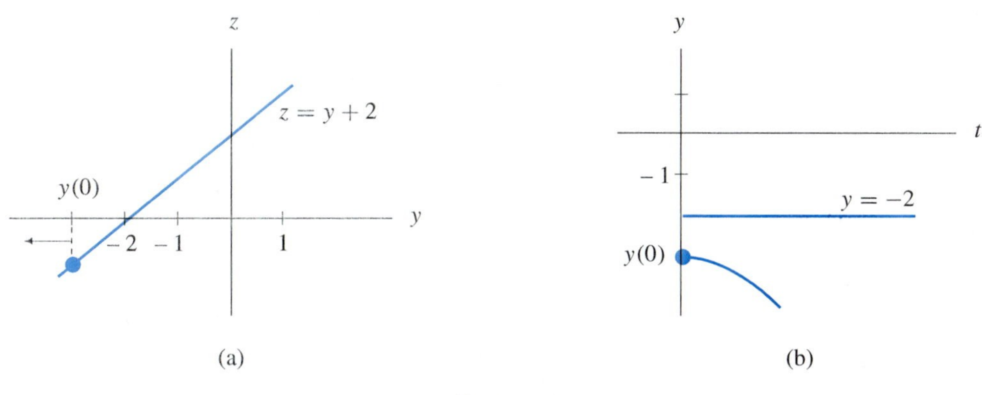

FIGURE 8

4. Determine whether the value of $g(y)$ is positive or negative when $y = y(0)$. This tells us whether the solution is increasing or decreasing. On the ty graph, indicate the direction of the solution through $y(0)$.
5. On the yz graph, indicate which direction y should move. (*Note*: If y is moving *down* on the ty graph, y moves to the *left* on the yz graph.) As y moves in the proper direction on the yz graph, determine whether $g(y)$ becomes more positive, less positive, more negative, or less negative. This tells us the concavity of the solution.
6. Beginning at $y(0)$ on the ty graph, sketch the solution, being guided by the principle that the solution will grow (positively or negatively) without bound unless it encounters a constant solution. In this case, it will approach the constant solution asymptotically.

EXAMPLE 3 Sketch the solutions to $y' = y^2 - 4y$ satisfying $y(0) = 4.5$ and $y(0) = 3$.

Solution Refer to Figure 9. Since $g(y) = y^2 - 4y = y(y-4)$, the zeros of $g(y)$ are 0 and 4; hence the constant solutions are $y = 0$ and $y = 4$. The solution satisfying

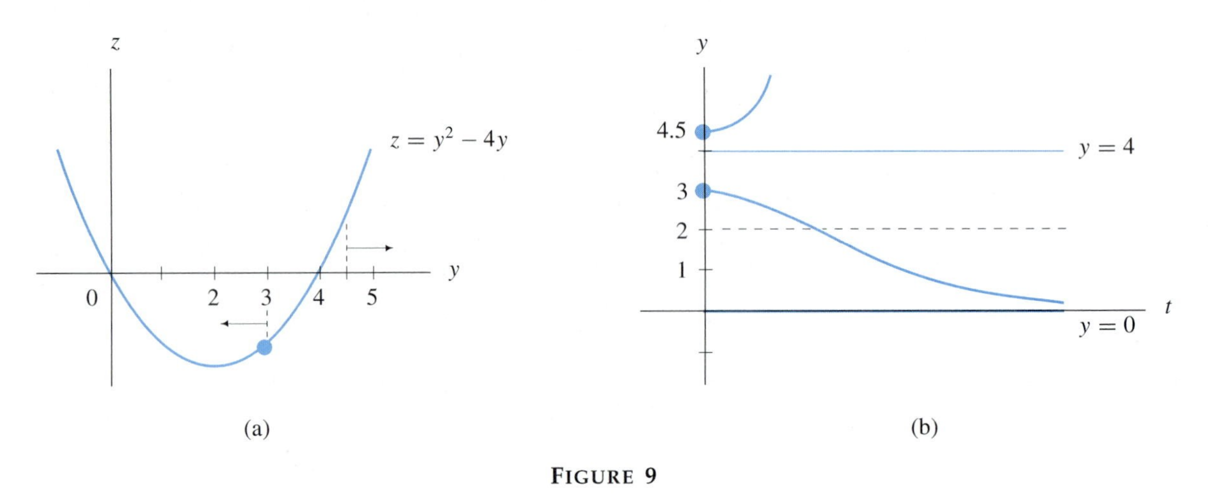

FIGURE 9

$y(0) = 4.5$ is increasing, because the z-coordinate is positive when $y = 4.5$ on the yz graph. This solution continues to increase without bound. The solution satisfying $y(0) = 3$ is decreasing because the z-coordinate is negative when $y = 3$ on the yz graph. This solution will decrease and asymptotically approach the constant solution $y = 0$.

An additional piece of information about the solution satisfying $y(0) = 3$ may be obtained from the graph of $z = g(y)$. We know that y decreases from 3 and approaches 0. From the graph of $z = g(y)$ in Figure 9(a) it appears that at first the z-coordinates become more negative until y reaches 2 and then become less negative as y moves on toward 0. Since these z-coordinates are slopes on the solution curve, we conclude that as the solution moves downward from its initial point on the ty-coordinate system, its slope becomes more negative until the y-coordinate is 2 and then the slope becomes less negative as the y-coordinate approaches 0. Hence the solution is concave down until $y = 2$ and then is concave up. Thus there is an inflection point at $y = 2$, where the concavity changes. ■

We saw in Example 3 that the inflection point at $y = 2$ was produced by the fact that $g(y)$ had a minimum at $y = 2$. A generalization (see below) of the argument in Example 3 shows that inflection points of solution curves occur at each value of y where $g(y)$ has a nonzero relative maximum or minimum point. Thus we may formulate an additional rule for sketching a solution of $y' = g(y)$.

7. On the ty-coordinate system draw dashed horizontal lines at all values of y at which $g(y)$ has a *nonzero* relative maximum or minimum point. A solution curve will have an inflection point whenever it crosses such a dashed line.

It is useful to note that when $g(y)$ is a quadratic function, as in Example 3, its maximum or minimum point occurs at a value of y halfway between the zeros of $g(y)$. This is because the graph of a quadratic function is a parabola, which is symmetric about a vertical line through its vertex.

EXAMPLE 4 Sketch a solution to $y' = e^{-y}$ with $y(0) > 0$.

Solution Refer to Figure 10. Since $g(y) = e^{-y}$ is always positive, there are no constant solutions to the differential equation and every solution will increase without bound. When drawing solutions that asymptotically approach a horizontal line, we have no choice as to whether to draw it concave up or concave down. This decision will be obvious from its increasing or decreasing nature and from knowledge of inflection points. However, for solutions that grow without bound, we must look at $g(y)$ in order to determine concavity. In this example, as t increases, the values of y increase. As y increases, $g(y)$ becomes less positive. Since $g(y) = y'$, we deduce that the slope of the solution curve becomes less positive; therefore, the solution curve is concave down. ■

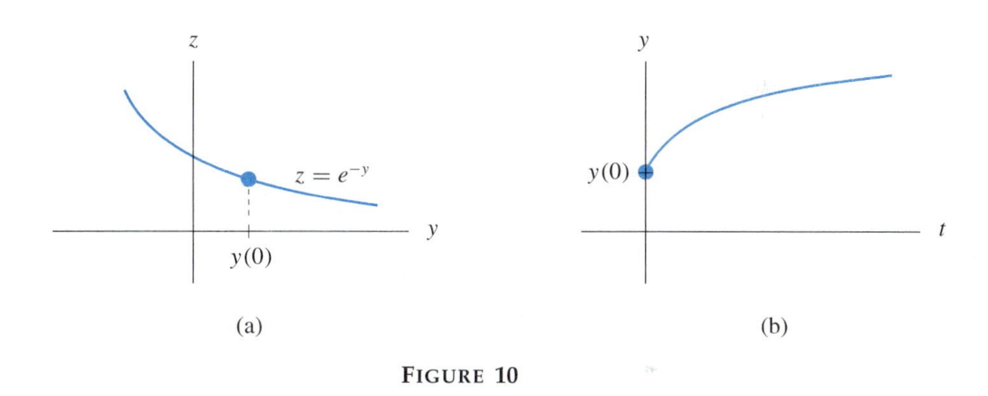

FIGURE 10

USING GRAPHING UTILITIES

Sketching solutions to an autonomous differential equation relies on a good graph of the function $z = g(y)$. A graphing utility can help make the graph accurate and determine its zeros and relative extreme points. The function must be entered with x as the variable.

PRACTICE PROBLEMS 6.3

Consider the differential equation $y' = g(y)$, where $g(y)$ is the function whose graph is drawn in Figure 11.

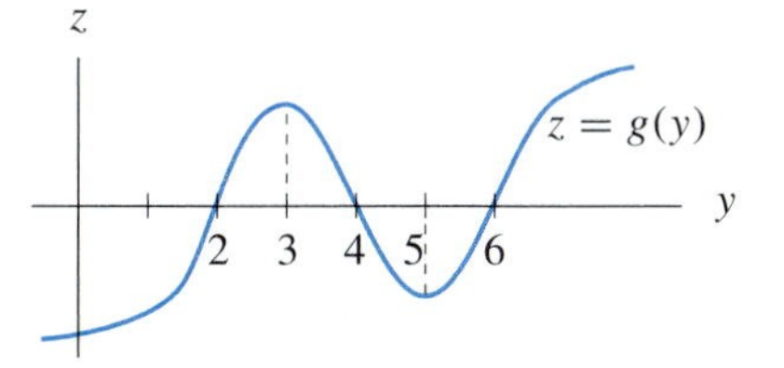

FIGURE 11

1. How many constant solutions are there to the differential equation $y' = g(y)$?
2. For what values $y(0)$ will the corresponding solution of the differential equation be an increasing function?
3. Is it true that if the initial value $y(0)$ is near 4, the corresponding solution will be asymptotic to the constant solution $y = 4$?
4. For what initial values $y(0)$ will the corresponding solution of the differential equation have an inflection point?

EXERCISES 6.3

One or more initial conditions are given for each differential equation below. Use the qualitative theory of autonomous differential equations to sketch the graphs of the corresponding solutions. Include a yz graph if one is not already provided. Always indicate the constant solutions on the ty graph whether they are mentioned or not.

1. $y' = 2y - 6$, $y(0) = 1$, $y(0) = 4$. [The graph of $z = g(y)$ is drawn in Figure 12.]
2. $y' = 5 - 2y$, $y(0) = 1$, $y(0) = 4$. (See Figure 13.)
3. $y' = 4 - y^2$, $y(0) = -3$, $y(0) = -1$, $y(0) = 3$. (See Figure 14.)
4. $y' = y^2 - 5$, $y(0) = -4$, $y(0) = 2$, $y(0) = 3$. (See Figure 15.)

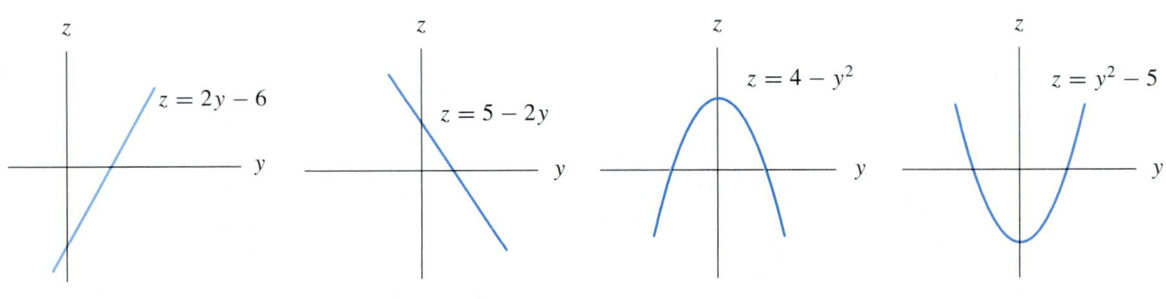

FIGURE 12

FIGURE 13

FIGURE 14

FIGURE 15

5. $y' = y^2 - 6y + 5$ or $y' = (y - 1)(y - 5)$, $y(0) = -1$, $y(0) = 2$, $y(0) = 4$, $y(0) = 6$. (See Figure 16.)
6. $y' = -\frac{1}{3}(y + 2)(y - 4)$, $y(0) = -3$, $y(0) = -1$, $y(0) = 6$. (See Figure 17.)
7. $y' = y^3 - 9y = y(y^2 - 9)$, $y(0) = -4$, $y(0) = -1$, $y(0) = 2$, $y(0) = 4$. (See Figure 18.)
8. $y' = 9y - y^3$, $y(0) = -4$, $y(0) = -1$, $y(0) = 2$, $y(0) = 4$. (See Figure 19.)

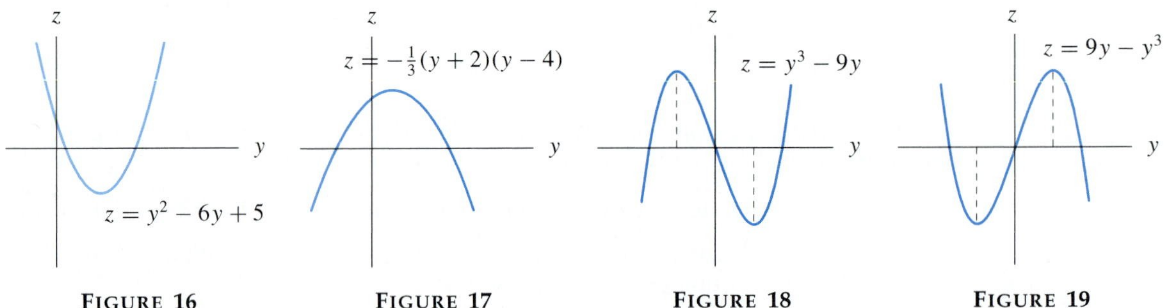

FIGURE 16

FIGURE 17

FIGURE 18

FIGURE 19

9. $y' = 3 - \frac{1}{2}y$, $y(0) = 2$, $y(0) = 8$.
10. $y' = 3y$, $y(0) = -2$, $y(0) = 2$.
11. $y' = 5y - y^2$, $y(0) = 1$, $y(0) = 7$.

12. $y' = -y^2 + 10y - 21$, $y(0) = 1$, $y(0) = 4$.
13. $y' = y^2 - 3y - 4$, $y(0) = 0$, $y(0) = 3$.
14. $y' = \frac{1}{2}y^2 - 3y$, $y(0) = 3$, $y(0) = 6$, $y(0) = 9$.
15. $y' = y^2 + 2$, $y(0) = -1$, $y(0) = 1$.
16. $y' = y - \frac{1}{4}y^2$, $y(0) = -1$, $y(0) = 1$.
17. $y' = 1/y$, $y(0) = -1$, $y(0) = 1$.
18. $y' = y^3$, $y(0) = -1$, $y(0) = 1$.
19. $y' = ky^2$, where k is a negative constant, $y(0) = -2$, $y(0) = 2$.
20. $y' = ky(M - y)$, where $k > 0$, $M > 10$, and $y(0) = 1$.
21. $y' = ky - A$, where k and A are positive constants. Sketch solutions where $0 < y(0) < A/k$ and $y(0) > A/k$.
22. $y' = k(y - A)$, where $k < 0$ and $A > 0$. Sketch solutions where $y(0) < A$ and $y(0) > A$.
23. Suppose that once a sunflower plant has started growing, the rate of growth at any time is proportional to the product of its height and the difference between its height at maturity and its current height. Give a differential equation that is satisfied by $f(t)$, the height at time t, and sketch the solution.
24. A parachutist has a terminal velocity of -176 feet per second. That is, no matter how long a person falls, his or her speed will not exceed 176 feet per second, but it will get arbitrarily close to that value. The velocity in feet per second, $v(t)$, after t seconds satisfies the differential equation $v'(t) = 32 - k \cdot v(t)$. What is the value of k?

In the remaining exercises, use a graphing utility to obtain the graphs of the functions $z = g(y)$ in the specified windows. Then sketch the graphs of the solutions.

25. $y' = -\frac{1}{10}y \ln \frac{y}{100}$, $y(0) = 10$ and $y(0) = 150$; $[0, 200]$ *by* $[-15, 5]$.
    (*Note*: Solutions to this differential equation are called Gompertz growth equations.)
26. $y' = -\frac{1}{2}y \ln \frac{y}{30}$, $y(0) = 1$, $y(0) = 5$, and $y(0) = 40$; $[0, 50]$ *by* $[-15, 10]$.
27. $y' = ye^{-.5y^2}$, $y(0) = -.5$; $[-4, 4]$ *by* $[-1, 1]$.
28. $y' = (y - 1)^2(6 - y)$, $y(0) = 5$, $y(0) = 7$; $[-1, 7]$ *by* $[-30, 30]$.
29. $y' = -.1(y^2 - 12y + 44)(y^2 - 12y + 20)$, $y(0) = 3$ and $y(0) = 9$; $[-2, 14]$ *by* $[-1, 15]$.
30. $y' = (y - 4)(y^3 - 4y^2 + y + 6)e^{.5y}$, $y(0) = -1.25$, $y(0) = 1$, and $y(0) = 2.25$; $[-2, 5]$ *by* $[-30, 15]$.
31. $y' = ye^{-y/5}$, $y(0) = -2$, $y(0) = 2$, $y(0) = 10$; $[-5, 20]$ *by* $[-10, 5]$.
32. $y' = -2ye^{-y^2/3}$, $y(0) = -2$, $y(0) = 1$; $[-5, 5]$ *by* $[-2.5, 2.5]$.

SOLUTIONS TO PRACTICE PROBLEMS 6.3

1. Three. The function $g(y)$ has zeros when y is 2, 4, and 6. Therefore, $y' = g(y)$ has the constant functions $y = 2$, $y = 4$, and $y = 6$ as solutions.
2. For $2 < y(0) < 4$ and $y(0) > 6$. Since nonconstant solutions are either strictly increasing or strictly decreasing, a solution is an increasing function provided that it is increasing at time $t = 0$. This is the case when the first derivative is positive at $t = 0$. When $t = 0$, $y' = g(y(0))$. Therefore, the solution corresponding to $y(0)$ is increasing whenever $g(y(0))$ is positive.
3. Yes. If $y(0)$ is slightly to the right of 4, then $g(y(0))$ is negative, so the corresponding solution will be a decreasing function with values moving to the left closer and closer to 4. If $y(0)$ is slightly to the left of 4, then $g(y(0))$ is positive, so the corresponding solution will be an increasing function with values moving to the right closer and closer to 4. (The constant solution $y = 4$ is referred to as a *stable* constant solution. The solution with initial value 4 stays at 4, and solutions with initial values near 4 move toward 4. The constant solution $y = 2$ is *unstable*. Solutions with initial values near 2 move away from 2.)
4. For $2 < y(0) < 3$ and $5 < y(0) < 6$. Inflection points of solutions correspond to nonzero relative maximum and relative minimum points of the function $g(y)$. If $2 < y(0) < 3$, the corresponding solution will be an increasing function. The values of y will move to the right (toward 4) and therefore will cross 3, a place at which $g(y)$ has a relative maximum point. Similarly, if $5 < y(0) < 6$, the corresponding solution will be decreasing. The values of y on the yz graph will move to the left and cross 5.

6.4 APPLICATIONS OF AUTONOMOUS DIFFERENTIAL EQUATIONS

Equations describing conditions in a physical process are often referred to as mathematical models. In this section we study real-life situations that may be modeled by an autonomous differential equation $y' = g(y)$. Here y will represent some quantity that is changing with time, and the equation $y' = g(y)$ will be obtained from a description of the rate of change of y.

For the differential equations considered in this section, the rate of change of y is *proportional* to some quantity. We have already encountered two such differential equations.

1. $y' = ky$: "the rate of change of y is proportional to y" (exponential growth or decay).
2. $y' = k(M - y)$: "the rate of change of y is proportional to the difference between M and y" (Newton's law of cooling).

A third situation is the *logistic differential equation*. It has the proportionality features of both 1 and 2.

3. $y' = ky(M - y)$: "the rate of change of y is proportional to the product of y and the difference between M and y."

Numerous applications of Type 1 differential equation were presented in Chapter 4. Let us now consider application of the other two types of differential equations.

APPLICATIONS OF DIFFERENTIAL EQUATIONS OF THE TYPE $y' = k(M - y)$, $y(0) = 0$

For the applications discussed in this section, both k and M are positive numbers. Figure 1 shows the qualitative analysis of this differential equation. As time progresses, the solution in Figure 1(b) gets asymptotically close to the line $y = M$.

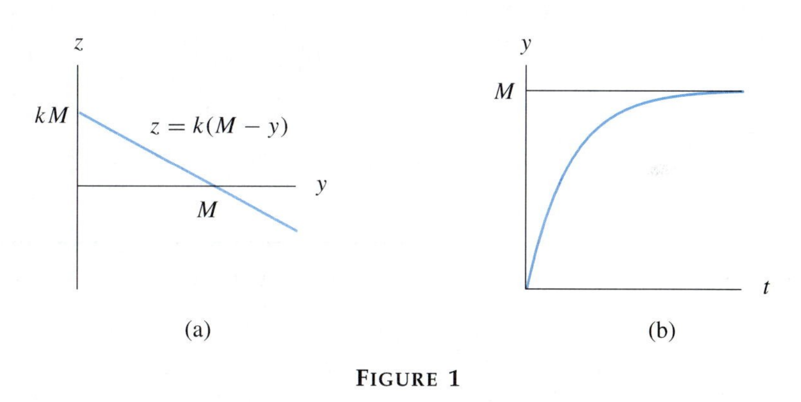

FIGURE 1

SPEED OF A SKYDIVER A skydiver, on jumping out of an airplane, falls at an increasing rate. However, the wind rushing past the skydiver's body creates an upward force that begins to counterbalance the downward force of gravity. This air friction finally becomes so great that the skydiver's speed reaches a limiting value called the *terminal speed*. Let $v(t)$ be the downward speed of the skydiver after t seconds of free fall. Then at any time, the rate of change of $v(t)$ is proportional to the difference between the terminal speed and the current speed. Also, the initial speed is zero. That is,

$$v'(t) = k\,[M - v(t)], \quad v(0) = 0,$$

where M is the terminal speed.

THE LEARNING CURVE Psychologists have found that in many learning situations a person's rate of learning is rapid at first and then slows down. Finally, as the task is mastered, the person's level of performance reaches a level above which it is almost physically impossible to rise. For example, within reasonable limits, each person seems to have a certain maximum capacity for memorizing a list of nonsense syllables. Suppose that a subject can memorize M syllables in a row if given sufficient time, say an hour, to study the list but cannot memorize $M+1$ syllables in a row even if allowed several hours of study. By giving the subject different lists of syllables and varying lengths of time to study the lists, the psychologists can determine an empirical relationship between $m(t)$, the number of nonsense syllables memorized accurately, and the number of minutes of study time. The *slope* of this learning curve at time t is approximately the number of additional syllables that can be memorized if the subject is given one more minute of study time. Thus the slope is a measure of the rate of learning. It turns out that if the subject is given a list of M nonsense syllables, then at any time the rate of learning is proportional to the number of syllables remaining to be memorized. Also, initially no syllables have been memorized. That is,

$$m'(t) = k[M - m(t)], \quad m(0) = 0.$$

DIFFUSION OF INFORMATION BY MASS MEDIA Sociologists have studied the spread of information that is propagated constantly by mass media, such as television or magazines.* They have concluded that at any time, the number of *newly informed* people per unit time, is proportional to the number of people who have not yet heard the news. Given a fixed population M, let $f(t)$ be the number of people who have already heard a certain piece of information by time t. Then $M - f(t)$ is the number who have not yet heard the information. The number of newly informed people per unit time is the rate of increase of the number of people who have heard the news. Also, initially no people have heard the news. Therefore,

$$f'(t) = k[M - f(t)], \quad f(0) = 0.$$

INTRAVENOUS INFUSION OF GLUCOSE The human body both manufactures and uses glucose ("blood sugar"). Usually, there is a balance in these two processes, so that the bloodstream has a certain "equilibrium level" of glucose. Suppose that a patient is given a single intravenous injection of glucose and let $A(t)$ be the amount of glucose in the bloodstream (in milligrams) above the equilibrium level. Then the body will start using up the excess glucose at a rate proportional to the amount of excess glucose; that is,

$$A'(t) = -\lambda A(t), \tag{1}$$

where λ is a positive constant called the velocity constant of elimination. This constant depends on how fast the patient's metabolic processes eliminate the excess glucose from the blood. Equation (1) describes a simple exponential decay process.

Now suppose that, instead of a single shot, the patient receives a continuous intravenous infusion of glucose. A bottle of glucose solution is suspended above the patient, and a small tube carries the glucose down to a needle that runs into a vein. In this case, there are two influences on the amount of excess glucose in the blood: the glucose being added steadily from the bottle and the glucose being removed from the blood by metabolic processes. Let r be the rate of infusion of glucose (often from 10 to 100 milligrams per minute). If the body did not remove any glucose, the excess glucose would increase at a constant rate of r milligrams

* Coleman, *Introduction to Mathematical Sociology* (New York: The Free Press, 1964), p. 43.

per minute; that is,

$$A'(t) = r. \tag{2}$$

Taking into account the two influences on $A'(t)$ described by (1) and (2), we can write

$$A'(t) = r - \lambda A(t). \tag{3}$$

If we let $M = r/\lambda$ and note that initially there is no excess glucose, then

$$A'(t) = \lambda(M - A(t)), \quad A(0) = 0.$$

From the solution of this differential equation in Figure 1(b), we conclude that the amount of excess glucose rises until it reaches a stable level.

Figure 1 showed a qualitatively obtained graph of the solution of the differential equation $y' = k(M - y)$, $y(0) = 0$. The following example yields the solution explicitly.

EXAMPLE 1 Show that the function $f(t) = M(1 - e^{-kt})$ is the solution to the differential equation $y' = k(M - y)$, $y(0) = 0$.

Solution $f(t) = M(1 - e^{-kt}) = M - Me^{-kt}$ and so $f'(t) = Mke^{-kt}$. However,

$$k[M - f(t)] = k\left[M - (M - Me^{-kt})\right] = kMe^{-kt},$$

so that the differential equation $y' = k(M - y)$ holds. Also,

$$f(0) = M - Me^0 = M - M = 0. \qquad \blacksquare$$

EXAMPLE 2 Suppose that a certain piece of news (such as the resignation of a public official) is broadcast frequently by radio and television stations in a community of 10,000 people. Also suppose that one-half of the residents of the community have heard the news within 3 hours of its initial release.

(a) Find the formula for the number of people who have heard the news after t hours.

(b) How many people will have heard the news after 7 hours?

(c) At what rate will the news be spreading after 7 hours?

(d) When will 95% of the people have heard the news?

(e) When will the news be spreading at the rate of 182 people per hour?

Solution Let $f(t)$ be the number of people who have heard the news after t hours. Then $f(t)$ satisfies a differential equation of the form $y' = k(10{,}000 - y)$, $y(0) = 0$, and therefore by Example 1, $f(t) = 10{,}000(1 - e^{-kt})$.

(a) We must find the value of k in $f(t) = 10{,}000(1 - e^{-kt})$. The problem states that $5000 = f(3)$. So

$$\begin{aligned} 5000 &= 10{,}000(1 - e^{-k \cdot 3}) \\ .5 &= 1 - e^{-3k} \\ e^{-3k} &= 1 - .5 = .5 \\ \ln e^{-3k} &= \ln .5 \\ -3k &= \ln .5 \\ k &= \frac{\ln .5}{-3} \approx .231. \end{aligned}$$

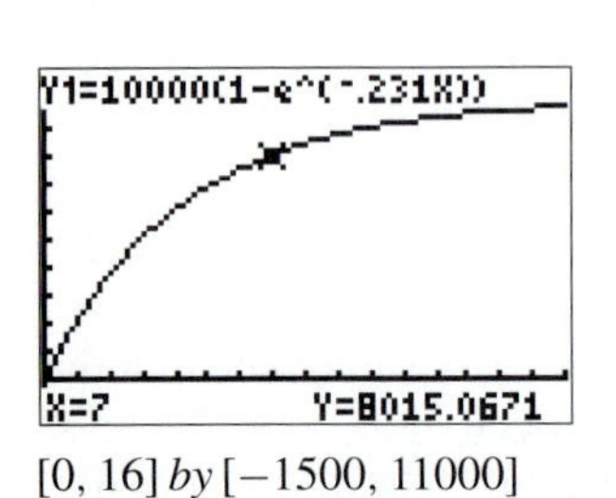

[0, 16] *by* [−1500, 11000]

Therefore, $f(t) = 10{,}000(1 - e^{-.231t})$.

(b) $f(7) = 10{,}000(1 - e^{-.231(7)}) \approx 8015$.

(c) The answer, $f'(7)$, can be found either with the differential equation, or the formula. With the differential equation,

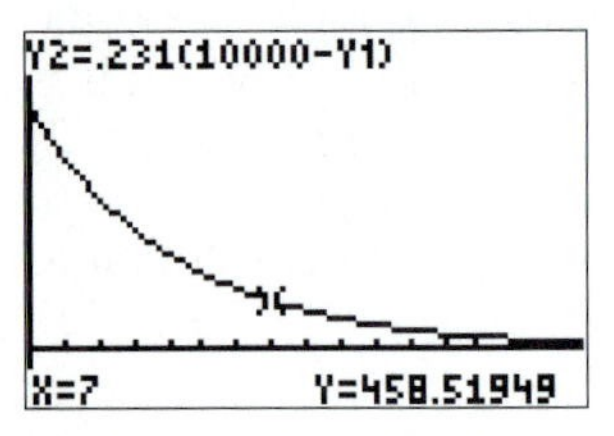

[0, 16] *by* [−500, 3000]

$$\begin{aligned} y' &= .231(10{,}000 - y), \\ f'(x) &= .231(10{,}000 - f(x)), \\ f'(7) &= .231(10{,}000 - f(7)) \\ &= .231(10{,}000 - 8015) \approx 459. \end{aligned}$$

With the formula,

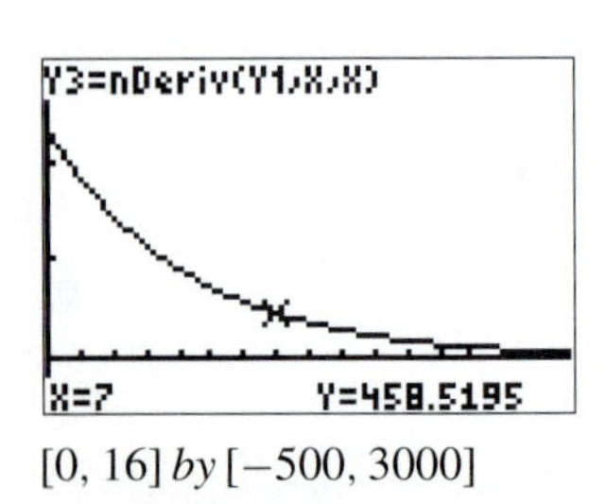

[0, 16] *by* [−500, 3000]

$$\begin{aligned} f(t) &= 10{,}000(1 - e^{-.231t}), \\ f'(t) &= 10{,}000(.231e^{-.231t}) = 2310e^{-.231t}, \\ f'(7) &= 2310e^{-.231(7)} \approx 459. \end{aligned}$$

Therefore, after 7 hours the news will be spreading at the rate of 459 people per hour.

(d) The problem asks for the solution to $f(t) = 9500$.

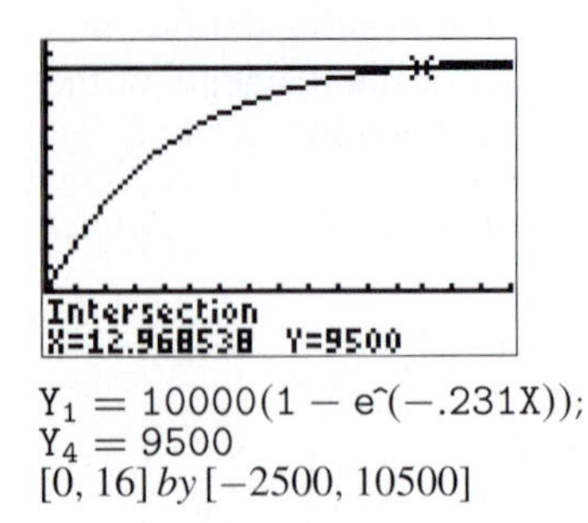

$Y_1 = 10000(1 - e\hat{\ }(-.231X))$;
$Y_4 = 9500$
[0, 16] *by* [−2500, 10500]

$$\begin{aligned} 9500 &= 10{,}000(1 - e^{-.231t}) \\ .95 &= 1 - e^{-.231t} \\ e^{-.231t} &= 1 - .95 = .05 \\ \ln e^{-.231t} &= \ln .05 \\ -.231t &= \ln .05 \\ t &= \frac{\ln .05}{-.231} \approx 13. \end{aligned}$$

Therefore, 95% of the population will have heard the news after about 13 hours.

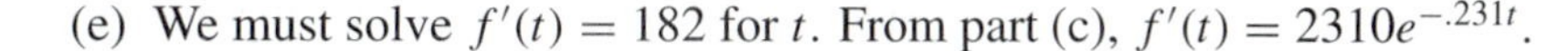

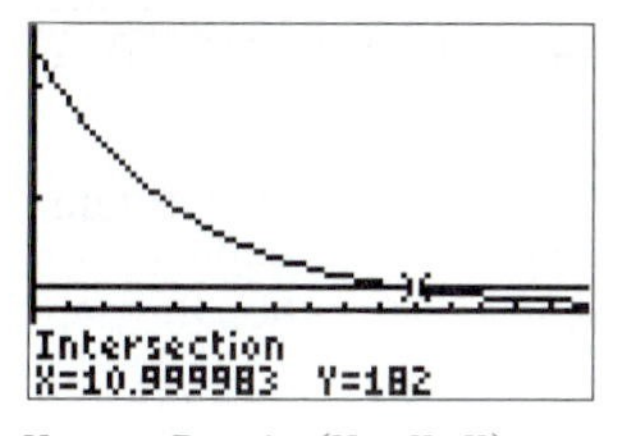

$Y_3 = \text{nDeriv}(Y_1, X, X)$;
$Y_4 = 182$
$[0, 16]$ *by* $[-750, 2500]$

(e) We must solve $f'(t) = 182$ for t. From part (c), $f'(t) = 2310e^{-.231t}$.

$$2310e^{-.231t} = 182$$

$$e^{-.231t} = \frac{182}{2310}.$$

Solving for t, we find $t \approx 11$. Therefore, after 11 hours the news is spreading at the rate of 182 people per hour. ■

Note: With Visual Calculus, all parts of Example 2 can be answered with the "Analyze a Function and Its Derivatives" routine.

APPLICATIONS OF DIFFERENTIAL EQUATIONS OF THE TYPE $y' = ky(M - y)$

For the applications discussed in this section, k and M are positive numbers. Figure 2 shows the qualitative analysis of this differential equation. The graph of the solution beginning at y_2 is called a *logistic curve*. As time progresses, it gets asymptotically close to the line $y = M$. The logistic curve has an inflection point at that value of t for which $f(t) = M/2$. The position of this inflection point is significant for applications of the logistic curve since it is the point at which the curve has greatest slope. In other words, the inflection point corresponds to the instant of fastest growth of the logistic curve.

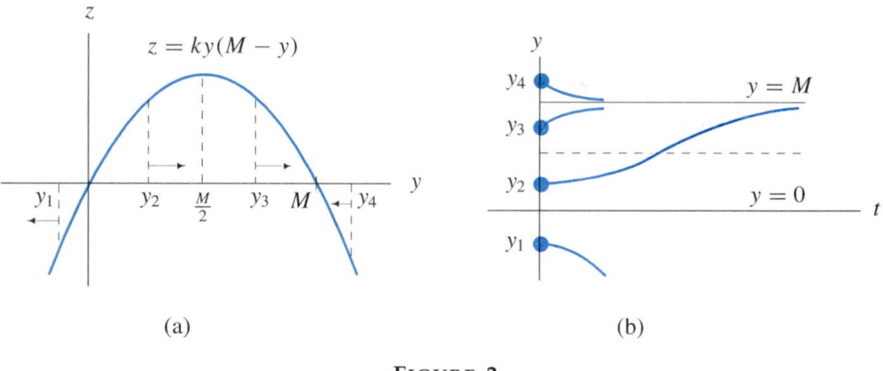

FIGURE 2

RESTRICTED POPULATION GROWTH The model for simple exponential growth discussed in Section 5.1 is adequate for describing the growth of many types of populations, but obviously a population cannot increase exponentially forever. The simple exponential growth model becomes inapplicable when the environment begins to inhibit the growth of the population. The logistic differential equation takes into account some of the effects of the environment on a population. For small values of t, the curve has the same basic shape as an exponential growth curve. Then when the population begins to suffer from overcrowding or lack of food, the growth rate (the slope of the population curve) begins to slow down. Eventually, the growth rate tapers off to zero as the population reaches the maximum size, M, that the environment will support. The constant M is called the *carrying capacity* of the environment. In the logistic differential equation $y' = ky(M - y)$, the factor y reflects the fact that the growth rate (y') depends in part on the size of the population. The factor $M - y$ reflects the fact that the growth rate also depends on how close y is to the carrying capacity M. The population is thriving best, that is, growing at the greatest rate, when the population is one-half of the carrying capacity.

SPREAD OF AN EPIDEMIC Consider the spread of a highly contagious disease among a population of M people who are susceptible to the disease. Assume that all infected individuals are contagious and circulate freely among the population. Let $f(t)$ be the number of infected people at time t. The growth rate of the disease, $f'(t)$, is small when either few people have the disease ($f(t)$ is small) or few people are free of the disease ($M - f(t)$ is small). Therefore, $f'(t)$ depends on both $f(t)$ and $M - f(t)$ and so the differential equation $y' = ky(M - y)$ is reasonable. Medical facilities are most strained when half the population has been infected, for then the disease is spreading at the greatest rate. Medical intervention to halt the spread of the disease is most effective if initiated before that time.

Figure 2 shows qualitatively obtained graphs of solutions of the logistic differential equation $y' = ky(M - y)$. Explicit solutions have the form

$$f(t) = \frac{M}{1 + Be^{-ct}},$$

where $c = kM$. See Exercise 12.

EXAMPLE 3 The Public Health Service monitors the spread of an epidemic of a particularly long-lasting strain of flu in a city of 500,000 people. At the beginning of the first week of monitoring, 200 cases have been reported; during the first week 300 new cases are reported.

(a) Find the differential equation and the formula for the number of cases after t weeks.

(b) How many cases will there be after 9 weeks?

(c) At what rate will the flu be spreading after 9 weeks?

(d) When is the disease spreading at the greatest rate?

Solution Let $f(t)$ be the number of cases at the end of t weeks. Then $f(t)$ satisfies a differential equation of the form $y' = ky(500{,}000 - y)$, $y(0) = 200$, and

$$f(t) = \frac{500{,}000}{1 + Be^{-ct}}.$$

(a) We must find the values of B and c in the formula for $f(t)$. Since $f(0) = 200$,

$$\begin{aligned}
200 &= \frac{500{,}000}{1 + Be^{0}} = \frac{500{,}000}{1 + B} \\
200(1 + B) &= 500{,}000 \\
1 + B &= \frac{500{,}000}{200} = 2500 \\
B &= 2499.
\end{aligned}$$

Therefore,

$$f(t) = \frac{500{,}000}{1 + 2499e^{-ct}}.$$

To find c, we use the fact that $f(1) = 300 + 200 = 500$.

$$\begin{aligned} 500 = f(1) &= \frac{500{,}000}{1 + 2499e^{-c(1)}} \\ 500(1 + 2499e^{-c}) &= 500{,}000 \\ (1 + 2499e^{-c}) &= 1000 \\ e^{-c} &= \frac{999}{2499}. \end{aligned}$$

Solving for c, we find $c \approx .917$. Since $c = kM$, $k = c/M = .917/500{,}000 \approx .000001834$. Therefore, the differential equation is

$$y' = .000001834y(500{,}000 - y), \quad y(0) = 200,$$

and the formula for $f(t)$ is

$$f(t) = \frac{500{,}000}{1 + 2499e^{-.917t}}.$$

(b) $f(9) = \dfrac{500{,}000}{1 + 2499e^{-.917(9)}} \approx 302{,}860.$

(c) The answer, $f'(9)$, can be found either with the differential equation or the formula. With the differential equation,

$$\begin{aligned} f'(9) &= .000001834f(9)(500{,}000 - f(9)) \\ &= .000001834(302{,}860)(500{,}000 - 302{,}860) \approx 109{,}501. \end{aligned}$$

With the formula,

$$\begin{aligned} f(t) &= \frac{500{,}000}{1 + 2499e^{-.917t}}, \\ f'(t) &= \frac{1{,}145{,}791{,}500e^{-.917t}}{(1 + 2499e^{-.917t})^2}, \\ f'(9) &= \frac{1{,}145{,}791{,}500e^{-.917(9)}}{(1 + 2499e^{-.917(9)})^2} \approx 109{,}501. \end{aligned}$$

(d) The disease is spreading at the greatest rate when $f(t) = 500{,}000/2 = 250{,}000$. The solution to

$$\frac{500{,}000}{1 + 2499e^{-.917t}} = 250{,}000$$

is approximately 8.5. Therefore, the disease is spreading fastest after eight and one-half weeks. ■

PRACTICE PROBLEM 6.4

1. A sociological study* was made to examine the process by which doctors decide to adopt a new drug. The doctors were divided into two groups of P doctors. The doctors in group A had little interaction with other doctors and so received most of their information via mass media. The doctors in group B had extensive interaction with other doctors and so received most of their information via word of mouth. For each group, let $f(t)$ be the number who have learned about a new drug after t months. Examine the appropriate differential equations to explain why the two graphs were of the types shown in Figure 3.

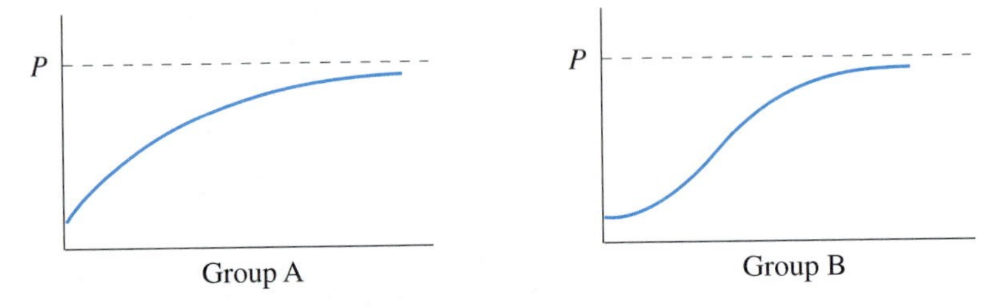

FIGURE 3 Results of a sociological study

EXERCISES 6.4

1. (Driving accidents) Let t represent the total number of hours that a truck driver spends during a year driving on a certain highway connecting two cities, and let $p(t)$ represent the probability that the driver will have at least one accident during these t hours. Then $0 \leq p(t) \leq 1$, and $1-p(t)$ represents the probability of not having an accident. Under ordinary conditions the rate of increase in the probability of an accident (as a function of t) is proportional to the probability of not having an accident. Construct and sketch the solution to a differential equation for this situation.

2. (Construction cost index) Some homeowner's insurance policies include automatic inflation coverage based on the U.S. Commerce Department's construction cost index (CCI). Each year the property insurance coverage is increased by an amount based on the change in the CCI. Let $f(t)$ be the CCI at time t years since January 1, 1990, and let $f(0) = 100$. Suppose the construction cost index is rising at a rate proportional to the CCI. Construct and sketch the solution to a differential equation satisfied by $f(t)$.

3. (Autocatalytic reaction) In an autocatalytic reaction, one substance is converted into a second substance in such a way that the second substance catalyzes its own formation. This is the process by which trypsinogen is converted into the enzyme trypsin. The reaction starts only in the presence of some trypsin, and each molecule of trypsinogen yields one molecule of trypsin. The rate of formation of trypsin is proportional to the product of the amounts of the two substances present. Set up the differential equation that is satisfied

* James S. Coleman, Elihu Katz, and Herbert Menzel, "The Diffusion of an Innovation Among Physicians," *Sociometry*, **20** (1957), 253–270.

by $y = f(t)$, the amount (number of molecules) of trypsin present at time t. Sketch the solution. For what value of y is the reaction proceeding the fastest? [*Note*: Letting M be the total amount of the two substances, the amount of trypsinogen present at time t is $M - f(t)$.]

4. (Movement of solutes through a cell membrane) Let c be the concentration of a solute outside a cell that we assume to be constant throughout the process—that is, unaffected by the small influx of the solute across the membrane due to a difference in concentration. The rate of change of the concentration of the solute inside the cell at any time t is proportional to the difference between the outside concentration and the inside concentration. Set up the differential equation whose solution is $y = f(t)$, the concentration of the solute inside the cell at time t. Sketch a solution.

5. (Capital investment model) In economic theory, the following model is used to describe a possible capital investment policy. Let $f(t)$ represent the total invested capital of a company at time t. Additional capital is invested whenever $f(t)$ is below a certain equilibrium value E, and capital is withdrawn whenever $f(t)$ exceeds E. The rate of investment is proportional to the difference between $f(t)$ and E. Construct a differential equation whose solution is $f(t)$ and sketch two or three typical solution curves.

6. (Evans price adjustment model) Consider a certain commodity that is produced by many companies and purchased by many other firms. Over a relatively short period there tends to be an equilibrium price M per unit of the commodity that balances the supply and the demand. Suppose that, for some reason, the price is different from the equilibrium price. The Evans price adjustment model says that the rate of change of price with respect to time is proportional to the difference between the actual market price p and the equilibrium price. Write a differential equation that expresses this relation and sketch a solution.

7. (Drying) A porous material dries outdoors at a rate that is proportional to the moisture content. Set up the differential equation whose solution is $y = f(t)$, the amount of water at time t in a towel on a clothesline. Sketch a solution.

8. (War fever) L. F. Richardson proposed the following model to describe the spread of war fever.* If $y = f(t)$ is the percent of the population advocating war at time t, then the rate of change of $f(t)$ at any time is proportional to the product of the percentage of the population advocating war and the percentage not advocating war. Set up a differential equation that is satisfied by $y = f(t)$ and sketch a solution.

9. (Chemical reaction) Suppose that substance A is converted into substance B at a rate that, at any time t, is proportional to the square of the amount of A. This situation occurs, for instance, when it is necessary for two molecules of A to collide in order to create one molecule of B. Set up the differential equation that is satisfied by $y = f(t)$, the amount of substance A at time t. Sketch a solution.

---

* See L. F. Richardson, "War Moods 1," *Psychometrica*, 1945, p. 13.

10. (Spread of a technological innovation) The by-product coke oven was first introduced into the iron and steel industry in 1894. It took about 30 years before all the major steel producers had adopted this innovation. Let $f(t)$ be the percentage of the producers that had installed the new coke ovens by time t. Then a reasonable model* for the way $f(t)$ increased is given by the assumption that the rate of change of $f(t)$ at time t was proportional to the product of $f(t)$ and the percentage of firms that had not yet installed the new coke ovens at time t. Write a differential equation that is satisfied by $f(t)$.

11. Refer to the discussion of the intravenous infusion of glucose. Determine the formula for the amount $A(t)$ of excess glucose in the bloodstream of a patient at time t. Describe what would happen if the rate r of infusion of glucose were doubled.

12. Verify that

$$f(t) = \frac{M}{1 + Be^{-ct}}$$

is a solution to the logistic differential equation $y' = ky(M - y)$, where $c = kM$.

In the remaining exercises, use algebra and/or a graphing utility to answer the questions.

13. A parachutist has a terminal speed of 160 feet per second. That is, no matter how long she falls, her speed will not exceed 160 feet per second, but it will get arbitrarily close to that value. Her speed after one second is 29 feet per second. Let $f(t)$ be her speed after t seconds.
    (a) Find the differential equation and the formula for $f(t)$.
    (b) Make a rough sketch of the graph of $f(t)$.
    (c) How fast will she be traveling after 15 seconds?
    (d) How fast will she be accelerating after 15 seconds?
    (e) Approximately when will she be traveling at the rate of 158 feet per second?
    (f) Approximately when will she be accelerating at the rate of one foot per second^2?

14. Suppose that a lake is stocked with 100 fish. After 3 months there are 250 fish. A study of the ecology of the lake predicts that the lake can support 1000 fish. Let $f(t)$ be the number of fish after t months.
    (a) Find the differential equation and the formula for $f(t)$.
    (b) Make a rough sketch of the graph of $f(t)$.
    (c) How many fish will there be after 20 months?
    (d) At what rate will the fish population be increasing after 3 months?
    (e) When is the fish population growing at the greatest rate?

* See E. Mansfield, "Technical Change and the Rate of Imitation," *Econometrica*, **29** (1961), 741–766.

15. The growth of the yellow nutsedge weed is described by a logistic differential equation. A typical weed has length 8 centimeters after 9 days, has length 48 centimeters after 25 days, and reaches length 55 centimeters at maturity.
    (a) Find the differential equation and formula for $f(t)$.
    (b) Make a rough sketch of the graph of $f(t)$.
    (c) How long will the weed be after 12 days?
    (d) At what rate will the weed be growing after 12 days?
    (e) When is the weed growing at the greatest rate?

16. A steel rod of temperature 125° is immersed in a large vat of water at temperature 75°. At any time, the rate of change in the temperature is proportional to the difference between the temperature of the water and the temperature of the rod. After 4 seconds, the rod has temperature 41.3°. Let $f(t)$ be the temperature after t seconds.
    (a) Find the differential equation and the formula for $f(t)$.
    (b) Make a rough sketch of the graph of $f(t)$.
    (c) What is the temperature of the rod after 10 seconds?
    (d) At what rate is the temperature changing after 10 seconds?
    (e) Approximately when is the temperature of the rod 100°?
    (f) Approximately when is the rod cooling at the rate of 2 degrees per second?

SOLUTIONS TO PRACTICE PROBLEM 6.4

1. The difference between transmission of information via mass media and via word of mouth is that in the second case the rate of transmission depends not only on the number of people who have not yet received the information, but also on the number of people who know the information and therefore are capable of spreading it. Therefore, for group A, $f'(t) = k[M - f(t)]$, and for group B, $f'(t) = kf(t)[M - f(t)]$. Note that the spread of information by word of mouth follows the same pattern as the spread of an epidemic.

REVIEW OF THE FUNDAMENTAL CONCEPTS OF CHAPTER 6

1. What does it mean to antidifferentiate a function?
2. What is a differential equation?
3. What does it mean for a function to be a solution to a differential equation?
4. What is a constant solution to a differential equation?
5. What is an autonomous differential equation?
6. Outline the procedure for sketching a solution of an autonomous differential equation.
7. Describe an application of the differential equation $y' = k(M - y)$.
8. Describe an application of the differential equation $y' = ky(M - y)$.

Chapter 7

THE DEFINITE INTEGRAL

There are two fundamental geometric problems of calculus. The first is to find the slope of a curve at a point, and the second is to find the area of a region under a curve. These problems are quite simple when the curve is a straight line, as in Figure 1. Both the slope of the line and the area of the shaded trapezoid can be calculated by geometric principles. When the graph consists of several line segments, as in Figure 2, the slope of each line segment can be computed separately, and the area of the region can be found by adding the areas of the regions under each line segment.

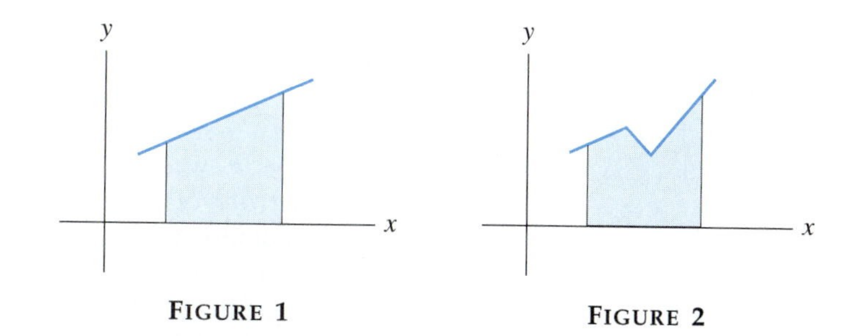

FIGURE 1 FIGURE 2

Calculus is needed when the curves are not straight lines. We have seen that the slope problem is resolved with the derivative of a function. In this chapter, we describe how the area problem is connected with the notion of the "integral" of a function. Both the slope problem and the area problem were studied by the ancient Greeks and solved in special cases, but it was not until the development of calculus in the seventeenth century that the intimate connection between the two problems was discovered. In this chapter we will discuss this connection as stated in the Fundamental Theorem of Calculus.

7.1 AREAS AND RIEMANN SUMS

This chapter reveals the important connection between antiderivatives and areas of regions under curves. Although the full story will have to wait until Section 7.5, we can give a hint now by describing how "area" is related to a problem solved in Section 6.2.

Example 2 in Section 6.2 concerned a company's marginal cost function, $f(x) = .015x^2 - 2x + 80$. A calculation with an antiderivative of $f(x)$ showed that if the production level is increased from $x = 30$ to $x = 60$ units per day, the change in total cost is \$645. As we'll see, this change in cost exactly equals the area of the region under the graph of the marginal cost curve in Figure 1 from $x = 30$ to $x = 60$. First, however, we need to learn how to find areas of such regions.

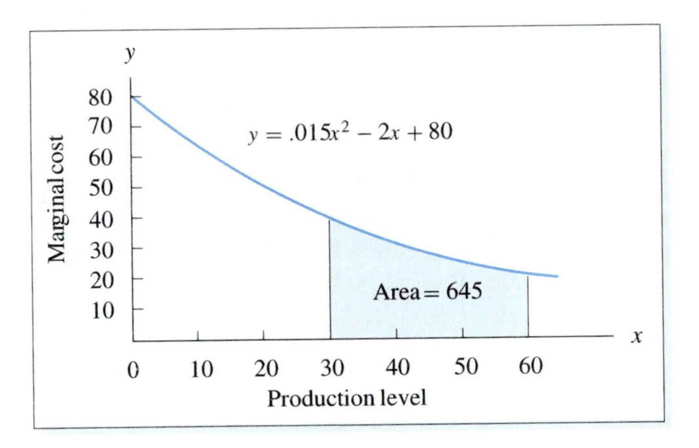

FIGURE 1 Area under a marginal cost curve

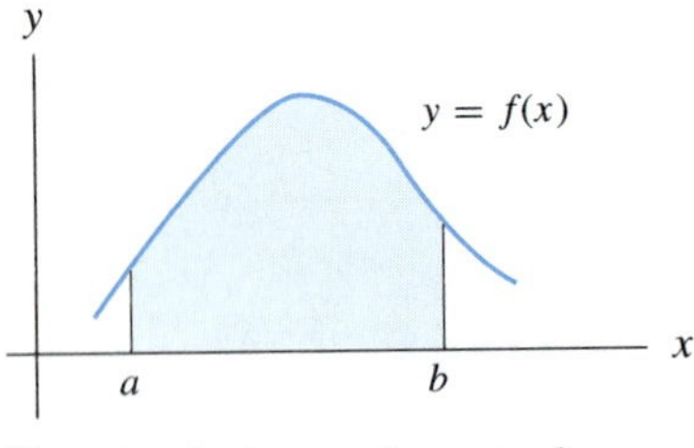

FIGURE 2 Area under a graph

AREA UNDER A GRAPH If $f(x)$ is a continuous nonnegative function on the interval $a \leq x \leq b$, we refer to the area of the region shown in Figure 2 as the *area under the graph of* $f(x)$ *from* a *to* b.

The computation of the area in Figure 2 is not a trivial matter when the top boundary of the region is curved. However, we can *estimate* the area to any desired degree of accuracy. The basic idea is to construct rectangles whose total area is approximately the same as the area to be computed. The area of each rectangle, of course, is easy to compute.

Figure 3 shows three rectangular approximations to the area under a graph. When the rectangles are thin, the mismatch between the rectangles and the region under the graph is quite small. In general, a rectangular approximation can be made as close as desired to the exact area simply by making the width of the rectangles sufficiently small.

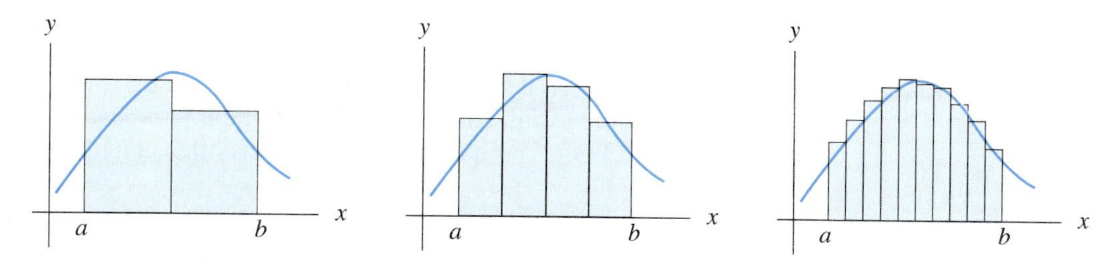

FIGURE 3 Approximating a region with rectangles

Given a continuous nonnegative function $f(x)$ on the interval $a \leq x \leq b$, divide this interval on the x-axis into n equal subintervals, where n represents some positive integer. Such a subdivision is called a *partition* of the interval from a to b. Since the entire interval is of width $b-a$, the width of each of the n subintervals is $(b-a)/n$. For brevity, denote this width by Δx. That is,

$$\Delta x = \frac{b-a}{n} \quad \text{(width of one subinterval).}$$

In each subinterval, select a point. (Any point in the subinterval will do.) Let x_1 be the point selected from the first subinterval, x_2 the point from the second subinterval, and so on. These points are used to form the rectangles that approximate the region under the graph of $f(x)$. Construct the first rectangle with height $f(x_1)$ and the first subinterval as base, as in Figure 4.

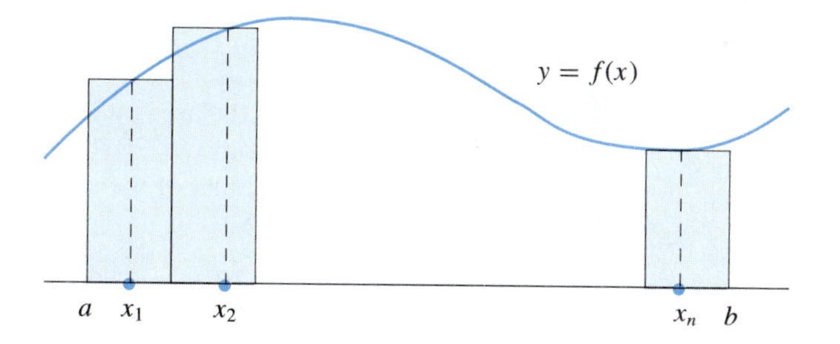

FIGURE 4 Rectangles with heights $f(x_1), \ldots, f(x_n)$

The top of the rectangle touches the graph directly above x_1. Notice that

$$[\text{area of first rectangle}] = [\text{height}]\,[\text{width}] = f(x_1)\Delta x.$$

The second rectangle rests on the second subinterval and has height $f(x_2)$. Thus

$$[\text{area of second rectangle}] = [\text{height}]\,[\text{width}] = f(x_2)\Delta x.$$

Continuing in this way, we construct n rectangles with a combined area of

$$f(x_1)\Delta x + f(x_2)\Delta x + \cdots + f(x_n)\Delta x. \tag{1}$$

A sum as in (1) is called a *Riemann sum*. It provides an approximation to the area under the graph of $f(x)$ when $f(x)$ is nonnegative and continuous. In fact, as the number of subintervals increases indefinitely, the Riemann sums (1) approach a limiting value, the area under the graph.*

For purposes of computation it is often convenient to rewrite the sum in (1) as

$$[f(x_1) + f(x_2) + \cdots + f(x_n)]\,\Delta x.$$

This calculation requires only one multiplication.

CALCULATING RIEMANN SUMS WITH A TI CALCULATOR

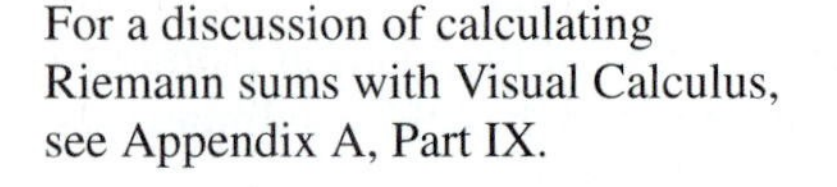

For a discussion of calculating Riemann sums with Visual Calculus, see Appendix A, Part IX.

When the points are selected as all midpoints, all left endpoints, or all right endpoints, the above sum is computed in the Home screen with

$$\texttt{sum(seq(Y}_1\texttt{, X, x}_1\texttt{, x}_n\texttt{, }\Delta\texttt{x))*}\Delta x$$

where "sum(seq(" is displayed on a TI-85 by pressing

[2nd] [LIST] [F5] [MORE] [F1] [(] [F3].

On a TI-82 or TI-83, it is displayed with the following four steps:

1. Press [2nd] [LIST] and cursor right to MATH.
2. Press **5** (and a left parenthesis, if necessary) to display sum(.
3. Press [2nd] [LIST] and place cursor on OPS.
4. Press **5** (and a left parenthesis, if necessary) to display seq(.

$\texttt{Y}_1$ can be either a function name or an expression for a function.

EXAMPLE 1 Estimate the area under the graph of the marginal cost function $f(x) = .015x^2 - 2x + 80$ from $x = 30$ to $x = 60$. Use partitions of 5, 30, and 100 subintervals. Use the midpoints of the subintervals as $x_1, x_2, \ldots, x_n$ to construct the rectangles. See Figure 5.

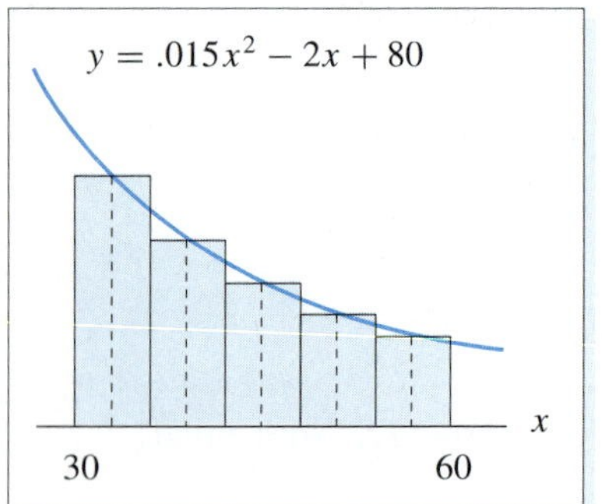

FIGURE 5 Estimating the area under a marginal cost curve

Solution The partition of $30 \le x \le 60$ with $n = 5$ is shown in Figure 5. The length of each subinterval is

$$\Delta x = \frac{60 - 30}{5} = 6.$$

Observe that the first midpoint is $\Delta x/2$ units from the left endpoint, and the midpoints themselves are Δx units apart. The first midpoint is $x_1 = 30 + \Delta x/2 = 30 + 3 = 33$. Subsequent midpoints are found by successively adding $\Delta x = 6$.

$$\text{midpoints:} \quad 33, 39, 45, 51, 57.$$

---

* Riemann sums are named after the nineteenth-century German mathematician G. B. Riemann (pronounced "Reemahn"), who used them extensively in his work on calculus. The concept of a Riemann sum has several uses: to approximate areas under curves, to construct mathematical models in applied problems, and to give a formal definition of area. In the next section, Riemann sums are used to define the definite integral of a function.

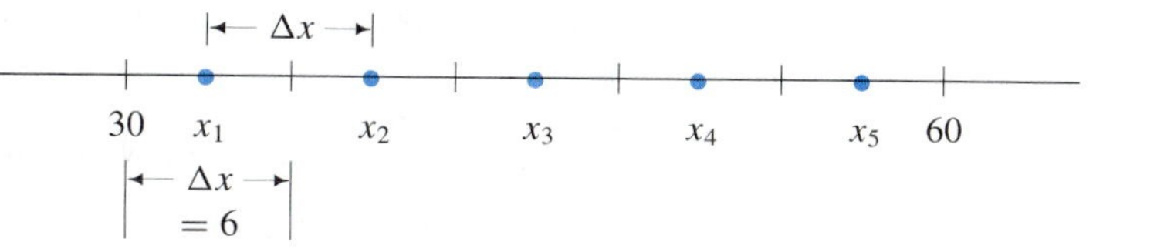

FIGURE 6 A partition of the interval $30 \le x \le 60$

```
sum(seq(Y1,X,33,
57,6))*6
             643.65
sum(seq(Y1,X,30.
5,59.5,1))*1
           644.9625
```

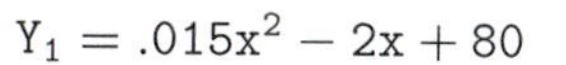

The corresponding estimate for the area under the graph of $f(x)$ is

$$
\begin{aligned}
f(33)\Delta x + f(39)\Delta x + f(45)\Delta x + f(51)\Delta x + f(57)\Delta x \\
&= [f(33) + f(39) + f(45) + f(51) + f(57)]\,\Delta x \\
&= [30.335 + 24.815 + 20.375 + 17.015 + 14.735] \cdot 6 \\
&= 107.275 \cdot 6 = 643.65.
\end{aligned}
$$

A similar calculation with 30 subintervals produces an area estimate of 644.9625. (See last 3 lines of calculator screen at left.) With 100 subintervals the estimate is 644.996625. ■

The approximations in Example 1 seem to confirm the claim made at the beginning of the section that the area under the marginal cost curve equals the change in total cost, \$645. The verification of this fact will be given in Section 7.5.

Although the midpoints of subintervals are often selected as the $x_1, x_2, \ldots, x_n$ in a Riemann sum, left endpoints and right endpoints are also convenient.

EXAMPLE 2 Use a Riemann sum with $n = 4$ to estimate the area under the graph of $f(x) = x^2$ from $x = 1$ to $x = 3$. Select the right endpoints of the subintervals as x_1, x_2, x_3, x_4.

Solution Here $\Delta x = (3 - 1)/4 = .5$. The right endpoint of the first subinterval is $1 + \Delta x = 1.5$. Subsequent right endpoints are obtained by successively adding .5, as shown below.

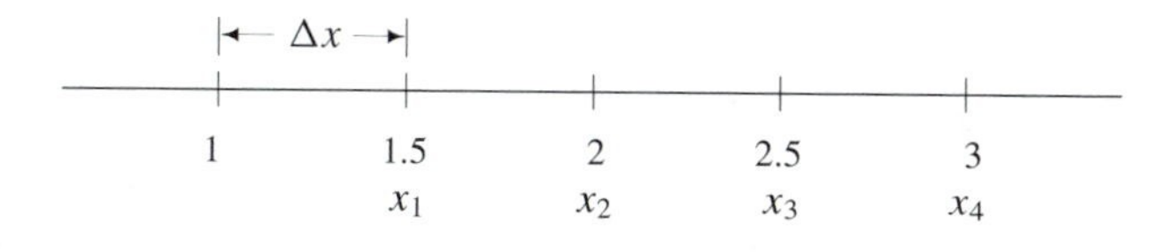


```
sum(seq(X²,X,1.5
,3,.5))*.5
              10.75
```

The corresponding Riemann sum is

$$
\begin{aligned}
f(x_1)\Delta x + f(x_2)\Delta x + f(x_3)\Delta x + f(x_4)\Delta x \\
&= [f(x_1) + f(x_2) + f(x_3) + f(x_4)]\,\Delta x \\
&= \left[(1.5)^2 + (2)^2 + (2.5)^2 + (3)^2\right] \cdot (.5) \\
&= 21.5 \cdot (.5) = 10.75.
\end{aligned}
$$

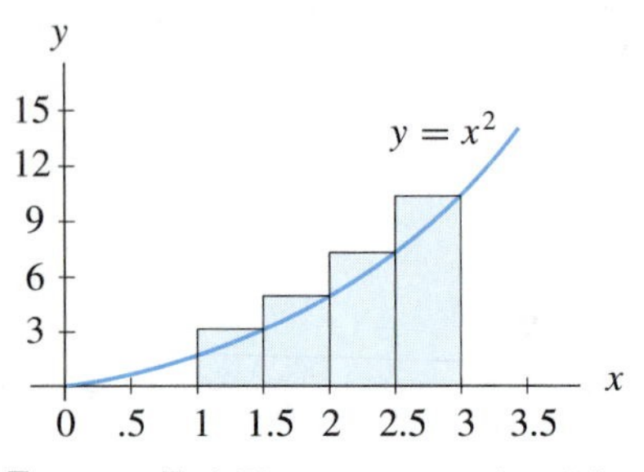

FIGURE 7 A Riemann sum using right endpoints

The rectangles used for this Riemann sum are shown in Figure 7. The right endpoints here give an area estimate that is obviously greater than the exact area. Midpoints would work better. But if the rectangles are sufficiently narrow, even a Riemann sum using right endpoints will be close to the exact area. ■

EXAMPLE 3 To estimate the area of a 100-foot-wide waterfront lot, a surveyor measured the distance from the street to the waterline at 20-foot intervals, starting 10 feet from one corner of the lot. Use the data to construct a Riemann sum approximation to the area of the lot. See Figure 8.

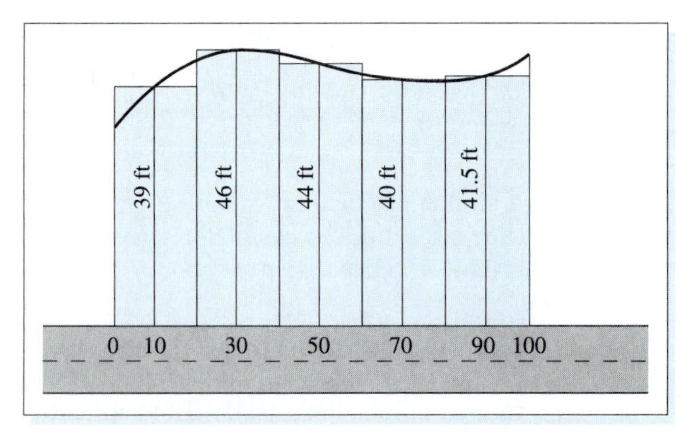

FIGURE 8 Survey of a waterfront property

Solution Treat the street as the x-axis, and consider the waterline along the property as the graph of a function $f(x)$ over the interval from 0 to 100. The five "vertical" distances give $f(x_1), \ldots, f(x_5)$, where $x_1 = 10, \ldots, x_5 = 90$. Since there are five points $x_1, \ldots, x_5$ spread across the interval $0 \le x \le 100$, we partition the interval into five subintervals, with $\Delta x = 100/5 = 20$. Fortunately, each subinterval contains one x_i. (In fact, each x_i is the midpoint of a subinterval.) Thus the area of the lot is approximated by the Riemann sum

$$\begin{aligned} f(x_1)\Delta x + \cdots + f(x_5)\Delta x &= [f(x_1) + f(x_2) + f(x_3) + f(x_4) + f(x_5)]\,\Delta x \\ &= [39 + 46 + 44 + 40 + 41.5] \cdot 20 \\ &= 210.5 \cdot 20 = 4210 \text{ square feet.} \end{aligned}$$

For a better estimate of the area, the surveyor will have to make more measurements from the street to the waterline. ■

The final example shows how Riemann sums arise in an application. Since the variable is time, t, we write Δt in place of Δx.

EXAMPLE 4 The velocity of a rocket at time t is $v(t)$ feet per second. Construct a Riemann sum that estimates how far the rocket travels in the first 10 seconds. What happens when the number of subintervals in the partition increases without bound?

Solution Partition the interval $0 \le t \le 10$ into n subintervals of width Δt, and select points $t_1, t_2, \ldots, t_n$ from these subintervals. Although the rocket's velocity is not constant, it does not change much during a small subinterval of time. So we may use $v(t_1)$ as an approximation of the rocket's velocity during the first subinterval. Over this time subinterval of length $\Delta t = (10 - 0)/n$,

$$[\text{distance traveled}] \approx [\text{velocity}] \cdot [\text{time}] \approx v(t_1)\Delta t. \tag{2}$$

The distance traveled during the second time subinterval is approximately $v(t_2)\Delta t$, and so on. Thus an estimate for the total distance traveled is

$$v(t_1)\Delta t + v(t_2)\Delta t + \cdots + v(t_n)\Delta t. \tag{3}$$

The sum in (3) is a Riemann sum for the velocity function on the interval $0 \le t \le 10$. As n increases, such a Riemann sum approaches the area under the graph of the velocity function. However, from our derivation of (3), it seems reasonable that as n increases, the sums become a better and better estimate of the total distance traveled. We conclude that

$$\begin{bmatrix} \text{total distance rocket travels} \\ \text{during the first 10 seconds} \end{bmatrix} = \begin{bmatrix} \text{area under graph of} \\ v(t) \text{ over } 0 \le t \le 10 \end{bmatrix}.$$

See Figure 9. ■

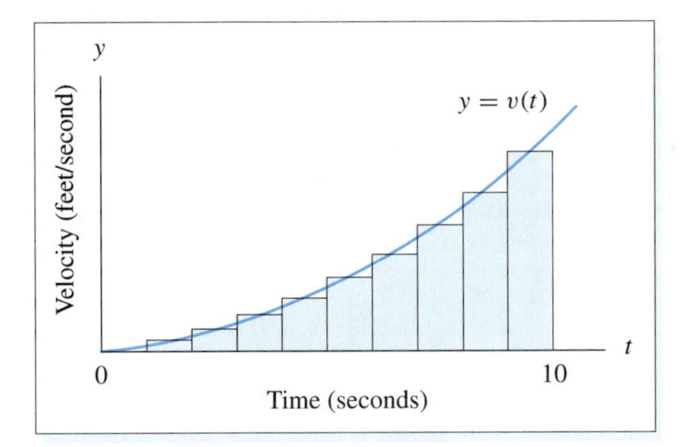

FIGURE 9 Area under a velocity curve

It is helpful to view Example 4 from a slightly different point of view. The height of the rocket is an increasing quantity and $v(t)$ is the *rate* of change of height at time t. The area under the graph of $v(t)$ from time $a = 0$ to time $b = 10$ is the *amount* of increase in the height during the first 10 seconds.

This connection between the rate of change of a function and the amount of increase of the function generalizes to a variety of situations.

If a quantity is increasing, then the area under the rate of change function from a to b is the amount of increase in the quantity from a to b.

Table 1 INTERPRETATION OF AREAS

Function	a to b	Area under the graph from a to b
Marginal cost at production level x	30 to 60	Additional cost when production is increased from 30 to 60 units
Rate of sulfur emissions from a power plant t years after 1990	1 to 3	Amount of sulfur released from 1991 to 1993
Birth rate t years after 1980 (in babies per year)	0 to 10	Number of babies born from 1980 to 1990
Rate of gas consumption t years after 1985	3 to 6	Amount of gas used from 1988 to 1991

Table 1 shows four instances of this principle. The result is justified in the next section. The first example in the table was discussed in Example 1.

PRACTICE PROBLEMS 7.1

1. Determine Δx and the midpoints of the subintervals formed by partitioning the interval $-2 \leq x \leq 2$ into five subintervals.
2. The graph in Figure 10 gives the rate at which new jobs were created (in millions of jobs per year) in the United States, where $t = 0$ corresponds to 1983. For instance, on January 1, 1986, jobs were being created at the rate of 2.4 million new jobs per year. Interpret the area of the shaded region.

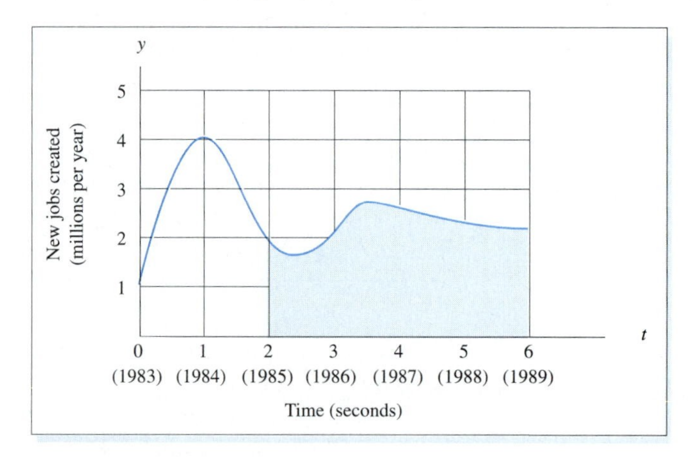

FIGURE 10 Rate of job creation

EXERCISES 7.1

Determine Δx and the midpoints of the subintervals formed by partitioning the given interval into n subintervals. (Hint: Decimals are sometimes easier to use than fractions.)

1. $0 \leq x \leq 2; n = 4$
2. $0 \leq x \leq 3; n = 6$
3. $1 \leq x \leq 4; n = 5$
4. $3 \leq x \leq 5; n = 5$

In Exercises 5–10, use a Riemann sum to approximate the area under the graph of $f(x)$ on the given interval, with selected points as specified.

5. $f(x) = x^2$; $1 \le x \le 3$, $n = 4$, midpoints of subintervals.
6. $f(x) = x^2$; $-2 \le x \le 2$, $n = 4$, midpoints of subintervals.
7. $f(x) = x^3$; $1 \le x \le 3$, $n = 5$, left endpoints.
8. $f(x) = x^3$; $0 \le x \le 1$, $n = 5$, right endpoints.
9. $f(x) = e^{-x}$; $2 \le x \le 3$, $n = 5$, right endpoints.
10. $f(x) = \ln x$; $2 \le x \le 4$, $n = 5$, left endpoints.

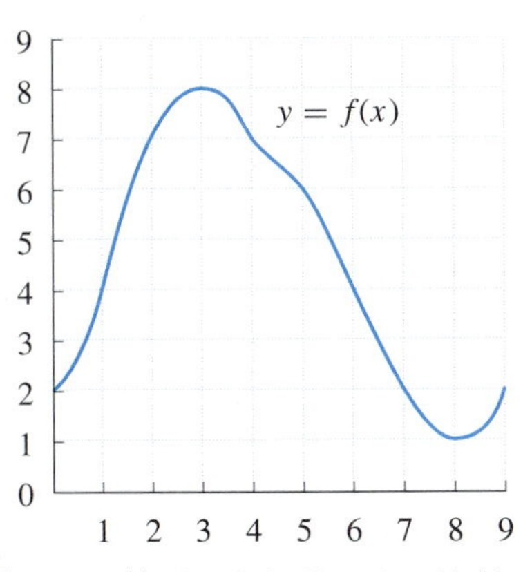

FIGURE 11 Graph for Exercises 11–14

In Exercises 11–14, use a Riemann sum to approximate the area under the graph of $f(x)$ in Figure 11 on the given interval, with selected points as specified. Draw the approximating rectangles.

11. $0 \le x \le 8$, $n = 4$, midpoints of subintervals
12. $3 \le x \le 7$, $n = 4$, left endpoints
13. $4 \le x \le 9$, $n = 5$, right endpoints
14. $0 \le x \le 9$, $n = 3$, midpoints of subintervals
15. Use a Riemann sum with $n = 4$ and left endpoints to estimate the area under the graph of $f(x) = 4 - x$ on the interval $1 \le x \le 4$. Then repeat with $n = 4$ and midpoints. Compare the answers with the exact answer, 4.5, which can be computed from the formula for the area of a triangle.
16. Use a Riemann sum with $n = 4$ and right endpoints to estimate the area under the graph of $f(x) = 2x - 4$ on the interval $2 \le x \le 3$. Then repeat with $n = 4$ and midpoints. Compare the answers with the exact answer, 1, which can be computed from the formula for the area of a triangle.

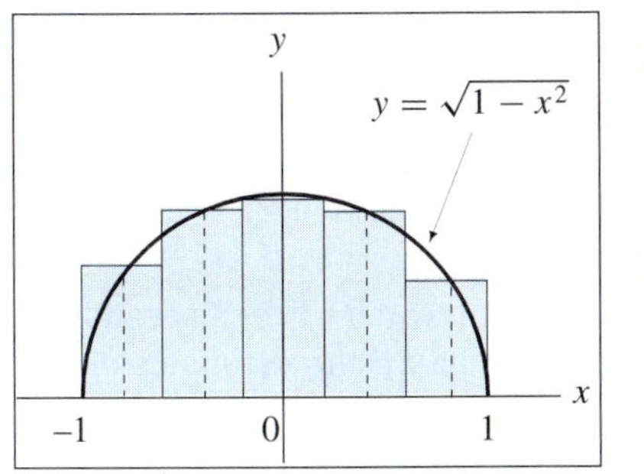

FIGURE 12(a)

17. The graph of the function $f(x) = \sqrt{1 - x^2}$ on the interval $-1 \le x \le 1$ is a semicircle. The area under the graph is $\frac{1}{2} \cdot \pi(1)^2 = \pi/2 = 1.57080$, to five decimal places. Use a Riemann sum with $n = 5$ and midpoints to estimate the area under the graph. See Figure 12(a). Carry out the calculations to five decimal places, and compute the error (the difference between the estimate and 1.57080).
18. Use a Riemann sum with $n = 5$ and midpoints to estimate the area under the graph of $f(x) = \sqrt{1 - x^2}$ on the interval $0 \le x \le 1$. The graph is a quarter-circle, and the area under the graph is .78540, to five decimal places. See Figure 12(b). Carry out the calculations to five decimal places, and compute the error. If you double the estimate from this exercise, will it be more accurate than the estimate computed in Exercise 17?

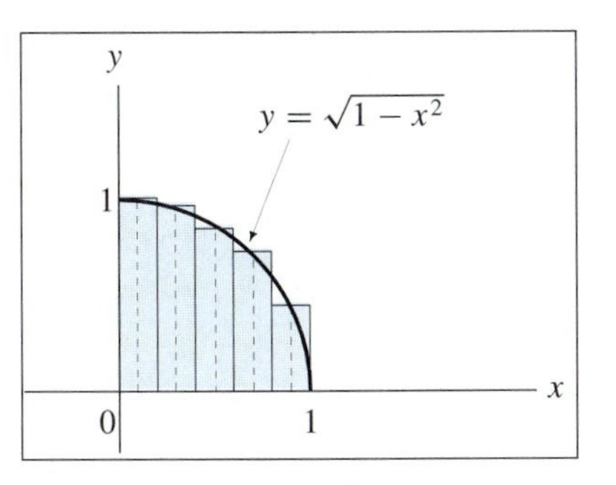

FIGURE 12(b)

19. Lung physiologists measure the velocity of air passing through a patient's throat by having the patient breathe into a pneumotachograph. This machine produces a graph that plots air-flow rate as a function of time. The graph in Figure 13 shows the flow rate while a patient is breathing out. The area under the graph gives the total volume of air during exhalation. Estimate this volume with a Riemann sum. Use $n = 5$ and midpoints.
20. Estimate the area (in square feet) of the piece of land shown in Figure 14.

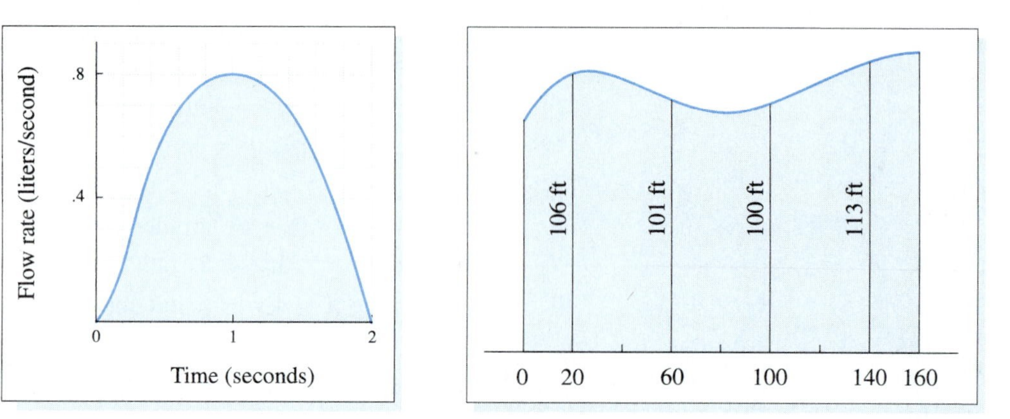

FIGURE 13 Data from a pneumotachograph

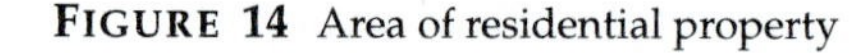

FIGURE 14 Area of residential property

21. The velocity of a car (in feet per second) is recorded from the speedometer every 10 seconds, beginning 5 seconds after the car starts to move. See Table 2. Use a Riemann sum to estimate the distance the car travels during the first 60 seconds.

Table 2 A CAR'S VELOCITY

Time	5	15	25	35	45	55
Velocity	20	44	32	39	65	80

22. Table 3 shows the velocity (in feet per second) at the end of each second for a person starting a morning jog. Make three Riemann sum estimates of the total distance jogged during the time from $t = 2$ to $t = 8$.

(a) $n = 6$, left endpoints.

(b) $n = 6$, right endpoints.

(c) $n = 3$, midpoints.

Table 3 A JOGGER'S VELOCITY

Time	0	1	2	3	4	5	6	7	8
Velocity	0	2	3	5	5	6	6	7	8

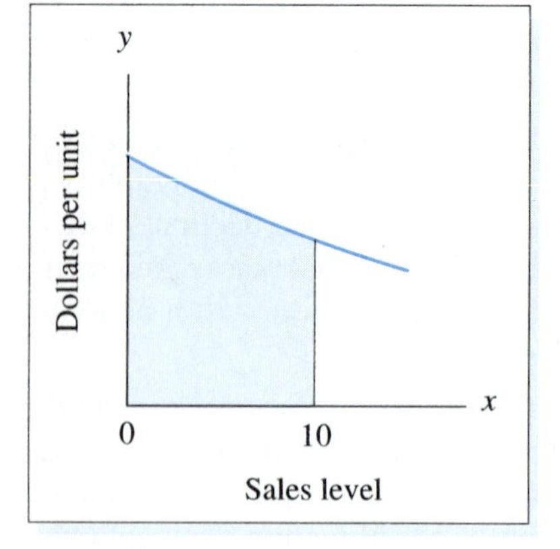

FIGURE 15 Marginal revenue

23. Complete the missing entries in Table 4.
24. Interpret the area of the shaded region in Figure 15.
25. Interpret the area of the shaded region in Figure 16.
26. Interpret the area of the shaded region in Figure 17.

Table 4 INTERPRETATION OF AREAS

Function	a to b	Area under the graph from a to b
Rate of growth of a population t years after 1900 (in millions of people per year)	10 to 50	
	5 to 7	Number of cigarettes smoked from 1990 to 1992
Marginal profit at production level x		Additional profit created by increasing production from 20 to 50 units

FIGURE 16 Rate of soil erosion

FIGURE 17 A child's growth rate

27. How does the area under $y = x^2$ from $x = 2$ to $x = 3$ compare with the area under $y = \frac{1}{5}x^2$ from 2 to 3? To answer this, write Riemann sums that approximate each area and compare them. Use $n = 5$ with midpoints. Write a brief paragraph about your conclusions.

28. How are the areas under $y = x^2$ and $y = 2 + x^3$ from $x = 1$ to $x = 2$ related to the area under the graph of $y = x^2 + 2 + x^3$? The answer is not obvious from an examination of the graphs in Figure 18. Examine Riemann sums for the three areas with $n = 5$ and midpoints, and write a short paragraph about

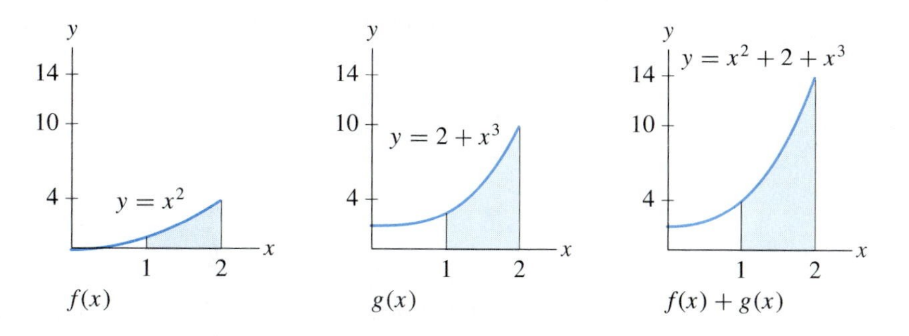

FIGURE 18

what you find.

29. Let $f(x)$ and $g(x)$ be nonnegative functions on $a \le x \le b$. Write a short paragraph about how you think the areas under the graphs of $f(x)$, $g(x)$, and $f(x) + g(x)$ are related. Use a general Riemann sum for $f(x) + g(x)$ in your discussion.

30. Let $f(x)$ be a nonnegative continuous function on $a \le x \le b$, and let k be a positive constant. Write a short paragraph about how you think the area under the graph of $k \cdot f(x)$ is related to the area under the graph of $f(x)$. Use a general Riemann sum for $k \cdot f(x)$ in your discussion.

In Exercises 31 and 32, let A be the area under the graph of $y = f(x)$ from $x = a$ to $x = b$.

31. Suppose that the function $f(x)$ is always increasing on the interval $a \le x \le b$. Explain why [approximation given by the Riemann sum with left endpoints] $< A <$ [approximation given by the Riemann sum with right endpoints].

32. Suppose that the function $f(x)$ is always decreasing on the interval $a \le x \le b$. Explain why [approximation given by the Riemann sum with left endpoints] $> A >$ [approximation given by the Riemann sum with right endpoints].

In Exercises 33–38, use a graphing utility to evaluate a Riemann sum to approximate the area under the graph of $f(x)$ on the given interval, with selected points as specified.

33. $f(x) = x\sqrt{1 + x^2}$; $1 \le x \le 3$, $n = 10$, midpoints of intervals.

34. $f(x) = \sqrt{1 - x^2}$; $-1 \le x \le 1$, $n = 10$, midpoints of intervals.

35. $f(x) = e^{-x^2}$; $-2 \le x \le 2$, $n = 10$, left endpoints.

36. $f(x) = x \ln x$; $2 \le x \le 6$, $n = 10$, right endpoints.

37. $f(x) = (\ln x)^2$; $2 \le x \le 3$, $n = 10$, right endpoints.

38. $f(x) = (1 + x)^2 e^{.4x}$; $2 \le x \le 4$, $n = 10$, left endpoints.

SOLUTIONS TO PRACTICE PROBLEMS 7.1

1. Since $n = 5$,

$$\Delta x = \frac{2 - (-2)}{5} = \frac{4}{5} = .8.$$

The first midpoint is $x_1 = -2 + .8/2 = -1.6$. Subsequent midpoints are found by successively adding .8, to obtain $x_2 = -.8$, $x_3 = 0$, $x_4 = .8$, and $x_5 = 1.6$.

2. The area under the curve from $t = 2$ to $t = 6$ is the number of new jobs created from 1985 to 1989.

7.2 THE DEFINITE INTEGRAL

In Section 7.1 we saw that the area under the graph of a continuous nonnegative function $f(x)$ from a to b is the limiting value of Riemann sums of the form

$$f(x_1)\Delta x + f(x_2)\Delta x + \cdots + f(x_n)\Delta x$$

as n increases without bound or, equivalently, as Δx approaches zero. (Recall that $x_1, x_2, \ldots, x_n$ are selected points from a partition of $a \le x \le b$ and Δx is the width of each of the n subintervals.) It can be shown that even if $f(x)$ has negative values, the Riemann sums still approach a limiting value as $\Delta x \to 0$. This number is called the *definite integral of $f(x)$ from a to b* and is denoted by

$$\int_a^b f(x)\,dx.$$

That is,

$$\int_a^b f(x)\,dx = \lim_{\Delta x \to 0} [f(x_1)\Delta x + f(x_2)\Delta x + \cdots + f(x_n)\Delta x]. \qquad (1)$$

If $f(x)$ is a nonnegative function, we know from Section 7.1 that the Riemann sum on the right side of (1) approaches the area under the graph of $f(x)$ from a to b. Thus, *the definite integral of a nonnegative function $f(x)$ equals the area under the graph of $f(x)$ from a to b.*

EXAMPLE 1 Calculate $\int_1^4 \left(\frac{1}{3}x + \frac{2}{3}\right) dx.$

Solution Figure 1 shows the graph of the function $f(x) = \frac{1}{3}x + \frac{2}{3}$. Since $f(x)$ is nonnegative for $1 \le x \le 4$, the definite integral of $f(x)$ equals the area of the shaded region in Figure 1. The region consists of a rectangle and a triangle. By geometry,

$$[\text{area of rectangle}] = [\text{width}] \cdot [\text{height}] = 3 \cdot 1 = 3,$$

$$[\text{area of triangle}] = \frac{1}{2}[\text{base}] \cdot [\text{height}] = \frac{1}{2} \cdot 3 \cdot 1 = \frac{3}{2}.$$

Thus the area under the graph is $4\frac{1}{2}$, and hence

$$\int_1^4 \left(\frac{1}{3}x + \frac{2}{3}\right) dx = 4\frac{1}{2}.$$

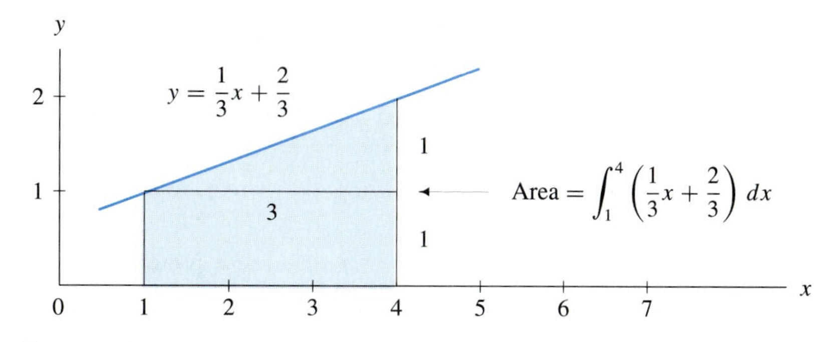

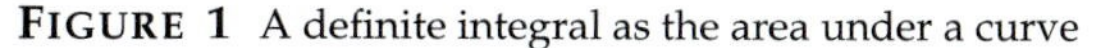
FIGURE 1 A definite integral as the area under a curve

In case $f(x)$ is negative at some points in the interval, we may also give a geometric interpretation of the definite integral. Consider the function $f(x)$ in Figure 2.

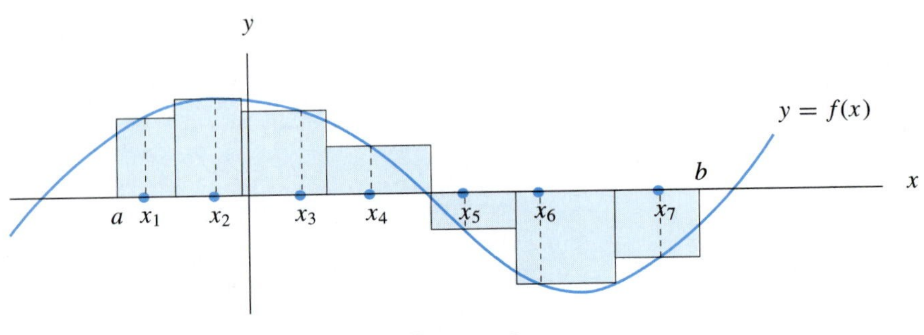

FIGURE 2

It shows a rectangular approximation of the region between the graph and the x-axis from a to b. Consider a typical rectangle located above or below the representative point x_i. If $f(x_i)$ is nonnegative, the area of the rectangle equals $f(x_i)\Delta x$. In case $f(x_i)$ is negative, the area of the rectangle equals $(-f(x_i))\Delta x$. So the expression $f(x_i)\Delta x$ equals either the area of the corresponding rectangle or the negative of the area, according to whether $f(x_i)$ is nonnegative or negative, respectively. In particular, the Riemann sum

$$f(x_1)\Delta x + f(x_2)\Delta x + \cdots + f(x_n)\Delta x$$

is equal to the area of the rectangles above the x-axis minus the area of the rectangle below the x-axis. Now take the limit as Δx approaches 0. On the one hand, the Riemann sum approaches the definite integral. On the other hand, the rectangular approximations approach the area bounded by the graph that is above the x-axis minus the area bounded by the graph that is below the x-axis. This gives us the following geometric interpretation of the definite integral.

Suppose that $f(x)$ is continuous on the interval $a \le x \le b$. Then

$$\int_a^b f(x)\,dx$$

is equal to the area above the x-axis bounded by the graph of $y = f(x)$ from $x = a$ to $x = b$ minus the corresponding area below the x-axis.

Referring to Figure 3, we have

$$\int_a^b f(x)\,dx = [\text{combined area of } B \text{ and } D] - [\text{combined area of } A \text{ and } C].$$

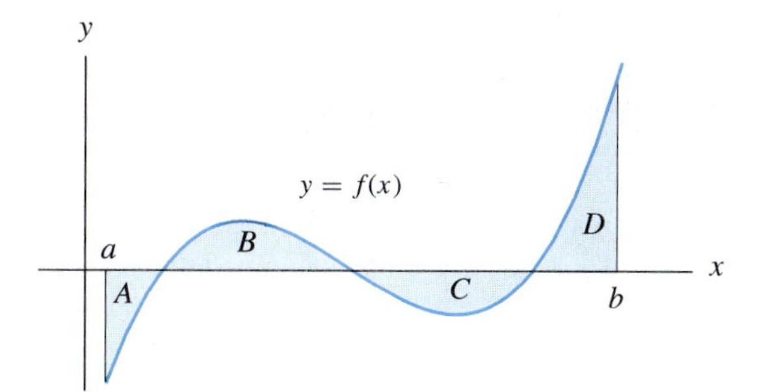

FIGURE 3 Regions above and below the x-axis

EXAMPLE 2 Calculate $\int_0^5 (2x - 4)\,dx$.

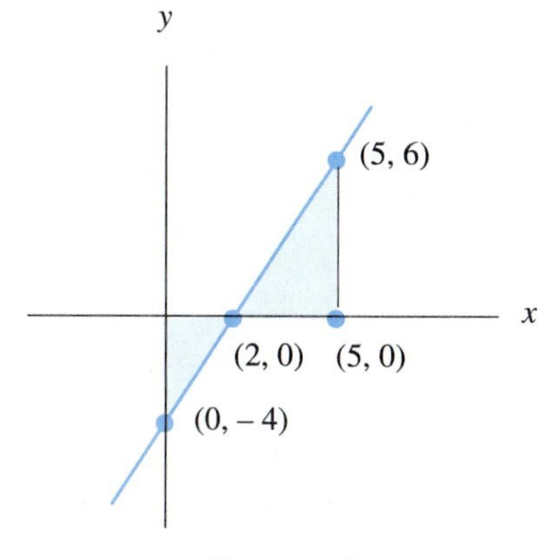

FIGURE 4

Solution Figure 4 shows the graph of the function $f(x) = 2x - 4$ on the interval $0 \le x \le 5$. The area of the triangle above the x-axis is 9 and the area of the triangle below the x-axis is 4. Therefore, from geometry, we find that

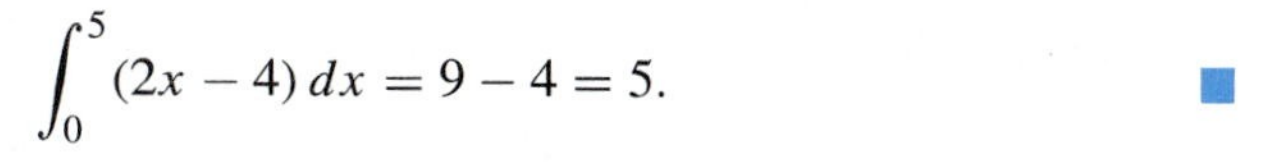

$$\int_0^5 (2x - 4)\,dx = 9 - 4 = 5.$$

The values of the definite integrals in Examples 1 and 2 follow from simple geometric formulas. For integrals of more complex functions, analogous area formulas are not available. Of course, Riemann sums may always be used to estimate the value of a definite integral to any desired degree of accuracy. Graphing utilities directly provide a very accurate approximation of the value of a definite integral. (Section 7.5 presents an algebraic method of evaluating a definite integral exactly.)

For a discussion of evaluating a definite integral with Visual Calculus, see Appendix A, Part X.

EVALUATING A DEFINITE INTEGRAL WITH A TI CALCULATOR

$\int_a^b f(x)\,dx$ is evaluated on the Home screen with the instruction

```
fnInt(Y1,X,a,b)
```

where Y_1 has been set to $f(x)$. For instance, the definite integral of Example 2 can be evaluated by setting Y_1 = 2X-4, typing fnInt(Y1,X,0,5), and pressing [ENTER]. (From the Home screen, "fnInt(" is invoked with [MATH] **9** on a TI-82 or TI-83, and with [2nd] [CALC] [F5] on a TI-85.)

With a TI-82 or TI-83, a definite integral can be calculated and the corresponding region of the graph shaded. Suppose Y_1 has been set to $f(x)$ and the graph of $f(x)$ appears on the screen. Press [2nd] [CALC] **7** and answer the questions. Reply to "Lower Bound (Limit)?" by moving the trace cursor to the point whose x-coordinate is a and pressing [ENTER]. Reply to "Upper Bound (Limit)?" by moving the trace cursor to the point whose x-coordinate is b and pressing [ENTER]. Figure 5(a) shows the resulting display. (*Note*: This task is usually easiest to accomplish when the difference between xMax and xMin is a multiple of

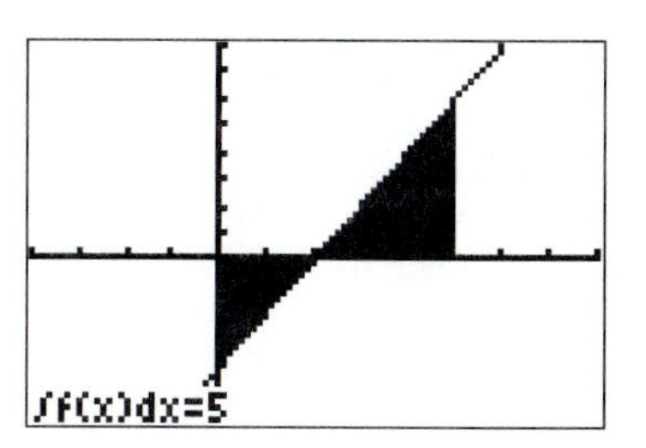

FIGURE 5(a) TI-82 or TI-83 $\int_0^5 (2x - 4)\,dx$

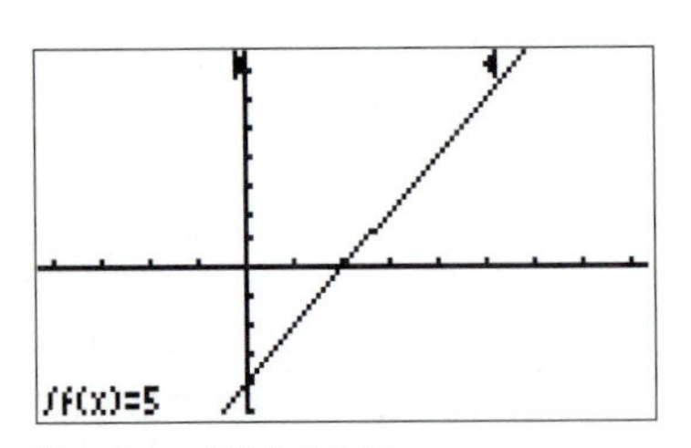

FIGURE 5(b) TI-85
$\int_0^5 (2x-4)\,dx$

9.4.) With a TI-83, the questions can be replied to by entering numbers. For instance, after pressing [2nd] [CALC] **7**, respond to "Lower Limit?" by typing in the number 0 and pressing [ENTER] and respond to "Upper Limit?" by typing in the number 5 and pressing [ENTER].

With a TI-85, a definite integral can be calculated from the graph, but the corresponding region of the graph will not be shaded. Suppose `y1` has been set to $f(x)$ and the graph of $f(x)$ appears on the screen. Press [GRAPH] [MORE] [F1] [F5]. Move the trace cursor to the point whose x-coordinate is a and press [ENTER]. Move the trace cursor to the point whose x-coordinate is b and press [ENTER]. Figure 5(b) shows the resulting display. (*Note*: This task is usually easiest to accomplish when the difference between xMax and xMin is a multiple of 12.6.)

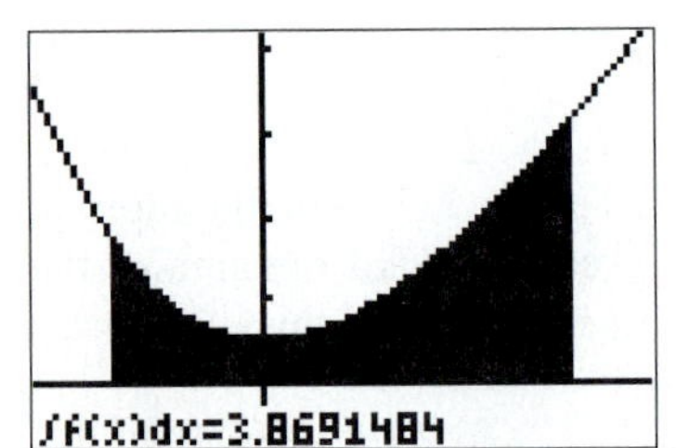

FIGURE 6

EXAMPLE 3 Compute the area under the curve $y = .5+x^2e^{-.2x}$ from $x = -1$ to $x = 2$.

Solution The graph of $f(x) = .5 + x^2e^{-.2x}$ is shown in Figure 6. Since $f(x)$ is nonnegative for $-1 \le x \le 2$, the area under the curve is given by the definite integral

$$\int_{-1}^{2} (.5 + x^2e^{-.2x})\,dx \approx 3.8691.$$

Many applications of the integral are developed by realizing that the quantity to be evaluated can be approximated arbitrarily closely by a Riemann sum with many terms. The quantity is then computed as the value of the associated definite integral. The following example illustrates the technique of progressing from a Riemann sum to the associated definite integral.

EXAMPLE 4 Use an appropriate definite integral to estimate the value of the following Riemann sum:

$$\left(e^2 - \frac{3}{2}\right)(.01) + \left(e^{2.01} - \frac{3}{2.01}\right)(.01)$$
$$+ \left(e^{2.02} - \frac{3}{2.02}\right)(.01) + \cdots + \left(e^{2.99} - \frac{3}{2.99}\right)(.01).$$

Solution This Riemann sum corresponds to the function

$$f(x) = e^x - \frac{3}{x}$$

and the one hundred points $x_1 = 2, x_2 = 2.01, x_3 = 2.02, \ldots, x_{100} = 2.99$. These points are the left endpoints when the interval $2 \le x \le 3$ is partitioned into 100 subintervals. The value of Δx is $\frac{3-2}{100} = .01$. The sum is close to

$$\int_2^3 \left(e^x - \frac{3}{x}\right) dx.$$

```
fnInt(e^(X)-3/X,
X,2,3)
            11.4800855
```

Therefore, the Riemann sum is approximately 11.48.

AREAS IN APPLICATIONS At the end of Section 7.1, we described how the area under a graph can represent the amount of change in a quantity. Since the area under a graph can be computed as a definite integral, this result can be restated in terms of a definite integral. Also, the earlier restriction that the quantity is increasing (equivalently, that the rate-of-change function is positive) can be dropped.

> If a quantity is changing, and $f(x)$ is its rate-of-change function, then the amount of change in the quantity from a to b is $\int_a^b f(x)\,dx$.

EXAMPLE 5 A rocket is fired vertically into the air. Its velocity at t seconds after liftoff is $v(t) = 6t + .5$ meters per second.

(a) Describe a region whose area represents the distance the rocket travels from time $t = 40$ to $t = 100$ seconds.

(b) Compute the distance in (a).

Solution (a) Let $s(t)$ be the position of the rocket at time t, measured from some reference point (such as the launch pad). This quantity is changing and its rate-of-change function is $v(t)$. The distance traveled during this time period is represented by the area under the velocity curve from 40 to 100. See Figure 7.

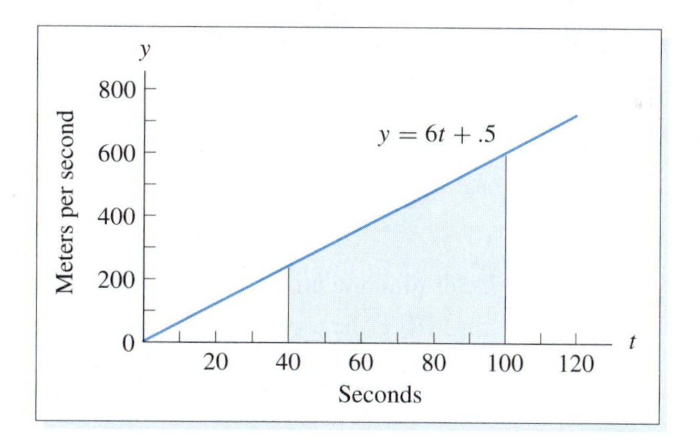

FIGURE 7 Area under a velocity curve is distance traveled

(b) With the aid of a graphing utility, the area under the curve is found to be

$$\int_{40}^{100} (6t + .5)\,dt = 25{,}230 \text{ meters.}$$

EXAMPLE 6 During the early 1970s, the annual worldwide rate of oil consumption was $R(t) = 16.1e^{.07t}$ billion barrels of oil per year, where t is the number of years since the beginning of 1970.

(a) Determine the amount of oil consumed from 1972 to 1974.

(b) Represent the answer to part (a) as an area.

Solution (a) Let $T(t)$ be the total amount of oil consumed t years since the beginning of 1970. This quantity is changing and its rate-of-change function is $R(t)$. Therefore, the amount of change from 1972 to 1974 is

$$\int_2^4 R(t)\,dt = \int_2^4 16.1e^{.07t}\,dt \approx 39.76 \text{ billion barrels of oil.}$$

A graphing utility was used to evaluate the definite integral.

(b) The amount of change in $T(t)$ is the area of the region under the graph of the rate function $R(t)$ for $t = 2$ to $t = 4$. See Figure 8.

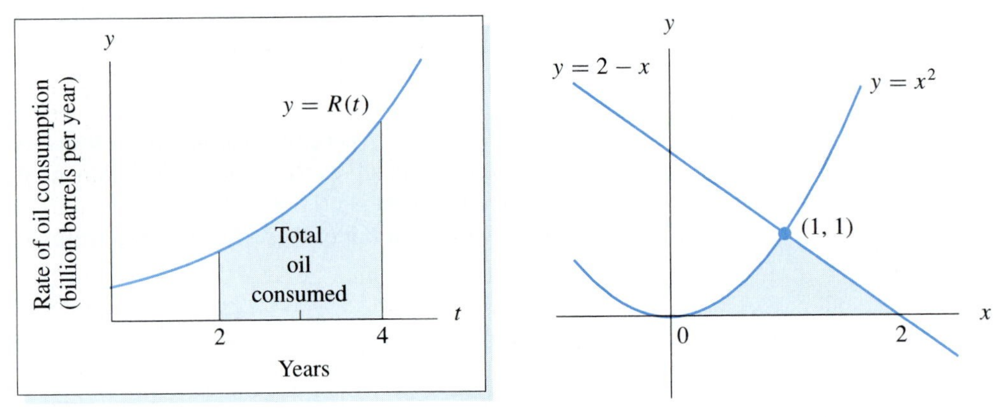

FIGURE 8 Total oil consumed

FIGURE 9

PRACTICE PROBLEMS 7.2

1. Determine the area of the shaded region in Figure 9.
2. Let $\mathrm{MR}(x)$ be a company's marginal revenue at production level x. Give an economic interpretation of the number $\int_{75}^{80} \mathrm{MR}(x)\,dx$.

EXERCISES 7.2

In Exercises 1–8, set up the definite integral(s) that gives the area of the shaded region, and use a graphing utility to evaluate the definite integral(s).

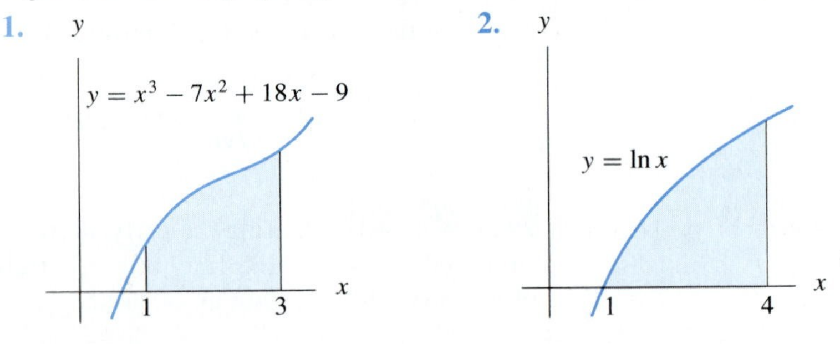

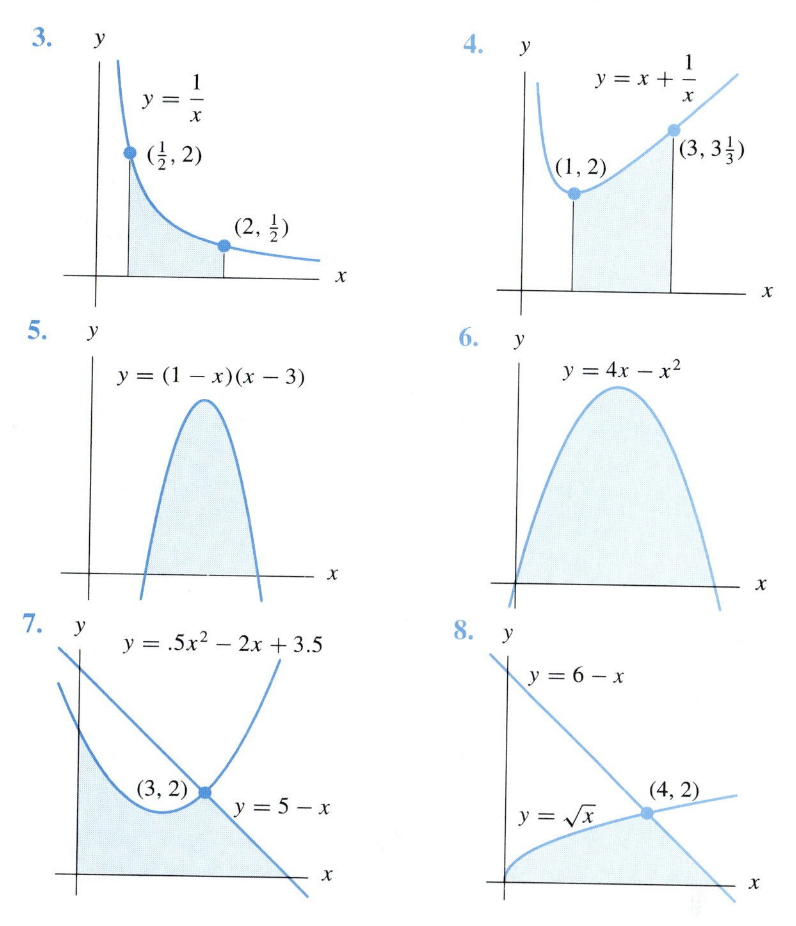

In Exercises 9–12, draw the region whose area is given by the definite integral.

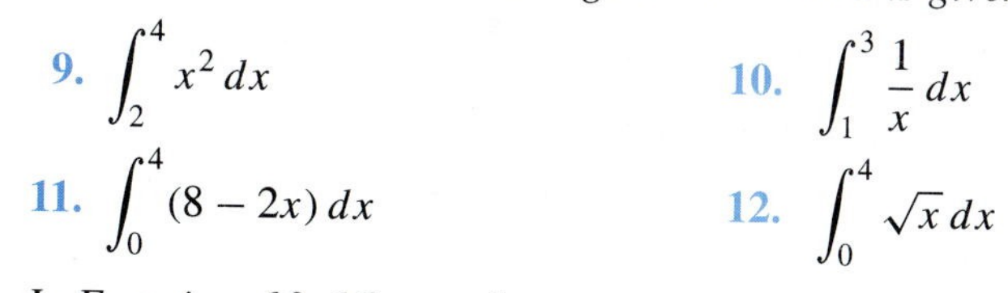

9. $\int_2^4 x^2\,dx$ **10.** $\int_1^3 \frac{1}{x}\,dx$

11. $\int_0^4 (8 - 2x)\,dx$ **12.** $\int_0^4 \sqrt{x}\,dx$

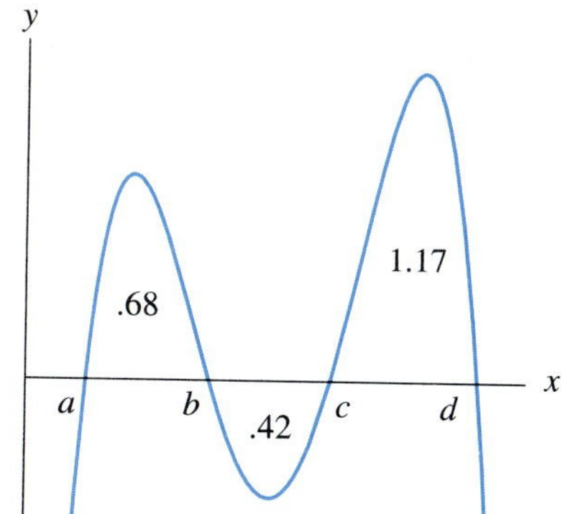

FIGURE 10 Graph for Exercises 17–20

In Exercises 13–16, use elementary geometry to find the value of the definite integral. Optionally, check your answer with a graphing utility.

13. $\int_2^6 5\,dx$ **14.** $\int_2^4 x\,dx$

15. $\int_0^3 2x\,dx$ **16.** $\int_0^3 (2x + 1)\,dx$

In Exercises 17–20, use Figure 10 to determine the value of the definite integral. Note: In Figure 10, three regions are labeled with their areas.

17. $\int_a^b f(x)\,dx$ **18.** $\int_a^c f(x)\,dx$

19. $\int_b^d f(x)\,dx$

20. $\int_a^d f(x)\,dx$

Exercises 21 and 22 refer to the graph of $f(x)$ shown in Figure 11. The graph consists of two straight line segments and two quarter-circles of radius 3. [Note: The area of a quarter-circle of radius r is $\frac{1}{4}\pi r^2$.]

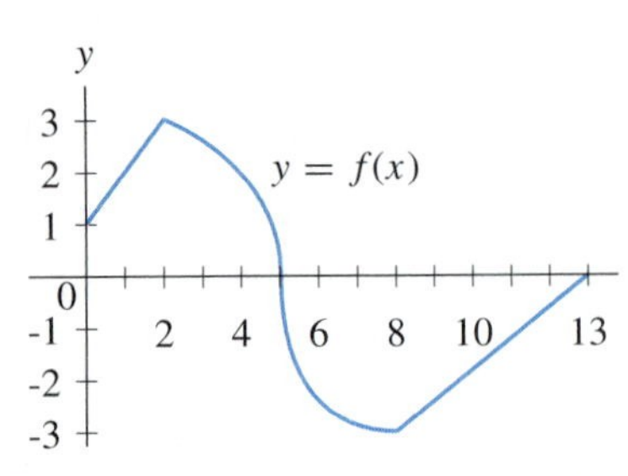

FIGURE 11 Graph for Exercises 21 and 22

21. Determine $\int_0^5 f(x)\,dx$.

22. Determine $\int_0^{13} f(x)\,dx$.

In Exercises 23–25, let $f(x) = 3x^2 + 2x + 1$ and $g(x) = x^2 - x + 3$ and use a graphing utility to verify the statement.

23. $\int_2^4 [f(x) + g(x)]\,dx = \int_2^4 f(x)\,dx + \int_2^4 g(x)\,dx$

24. $\int_2^4 [f(x) - g(x)]\,dx = \int_2^4 f(x)\,dx - \int_2^4 g(x)\,dx$

25. $\int_2^4 3 \cdot f(x)\,dx = 3 \cdot \int_2^4 f(x)\,dx$

26. Determine the value of $\int_{-2}^{2} x^3\,dx$ by looking at the graph of $y = x^3$.

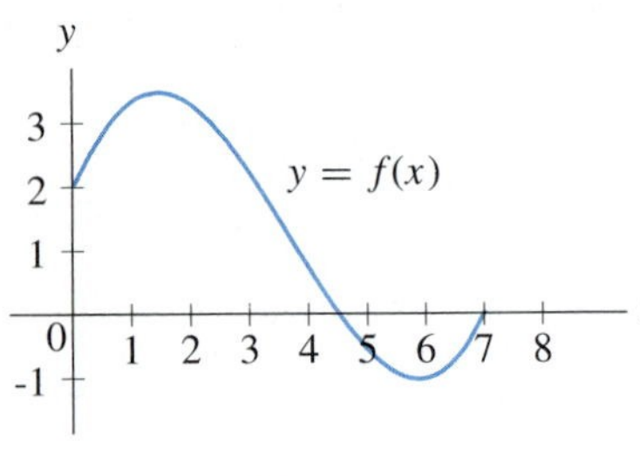

27. Let $f(x)$ be the function pictured in Figure 12. Determine whether $\int_0^7 f(x)\,dx$ is positive, negative, or zero.

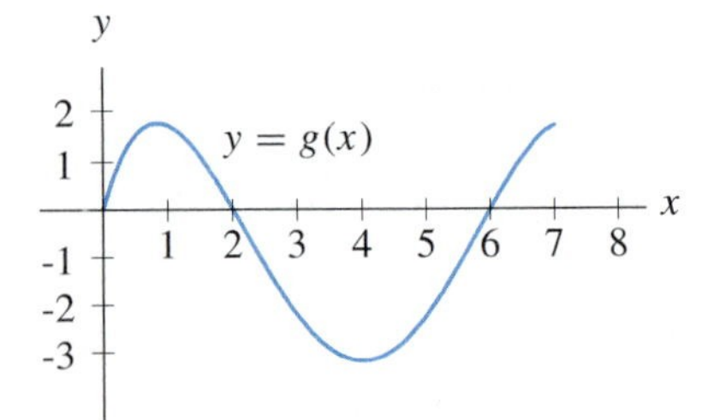

FIGURE 13

28. Let $g(x)$ be the function pictured in Figure 13. Determine whether $\int_0^7 g(x)\,dx$ is positive, negative, or zero.

In Exercises 29–32, use a graphing utility to graph the function and find the approximate coordinates of its two x-intercepts on the positive part of the x-axis. Then determine the area of the region under the graph from one x-intercept to the other.

29. $f(x) = 10 - x - \dfrac{2}{x}$

30. $f(x) = 10x - x^2 - 20$

31. $f(x) = (5 + x - x^2)\ln x$

32. $f(x) = -x^3 + 9x^2 - 15x - 1$

In Exercises 33–38, determine a definite integral that has the given expression as a Riemann sum.

33. $\left[(8.25)^3 + (8.75)^3 + (9.25)^3 + (9.75)^3\right](.5)$

34. $[\ln 4.2 + \ln 4.4 + \ln 4.6 + \ln 4.8 + \ln 5]\,(.2)$

35. $\left[(1 + e^{-1}) + (1.1 + e^{-1.1}) + \cdots + (1.9 + e^{-1.9})\right](.1)$

36. $\left[(3.1^2 + e^{3.1}) + (3.3^2 + e^{3.3}) + \cdots + (3.9^2 + e^{3.9})\right](.2)$

37. $\left[(3 - x_1)^2 + (3 - x_2)^2 + \cdots + (3 - x_{100})^2\right]\Delta x$, where the interval $0 \le x \le 3$ is divided into 100 subintervals of width $\Delta x = .03$ and $x_1, x_2, \ldots, x_{100}$ are points in these subintervals

38. $\left[5e^{.02(9-x_1)} + 5e^{.02(9-x_2)} + \cdots + 5e^{.02(9-x_{50})}\right]\Delta x$, where the interval $1 \le x \le 5$ is divided into 50 subintervals of width $\Delta x = .08$ and $x_1, x_2, \ldots, x_{50}$ are points in these subintervals

39. The worldwide rate of cigarette consumption (in trillions of cigarettes per year) since 1960 is given approximately by the function $c(t) = .1t + 2.4$, where $t = 0$ corresponds to 1960. Determine the number of cigarettes sold from 1980 to 1990.

40. Suppose $p(t)$ is the rate (in tons per year) at which pollutants are discharged into a lake, where t is the number of years since 1985. Interpret $\int_5^7 p(t)\,dt$.

41. A helicopter is rising straight up in the air. Its velocity at time t is $2t + 1$ feet per second.

    (a) How high does the helicopter rise during the first 5 seconds?

    (b) Represent the answer to part (a) as an area.

42. After t hours of operation, an assembly line is producing power lawn mowers at the rate of $21 - \frac{4}{5}t$ mowers per hour.

    (a) How many mowers are produced during the time from $t = 2$ to $t = 5$ hours?

    (b) Represent the answer to part (a) as an area.

43. Suppose that the marginal cost function of a handbag manufacturer is $\frac{3}{32}x^2 - x + 200$ dollars per unit at production level x (where x is measured in units of 100 handbags).

    (a) Find the total cost of producing 6 additional units if 2 units are currently being produced.

    (b) Describe the answer to part (a) as an area. (Give a written description rather than a sketch.)

44. Suppose that the marginal profit function for a company is $100 + 50x - 3x^2$ at production level x.

    (a) Find the extra profit earned from the sale of 3 additional units if 5 units are currently being produced.

    (b) Describe the answer to part (a) as an area. (Do not make a sketch.)

45. Let MP(x) be a company's marginal profit at production level x. Give an economic interpretation of the number $\int_{44}^{48} \text{MP}(x)\,dx$.

46. Let MC(x) be a company's marginal cost at production level x. Give an economic interpretation of the number $\int_0^{100} \text{MC}(x)\,dx$. (*Note*: At any production level, the total cost equals the fixed cost plus the total variable cost.)

47. Some food is placed in a freezer. After t hours the temperature of the food is dropping at the rate of $r(t)$ degrees Fahrenheit per hour, where $r(t) = 12 + 4/(t+3)^2$.

    (a) Compute the area under the graph of $y = r(t)$ over the interval $0 \le t \le 2$.

    (b) What does the area in part (a) represent?

48. Suppose that the velocity of a car at time t is $40 + 8/(t+1)^2$ kilometers per hour.

    (a) Compute the area under the velocity curve from $t = 1$ to $t = 9$.

    (b) What does the area in part (a) represent?

49. Deforestation is one of the major problems facing sub-Saharan Africa. Although the clearing of land for farming has been the major cause, the steadily increasing demand for fuel wood has become a significant factor. Figure 14 summarizes projections of the World Bank. The rate of fuel-wood consumption (in millions of cubic meters per year) in Sudan t years after 1980 is given approximately by the function $c(t) = 76.2e^{.03t}$. Determine the amount of fuel wood that will be consumed from 1980 to 2000.

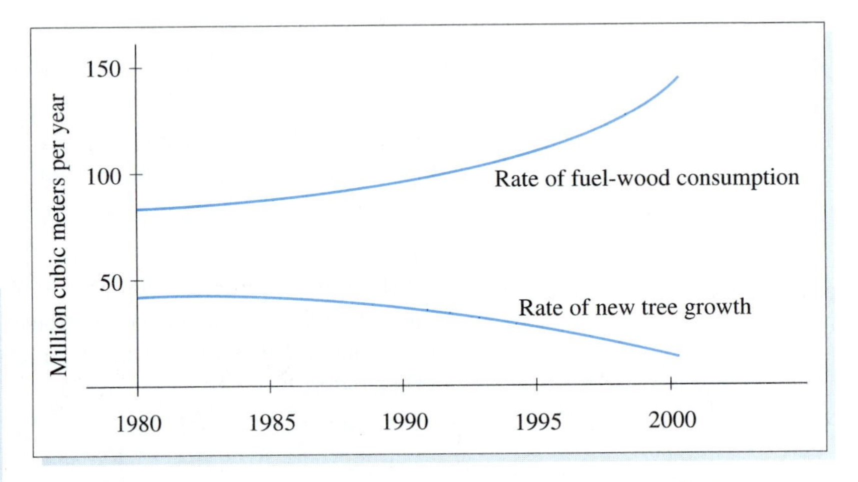

FIGURE 14 Data from "Sudan and Options in the Energy Sector," World Bank/UNDP, July 1983

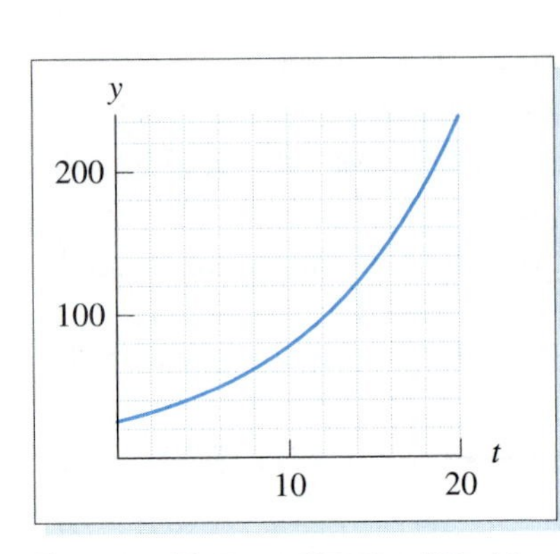

FIGURE 15 Rate of National Health Expenditures: 1960 to 1980

50. The graph in Figure 15 gives the annual rate of national health expenses (in billions of dollars) t years after 1960, for $0 \le t \le 20$. For instance, at the beginning of 1964, money was being spent at the rate of \$40 billion per year. Which of the following numbers is closest to the total national health expenditures for the years 1960 through 1980?

(a) \$1800 billion (b) \$3600 billion (c) \$900 billion

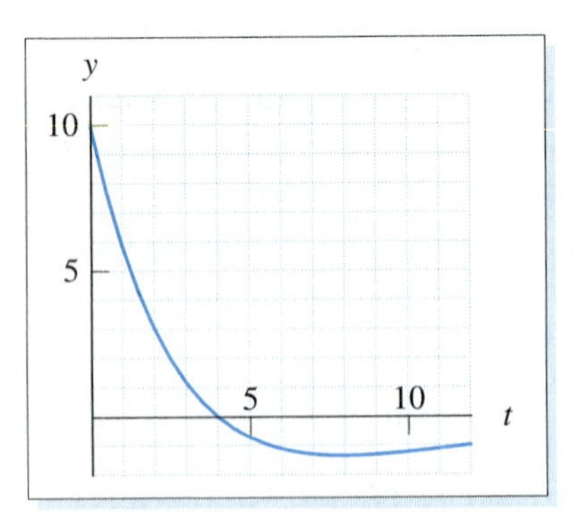

FIGURE 16

51. When a drug is injected intramuscularly (into a muscle), the amount of the drug in the bloodstream initially increases and then decreases. Figure 16 shows the rate at which the drug is entering the bloodstream t hours after the drug has been injected.

(a) When is there the greatest amount of the drug in the bloodstream?

(b) Which of the following numbers is closest to the greatest number of units of the drug in the bloodstream?

(i) 7 (ii) 15 (iii) 30

(c) Which of the following numbers is closest to the number of units of the drug in the bloodstream after 6 hours?

(i) 4 (ii) 16 (iii) 8

SOLUTIONS TO PRACTICE PROBLEMS 7.2

1. The simplest way to calculate the area is to split the region into two pieces by drawing the line $x = 1$. The shaded region is bounded above by the graph of $y = x^2$ for $x \leq 1$ and is bounded above by the graph of $y = 2 - x$ for $x \geq 1$. Therefore, the total area is given by

$$\int_0^1 x^2\,dx + \int_1^2 (2 - x)\,dx \approx .33333 + .5 = .83333.$$

2. Since $\mathrm{MR}(x)$ gives the rate of change of revenue, the number $\int_{75}^{80} \mathrm{MR}(x)\,dx$ is the amount of additional revenue received when the production level is raised from $x = 75$ to $x = 80$ units.

7.3 AREAS IN THE xy-PLANE

In this section we show how to use the definite integral to compute the area of a region that lies between the graphs of two or more functions. Three simple but important properties of the definite integral follow from the definition of the definite integral. These properties will be proven in Section 7.5.

Let $f(x)$ and $g(x)$ be functions and a, b, k any constants. Then

$$\int_a^b f(x)\,dx + \int_a^b g(x)\,dx = \int_a^b [f(x) + g(x)]\,dx, \tag{1}$$

$$\int_a^b f(x)\,dx - \int_a^b g(x)\,dx = \int_a^b [f(x) - g(x)]\,dx, \tag{2}$$

$$\int_a^b kf(x)\,dx = k\int_a^b f(x)\,dx. \tag{3}$$

Let us now consider regions that are bounded both above and below by graphs of functions. Referring to Figure 1, we would like to find a simple expression for the area of the shaded region under the graph of $y = f(x)$ and above the graph of $y = g(x)$ from $x = a$ to $x = b$. It is the region under the graph of $f(x)$ with the region under the graph of $g(x)$ taken away. Therefore,

$$\begin{aligned}[\text{area of shaded region}] &= [\text{area under } f(x)] - [\text{area under } g(x)]\\ &= \int_a^b f(x)\,dx - \int_a^b g(x)\,dx\\ &= \int_a^b [f(x) - g(x)]\,dx \qquad [\text{by property (2)}].\end{aligned}$$

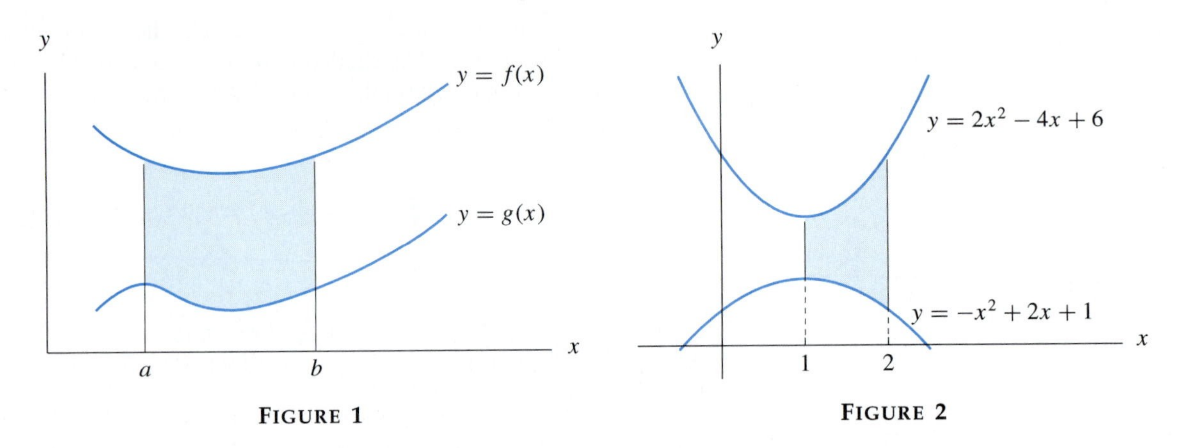

FIGURE 1

FIGURE 2

> *Area Between Two Curves* If $y = f(x)$ lies above $y = g(x)$ from $x = a$ to $x = b$, the area of the region between $f(x)$ and $g(x)$ from $x = a$ to $x = b$ is
>
> $$\int_a^b [f(x) - g(x)]\, dx.$$

EXAMPLE 1 Find the area of the region between $y = 2x^2 - 4x + 6$ and $y = -x^2 + 2x + 1$ from $x = 1$ to $x = 2$.

Solution Upon sketching the two graphs (Figure 2), we see that $f(x) = 2x^2 - 4x + 6$ lies above $g(x) = -x^2 + 2x + 1$ for $1 \le x \le 2$. Therefore, our formula gives the area of the shaded region as

$$\int_1^2 \left[(2x^2 - 4x + 6) - (-x^2 + 2x + 1)\right] dx = \int_1^2 (3x^2 - 6x + 5)\, dx = 3 \quad \blacksquare$$

EXAMPLE 2 Find the area of the region between $y = x^2$ and $y = (x - 2)^2 = x^2 - 4x + 4$ from $x = 0$ to $x = 3$.

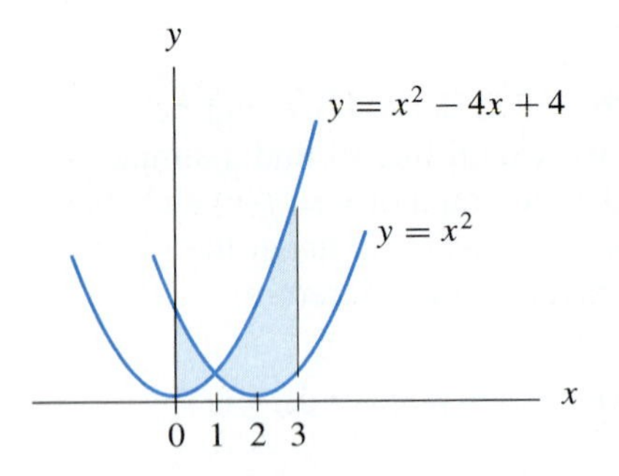

FIGURE 3

Solution Upon sketching the graphs (Figure 3), we see that the two graphs cross; by setting $x^2 = x^2 - 4x + 4$, we find that they cross when $x = 1$. Thus one graph does not always lie above the other from $x = 0$ to $x = 3$, so that we cannot directly apply our rule for finding the area between two curves. However, the difficulty is easily surmounted if we break the region into two parts, namely, the area from $x = 0$ to $x = 1$ and the area from $x = 1$ to $x = 3$. For from $x = 0$ to $x = 1$, $y = x^2 - 4x + 4$ is on top; and from $x = 1$ to $x = 3$, $y = x^2$ is on top. Consequently,

$$[\text{area from } x = 0 \text{ to } x = 1] = \int_0^1 \left[(x^2 - 4x + 4) - (x^2)\right] dx$$

$$= \int_0^1 (-4x + 4)\, dx = 2.$$

$$[\text{area from } x = 1 \text{ to } x = 3] = \int_1^3 \left[(x^2) - (x^2 - 4x + 4)\right] dx$$
$$= \int_1^3 (4x - 4)\, dx = 8.$$

Thus the total area is $2 + 8 = 10$. ■

Note: In Example 2, the intersection point of the two graphs was found algebraically. It also can be found on a calculator with the "intersect" or "ISECT" command and in Visual Calculus with the "Solve" command.

In our derivation of the formula for the area between two curves, we examined functions that are nonnegative. However, the statement of the rule does not contain this stipulation, and rightly so. Consider the case where $f(x)$ and $g(x)$ are not always positive. Let us determine the area of the shaded region in Figure 4(a). Select some constant c such that the graphs of the functions $f(x) + c$ and $g(x) + c$ lie completely above the x-axis [Figure 4(b)]. The region between them will have the same area as the original region. Using the rule as applied to nonnegative functions, we have

$$[\text{area of the region}] = \int_a^b [(f(x) + c) - (g(x) + c)]\, dx$$
$$= \int_a^b [f(x) - g(x)]\, dx.$$

Therefore, we see that our rule is valid for any functions $f(x)$ and $g(x)$ as long as the graph of $f(x)$ lies above the graph of $g(x)$ for all x from $x = a$ to $x = b$.

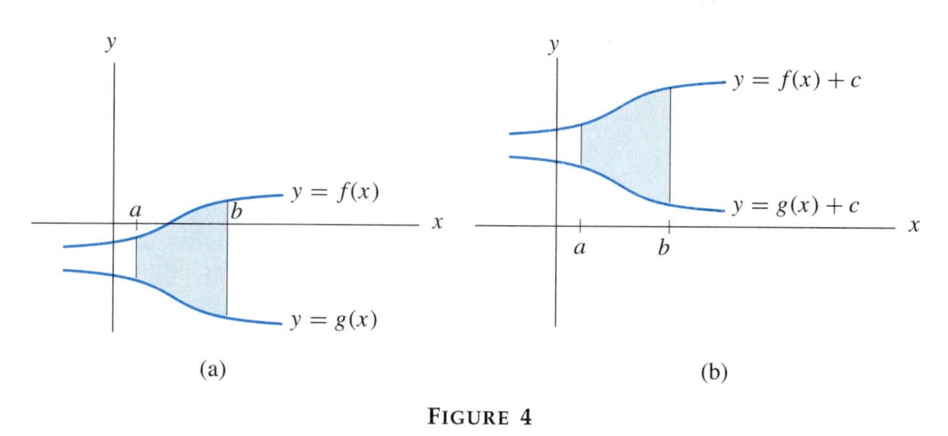

FIGURE 4

EXAMPLE 3 Set up the integral that gives the area between the curves $y = x^2 - 2x$ and $y = -e^x$ from $x = -1$ to $x = 2$.

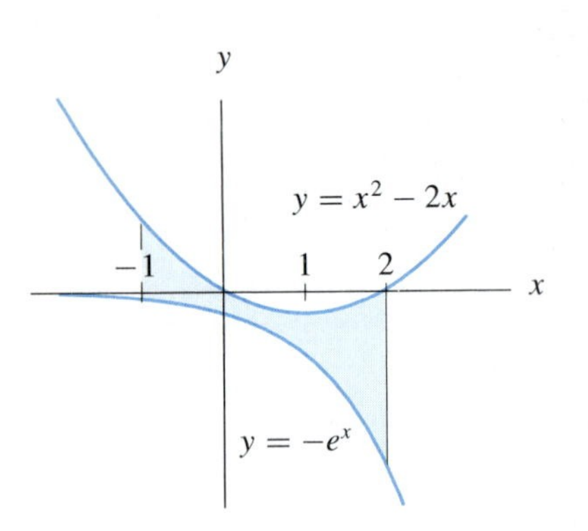

FIGURE 5

Solution Since $y = x^2 - 2x$ lies above $y = -e^x$ (Figure 5), the rule for finding the area between two curves can be applied directly. The area between the curves is

$$\int_{-1}^{2} (x^2 - 2x + e^x)\, dx. \qquad \blacksquare$$

Sometimes we are asked to find the area between two curves without being given the values of a and b. In these cases there is a region that is completely enclosed by the two curves. As the next examples illustrate, we must first find the points of intersection of the two curves in order to obtain the values of a and b. In such problems careful curve sketching is especially important.

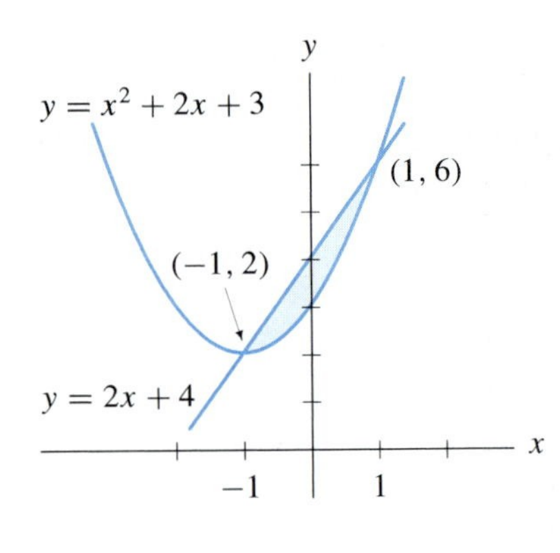

FIGURE 6

EXAMPLE 4 Set up the integral that gives the area bounded by the curves $y = x^2 + 2x + 3$ and $y = 2x + 4$.

Solution The two curves are sketched in Figure 6, and the region bounded by them is shaded. In order to find the points of intersection, we set $x^2 + 2x + 3 = 2x + 4$ and solve for x. We obtain $x^2 = 1$, or $x = -1$ and $x = +1$. When $x = -1$, $2x + 4 = 2(-1) + 4 = 2$. When $x = 1$, $2x + 4 = 2(1) + 4 = 6$. Thus the curves intersect at the points $(1, 6)$ and $(-1, 2)$.

Since $y = 2x + 4$ lies above $y = x^2 + 2x + 3$ from $x = -1$ to $x = 1$, the area between the curves is given by

$$\int_{-1}^{1} \left[(2x + 4) - (x^2 + 2x + 3)\right] dx = \int_{-1}^{1} (1 - x^2)\, dx. \qquad \blacksquare$$

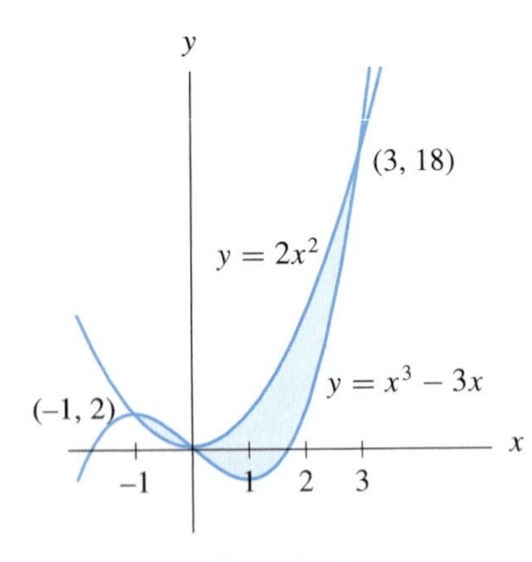

FIGURE 7

EXAMPLE 5 Set up the integral that gives the area bounded by the two curves $y = 2x^2$ and $y = x^3 - 3x$.

Solution First we make a rough sketch of the two curves, as in Figure 7. The curves intersect where $x^3 - 3x = 2x^2$ or $x^3 - 2x^2 - 3x = 0$. Note that

$$x^3 - 2x^2 - 3x = x(x^2 - 2x - 3) = x(x - 3)(x + 1).$$

So the solutions to $x^3 - 2x^2 - 3x = 0$ are $x = 0, 3, -1$, and the curves intersect at $(-1, 2)$, $(0, 0)$, and $(3, 18)$. From $x = -1$ to $x = 0$, the curve $y = x^3 - 3x$ lies above $y = 2x^2$. But from $x = 0$ to $x = 3$, the reverse is true. Thus the area between the curves is given by

$$\int_{-1}^{0} (x^3 - 3x - 2x^2)\, dx + \int_{0}^{3} (2x^2 - x^3 + 3x)\, dx. \qquad \blacksquare$$

EXAMPLE 6 Beginning in 1974, with the advent of dramatically higher oil prices, the exponential rate of growth of world oil consumption slowed down from a growth constant of 7% to a growth constant of 4% per year. A fairly good model for the annual rate of oil consumption since 1974 is given by

$$R_1(t) = 21.3e^{.04(t-4)}, \quad t \geq 4,$$

where $t = 0$ corresponds to 1970. Determine the total amount of oil saved between 1976 and 1980 by not consuming oil at the rate predicted by the model of Example 1, Section 6.1, namely,

$$R(t) = 16.1e^{.07t}, \quad t \geq 0.$$

Solution If oil consumption had continued to grow as it did prior to 1974, then the total oil consumed between 1976 and 1980 would have been

$$\int_6^{10} R(t)\,dt. \tag{4}$$

However, taking into account the slower increase in the rate of oil consumption since 1974, we find that the total oil consumed between 1976 and 1980 was approximately

$$\int_6^{10} R_1(t)\,dt. \tag{5}$$

The integrals in (4) and (5) may be interpreted as the areas under the curves $y = R(t)$ and $y = R_1(t)$, respectively, from $t = 6$ to $t = 10$. (See Figure 8.) By superimposing the two curves we see that the area between them from $t = 6$ to $t = 10$ represents the total oil that was saved by consuming oil at the rate given by $R_1(t)$ instead of $R(t)$. (See Figure 9.) The area between the two curves equals

$$\int_6^{10} [R(t) - R_1(t)]\,dt = \int_6^{10} \left[16.1e^{.07t} - 21.3e^{.04(t-4)}\right] dt \approx 13.02.$$

Thus about 13 billion barrels of oil were saved between 1976 and 1980. ■

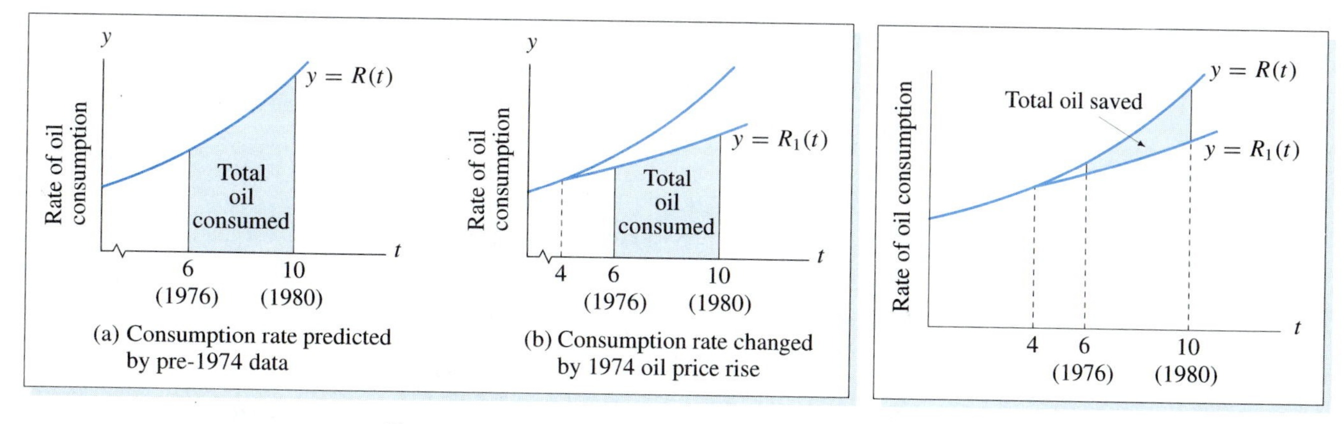

FIGURE 8

FIGURE 9

PRACTICE PROBLEMS 7.3

1. Find the area between the curves $y = x+3$ and $y = \frac{1}{2}x^2+x-7$ from $x = -2$ to $x = 1$.
2. A company plans to increase its production from 10 to 15 units per day. The present marginal cost function is $MC_1(x) = x^2 - 20x + 108$. By redesigning the production process and purchasing new equipment, the company can change the marginal cost function to $MC_2(x) = \frac{1}{2}x^2 - 12x + 75$. Determine the area between the graphs of the two marginal cost curves from $x = 10$ to $x = 15$. Interpret this area in economic terms.

EXERCISES 7.3

1. Write down a definite integral or sum of definite integrals that gives the area of the shaded portion of Figure 10.
2. Write down a definite integral or sum of definite integrals that gives the area of the shaded portion of Figure 11.

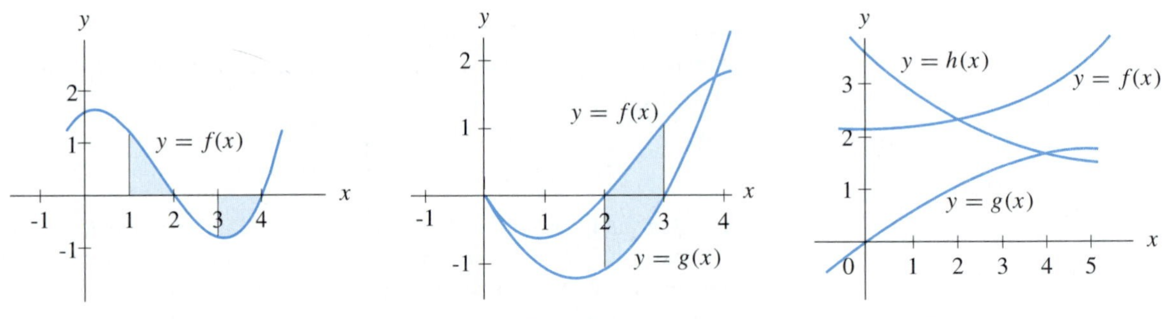

FIGURE 10 FIGURE 11 FIGURE 12

3. Shade the portion of Figure 12 whose area is given by the integral

$$\int_0^2 [f(x) - g(x)]\,dx + \int_2^4 [h(x) - g(x)]\,dx.$$

4. Shade the portion of Figure 13 whose area is given by the integral

$$\int_0^1 [f(x) - g(x)]\,dx + \int_1^2 [g(x) - f(x)]\,dx.$$

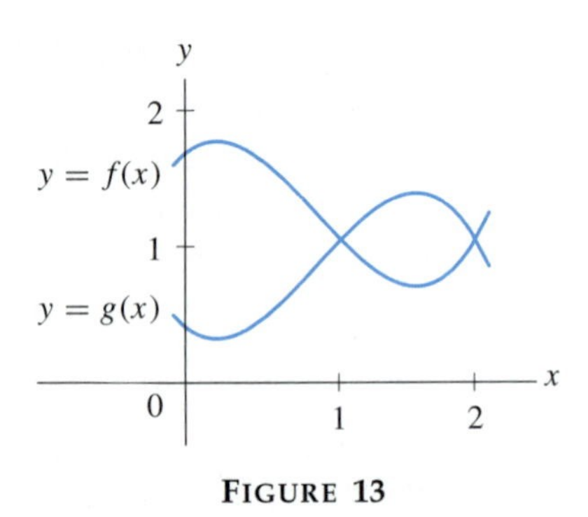

FIGURE 13

Find the area of the region between the curves.

5. $y = 2x^2$ and $y = 8$ (a horizontal line) from $x = -2$ to $x = 2$.
6. $y = 13 - 3x^2$ and $y = 1$ from $x = -2$ to $x = 2$.
7. $y = x^2 - 6x + 12$ and $y = 1$ from $x = 0$ to $x = 4$.
8. $y = x(2 - x)$ and $y = 4$ from $x = 0$ to $x = 2$.
9. $y = 3x^2$ and $y = -3x^2$ from $x = -1$ to $x = 2$.
10. $y = e^{2x}$ and $y = -e^{2x}$ from $x = -1$ to $x = 1$.

Find the area of the region bounded by the curves.

11. $y = x^2 + x$ and $y = 3 - x$.

12. $y = 3x - x^2$ and $y = 4 - 2x$.

13. $y = -x^2 + 6x - 5$ and $y = 2x - 5$.

14. $y = 2x^2 + x - 7$ and $y = x + 1$.

15. $y = x^2$ and $y = 18 - x^2$.

16. $y = 4x^2 - 24x + 20$ and $y = 2 - 2x^2$.

17. $y = x^2 - 6x - 7$ and $y = -2x^2 + 6x + 8$.

18. $y = 2x^2 - 8x + 8$ and $y = -x^2 + 7x + 8$.

19. Find the area of the region between $y = x^2 - 3x$ and the x-axis
    (a) from $x = 0$ to $x = 3$
    (b) from $x = 0$ to $x = 4$
    (c) from $x = -2$ to $x = 3$

20. Find the area of the region between $y = x^2$ and $y = 1/x$
    (a) from $x = 1$ to $x = 4$
    (b) from $x = \frac{1}{2}$ to $x = 4$

21. Find the area of the region bounded by $y = 1/x^2$, $y = x$, and $y = 8x$, for $x \geq 0$. [See Figure 14(a).]

22. Find the area of the region bounded by $y = 1/x$, $y = 4x$, and $y = x/2$, for $x \geq 0$. [The region resembles the shaded region in Figure 14(a).]

23. Find the area of the shaded region in Figure 14(b) bounded by $y = 12/x$, $y = \frac{3}{2}\sqrt{x}$, and $y = x/3$.

24. Find the area of the shaded region in Figure 14(c) bounded by $y = 12 - x^2$, $y = 4x$, and $y = x$.

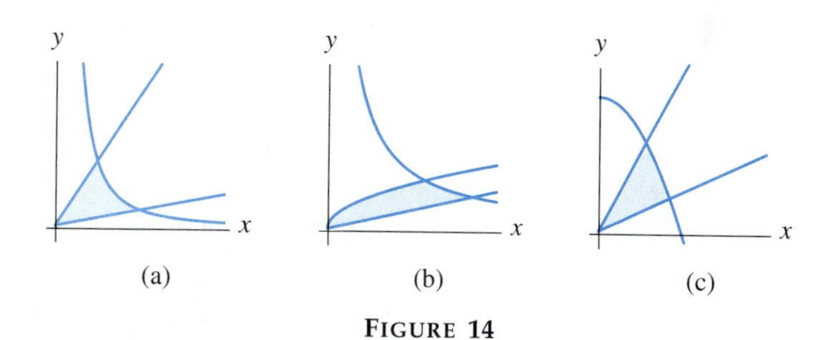

FIGURE 14

25. Refer to Exercise 49 of Section 7.2. The rate of new tree growth (in millions of cubic meters per year) in Sudan t years after 1980 is given approximately by the function $g(t) = 50 - 6.03e^{.09t}$. Set up the definite integral giving the amount of depletion of the forests due to the excess of fuel-wood consumption over new growth from 1980 to 2000.

26. Refer to the oil-consumption data in Example 1, Section 6.1. Suppose that in 1970 the growth constant for the annual rate of oil consumption had been held to .04. What effect would this action have had on oil consumption from 1970 to 1974?

27. The marginal profit for a certain company is $MP_1(x) = -x^2 + 14x - 24$. The company expects the daily production level to rise from $x = 6$ to $x = 8$ units. The management is considering a plan that would have the effect of changing the marginal profit to $MP_2(x) = -x^2 + 12x - 20$. Should the company adopt the plan? Determine the area between the graphs of the two marginal profit functions from $x = 6$ to $x = 8$. Interpret this area in economic terms.

28. Two rockets are fired simultaneously straight up into the air. Their velocities (in meters per second) are $v_1(t)$ and $v_2(t)$, respectively, and $v_1(t) \geq v_2(t)$ for $t \geq 0$. Let A denote the area of the region between the graphs of $y = v_1(t)$ and $y = v_2(t)$ for $0 \leq t \leq 10$. What physical interpretation may be given to the value of A?

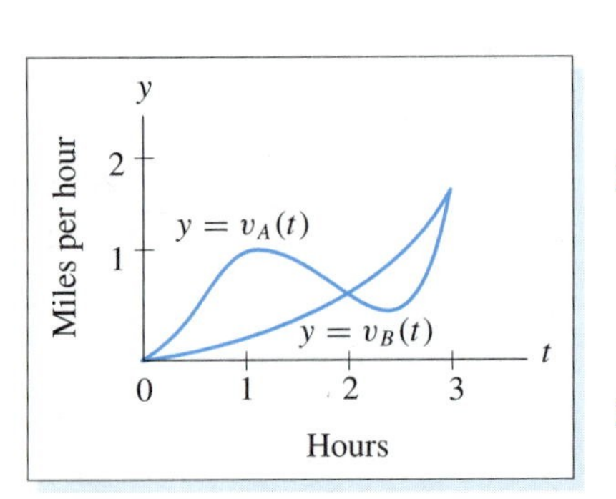

FIGURE 15 Velocity functions of two cars

29. Cars A and B start at the same place and travel in the same direction, with velocities after t hours given by the functions $v_A(t)$ and $v_B(t)$ in Figure 15.

(a) What does the area between the two curves from $t = 0$ to $t = 1$ represent?

(b) At what time will the distance between the cars be greatest?

SOLUTIONS TO PRACTICE PROBLEMS 7.3

1. First graph the two curves, as shown in Figure 16. The curve $y = x + 3$ lies on top. So the area between the curves is

$$\int_{-2}^{1} \left[(x+3) - \left(\tfrac{1}{2}x^2 + x - 7\right)\right] dx = \int_{-2}^{1} \left(-\tfrac{1}{2}x^2 + 10\right) dx = 28.5.$$

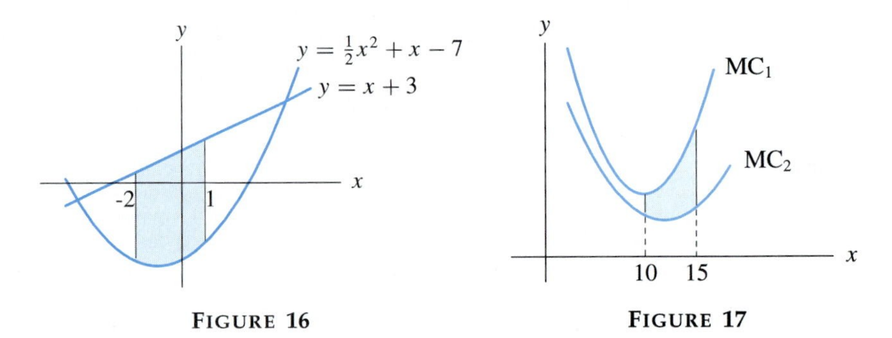

FIGURE 16 **FIGURE 17**

2. Graphing the two marginal cost functions yields the results shown in Figure 17. So the area between the curves equals

$$\int_{10}^{15} [\text{MC}_1(x) - \text{MC}_2(x)]\, dx$$

$$= \int_{10}^{15} \left[(x^2 - 20x + 108) - \left(\tfrac{1}{2}x^2 - 12x + 75\right)\right] dx$$

$$= \int_{10}^{15} \left[\tfrac{1}{2}x^2 - 8x + 33\right] dx = 60.83333.$$

This amount represents the cost savings on the increased production (from 10 to 15) provided the new production process is used.

7.4 APPLICATIONS OF THE DEFINITE INTEGRAL

The applications in this section have two features in common. First, each example contains a quantity that is computed by evaluating a definite integral. Second, the formula for the definite integral is derived by looking at Riemann sums.

In each application, we show that a certain quantity, call it Q, may be approximated by subdividing an interval into equal subintervals and forming an approximate sum. We then observe that this sum is a Riemann sum for some function $f(x)$. [Sometimes $f(x)$ is not given beforehand.] The Riemann sums approach Q as the number of subintervals becomes large. Since the Riemann sums also approach a definite integral, we conclude that the value of Q is given by the definite integral.

Once we have expressed Q as a definite integral, we may determine its value with a graphing utility. It is not necessary to calculate the value of any Riemann sum. Rather, we use the Riemann sums only as a device to express Q as a definite integral. The key step is to recognize a given sum as a Riemann sum and to determine the corresponding definite integral. Example 1 illustrates this step.

EXAMPLE 1 Suppose that the interval $1 \le x \le 2$ is divided into 50 subintervals, each of length Δx. Let $x_1, x_2, \ldots, x_{50}$ denote representative points selected from these subintervals. Find an approximate value for the sum

$$(8x_1^7 + 6x_1)\Delta x + (8x_2^7 + 6x_2)\Delta x + \cdots + (8x_{50}^7 + 6x_{50})\Delta x.$$

Solution The sum is clearly a Riemann sum for the function $f(x) = 8x^7 + 6x$ on the interval $1 \le x \le 2$. Therefore, an approximation to the sum is given by the integral

$$\int_1^2 (8x^7 + 6x)\,dx.$$

A graphing utility will calculate the value of this integral to be 264. Therefore, the sum is approximately equal to 264. ■

THE AVERAGE VALUE OF A FUNCTION Let $f(x)$ be a continuous function on the interval $a \le x \le b$. The definite integral may be used to define the *average value* of $f(x)$ on this interval. To calculate the average of a collection of numbers $y_1, y_2, \ldots, y_n$, we add the numbers and divide by n to obtain

$$\frac{y_1 + y_2 + \cdots + y_n}{n}.$$

To determine the average value of $f(x)$, we proceed similarly. Choose n values of x, say $x_1, x_2, \ldots, x_n$, and calculate the corresponding function values $f(x_1)$, $f(x_2), \ldots, f(x_n)$. The average of these values is

$$\frac{f(x_1) + f(x_2) + \cdots + f(x_n)}{n}. \tag{1}$$

Our goal now is to obtain a reasonable definition of the average of all the values of $f(x)$ on the interval $a \leq x \leq b$. If the points $x_1, x_2, \ldots, x_n$ are spread "evenly" throughout the interval, then the average (1) should be a good approximation to our intuitive concept of the average value of $f(x)$. In fact, as n becomes large, the average (1) should approximate the average value of $f(x)$ to any arbitrary degree of accuracy. To guarantee that the points $x_1, x_2, \ldots, x_n$ are "evenly" spread out from a to b, let us divide the interval from $x = a$ to $x = b$ into n subintervals of equal length $\Delta x = (b - a)/n$. Then choose x_1 from the first subinterval, x_2 from the second, and so forth. The average (1) that corresponds to these points may be arranged in the form of a Riemann sum as follows:

$$\begin{aligned}\frac{f(x_1) + f(x_2) + \cdots + f(x_n)}{n} &\\ &= f(x_1) \cdot \frac{1}{n} + f(x_2) \cdot \frac{1}{n} + \cdots + f(x_n) \cdot \frac{1}{n}\\ &= \frac{1}{b-a}\left[f(x_1) \cdot \frac{b-a}{n} + f(x_2) \cdot \frac{b-a}{n} + \cdots + f(x_n) \cdot \frac{b-a}{n}\right]\\ &= \frac{1}{b-a}[f(x_1)\Delta x + f(x_2)\Delta x + \cdots + f(x_n)\Delta x].\end{aligned}$$

The sum inside the brackets is a Riemann sum for the integral of $f(x)$. Thus we see that for a large number of points x_i, the average in (1) approaches the quantity

$$\frac{1}{b-a}\int_a^b f(x)\,dx.$$

This argument motivates the following definition.

> The *average value* of a continuous function $f(x)$ over the interval $a \leq x \leq b$ is defined as the quantity
>
> $$\frac{1}{b-a}\int_a^b f(x)\,dx. \qquad (2)$$

EXAMPLE 2 Compute the average value of $f(x) = \sqrt{x}$ over the interval $0 \leq x \leq 9$.

Solution Using (2) with $a = 0$ and $b = 9$, the average value of $f(x) = \sqrt{x}$ over the interval $0 \leq x \leq 9$ is equal to

$$\frac{1}{9-0}\int_0^9 \sqrt{x}\,dx = 2,$$

so the average value of $\sqrt{x}$ over the interval $0 \leq x \leq 9$ is 2. The area of the shaded region is the same as the area of the rectangle of height 2 pictured in Figure 1.

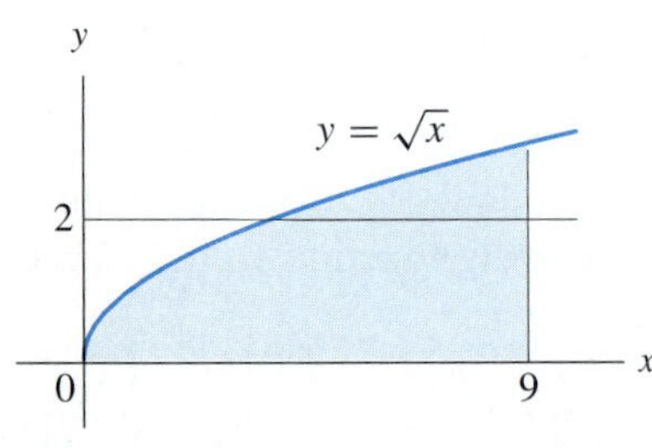

FIGURE 1 Average value of a function

EXAMPLE 3 The current world population is about 5.8 billion, and the population t years from now is projected to be given by the exponential growth law

$$P(t) = 5.8e^{.014t}.$$

Determine the average population of the earth during the next 30 years. (This average is important in long-range planning for agricultural production and the allocation of goods and services.)

Solution The average value of the population $P(t)$ from $t = 0$ to $t = 30$ is

$$\frac{1}{30-0}\int_0^{30} P(t)\,dt = \frac{1}{30}\int_0^{30} 5.8e^{.014t}\,dt \approx 7.208 \text{ billion.}$$

CONSUMERS' SURPLUS Using a demand curve from economics, we can derive a formula showing the amount that consumers benefit from an open system that has no price discrimination. Figure 2(a) is a *demand curve* for a commodity. It is determined by complex economic factors and gives a relationship between the quantity sold and the unit price of a commodity. Specifically, it says that in order to sell x units, the price must be set at $f(x)$ dollars per unit. Since, for most commodities, selling larger quantities requires a lowering of the price, demand functions are usually decreasing. Interactions between supply and demand determine the amount of a quantity available. Let A designate the amount of the commodity currently available and $B = f(A)$ the current selling price.

Divide the interval from 0 to A into n subintervals, each of length $\Delta x = (A - 0)/n$, and take x_i to be the right-hand endpoint of the ith subinterval. Consider the first subinterval, from 0 to x_1. [See Figure 2(b).] Suppose that only x_1 units had been available. Then the price per unit could have been set at $f(x_1)$ dollars and these x_1 units sold. Of course, at this price we could not have sold any more units. However, those people who paid $f(x_1)$ dollars had a great demand for the commodity. It was extremely valuable to them, and there was no advantage in substituting another commodity at that price. They were actually paying what the commodity was worth to them. In theory, then, the first x_1 units of the commodity could be sold to these people at $f(x_1)$ dollars per unit, yielding (price per unit) $\cdot$ (number of units) $= f(x_1)\cdot(x_1) = f(x_1)\cdot\Delta x$ dollars.

After selling the first x_1 units, suppose that more units become available, so that now a total of x_2 units has been produced. Setting the price at $f(x_2)$, the remaining $x_2 - x_1 = \Delta x$ units can be sold, yielding $f(x_2)\cdot\Delta x$ dollars. Here, again, the second group of buyers would have paid as much for the commodity as it was worth to them. Continuing this process of price discrimination, the amount of money paid by the consumers would be

$$f(x_1)\Delta x + f(x_2)\Delta x + \cdots + f(x_n)\Delta x.$$

Taking n large, we see that this Riemann sum approaches $\int_0^A f(x)\,dx$. Since $f(x)$ is positive, this integral equals the area under the graph of $f(x)$ from $x = 0$ to $x = A$.

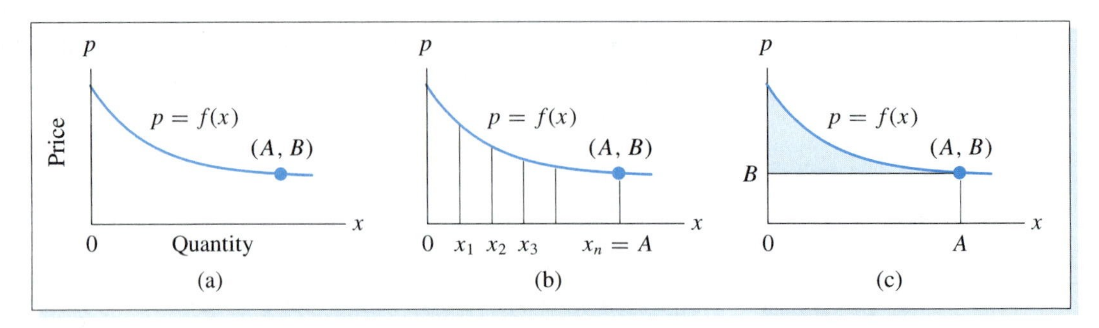

FIGURE 2 Consumers' surplus

Of course, in our open system, everyone pays the same price, B, so the total amount paid is [price per unit] $\cdot$ [number of units] $= BA$. Since BA is the area of the rectangle under the graph of the line $p = B$ from $x = 0$ to $x = A$, the amount of money saved by consumers is the area of the shaded region in Figure 2(c). That is, the area between the curves $p = f(x)$ and $p = B$ gives a numerical value to one benefit of a modern efficient economy.

> The *consumers' surplus* for a commodity having demand curve $p = f(x)$ is
>
> $$\int_0^A [f(x) - B]\,dx,$$
>
> where the quantity demanded is A and the price is $B = f(A)$.

EXAMPLE 4 Find the consumers' surplus for the demand curve $p = 50 - .06x^2$ at the sales level 20.

Solution Since 20 units are sold, the price must be

$$B = 50 - .06(20)^2 = 50 - 24 = 26.$$

Therefore, the consumers' surplus is

$$\int_0^{20} [(50 - .06x^2) - 26]\,dx = \int_0^{20} (24 - .06x^2)\,dx = 320.$$

That is, the consumers' surplus is \$320. ■

FUTURE VALUE OF AN INCOME STREAM The next example shows how the definite integral can be used to approximate the sum of a large number of terms.

EXAMPLE 5 Suppose that money is deposited daily into a savings account at an annual rate of \$1000. The account pays 6% interest compounded continuously. Approximate the amount of money in the account at the end of 5 years.

Solution Divide the time interval from 0 to 5 years into daily subintervals. Each subinterval is then of duration $\Delta t = \frac{1}{365}$ years. Let $t_1, t_2, \ldots, t_n$ be points chosen from these subintervals. Since we deposit money at an annual rate of $1000, the amount deposited during one of the subintervals is $1000\Delta t$ dollars. If this amount is deposited at time t_i, the $1000\Delta t$ dollars will earn interest for the remaining $5-t_i$ years. The total amount resulting from this one deposit at time t_i is then

$$1000\Delta t e^{.06(5-t_i)}.$$

Add the effects of the deposits at times $t_1, t_2, \ldots, t_n$ to arrive at the total balance in the account:

$$A = 1000e^{.06(5-t_1)}\Delta t + 1000e^{.06(5-t_2)}\Delta t + \cdots + 1000e^{.06(5-t_n)}\Delta t.$$

This is a Riemann sum for the function $f(t) = 1000e^{.06(5-t)}$ on the interval $0 \leq t \leq 5$. Since Δt is very small when compared with the interval, the total amount in the account, A, is approximately

$$A = \int_0^5 1000e^{.06(5-t)}\, dt \approx 5831.$$

That is, the approximate balance in the account at the end of 5 years is $5831. ■

In Example 5, money was deposited into the account. If the money had been deposited several times a day, the definite integral would have given an even better approximation to the balance. Actually, the more frequently the money is deposited, the better the approximation. Economists consider a hypothetical situation where money is deposited steadily throughout the year. This flow of money is called a *continuous income stream*, and the balance in the account is given exactly by the definite integral.

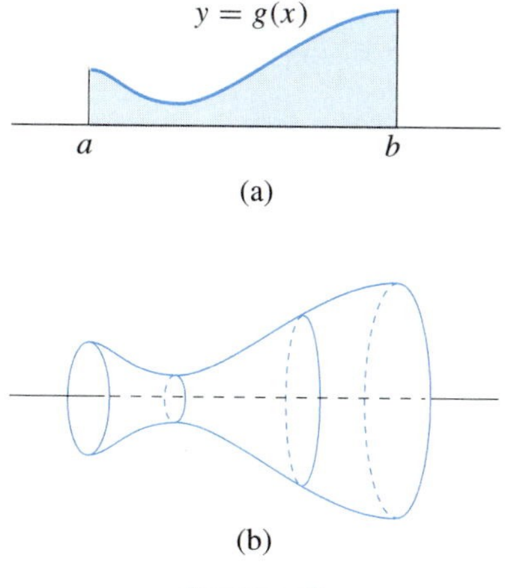

FIGURE 3

> The *future value of a continuous stream* of P dollars per year for N years at interest rate r compounded continuously is
>
> $$\int_0^N Pe^{r(N-t)}\, dt.$$

SOLIDS OF REVOLUTION When the region of Figure 3(a) is revolved about the x-axis, it sweeps out a solid [Figure 3(b)]. Riemann sums can be used to derive a formula for the volume of this *solid of revolution*. Let us break the x-axis between a and b into a large number n of equal subintervals, each of length $\Delta x = (b - a)/n$. Using each subinterval as a base, we can divide the region into strips (see Figure 4).

Let x_i be a point in the ith subinterval. Then the volume swept out by revolving the ith strip is approximately the same as the volume of the cylinder swept out

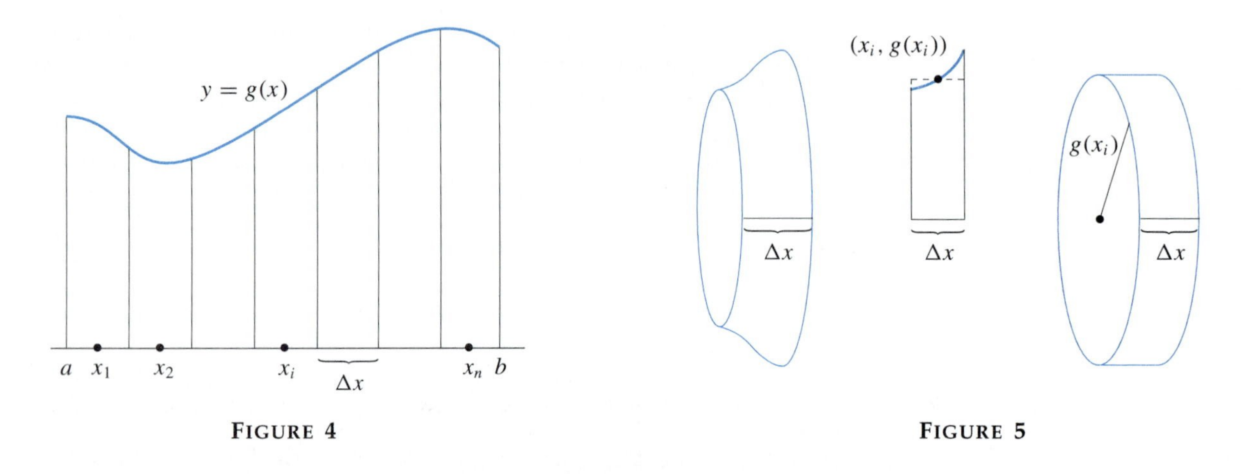

FIGURE 4

FIGURE 5

by revolving the rectangle of height $g(x_i)$ and base Δx around the x-axis (Figure 5). The volume of the cylinder is

$$[\text{area of circular side}] \cdot [\text{width}] = \pi[g(x_i)]^2 \cdot \Delta x.$$

The total volume swept out by all the strips is approximated by the total volume swept out by the rectangles, which is

$$[\text{volume}] \approx \pi[g(x_1)]^2\Delta x + \pi[g(x_2)]^2\Delta x + \cdots + \pi[g(x_n)]^2\Delta x.$$

As n gets larger and larger, this approximation becomes arbitrarily close to the true volume. The expression on the right is a Riemann sum for the definite integral of $f(x) = \pi[g(x)]^2$. Therefore, the volume of the solid equals the value of the definite integral.

> The volume of the *solid of revolution* obtained from revolving the region below the graph of $f = g(x)$ from $x = a$ to $x = b$ about the x-axis is
>
> 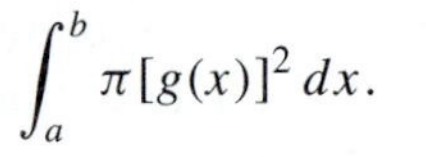
>
> $$\int_a^b \pi[g(x)]^2\,dx.$$

EXAMPLE 6 Find the volume of the solid of revolution obtained by revolving the region of Figure 6 about the x-axis.

Solution Here $g(x) = e^{.5x}$, and

$$[\text{volume}] = \int_0^1 \pi\left(e^{.5x}\right)^2 dx = \int_0^1 \pi e^x\,dx \approx 5.3981.$$ ■

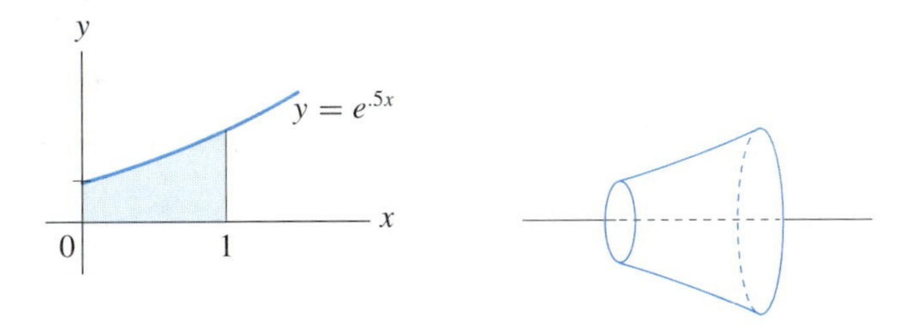

FIGURE 6

PRACTICE PROBLEMS 7.4

1. A rock dropped from a bridge has a velocity of $-32t$ feet per second after t seconds. Find the average velocity of the rock during the first three seconds.
2. An investment steadily yields income at an annual rate of \$300. (Assume an interest rate of 10% compounded continuously.) What is the (future) value of this income stream at the end of 10 years?

EXERCISES 7.4

Determine the average value of $f(x)$ over the interval from $x = a$ to $x = b$, where:

1. $f(x) = x^2$; $a = 0, b = 3$
2. $f(x) = e^{x/3}$; $a = 0, b = 3$
3. $f(x) = x^3$; $a = -1, b = 1$
4. $f(x) = 5$; $a = 1, b = 10$
5. $f(x) = 1/x^2$; $a = \frac{1}{4}, b = \frac{1}{2}$
6. $f(x) = 2x - 6$; $a = 2, b = 4$
7. During a certain 12-hour period the temperature at time t (measured in hours from the start of the period) was $47 + 4t - \frac{1}{3}t^2$ degrees. What was the average temperature during that period?
8. Assuming that a country's population is now 3 million and growing exponentially with growth constant .02, what will be the average population during the next 50 years?
9. One hundred grams of radioactive radium having a half-life of 1690 years are placed in a concrete vault. What will be the average amount of radium in the vault during the next 1000 years?
10. One hundred dollars are deposited in the bank at 5% interest compounded continuously. What will be the average value of the money in the account during the next 20 years?

Find the consumers' surplus for each of the following demand curves at the given sales level x.

11. $p = (10 - .02x)^2$; $x = 100$
12. $p = 20\sqrt[3]{\dfrac{1}{x+1}}$; $x = 63$
13. $p = \sqrt{500 - 2x}$; $x = 178$
14. $p = .2\ln\left(\dfrac{300}{x+1}\right)$; $x = 66$

Figure 7 shows a supply curve for a commodity. It gives the relationship between the selling price of the commodity and the quantity that producers will manufacture. At a higher selling price, a greater quantity will be produced. Therefore, the

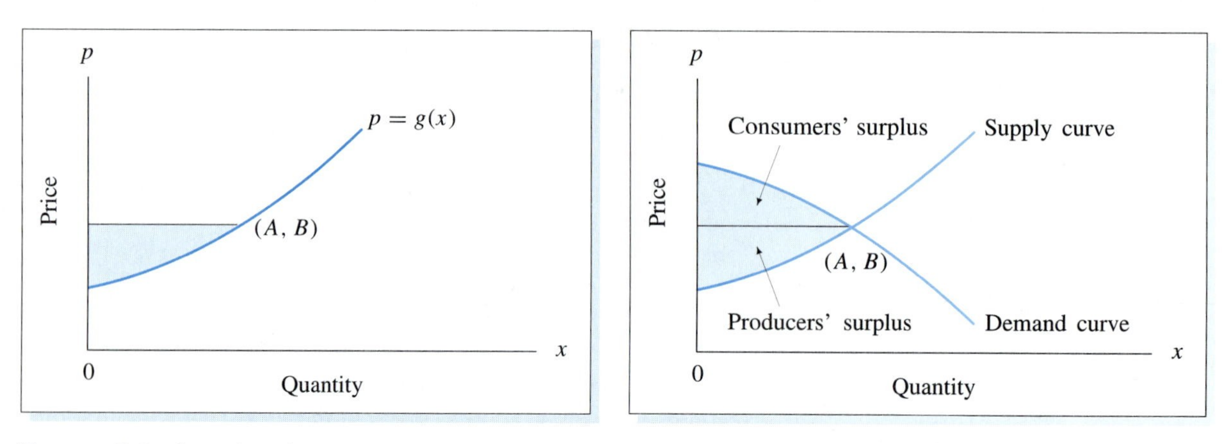

FIGURE 7 Producers' surplus

FIGURE 8

curve is increasing. If (A, B) is a point on the curve, then, in order to stimulate the production of A units of the commodity, the price per unit must be B dollars. Of course, some producers will be willing to produce the commodity even with a lower selling price. Since everyone receives the same price in an open, efficient economy, most producers are receiving more than their minimal required price. The excess is called the producers' surplus. *Using an argument analogous to that of the* consumers' surplus, *one can show that the total producers' surplus when the price is B is the area of the shaded region in Figure 8. Find the producers' surplus for each of the following supply curves at the given sales level x.*

15. $p = .01x + 3;\ x = 200$

16. $p = \dfrac{x^2}{9} + 1;\ x = 3$

17. $p = \dfrac{x}{2} + 7;\ x = 10$

18. $p = 1 + \frac{1}{2}\sqrt{x};\ x = 36$

For a particular commodity, the quantity produced and the unit price are given by the coordinates of the point where the supply and demand curves intersect. For each pair of supply and demand curves, determine the point of intersection (A, B) and the consumers' and producers' surplus. (See Figure 8.)

19. Demand curve: $p = 12 - (x/50)$; supply curve: $p = (x/20) + 5$.

20. Demand curve: $p = \sqrt{25 - .1x}$; supply curve: $p = \sqrt{.1x + 9} - 2$.

21. Suppose that money is deposited daily into a savings account at an annual rate of \$1000. If the account pays 5% interest compounded continuously, estimate the balance in the account at the end of 3 years.

22. Suppose that money is deposited daily into a savings account at an annual rate of \$2000. If the account pays 6% interest compounded continuously, approximately how much will be in the account at the end of 2 years?

23. Suppose that money is deposited steadily into a savings account at the rate of \$16,000 per year. Determine the balance at the end of 4 years if the account pays 8% interest compounded continuously.

24. Suppose that money is deposited steadily into a savings account at the rate of \$14,000 per year. Determine the balance at the end of 6 years if the account pays 7% interest compounded continuously.

25. Mary deposits money steadily into a savings account paying 4% interest compounded continuously at the rate of \$5000 per year. Beth makes a single deposit of \$35,000 into a savings account. What rate of interest must Beth earn in order to have the same amount of money as Mary after 10 years?

26. Suppose money can be invested at 5% interest compounded annually. Bob deposits money steadily into a savings account at the rate of \$2000 per year. Tom makes a single deposit. How much must Tom deposit in order to have the same amount as Bob after three years?

27. Suppose money is to be deposited daily for 5 years into a savings account at an annual rate of \$1000 and the account pays 6% interest compounded continuously. Let the interval from 0 to 5 be divided into daily subintervals, with each subinterval of duration $\Delta t = \frac{1}{365}$ years. Let $t_1, \ldots, t_n$ be points chosen from the subintervals.
   (a) Show that the present value of a daily deposit at time t_i is $1000\Delta t e^{-.06t_i}$.
   (b) Find the Riemann sum corresponding to the sum of the present values of all the deposits.
   (c) What is the function and interval corresponding to the Riemann sum in (b)?
   (d) Give the definite integral that approximates the Riemann sum in (b).
   (e) Evaluate the definite integral in (d). This number is the *present value of a continuous income stream.*

28. Use the result of Exercise 27 to calculate the present value of a continuous income stream of \$5000 per year for 10 years at an interest rate of 5% compounded continuously.

29. Suppose money can be invested at 5% interest compounded continuously. Would you rather receive \$100,000 now or receive funds steadily at the rate of \$20,000 a year for six years? (See Exercise 27.)

30. Suppose money can be invested at 4% interest compounded continuously. Would you rather receive \$100,000 now or receive funds steadily at the rate of \$20,000 a year for six years? (See Exercise 27.)

In Exercises 31–38, find the volume of the solid of revolution generated by revolving about the x-axis the region under each of the following curves.

31. $y = \sqrt{100 - x^2}$ from $x = -10$ to $x = 10$ (generates a sphere of radius 10)
32. $y = 2x$ from $x = 0$ to $x = 5$ (generates a cone)
33. $y = x^2$ from $x = 1$ to $x = 2$
34. $y = 1/x$ from $x = 1$ to $x = 100$
35. $y = \sqrt{x}$ from $x = 0$ to $x = 4$ (The solid generated is called a *paraboloid*.)
36. $y = 2x - x^2$ from $x = 0$ to $x = 2$
37. $y = e^{-x}$ from $x = 0$ to $x = 1$

38. $y = 2x + 1$ from $x = 0$ to $x = 1$ (The solid generated is called a *truncated cone*.)

SOLUTIONS TO PRACTICE PROBLEMS 7.4

1. By definition, the average value of the function $v(t) = -32t$ for $t = 0$ to $t = 3$ is

$$\frac{1}{3-0}\int_0^3 -32t\,dt = -48 \text{ feet per second.}$$

Note: There is another way to approach this problem:

$$[\text{average velocity}] = \frac{[\text{distance traveled}]}{[\text{time elaspsed}]}.$$

As we discussed in Section 6.2, distance traveled equals the area under the velocity curve. Therefore,

$$[\text{average velocity}] = \frac{\int_0^3 -32t\,dt}{3}.$$

2. According to the formula developed in the text, the future value of the income stream after 10 years is equal to

$$\int_0^{10} 300e^{.1(10-t)}\,dt \approx \$5154.85.$$

7.5 THE FUNDAMENTAL THEOREM OF CALCULUS

The fundamental theorem of calculus provides an algebraic method of evaluating a definite integral.

> *Fundamental Theorem of Calculus* Suppose that $f(x)$ is continuous on the interval $a \le x \le b$, and let $F(x)$ be an antiderivative of $f(x)$. Then
>
> $$\int_a^b f(x)\,dx = F(b) - F(a). \tag{1}$$

This theorem connects the two key concepts of calculus—the integral and the derivative. An explanation of why the theorem is true is given later. First we show how to use the theorem to evaluate definite integrals.

EXAMPLE 1 Use the fundamental theorem of calculus to evaluate the following definite integrals.

(a) $\displaystyle\int_1^4 \left(\frac{1}{3}x + \frac{2}{3}\right) dx$ (b) $\displaystyle\int_0^5 (2x - 4)\,dx$

Solution (a) An antiderivative of $\frac{1}{3}x + \frac{2}{3}$ is $F(x) = \frac{1}{6}x^2 + \frac{2}{3}x$. Therefore, by the fundamental theorem,

$$\begin{aligned}\int_1^4 \left(\frac{1}{3}x + \frac{2}{3}\right) dx &= F(4) - F(1) \\ &= \left[\frac{1}{6}(4)^2 + \frac{2}{3}(4)\right] - \left[\frac{1}{6}(1)^2 + \frac{2}{3}(1)\right] \\ &= \left[\frac{16}{6} + \frac{8}{3}\right] - \left[\frac{1}{6} + \frac{2}{3}\right] = 4\frac{1}{2}.\end{aligned}$$

This result is the same as in Example 1 of Section 7.2

(b) An antiderivative of $2x - 4$ is $F(x) = x^2 - 4x$. Therefore,

$$\int_0^5 (2x - 4)\, dx = F(5) - F(0) = \left[5^2 - 4(5)\right] - \left[0^2 - 4(0)\right] = 5.$$

This result is the same as in Example 2 of Section 7.2. ■

The next example illustrates the fact that when computing the definite integral of a function, we may use *any* antiderivative of the function.

EXAMPLE 2 Evaluate $\displaystyle\int_2^5 3x^2\, dx$.

Solution An antiderivative of $f(x) = 3x^2$ is $F(x) = x^3 + C$, where C is any constant. Then

$$\begin{aligned}\int_2^5 3x^2\, dx &= F(5) - F(2) = \left[5^3 + C\right] - \left[2^3 + C\right] \\ &= 5^3 + C - 2^3 - C = 117.\end{aligned}$$

Notice how the C in $F(2)$ is subtracted from the C in $F(5)$. Thus the value of the definite integral does not depend on the choice of the constant C. For convenience, we may take $C = 0$ when evaluating a definite integral. ■

The quantity $F(b) - F(a)$ is called the *net change of* $F(x)$ *from* $x = a$ *to* $x = b$. It is abbreviated by the symbol $F(x)\big|_a^b$. For instance, the net change of $F(x) = \frac{1}{3}e^{3x}$ from $x = 0$ to $x = 2$ is written as $\frac{1}{3}e^{3x}\big|_0^2$ and is evaluated as $F(2) - F(0)$.

EXAMPLE 3 Evaluate $\displaystyle\int_0^2 e^{3x}\, dx$.

Solution An antiderivative of e^{3x} is $\frac{1}{3}e^{3x}$. Therefore,

$$\int_0^2 e^{3x}\, dx = \frac{1}{3}e^{3x}\bigg|_0^2 = \frac{1}{3}e^{3(2)} - \frac{1}{3}e^{3(0)} = \frac{1}{3}e^6 - \frac{1}{3}.$$ ■

EXAMPLE 4 Compute the area under the curve $y = x^2 - 4x + 5$ from $x = -1$ to $x = 3$.

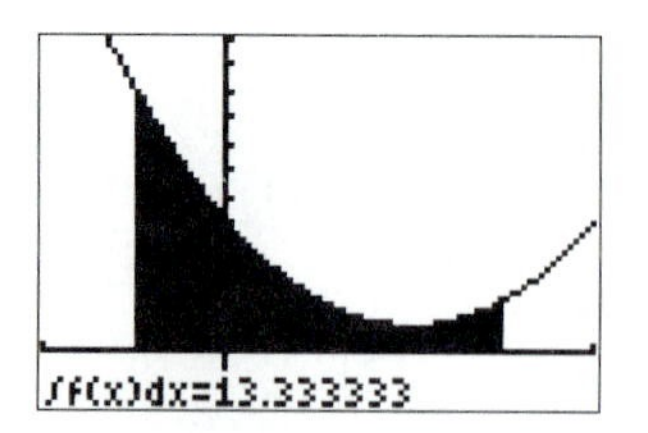

FIGURE 1

Solution The graph of $f(x) = x^2 - 4x + 5$ is shown in Figure 1. Since $f(x)$ is nonnegative for $-1 \le x \le 3$, the area under the curve is given by the definite integral

$$\int_{-1}^{3} (x^2 - 4x + 5)\, dx = \left(\frac{x^3}{3} - 2x^2 + 5x\right)\Bigg|_{-1}^{3}$$

$$= \left[\frac{(3)^3}{3} - 2(3)^2 + 5(3)\right] - \left[\frac{(-1)^3}{3} - 2(-1)^2 + 5(-1)\right]$$

$$= [9 - 18 + 15] - \left[-\frac{1}{3} - 2 - 5\right]$$

$$= [6] - \left[-\frac{22}{3}\right] = \frac{18}{3} + \frac{22}{3} = \frac{40}{3}.$$

Study this solution carefully. The calculations show how to use parentheses to avoid errors in arithmetic. It is particularly important to include the outside brackets around the value of the antiderivative at -1.

EXAMPLE 5 Use the fundamental theorem of calculus to evaluate

$$\int_{2}^{4} 16.1e^{.07t}\, dt.$$

This is the integral considered earlier that gives the amount of oil consumed from 1972 to 1974.

Solution The antidifferentiation requires the fact that $\frac{1}{k}e^{kt}$ is an antiderivative of e^{kt}.

$$\int_{2}^{4} 16.1e^{.07t}\, dt = 16.1 \cdot \frac{1}{.07}e^{.07t}\Bigg|_{2}^{4}$$

$$= 230e^{.07t}\Big|_{2}^{4} \qquad \left(\text{since } 16.1 \cdot \tfrac{1}{.07} = 230\right)$$

$$= 230e^{.07(4)} - 230e^{.07(2)}$$

$$= 230\left(e^{.28} - e^{.14}\right) \approx 39.76.$$

The definite integrals in the previous examples could have easily been evaluated with a graphing utility. However, in the next example the definite integral involves an unknown value N and therefore cannot be easily evaluated with a graphing utility.

EXAMPLE 6 A savings account pays 5% interest compounded continuously. If money is deposited steadily at the rate of \$4000 per year, how much time is required until the balance reaches \$40,000?

Solution Let N be the number of years required. The formula for the future value of a continuous income stream gives

$$\int_0^N 4000e^{.05(N-t)}\,dt = 40{,}000.$$

Notice that an antiderivative of $e^{.05(N-t)}$ is $\frac{-1}{.05}e^{.05(N-t)}$. (Check by differentiating the second function.) Therefore,

$$\begin{aligned}
4000 \cdot \frac{-1}{.05}e^{.05(N-t)}\Big|_0^N &= 40{,}000\\
-80{,}000e^{.05(N-t)}\Big|_0^N &= 40{,}000\\
-80{,}000e^{.05(0)} - \left(-80{,}000e^{.05N}\right) &= 40{,}000\\
-80{,}000 + 80{,}000e^{.05N} &= 40{,}000\\
80{,}000e^{.05N} &= 120{,}000\\
e^{.05N} &= 1.5\\
\ln e^{.05N} &= \ln 1.5\\
.05N &= \ln 1.5\\
N &= \frac{\ln 1.5}{.05} \approx 8.1 \text{ years.}
\end{aligned}$$

VERIFICATION OF THE FUNDAMENTAL THEOREM OF CALCULUS The following explanation of the fundamental theorem of calculus is based on the Riemann sum definition of the definite integral. The fundamental theorem may be written in the form

$$\int_a^b F'(x)\,dx = F(b) - F(a), \tag{2}$$

where $F(x)$ is any function with a continuous derivative on $a \le x \le b$. There are three key ideas in an explanation of why (2) is true.

I. If the interval $a \le x \le b$ is partitioned into n subintervals of width $\Delta x = (b-a)/n$, then the net change in $F(x)$ over $a \le x \le b$ is the sum of the net change in $F(x)$ over each subinterval.

For example, partition $a \le x \le b$ into three subintervals and denote the left endpoints by x_1, x_2, x_3, as shown below.

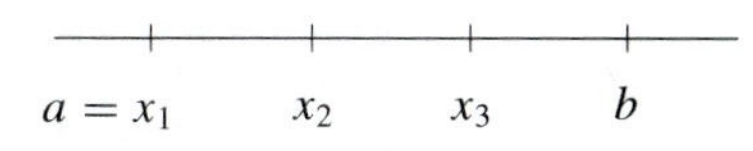

Then

$$[\text{change in } F(x) \text{ over 1st interval}] = F(x_2) - F(a),$$
$$[\text{change in } F(x) \text{ over 2nd interval}] = F(x_3) - F(x_2),$$
$$[\text{change in } F(x) \text{ over 3rd interval}] = F(b) - F(x_3).$$

When these changes are summed, the intermediate terms cancel:

$$F(b) - F(x_3) + F(x_3) - F(x_2) + F(x_2) - F(a) = F(b) - F(a).$$

II. If Δx is small, then the change in $F(x)$ over the ith subinterval is approximately $F'(x_i)\Delta x$.

This is the approximation of the change in a function discussed in Section 2.2. Here x_i is the left endpoint of the ith subinterval, as shown below.

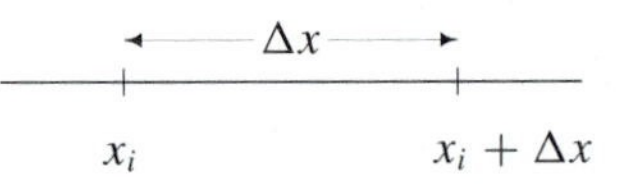

III. The sum of the approximations in (II) is a Riemann sum for the definite integral $\int_a^b F'(x)\,dx$.

That is,

$$\begin{aligned} F(b) - F(a) & \\ &= \begin{bmatrix} \text{change over} \\ \text{1st subinterval} \end{bmatrix} + \begin{bmatrix} \text{change over} \\ \text{2nd subinterval} \end{bmatrix} + \cdots + \begin{bmatrix} \text{change over} \\ n\text{th subinterval} \end{bmatrix} \\ &\approx F'(x_1)\Delta x + F'(x_2)\Delta x + \cdots + F'(x_n)\Delta x. \end{aligned}$$

As $\Delta x \to 0$, these approximations improve. Thus, in the limit

$$F(b) - F(a) = \int_a^b F'(x)\,dx.$$

To summarize, the derivative of $F(x)$ determines the approximate change of $F(x)$ on small subintervals. The definite integral sums these approximate changes and in the limit gives the exact change of $F(x)$ over the entire interval $a \le x \le b$.

If you look back at the solution of Example 4 of Section 7.1, you will see essentially the same argument. The distance a rocket travels is the sum of the distances it travels over small intervals of time. On each subinterval, this distance is approximately the velocity (the derivative) times Δt.

VERIFICATION OF BASIC PROPERTIES OF DEFINITE INTEGRALS Three properties of definite integrals were stated at the beginning of Section 7.3. These properties involved the definite integral of the sum, difference, and scalar multiples of functions. The first property was

$$\int_a^b f(x)\,dx + \int_a^b g(x)\,dx = \int_a^b [f(x) + g(x)]\,dx.$$

To verify this property, let $F(x)$ and $G(x)$ be antiderivatives of $f(x)$ and $g(x)$, respectively. Then $F(x) + G(x)$ is an antiderivative of $f(x) + g(x)$. By the fundamental theorem of calculus,

$$\begin{aligned}\int_a^b [f(x) + g(x)]\,dx &= [F(x) + G(x)]\Big|_a^b \\ &= [F(b) + G(b)] - [F(a) + G(a)] \\ &= [F(b) - F(a)] + [G(b) - G(a)] \\ &= \int_a^b f(x)\,dx + \int_a^b g(x)\,dx.\end{aligned}$$

The verification of the other two properties are similar and use the facts that $F(x) - G(x)$ is an antiderivative of $f(x) - g(x)$ and $kF(x)$ is an antiderivative of $kf(x)$.

PRACTICE PROBLEMS 7.5

1. Evaluate $\int_0^6 1000e^{k(6-t)}\,dx$.

2. Money is deposited steadily at the rate of \$1000 per year into a savings account. After 6 years the balance is \$6781.20. What interest rate, with interest compounded continuously, did the money earn?

EXERCISES 7.5

In Exercises 1–8, set up the definite integral that gives the area of the shaded region and use the fundamental theorem of calculus to evaluate the integral(s).

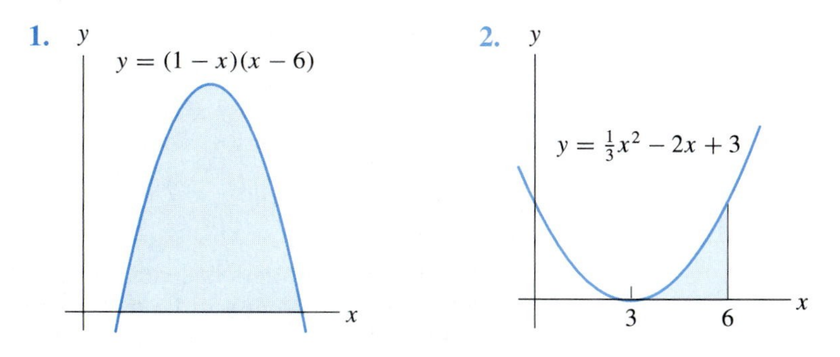

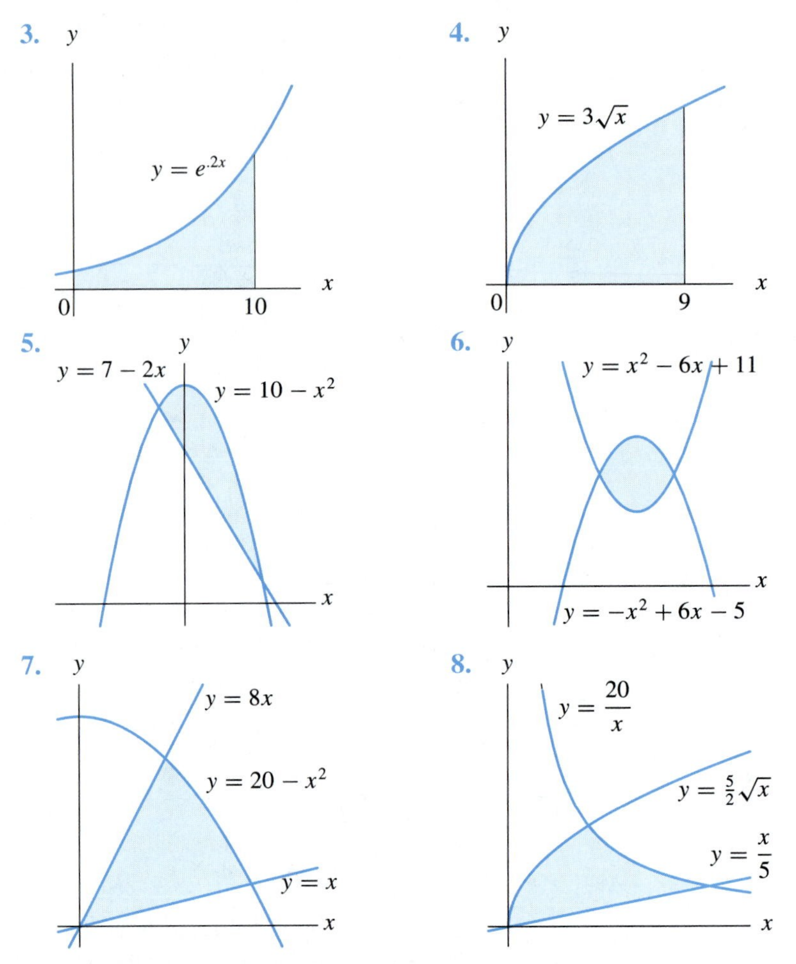

In Exercises 9–16, use the fundamental theorem of calculus to evaluate the definite integrals.

9. $\int_1^2 x^3\,dx$
10. $\int_0^1 e^{x/3}\,dx$
11. $\int_0^1 4e^{-3x}\,dx$
12. $\int_3^6 x^{-1}\,dx$
13. $\int_1^3 (1-t)^3\,dt$
14. $\int_2^3 (5-t)^4\,dt$
15. $\int_0^2 5\,dx$
16. $\int_0^4 x\,dx$

In Exercises 17–20, use the fundamental theorem of calculus to find the area of the region bounded by the curves.

17. $y = x^2 - 3x + 3$ and $y = 2x - 1$
18. $y = 8x - x^2$ and $y = 2x$
19. $y = x^2 - 4x + 8$ and $y = 2 + 4x - x^2$
20. $y = 8 - x^2$ and $y = x^2$
21. One thousand dollars is deposited into the bank at 4% interest compounded continuously. What will be the average value of the money in the account during the next five years?

22. The temperature at an airport t hours after midnight was $.5t^2 - 6t + 4$ degrees, for $0 \leq t \leq 14$. What was the average temperature from midnight to noon?

23. Refer to Example 3 (world population) of Section 7.4. When will the average world population from now until then be 7 billion people?

24. Five grams of a radioactive element with decay constant .02 (time measured in years) are placed in a steel container. In how many years will the average amount of the element present from now until then be 3 grams?

In Exercises 25–28, find the consumers' surplus for the demand curves at the given sales level x.

25. $p = 3 - \frac{x}{10}$; $x = 20$

26. $p = \frac{x^2}{200} - x + 50$; $x = 20$

27. $p = \frac{500}{x + 10}$; $x = 40$

28. $p = \sqrt{16 - .02x}$; $x = 350$

29. One thousand dollars is to be deposited into a savings account and the interest allowed to accumulate for 3 years. For what interest rate (compounded continuously) will the average amount of money in the account be \$1070.60?

30. One thousand dollars is to be deposited into a savings account and the interest allowed to accumulate at the rate of 5% compounded continuously for several years. Over the course of what number of years will the average amount of money in the account be \$1166.20?

31. A savings account pays 4% interest compounded continuously. At what rate per year must money be deposited steadily into the account in order to accumulate a balance of \$50,000 after 10 years?

32. Money is deposited steadily at the rate of \$1000 per year into a savings account. After 5 years the balance is \$5607.20. What interest rate, with interest compounded continuously, did the money earn?

33. Money is deposited steadily at the rate of \$3000 per year into a savings account. After 10 years the balance is \$36,887. What interest rate, with interest compounded continuously, did the money earn?

34. A savings account pays 3.5% interest compounded continuously. If money is deposited steadily at the rate of \$2000 per year, how much time is required until the balance reaches \$10,000?

35. The right circular cone in Figure 2(a) is the solid of revolution swept out by the region under the graph in Figure 2(b). Find the volume of the cone in Figure 2(a).

36. A sphere of radius r is the solid of revolution swept out by the semicircular region under the graph of $y = \sqrt{r^2 - x^2}$ from $x = -r$ to $x = r$. Find the volume of a sphere of radius r.

37. Use the fundamental theorem of calculus to show that

$$\int_a^b f(x)\,dx - \int_a^b g(x)\,dx = \int_a^b [f(x) - g(x)]\,dx.$$

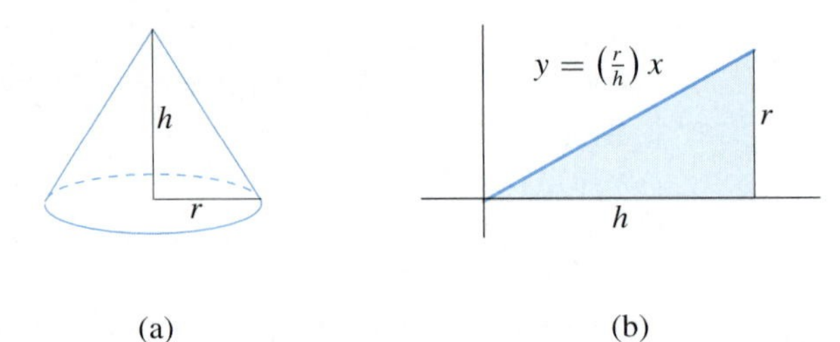

FIGURE 2 A right circular cone and its generating region

38. Use the fundamental theorem of calculus to show that

$$\int_a^b k \cdot f(x)\,dx = k \int_a^b f(x)\,dx.$$

SOLUTIONS TO PRACTICE PROBLEMS 7.5

1. The tricky part here is antidifferentiating $e^{k(6-t)}$. The antiderivative is $\frac{-1}{k}e^{k(6-t)}$. The simplest method to arrive at this result is to make a rough guess and then modify the guess. Let's guess that the function is its own derivative. Differentiating,

$$\frac{d}{dt}e^{k(6-t)} = e^{k(6-t)} \cdot \frac{d}{dt}k(6-t) = e^{k(6-t)} \cdot (-k) = -ke^{k(6-t)}.$$

The derivative is not what we want; it has an extra factor of $-k$. Therefore, if we multiply our *guess* function by $\frac{1}{-k}$ (that is, by $\frac{-1}{k}$) the derivative will turn out just right. Now,

$$\begin{aligned}\int_0^6 1000e^{k(6-t)}\,dt &= \left.\frac{-1000}{k}e^{k(6-t)}\right|_0^6 \\ &= \frac{-1000}{k}e^{k(0)} - \left(\frac{-1000}{k}e^{k(6)}\right) \\ &= \frac{-1000}{k}\left(1 - e^{6k}\right).\end{aligned}$$

2. By the formula for the future value of a continuous income stream, we must solve

$$\int_0^6 1000e^{k(6-t)}\,dt = 6781.20,$$

for k. That is,

$$\frac{-1000}{k}\left(1 - e^{6k}\right) = 6781.20.$$

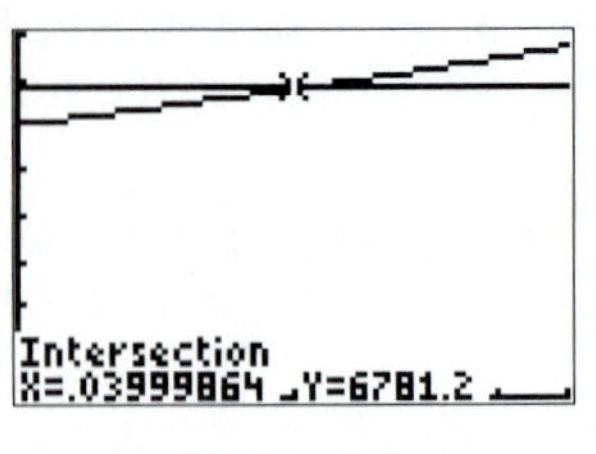

FIGURE 3

This equation is difficult to solve algebraically, so let's use a graphing utility to graph

$$y_1 = \frac{-1000}{x}\left(1 - e^{6x}\right) \quad \text{and} \quad y_2 = 6781.2$$

together in the window [0, .08] *by* [0, 8000] and find their point of intersection. (See Figure 3.) They intersect when $x \approx .04$. Therefore, the interest rate is 4%.

7.6 TECHNIQUES OF INTEGRATION

This section presents two techniques of antidifferentiation that will greatly expand our ability to find antiderivatives. Like the other rules of antidifferentiation discussed in Section 6.1, these new methods are based on corresponding rules of differentiation: the chain rule and the product rule.

INTEGRATION BY SUBSTITUTION Many integrals that appear complex have a simple form that becomes evident after making an appropriate change of variable. Integration by substitution is a technique for making such changes of variables.

Let $f(x)$ and $g(x)$ be two given functions, and let $F(x)$ be an antiderivative of $f(x)$. The chain rule asserts that

$$\frac{d}{dx}[F(g(x))] = F'(g(x))g'(x) = f(g(x))g'(x) \qquad [\text{since } F'(x) = f(x)].$$

Turning this formula into an integration formula, we have

$$\int f(g(x))g'(x)\,dx = F(g(x)) + C, \tag{1}$$

where C is any constant. One way to use this formula is illustrated in the following example.

EXAMPLE 1 Determine $\int (x^2 + 1)^3 \cdot 2x\,dx$.

Solution If we set $f(x) = x^3$ and $g(x) = x^2 + 1$, then $f(g(x)) = (x^2 + 1)^3$ and $g'(x) = 2x$. Therefore, we can apply formula (1). An antiderivative $F(x)$ of $f(x)$ is given by

$$F(x) = \frac{1}{4}x^4,$$

so that, by formula (1), we have

$$\int (x^2 + 1)^3 \cdot 2x\,dx = F(g(x)) + C = \frac{1}{4}(x^2 + 1)^4 + C.$$ ■

Formula (1) can be elevated from the status of a sometimes useful formula to a technique of integration by the introduction of a simple mnemonic device. Suppose that we are faced with integrating a function of the form $f(g(x))g'(x)$. Of course, we know the answer from formula (1). However, let us proceed somewhat differently. Replace the expression $g(x)$ by a new variable u and replace $g'(x)\,dx$

by du. Such a substitution has the advantage that it reduces the generally complex expression $f(g(x))$ to the simpler form $f(u)$. In terms of u, the integration problem may be written

$$\int f(g(x))g'(x)\,dx = \int f(u)\,du.$$

However, the integral on the right is easy to evaluate, since

$$\int f(u)\,du = F(u) + C.$$

Since $u = g(x)$, we then obtain

$$\int f(g(x))g'(x)\,dx = F(u) + C = F(g(x)) + C,$$

which is the correct answer by (1). Remember, however, that replacing $g'(x)\,dx$ by du has status as a correct mathematical statement only because doing so leads to the correct answers. We do not, in this book, seek to explain in any deeper way what this replacement means.

Let us rework Example 1 using this method.

Second Solution of Example 1 Set $u = x^2 + 1$. Then

$$du = \frac{d}{dx}(x^2 + 1)\,dx = 2x\,dx,$$

and

$$\begin{aligned}\int (x^2+1)^3 \cdot 2x\,dx &= \int u^3\,du\\ &= \frac{1}{4}u^4 + C\\ &= \frac{1}{4}(x^2+1)^4 + C \qquad \text{(since } u = x^2 + 1\text{).}\end{aligned}$$

EXAMPLE 2 Evaluate $\int 2xe^{x^2}\,dx$.

Solution Let $u = x^2$, so that $du = \frac{d}{dx}(x^2)\,dx = 2x\,dx$. Therefore,

$$\int 2xe^{x^2}\,dx = \int e^{x^2} \cdot 2x\,dx = \int e^u\,du = e^u + C = e^{x^2} + C.$$

From Examples 1 and 2 we can deduce the following method for integration of functions of the form $f(g(x))g'(x)$.

Integration by Substitution

1. Define a new variable $u = g(x)$, where $g(x)$ is chosen in such a way that, when written in terms of u, the integrand (the function to be integrated) is simpler than when written in terms of x.
2. Transform the integral with respect to x into an integral with respect to u by replacing $g(x)$ everywhere by u and $g'(x)\,dx$ by du.
3. Integrate the resulting function of u.
4. Rewrite the answer in terms of x by replacing u with $g(x)$.

Let us try a few more examples.

EXAMPLE 3 Evaluate $\int 3x^2\sqrt{x^3+1}\,dx$.

Solution The first problem facing us is to find an appropriate substitution that will simplify the integral. An immediate possibility is offered by setting $u = x^3 + 1$. Then $\sqrt{x^3+1}$ will become $\sqrt{u}$, a significant simplification. If $u = x^3 + 1$, then

$$du = \frac{d}{dx}(x^3+1)\,dx = 3x^2\,dx,$$

so that

$$\int 3x^2\sqrt{x^3+1}\,dx = \int \sqrt{u}\,du = \frac{2}{3}u^{3/2} + C = \frac{2}{3}(x^3+1)^{3/2} + C.$$

EXAMPLE 4 Find $\int \frac{(\ln x)^2}{x}\,dx$.

Solution Let $u = \ln x$. Then $du = (1/x)\,dx$ and

$$\begin{aligned}\int \frac{(\ln x)^2}{x}\,dx &= \int (\ln x)^2 \cdot \frac{1}{x}\,dx \\ &= \int u^2\,du \\ &= \frac{u^3}{3} + C \\ &= \frac{(\ln x)^3}{3} + C \qquad \text{(since } u = \ln x\text{).}\end{aligned}$$

Knowing the correct substitution to make is a skill that develops through practice. Basically, we look for an occurrence of function composition, $f(g(x))$, where $f(x)$ is a function that we know how to integrate and where $g'(x)$ also appears in the integrand. Sometimes $g'(x)$ does not appear exactly but can be obtained by multiplying by a constant. Such a shortcoming is easily remedied, as illustrated in Examples 5, 6, and 7.

EXAMPLE 5 Find $\int x^2 e^{x^3}\, dx$.

Solution Let $u = x^3$; then $du = 3x^2\, dx$. A simple trick allows us to introduce the needed factor of 3 into the integrand. Note that

$$\int x^2 e^{x^3}\, dx = \int \frac{1}{3} \cdot 3x^2 e^{x^3}\, dx = \frac{1}{3} \int 3x^2 e^{x^3}\, dx.$$

Substituting, we obtain

$$\begin{aligned} \int x^2 e^{x^3}\, dx &= \frac{1}{3} \int e^u\, du \\ &= \frac{1}{3} e^u + C \\ &= \frac{1}{3} e^{x^3} + C \qquad \text{(since } u = x^3\text{)}. \end{aligned}$$

Another way to handle the missing factor 3 is to write

$$u = x^3, \quad du = 3x^2\, dx, \quad \text{and} \quad \frac{1}{3}\, du = x^2\, dx.$$

Then substitution yields

$$\begin{aligned} \int x^2 e^{x^3}\, dx &= \int e^{x^3} \cdot x^2\, dx = \int e^u \cdot \frac{1}{3}\, du = \frac{1}{3} \int e^u\, du \\ &= \frac{1}{3} e^u + C = \frac{1}{3} e^{x^3} + C. \end{aligned}$$

■

EXAMPLE 6 Find $\int \frac{2 - x}{\sqrt{2x^2 - 8x + 1}}\, dx$.

Solution Let $u = 2x^2 - 8x + 1$; then $du = (4x - 8)\, dx$. Observe that $4x - 8 = -4(2 - x)$. So we multiply the integrand by -4 and compensate by placing a factor of $-\frac{1}{4}$ in front of the integral:

$$\begin{aligned} \int \frac{1}{\sqrt{2x^2 - 8x + 1}} \cdot (2 - x)\, dx &= -\frac{1}{4} \int \frac{1}{\sqrt{2x^2 - 8x + 1}} \cdot (-4)(2 - x)\, dx \\ &= -\frac{1}{4} \int \frac{1}{\sqrt{u}}\, du = -\frac{1}{4} \int u^{-1/2}\, du \\ &= -\frac{1}{4} \cdot 2u^{1/2} + C = -\frac{1}{2} u^{1/2} + C \\ &= -\frac{1}{2} (2x^2 - 8x + 1)^{1/2} + C. \end{aligned}$$

■

EXAMPLE 7 Find $\int \frac{x}{x^2+1}\,dx$.

Solution If $u = x^2 + 1$, then $du = 2x\,dx$, and

$$\begin{aligned}\int \frac{x}{x^2+1}\,dx &= \int \frac{1}{2}\cdot\frac{1}{x^2+1}\cdot 2x\,dx\\ &= \frac{1}{2}\int \frac{1}{x^2+1}\cdot 2x\,dx = \frac{1}{2}\int \frac{1}{u}\,du\\ &= \frac{1}{2}\ln|u| + C = \frac{1}{2}\ln\left|x^2+1\right| + C.\end{aligned}$$

■

INTEGRATION BY PARTS Integration by parts is a technique of integration that exchanges a given integral for another (less complicated) integral. Let $f(x)$ and $g(x)$ be given functions, and let $G(x)$ be an antiderivative of $g(x)$. The product rule asserts that

$$\begin{aligned}\frac{d}{dx}[f(x)G(x)] &= f(x)G'(x) + G(x)f'(x)\\ &= f(x)g(x) + f'(x)G(x) \qquad [\text{since } G'(x) = g(x)].\end{aligned}$$

Therefore,

$$f(x)G(x) = \int f(x)g(x)\,dx + \int f'(x)G(x)\,dx.$$

This last formula may be rewritten in the following more useful form:

$$\int f(x)g(x)\,dx = f(x)G(x) - \int f'(x)G(x)\,dx. \tag{2}$$

Equation (2) is the basis of *integration by parts*, one of the most important techniques of integration. The next four examples illustrate this technique.

EXAMPLE 8 Evaluate $\int xe^x\,dx$.

Solution Set $f(x) = x$, $g(x) = e^x$. Then $f'(x) = 1$, $G(x) = e^x$, and (2) yields

$$\int xe^x\,dx = xe^x - \int 1\cdot e^x\,dx = xe^x - e^x + C.$$

■

The following principles underlie Example 8 and also illustrate general features of situations to which integration by parts may be applied:

1. The integrand is the product of two functions $f(x)$ and $g(x)$.
2. It is easy to compute $f'(x)$ and $G(x)$. That is, we can differentiate $f(x)$ and integrate $g(x)$.

3. The integral $\int f'(x)G(x)\,dx$ can be calculated.

Let us consider another example in order to see how these three principles work.

EXAMPLE 9 Evaluate $\displaystyle\int x(x+5)^8\,dx$.

Solution Our calculations can be set up as follows:

$$f(x) = x, \qquad g(x) = (x+5)^8,$$
$$f'(x) = 1, \qquad G(x) = \frac{1}{9}(x+5)^9.$$

Then

$$\begin{aligned}\int x(x+5)^8\,dx &= x\cdot\frac{1}{9}(x+5)^9 - \int 1\cdot\frac{1}{9}(x+5)^9\,dx\\ &= \frac{1}{9}x(x+5)^9 - \frac{1}{9}\int (x+5)^9\,dx\\ &= \frac{1}{9}x(x+5)^9 - \frac{1}{9}\cdot\frac{1}{10}(x+5)^{10} + C\\ &= \frac{1}{9}x(x+5)^9 - \frac{1}{90}(x+5)^{10} + C.\end{aligned}$$

We were led to try integration by parts because our integrand is the product of two functions. We were led to choose $f(x) = x$ [and not $(x+5)^9$] because $f'(x) = 1$, so that the factor x in the integrand is made to disappear, thereby simplifying the integral. ■

EXAMPLE 10 Evaluate $\displaystyle\int x^2\ln x\,dx$.

Solution Set

$$f(x) = \ln x, \qquad g(x) = x^2,$$
$$f'(x) = \frac{1}{x}, \qquad G(x) = \frac{x^3}{3}.$$

Then

$$\begin{aligned}\int x^2\ln x\,dx &= \frac{x^3}{3}\ln x - \int \frac{1}{x}\cdot\frac{x^3}{3}\,dx\\ &= \frac{x^3}{3}\ln x - \frac{1}{3}\int x^2\,dx\\ &= \frac{x^3}{3}\ln x - \frac{1}{9}x^3 + C.\end{aligned}$$

■

EXAMPLE 11 Evaluate $\displaystyle\int \ln x\,dx$.

Solution Since $\ln x = 1 \cdot \ln x$, we may view $\ln x$ as a product $f(x)g(x)$, where $f(x) = \ln x$, $g(x) = 1$. Then

$$f'(x) = \frac{1}{x}, \qquad G(x) = x.$$

Finally,

$$\int \ln x \, dx = x \ln x - \int \frac{1}{x} \cdot x \, dx$$
$$= x \ln x - \int 1 \, dx = x \ln x - x + C.$$

PRESENT VALUE OF AN INCOME STREAM Let us illustrate how an integration technique such as integration by parts can arise in a fairly simple business situation. Recall from our discussion of compound interest that the present value of A dollars to be received t years from now is given by

$$P = Ae^{-rt},$$

where r is a specified annual rate of interest, with interest compounded continuously. Now suppose that a sum of money is to be received in a series of frequent payments over a period of time instead of in one lump sum at the end of the period. Such a series of payments is often viewed as a "continuous stream of income." If $K(t)$ is the annual rate of income at time t, and if the income is to be received over the next N years, then the *present value P of the stream of income* at interest rate r is defined by the integral

$$P = \int_0^N K(t)e^{-rt} \, dt. \tag{3}$$

A Riemann sum argument can be used to show that it takes a fund of P dollars available today in order to create a continuous stream of income of $K(t)$ dollars (annual rate) at time t. (See Exercise 27 in Section 7.4.)

The concept of the present value of a continuous stream of income is an important tool in management decision processes involving the selection or replacement of equipment. Even when $K(t)$ is a simple function, the evaluation of the integral in (3) usually requires special techniques such as integration by parts, as we see in the following example.

EXAMPLE 12 A printing company estimates that the rate of revenue generated by one of its printing presses at time t will be $80 - 2t$ thousand dollars per year. Find the present value of this continuous stream of income over the next 4 years at a 10% interest rate.

Solution We use (3) with $K(t) = 80 - 2t$, $N = 4$, and $r = .1$. Our first task is to find an antiderivative of $K(t)e^{-rt} = (80 - 2t)e^{-.1t}$. We integrate by parts, with

$f(t) = 80 - 2t$, $g(t) = e^{-.1t}$, $f'(t) = -2$, and $G(t) = -10e^{-.1t}$.

$$\begin{aligned}\int (80 - 2t)e^{-.1t}\,dt &= (80 - 2t)(-10e^{-.1t}) - \int 20e^{-.1t}\,dt \\ &= -800e^{-.1t} + 20te^{-.1t} + 200e^{-.1t} + C \\ &= 20te^{-.1t} - 600e^{-.1t} + C.\end{aligned}$$

Then

$$\begin{aligned}P = \int_0^4 (80 - 2t)e^{-.1t}\,dt &= (20te^{-.1t} - 600e^{-.1t})\Big|_0^4 \\ &= (80e^{-.4} - 600e^{-.4}) - (0 - 600e^0) \\ &= 600 - 520e^{-.4} \\ &\approx 251.\end{aligned}$$

Thus the present value of the machine's earnings is approximately $251,000. ■

PRACTICE PROBLEMS 7.6

Calculate the following integrals.

1. $\displaystyle\int \frac{e^{5x} + x^4}{(e^{5x} + x^5)^2}\,dx$ (*Hint*: Let $u = e^{5x} + x^5$.)

2. $\displaystyle\int \frac{2x - 1}{(x + 4)^{1/3}}\,dx$ [*Hint*: Let $f(x) = 2x - 1$, $g(x) = (x + 4)^{-1/3}$.]

3. $\displaystyle\int \ln\sqrt{x}\,dx$

EXERCISES 7.6

Calculate each of the following indefinite integrals.

1. $\displaystyle\int 2x(x^2 + 4)^5\,dx$

2. $\displaystyle\int 3x^2(x^3 + 1)^2\,dx$

3. $\displaystyle\int (x^2 - 5x)^3(2x - 5)\,dx$

4. $\displaystyle\int 2x\sqrt{x^2 + 3}\,dx$

5. $\displaystyle\int 5e^{5x-3}\,dx$

6. $\displaystyle\int 2xe^{-x^2}\,dx$

7. $\displaystyle\int \frac{3x^2}{x^3 - 1}\,dx$

8. $\displaystyle\int \frac{2x + 1}{(x^2 + x + 3)^6}\,dx$

Calculate the following integrals, making the indicated substitutions.

9. $\displaystyle\int \frac{x^2}{\sqrt{x^3 - 1}}\,dx;\ u = x^3 - 1$

10. $\displaystyle\int \frac{1}{\sqrt{2x + 1}}\,dx;\ u = 2x + 1$

11. $\displaystyle\int \frac{e^{1/x}}{x^2}\,dx;\ u = \frac{1}{x}$

12. $\displaystyle\int \frac{e^{3x}}{e^{3x} + 1}\,dx;\ u = e^{3x} + 1$

13. $\displaystyle\int \frac{x^2 + 1}{x^3 + 3x + 2}\,dx;\ u = x^3 + 3x + 2$

14. $\int \frac{\ln x}{x}\,dx;\ u = \ln x$

Determine the following integrals by making an appropriate substitution.

15. $\int (x^5 - 2x + 1)^{10}(5x^4 - 2)\,dx$

16. $\int (x + 1)e^{x^2+2x+4}\,dx$

17. $\int \frac{3}{2x - 4}\,dx$

18. $\int \frac{e^{\sqrt{x}}}{\sqrt{x}}\,dx$

19. $\int \frac{x}{e^{x^2}}\,dx$

20. $\int \frac{3x - x^3}{x^4 - 6x^2 + 5}\,dx$

Use integration by parts to determine the following integrals.

21. $\int xe^{5x}\,dx$

22. $\int xe^{-x/2}\,dx$

23. $\int x(2x + 1)^4\,dx$

24. $\int (x + 1)e^x\,dx$

25. $\int x\sqrt{x + 1}\,dx$

26. $\int x(x + 5)^{-3}\,dx$

27. $\int \frac{3x}{e^x}\,dx$

28. $\int x^3 \ln x\,dx$

29. $\int \ln 3x\,dx$ [Let $f(x) = \ln x$, $g(x) = 1$.]

30. $\int \frac{x}{\sqrt{3 + 2x}}\,dx$ [Let $f(x) = x$, $g(x) = (3 + 2x)^{-1/2}$.]

Determine the following indefinite integrals.

31. $\int xe^{2x}\,dx$

32. $\int x\sqrt{x^2 - 1}\,dx$

33. $\int xe^{x^2}\,dx$

34. $\int x\sqrt{x - 1}\,dx$

35. $\int \frac{2x - 1}{\sqrt{3x - 3}}\,dx$

36. $\int \frac{2x - 1}{3x^2 - 3x + 1}\,dx$

37. $\int (4x + 3)(6x^2 + 9x)^{-7}\,dx$

38. $\int (4x + 3)(6x + 9)^{-7}\,dx$

39. $\int \frac{\ln x}{\sqrt{x}}\,dx$

40. $\int \frac{x + 4}{e^{4x}}\,dx$

41. $\int \frac{1}{x \ln 5x}\,dx$

42. $\int x \ln 5x\,dx$

43. $\int \ln x^4\,dx$

44. $\int \frac{(\ln x)^4}{x}\,dx$

45. $\int \frac{\ln \sqrt{x}}{x}\,dx$

46. $\int \sqrt{x} \ln \sqrt{x}\,dx$

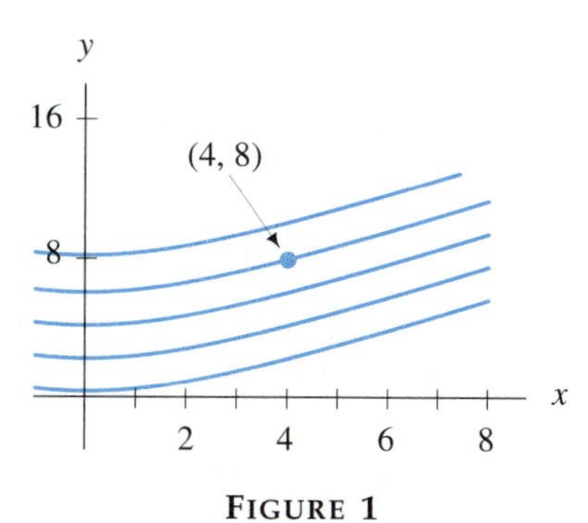

FIGURE 1

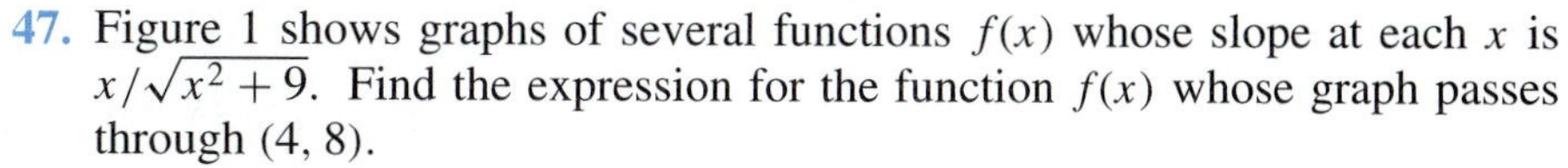
47. Figure 1 shows graphs of several functions $f(x)$ whose slope at each x is $x/\sqrt{x^2 + 9}$. Find the expression for the function $f(x)$ whose graph passes through $(4, 8)$.

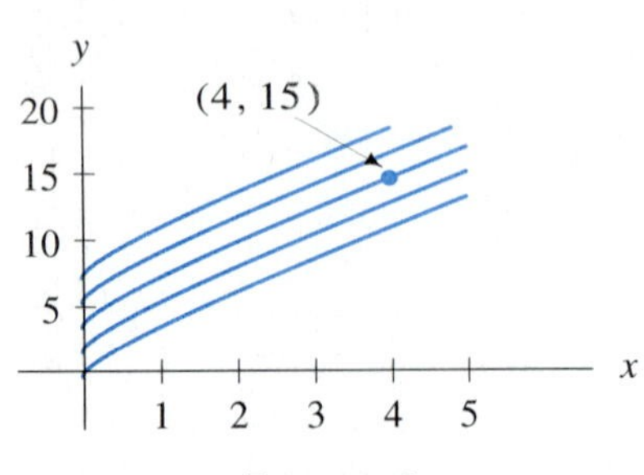

FIGURE 2

48. Figure 2 shows graphs of several functions $f(x)$ whose slope at each x is $(2\sqrt{x}+1)/\sqrt{x}$. Find the expression for the function $f(x)$ whose graph passes through (4, 15).

49. Suppose that an investment produces a continuous stream of income at the rate of $300t + 500$ dollars per year at time t. Find the present value of the investment income over the next 5 years, using a 10% interest rate.

50. Recompute the present value of the stream of income in Exercise 49 over 5 years using a 5% interest rate.

In the remaining exercises, use a graphing utility to obtain the graphs of the functions. In Exercises 51–54, find an antiderivative of $f(x)$, call it $F(x)$, and then compare the graphs of $F(x)$ and $f(x)$ in the given window to check that the expression for $F(x)$ is reasonable. (That is, determine whether the two graphs are consistent. When $F(x)$ has a relative extreme point, $f(x)$ should be zero. When $F(x)$ is increasing, $f(x)$ should be positive. And so on.)

51. $f(x) = \dfrac{90 - 180x}{(x^2 - x + 5)^3}$; $[-4, 4]$ *by* $[-1, 2]$

52. $f(x) = (3 - x)(x^2 - 6x)^4$; $[-4, 9]$ *by* $[-6500, 6500]$

53. $f(x) = x(x + 5)^3$; $[-7, 2]$ *by* $[-200, 200]$

54. $f(x) = x^2e^x$; $[-6, 2]$ *by* $[0, 3]$

SOLUTIONS TO PRACTICE PROBLEMS 7.6

1. Let $u = e^{5x} + x^5$ and $du = (5e^{5x} + 5x^4)\,dx$, and note that $5e^{5x} + 5x^4 = 5(e^{5x} + x^4)$. Then

$$\begin{aligned}\int \frac{e^{5x} + x^4}{(e^{5x} + x^5)^2}\,dx &= \int \frac{1}{5} \cdot \frac{5e^{5x} + 5x^4}{(e^{5x} + x^5)^2}\,dx \\ &= \frac{1}{5}\int \frac{1}{(e^{5x} + x^5)^2} \cdot (5e^{5x} + 5x^4)\,dx \\ &= \frac{1}{5}\int \frac{1}{u^2}\,du \\ &= \frac{1}{5}(-u^{-1}) + C \\ &= -\frac{1}{5}(e^{5x} + x^5)^{-1} + C.\end{aligned}$$

2. Use integration by parts with $f(x) = 2x - 1$, $g(x) = (x + 4)^{-1/3}$, $f'(x) = 2$, and $G(x) = \frac{3}{2}(x + 4)^{2/3}$.

$$\begin{aligned}\int \frac{2x - 1}{(x + 4)^{1/3}}\,dx &= \int (2x - 1)(x + 4)^{-1/3}\,dx \\ &= (2x - 1) \cdot \frac{3}{2}(x + 4)^{2/3} - \int 2 \cdot \frac{3}{2}(x + 4)^{2/3}\,dx \\ &= \frac{3}{2}(2x - 1)(x + 4)^{2/3} - 3\int (x + 4)^{2/3}\,dx \\ &= \frac{3}{2}(2x - 1)(x + 4)^{2/3} - \frac{9}{5}(x + 4)^{5/3} + C.\end{aligned}$$

3. This problem is similar to Example 11, which asks for $\int \ln x\, dx$, and can be approached in the same way by letting $f(x) = \ln\sqrt{x}$ and $g(x) = 1$. Another approach is to use a property of logarithms to simplify the integrand.

$$\int \ln\sqrt{x}\, dx = \int \ln\left(x^{1/2}\right) dx = \int \frac{1}{2}\ln x\, dx = \frac{1}{2}\int \ln x\, dx.$$

Since we know $\int \ln x\, dx$ from Example 11,

$$\int \ln\sqrt{x}\, dx = \frac{1}{2}\int \ln x\, dx = \frac{1}{2}(x\ln x - x) + C.$$

7.7 IMPROPER INTEGRALS

In applications of calculus, especially to statistics, it is often necessary to consider the area of a region that extends far to the right or left along the x-axis. We have drawn several such regions in Figure 1. The areas of such "infinite" regions may be computed using *improper integrals*.

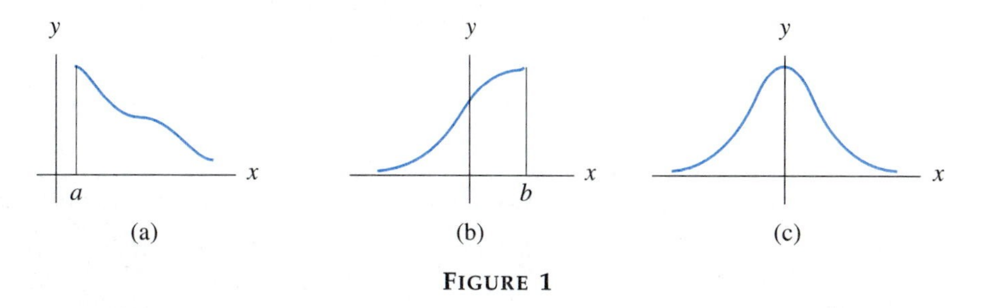

FIGURE 1

In order to motivate the idea of an improper integral, let us attempt to calculate the area under the curve $y = 3/x^2$ to the right of $x = 1$ (Figure 2).

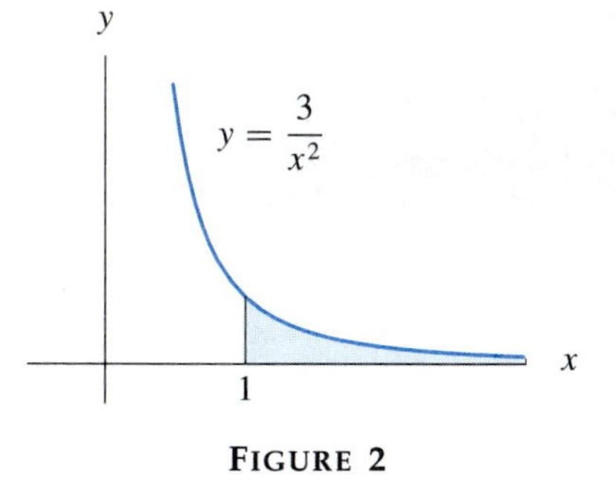

FIGURE 2

First, we shall compute the area under the graph of this function from $x = 1$ to $x = b$, where b is some number greater than 1. [See Figure 3(a).] Then we shall examine how the area increases as we let b get larger, as in Figure 3(b) and (c).

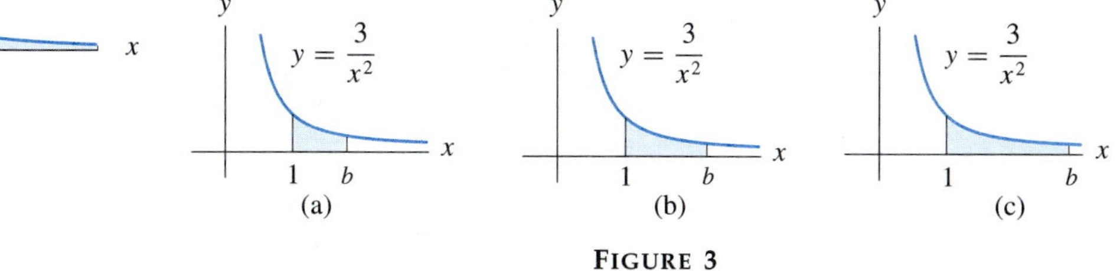

FIGURE 3

The area from 1 to b is given by

$$\int_1^b \frac{3}{x^2}\, dx = -\frac{3}{x}\bigg|_1^b = \left(-\frac{3}{b}\right) - \left(-\frac{3}{1}\right) = 3 - \frac{3}{b}.$$

When b is large, $3/b$ is small and the integral nearly equals 3. That is, the area under the curve from 1 to b nearly equals 3. (See Table 1.) In fact, the area gets arbitrarily close to 3 as b gets larger. Thus it is reasonable to say that the region under the curve $y = 3/x^2$ for $x \geq 1$ has area 3.

Table 1

b	AREA $= \int_1^b \frac{3}{x^2}\,dx = 3 - \frac{3}{b}$
10	2.7000
100	2.9700
1,000	2.9970
10,000	2.9997

Recall from Chapter 2 that we write $b \to \infty$ as a shorthand for "b gets arbitrarily large, without bound." Then, to express the fact that the value of $\int_1^b \frac{3}{x^2}\,dx$ approaches 3 as $b \to \infty$, we write

$$\int_1^\infty \frac{3}{x^2}\,dx = \lim_{b\to\infty} \int_1^b \frac{3}{x^2}\,dx = 3.$$

We call $\int_1^\infty \frac{3}{x^2}\,dx$ an *improper* integral because the upper limit of the integral is ∞ (infinity) rather than a finite number.

Definition Let a be fixed and suppose that $f(x)$ is a nonnegative function for $x \geq a$. If $\lim_{b\to\infty} \int_a^b f(x)\,dx = L$, we define

$$\int_a^\infty f(x)\,dx = \lim_{b\to\infty} \int_a^b f(x)\,dx = L.$$

We say that the improper integral $\int_a^\infty f(x)\,dx$ is *convergent* and that the region under the curve $y = f(x)$ for $x \geq a$ has area L. (See Figure 4.)

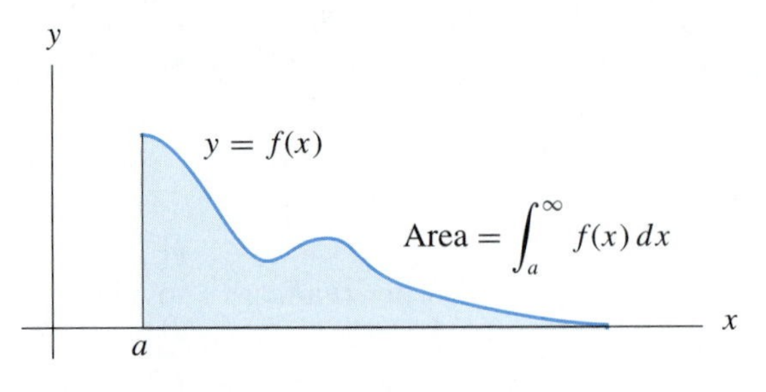

FIGURE 4

It is possible to consider improper integrals in which $f(x)$ is both positive and negative. However, we shall consider only nonnegative functions, since this is the case occurring in most applications.

EXAMPLE 1 Find the area under the curve $y = e^{-x}$ for $x \geq 0$ (Figure 5).

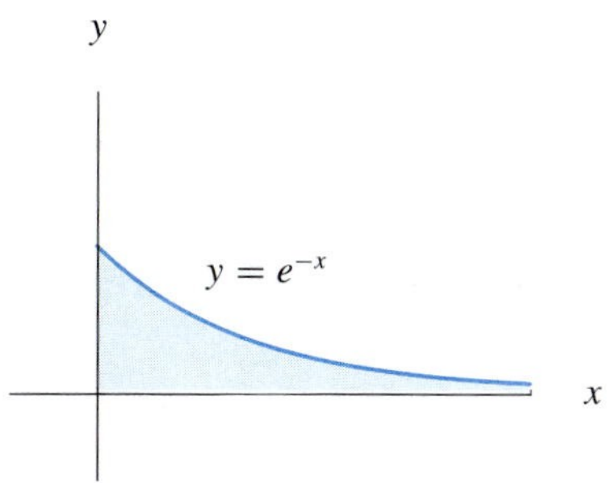

y

$y = e^{-x}$

x

FIGURE 5

Solution We must calculate the improper integral

$$\int_0^\infty e^{-x}\,dx.$$

We take $b > 0$ and compute

$$\int_0^b e^{-x}\,dx = -e^{-x}\Big|_0^b = (-e^{-b}) - (-e^0) = 1 - e^{-b} = 1 - \frac{1}{e^b}.$$

We now consider the limit as $b \to \infty$ and note that $1/e^b$ approaches zero. Thus

$$\int_0^\infty e^{-x}\,dx = \lim_{b\to\infty}\int_0^b e^{-x}\,dx = \lim_{b\to\infty}\left(1 - \frac{1}{e^b}\right) = 1.$$

Therefore, the region in Figure 5 has area 1. ■

EXAMPLE 2 Evaluate the improper integral $\displaystyle\int_7^\infty \frac{1}{(x-5)^2}\,dx$.

Solution $$\int_7^b \frac{1}{(x-5)^2}\,dx = -\frac{1}{x-5}\Big|_7^b = -\frac{1}{b-5} - \left(-\frac{1}{7-5}\right) = \frac{1}{2} - \frac{1}{b-5}.$$

As $b \to \infty$, the fraction $1/(b-5)$ approaches zero, so

$$\int_7^\infty \frac{1}{(x-5)^2}\,dx = \lim_{b\to\infty}\int_7^b \frac{1}{(x-5)^2}\,dx = \lim_{b\to\infty}\left(\frac{1}{2} - \frac{1}{b-5}\right) = \frac{1}{2}.$$ ■

Not every improper integral is convergent. If the value of $\int_a^b f(x)\,dx$ does not have a limit as $b \to \infty$, we cannot assign any numerical value to $\int_a^\infty f(x)\,dx$, and we say that the improper integral $\int_a^\infty f(x)\,dx$ is *divergent*.

EXAMPLE 3 Show that $\displaystyle\int_1^\infty \frac{1}{\sqrt{x}}\,dx$ is divergent.

Solution For $b > 1$ we have

$$\int_1^b \frac{1}{\sqrt{x}}\,dx = 2\sqrt{x}\Big|_1^b = 2\sqrt{b} - 2. \tag{1}$$

As $b \to \infty$, the quantity $2\sqrt{b} - 2$ increases without bound. That is, $2\sqrt{b} - 2$ can be made larger than any specific number. Therefore,

$$\int_1^b \frac{1}{\sqrt{x}}\,dx$$

has no limit as $b \to \infty$, so $\displaystyle\int_1^\infty \frac{1}{\sqrt{x}}\,dx$ is divergent. ■

In some cases it is necessary to consider improper integrals of the form

$$\int_{-\infty}^b f(x)\,dx.$$

Let b be fixed and examine the value of $\int_a^b f(x)\,dx$ as $a \to -\infty$ — that is, as a moves arbitrarily far to the left on the number line. If $\lim_{a\to-\infty} \int_a^b f(x)\,dx = L$, we say that the improper integral $\int_{-\infty}^b f(x)\,dx$ is *convergent* and we write

$$\int_{-\infty}^b f(x)\,dx = L.$$

Otherwise, the improper integral is divergent. An integral of the form $\int_{-\infty}^b f(x)\,dx$ may be used to compute the area of a region such as that shown in Figure 1(b).

EXAMPLE 4 Determine if $\int_{-\infty}^0 e^{5x}\,dx$ is convergent. If convergent, find its value.

Solution $\displaystyle\int_{-\infty}^0 e^{5x}\,dx = \lim_{a\to-\infty}\int_a^0 e^{5x}\,dx = \lim_{a\to-\infty}\frac{1}{5}e^{5x}\Big|_a^0 = \lim_{a\to-\infty}\left(\frac{1}{5} - \frac{1}{5}e^{5a}\right).$

As $a \to -\infty$, e^{5a} approaches 0 so that $\frac{1}{5} - \frac{1}{5}e^{5a}$ approaches $\frac{1}{5}$. Thus the improper integral converges and has value $\frac{1}{5}$. ■

Areas of regions that extend infinitely far to the left *and* right, such as the region in Figure 1(c), are calculated using improper integrals of the form

$$\int_{-\infty}^\infty f(x)\,dx.$$

We define such an integral to have the value

$$\lim_{b\to\infty}\int_{-b}^b f(x)\,dx,$$

provided this limit exists.

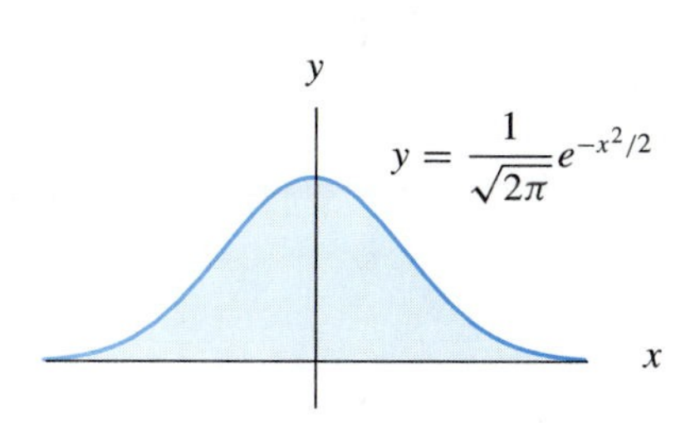

FIGURE 6

An important area that arises in probability theory is the area under the so-called *normal curve*, whose equation is

$$y = \frac{1}{\sqrt{2\pi}}e^{-x^2/2}.$$

(See Figure 6.) It is of fundamental importance for probability theory that this area is 1. In terms of an improper integral, this fact may be written

$$\int_{-\infty}^{\infty} \frac{1}{\sqrt{2\pi}}e^{-x^2/2}\,dx = 1.$$

The proof of this result is beyond the scope of this book. A graphing utility shows that the integral from $x = -4$ to $x = 4$ is .9999367 and that the integral from $x = -5$ to $x = 5$ is .9999994.

PRACTICE PROBLEMS 7.7

1. Does $1 - 2(1 - 3b)^{-4}$ approach a limit as $b \to \infty$?
2. Evaluate $\displaystyle\int_{1}^{\infty} \frac{x^2}{x^3 + 8}\,dx$.
3. Evaluate $\displaystyle\int_{-\infty}^{-2} \frac{1}{x^4}\,dx$.

EXERCISES 7.7

In Exercises 1–14, determine if the expression approaches a limit as $b \to \infty$, and find the number when it exists. (Note: To check your answer, graph the function with a graphing utility to see what happens for large values of b. Replace b by x and graph the function for say $2 \le x \le 100$.)

1. $\frac{5}{b}$
2. b^2
3. $-3e^{2b}$
4. $\frac{1}{b} + \frac{1}{3}$
5. $\frac{1}{4} - \frac{1}{b^2}$
6. $\frac{1}{2}\sqrt{b}$
7. $2 - (b + 1)^{-1/2}$
8. $\frac{2}{b} - \frac{3}{b^{3/2}}$
9. $5 - \frac{1}{b - 1}$
10. $5(b^2 + 3)^{-1}$
11. $6 - 3b^{-2}$
12. $e^{-b/2} + 5$
13. $2(1 - e^{-3b})$
14. $4(1 - b^{-3/4})$
15. Find the area under the graph of $y = 1/x^2$ for $x \ge 2$.
16. Find the area under the graph of $y = (x + 1)^{-2}$ for $x \ge 0$.
17. Find the area under the graph of $y = e^{-x/2}$ for $x \ge 0$.
18. Find the area under the graph of $y = 4e^{-4x}$ for $x \ge 0$.
19. Find the area under the graph of $y = (x + 1)^{-3/2}$ for $x \ge 3$.
20. Find the area under the graph of $y = (2x + 6)^{-4/3}$ for $x \ge 1$. (See Figure 7.)
21. Show that the region under the graph of $y = (14x + 18)^{-4/5}$ for $x \ge 1$ cannot be assigned any finite number as its area. (See Figure 8.)
22. Show that the region under the graph of $y = (x - 1)^{-1/3}$ for $x \ge 2$ cannot be assigned any finite number as its area.

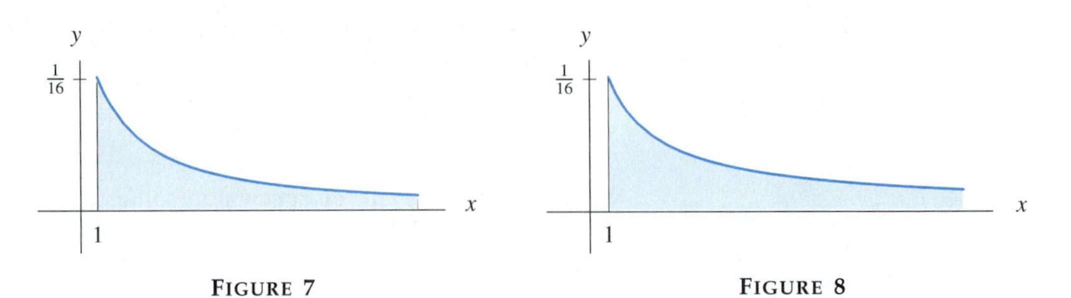

FIGURE 7

FIGURE 8

Evaluate the following integrals whenever they are convergent. Confirm your answer by using a graphing utility to evaluate the integral with ∞ replaced by a large number such as 100.

23. $\int_1^\infty \frac{1}{x^3}\,dx$ **24.** $\int_1^\infty \frac{2}{x^{3/2}}\,dx$ **25.** $\int_0^\infty e^{2x}\,dx$

26. $\int_0^\infty (x^2+1)\,dx$ **27.** $\int_0^\infty \frac{1}{(2x+3)^2}\,dx$ **28.** $\int_0^\infty e^{-3x}\,dx$

29. $\int_2^\infty \frac{1}{(x-1)^{5/2}}\,dx$ **30.** $\int_2^\infty e^{2-x}\,dx$ **31.** $\int_0^\infty .01e^{-.01x}\,dx$

32. $\int_0^\infty \frac{4}{(2x+1)^3}\,dx$ **33.** $\int_0^\infty 6e^{1-3x}\,dx$ **34.** $\int_1^\infty e^{-.2x}\,dx$

35. $\int_3^\infty \frac{x^2}{\sqrt{x^3-1}}\,dx$ **36.** $\int_2^\infty \frac{1}{x\ln x}\,dx$ **37.** $\int_0^\infty xe^{-x^2}\,dx$

38. $\int_0^\infty \frac{x}{x^2+1}\,dx$ **39.** $\int_0^\infty 2x(x^2+1)^{-3/2}\,dx$

40. $\int_1^\infty (5x+1)^{-4}\,dx$ **41.** $\int_{-\infty}^0 e^{4x}\,dx$ **42.** $\int_{-\infty}^0 \frac{8}{(x-5)^2}\,dx$

43. $\int_{-\infty}^0 \frac{6}{(1-3x)^2}\,dx$ **44.** $\int_{-\infty}^0 \frac{1}{\sqrt{4-x}}\,dx$ **45.** $\int_0^\infty \frac{e^{-x}}{(e^{-x}+2)^2}\,dx$

46. $\int_{-\infty}^\infty \frac{e^{-x}}{(e^{-x}+2)^2}\,dx$

47. If $k > 0$, show that $\int_0^\infty ke^{-kx}\,dx = 1$.

48. If $k > 0$, show that $\int_1^\infty \frac{k}{x^{k+1}}\,dx = 1$.

49. The *capital value* of an asset such as a machine is sometimes defined as the present value of all future net earnings of the asset. (See Section 7.6.) The actual lifetime of the asset may not be known, and since some assets may last indefinitely, the capital value of the asset may be written in the form

$$\int_0^\infty K(t)e^{-rt}\,dt,$$

where $K(t)$ is the annual rate of earnings produced by the asset at time t, and where r is the annual rate of interest, compounded continuously. Find the capital value of an asset that generates income at the rate of \$5000 per year, assuming an interest rate of 10%.

50. Find the capital value of an asset that at time t is producing income at the rate of $6000e^{.04t}$ dollars per year, assuming an interest rate of 16%.

SOLUTIONS TO PRACTICE PROBLEMS 7.7

1. The expression $1 - 2(1 - 3b)^{-4}$ may also be written in the form

$$1 - \frac{2}{(1 - 3b)^4}.$$

When b is large, $(1 - 3b)^4$ is very large, and so $2/(1 - 3b)^4$ is very small. Thus $1 - 2(1 - 3b)^{-4}$ approaches 1 as $b \to \infty$.

2. The first step is to find an antiderivative of $x^2/(x^3+8)$. Using the substitution $u = x^3 + 8$, $du = 3x^2\,dx$, we obtain

$$\int \frac{x^2}{x^3 + 8}\,dx = \frac{1}{3}\int \frac{1}{u}\,du = \frac{1}{3}\ln|u| + C = \frac{1}{3}\ln\left|x^3 + 8\right| + C.$$

Now,

$$\int_1^b \frac{x^2}{x^3 + 8}\,dx = \frac{1}{3}\left|x^3 + 8\right|\Big|_1^b = \frac{1}{3}\ln\left(b^3 + 8\right) - \frac{1}{3}\ln 9.$$

Finally, we examine what happens as $b \to \infty$. Certainly, $b^3 + 8$ gets arbitrarily large, and so $\ln(b^3 + 8)$ must also get arbitrarily large. Hence

$$\int_1^b \frac{x^2}{x^3 + 8}\,dx$$

has no limit as $b \to \infty$, so the improper integral

$$\int_1^\infty \frac{x^2}{x^3 + 8}\,dx$$

is divergent.

3.
$$\int_a^{-2} \frac{1}{x^4}\,dx = \int_a^{-2} x^{-4}\,dx = \frac{x^{-3}}{-3}\Big|_a^{-2} = \frac{1}{-3x^3}\Big|_a^{-2}$$

$$= \frac{1}{-3(-2)^3} - \left(\frac{1}{-3 \cdot a^3}\right)$$

$$= \frac{1}{24} + \frac{1}{3a^3}.$$

$$\int_{-\infty}^{-2} \frac{1}{x^4}\,dx = \lim_{a \to -\infty} \int_a^{-2} \frac{1}{x^4}\,dx = \lim_{a \to -\infty}\left(\frac{1}{24} + \frac{1}{3a^3}\right) = \frac{1}{24}.$$

7.8 APPLICATIONS OF CALCULUS TO PROBABILITY

Consider a cell population that is growing vigorously. Suppose that when a cell is 3 days old it divides and forms two new "daughter" cells. If the population is sufficiently large, it will contain cells of various ages between 0 and 3, and it will turn out that the proportion of cells of various ages remains constant. That is, if a and b are any two numbers between 0 and 3, with $a < b$, then the proportion of cells whose ages lie between a and b will be essentially constant from one moment to the next, even though individual cells are aging and new cells are being formed all the time. In fact, biologists have found that under the ideal circumstances described, the proportion of cells whose ages are between a and b is given by the area under the graph of the function $f(x) = 2ke^{-kx}$ from $x = a$ to $x = b$, where $k = (\ln 2)/3$.* (See Figure 1.)

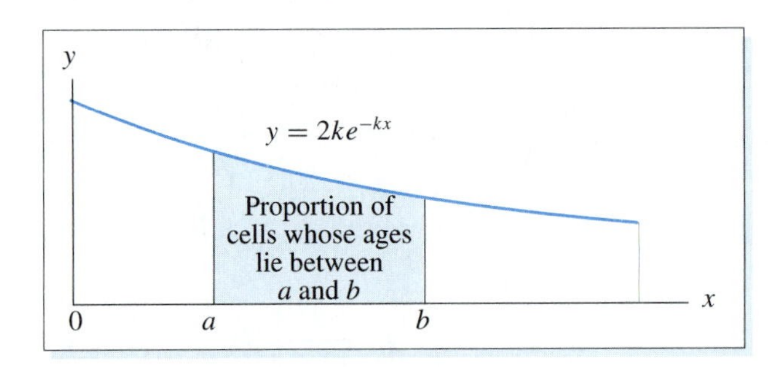

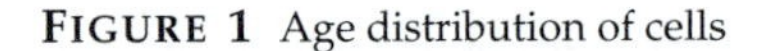

FIGURE 1 Age distribution of cells

Now consider an experiment where we select a cell at random from the population and observe its age, X. Then the probability or likelihood** that X lies between a and b is given by the area under the graph of $f(x) = 2ke^{-kx}$ from a to b, as in Figure 1. Let us denote this probability by $\Pr(a \le X \le b)$. Using the fact that the area under the graph of $f(x)$ is given by a definite integral, we have

$$\Pr(a \le X \le b) = \int_a^b f(x)\,dx = \int_a^b 2ke^{-kx}\,dx. \tag{1}$$

The function $f(x)$ that determines the probability in (1) for each a and b is called the *probability density function* of X (or of the experiment whose outcome is X).

The general situation we wish to describe in this section concerns an experiment whose outcome is a number X in a certain interval, say between A and B. For the cell population above, $A = 0$ and $B = 3$. Another typical experiment might consist of choosing a decimal X at random between $A = 5$ and $B = 6$. Or, one could select a random telephone call at some telephone switchboard and

* See J. R. Cook and T. W. James, "Age Distribution of Cells in Logarithmically Growing Cell Populations," in *Synchrony in Cell Division and Growth*, Erik Zeuthen, ed. (New York: John Wiley & Sons, 1964), pp. 485–495.

** For our purposes, it is sufficient to think of probability in the following intuitive terms. Suppose than an experiment with observed outcome X is repeated very often. Then the probability $\Pr(a \le X \le b)$ is given (approximately) as the fraction of repetitions in which X was between a and b.

observe its duration, X. If we have no way of knowing how long a call might last, then X might be any nonnegative number. In this case it is convenient to say that X lies between 0 and ∞ and to take $A = 0$ and $B = \infty$. A similar situation arise in reliability studies where one measures the lifetime X of a transistor selected at random from a manufacturer's production line. Again, the possible values of X lie between 0 and ∞.

When we are dealing with experiments such as those described above, many questions can be reduced to calculating the probability that the outcome X lies between two specified numbers, say a and b. This probability, $\Pr(a \le X \le b)$, is a measure of the likelihood that an outcome of the experiment will lie between a and b. If the experiment is repeated many times, then the proportion of times X has a value between a and b will be close to $\Pr(a \le X \le b)$. In experiments of practical interest, it is often possible to find a density function $f(x)$ such that

$$\Pr(a \le X \le b) = \int_a^b f(x)\, dx, \tag{2}$$

for all a and b in the range of possible values of X.

Any function $f(x)$ with the following two properties is said to be a *(probability) density function*:

I. $f(x) \ge 0,\ A \le x \le B$;

II. $\displaystyle\int_A^B f(x)\, dx = 1.$

The additional condition (2) is what relates a density function to the outcome X of a specific experiment. Graphically, properties I and II mean that for x between A and B, the graph of $f(x)$ must lie above or on the x-axis and the area under the graph must equal 1. Property II simply says that there is probability 1 (certainty) that X has a value between A and B. Of course, if $B = \infty$, then the integral in property II is an improper integral.

EXAMPLE 1 Consider the cell population described earlier. Let $f(x) = 2ke^{-kx}$, where $k = (\ln 2)/3$. Show that $f(x)$ is indeed a probability density function on the interval from $x = 0$ to $x = 3$.

Solution Clearly, $f(x) \ge 0$, since the exponential function is never negative. Thus property I is satisfied. For property II, we check that

$$\begin{aligned}\int_0^3 f(x)\, dx &= \int_0^3 2ke^{-kx}\, dx = -2e^{-kx}\Big|_0^3 = -2e^{-k\cdot 3} + 2e^0\\ &= 2 - 2e^{-[(\ln 2)/3]3} = 2 - 2e^{-\ln 2}\\ &= 2 - 2\left(e^{\ln 2}\right)^{-1} = 2 - 2(2)^{-1} = 2 - 1 = 1.\end{aligned}$$

The simplest probability density function is that which assumes a constant value for $A \le x \le B$. In order for property II to hold, this constant value must be

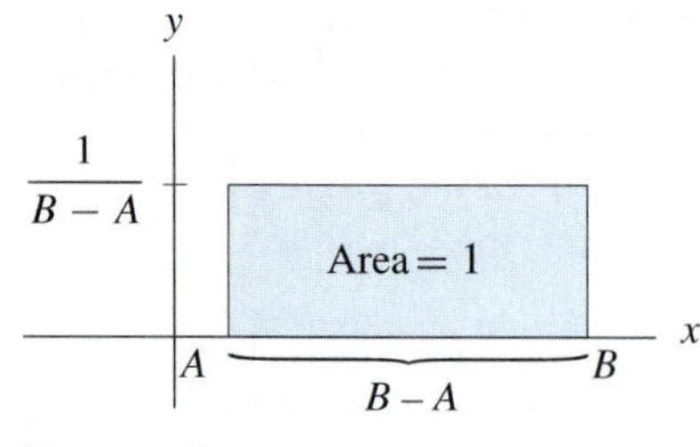

FIGURE 2 A uniform density function

$1/(B-A)$; that is

$$f(x) = \frac{1}{B-A}, \qquad A \le x \le B,$$

for in this case the area under the graph of $f(x)$ in Figure 2 is 1. Such a probability density function is said to be *uniform*. Experiments having uniform density functions generalize those experiments with a finite number of equally likely outcomes.

EXAMPLE 2 Suppose that a subway train leaves the station every 15 minutes. A person who arrives at a random time during the day must wait between 0 and 15 minutes for the next train. What are the chances the person must wait at least 10 minutes?

Solution Let X be the number of minutes the person must wait. The statement that the person arrives at a "random" time is usually interpreted to mean that X has a uniform probability density function. Since the possible values of X lie between 0 and 15, we take $f(x) = \frac{1}{15}$. Then the probability that the person waits at least 10 minutes is given by

$$\Pr(10 \le X \le 15) = \int_{10}^{15} \frac{1}{15}\,dx = \frac{1}{15}x\Big|_{10}^{15} = \frac{15}{15} - \frac{10}{15} = \frac{1}{3}.$$

(See Figure 3.)

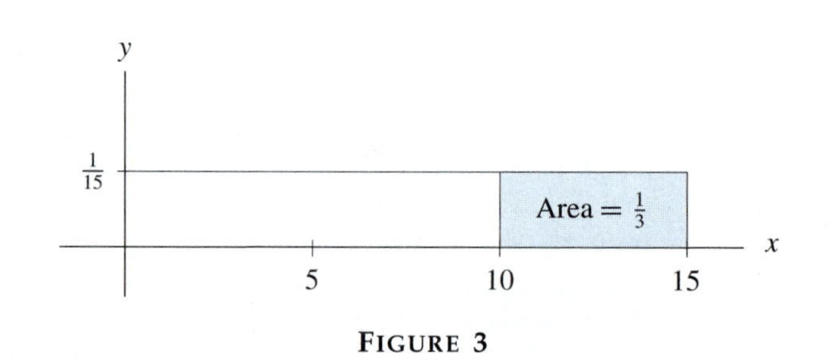

FIGURE 3

As we mentioned earlier, for some experiments the possible outcomes are the numbers between 0 and ∞. In such a case the probability density function $f(x)$ is defined for all $x \ge 0$, and property II is written as

$$\int_0^\infty f(x)\,dx = 1.$$

The most important function of this type has the form $f(x) = \lambda e^{-\lambda x}$, where λ is a positive constant. Just as in Example 1 of Section 7.7, one easily verifies that

$$\int_0^\infty \lambda e^{-\lambda x}\,dx = 1.$$

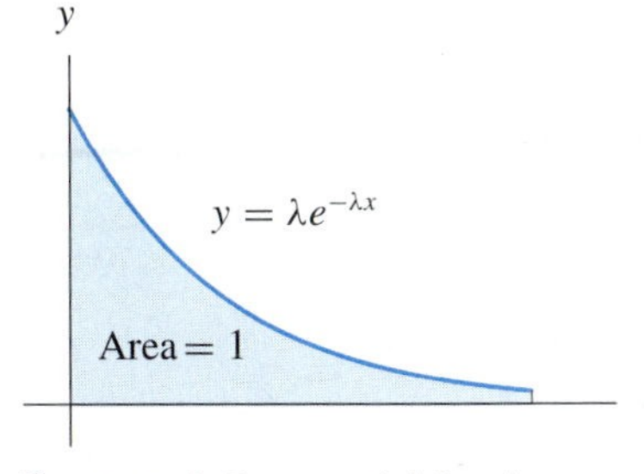

FIGURE 4 Exponential density function

(See Figure 4.) If the outcome X of an experiment has such a probability density function, the experiment is said to be *exponential* (or *exponentially distributed*). It can be shown that the constant λ may be interpreted as

$$\lambda = \frac{1}{a}, \quad \text{where } a = \text{average value of } X.$$

Typical uses of exponential probability density functions are given in the next two examples.

EXAMPLE 3 Experiment has shown that the lifetime of a light bulb is exponentially distributed. Let X be the lifetime of a light bulb selected at random from the production line of a light bulb manufacturer. For simplicity, let us measure the lifetime in years rather than hours and suppose that the average light bulb produced by the manufacturer burns out in $\frac{1}{4}$ year of continuous use.

(a) What proportion of the light bulbs will burn out within $\frac{1}{2}$ year?

(b) What proportion will continue to burn for at least 1 year?

Solution The average value of X is $\frac{1}{4}$, so we let $\lambda = 4$ and $f(x) = 4e^{-4x}$.

(a) The proportion of light bulbs where X is less than or equal to $\frac{1}{2}$ is

$$\begin{aligned}\Pr(0 \le X \le \tfrac{1}{2}) &= \int_0^{1/2} 4e^{-4x}\,dx = -e^{-4x}\Big|_0^{1/2} = -e^{-2} + 1 \\ &= 1 - .13534 = .86466.\end{aligned}$$

(b) The proportion of light bulbs that do not burn out for at least 1 year is

$$\Pr(1 \le X \le \infty) = \int_1^{\infty} 4e^{-4x}\,dx.$$

To evaluate this improper integral, we compute

$$\int_1^b 4e^{-4x}\,dx = -e^{-4x}\Big|_1^b = -e^{-4b} + e^{-4} \to e^{-4}$$

as $b \to \infty$. Hence $\Pr(1 \le X \le \infty) = e^{-4} = .01832$. ■

Note: The probability in Part (b) of Example 3 also can be calculated as $1 - \Pr(0) \le X \le 1)$. This calculation avoids the evaluation of an improper integral.

EXAMPLE 4 A company makes a survey of the duration of telephone calls made by its employees. It finds that the lengths of calls are exponentially distributed, with the average call lasting 5 minutes. What is the probability that a randomly chosen call will last between 5 and 10 minutes?

Solution Let X be the length of the call. Since the average value of X is 5, we take $\lambda = \frac{1}{5} = .2$ and $f(x) = .2e^{-.2x}$. The desired probability is

$$\begin{aligned}\Pr(5 \le X \le 10) &= \int_5^{10} .2e^{-.2x}\,dx = -e^{-.2x}\Big|_5^{10} \\ &= -e^{-2} + e^{-1} \\ &= -.13534 + .36788 = .23254.\end{aligned}$$

PRACTICE PROBLEMS 7.8

1. An experimenter determines that the probability density function for a certain experiment with outcomes between 0 and 1 is given by $f(x) = x$. Why might you doubt his conclusion?
2. An experiment with outcomes between 0 and 1 has probability density function $f(x) = 4x^3$. What is the probability of an outcome larger than $\frac{1}{2}$?

EXERCISES 7.8

1. An experiment has the probability density function $f(x) = 6(x - x^2)$ and outcomes lying between 0 and 1. Determine the probability that an outcome
   (a) lies between $\frac{1}{4}$ and $\frac{1}{2}$ (b) lies between 0 and $\frac{1}{3}$
   (c) is at least $\frac{1}{4}$ (d) is at most $\frac{3}{4}$
2. Suppose that the outcome X of an experiment lies between 0 and 4, and the probability density function for X is $f(x) = \frac{1}{8}x$. Find each value
   (a) $\Pr(X \le 1)$ (b) $\Pr(2 \le X \le 2.5)$ (c) $\Pr(3.5 \le X)$
3. If $f(x) = kx^2$, determine the value of k that makes $f(x)$ a probability density function on $0 \le x \le 2$.
4. If $f(x) = k/\sqrt{x}$, determine the value of k that makes $f(x)$ a probability density function on $1 \le x \le 4$.
5. Suppose that the outcomes X of an experiment lie between 0 and ∞, and x has an exponential density function $f(x) = 2e^{-2x}$. Find each value.
   (a) $\Pr(X \le .1)$ (b) $\Pr(.1 \le X \le .5)$
   (c) $\Pr(1 \le X)$ (d) the average value of X
6. Suppose that the outcomes X of an experiment are exponentially distributed, with density function $f(x) = .25e^{-.25x}$. Find each value.
   (a) $\Pr(1 \le X \le 2)$ (b) $\Pr(X \le 3)$
   (c) $\Pr(4 \le X)$ (d) the average value of X
7. An automated machine produces an automobile part every 3 minutes. An inspector arrives at a random time and must wait X minutes for a part.
   (a) Find the probability density function for X.
   (b) Find the probability that the inspector must wait at least 1 minute.
   (c) Find the probability that the inspector must wait no more than 1 minute.

8. The annual income of the households in a certain community ranges between \$5000 and \$25,000. Let X represent the annual income (in thousands of dollars) of a household chosen at random in this community, and suppose that the probability density function for X is $f(x) = kx$, $5 \le x \le 25$.

   (a) Find the value of k that makes $f(x)$ a density function.

   (b) Find the fraction of the households that have an annual income between \$5000 and \$10,000.

   (c) What fraction of the households have an income exceeding \$20,000?

9. Suppose that in a certain farming region, and in a certain year, the number X of bushels of wheat produced on a given acre has a probability density function $f(x) = (x - 30)/50$, $30 \le x \le 40$.

   (a) What is the probability that an acre selected at random produced less than 35 bushels of wheat?

   (b) If the farming region had 20,000 acres of wheat, how many acres produced less than 35 bushels of wheat?

10. The parent corporation for a franchised chain of fast-food restaurants claims that the fraction X of their new restaurants that make a profit during their first year of operation has the probability density function $f(x) = 12x^2 - 12x^3$, $0 \le x \le 1$.

    (a) What is the likelihood that less than 40% of the restaurants opened this year will make a profit during their first year of operation?

    (b) What is the likelihood that more than 50% of the restaurants will make a profit during their first year of operation?

11. Suppose that at a certain supermarket, the amount of time X one must wait at the express lane has the probability density function $f(x) = \frac{11}{10}(x + 1)^{-2}$, $0 \le x \le 10$. Find the probability of having to wait less than 4 minutes at the express lane.

12. Suppose that in a certain cell population, cells divide every 10 days and the age X of a cell selected at random has the probability density function $f(x) = 2ke^{-kx}$, $0 \le x \le 10$, where $k = (.1)\ln 2$.

    (a) Find the probability that a cell is at most 5 days old.

    (b) Upon examining a slide, a microbiologist finds that 10% of the cells are undergoing mitosis (a change in the cell leading to division). Compute the length of time required for mitosis; that is, find the number M such that

    $$\int_{10-M}^{10} 2ke^{-kx}\,dx = .10.$$

13. At a certain gas station, it takes an average of 4 minutes to get serviced. Suppose that the service time X for a car has an exponential probability density function.

    (a) What fraction of the cars are serviced within 2 minutes?

    (b) What is the probability that a car will have to wait at least 4 minutes?

14. The emergency flasher on an automobile is guaranteed for the first 12,000 miles that the car is driven. On the average, the flashers last about 50,000 miles. Let X be the time of failure of the flasher (measured in thousands of miles), and suppose X has an exponential probability density function. What percentage of the emergency flashers will have to be replaced during the warranty period?

15. Let X be the number of seconds between successive cars at a toll booth on the Ohio Turnpike on a typical Saturday afternoon. It can be shown that X has an exponential density function. If the average interarrival time is 2 seconds, find the probability that X is at least 3 seconds.

16. Let X be the relief time (in seconds) of an arthritic patient who has taken an analgesic for pain. Suppose that a certain analgesic provides relief within 4 minutes for 75% of a large group of patients, and suppose the density function for X is $f(x) = ke^{-kx}$. (This model has been used by some medical researchers.) Then one estimates that $\Pr(X \le 4) = .75$. Use this estimate to find an approximate value for k. [*Hint*: First show that $\Pr(X \le 4) = 1 - e^{-4k}$.]

*Upon studying the vacancies occurring in the U.S. Supreme Court, it has been determined that the time elapsed between successive resignations is exponentially distributed with average value 2 years.**

17. A new president takes office at the same time a justice retires. Find the probability that the next vacancy will take place during his 4-year term.

18. Find the probability that the composition of the U.S. Supreme Court will remain unchanged for a period of 5 years or more.

19. Consider a group of patients that have been treated for an acute disease such as cancer, and let X be the number of years a person lives after receiving the treatment (the "survival time"). Under suitable conditions, the density function for X will be $f(x) = ke^{-kx}$ for some constant k. The *survival function* $S(x)$ is the probability that a person chosen at random from the group of patients survives until at least time x. Suppose that the probability is .90 that a patient will survive at least 5 years [i.e., $S(5) = .90$]. Find the constant k in the exponential density function $f(x)$.

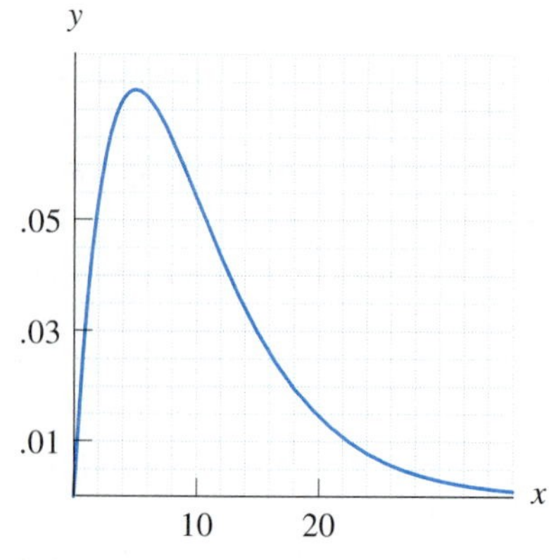

FIGURE 5 Probability density function for Exercises 20–23

Exercises 20–23 refer to the probability density function in Figure 5, where $0 \le X < \infty$. [*Note*: $\Pr(X \le 2) \approx .06$.]

20. Estimate $\Pr(14 \le X \le 16)$.

21. Estimate $\Pr(16 \le X \le 20)$.

22. Estimate $\Pr(4 \le X)$.

23. Estimate $\Pr(6 \le X)$.

SOLUTIONS TO PRACTICE PROBLEMS 7.8

1. Property II is not satisfied: $\int_0^1 f(x)\,dx = \int_0^1 x\,dx = \frac{1}{2}x^2\Big|_0^1 = \frac{1}{2}$. Property II says that this integral should be 1.

2. $\int_{1/2}^1 4x^3\,dx = x^4\Big|_{1/2}^1 = 1 - \frac{1}{16} = \frac{15}{16}$.

* See W. A. Wallis, "The Poisson Distribution and the Supreme Court," *Journal of the American Statistical Association*, **31** (1936), 376–380.

REVIEW OF THE FUNDAMENTAL CONCEPTS OF CHAPTER 7

1. What is a Riemann sum?
2. Give an interpretation of the area under a rate of change function. Give a concrete example.
3. What is a definite integral?
4. What is the difference between a definite integral and an indefinite integral?
5. Outline a procedure for finding the area of a region bounded by two curves.
6. State the formula for each of the following quantities.
   (a) average value of a function
   (b) consumers' surplus
   (c) future value of an income stream
   (d) volume of a solid of revolution
7. State the fundamental theorem of calculus.
8. Describe integration by substitution in your own words.
9. Describe integration by parts in your own words.
10. What is an improper integral and how is it evaluated?
11. What is a probability density function and how is it used?
12. What is an exponential density function? Give an example.

Chapter 8

Functions of Several Variables

Until now, most of our applications of calculus have involved functions of one variable. In real life, however, a quantity of interest often depends on more than one variable. For instance, the sales level of a product may depend not only on its price but also on the prices of competing products, the amount spent on advertising, and perhaps the time of year. The total cost of manufacturing the product depends on the cost of raw materials, labor, plant maintenance, and so on.

This chapter introduces the basic ideas of calculus for functions of more than one variable. Section 8.1 presents several examples that will be used throughout the chapter. Derivatives are treated in Section 8.2 and then used in Sections 8.3 and 8.4 to solve optimization problems more general than those in Chapter 3.

8.1 Examples of Functions of Several Variables

A function $f(x, y)$ of the two variables x and y is a rule that assigns a number to each of values for the variables; for instance,

$$f(x, y) = e^x(x^2 + 2y).$$

An example of a function of three variables is

$$f(x, y, z) = 5xy^2z.$$

Example 1 A store sells butter at \$2.50 per pound and margarine at \$1.40

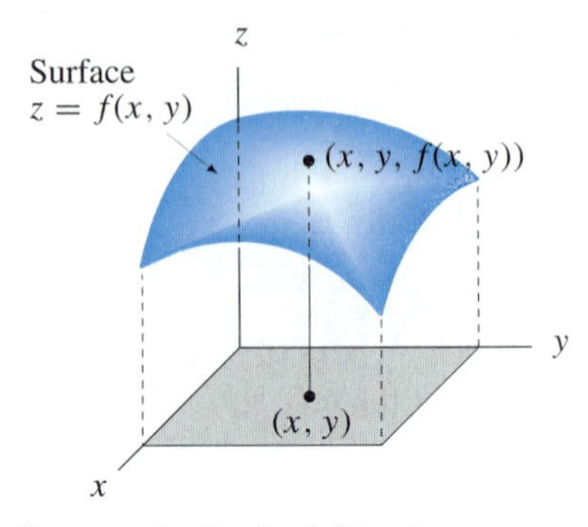

FIGURE 1 Graph of $f(x, y)$

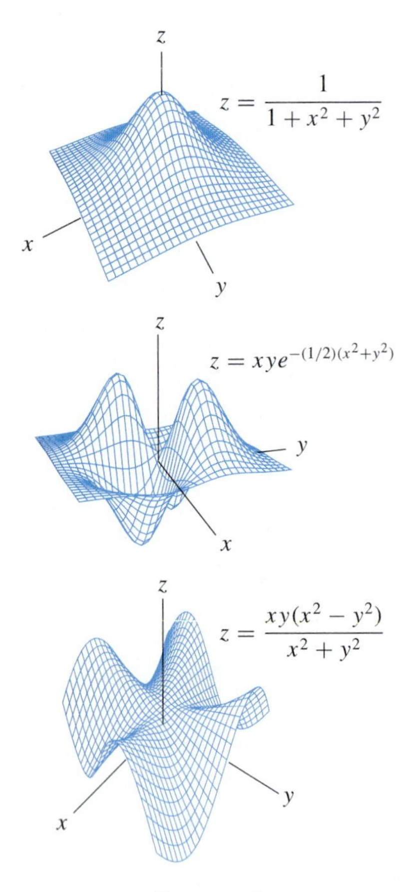

FIGURE 2

per pound. The revenue from the sale of x pounds of butter and y pounds of margarine is given by the function

$$f(x, y) = 2.50x + 1.40y.$$

Determine and interpret $f(200, 300)$.

Solution $f(200, 300) = 2.50(200) + 1.40(300) = 500 + 420 = 920$. The revenue from the sale of 200 pounds of butter and 300 pounds of margarine is \$920.

A function $f(x, y)$ of two variables may be graphed in a manner analogous to that for functions of one variable. It is necessary to use a three-dimensional coordinate system, where each point is identified by three coordinates (x, y, z). For each choice of x, y, the graph of $f(x, y)$ includes the point $(x, y, f(x, y))$. This graph is a surface in three-dimensional space, with equation $z = f(x, y)$. See Figure 1. Three graphs of specific functions are shown in Figure 2.

We do not expect the reader to make three-dimensional drawings, but we will use selected figures to provide geometric descriptions of key concepts.

The next two examples recur throughout the chapter to illustrate various concepts about functions of several variables.

APPLICATION TO ARCHITECTURAL DESIGN When designing a building, it is necessary to know, at least approximately, how much heat the building loses per day. The heat loss affects many aspects of the design, such as the size of the heating plant, the size and location of duct work, and so on. A building loses heat through its sides, roof, and floor. How much heat is lost will generally differ for each face of the building and will depend on such factors as insulation, materials used in construction, exposure (north, south, east, or west), and climate. It is possible to estimate how much heat is lost per square foot of each face. Using this data, one can construct a heat-loss function as in the following example.

EXAMPLE 2 A rectangular industrial building of dimensions x, y, and z is shown in Figure 3(a). In Figure 3(b) we give the amount of heat lost per day by each side of the building, measured in suitable units of heat per square foot. Let $f(x, y, z)$ be the total daily heat loss for such a building.

(a) Find a formula for $f(x, y, z)$.

(b) Find the total daily heat loss if the building has length 100 feet, width 70 feet, and height 50 feet.

Solution (a) The total heat loss is the sum of the amount of heat loss through each side of the building. The heat loss through the roof is

$$[\text{heat loss per square foot of roof}] \cdot [\text{area of roof in square feet}] = 10xy.$$

Similarly, the heat loss through the east side is $8yz$. Continuing in this way, we see that the total daily heat loss is

$$f(x, y, z) = 10xy + 8yz + 6yz + 10xz + 5xz + 1 \cdot xy.$$

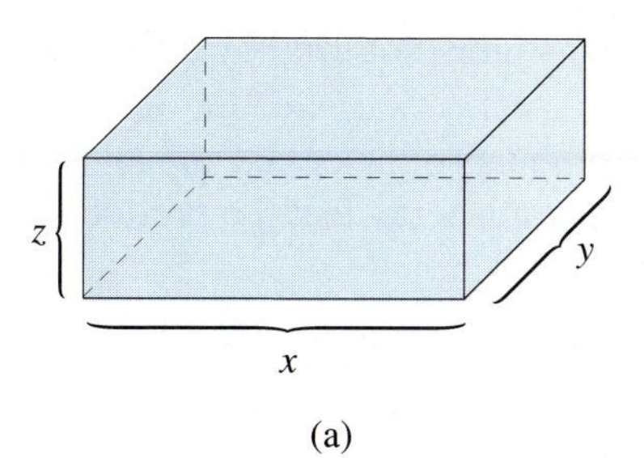

(a)

	Roof	East side	West side	North side	South side	Floor
Heat loss (per sq ft)	10	8	6	10	5	1
Area (sq ft)	xy	yz	yz	xz	xz	xy

(b)

FIGURE 3 Heat loss from an industrial building

We collect terms to obtain

$$f(x, y, z) = 11xy + 14yz + 15xz.$$

The amount of heat loss when $x = 100$, $y = 70$, and $z = 50$ is given by $f(100, 70, 50)$, which equals

$$\begin{aligned} f(100, 70, 50) &= 11(100)(70) + 14(70)(50) + 15(100)(50) \\ &= 77{,}000 + 49{,}000 + 75{,}000 = 201{,}000. \end{aligned}$$

■

In Section 8.3 we will determine the dimensions x, y, z that minimize the heat loss for a building of specified volume.

PRODUCTION FUNCTIONS IN ECONOMICS The costs of a manufacturing process can generally be classified as one of two types: cost of labor and cost of capital. The meaning of the cost of labor is clear. By the cost of capital we mean the cost of buildings, tools, machines, and similar items used in the production process. A manufacturer usually has some control over the relative portions of labor and capital utilized in his production process. He can completely automate production so that labor is at a minimum, or he can utilize mostly labor and little capital. Suppose that x units of labor and y units of capital are used.* Let $f(x, y)$ denote the number of units of finished product that are manufactured. Economists have found that $f(x, y)$ is often a function of the form

$$f(x, y) = Cx^A y^{1-A},$$

where A and C are constants, $0 < A < 1$. Such a function is called a *Cobb-Douglas production function*. An example is $f(x, y) = 60x^{3/4}y^{1/4}$.

EXAMPLE 3 (*Production in a firm*) Suppose that during a certain time period the number of units of goods produced when utilizing x units of labor and y units of capital is $f(x, y) = 60x^{3/4}y^{1/4}$.

---

* Economists normally use L and K, respectively, for labor and capital. However, for simplicity, we use x and y.

(a) How many units of goods will be produced by using 81 units of labor and 16 units of capital?

(b) Show that whenever the amounts of labor and capital being used are doubled, so is the production. (Economists say that the production function has "constant returns to scale.")

Solution (a) $f(81, 16) = 60(81)^{3/4} \cdot (16)^{1/4} = 60 \cdot 27 \cdot 2 = 3240$. There will be 3240 units of goods produced.

(b) Utilization of a units of labor and b units of capital results in the production of $f(a, b) = 60a^{3/4}b^{1/4}$ units of goods. Utilizing $2a$ and $2b$ units of labor and capital, respectively, results in $f(2a, 2b)$ units produced. Set $x = 2a$ and $y = 2b$. Then we see that

$$\begin{aligned} f(2a, 2b) &= 60(2a)^{3/4}(2b)^{1/4} \\ &= 60 \cdot 2^{3/4} \cdot a^{3/4} \cdot 2^{1/4} \cdot b^{1/4} \\ &= 60 \cdot 2^{(3/4+1/4)} \cdot a^{3/4}b^{1/4} \\ &= 2^1 \cdot 60a^{3/4}b^{1/4} \\ &= 2f(a, b). \end{aligned}$$

■

LEVEL CURVES It is possible graphically to depict a function $f(x, y)$ of two variables using a family of curves called level curves. Let c be any number. Then the graph of the equation $f(x, y) = c$ is a curve in the xy-plane called the *level curve of height c*. This curve describes all points of height c on the graph of the function $f(x, y)$. As c varies, we have a family of level curves indicating the sets of points on which $f(x, y)$ assumes various values c. In Figure 4, we have drawn the graph and various level curves for the function $f(x, y) = x^2 + y^2$.

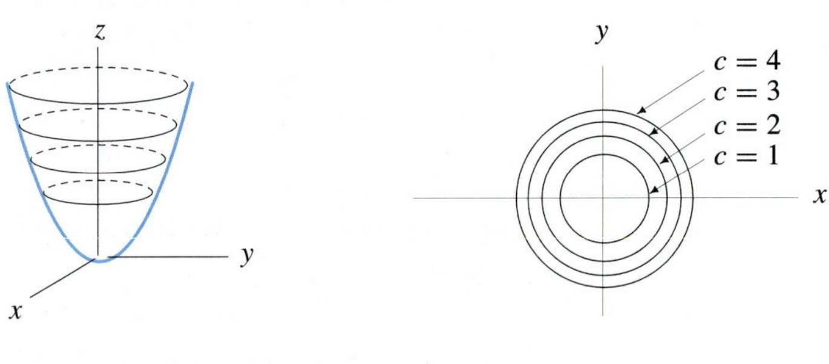

FIGURE 4 Level curves

Level curves often have interesting physical interpretations. For example, surveyors draw *topographic maps* that use level curves to represent points having equal altitude. Here $f(x, y) =$ the altitude at point (x, y). Figure 5(a) shows the graph of $f(x, y)$ for a typical hilly region. Figure 5(b) shows the level curves corresponding to various altitudes.

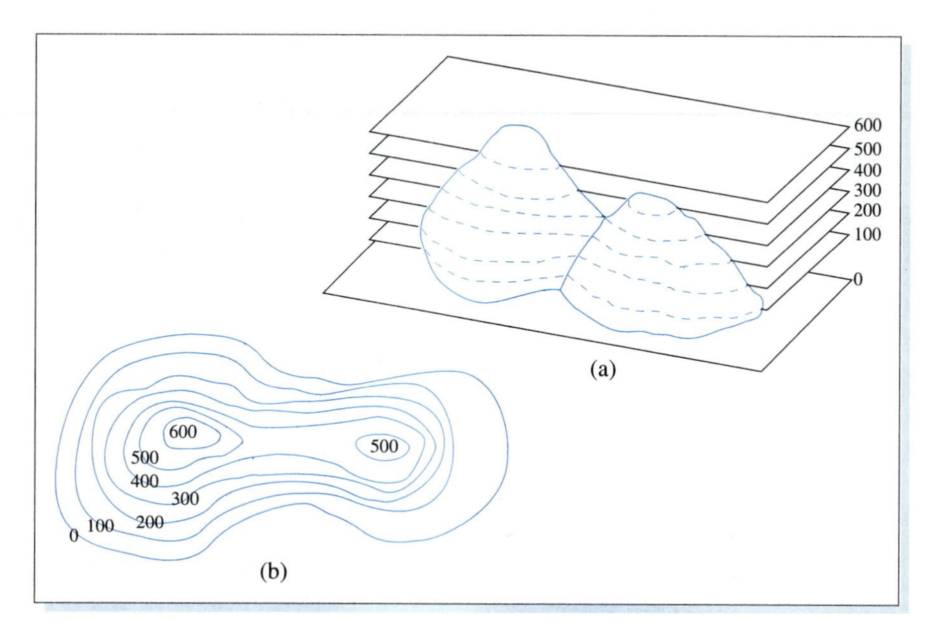

FIGURE 5 Level curves

EXAMPLE 4 Determine the level curve at height 600 for the production function $f(x, y) = 60x^{3/4}y^{1/4}$ of Example 3.

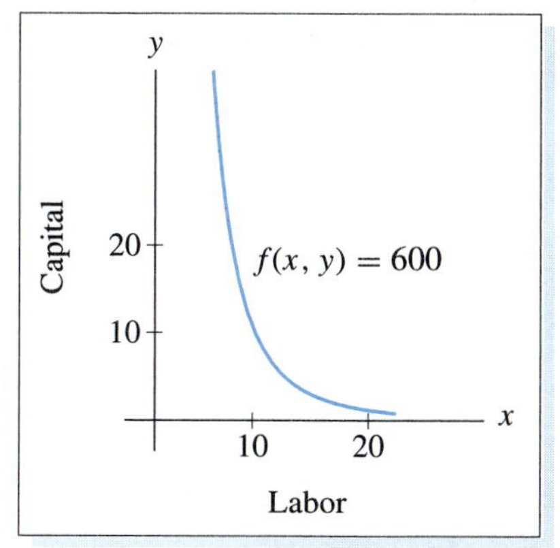

FIGURE 6 Isoquant of a production function

Solution The level curve is the graph of $f(x, y) = 600$, or

$$60x^{3/4}y^{1/4} = 600$$
$$y^{1/4} = \frac{10}{x^{3/4}}$$
$$y = \frac{10{,}000}{x^3}.$$

Of course, since x and y represent quantities of labor and capital, they must both be positive. We have sketched the graph of the level curve in Figure 6. The points on the curve are precisely those combinations of capital and labor that yield 600 units of production. Since all points on the curve yield the same amount of production, the level curve is also called an *isoquant* ("iso" means "equal," "quant" means "amount").

EVALUATING A FUNCTION OF SEVERAL VARIABLES WITH A TI CALCULATOR

To be specific, let's evaluate $f(x, y) = 60x^{3/4}y^{1/4}$ at the point (81, 16).

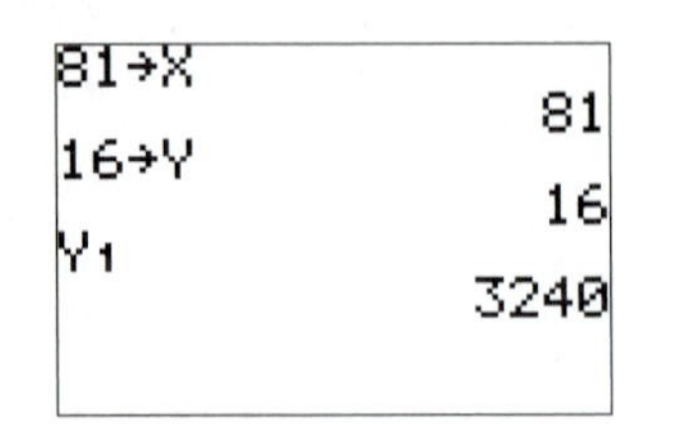

1. Set `Y1=60X^(3/4)Y^(1/4)`. (*Note*: To obtain `Y` press [ALPHA] **Y**.)
2. Press [2nd] [QUIT] to invoke the Home screen.
3. Assign the value 81 to the variable X with [STO ▶].
4. Assign the value 16 to the variable Y with [STO ▶].
5. Place `Y1` on the screen and press [ENTER] to display the value of $f(81, 16)$. See screen at left.

For a discussion of evaluating a function of two variables with Visual Calculus, see Appendix A, Part XI.

PRACTICE PROBLEMS 8.1

1. Let $f(x, y, z) = x^2 + y/(x - z) - 4$. Compute $f(3, 5, 2)$.
2. Suppose that in a certain country the daily demand for coffee is given by $f(p_1, p_2) = 16p_1/p_2$ thousand pounds, where p_1 and p_2 are the respective prices of tea and coffee per pound. Compute and interpret $f(3, 4)$.

EXERCISES 8.1

1. Let $f(x, y) = x^2 + 8y$. Compute $f(1, 0)$, $f(0, 1)$, and $f(3, 2)$.
2. Let $g(x, y) = 3xe^y$. Compute $g(2, 1)$, $g(1, 0)$, and $g(0, 0)$.
3. Let $f(L, K) = 3\sqrt{LK}$. Compute $f(0, 1)$, $f(3, 12)$, and $f(a, b)$.
4. Let $f(p, q) = pe^{q/p}$. Compute $f(1, 0)$, $f(3, 12)$, and $f(a, b)$.
5. Let $f(x, y, z) = x/(y - z)$. Compute $f(2, 3, 4)$ and $f(7, 46, 44)$.
6. Let $f(x, y, z) = x^2e^{\sqrt{y^2+z^2}}$. Compute $f(1, 0, 1)$ and $f(5, 2, 3)$.
7. Let $f(x, y) = xy$. Show that $f(2 + h, 3) - f(2, 3) = 3h$.
8. Let $f(x, y) = xy$. Show that $f(2, 3 + K) - f(2, 3) = 2K$.
9. Let
$$f(x, y) = \frac{x^2 + 3xy + 3y^2}{x + y}.$$
Show that $f(2a, 2b) = 2f(a, b)$.
10. Let $f(x, y) = 75x^Ay^{1-A}$, where $0 < A < 1$. Show that $f(2a, 2b) = 2f(a, b)$.
11. The present value of A dollars to be paid t years in the future (assuming a 5% continuous interest rate) is $P(A, t) = Ae^{-.05t}$. Find and interpret $P(100, 13.8)$.
12. Refer to Example 3. Suppose that labor costs \$100 per unit and capital costs \$200 per unit. Express as a function of two variables, $C(x, y)$, the cost of utilizing x units of labor and y units of capital.
13. The value of residential property for tax purposes is usually much lower than its actual market value. If v is the market value, then the *assessed value* for

real estate taxes might be only 40% of v. Suppose the property tax, T, in a community is given by the function

$$T = f(r, v, x) = \frac{r}{100}(.40v - x),$$

where v is the estimated market value of a property (in dollars), x is a *homeowner's exemption* (a number of dollars depending on the type of property), and r is the tax rate (stated in dollars per hundred dollars) of net assessed value.

(a) Determine the real estate tax on a property valued at \$200,000 with a homeowner's exemption of \$5000, assuming a tax rate of \$2.50 per hundred dollars of net assessed value.

(b) Determine the tax due if the tax rate increases by 20% to \$3.00 per hundred dollars of net assessed value. Assume the same property value and homeowner's exemption. Does the tax due also increase by 20%? Explain your reasoning in a complete sentence.

14. Let $f(r, v, x)$ be the real estate tax function of Exercise 13.

(a) Determine the real estate tax on a property valued at \$100,000 with a homeowner's exemption of \$5000, assuming a tax rate of \$2.20 per hundred dollars of net assessed value.

(b) Determine the real estate tax when the market value rises 20% to \$120,000. Assume the same homeowner's exemption and a tax rate of \$2.20 per hundred dollars of net assessed valued. Does the tax due also increase by 20%?

Draw the level curves of heights 0, 1, and 2 for the functions in Exercises 15 and 16.

15. $f(x, y) = 2x + y$

16. $f(x, y) = -x^2 + y$

17. Draw the level curve of the function $f(x, y) = x - y$ containing the point $(5, 3)$.

18. Draw the level curve of the function $f(x, y) = xy$ containing the point $(2, \frac{1}{2})$.

19. Find a function $f(x, y)$ that has the line $y = 3x - 4$ as a level curve.

20. Find a function $f(x, y)$ that has the curve $y = 3/x^2$ as a level curve.

21. Suppose that a topographic map is viewed as the graph of a certain function $f(x, y)$. What are the level curves?

22. A certain production process uses labor and capital. If the quantities of these commodities are x and y, respectively, then the total cost is $100x + 200y$ dollars. Draw the level curves of height 600, 800, and 1000 for this function. Explain the significance of these curves. (Economists frequently refer to these lines as *budget lines* or *isocost lines*.)

Match the graphs of the functions in Exercises 23–26 to the systems of level curves shown in Figures 7(a)–(d).

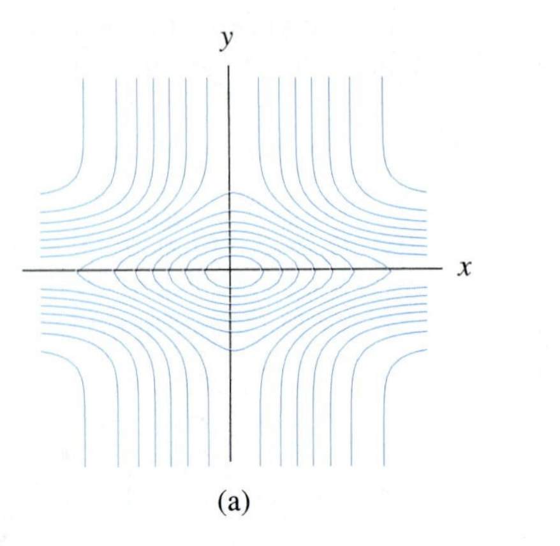

(a)

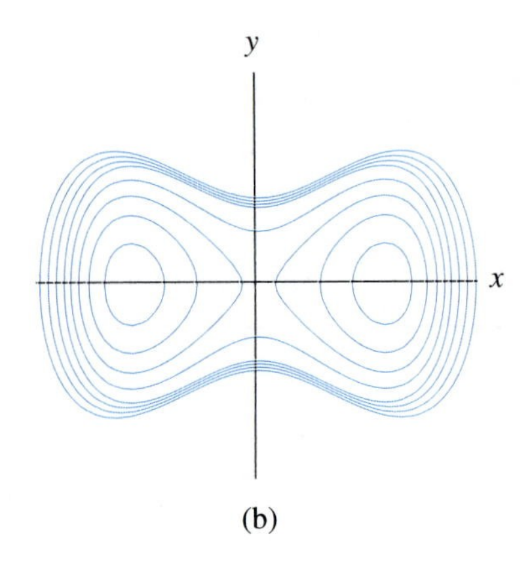

(b)

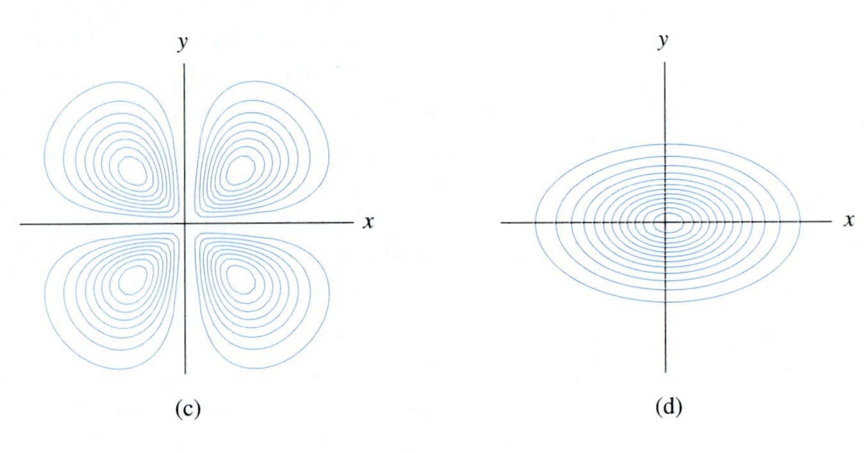

(c)

(d)

FIGURE 7

23.

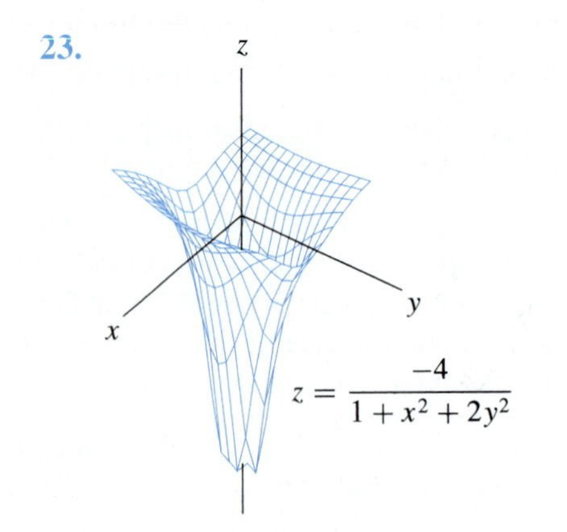

24.

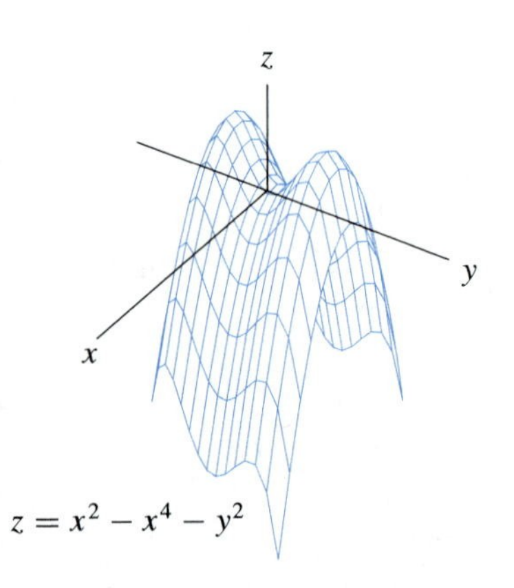

25. **26.**

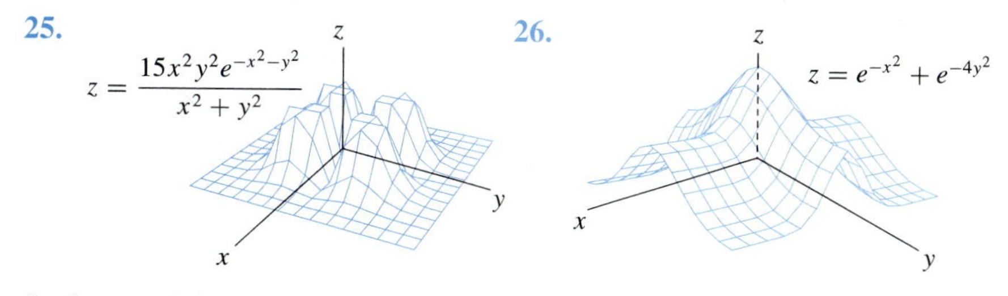

In the remaining exercises, use a graphing utility to evaluate the function.

27. Example 3 determined that $f(81, 16) = 3240$, where $f(x, y) = 60x^{3/4}y^{1/4}$. Show that $f(162, 32) = 2(3240)$ as predicted by part (b) of the example. Also show that $f(243, 48) = 3(3240)$.

28. Let $f(x, y) = xe^{xy}$. Evaluate $f(2, .6)$.

SOLUTIONS TO PRACTICE PROBLEMS 8.1

1. Substitute 3 for x, 5 for y, and 2 for z.

$$f(3, 5, 2) = 3^2 + \frac{5}{3-2} - 4 = 10.$$

2. To compute $f(3, 4)$, substitute 3 for p_1 and 4 for p_2 into $f(p_1, p_2) = 16p_1/p_2$. Thus,

$$f(3, 4) = 16 \cdot \frac{3}{4} = 12.$$

Therefore, if the price of tea is \$3 per pound and the price of coffee is \$4 per pound, then 12,000 pounds of coffee will be sold each day. (Notice that as the price of coffee increases, the demand decreases.)

8.2 PARTIAL DERIVATIVES

In Chapter 2 we introduced the notion of a derivative to measure the rate at which a function $f(x)$ is changing with respect to changes in the variable x. Let us now study the analogue of the derivative for functions of two (or more) variables.

Let $f(x, y)$ be a function of the two variables x and y. Since we want to know how $f(x, y)$ changes with respect to changes in the variable x and changes in the variable y, we shall define two derivatives of $f(x, y)$ (to be called "partial derivatives"), one with respect to each of the variables. The *partial derivative of $f(x, y)$ with respect to x*, written $\frac{\partial f}{\partial x}$, is the derivative of $f(x, y)$, where y is treated as a constant and $f(x, y)$ is considered as a function of x alone. The *partial derivative of $f(x, y)$ with respect to y*, written $\frac{\partial f}{\partial y}$, is the derivative of $f(x, y)$, where x is treated as a constant.

EXAMPLE 1 Let $f(x, y) = 5x^3y^2$. Compute $\frac{\partial f}{\partial x}$ and $\frac{\partial f}{\partial y}$.

Solution To compute $\frac{\partial f}{\partial x}$, we think of $f(x, y)$ written as

$$f(x, y) = [5y^2]x^3,$$

where the brackets emphasize that $5y^2$ is to be treated as a constant. Therefore, when differentiating with respect to x, $f(x, y)$ is just a constant times x^3. Recall that if k is any constant, then

$$\frac{d}{dx}(kx^3) = 3 \cdot k \cdot x^2.$$

Thus

$$\frac{\partial f}{\partial x} = 3 \cdot [5y^2] \cdot x^2 = 15x^2y^2.$$

After some practice, it is unnecessary to place the y^2 in front of the x^3 before differentiating.

Now, in order to compute $\frac{\partial f}{\partial y}$, we think of

$$f(x, y) = [5x^3]y^2.$$

When differentiating with respect to y, $f(x, y)$ is simply a constant (namely, $5x^3$) times y^2. Hence

$$\frac{\partial f}{\partial y} = 2 \cdot [5x^3] \cdot y = 10x^3y.$$ ■

EXAMPLE 2 Let $f(x, y) = 3x^2 + 2xy + 5y$. Compute $\frac{\partial f}{\partial x}$ and $\frac{\partial f}{\partial y}$.

Solution To compute $\frac{\partial f}{\partial x}$, we think of

$$f(x, y) = 3x^2 + [2y]x + [5y].$$

Now we differentiate $f(x, y)$ as if it were a quadratic polynomial in x:

$$\frac{\partial f}{\partial x} = 6x + [2y] + 0 = 6x + 2y.$$

Note that $5y$ is treated as a constant when differentiating with respect to x, so the partial derivative of $5y$ with respect to x is zero.

To compute $\frac{\partial f}{\partial y}$, we think of

$$f(x, y) = [3x^2] + [2x]y + 5y.$$

Then

$$\frac{\partial f}{\partial y} = 0 + [2x] + 5 = 2x + 5.$$

Note that $3x^2$ is treated as a constant when differentiating with respect to y, so the partial derivative of $3x^2$ with respect to y is zero. ■

EXAMPLE 3 Compute $\frac{\partial f}{\partial x}$ and $\frac{\partial f}{\partial y}$ for each of the following:

(a) $f(x, y) = (4x + 3y - 5)^8$

(b) $f(x, y) = e^{xy^2}$

(c) $f(x, y) = y/(x + 3y)$

Solution (a) To compute $\frac{\partial f}{\partial x}$, we think of

$$f(x, y) = (4x + [3y - 5])^8.$$

By the general power rule,

$$\frac{\partial f}{\partial x} = 8 \cdot (4x + [3y - 5])^7 \cdot 4 = 32(4x + 3y - 5)^7.$$

Here we used the fact that the derivative of $4x + 3y - 5$ with respect to x is just 4.

To compute $\frac{\partial f}{\partial y}$, we think of

$$f(x, y) = ([4x] + 3y - 5)^8.$$

Then

$$\frac{\partial f}{\partial y} = 8 \cdot ([4x] + 3y - 5)^7 \cdot 3 = 24(4x + 3y - 5)^7.$$

(b) To compute $\frac{\partial f}{\partial x}$, we observe that

$$f(x, y) = e^{x[y^2]},$$

so that

$$\frac{\partial f}{\partial x} = [y^2]e^{x[y^2]} = y^2 e^{xy^2}.$$

To compute $\frac{\partial f}{\partial y}$, we think of

$$f(x, y) = e^{[x]y^2}.$$

Thus

$$\frac{\partial f}{\partial y} = e^{[x]y^2} \cdot 2[x]y = 2xye^{xy^2}.$$

(c) To compute $\frac{\partial f}{\partial x}$, we use the general power rule to differentiate $[y](x+[3y])^{-1}$ with respect to x:

$$\frac{\partial f}{\partial x} = (-1) \cdot [y](x + [3y])^{-2} \cdot 1 = -\frac{y}{(x + 3y)^2}.$$

To compute $\frac{\partial f}{\partial y}$, we use the quotient rule to differentiate

$$f(x, y) = \frac{y}{[x] + 3y}$$

with respect to y. We find that

$$\frac{\partial f}{\partial y} = \frac{([x] + 3y) \cdot 1 - y \cdot 3}{([x] + 3y)^2} = \frac{x}{(x + 3y)^2}.$$

The use of brackets to highlight constants is helpful initially in order to compute partial derivatives. From now on we shall merely form a mental picture of those terms to be treated as constants and dispense with brackets. ■

A partial derivative of a function of several variables is also a function of several variables and hence can be evaluated at specific values of the variables. We write

$$\frac{\partial f}{\partial x}(a, b)$$

for $\frac{\partial f}{\partial x}$ evaluated at $x = a$, $y = b$. Similarly,

$$\frac{\partial f}{\partial y}(a, b)$$

denotes the function $\frac{\partial f}{\partial y}$ evaluated at $x = a$, $y = b$.

EXAMPLE 4 Let $f(x, y) = 3x^2 + 2xy + 5y$.

(a) Calculate $\frac{\partial f}{\partial x}(1, 4)$. (b) Evaluate $\frac{\partial f}{\partial y}$ at $(x, y) = (1, 4)$.

Solution (a) $\frac{\partial f}{\partial x} = 6x + 2y$, $\frac{\partial f}{\partial x}(1, 4) = 6 \cdot 1 + 2 \cdot 4 = 14$.

(b) $\frac{\partial f}{\partial y} = 2x + 5$, $\frac{\partial f}{\partial y}(1, 4) = 2 \cdot 1 + 5 = 7$. ■

EVALUATING PARTIAL DERIVATIVES WITH A TI CALCULATOR

To be specific, let's evaluate the partial derivatives of Example 4.

1. Set Y_1=3X^3+2XY+5Y. (*Notes*: To obtain Y, press ALPHA [Y]. With a TI-85, set y1=3x^2+2x * y+5y.)

2. Set Y_2=nDeriv(Y_1,X,X) or y2=der1(y1,x,x). Y_2 will be $\frac{\partial f}{\partial x}$.

3. Set Y_3=nDeriv(Y_1,Y,Y) or y3=der1(y1,y,y).

4. Press [2nd][QUIT] to invoke the Home screen.

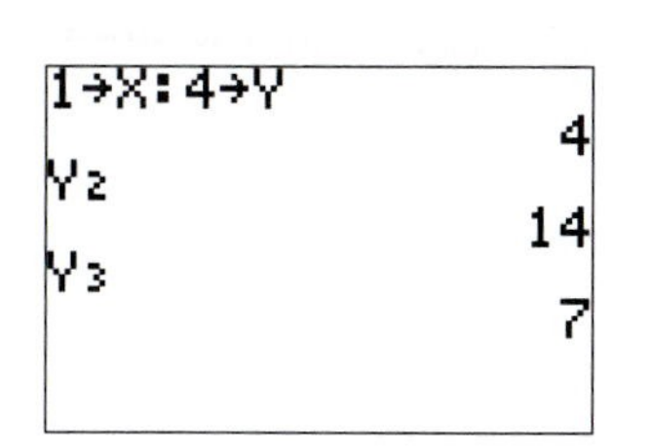

5. Assign the value 1 to the variable X with [STO ▶].
6. Assign the value 4 to the variable Y with [STO ▶].
7. Place Y_2 on the screen and press [ENTER] to display the value of $\frac{\partial f}{\partial x}(1, 4)$.
8. Place Y_3 on the screen and press [ENTER] to display the value of $\frac{\partial f}{\partial y}(1, 4)$. See figure at left.

With Visual Calculus, when a table of values for a function of two variables is formed, the last two columns contain values of $\frac{\partial f}{\partial x}$ and $\frac{\partial f}{\partial y}$ for various values of x and y.

GEOMETRIC INTERPRETATION OF PARTIAL DERIVATIVES Consider the three-dimensional surface $z = f(x, y)$ in Figure 1. If y is held constant at b and x is allowed to vary, the equation

$$z = f(x, b) \quad \text{(} b \text{ constant)}$$

describes a curve on the surface. (The curve is formed by cutting the surface $z = f(x, y)$ with a vertical plane parallel to the xz-plane.) The value of $\frac{\partial f}{\partial x}(a, b)$ is the slope of the tangent line to the curve at the point where $x = a$ and $y = b$.

Likewise, if x is held constant at a and y is a allowed to vary, the equation

$$z = f(a, y) \quad \text{(} a \text{ constant)}$$

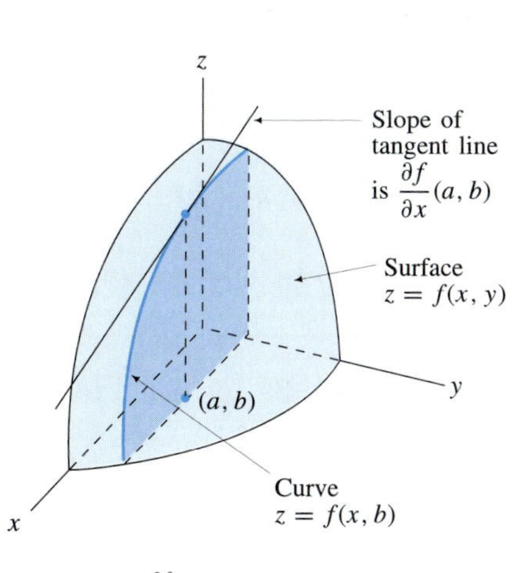

FIGURE 1 $\frac{\partial f}{\partial x}$ gives the slope of a curve formed by holding y constant

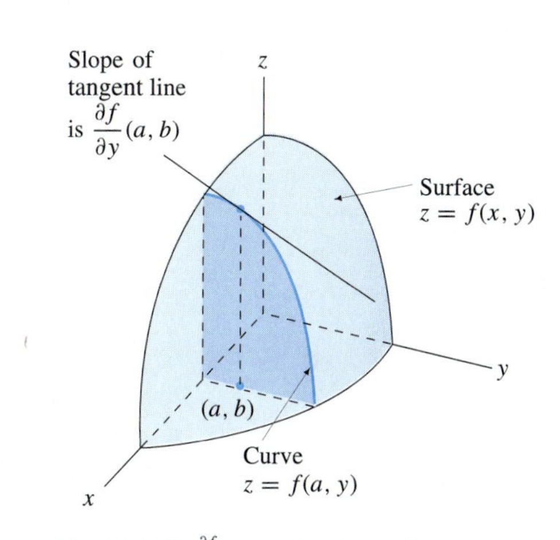

FIGURE 2 $\frac{\partial f}{\partial y}$ gives the slope of a curve formed by holding x constant

describes the curve on the surface $z = f(x, y)$ shown in Figure 2. The value of the partial derivative $\frac{\partial f}{\partial y}(a, b)$ is the slope of this curve at the point where $x = a$ and $y = b$.

EXAMPLE 5 Interpret the partial derivatives of $f(x, y) = 3x^2 + 2xy + 5y$ calculated in Example 4.

Solution We showed in Example 4 that

$$\frac{\partial f}{\partial x}(1, 4) = 14, \quad \frac{\partial f}{\partial y}(1, 4) = 7.$$

The fact that $\frac{\partial f}{\partial x}(1, 4) = 14$ means that if y is kept constant at 4 and x is allowed to vary near 1, then $f(x, y)$ changes at a rate of 14 times the change in x. That is, if x increases by one small unit, then $f(x, y)$ increases by approximately 14 units. If x increases by h units (where h is small), then $f(x, y)$ increases by approximately $14 \cdot h$ units. That is,

$$f(1 + h, 4) - f(1, 4) \approx 14 \cdot h.$$

Similarly, the fact that $\frac{\partial f}{\partial y}(1, 4) = 7$ means that if we keep x constant at 1 and let y vary near 4, then $f(x, y)$ changes at a rate equal to seven times the change in y. So for a small value of k, we have

$$f(1, 4 + k) - f(1, 4) \approx 7 \cdot k.$$

■

We can generalize the interpretations of $\frac{\partial f}{\partial x}$ and $\frac{\partial f}{\partial y}$ given in Example 5 to yield the following general fact.

> Let $f(x, y)$ be a function of two variables. Then if h and k are small, we have
>
> $$f(a + h, b) - f(a, b) \approx \frac{\partial f}{\partial x}(a, b) \cdot h,$$
>
> $$f(a, b + k) - f(a, b) \approx \frac{\partial f}{\partial y}(a, b) \cdot k.$$

Partial derivatives can be computed for functions of any number of variables. When taking the partial derivative with respect to one variable, we treat the other variables as constant.

EXAMPLE 6 Let $f(x, y, z) = x^2yz - 3z$.

(a) Compute $\frac{\partial f}{\partial x}$, $\frac{\partial f}{\partial y}$, and $\frac{\partial f}{\partial z}$. (b) Calculate $\frac{\partial f}{\partial z}(2, 3, 1)$.

Solution (a) $\frac{\partial f}{\partial x} = 2xyz$, $\frac{\partial f}{\partial y} = x^2z$, $\frac{\partial f}{\partial z} = x^2y - 3$.

(b) $\frac{\partial f}{\partial z}(2, 3, 1) = 2^2 \cdot 3 - 3 = 12 - 3 = 9$.

EXAMPLE 7 Let $f(x, y, z)$ be the heat-loss function computed in Example 2 of Section 8.1. That is, $f(x, y, z) = 11xy + 14yz + 15xz$. Calculate and interpret $\frac{\partial f}{\partial x}(10, 7, 5)$.

Solution We have

$$\frac{\partial f}{\partial x} = 11y + 15z$$

$$\frac{\partial f}{\partial x}(10, 7, 5) = 11 \cdot 7 + 15 \cdot 5 = 77 + 75 = 152.$$

The quantity $\frac{\partial f}{\partial x}$ is commonly referred to as the *marginal heat loss with respect to change in* x. Specifically, if x is changed from 10 by h units (where h is small) and the values of y and z remain fixed at 7 and 5, then the amount of heat loss will change by approximately $152 \cdot h$ units.

EXAMPLE 8 (*Production*) Consider the production function $f(x, y) = 60x^{3/4}y^{1/4}$, which gives the number of units of goods produced when utilizing x units of labor and y units of capital.

(a) Find $\frac{\partial f}{\partial x}$ and $\frac{\partial f}{\partial y}$.

(b) Evaluate $\frac{\partial f}{\partial x}$ and $\frac{\partial f}{\partial y}$ at $x = 81$, $y = 16$.

(c) Interpret the numbers computed in part (b).

Solution (a) $\frac{\partial f}{\partial x} = 60 \cdot \frac{3}{4}x^{-1/4} \cdot y^{1/4} = 45x^{-1/4}y^{1/4} = 45\frac{y^{1/4}}{x^{1/4}}$.

$\frac{\partial f}{\partial y} = 60 \cdot \frac{1}{4}x^{3/4}y^{-3/4} = 15x^{3/4}y^{-3/4} = 15\frac{x^{3/4}}{y^{3/4}}$.

(b) $\frac{\partial f}{\partial x}(81, 16) = 45 \cdot \frac{16^{1/4}}{81^{1/4}} = 45 \cdot \frac{2}{3} = 30$.

$\frac{\partial f}{\partial y}(81, 16) = 15 \cdot \frac{81^{3/4}}{16^{3/4}} = 15 \cdot \frac{27}{8} = \frac{405}{8} = 50\frac{5}{8}$.

(c) The quantities $\frac{\partial f}{\partial x}$ and $\frac{\partial f}{\partial y}$ are referred to as the *marginal productivity of labor* and the *marginal productivity of capital*, respectively. If the amount of capital is held fixed at $y = 16$ and the amount of labor increases by 1 unit, then the quantity of goods produced will increase by approximately 30 units. Similarly, an increase in capital of 1 unit (with labor fixed at 81 units) results in an increase in production of approximately $50\frac{5}{8}$ units of goods.

Just as we formed second derivatives in the case of one variable, we can form second partial derivatives of a function $f(x, y)$ of two variables. Since $\frac{\partial f}{\partial x}$ is a function of x and y, we can differentiate it with respect to x or y. The partial derivative of $\frac{\partial f}{\partial x}$ with respect to x is denoted by $\frac{\partial^2 f}{\partial x^2}$. The partial derivative of $\frac{\partial f}{\partial x}$ with respect to y is denoted by $\frac{\partial^2 f}{\partial y \partial x}$. Similarly, the partial derivative of $\frac{\partial f}{\partial y}$ with respect to x is denoted by $\frac{\partial^2 f}{\partial x \partial y}$, and the partial derivative of $\frac{\partial f}{\partial y}$ with respect to y is denoted by $\frac{\partial^2 f}{\partial y^2}$. Almost all functions $f(x, y)$ encountered in applications [and all functions $f(x, y)$ in this text] have the property that

$$\frac{\partial^2 f}{\partial y \partial x} = \frac{\partial^2 f}{\partial x \partial y}.$$

EXAMPLE 9 Let $f(x, y) = x^2 + 3xy + 2y^2$. Calculate $\frac{\partial^2 f}{\partial x^2}, \frac{\partial^2 f}{\partial y^2}, \frac{\partial^2 f}{\partial x \partial y}$, and $\frac{\partial^2 f}{\partial y \partial x}$.

Solution First we compute $\frac{\partial f}{\partial x}$ and $\frac{\partial f}{\partial y}$.

$$\frac{\partial f}{\partial x} = 2x + 3y, \qquad \frac{\partial f}{\partial y} = 3x + 4y.$$

To compute $\frac{\partial^2 f}{\partial x^2}$, we differentiate $\frac{\partial f}{\partial x}$ with respect to x:

$$\frac{\partial^2 f}{\partial x^2} = 2.$$

Similarly, to compute $\frac{\partial^2 f}{\partial y^2}$, we differentiate $\frac{\partial f}{\partial y}$ with respect to y:

$$\frac{\partial^2 f}{\partial y^2} = 4.$$

To compute $\frac{\partial^2 f}{\partial x \partial y}$, we differentiate $\frac{\partial f}{\partial y}$ with respect to x:

$$\frac{\partial^2 f}{\partial x \partial y} = 3.$$

Finally, to compute $\frac{\partial^2 f}{\partial y \partial x}$, we differentiate $\frac{\partial f}{\partial x}$ with respect to y:

$$\frac{\partial^2 f}{\partial y \partial x} = 3.$$

■

EVALUATING HIGHER PARTIAL DERIVATIVES WITH A TI-82 OR TI-83

The following steps produce the higher partial derivatives of $f(x, y)$. Any one of these functions can be evaluated by assigning values to X and Y and then invoking the function.

1. Set $\mathtt{Y}_1 = f(x, y)$.
2. Set `Y2=nDeriv(Y1,X,X)`. $\mathtt{Y}_2$ will be $\frac{\partial f}{\partial x}$.
3. Set `Y3=nDeriv(Y1,Y,Y)`. $\mathtt{Y}_3$ will be $\frac{\partial f}{\partial y}$.
4. Set `Y4=nDeriv(Y2,X,X)`. $\mathtt{Y}_4$ will be $\frac{\partial^2 f}{\partial x^2}$.
5. Set `Y5=nDeriv(Y3,Y,Y)`. $\mathtt{Y}_5$ will be $\frac{\partial^2 f}{\partial y^2}$.
6. Set `Y6=nDeriv(Y3,X,X)`. $\mathtt{Y}_6$ will be $\frac{\partial^2 f}{\partial x \partial y}$.

EVALUATING HIGHER PARTIAL DERIVATIVES WITH A TI-85

The following steps produce the higher partial derivatives of $f(x, y)$. Any one of these functions can be evaluated by assigning values to x and y and then invoking the function.

1. Set `y1=`$f(x, y)$
2. Set `y2=der1(y1,x,x)`. `y2` will be $\frac{\partial f}{\partial x}$.
3. Set `y3=der1(y1,y,y)`. `y3` will be $\frac{\partial f}{\partial y}$.
4. Set `y4=der1(y2,x,x)`. `y4` will be $\frac{\partial^2 f}{\partial x^2}$.
5. Set `y5=der1(y2,y,y)`. `y5` will be $\frac{\partial^2 f}{\partial y^2}$.
6. Set `y6=der1(y3,x,x)`. `y6` will be $\frac{\partial^2 f}{\partial x \partial y}$.

PRACTICE PROBLEMS 8.2

1. The number of TV sets sold per week by an appliance store is given by a function of two variables, $f(x, y)$, where x is the price per TV sets and y is the amount of money spent weekly on advertising. Suppose that the current price is \$400 per set and that currently \$2000 per week is being spent for advertising.

   (a) Would you expect $\frac{\partial f}{\partial x}(400, 2000)$ to be positive or negative?

   (b) Would you expect $\frac{\partial f}{\partial y}(400, 2000)$ to be positive or negative?

2. The monthly mortgage payment for a house is a function of two variables, $f(A, r)$, where A is the amount of the mortgage and the interest rate is $r\%$. For a 30-year mortgage, $f(92{,}000, 9) = 740.25$ and $\frac{\partial f}{\partial r}(92{,}000, 9) = 66.20$. What is the significance of the number 66.20?

EXERCISES 8.2

Find $\frac{\partial f}{\partial x}$ *and* $\frac{\partial f}{\partial y}$ *for each of the following functions.*

1. $f(x, y) = 5xy$
2. $f(x, y) = 3x^2 + 2y + 1$
3. $f(x, y) = 2x^2e^y$
4. $f(x, y) = x + e^{xy}$
5. $f(x, y) = \frac{y^2}{x}$
6. $f(x, y) = \frac{x}{1 + e^y}$
7. $f(x, y) = (2x - y + 5)^2$
8. $f(x, y) = (9x^2y + 3x)^{12}$
9. $f(x, y) = x^2e^{3x} \ln y$
10. $f(x, y) = (x - \ln y)e^{xy}$
11. $f(x, y) = \frac{x - y}{x + y}$
12. $f(x, y) = \frac{2xy}{e^x}$
13. Let $f(L, K) = 3\sqrt{LK}$. Compute $\frac{\partial f}{\partial L}$.
14. Let $f(p, q) = e^{q/p}$. Compute $\frac{\partial f}{\partial q}$ and $\frac{\partial f}{\partial p}$.
15. Let $f(x, y, z) = (1 + x^2y)/z$. Compute $\frac{\partial f}{\partial x}$, $\frac{\partial f}{\partial y}$, and $\frac{\partial f}{\partial z}$.
16. Let $f(x, y, z) = x^2y + 3yz - z^2$. Compute $\frac{\partial f}{\partial x}$, $\frac{\partial f}{\partial y}$, and $\frac{\partial f}{\partial z}$.
17. Let $f(x, y, z) = xze^{yz}$. Find $\frac{\partial f}{\partial x}$, $\frac{\partial f}{\partial y}$, and $\frac{\partial f}{\partial z}$.
18. Let $f(x, y, z) = ze^{z/xy}$. Find $\frac{\partial f}{\partial x}$, $\frac{\partial f}{\partial y}$, and $\frac{\partial f}{\partial z}$.
19. Let $f(x, y) = x^2 + 2xy + y^2 + 3x + 5y$. Compute $\frac{\partial f}{\partial x}(2, -3)$ and $\frac{\partial f}{\partial y}(2, -3)$.
20. Let $f(x, y) = xye^{2x-y}$. Evaluate $\frac{\partial f}{\partial x}$ and $\frac{\partial f}{\partial y}$ at $(x, y) = (1, 2)$.
21. Let $f(x, y, z) = xy^2z + 5$. Evaluate $\frac{\partial f}{\partial y}$ at $(x, y, z) = (2, -1, 3)$.
22. Let $f(x, y, z) = \frac{x}{y - z}$. Compute $\frac{\partial f}{\partial y}(2, -1, 3)$.
23. Let $f(x, y) = x^3y + 2xy^2$. Find $\frac{\partial^2 f}{\partial x^2}$, $\frac{\partial^2 f}{\partial y^2}$, $\frac{\partial^2 f}{\partial x \partial y}$, and $\frac{\partial^2 f}{\partial y \partial x}$.
24. Let $f(x, y) = xe^y + x^4y + y^3$. Find $\frac{\partial^2 f}{\partial x^2}$, $\frac{\partial^2 f}{\partial y^2}$, $\frac{\partial^2 f}{\partial x \partial y}$, and $\frac{\partial^2 f}{\partial y \partial x}$.

25. A farmer can produce $f(x, y) = 200\sqrt{6x^2 + y^2}$ units of produce by utilizing x units of labor and y units of capital. (The capital is used to rent or purchase land, materials, and equipment.)
   (a) Calculate the marginal productivities of labor and capital when $x = 10$ and $y = 5$.
   (b) Use the result of part (a) to determine the approximate effect on production of utilizing 5 units of capital but cutting back to $9\frac{1}{2}$ units of labor.
26. The productivity of a country is given by $f(x, y) = 300x^{2/3}y^{1/3}$, where x and y are the amounts of labor and capital.
   (a) Compute the marginal productivities of labor and capital when $x = 125$ and $y = 64$.
   (b) What would be the approximate effect of utilizing 125 units of labor but cutting back to 62 units of capital?
27. In a certain suburban community commuters have the choice of getting into the city by bus or train. The demand for these modes of transportation varies with their cost. Let $f(p_1, p_2)$ be the number of people who will take the bus when p_1 is the price of the bus ride and p_2 is the price of the train ride. For example, if $f(4.50, 6) = 7000$, then 7000 commuters will take the bus when the price of a bus ticket is \$4.50 and the price of a train ticket is \$6.00. Explain why $\frac{\partial f}{\partial p_1} < 0$ and $\frac{\partial f}{\partial p_2} > 0$.
28. Refer to Exercise 27. Let $g(p_1, p_2)$ be the number of people who will take the train when p_1 is the price of the bus ride and p_2 is the price of the train ride. Would you expect $\frac{\partial g}{\partial p_1}$ to be positive or negative? How about $\frac{\partial g}{\partial p_2}$?
29. Let p_1 be the average price of VCRs, p_2 the average price of a video tape, $f(p_1, p_2)$ the demand for VCRs, and $g(p_1, p_2)$ the demand for video tape. Explain why $\frac{\partial f}{\partial p_2} < 0$ and $\frac{\partial g}{\partial p_1} < 0$.
30. The demand for a certain gas-guzzling car is given by $f(p_1, p_2)$, where p_1 is the price of the car and p_2 is the price of gasoline. Explain why $\frac{\partial f}{\partial p_1} < 0$ and $\frac{\partial f}{\partial p_2} < 0$.
31. The volume (V) of a certain amount of a gas is determined by the temperature (T) and the pressure (P) by the formula $V = .08(T/P)$. Calculate and interpret $\frac{\partial V}{\partial P}$ and $\frac{\partial V}{\partial T}$ when $P = 20$, $T = 300$.
32. Using data collected from 1929–1941, Richard Stone* determined that the yearly quantity Q of beer consumed in the United Kingdom was approximately given by the formula $Q = f(m, p, r, s)$ where

$$f(m, p, r, s) = (1.058)m^{.136}p^{-.727}r^{.914}s^{.816},$$

* Richard Stone, "The Analysis of Market Demand," *Journal of the Royal Statistical Society*, **CVIII** (1945), 286–391.

m is the aggregate real income (personal income after taxes, adjusted for retail price changes), p is the average retail price of this commodity (beer), r is the average retail price level of all other consumer goods and services, and s is a measure of the strength of the beer. Determine which partial derivatives are positive and which are negative and give interpretations. (For example, since $\frac{\partial f}{\partial r} > 0$, people buy more beer when the prices of other goods increase and the other factors remain constant.)

33. Richard Stone (see Exercise 32) determined that the yearly consumption of food in the United States was given by

$$f(m, p, r) = (2.186)m^{.595}p^{-.543}r^{.922}.$$

Determine which partial derivatives are positive and which are negative and give interpretations of these facts.

34. For the production function $f(x, y) = 60x^{3/4}y^{1/4}$ considered in Example 8, think of $f(x, y)$ as the revenue when utilizing x units of labor and y units of capital. Under actual operating conditions, say $x = a$ and $y = b$, $\frac{\partial f}{\partial x}(a, b)$ is referred to as the *wage per unit of labor* and $\frac{\partial f}{\partial y}(a, b)$ is referred to as the *wage per unit of capital*. Show that

$$f(a, b) = a \cdot \left[\frac{\partial f}{\partial x}(a, b)\right] + b \cdot \left[\frac{\partial f}{\partial y}(a, b)\right].$$

(This equation shows how the revenue is distributed between labor and capital.)

35. Compute $\frac{\partial^2 f}{\partial x^2}$, where $f(x, y) = 60x^{3/4}y^{1/4}$, a production function (where x is units of labor). Explain why $\frac{\partial^2 f}{\partial x^2}$ is always negative.

36. Compute $\frac{\partial^2 f}{\partial y^2}$, where $f(x, y) = 60x^{3/4}y^{1/4}$, a production function (where y is units of capital). Explain why $\frac{\partial^2 f}{\partial y^2}$ is always negative.

37. Let $f(x, y) = 3x^2 + 2xy + 5y$, as in Example 5. Show that

$$f(1 + h, 4) - f(1, 4) = 14h + 3h^2.$$

Thus the error in approximating $f(1 + h, 4) - f(1, 4)$ by $14h$ is $3h^2$. (If $h = .01$, for instance, the error is only .0003.)

38. Physicians, particularly pediatricians, sometimes need to know the body surface area of a patient. For instance, the surface area is used to adjust the results of certain tests of kidney performance. Tables are available that give the approximate body surface A in square meters of a person who weighs W

kilograms and is H centimeters tall. The following empirical formula* is also used:

$$A = .007W^{.425}H^{.725}.$$

Evaluate $\frac{\partial A}{\partial W}$ and $\frac{\partial A}{\partial H}$ when $W = 54$, $H = 165$, and give a physical interpretation of your answers.

In the remaining exercises, use a graphing utility to answer the questions.

39. Example 8, part (b) determined that $\frac{\partial f}{\partial x}(81, 16) = 30$, where $f(x, y) = 60x^{3/4}y^{1/4}$. Previously, $f(81, 16)$ was found to be 3240. Find $f(82, 16)$. How close is it to $f(81, 16)+30$? Also, how close is $f(80, 16)$ to $f(81, 16)-30$?

40. Example 8, part (c) determined that $\frac{\partial f}{\partial y}(81, 16) = 50\frac{5}{8}$, where $f(x, y) = 60x^{3/4}y^{1/4}$. Previously, $f(81, 16)$ was found to be 3240. Find $f(81, 17)$. How close is it to $f(81, 16)+50\frac{5}{8}$? Also, how close is $f(81, 15)$ to $f(81, 16)-50\frac{5}{8}$?

41. Rework Exercise 19 with a graphing utility.

42. Rework Exercise 20 with a graphing utility.

SOLUTIONS TO PRACTICE PROBLEMS 8.2

1. (a) Negative. $\frac{\partial f}{\partial x}(400, 2000)$ is approximately the change in sales due to a \$1 increase in x (price). Since raising prices lowers sales, we would expect $\frac{\partial f}{\partial x}(400, 2000)$ to be negative.

   (b) Positive. $\frac{\partial f}{\partial y}(400, 2000)$ is approximately the change in sales due to a \$1 increase in advertising. Since spending more money on advertising brings in more customers, we would expect sales to increase; that is, $\frac{\partial f}{\partial y}(400, 2000)$ is most likely positive.

2. If the interest rate is raised from 9% to 10%, then the monthly payment will increase by about \$66.20. [An increase to $9\frac{1}{2}\%$ causes an increase in the monthly payment of about $\frac{1}{2} \cdot (66.20)$ or \$33.10, and so on.]

8.3 MAXIMA AND MINIMA OF FUNCTIONS OF SEVERAL VARIABLES

Previously, we studied how to determine the maxima and minima of functions of a single variable.

If $f(x, y)$ is a function of two variables, then we say that $f(x, y)$ has a *relative maximum* when $x = a$, $y = b$ if $f(x, y)$ is at most equal to $f(a, b)$ whenever

* See J. Routh, *Mathematical Preparation for Laboratory Technicians* (Philadelphia; W. B. Saunders Co., 1971), p. 92.

x is near a and y is near b. Geometrically, the graph of $f(x, y)$ has a peak at the point (a, b). [See Figure 1(a).] Similarly, we say that $f(x, y)$ has a *relative minimum* when $x = a$, $y = b$ if $f(x, y)$ is at least equal to $f(a, b)$ whenever x is near a and y is near b. Geometrically, the graph of $f(x, y)$ has a pit with bottom at the point (a, b). [See Figure 1(b).]

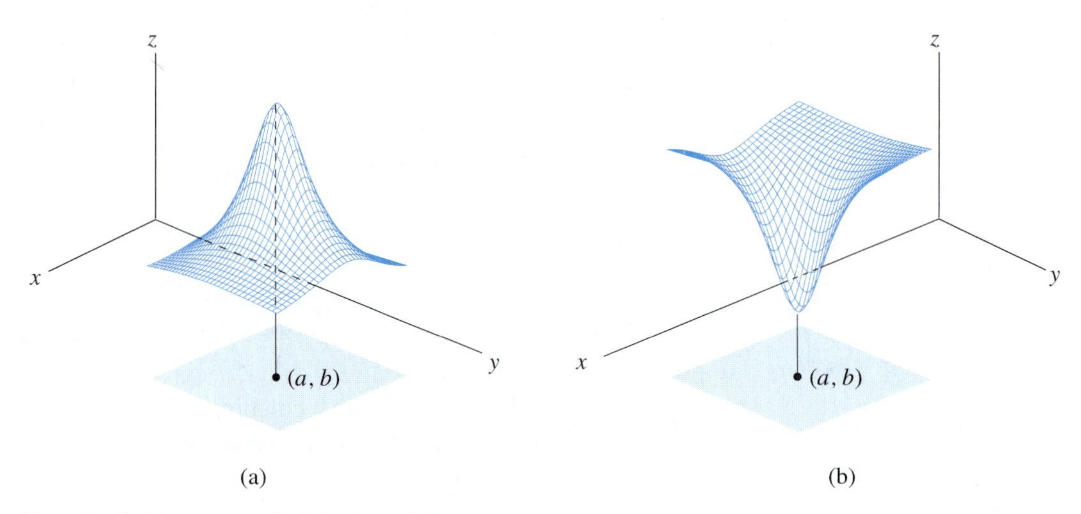

FIGURE 1 Maximum and minimum points

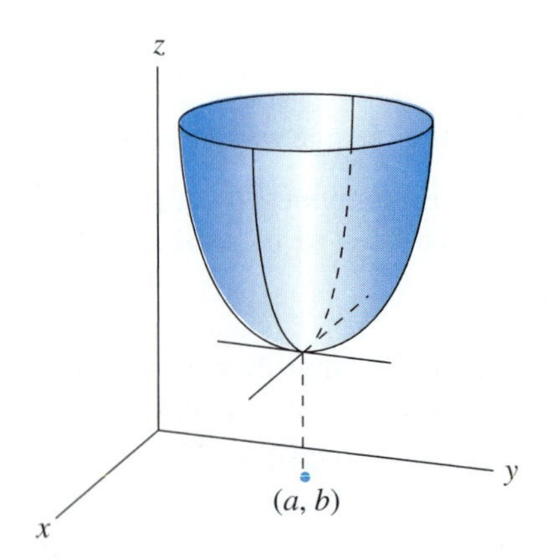

FIGURE 2 Horizontal tangent lines at a relative minimum

Suppose the function $f(x, y)$ has a relative minimum at $(x, y) = (a, b)$, as in Figure 2. When y is held constant at b, $f(x, y)$ is a function of x with a relative minimum at $x = a$. Therefore, the tangent line to the curve $z = f(x, b)$ is horizontal at $x = a$ and hence has slope 0. That is,

$$\frac{\partial f}{\partial x}(a, b) = 0.$$

Likewise, when x is held constant at a, $f(x, y)$ is a function of y with a relative minimum at $y = b$. Therefore, its derivative with respect to y is zero at $y = b$. That is,

$$\frac{\partial f}{\partial y}(a, b) = 0.$$

Similar considerations apply when $f(x, y)$ has a relative maximum at $(x, y) = (a, b)$.

First-Derivative Test for Functions of Two Variables If $f(x, y)$ has either a relative maximum or minimum at $(x, y) = (a, b)$, then

$$\frac{\partial f}{\partial x}(a, b) = 0$$

and

$$\frac{\partial f}{\partial y}(a, b) = 0.$$

A relative maximum or minimum may or may not be an absolute maximum or minimum. However, to simplify matters in this text, the examples and exercises have been chosen so that if an absolute extremum of $f(x, y)$ exists, it will occur at a point where $f(x, y)$ has a relative extremum.

EXAMPLE 1 The function $f(x, y) = 3x^2 - 4xy + 3y^2 + 8x - 17y + 30$ has the graph pictured in Figure 2. Find the point (a, b) at which $f(x, y)$ attains its minimum.

Solution We look for those values of x and y at which both partial derivatives are zero. The partial derivatives are

$$\frac{\partial f}{\partial x} = 6x - 4y + 8$$
$$\frac{\partial f}{\partial y} = -4x + 6y - 17.$$

Setting $\frac{\partial f}{\partial x} = 0$ and $\frac{\partial f}{\partial y} = 0$, we obtain

$$6x - 4y + 8 = 0 \quad \text{or} \quad y = \frac{6x + 8}{4}$$
$$-4x + 6y - 17 = 0 \quad \text{or} \quad y = \frac{4x + 17}{6}.$$

By equating these two expressions for y, we have

$$\frac{6x + 8}{4} = \frac{4x + 17}{6}.$$

Cross-multiplying, we see that

$$36x + 48 = 16x + 68$$
$$20x = 20$$
$$x = 1$$

When we substitute this value for x into our first equation for y in terms of x, we obtain

$$y = \frac{6x + 8}{4} = \frac{6 \cdot 1 + 8}{4} = \frac{7}{2}.$$

If $f(x, y)$ has a minimum, it must occur where $\frac{\partial f}{\partial x} = 0$ and $\frac{\partial f}{\partial y} = 0$. We have determined that the partial derivatives are zero only when $x = 1$, $y = \frac{7}{2}$. From Figure 2 we know that $f(x, y)$ has a minimum, so it must be at $(x, y) = (1, \frac{7}{2})$. ■

EXAMPLE 2 (*Price discrimination*) A monopolist markets his product in two countries and can charge different amounts in each country. Let x be the number of units to be sold in the first country and y the number of units to be sold in the second country. Due to the laws of demand, the monopolist must set the price at $97 - (x/10)$ dollars in the first country and $83 - (y/20)$ dollars in the second country in order to sell all the units. The cost of producing these units is $20{,}000 + 3(x + y)$. Find the values of x and y that maximize the profit.

Solution Let $f(x, y)$ be the profit derived from selling x units in the first country and y in the second. Then

$$\begin{aligned} f(x, y) &= [\text{revenue from first country}] \\ &\qquad + [\text{revenue from second country}] - [\text{cost}] \\ &= \left(97 - \frac{x}{10}\right)x + \left(83 - \frac{y}{20}\right)y - [20{,}000 + 3(x + y)] \\ &= 97x - \frac{x^2}{10} + 83y - \frac{y^2}{20} - 20{,}000 - 3x - 3y \\ &= 94x - \frac{x^2}{10} + 80y - \frac{y^2}{20} - 20{,}000 \end{aligned}$$

To find where $f(x, y)$ has its maximum value, we look for those values of x and y at which both partial derivatives are zero:

$$\frac{\partial f}{\partial x} = 94 - \frac{x}{5},$$
$$\frac{\partial f}{\partial y} = 80 - \frac{y}{10}.$$

We set $\dfrac{\partial f}{\partial x} = 0$ and $\dfrac{\partial f}{\partial y} = 0$ to obtain

$$94 - \frac{x}{5} = 0 \quad \text{or} \quad x = 470,$$
$$80 - \frac{y}{10} = 0 \quad \text{or} \quad y = 800.$$

Therefore, the firm should adjust its prices to levels where it will sell 470 units in the first country and 800 units in the second country. ■

EXAMPLE 3 Suppose that we want to design a rectangular building having volume 147,840 cubic feet. Assuming that the daily loss of heat is given by

$$11xy + 14yz + 15xz,$$

where x, y, and z are, respectively, the length, width, and height of the building, find the dimensions of the building for which the daily heat loss is minimal.

Solution We must minimize the function

$$11xy + 14yz + 15xz, \tag{1}$$

where x, y, z satisfy the constraint equation

$$xyz = 147{,}840.$$

For simplicity, let us denote 147,840 by V. Then $xyz = V$, so that $z = V/xy$. We substitute this expression for z into the objective function (1) to obtain a heat-loss function $g(x, y)$ of two variables—namely,

$$g(x, y) = 11xy + 14y\frac{V}{xy} + 15x\frac{V}{xy} = 11xy + \frac{14V}{x} + \frac{15V}{y}.$$

To minimize this function, we first compute the partial derivatives with respect to x and y; then equate them to zero.

$$\frac{\partial g}{\partial x} = 11y - \frac{14V}{x^2} = 0,$$
$$\frac{\partial g}{\partial y} = 11x - \frac{15V}{y^2} = 0.$$

These two equations yield

$$y = \frac{14V}{11x^2}, \quad (2)$$
$$11xy^2 = 15V. \quad (3)$$

If we substitute the value of y from (2) into (3), we see that

$$11x\left(\frac{14V}{11x^2}\right)^2 = 15V$$
$$\frac{14^2V^2}{11x^3} = 15V$$
$$x^3 = \frac{14^2 \cdot V^2}{11 \cdot 15 \cdot V} = \frac{14^2 \cdot V}{11 \cdot 15}$$
$$= \frac{14^2 \cdot 147{,}840}{11 \cdot 15}$$
$$= 175{,}616.$$

Therefore, we see that (by taking cube roots)

$$x = 56.$$

From equation (2) we find that

$$y = \frac{14 \cdot V}{11x^2} = \frac{14 \cdot 147{,}840}{11 \cdot 56^2} = 60.$$

Finally,

$$z = \frac{V}{xy} = \frac{147{,}840}{56 \cdot 60} = 44.$$

Thus the building should be 56 feet long, 60 feet wide, and 44 feet high in order to minimize the heat loss.*

When considering a function of two variables, we find points (x, y) at which $f(x, y)$ has a potential relative maximum or minimum by setting $\frac{\partial f}{\partial x}$ and $\frac{\partial f}{\partial y}$ equal to zero and solving for x and y. However, if we are given no additional information about $f(x, y)$, it may be difficult to determine whether we have found a maximum or a minimum (or neither). In the case of functions of one variable, we studied concavity and deduced the second-derivative test. There is an analogue of the second derivative test for functions of two variables, but it is much more complicated than the one-variable test. We state it without proof.

Second-Derivative Test for Functions of Two Variables Suppose $f(x, y)$ is a function and (a, b) is a point at which

$$\frac{\partial f}{\partial x}(a, b) = 0 \quad \text{and} \quad \frac{\partial f}{\partial y}(a, b) = 0,$$

and let

$$D(x, y) = \frac{\partial^2 f}{\partial x^2} \cdot \frac{\partial^2 f}{\partial y^2} - \left(\frac{\partial^2 f}{\partial x \partial y}\right)^2.$$

1. If

$$D(a, b) > 0 \quad \text{and} \quad \frac{\partial^2 f}{\partial x^2}(a, b) > 0,$$

then $f(x, y)$ has a relative minimum at (a, b).

2. If

$$D(a, b) > 0 \quad \text{and} \quad \frac{\partial^2 f}{\partial x^2}(a, b) < 0,$$

then $f(x, y)$ has relative maximum at (a, b).

3. If

$$D(a, b) < 0,$$

then $f(x, y)$ has neither a relative maximum nor a relative minimum at (a, b).

4. If $D(a, b) = 0$, then no conclusion can be drawn from this test.

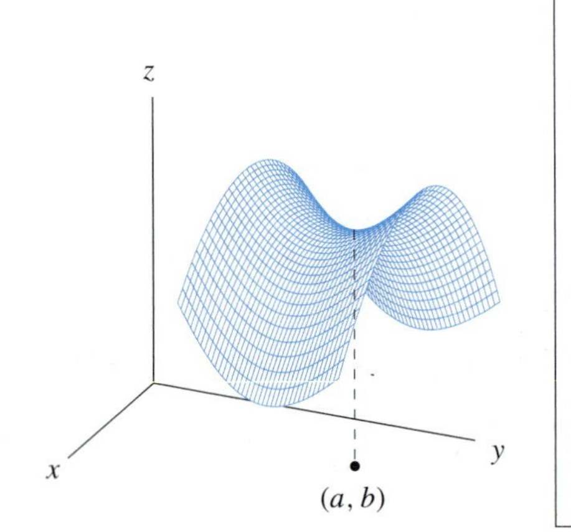

FIGURE 3

The saddle-shaped graph in Figure 3 illustrates a function $f(x, y)$ for which $D(a, b) < 0$. Both partial derivatives are zero at $(x, y) = (a, b)$ and yet the function has neither a relative maximum nor a relative minimum there. (Observe that

* For further discussion of this heat-loss problem, as well as other examples of optimization in architectural design, see L. March, "Elementary Models of Built Forms," Chapter 3 in *Urban Space and Structures*, L. Martin and L. March, eds. (Cambridge University Press, 1972).

the function has a relative maximum with respect to x when y is held constant and a relative minimum with respect to y when x is held constant.) The point (a, b) is called a *saddle point*.

EXAMPLE 4 Let $f(x, y) = y^3 - x^2 + 6x - 12y + 5$. Find all possible relative maximum and minimum points of $f(x, y)$. Use the second-derivative test to determine the nature of each such point.

Solution Since

$$\frac{\partial f}{\partial x} = -2x + 6, \qquad \frac{\partial f}{\partial y} = 3y^2 - 12,$$

we find that $f(x, y)$ has a potential relative extreme point when

$$\begin{aligned} -2x + 6 &= 0 \\ 3y^2 - 12 &= 0. \end{aligned}$$

The first equation implies that $x = 3$. From the second equation we have

$$\begin{aligned} 3y^2 &= 12 \\ y^2 &= 4 \\ y &= \pm 2. \end{aligned}$$

Thus $\dfrac{\partial f}{\partial x}$ and $\dfrac{\partial f}{\partial y}$ are both zero when $(x, y) = (3, 2)$ and when $(x, y) = (3, -2)$. To apply the second-derivative test, we compute

$$\frac{\partial^2 f}{\partial x^2} = -2, \qquad \frac{\partial^2 f}{\partial y^2} = 6y, \qquad \frac{\partial^2 f}{\partial x \partial y} = 0,$$

and

$$D(x, y) = \frac{\partial^2 f}{\partial x^2} \cdot \frac{\partial^2 f}{\partial y^2} - \left(\frac{\partial^2 f}{\partial x \partial y}\right)^2 = (-2)(6y) - 0 = -12y. \tag{4}$$

Since $D(3, 2) = -24$ is negative, case 3 of the second-derivative test says that $f(x, y)$ has neither a relative maximum nor a relative minimum at $(3, 2)$. Next, note that

$$D(3, -2) = 24 > 0, \qquad \frac{\partial^2 f}{\partial x^2}(3, -2) = -2 < 0.$$

Thus, by case 2 of the second-derivative test, the function $f(x, y)$ has a relative maximum at $(3, -2)$. ■

OBTAINING $D(x, y)$ WITH A TI CALCULATOR

Refer to the discussion "Evaluating Higher Partial Derivatives with a TI Calculator" in Section 8.2. After the following step is added to steps 1–6, Y_7 will be $D(x, y)$.

7. Set `Y7=Y4Y5-Y6^2`.

In this section we have restricted ourselves to functions of two variables, but the case of three or more variables is handled in a similar fashion. For instance, here is the first-derivative test for a function of three variables.

> If $f(x, y)$ has a relative maximum or minimum at $(x, y, z) = (a, b, c)$, then
>
> $$\frac{\partial f}{\partial x}(a, b, c) = 0,$$
> $$\frac{\partial f}{\partial y}(a, b, c) = 0,$$
> $$\frac{\partial f}{\partial z}(a, b, c) = 0.$$

PRACTICE PROBLEMS 8.3

1. Find all points (x, y) where $f(x, y) = x^3 - 3xy + \frac{1}{2}y^2 + 8$ has a possible relative maximum or minimum.
2. Apply the second-derivative test to the function $g(x, y)$ of Example 3 to confirm that a relative minimum actually occurs when $x = 56$ and $y = 60$.

EXERCISES 8.3

Find all points (x, y) where $f(x, y)$ has a possible relative maximum or minimum.

1. $f(x, y) = x^2 - 3y^2 + 4x + 6y + 8$
2. $f(x, y) = \frac{1}{2}x^2 + y^2 - 3x + 2y - 5$
3. $f(x, y) = x^2 - 5xy + 6y^2 + 3x - 2y + 4$
4. $f(x, y) = -3x^2 + 7xy - 4y^2 + x + y$
5. $f(x, y) = x^3 + y^2 - 3x + 6y$
6. $f(x, y) = x^2 - y^3 + 5x + 12y + 1$
7. $f(x, y) = \frac{1}{3}x^3 - 2y^3 - 5x + 6y - 5$
8. $f(x, y) = x^4 - 8xy + 2y^2 - 3$
9. The function $f(x, y) = 2x + 3y + 9 - x^2 - xy - y^2$ has a maximum at some point (x, y). Find the values of x and y where this maximum occurs.
10. The function $f(x, y) = \frac{1}{2}x^2 + 2xy + 3y^2 - x + 2y$ has a minimum at some point (x, y). Find the values of x and y where this minimum occurs.

In Exercises 11–16, both first partial derivatives of the function $f(x, y)$ are zero at the given points. Use the second-derivative test to determine the nature of $f(x, y)$ at each of these points. If the second-derivative test is inconclusive, so state.

11. $f(x, y) = 3x^2 - 6xy + y^3 - 9y$; $(3, 3)$, $(-1, -1)$
12. $f(x, y) = 6xy^2 - 2x^3 - 3y^4$; $(0, 0)$, $(1, 1)$, $(1, -1)$
13. $f(x, y) = 2x^2 - x^4 - y^2$; $(-1, 0)$, $(0, 0)$, $(1, 0)$

14. $f(x, y) = x^4 - 4xy + y^4$; $(0, 0)$, $(1, 1)$, $(-1, -1)$
15. $f(x, y) = ye^x - 3x - y + 5$; $(0, 3)$
16. $f(x, y) = \dfrac{1}{x} + \dfrac{1}{y} + xy$; $(1, 1)$

Find all points (x, y) where $f(x, y)$ has a possible relative maximum or minimum. Then use the second-derivative test to determine, if possible, the nature of $f(x, y)$ at each of these points. If the second-derivative test is inconclusive, so state.

17. $f(x, y) = x^2 - 2xy + 4y^2$
18. $f(x, y) = 2x^2 + 3xy + 5y^2$
19. $f(x, y) = -2x^2 + 2xy - y^2 + 4x - 6y + 5$
20. $f(x, y) = -x^2 - 8xy - y^2$
21. $f(x, y) = x^2 + 2xy + 5y^2 + 2x + 10y - 3$
22. $f(x, y) = x^2 - 2xy + 3y^2 + 4x - 16y + 22$
23. $f(x, y) = x^3 - y^2 - 3x + 4y$
24. $f(x, y) = x^3 - 2xy + 4y$
25. $f(x, y) = 2x^2 + y^3 - x - 12y + 7$
26. $f(x, y) = x^2 + 4xy + 2y^4$
27. Find the possible values of x, y, z at which

$$f(x, y, z) = 2x^2 + 3y^2 + z^2 - 2x - y - z$$

assumes its minimum value.

28. Find the possible values of x, y, z at which

$$f(x, y, z) = 5 + 8x - 4y + x^2 + y^2 + z^2$$

assumes its minimum value.

29. United States postal rules require that the length plus the girth of a package cannot exceed 84 inches in order to be mailed. Find the dimensions of the rectangular package of greatest volume that can be mailed. [*Note*: From Figure 4 we see that $84 = (\text{length}) + (\text{girth}) = l + (2x + 2y)$.]

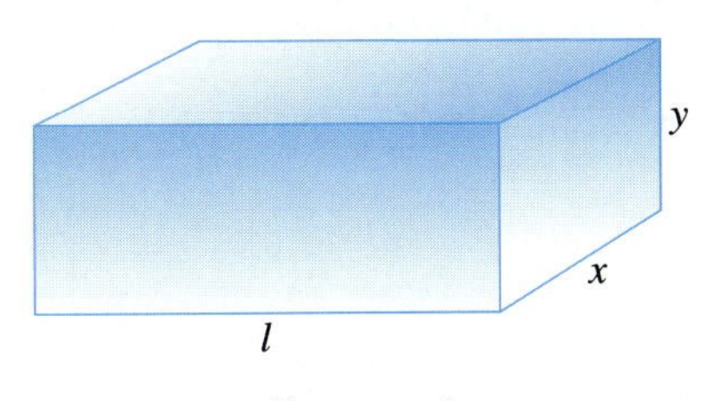

FIGURE 4

30. Find the dimensions of the rectangular box of least surface area that has a volume of 1000 cubic inches.

31. A company manufactures and sells two products, call them I and II, that sell for \$10 and \$9 per unit, respectively. The cost of producing x units of product I and y units of product II is

$$400 + 2x + 3y + .01(3x^2 + xy + 3y^2).$$

Find the values of x and y that maximize the company's profit. (*Note*: profit = revenue − cost.)

32. A monopolist manufactures and sells two competing products, call them I and II, that cost \$30 and \$20 per unit, respectively, to produce. The revenue from marketing x units of product I and y units of product II is $98x+112y-.04xy-.1x^2-.2y^2$. Find the values of x and y that maximize the monopolist's profits.

33. A company manufactures and sells two products, call them I and II, that sell for $\$p_{\mathrm{I}}$ and $\$p_{\mathrm{II}}$ per unit, respectively. Let $C(x, y)$ be the cost of producing x units of product I and y units of product II. Show that if the company's profit is maximized when $x = a, y = b$, then

$$\frac{\partial C}{\partial x}(a, b) = p_{\mathrm{I}} \quad \text{and} \quad \frac{\partial C}{\partial y}(a, b) = p_{\mathrm{II}}.$$

34. A monopolist manufactures and sells two competing products, call them I and II, that cost $\$p_{\mathrm{I}}$ and $\$p_{\mathrm{II}}$ per unit, respectively, to produce. Let $R(x, y)$ be the revenue from marketing x units of product I and y units of product II. Show that if the monopolist's profit is maximized when $x = a, y = b$, then

$$\frac{\partial R}{\partial x}(a, b) = p_{\mathrm{I}} \quad \text{and} \quad \frac{\partial R}{\partial y}(a, b) = p_{\mathrm{II}}.$$

In the remaining exercises, use a graphing utility to answer the questions. In Exercises 35–40, each point P is either a relative maximum point, relative minimum point, or saddle point for the given function. Evaluate the function at points near P and make your best guess of the nature of the function at P. (Evaluate the function at P and at points to the left of, to the right of, above, and below P.)

35. $f(x, y) = 2xy^2 - x^2y + 4xy$; $P = (4/3, -2/3)$

36. $f(x, y) = 3x - x^3 - 3xy^2$; $P = (1, 0)$, $P = (-1, 0)$

37. $f(x, y) = \frac{1}{3}x^3 - \frac{2}{3}y^3 + \frac{1}{2}x^2 - 6x + 32y$; $P = (2, -4)$, $P = (-3, 4)$, $P = (2, 4)$

38. $f(x, y) = y^3 - x^2 + 6x - 12y + 5$; $P = (3, 2)$, $P = (3, -2)$

39. $f(x, y) = x^2 - 2xy + \frac{1}{3}y^3 - 3y$; $P = (3, 3)$, $P = (-1, -1)$

40. $f(x, y) = x^2y - 2xy + 2y^2 - 15y$; $P = (1, 4)$

41. Rework Exercise 11 with a TI calculator.

42. Rework Exercise 12 with a TI calculator.

43. Rework Exercise 13 with a TI calculator.

44. Rework Exercise 14 with a TI calculator.

Solutions to Practice Problems 8.3

1. Compute the first partial derivatives of $f(x, y)$ and solve the system of equations that results from setting the partials equal to zero.

$$\frac{\partial f}{\partial x} = 3x^2 - 3y = 0$$
$$\frac{\partial f}{\partial y} = -3x + y = 0.$$

Solve each equation for y in terms of x.

$$\begin{cases} y = x^2 \\ y = 3x. \end{cases}$$

Equate expressions for y and solve for x.

$$\begin{aligned} x^2 &= 3x \\ x^2 - 3x &= 0 \\ x(x-3) &= 0 \\ x = 0 \quad \text{or} \quad x &= 3. \end{aligned}$$

When $x = 0$, $y = 0^2 = 0$. When $x = 3$, $y = 3^2 = 9$. Therefore, the possible relative maximum or minimum points are $(0, 0)$ and $(3, 9)$.

2. We have

$$g(x, y) = 11xy + \frac{14V}{x} + \frac{15V}{y},$$

$$\frac{\partial g}{\partial x} = 11y - \frac{14V}{x^2}, \quad \text{and} \quad \frac{\partial g}{\partial y} = 11x - \frac{15V}{y^2}.$$

Now,

$$\frac{\partial^2 g}{\partial x^2} = \frac{28V}{x^3}, \quad \frac{\partial^2 g}{\partial y^2} = \frac{30V}{y^3}, \quad \text{and} \quad \frac{\partial^2 g}{\partial x \partial y} = 11.$$

Therefore,

$$\begin{aligned} D(x, y) &= \frac{28V}{x^3} \cdot \frac{30V}{y^3} - (11)^2 \\ D(56, 60) &= \frac{28(147{,}840)}{(56)^3} \cdot \frac{30(147{,}840)}{(60)^3} - 121 \\ &= 484 - 121 = 363 > 0. \end{aligned}$$

and

$$\frac{\partial^2 g}{\partial x^2}(56, 60) = \frac{28(147{,}840)}{(56)^3} > 0.$$

It follows that $g(x, y)$ has a relative minimum at $x = 56$, $y = 60$.

8.4 LAGRANGE MULTIPLIERS AND CONSTRAINED OPTIMIZATION

We have see a number of optimization problems in which we were required to minimize (or maximize) an objective function where the variables were subject to a constraint equation. For instance, in Example 4 of Section 3.5, we minimized the cost of a rectangular enclosure by minimizing the objective function $21x + 14y$, where x and y were subject to the constraint equation $600 - xy = 0$. In the preceding section (Example 3) we minimized the daily heat loss from a building by minimizing the objective function $11xy+14yz+15xz$, subject to the constraint equation $147{,}840 - xyz = 0$.

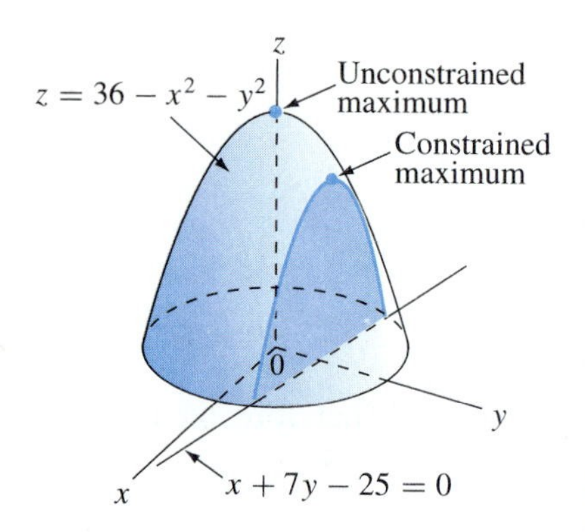

FIGURE 1 A constrained optimization problem

Figure 1 gives a graphical illustration of what happens when an objective function is maximized subject to a constraint. The graph of the objective function is the cone-shaped surface $z = 36-x^2-y^2$, and the colored curve on that surface consists of those points whose x- and y-coordinates satisfy the constraint equation $x+7y-25 = 0$. The constrained maximum is at the highest point on this curve. Of course, the surface itself has a higher "unconstrained maximum" at $(x, y, z) = (0, 0, 36)$, but these values of x and y do not satisfy the constraint equation.

In this section we introduce a powerful technique for solving problems of this type. Let us begin with the following general problem, which involves two variables

Problem Let $f(x, y)$ and $g(x, y)$ be functions of two variables. Find values of x and y that maximize (or minimize) the objective function $f(x, y)$ and that also satisfy the constraint equation $g(x, y) = 0$.

Of course, if we can solve the equation $g(x, y) = 0$ for one variable in terms of the other and substitute the resulting expression into $f(x, y)$, we arrive at a function of a single variable that can be maximized (or minimized) by using the methods of Chapter 3. However, this technique can be unsatisfactory for two reasons. First, it may be difficult to solve the equation $g(x, y) = 0$ for x or for y. For example, if $g(x, y) = x^4+5x^3y+7x^2y^3+y^5-17 = 0$, then it is difficult to write y as a function of x or x as a function of y. Second, even if $g(x, y) = 0$ can be solved for one variable in terms of the other, substitution of the result into $f(x, y)$ may yield a complicated function.

One clever idea for handling the preceding problem was discovered by the eighteenth-century mathematician Lagrange, and the technique that he pioneered today bears his name—the method of *Lagrange multipliers*. The basic idea of this method is to replace $f(x, y)$ by an auxiliary function of three variables $F(x, y, \lambda)$, defined as

$$F(x, y, \lambda) = f(x, y) - \lambda g(x, y).$$

The new variable λ (lambda) is called a *Lagrange multiplier* and always multiplies the constraint function $g(x, y)$. The following theorem is stated without proof.

> *Theorem* Suppose that, subject to the constraint $g(x, y) = 0$, the function $f(x, y)$ has a relative maximum or minimum at $(x, y) = (a, b)$. Then there is a value of λ, say $\lambda = c$, such that the partial derivatives of $F(x, y, \lambda)$ all equal zero at $(x, y, \lambda) = (a, b, c)$.

The theorem implies that if we locate all points (x, y, λ) where the partial derivatives of $F(x, y, \lambda)$ are all zero, then among the corresponding points (x, y) we will find all possible places where $f(x, y)$ may have a constrained relative maximum or minimum. Thus the first step in the method of Lagrange multipliers is to set the partial derivatives of $F(x, y, \lambda)$ equal to zero and solve for x, y, and λ:

$$\frac{\partial F}{\partial x} = 0, \tag{L-1}$$

$$\frac{\partial F}{\partial y} = 0, \tag{L-2}$$

$$\frac{\partial F}{\partial \lambda} = 0. \tag{L-3}$$

From the definition of $F(x, y, \lambda)$, we see that $\frac{\partial F}{\partial \lambda} = g(x, y)$. Thus the third equation (L-3) is just the original constraint equation $g(x, y) = 0$. So when we find a point (x, y, λ) that satisfies (L-1), (L-2), and (L-3), the coordinates x and y will automatically satisfy the constraint equation.

The first example applies this method to the problem described in Figure 1.

EXAMPLE 1 Maximize $36 - x^2 - y^2$ subject to the constraint $x + 7y - 25 = 0$.

Solution Here $f(x, y) = 36 - x^2 - y^2$, $g(x, y) = x + 7y - 25$, and

$$F(x, y, \lambda) = 36 - x^2 - y^2 + \lambda(x + 7y - 25).$$

Equations (L-1) to (L-3) read

$$\frac{\partial F}{\partial x} = -2x + \lambda = 0 \tag{1}$$

$$\frac{\partial F}{\partial y} = -2y + 7\lambda = 0 \tag{2}$$

$$\frac{\partial F}{\partial \lambda} = x + 7y - 25 = 0. \tag{3}$$

We solve the first two equations for λ:

$$\lambda = 2x \tag{4}$$

$$\lambda = \frac{2}{7}y.$$

If we equate these two expressions for λ, we obtain

$$\begin{aligned} 2x &= \frac{2}{7}y \\ x &= \frac{1}{7}y. \end{aligned} \tag{5}$$

Substituting this expression for x into equation (3), we have

$$\begin{aligned} \frac{1}{7}y + 7y - 25 &= 0 \\ \frac{50}{7}y &= 25 \\ y &= \frac{7}{2}. \end{aligned}$$

With this value for y, equations (5) and (4) produce the values of x and λ:

$$\begin{aligned} x &= \frac{1}{7}y = \frac{1}{7}\left(\frac{7}{2}\right) = \frac{1}{2}, \\ \lambda &= 2x = 1. \end{aligned}$$

Therefore, the partial derivatives of $F(x, y, \lambda)$ are zero when $x = \frac{1}{2}$, $y = \frac{7}{2}$, and $\lambda = 1$. So the minimum value of $36-x^2-y^2$ subject to the constraint $x+7y-25 = 0$ is

$$36 - \left(\frac{1}{2}\right)^2 - \left(\frac{7}{2}\right)^2 = \frac{47}{2}.$$ ■

The preceding technique for solving three equations in the three variables x, y, and λ can usually be applied to solve Lagrange multiplier problems. Here is the basic procedure.

1. Solve (L-1) and (L-2) for λ in terms of x and y; then equate the resulting expressions for λ.

2. Solve the resulting equation for one of the variables.

3. Substitute the expression so derived into equation (L-3) and solve the resulting equation of one variable.

4. Use the one known variable and the equations of steps 1 and 2 to determine the other two variables.

In most applications we know that an absolute (constrained) maximum or minimum exists. In the event that the method of Lagrange multipliers produces exactly one possible relative extreme value, we will assume it is indeed the sought-after absolute extreme value. For instance, the statement of Example 1 is meant to imply that there is an absolute maximum value. Since we determined that there was just one possible relative extreme value, we concluded that it was the absolute minimum value.

EXAMPLE 2 Using Lagrange multipliers, minimize $42x+28y$, subject to the constraint $600 - xy = 0$, where x and y are restricted to positive values. (This problem arose in Example 4 of Section 3.5, where $42x+28y$ was the cost of building a 600-square-foot enclosure having dimensions x and y.)

Solution We have $f(x, y) = 42x + 28y$, $g(x, y) = 600 - xy$, and

$$F(x, y, \lambda) = 42x + 28y + \lambda(600 - xy).$$

The equations (L-1) to (L-3), in this case, are

$$\frac{\partial F}{\partial x} = 42 - \lambda y = 0,$$
$$\frac{\partial F}{\partial y} = 28 - \lambda x = 0,$$
$$\frac{\partial F}{\partial \lambda} = 600 - xy = 0.$$

From the first two equations we see that

$$\lambda = \frac{42}{y} = \frac{28}{x}. \qquad \text{(step 1)}$$

Therefore,

$$42x = 28y$$

and

$$x = \frac{2}{3}y. \qquad \text{(step 2)}$$

Substituting this expression for x into the third equation, we derive

$$600 - \left(\frac{2}{3}y\right)y = 0$$
$$y^2 = \frac{3}{2} \cdot 600 = 900$$
$$y = \pm 30. \qquad \text{(step 3)}$$

We discard the case $y = -30$ because we are interested only in positive values of x and y. Using $y = 30$, we find that

$$x = \frac{2}{3}(30) = 20$$
$$\lambda = \frac{14}{20} = \frac{7}{10}. \qquad \text{(step 4)}$$

So the minimum value of $42x + 28y$ with x and y subject to the constraints occurs when $x = 20$, $y = 30$, and $\lambda = \frac{7}{10}$. That minimum value is

$$42 \cdot (20) + 28 \cdot (30) = 1680.$$

EXAMPLE 3 (*Production*) Suppose that x units of labor and y units of capital can produce $f(x, y) = 60x^{3/4}y^{1/4}$ units of a certain product. Also suppose that each unit of labor costs \$100, whereas each unit of capital costs \$200. Assume that \$30,000 is available to spend on production. How many units of labor and how many of capital should be utilized in order to maximize production?

Solution The cost of x units of labor and y units of capital equals $100x + 200y$. Therefore, since we want to use all the available money (\$30,000), we must satisfy the constraint equation

$$100x + 200y = 30{,}000$$

or

$$g(x, y) = 30{,}000 - 100x - 200y = 0.$$

Our objective function is $f(x, y) = 60x^{3/4}y^{1/4}$. In this case, we have

$$F(x, y, \lambda) = 60x^{3/4}y^{1/4} + \lambda(30{,}000 - 100x - 200y).$$

The equations (L-1) to (L-3) read

$$\frac{\partial F}{\partial x} = 45x^{-1/4}y^{1/4} - 100\lambda = 0 \tag{L-1}$$

$$\frac{\partial F}{\partial y} = 15x^{3/4}y^{-3/4} - 200\lambda = 0 \tag{L-2}$$

$$\frac{\partial F}{\partial \lambda} = 30{,}000 - 100x - 200y = 0 \tag{L-3}$$

By solving the first two equations for λ, we see that

$$\lambda = \frac{45}{100}x^{-1/4}y^{1/4} = \frac{9}{20}x^{-1/4}y^{1/4}$$

$$\lambda = \frac{15}{200}x^{3/4}y^{-3/4} = \frac{3}{40}x^{3/4}y^{-3/4}.$$

Therefore, we must have

$$\frac{9}{20}x^{-1/4}y^{1/4} = \frac{3}{40}x^{3/4}y^{-3/4}.$$

To solve for y in terms of x, let us multiply both sides of this equation by $x^{1/4}y^{3/4}$:

$$\frac{9}{20}y = \frac{3}{40}x$$

or

$$y = \frac{1}{6}x.$$

Inserting this result in (L-3), we find that

$$\begin{aligned} 100x + 200\left(\frac{1}{6}x\right) &= 30{,}000 \\ \frac{400x}{3} &= 30{,}000 \\ x &= 225. \end{aligned}$$

Hence

$$y = \frac{225}{6} = 37.5.$$

So maximum production is achieved by using 225 units of labor and 37.5 units of capital. ■

In Example 3 it turns out that, at the optimum value of x and y,

$$\lambda = \frac{9}{20}x^{-1/4}y^{1/4} = \frac{9}{20}(225)^{-1/4}(37.5)^{1/4} \approx .2875$$

$$\frac{\partial f}{\partial x} = 45x^{-1/4}y^{1/4} = 45(225)^{-1/4}(37.5)^{1/4} \qquad (6)$$

$$\frac{\partial f}{\partial y} = 15x^{3/4}y^{-3/4} = 15(225)^{3/4}(37.5)^{-3/4}. \qquad (7)$$

It can be shown that the Lagrange multiplier λ can be interpreted as the marginal productivity of money. That is, if one additional dollar is available, then approximately .2875 additional units of the product can be produced.

Recall that the partial derivatives $\frac{\partial f}{\partial x}$ and $\frac{\partial f}{\partial y}$ are called the marginal productivity of labor and capital, respectively. From (6) and (7) we have

$$\begin{aligned} \frac{[\text{marginal productivity of labor}]}{[\text{marginal productivity of capital}]} &= \frac{45(225)^{-1/4}(37.5)^{1/4}}{15(225)^{3/4}(37.5)^{-3/4}} \\ &= \frac{45}{15}(225)^{-1}(37.5)^{1} \\ &= \frac{3(37.5)}{225} = \frac{37.5}{75} = \frac{1}{2}. \end{aligned}$$

On the other hand,

$$\frac{[\text{cost per unit of labor}]}{[\text{cost per unit of capital}]} = \frac{100}{200} = \frac{1}{2}.$$

This result illustrates the following law of economics. *If labor and capital are at their optimal levels, then the ratio of their marginal productivities equals the ratio of their unit costs.*

The method of Lagrange multipliers generalizes to functions of any number of variables. For instance, we can maximize $f(x, y, z)$, subject to the constraint equation $g(x, y, z) = 0$, by considering the Lagrange function

$$F(x, y, z, \lambda) = f(x, y, z) + \lambda g(x, y, z).$$

The analogues of equations (L-1) to (L-3) are

$$\frac{\partial F}{\partial x} = 0$$
$$\frac{\partial F}{\partial y} = 0$$
$$\frac{\partial F}{\partial z} = 0$$
$$\frac{\partial F}{\partial \lambda} = 0.$$

Let us now show how we can solve the heat-loss problem of Section 8.3 by using this method.

EXAMPLE 4 Use Lagrange multipliers to find the values of x, y, z, that minimize the objective function

$$f(x, y, z) = 11xy + 14yz + 15xz,$$

subject to the constraint

$$xyz = 147{,}840.$$

Solution The Lagrange function is

$$F(x, y, z, \lambda) = 11xy + 14yz + 15xz + \lambda(147{,}840 - xyz).$$

The conditions for a relative minimum are

$$\begin{aligned} \frac{\partial F}{\partial x} &= 11y + 15z - \lambda yz = 0, \\ \frac{\partial F}{\partial y} &= 11x + 14z - \lambda xz = 0, \\ \frac{\partial F}{\partial z} &= 14y + 15x - \lambda xy = 0, \\ \frac{\partial F}{\partial \lambda} &= 147{,}840 - xyz = 0. \end{aligned} \tag{8}$$

From the first three equations we have

$$\left.\begin{aligned} \lambda &= \frac{11y + 15z}{yz} = \frac{11}{z} + \frac{15}{y} \\ \lambda &= \frac{11x + 14z}{xz} = \frac{11}{z} + \frac{14}{x} \\ \lambda &= \frac{14y + 15x}{xy} = \frac{14}{x} + \frac{15}{y} \end{aligned}\right\} \tag{9}$$

Let us equate the first two expressions for λ:

$$\begin{aligned}\frac{11}{z}+\frac{15}{y}&=\frac{11}{z}+\frac{14}{x}\\ \frac{15}{y}&=\frac{14}{x}\\ x&=\frac{14}{15}y.\end{aligned}$$

Next, we equate the second and third expressions for λ in (9):

$$\begin{aligned}\frac{11}{z}+\frac{14}{x}&=\frac{14}{x}+\frac{15}{y}\\ \frac{11}{z}&=\frac{15}{y}\\ z&=\frac{11}{15}y.\end{aligned}$$

We now substitute the expressions for x and z into the constraint equation (8) and obtain

$$\begin{aligned}\frac{14}{15}y\cdot y\cdot\frac{11}{15}y&=147{,}840\\ y^3&=\frac{(147{,}840)(15)^2}{(14)(11)}=216{,}000\\ y&=60.\end{aligned}$$

From this, we find that

$$x=\frac{14}{15}(60)=56 \quad\text{and}\quad z=\frac{11}{15}(60)=44.$$

We conclude that the heat loss equation is minimized when $x=56$, $y=60$, and $z=44$. ■

In the solution of Example 4, we found that at the optimal values of x, y, and z,

$$\frac{14}{x}=\frac{15}{y}=\frac{11}{z}.$$

Referring to Example 2 of Section 8.1, we see that 14 is the combined heat loss through the east and west sides of the building, 15 is the heat loss through the north and south sides of the building, and 11 is the heat loss through the floor and roof. Thus we have that under optimal conditions

$$\begin{aligned}&\frac{[\text{heat loss through east and west sides}]}{[\text{distance between east and west sides}]}\\ &\qquad=\frac{[\text{heat loss through north and south sides}]}{[\text{distance between north and sides sides}]}\\ &\qquad=\frac{[\text{heat loss through floor and roof}]}{[\text{distance between floor and roof}]}.\end{aligned}$$

This is a principle of optimal design: Minimal heat loss occurs when the distance between each pair of opposite sides is some fixed constant times the heat loss from the pair of sides.

The value of λ in Example 4 corresponding to the optimal values of x, y and z is

$$\lambda = \frac{11}{z} + \frac{15}{y} = \frac{11}{44} + \frac{15}{60} = \frac{1}{2}.$$

One can show that the Lagrange multiplier λ is the marginal heat loss with respect to volume. That is, if a building of volume slightly more than 147,840 cubic feet is optimally designed, then $\frac{1}{2}$ unit of additional heat will be lost for each additional cubic foot of volume.

PRACTICE PROBLEMS 8.4

1. Let $F(x, y, \lambda) = 2x + 3y + \lambda(90 - 6x^{1/3}y^{2/3})$. Find $\frac{\partial F}{\partial x}$.
2. Refer to Exercise 29 of Section 8.3. What is the function $F(x, y, l, \lambda)$ when the exercise is solved using the method of Lagrange multipliers?

EXERCISES 8.4

Solve the following exercises by the method of Lagrange multipliers.

1. Minimize the function $x^2 + 3y^2 + 10$, subject to the constraint $8 - x - y = 0$.
2. Maximize the function $x^2 - y^2$, subject to the constraint $2x + y - 3 = 0$.
3. Maximize $x^2 + xy - 3y^2$, subject to the constraint $2 - x - 2y = 0$.
4. Minimize $\frac{1}{2}x^2 - 3xy + y^2 + \frac{1}{2}$, subject to the constraint $3x - y - 1 = 0$.
5. Find the values of x, y that maximize the function

$$-2x^2 - 2xy - \frac{3}{2}y^2 + x + 2y,$$

subject to the constraint $x + y - \frac{5}{2} = 0$.

6. Find the values of x, y that minimize the function

$$x^2 + xy + y^2 - 2x - 5y,$$

subject to the constraint $1 - x + y = 0$.

7. Find the two positive numbers whose product is 25 and whose sum is as small as possible.
8. Four hundred eighty dollars are available to fence in a rectangular garden. The fencing for the north and south sides of the garden costs \$10 per foot and the fencing for the east and west sides costs \$15 per foot. Find the dimensions of the largest possible garden.
9. Three hundred square inches of material are available to construct an open rectangular box with a square base. Find the dimensions of the box that maximize the volume.

10. The amount of space required by a particular firm is

$$f(x, y) = 1000\sqrt{6x^2 + y^2},$$

where x and y are, respectively, the number of units of labor and capital utilized. Suppose that labor costs \$480 per unit and capital costs \$40 per unit and that the firm has \$5000 to spend. Determine the amounts of labor and capital that should be utilized in order to minimize the amount of space required.

11. Find the dimensions of the rectangle of maximum area that can be inscribed in the unit circle. [See Figure 2(a).]

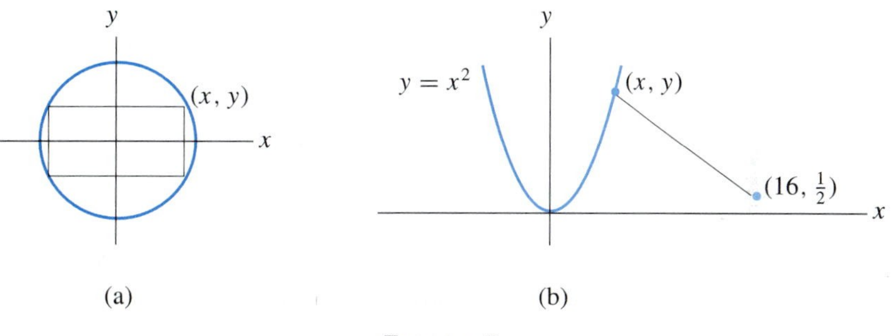

FIGURE 2

12. Find the point on the parabola $y = x^2$ that has minimal distance from the point $(16, \frac{1}{2})$. [See Figure 2(b).] [*Suggestion*: If d denotes the distance from (x, y) to $(16, \frac{1}{2})$, then $d^2 = (x - 16)^2 + (y - \frac{1}{2})^2$. If d^2 is minimized, then d will be minimized.]

13. Suppose that a firm makes two products A and B that use the same raw materials. Given a fixed amount of raw materials and a fixed amount of manpower, the firm must decide how much of its resources should be allocated to the production of A and how much to B. If x units of A and y units of B are produced, suppose that x and y must satisfy

$$9x^2 + 4y^2 = 18{,}000.$$

The graph of this equation (for $x \geq 0$, $y \geq 0$) is called a *production possibilities curve* (Figure 3). A point (x, y) on this curve represents a *production schedule* for the firm, committing it to produce x units of A and y units of B. The reason for the relationship between x and y involves the limitations on personnel and raw materials available to the firm. Suppose that each unit of A yields a \$3 profit, whereas each unit of B yields a \$4 profit. Then the profit of the firm is

$$P(x, y) = 3x + 4y.$$

Find the production schedule that maximizes the profit function $P(x, y)$.

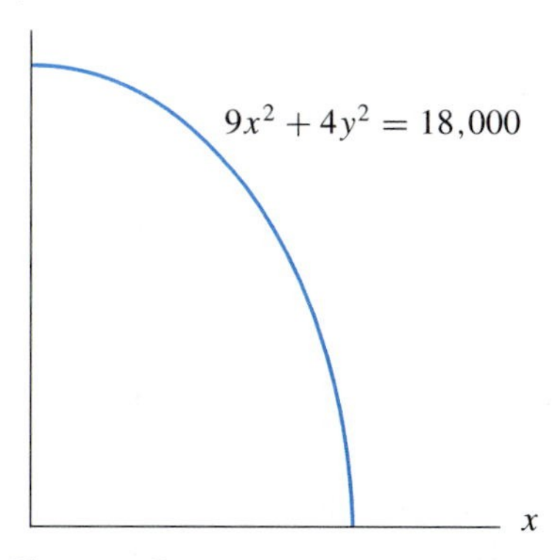

FIGURE 3 A production possibilities curve

14. A firm makes x units of product A and y units of product B and has a production possibilities curve given by the equation $4x^2 + 25y^2 = 50{,}000$ for $x \geq 0$,

$y \geq 0$. (See Exercise 13.) Suppose profits are \$2 per unit for product A and \$10 per unit for product B. Find the production schedule that maximizes the total profit.

15. The production function for a firm is $f(x, y) = 64x^{3/4}y^{1/4}$, where x and y are the number of units of labor and capital utilized. Suppose that labor costs \$96 per unit and capital costs \$162 per unit and that the firm decides to produce 3456 units of goods.
    (a) Determine the amounts of labor and capital that should be utilized in order to minimize the cost. That is, find the values of x, y that minimize $96x + 162y$, subject to the constraint $3456 - 64x^{3/4}y^{1/4} = 0$.
    (b) Find the value of λ at the optimal level of production.
    (c) Show that, at the optimal level of production, we have

$$\frac{[\text{marginal productivity of labor}]}{[\text{marginal productivity of capital}]} = \frac{[\text{unit price of labor}]}{[\text{unit price of capital}]}.$$

16. Consider the monopolist of Example 2, Section 8.3, who sells his goods in two countries. Suppose that he must set the same price in each country. That is, $97 - (x/10) = 83 - (y/20)$. Find the values of x and y that maximize profits under this new restriction.

17. Find the values of x, y, and z that maximize the function xyz subject to the constraint $36 - x - 6y - 3z = 0$.

18. Find the values of x, y, and z that maximize the function $xy + 3xz + 3yz$ subject to the constraint $9 - xyz = 0$.

19. Find the values of x, y, z that maximize the function

$$3x + 5y + z - x^2 - y^2 - z^2,$$

subject to the constraint $6 - x - y - z = 0$.

20. Find the values of x, y, z that minimize the function

$$x^2 + y^2 + z^2 - 3x - 5y - z,$$

subject to the constraint $20 - 2x - y - z = 0$.

21. The material for a rectangular box costs \$2 per square foot for the top and \$1 per square foot for the sides and bottom. Using Lagrange multipliers, find the dimensions for which the volume of the box is 12 cubic feet and cost of the materials is minimized. [Referring to Figure 4(a), the cost will be $3xy + 2xz + 2yz$.]

22. Use Lagrange multipliers to find the three positive numbers whose sum is 15 and whose product is as large as possible.

23. Find the dimensions of an open rectangular glass tank of volume 32 cubic feet for which the amount of material needed to construct the tank is minimized. [See Figure 4(a).]

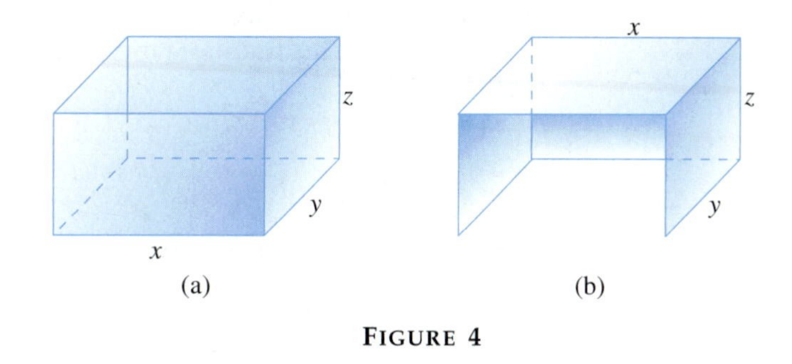

FIGURE 4

24. A shelter for use at the beach has a back, two sides, and a top made of canvas. [See Figure 4(b).] Find the dimensions that maximize the volume and require 96 square feet of canvas.

25. Let $f(x, y)$ be any production function where x represents labor (costing \$$a$ per unit) and y represents capital (costing \$$b$ per unit). Assuming that \$$c$ is available, show that, at the values of x, y that maximize production,

$$\frac{\frac{\partial f}{\partial x}}{\frac{\partial f}{\partial y}} = \frac{a}{b}.$$

[*Note*: Let $F(x, y, \lambda) = f(x, y) + \lambda(c - ax - by)$. The result follows from (L-1) and (L-2).]

26. By applying the result in Exercise 25 to the production function $f(x, y) = kx^{\alpha}y^{\beta}$, show that, for the values of x, y that maximize production, we have

$$\frac{y}{x} = \frac{a\beta}{b\alpha}.$$

(This tells us that the ratio of capital to labor does not depend on the amount of money available nor on the level of production but only on the numbers a, b, α, and β.)

SOLUTIONS TO PRACTICE PROBLEMS 8.4

1. The function can be written as

$$F(x, y, \lambda) = 2x + 3y + \lambda \cdot 90 - \lambda \cdot 6x^{1/3}y^{2/3}.$$

When differentiating with respect to x, both y and λ should be treated as constants (so $\lambda \cdot 90$ and $\lambda \cdot 6$ are also regarded as constants).

$$\frac{\partial F}{\partial x} = 2 - \lambda \cdot 6 \cdot \frac{1}{3}x^{-2/3} \cdot y^{2/3} = 2 - 2\lambda x^{-2/3}y^{2/3}.$$

(*Note*: It is not necessary to write out the multiplication by λ as we did. Most people just do this mentally and then differentiate.)

2. The quantity to be maximized is the volume xyl. The constraint is that length plus girth is 84. This translates to $84 = l + 2x + 2y$ or $84 - l - 2x - 2y = 0$. Therefore,

$$F(x, y, l, \lambda) = xyl + \lambda(84 - l - 2x - 2y).$$

REVIEW OF THE FUNDAMENTAL CONCEPTS OF CHAPTER 8

1. What is a function of two variables?
2. Give two examples of Cobb-Douglas production functions.
3. What is a level curve of a function of two variables?
4. Explain how to find a first partial derivative of a function of two variables.
5. Explain how to find a second partial derivative of a function of two variables.
6. What expression involving a partial derivative gives an approximation to $f(a + h, b) - f(a, b)$?
7. Interpret $\frac{\partial f}{\partial y}(2, 3)$ as a rate of change.
8. Explain the marginal productivities of labor and capital in your own words.
9. Explain how to find possible relative extreme points for a function of several variables.
10. State the second-derivative test for functions of two variables.
11. Explain how the method of Lagrange multipliers is used to solve an optimization problem.

APPENDICES

APPENDIX A USING VISUAL CALCULUS SOFTWARE

PART I AN INTRODUCTION TO VISUAL CALCULUS

Visual Calculus is invoked from DOS by typing VC and pressing the Enter key. Initially, we will work with the first routine in the main menu, "Graphs of Functions."

To invoke a routine, do one of the following:

1. Press the first letter of its name.
2. Cursor down to the name and press the Enter key.
3. Double-click on the name.

Press **G** to invoke the "Graphs of Functions" routine. The screen that appears is called an *input screen*. The blank spaces following `"f(x)="`, `"g(x)="`, and `"h(x)="` and the blank spaces preceding and following `"≤x≤"` and `"≤y≤"` are called *input boxes*. The remaining objects on the screen, such as "**D**raw Graph(s)", "Clear **f**(x)", and so on, are called *buttons*. Input boxes are used to enter information, such as expressions for functions and ranges for variables. Buttons are used to carry out actions, such as drawing the graph of a function or clearing input boxes.

Pressing the Tab key causes successive input boxes and buttons to be highlighted. (Pressing Shift+Tab reverses the direction in which items are highlighted.) When an input box is highlighted, the cursor is in that box and expressions or numbers can be typed into that box. When a button is highlighted, pressing the Enter key causes the action listed on the button to be carried out. With

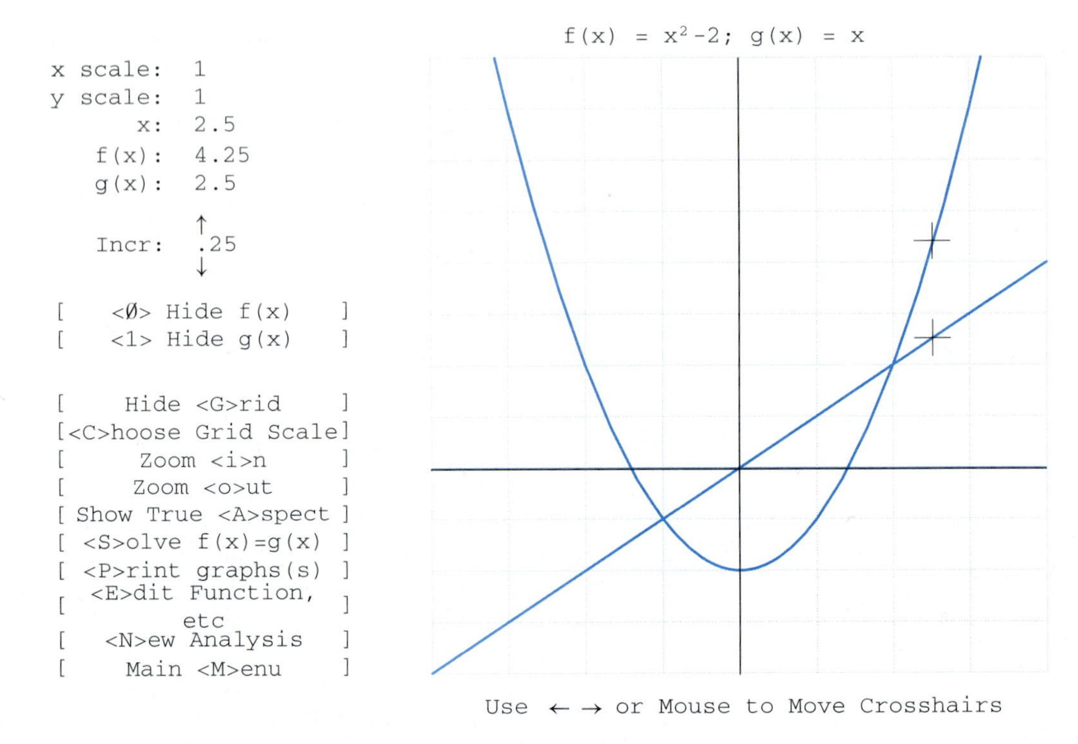

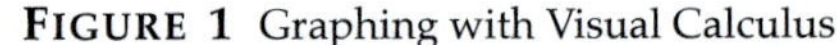
FIGURE 1 Graphing with Visual Calculus

a mouse, any input box can be highlighted and any button can be activated by clicking on it.

To graph one or more functions, type the expressions into the function input boxes, type in a range of values for x, and then activate the Draw Graph(s) button. (Optionally you can specify a range for y. Otherwise, the computer automatically will select the smallest range for y that displays all the graphs.)

The graphs appear on a screen such as the one in Figure 1. The coordinates of the points at the crosshairs (+) are shown in the left column. Press the right- and left-arrow keys to move the crosshairs. Press the down- or up-arrow keys to half or double the distance moved with each keypress. Alternately, you can move the crosshairs to a point on a graph by clicking on the point with the mouse.

Press **G** to superimpose a grid on the screen. (The computer automatically selects the scale for the grid.) If you would like to choose your own scale, press **C** and make your selections.

Any displayed function can be hidden and then redisplayed. (The keys **0**, **1**, and **2** toggle hiding and showing $f(x)$, $g(x)$, and $h(x)$.) When only one graph is displayed, all zeros are obtained by pressing **S**. When two graphs are displayed, all intersection points are displayed by pressing **S**. To return to the input screen, press **E**.

COMMENTS

1. When entering functions on the input screen, use ^ for exponentiation, and `sqr` for $\sqrt{\ }$. Be generous with parentheses. For instance, 2^{3x} should be entered as `2^(3x)`.

2. When entering functions on the input screen, you have the choice of typing in functions or pressing the F4 key to select functions from the text.

3. Visual Calculus works best when the length of the range for the values is a power of 2. For instance, some good ranges are $-4 \le x \le 4$, $-8 \le x \le 8$, and $-2 \le x \le 14$. Such ranges produce nice x values when the graphs are traced.

4. Each button on the input screen has a single white letter. The button can be activated by pressing Alt+<white letter>. For instance, to return to the main menu press Alt+M. That is, hold down the Alt key and press M.

5. Any instruction in the menu on the left side of Figure 1 can be initiated either by clicking on it with the mouse or by pressing the letter that is surrounded by $<$ and $>$. "Zoom in" enlarges a square portion of the screen and "Zoom out" undoes the most recent zoom. "Show True Aspect" changes either the x-range or the y-range so that a unit on the x-axis has the same length as a unit on the y-axis. "Print graph(s)" prints a smooth copy of the graph on either a dot matrix or laser printer. With a laser printer, the graph can be printed in four different sizes. The smallest size prints in less than half a minute, but the largest size requires a few minutes.

6. When graphing a function that assumes arbitrarily large values (such as $f(x) = 1/x$), specify a range for y (such as $-10 \le x \le 10$). Otherwise, the range selected by Visual Calculus will be too large to display the graph clearly.

7. A function also can be evaluated with the routine "1:Compute Values of a Function of one variable."

PART II OBTAINING THE LEAST-SQUARES LINE

The following steps find the straight line that minimizes the least-squares error for the points $(1, 4)$, $(2, 5)$, $(3, 8)$.

1. From the main menu press **L** to invoke the least squares routine.

2. Type in the three ordered pairs. (You can move from location to location with Enter, Tab, and Shift+Tab or by clicking with the mouse.)

3. Press Alt+C to obtain the slope and y-intercept of the line. The least squares line is $y = 2x + \frac{5}{3}$.

Note: As an alternate to Step 3, press Alt+P and then use the cursor-moving keys to try to obtain the least-squares line by experimentation. Press **B** to see the actual least-squares line and obtain its slope and x-intercept.

PART III FORMING COMBINATIONS OF FUNCTIONS

Consider the Graphs of Functions routine. After $f(x)$ and $g(x)$ have been specified, the expression for $h(x)$ can be *literally* specified as $f(x)+g(x)$, $f(x)-g(x)$, $f(x)g(x)$, $f(x)/g(x)$, or $f(g(x))$. After the functions are graphed, values of $h(x)$ can be obtained by moving the crosshairs.

PART IV SLOPES AND TANGENT LINES

The following steps graph a function, display the values of $f(x)$ and $f'(x)$ at any point on the graph, and draw tangent lines through selected points.

1. From the main menu, press **A** to invoke the routine "Analyze a Function and Its Derivatives."
2. Type the function into the box labeled `"f(x)="`.
3. Type the lower and upper values of x into the boxes preceding and following `"≤x≤"`.
4. Optionally type the lower and upper values of y into the boxes preceding and following `"≤y≤"`. For instance, the screen might read

```
f(x)=x^2
-4 ≤ x ≤ 4
-2 ≤ y ≤ 6
```

5. Highlight the [Graph ···] bar and press the Enter key, click on the [Graph ···] bar with the mouse, or press Alt+G.
6. Press the right- and left-arrow keys to move the cursor along the graph. At any time, the values of x, $f(x)$ and $f'(x)$ are displayed on the left, where x is the first coordinate of the point at the cursor.
7. Move the cursor to the point through which you would like to draw a tangent line, and then press **T** or click on `"Draw Tangent"` with the mouse. (*Note*: You can display as many tangent lines as you like, To clear all tangent lines, press **F** twice.)

The following steps calculate $f(x)$ and $f'(x)$ for any value of x in the domain of the function.

1. From the main menu, press **1** to invoke the routine "1:Compute Values of a Function of one variable."
2. Type the function into the box labeled `"f(x)="`.
3. Type the value of x into the box labeled `"x="`.
4. Press Alt+E to display the values of x, $f(x)$, and $f'(x)$. (Alternately, highlight the [Evaluate] bar and press the Enter key, or click on the [Evaluate] bar with the mouse.) Subsequent pressings of the bar increment the value of x and display new values of x, $f(x)$, and $f'(x)$.

PART V GRAPHING $f'(x)$ AND $f''(x)$

To obtain graphs of $f(x)$ and $f'(x)$:

1. Invoke the routine "Analyze a Function and Its Derivatives."

2. Graph the function $f(x)$.

3. Press [1] to show the graph of $f'(x)$ along with the graph of $f(x)$.

   If you so desire, you can obtain the graph of $f'(x)$ alone by pressing [0]. This is necessary of you want to use "Solve $f'(x) = 0$ to determine where the derivative function crosses the x-axis. *Tip*: If you leave the bounds for y blank and let the computer specify bounds, Visual Calculus will automatically adjust the bounds when you press [1] so that both $f(x)$ and $f'(x)$ will be fully displayed. When you press [0] to hide $f(x)$, the bounds automatically will adjust again to provide a good window for $f'(x)$.

4. Press [2] to show the graph of $f''(x)$. The comment following Step 3 also applies to $f''(x)$.

PART VI USING e^x

The function e^x is entered as `e^x` or as `exp(x)`. A function such as e^{3x} must be entered as either `e^(3x)` or `exp(3x)`. In each case, the parentheses are essential.

PART VII USING $\ln x$

The function $\ln x$ is entered as `lnx` or as `ln(x)`. A function such as $\ln 3x$ must be entered as `ln(3x)`. The parentheses are essential.

PART VIII GRAPHING ANTIDERIVATIVES

When solving differential equations, Visual Calculus uses t as the variable. The following steps graph the solution to the differential equation in $y' = x^2 - 2$, $y(1) = \frac{4}{3}$.

1. From the main menu, press **E** to invoke the routine "Examine Solutions of Differential Equations."

2. Type `t^2-2` into the box labeled `"g(t,y)="`.

3. In the next line of the screen, type in the point $(1, 4/3)$.

4. Specify $-2.25 \leq t \leq 2.25$ and $-6 \leq y \leq 6$.

5. Highlight the [Graph Solution] bar and press the Enter key, click on the [Graph Solution] bar with the mouse, or press Alt+S.

Continuing, the following steps show other solutions to $y' = t^2 - 2$ as in Figure 2 of Section 6.1, Example 6.

1. Press **O** to invoke "Other Solutions."

2. Use the arrow keys or click the mouse to move the crosshairs to any point in the plane.

3. Press **P** to plot the solution that passes through the crosshairs.

4. Repeat Steps 2 and 3 as often as you like. The resulting screen will resemble Figure 2 of Section 6.1

In general, the following steps graph the solution to the differential equation $y' = g(t)$, $y(a) = b$.

1. Type the expression for $g(t)$ into the box labeled `"g(t,y)="`.

2. In the next line of the screen, type in the point (a, b).

3. Specify an appropriate range for t. Optionally specify a range for y.

4. Highlight the [Graph Solution] bar and press the Enter key, or click on the [Graph Solution] bar with the mouse.

PART IX CALCULATING RIEMANN SUMS

Lets calculate the Riemann sum from Example 1 of Section 7.1.

1. From the main menu, press **I** to invoke the routine "Integral, Definite."

2. Type `.015x^2-2x+80` into the box labeled "Enter function or ..." and then press [Enter].

3. Type 30 into the box below the integral sign and then press [Enter].

4. Type 60 into the box above the integral sign.

5. Press the Tab key three times to highlight the small box following `"N="`. (Alternately, click on the box with the mouse.)

6. Type 5 into the small box and then press [Enter] twice.

7. The left column in the display in the center of the screen shows the values of the Riemann sums obtained with left endpoints, midpoints, and right endpoints.

Note: To visualize the Riemann sum approximations to the area under the curve, highlight the [Graph Approximations] bar and press [Enter]. (Initially the midpoint rule approximation with $n = 4$ is displayed.) The value of n can be doubled with the press of the up–arrow key. Press **L** or **R** to view the left and right endpoint approximations.

PART X EVALUATING A DEFINITE INTEGRAL

The following steps evaluate the definite integral of Example 2 of Section 7.2.

1. From the main menu, press **I** to invoke the routine "Integral, Definite."

2. Type `2x-4` into the box labeled "Enter function or ..." and then press Enter.

3. Type `0` into the box below the integral sign and then press Enter.

4. Type `5` into the box above the integral sign and then press Enter.

5. The Compute Definite Integral bar is highlighted. Press Enter to carry out the computation. The value 5 is shown.

PART XI EVALUATING A FUNCTION OF TWO VARIABLES

To be specific, lets evaluate $f(x, y) = 60x^{3/4}y^{1/4}$ at the point $(81, 16)$.

1. Invoke the routine "2: Compute Values of a Function of two variables."

2. Set `f(x,y)=60x^(3/4)y^(1/4)`

3. Set `x=81` and set `y=16`. (*Note*: The increment can be left alone or set to anything you like. Let set the increment for x to 1 and the increment for y to 0.)

4. Highlight the Evaluate bar and press the enter key, or click on the bar with the mouse. The values of x, y, and $f(x, y)$ appear in the first three columns in the table on the right. The last two columns contain the values of $\frac{\partial f}{\partial x}$ and $\frac{\partial f}{\partial y}$ which are discussed in Section 8.2.

5. Repeatedly press the Enter key to see new values. At each press, x and y are incremented by the increments given above. (At any time, the values for x and/or y and their increments can be altered to evaluate the function anywhere we like.)

APPENDIX B SOLVING QUADRATIC EQUATIONS

A zero of a function $f(x)$ is a value of x for which $f(x) = 0$. For instance, the function $f(x)$ whose graph is shown in Figure 1 has $x = -3$, $x = 3$, and $x = 7$ as zeros. Throughout this book we shall need to determine zeros of functions or, what amounts to the same thing, to solve the equation $f(x) = 0$.

Zeros of quadratic equations can be found algebraically by either the quadratic formula or by factoring. In addition, they can be found graphically with a graphing utility.

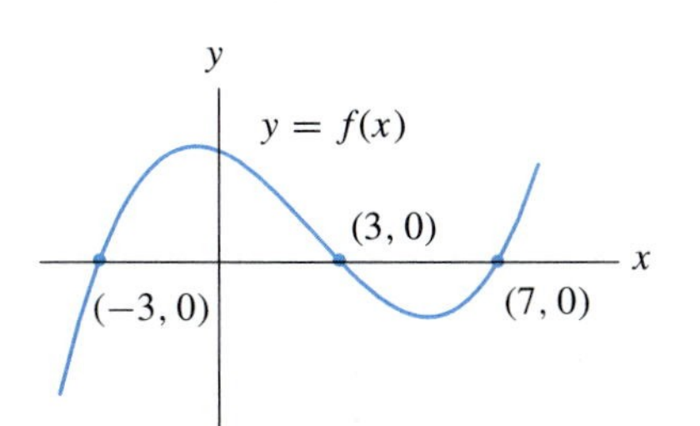

FIGURE 1 Zeros of a function

THE QUADRATIC FORMULA Consider the quadratic function $f(x) = ax^2 + bx+c$, $a \neq 0$. The zeros of this function are precisely the solutions of the quadratic equation

$$ax^2 + bx + c = 0.$$

One way of solving such an equation is via the *quadratic formula.*

> The solutions of the equation $ax^2 + bx + c = 0$ are
> $$x = \frac{-b \pm \sqrt{b^2 - 4ac}}{2a}.$$

The $\pm$ sign tells us to form two expressions, one with + and one with −. The quadratic formula is derived at the end of the section.

EXAMPLE 1 Solve the quadratic equation $3x^2 - 6x + 2 = 0$.

Solution Here $a = 3$, $b = -6$, and $c = 2$. Substituting these values into the quadratic formula, we find that

$$\sqrt{b^2 - 4ac} = \sqrt{(-6)^2 - 4(3)(2)} = \sqrt{36 - 24} = \sqrt{12} = \sqrt{4 \cdot 3} = 2\sqrt{3}$$

and

$$x = \frac{-b \pm \sqrt{b^2 - 4ac}}{2a} = \frac{-(-6) \pm 2\sqrt{3}}{2(3)} = \frac{6 \pm 2\sqrt{3}}{6} = 1 \pm \frac{\sqrt{3}}{3}.$$

The solutions of the equation are $1 + \sqrt{3}/3$ and $1 - \sqrt{3}/3$. (See Figure 2.) ■

FIGURE 2

EXAMPLE 2 Find the zeros of the following quadratic functions.

(a) $f(x) = 4x^2 - 4x + 1$ (b) $f(x) = \frac{1}{2}x^2 - 3x + 5$

Solution (a) We must solve $4x^2 - 4x + 1 = 0$. Here $a = 4$, $b = -4$, and $c = 1$, so that

$$\sqrt{b^2 - 4ac} = \sqrt{(-4)^2 - 4(4)(1)} = \sqrt{0} = 0.$$

There is only one zero, namely

$$x = \frac{-(-4) \pm 0}{2(4)} = \frac{4}{8} = \frac{1}{2}.$$

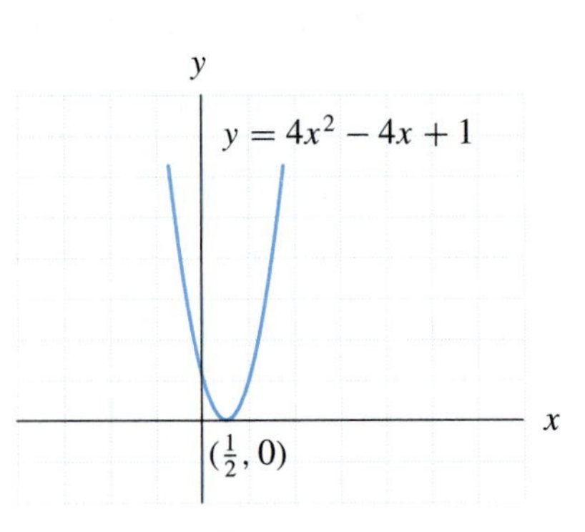

FIGURE 3

The graph of $f(x)$ is sketched in Figure 3.

(b) We must solve $\frac{1}{2}x^2 - 3x + 5 = 0$. Here $a = \frac{1}{2}$, $b = -3$, and $c = 5$, so that

$$\sqrt{b^2 - 4ac} = \sqrt{(-3)^2 - 4(\tfrac{1}{2})(5)} = \sqrt{9 - 10} = \sqrt{-1}.$$

The square root of a negative number is undefined, so we conclude that $f(x)$ has no zeros. The reason for this is clear from Figure 4. The graph of $f(x)$ lies entirely above the x-axis and has no x-intercepts. ■

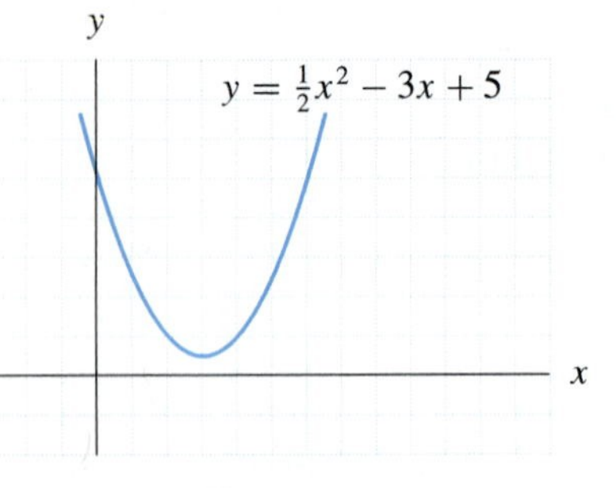

FIGURE 4

FACTORING If $f(x)$ is a polynomial, we can often write $f(x)$ as a product of linear factors (i.e., factors of the form $ax + b$). If this can be done, then the zeros of $f(x)$ can be determined by setting each of the linear factors equal to zero and solving for x. (The reason is that the product of numbers can be zero only when one of the numbers is zero.)

EXAMPLE 3 Factor the following quadratic functions.
(a) $x^2 + 7x + 12$ (b) $x^2 - 13x + 12$ (c) $x^2 - 4x - 12$ (d) $x^2 + 4x - 12$

Solution Note first that for any numbers c and d,

$$(x + c)(x + d) = x^2 + (c + d)x + cd.$$

In the quadratic on the right, the constant term is the product cd, whereas the coefficient of x is the sum $c + d$.

(a) Think of all integers c and d such that $cd = 12$. Then choose the pair that satisfies $c + d = 7$; that is, take $c = 3$, $d = 4$. Thus,

$$x^2 + 7x + 12 = (x + 3)(x + 4).$$

(b) We want $cd = 12$. Since 12 is positive, c and d must be both positive or both negative. We must also have $c + d = -13$. These facts lead us to

$$x^2 - 13x + 12 = (x - 12)(x - 1).$$

(c) We want $cd = -12$. Since -12 is negative, c and d must have opposite signs. Also, they must sum to give -4. We find that

$$x^2 - 4x - 12 = (x - 6)(x + 2).$$

(d) This is almost the same as part (c).

$$x^2 + 4x - 12 = (x + 6)(x - 2).$$ ■

EXAMPLE 4 Factor the following polynomials.
(a) $x^2 - 6x + 9$ (b) $x^2 - 25$ (c) $3x^2 - 21x + 30$ (d) $20 + 8x - x^2$

Solution (a) We look for $cd = 9$ and $c + d = -6$. The solution is $c = d = -3$, and

$$x^2 - 6x + 9 = (x - 3)(x - 3) = (x - 3)^2.$$

In general,

$$x^2 - 2cx + c^2 = (x - c)(x - c) = (x - c)^2.$$

(b) We use the identity

$$x^2 - c^2 = (x + c)(x - c).$$

Hence

$$x^2 - 25 = (x + 5)(x - 5).$$

(c) We first factor out a common factor of 3 and then use the method of Example 3.

$$3x^2 - 21x + 30 = 3(x^2 - 7x + 10) = 3(x - 5)(x - 2).$$

(d) We first factor out a -1 in order to make the coefficient of x^2 equal to $+1$.

$$20 + 8x - x^2 = (-1)(x^2 - 8x - 20) = (-1)(x - 10)(x + 2).$$ ■

EXAMPLE 5 Factor the following polynomials.

(a) $x^2 - 8x$ (b) $x^3 + 3x^2 - 18x$ (c) $x^3 - 10x$

Solution In each case we first factor out a common factor of x.

(a) $x^2 - 8x = x(x - 8)$.

(b) $x^3 + 3x^2 - 18x = x(x^2 + 3x - 18) = x(x + 6)(x - 3)$.

(c) $x^3 - 10x = x(x^2 - 10)$. To factor $x^2 - 10$, we use the identity $x^2 - c^2 = (x + c)(x - c)$, where $c^2 = 10$ and $c = \sqrt{10}$. Thus

$$x^3 - 10x = x(x^2 - 10) = x(x + \sqrt{10})(x - \sqrt{10}).$$ ■

The main use of factoring in this text will be to solve equations.

EXAMPLE 6 Solve the following equations.

(a) $x^2 - 2x - 15 = 0$ (b) $x^2 - 20 = x$ (c) $\dfrac{x^2 + 10x + 25}{x + 1} = 0$

Solution (a) The equation $x^2 - 2x - 15 = 0$ may be written in the form

$$(x - 5)(x + 3) = 0.$$

The product of two numbers is zero if one or the other of the numbers (or both) is zero. Hence

$$x - 5 = 0 \quad \text{or} \quad x + 3 = 0.$$

That is,

$$x = 5 \quad \text{or} \quad x = -3.$$

(b) First we must rewrite the equation $x^2 - 20 = x$ in the form $ax^2 + bx + c = 0$; that is,

$$x^2 - x - 20 = 0$$
$$(x - 5)(x + 4) = 0.$$

We conclude that

$$x - 5 = 0 \quad \text{or} \quad x + 4 = 0;$$

that is,

$$x = 5 \quad \text{or} \quad x = -4.$$

(c) A rational function will be zero only if the numerator is zero. Thus

$$x^2 + 10x + 25 = 0$$
$$(x + 5)^2 = 0$$
$$x + 5 = 0.$$

That is, $x = -5$. Since the denominator is not 0 at $x = -5$, we conclude that $x = -5$ is the solution. ■

SOLVING QUADRATIC EQUATIONS WITH A GRAPHING UTILITY To solve the quadratic equation

$$ax^2 + bx + c = 0$$

with a graphing utility, draw the graph of $y = ax^2 + bx + c$ with a graphing utility and then determine where the graph crosses the x-axis. This task is discussed in Section 1.2 for TI calculators and is discussed in Appendix A, Part I for Visual Calculus.

EXAMPLE 7 Solve the quadratic equation $x^2 - 1.5x - 7 = 0$ with a graphing calculator.

Solution First graph the function. In some cases you will have to experiment with the range of x-values until you obtain a range for which the graph touches or crosses the x-axis. (If experimentation leads you to suspect that the equation might have no solutions, calculate $b^2 - 4ac$ to determine whether there are indeed any solutions.) See Figure 5(a).

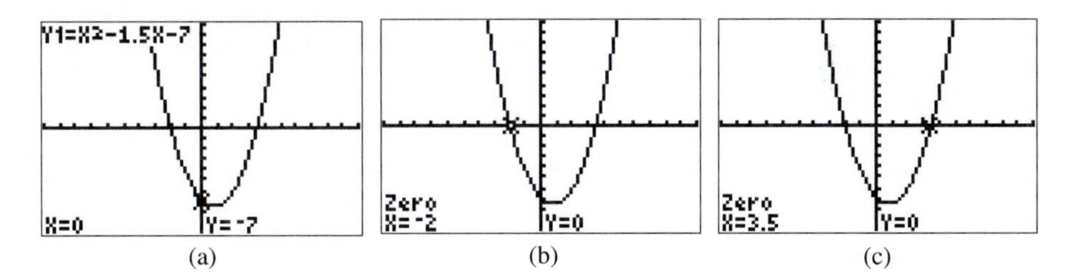

FIGURE 5 Solution to Example 7

Use the built-in capabilities of the calculator to find the x-intercepts of the graph as follows.

TI-82: select `root` from the CALC menu.

TI-83: Select `zero` from the CALC menu.

TI-85: Select `root` from the GRAPH MATH menu.

From Figures 5(b) and (c), the solutions are seen to be $x = -2$ and $x = 3.5$. ■

PRACTICE PROBLEMS APPENDIX B

1. Solve the equation $x - 14/x = 5$.
2. Use the quadratic formula to solve $7x^2 - 35x + 35 = 0$.

EXERCISES APPENDIX B

Use the quadratic formula to find the zeros of the functions in Exercises 1–6.

1. $f(x) = 2x^2 - 7x + 6$
2. $f(x) = 3x^2 + 2x - 1$
3. $f(x) = 4x^2 - 12x + 9$
4. $f(x) = \frac{1}{4}x^2 + x - 1$
5. $f(x) = -2x^2 + 3x - 4$
6. $f(x) = 11x^2 - 7x + 1$

Use the quadratic formula to solve the equations in Exercises 7–12.

7. $5x^2 - 4x - 1 = 0$
8. $x^2 - 4x + 5 = 0$
9. $15x^2 - 135x + 300 = 0$
10. $x^2 - \sqrt{2}x - \frac{5}{4} = 0$
11. $\frac{3}{2}x^2 - 6x + 5 = 0$
12. $9x^2 - 12x + 4 = 0$

Factor the polynomials in Exercises 13–24.

13. $x^2 + 8x + 15$
14. $x^2 - 10x + 16$
15. $x^2 - 16$
16. $x^2 - 1$
17. $3x^2 + 12x + 12$
18. $2x^2 - 12x + 18$
19. $30 - 4x - 2x^2$
20. $15 + 12x - 3x^2$
21. $3x - x^2$
22. $4x^2 - 1$
23. $6x - 2x^3$
24. $16x + 6x^2 - x^3$

In Exercises 25–30, solve the equation.

25. $\frac{21}{x} - x = 4$
26. $x + \frac{2}{x-6} = 3$
27. $x + \frac{14}{x+4} = 5$
28. $1 = \frac{5}{x} + \frac{6}{x^2}$
29. $\frac{x^2 + 14x + 49}{x^2 + 1} = 0$
30. $\frac{x^2 - 8x + 16}{1 + \sqrt{x}} = 0$

In the remaining exercises, use a graphing utility to solve the quadratic equation.

31. $x^2 + .4x - 3.57 = 0$
32. $x^2 - .8x - 4.32 = 0$
33. $x^2 + 5.5x - 78 = 0$
34. $x^2 - 14.5x - 110 = 0$
35. $-x^2 + 9x - 21 = 0$
36. $x^2 + 7x + 14 = 0$
37. $x^2 + x - 462 = 0$
38. $x^2 - x + 306 = 0$

SOLUTIONS TO PRACTICE PROBLEMS APPENDIX B

1. Multiply both sides of the equation by x. Then

$$x^2 - 14 = 5x.$$

Now, take the term $5x$ to the left side of the equation and solve by factoring.

$$\begin{aligned} x^2 - 5x - 14 &= 0 \\ (x - 7)(x + 2) &= 0 \\ x = 7 \quad \text{or} \quad x &= -2. \end{aligned}$$

2. In this case, each coefficient is a multiple of 7. To simplify the arithmetic, we divide both sides of the equation by 7 before using the quadratic formula.

$$x^2 - 5x + 5 = 0$$

$$\sqrt{b^2 - 4ac} = \sqrt{(-5)^2 - 4(1)(5)} = \sqrt{5}$$

$$x = \frac{-b \pm \sqrt{b^2 - 4ac}}{2a} = \frac{5 \pm \sqrt{5}}{2 \cdot 1} = \frac{5}{2} \pm \frac{1}{2}\sqrt{5}.$$

APPENDIX C USING THE TI-82 GRAPHING CALCULATOR: A SUMMARY

HOW TO USE FUNCTIONS

A. Define a function

1. Press [Y=] to obtain a list of functions.
2. Cursor down to the function to be defined. (Press [CLEAR] if the function is to be redefined.)
3. Type in the expression for the function. (*Note*: Press [X, T, θ] to display X.)

B. Select or deselect a function

(Functions with highlighted equal signs are said to be *selected*. Pressing [GRAPH] instructs the calculator to graph all selected functions and pressing [2nd] [TABLE] creates a column of the table for each selected function.)

1. Press [Y=] to obtain a list of functions.
2. Cursor down to the function to be selected or deselected.
3. Move the cursor to the equal sign and press [ENTER] to toggle the state of the function.

C. Display a function name, that is $Y_1, Y_2, Y_3, \ldots$.

1. Press [2nd] [Y-VARS] **1** to invoke the list of function names.
2. Press the number of the function name.

or

1. Press [2nd] [Y-VARS] [ENTER] to invoke the list of function names.
2. Cursor down to the desired function.
3. Press [ENTER] to display the selected function name.

D. Combining functions.

Suppose Y_1 is $f(x)$ and Y_2 is $g(x)$.

1. If $Y_3 = Y_1 + Y_2$, then Y_2 is $f(x) + g(x)$. (Similarly for $-$, $\times$, and $\div$.)
2. If $Y_3 = Y_1(Y_2)$, then Y_3 is $f(g(x))$.

HOW TO SPECIFY WINDOW SETTINGS

A. Customize a window

1. Press [WINDOW] to invoke the window-setting screen.
2. Set Xmin to the leftmost value on the x-axis.
3. Set Xmax to be the rightmost value on the x-axis.
4. Set Xscl to the distance between tick marks on the x-axis.
5. Set Ymin to be the bottom value on the y-axis.
6. Set Ymax to be the top value on the y-axis.
7. Set Yscl to the distance between tick marks on the y-axis.

Note 1: The notation $[a, b]$ *by* $[c, d]$ stands for the window settings Xmin $= a$, Xmax $= b$, Ymin $= c$, Ymax $= d$.

Note 2: The default values of Xscl and Yscl are 1. The value of Xscl should be made large (small) if the difference between Xmax and Xmin is large (small). For instance, with the window setting $[0, 100]$ *by* $[-1, 1]$, good scale settings are Xscl $= 10$ and Yscl $= .1$.

B. Use a predefined window setting

1. Press [ZOOM] to display the list of predefined settings.
2. Either press a number or move the cursor down to an item and press [ENTER].

   ZStandard produces: $[-10, 10]$ *by* $[-10, 10]$, Xscl $=$ Yscl $= 1$

   ZDecimal produces: $[-4.7, 4.7]$ *by* $[-3.1, 3.1]$, Xscl $=$ Yscl $= 1$

C. Some nice window setting

(With these settings, one unit on the x-axis has the same length as one unit on the y-axis, and tracing progresses over simple values.)

1. $[-4.7, 4.7]$ *by* $[-3.1, 3.1]$ (ZDecimal)
2. $[-2.35, 2.35]$ *by* $[-1.55, 1.55]$
3. $[-7.05, 7.05]$ *by* $[-4.65, 4.65]$
4. $[-9.4, 9.4]$ *by* $[-6.2, 6.2]$
5. $[0, 9.4]$ *by* $[0, 6.2]$
6. $[0, 18.8]$ *by* $[0, 12.4]$
7. $[0, 47]$ *by* $[0, 31]$
8. $[0, 94]$ *by* $[0, 62]$

General principle: (Xmax − Xmin) should be a number of the form $k \cdot 9.4$, where k is a whole number or $\frac{1}{2}, \frac{3}{2}, \frac{5}{2}, \ldots$. Then (Ymax − Ymin) should be .6595744(Xmax − Xmin).

HOW TO FIND TANGENT LINES AND SLOPES FOR A FUNCTION

A. Compute $f'(a)$ from the Home screen.

Evaluate `nDeriv(f(x),X,a)` as follows

1. Press [MATH] **8** to display `nDeriv(`.
2. Enter either $Y_1, Y_2, \ldots$ or an expression for $f(x)$.

3. Type in the remaining items and press ENTER.

B. Compute the slope of a point on a graph.

1. Press 2nd [CALC] **6**.
2. Use the arrow keys to move to the point of the graph.
3. Press ENTER.

Note: This process usually works best with a nice window setting.

C. Draw the tangent line to a point on a graph.

1. Press 2nd [DRAW] **5**.
2. Use the arrow keys to move to a point of the graph.
3. Press ENTER.

Note 1: This process usually works best with a nice window setting.

Note 2: To remove all tangent lines, press 2nd [DRAW] **1**.

HOW TO FIND THE COORDINATES OF SPECIAL POINTS ON THE GRAPH OF Y_1

A. Find a point of intersection with the graph of another function, say Y_2.

(From the Home screen)

Evaluate `solve(Y1 − Y2,X,g)`, where `g` is a guess, as follows:

1. Press MATH **0** to display `solve(`.
2. Enter either $Y_1 - Y_2$ or an expression for the difference of the two functions.
3. Type in the remaining items, where g is (hopefully) close to a value of x for which $Y_1 = Y_2$.
4. A value of x for which $Y_1 = Y_2$ will be displayed.

(With the graphs displayed.)

1. Press GRAPH to display the graphs of all selected functions.
2. Press 2nd [CALC].
3. Either press **5** or move the cursor down to the entry "5:intersect" and press ENTER.
4. Reply to "First curve?" by using the up- or down-arrow keys to move the cursor to one of the two curves and then pressing ENTER.
5. Reply to "Second curve?" by using the up- or down-arrow keys to move the cursor to the other curve and then pressing ENTER.
6. Reply to "Guess?" by moving the cursor near the point of intersection and pressing ENTER.

B. Find the second coordinate of point with first coordinate specified, call it a.

(From the Home screen)

1. Display $Y_1(a)$ and press [ENTER].

   or

1. Press a [STO ▶] [X, T, θ] [ENTER] to assign the value a to the variable X.
2. Display Y_1 and press [ENTER].

(From the Home screen or with the graph displayed)

1. Press [2nd] [CALC].
2. Press [ENTER] or **1**.
3. Type in the value of a and press [ENTER].
4. If desired, press the up-arrow key to move to points on other selected functions.

(With the graph displayed)

1. Press [TRACE].
2. Move the cursor with [▶] and/or [◀] until the x-coordinate of the cursor is as close as possible to a. (*Note*: This process usually works best if one of the nice window settings discussed above is used.)

C. Find the first coordinate of point with second coordinate specified, call it b.

(From the Home screen)

1. Press [MATH].
2. Press **0** or move the cursor down to the last entry "0:solve(" and press [ENTER].
3. Continue typing to obtain `solve(`$Y_1 -$ `b,X,c)` where `c` is a guess for the value of the first coordinate. (*Note*: The expression Y_1 can be used in place of Y_1.)
4. Press [ENTER]

(With the graph displayed)

1. Set $Y_2 = b$
2. Find the point of intersection of Y_1 and Y_2 as in Part A above.

D. Find an x-intercept

1. Graph Y_1.
2. Press [2nd] [CALC] **2** to select "root".
3. Move the cursor to a point just to the left of a root and press [ENTER].
4. Move the cursor to a point just to the right of the root and press [ENTER].

5. Move the cursor to a point near the root and press [ENTER].

E. Find a relative extreme point

1. Set `Y2=nDeriv(Y1,X,X)` or set Y_2 equal to the exact expression for the derivative of Y_1.
2. Select Y_2 and deselect all other functions.
3. Graph Y_2.
4. Find an x-intercept of Y_2, call it r, at which the graph of Y_2 crosses the x-axis.
5. The point $(r, Y_1(r))$ will be a possible relative extreme point of Y_1.

F. Find an inflection point

1. Set `Y2=nDeriv(nDeriv(Y1,X,X),X,X)` or set Y_2 equal to the exact expression for the second derivative of Y_1.
2. Select Y_2 and deselect all other functions.
3. Graph Y_2.
4. Find an x-intercept of Y_2, call it r, at which the graph of Y_2 crosses the x-axis.
5. The point $(r, Y_1(r))$ will be a possible inflection point of Y_1.

HOW TO USE TABLES

A. Display values of $f(x)$ for evenly spaced values of x.

1. Press [Y=], assign the function $f(x)$ to Y_1, and deselect all other functions.
2. Press [2nd] [TblSet].
3. Set `TblMin` = *first value of* x.
4. Set Δ`Tbl` = *increment for values of* x.
5. Set both `Indpnt` and `Depend` to `Auto`.
6. Press [2nd] [TABLE].

Note 1: You can use the down- and up-arrow keys to look at function values for other values of x.

Note 2: The table can display values of more than one function. For example, in Step 1 you can assign the function $g(x)$ to Y_2 and also select Y_2 to obtain a table with columns for X, Y_1, and Y_2.

B. Display values of $f(x)$ for arbitrary values of x.

1. Press [Y=], assign the function $f(x)$ to Y_1, and deselect all other functions.
2. Press [2nd] [TblSet].
3. set `Indpnt` to `Ask` by moving the cursor to Ask and pressing [ENTER].

4. Leave `Depend` set to `Auto`.
5. Press [2nd] [TABLE].
6. Type in any value for X and press [ENTER].
7. Repeat Step 6 for as many values as you like.

How to Obtain Riemann Sums, Definite Integrals, and Antiderivatives

A. Compute $[f(x_1) + f(x_2) + \cdots + f(x_n)] \cdot \Delta x$.

On the Home screen, evaluate $\text{sum}(\text{seq}(f(x), \text{X}, x_1, x_n, \Delta x)) * \Delta x$ as follows:

1. Press [2nd] [LIST] and move the cursor right to MATH.
2. Press **5** [(] to display `sum(`.
3. Press [2nd] [LIST] **5** to display `seq(`.
4. Enter either Y_1 or an expression for $f(x)$.
5. Type in the remaining items and press [ENTER].

B. Compute $\int_a^b f(x)\,dx$.

On the Home screen, evaluate `fnInt(`$f(x)$`,X,`a`,`b`)` as follows:

1. Press [MATH] **9** to display `fnInt`.
2. Enter either Y_1 or an expression for $f(x)$.
3. Type in the remaining items and press [ENTER].

C. Shade a region under the graph of $\text{Y}_1 = f(x)$ and find its area.

(Assume the graph is displayed.)

1. Press [2nd] [CALC] **7**.
2. Move the cursor to the left endpoint of the region and press [ENTER].
3. Move the cursor to the right endpoint of the region and press [ENTER].

Note 1: This process often works best with a nice window setting.

Note 2: To remove the shading, press [2nd] [DRAW] **1**.

D. Obtain the solution to the differential equation $y' = g(x)$, $y(a) = b$.

1. Set `Y`$_1$ `= g(X)`.
2. Set `Y`$_2$`=fnInt(Y`$_1$`,X,a,X)+b`. (Press [MATH] **9** to display `fnInt`.) The function Y_2 is an antiderivative of $g(x)$ and can be evaluated and graphed.

E. Shade the region between two curves.

Suppose the graph of Y_1 lies below the graph of Y_2 for $a \le x \le b$ and both functions have been selected. To shade the region between these two curves execute the instruction `Shade(Y`$_1$`,Y`$_2$`,1,a,b)` as follows:

1. Press [2nd] [DRAW] **7** to display `Shade(`.
2. Type in the remaining itmes and press [ENTER].

Note 1: Replace 1 by 2 for a striped shading.

Note 2: To remove the shading, press [2nd] [DRAW] **1**.

HOW TO FIND THE BEST LEAST-SQUARES APPROXIMATION

A. Obtain the equation of the line. [Assume the points are $(x_1, y_1), \ldots, (x_n, y_n)$.]

1. Press [STAT] **1** to obtain a table used for entering the data.
2. If there is no data in columns labeled L_1 and L_2, proceed to Step 4.
3. Move the cursor up to L_1 and press [CLEAR] [ENTER] to delete all data in L_1's column. Move the cursor right and up to L_2 and press [CLEAR] [ENTER] to delete all data in L_2's column.
4. If necessary, move the cursor left to the first blank row of the L_1 column. Type in the value of x_1 and press [ENTER], . . . , type in the value of x_2 and press [ENTER] to place the x-coordinates of the points into the L_1 column.
5. Move the cursor right to the L_2 column and place the y-coordinates of the points into the L_2 column.
6. Press [STAT] [▶] **5** [ENTER] to execute `LinReg(ax+b)` and obtain the slope and y-intercept of the least-squares line.

B. Assign the least-squares line to a function.

1. Press [Y=], move the cursor to the function, and press [CLEAR] to erase the current expression for the function.
2. Press [VARS] **5** [▶] [▶] **7** to assign the equation to the function.

C. Display the points from Part A.

1. Press [Y=] and deselect all functions.
2. Press [2nd] [STAT PLOT] [ENTER] [ENTER] [GRAPH].

Note: To turn off the point-plotting feature, press [2nd] [STAT PLOT] [ENTER] [▶] [ENTER].

D. Display the line and the points from Part A.

1. Press [Y=] and deselect all functions except for the function containing the equation of the least-squares line.
2. Press [2nd] [STAT PLOT] [ENTER] [ENTER] [GRAPH].

Note 1: To turn off the point-plotting feature, press [2nd] [STAT PLOT] [ENTER] [▶] [ENTER].

Note 2: Make sure the current window setting is large enough to display the points.

APPENDIX D USING THE TI-83 GRAPHING CALCULATOR: A SUMMARY

HOW TO USE FUNCTIONS

A. Define a function

1. Press [Y=] to obtain a list of functions.
2. Cursor down to the function to be defined. (Press [CLEAR] if the function is to be redefined.)
3. Type in the expression for the function. (*Note*: Press [X, T, θ, n] to display X.)

B. Select or deselect a function

(Functions with highlighted equal signs are said to be *selected*. Pressing [GRAPH] instructs the calculator to graph all selected functions and pressing [2nd] [TABLE] creates a column of the table for each selected function.)

1. Press [Y=] to obtain a list of functions.
2. Cursor down to the function to be selected or deselected.
3. Move the cursor to the equal sign and press [ENTER] to toggle the state of the function.

C. Display a function name, that is Y_1, Y_2, Y_3, ...

1. Press [VARS] [▶] **1** to invoke the list of function names.
2. Press the number of the function name.

   or

1. Press [VARS] [▶] [ENTER] to invoke the list of function names.
2. Cursor down to the desired function.
3. Press [ENTER] to display the selected function name.

D. Combining functions.

Suppose Y_1 is $f(x)$ and Y_2 is $g(x)$.

1. If $Y_3 = Y_1 + Y_2$, then Y_2 is $f(x) + g(x)$. (Similarly for $-$, $\times$, and $\div$.)
2. If $Y_3 = Y_1(Y_2)$, then Y_3 is $f(g(x))$.

HOW TO SPECIFY WINDOW SETTINGS

A. Customize a window

1. Press [WINDOW] to invoke the window-setting screen.
2. Set Xmin to the leftmost value on the x-axis.
3. Set Xmax to be the rightmost value on the x-axis.
4. Set Xscl to the distance between tick marks on the x-axis.

5. Set Ymin to be the bottom value on the y-axis.
6. Set Ymax to be the top value on the y-axis.
7. Set Yscl to the distance between tick marks on the y-axis.

Note 1: The notation $[a, b]$ *by* $[c, d]$ stands for the window settings Xmin $= a$, Xmax $= b$, Ymin $= c$, Ymax $= d$.

Note 2: The default values of Xscl and Yscl are 1. The value of Xscl should be made large (small) if the difference between Xmax and Xmin is large (small). For instance, with the window setting $[0, 100]$ *by* $[-1, 1]$, good scale settings are Xscl $= 10$ and Yscl $= .1$.

Note 3: Xres controls the pixel resolution (from 1 to 8) and is normally set to 1. Higher settings are used to speed up graphing.

B. Use a predefined window setting

1. Press [ZOOM] to display the list of predefined settings.
2. Either press a number or move the cursor down to an item and press [ENTER].
   ZStandard produces: $[-10, 10]$ *by* $[-10, 10]$, Xscl = Yscl = 1
   ZDecimal produces: $[-4.7, 4.7]$ *by* $[-3.1, 3.1]$, Xscl = Yscl = 1

C. Some nice window setting

(With these settings, one unit on the x-axis has the same length as one unit on the y-axis, and tracing progresses over simple values.)

1. $[-4.7, 4.7]$ *by* $[-3.1, 3.1]$ (ZDecimal)
2. $[-2.35, 2.35]$ *by* $[-1.55, 1.55]$
3. $[-7.05, 7.05]$ *by* $[-4.65, 4.65]$
4. $[-9.4, 9.4]$ *by* $[-6.2, 6.2]$
5. $[0, 9.4]$ *by* $[0, 6.2]$
6. $[0, 18.8]$ *by* $[0, 12.4]$
7. $[0, 47]$ *by* $[0, 31]$
8. $[0, 94]$ *by* $[0, 62]$

General principle: (Xmax − Xmin) should be a number of the form $k \cdot 9.4$, where k is a whole number or $\frac{1}{2}, \frac{3}{2}, \frac{5}{2}, \ldots$. Then (Ymax − Ymin) should be .6595744(Xmax − Xmin).

HOW TO FIND TANGENT LINES AND SLOPES FOR A FUNCTION

A. Compute $f'(a)$ from the Home screen.

Evaluate `nDeriv(f(x),X,a)` as follows

1. Press [MATH] **8** to display `nDeriv(`.
2. Enter either $Y_1, Y_2, \ldots$ or an expression for $f(x)$.
3. Type in the remaining items and press [ENTER].

B. Compute the slope of a point on a graph.

1. Press [2nd] [CALC] **6**.

2. Use the arrow keys to move to the point of the graph or type in the x-coordinate of a point.
3. Press [ENTER].

C. Draw the tangent line to a point on a graph.

1. Press [2nd] [DRAW] **5**.
2. Use the arrow keys to move to a point of the graph or type in the x-coordinate of a point.
3. Press [ENTER].

Note: To remove all tangent lines, press [2nd] [DRAW] **1**.

HOW TO FIND THE COORDINATES OF SPECIAL POINTS ON THE GRAPH OF Y_1

A. Find a point of intersection with the graph of another function, say Y_2.

(From the Home screen)

1. Press [MATH] **0** to invoke the solver.
2. Press [▲] [CLEAR] to invoke and clear the EQUATION SOLVER.
3. Enter either $Y_1 - Y_2$ or an expression for the difference of the two functions to the right of "`eqn:0=`".
4. Press [▼] [CLEAR]. The equation to be solved will be on the first line of the screen and the cursor will be just to the right of "`X=`".
5. Type in a guess for the x-coordinate of a point of intersection and then press [ALPHA] [SOLVE]. After a little delay, the value you typed in will be replaced by the value of the x-coordinate of a point of intersection. (*Note*: You needn't be concerned with the last two lines displayed.)
6. You can now insert a different guess to the right of "`X=`" and press [ALPHA] [SOLVE] again to obtain another point of intersection.

(With the graphs displayed.)

1. Press [GRAPH] to display the graphs of all selected functions.
2. Press [2nd] [CALC].
3. Either press **5** or move the cursor down to the entry "5:intersect" and press [ENTER].
4. Reply to "First curve?" by using the up- or down-arrow keys to move the cursor to one of the two curves and then pressing [ENTER].
5. Reply to "Second curve?" by using the up- or down-arrow keys to move the cursor to the other curve and then pressing [ENTER].
6. Reply to "Guess?" by moving the cursor near the point of intersection (or typing in an approximate value of the x-coordinate of the point of intersection) and pressing [ENTER].

B. Find the second coordinate of point with first coordinate specified, call it a.

(From the Home screen)

1. Display $Y_1(a)$ and press [ENTER].

   or

1. Press a [STO ▶] [X, T, θ, n] [ENTER] to assign the value a to the variable X.
2. Display Y_1 and press [ENTER].

(From the Home screen or with the graph displayed)

1. Press [2nd] [CALC].
2. Press [ENTER] or **1**.
3. Type in the value of a and press [ENTER].
4. If desired, press the up-arrow key to move to points on other selected functions.

(With the graph displayed)

1. Press [TRACE].
2. Type in the value of a and press [ENTER].

C. Find the first coordinate of point with second coordinate specified, call it b.

(From the Home screen)

1. Proceed as in Part A with $Y_2 = b$.

(With the graph displayed)

1. Set $Y_2 = b$
2. Find the point of intersection of Y_1 and Y_2 as in Part A above.

D. Find an x-intercept

1. Graph Y_1.
2. Press [2nd] [CALC] **2** to select `"zero"`.
3. Move the cursor to a point just to the left of a zero (or type in a number less than a zero) and press [ENTER].
4. Move the cursor to a point just to the right of the zero (or type in a number greater than a zero) and press [ENTER].
5. Move the cursor to a point near the zero (or type in a number near the zero) and press [ENTER].

E. Find a relative extreme point

1. Set `Y2=nDeriv(Y1,X,X)` or set Y_2 equal to the exact expression for the derivative of Y_1.

2. Select Y_2 and deselect all other functions.
3. Graph Y_2.
4. Find an x-intercept of Y_2, call it r, at which the graph of Y_2 crosses the x-axis.
5. The point $(r, Y_1(r))$ will be a possible relative extreme point of Y_1.

F. Find an inflection point

1. Set `Y2=nDeriv(nDeriv(Y1,X,X),X,X)` or set Y_2 equal to the exact expression for the second derivative of Y_1.
2. Select Y_2 and deselect all other functions.
3. Graph Y_2.
4. Find an x-intercept of Y_2, call it r, at which the graph of Y_2 crosses the x-axis.
5. The point $(r, Y_1(r))$ will be a possible inflection point of Y_1.

HOW TO USE TABLES

A. Display values of $f(x)$ for evenly spaced values of x.

1. Press [Y=], assign the function $f(x)$ to Y_1, and deselect all other functions.
2. Press [2nd] [TblSet].
3. Set `TblMin` = *first value of* x.
4. Set `ΔTbl` = *increment for values of* x.
5. Set both `Indpnt` and `Depend` to `Auto`.
6. Press [2nd] [TABLE].

Note 1: You can use the down- and up-arrow keys to look at function values for other values of x.

Note 2: The table can display values of more than one function. For example, in Step 1 you can assign the function $g(x)$ to Y_2 and also select Y_2 to obtain a table with columns for X, Y_1, and Y_2.

B. Display values of $f(x)$ for arbitrary values of x.

1. Press [Y=], assign the function $f(x)$ to Y_1, and deselect all other functions.
2. Press [2nd] [TblSet].
3. Set `Indpnt` to `Ask` by moving the cursor to Ask and pressing [ENTER].
4. Leave `Depend` set to `Auto`.
5. Press [2nd] [TABLE].
6. Type in any value for X and press [ENTER].
7. Repeat Step 6 for as many values as you like.

HOW TO OBTAIN RIEMANN SUMS, DEFINITE INTEGRALS, AND ANTIDERIVATIVES

A. Compute $[f(x_1) + f(x_2) + \cdots + f(x_n)] \cdot \Delta x$.

On the Home screen, evaluate sum(seq($f(x)$, X, x_1, x_n, Δx)) $*$ Δx as follows:

1. Press [2nd] [LIST] and move the cursor right to MATH.
2. Press **5** [(] to display `sum(`.
3. Press [2nd] [LIST] **5** to display `seq(`.
4. Enter either Y_1 or an expression for $f(x)$.
5. Type in the remaining items and press [ENTER].

B. Compute $\int_a^b f(x)\,dx$.

On the Home screen, evaluate `fnInt(`$f(x)$`,X,`a`,`b`)` as follows:

1. Press [MATH] **9** to display `fnInt`.
2. Enter either Y_1 or an expression for $f(x)$.
3. Type in the remaining items and press [ENTER].

C. Shade a region under the graph of $Y_1 = f(x)$ and find its area.

(Assume the graph is displayed.)

1. Press [2nd] [CALC] **7**.
2. Move the cursor to the left endpoint of the region (or type in the value of a) and press [ENTER].
3. Move the cursor to the right endpoint of the region (or type in the value of b) and press [ENTER].

Note: To remove the shading, press [2nd] [DRAW] **1**.

D. Obtain the solution to the differential equation $y' = g(x)$, $y(a) = b$.

1. Set Y_1 = `g(X)`.
2. Set Y_2=`fnInt(`Y_1`,X,a,X)+b`. (Press [MATH] **9** to display `fnInt`.) The function Y_2 is an antiderivative of $g(x)$ and can be evaluated and graphed.

E. Shade the region between two curves.

Suppose the graph of Y_1 lies below the graph of Y_2 for $a \le x \le b$ and both functions have been selected. To shade the region between these two curves execute the instruction `Shade(`Y_1`,`Y_2`,1,a,b)` as follows:

1. Press [2nd] [DRAW] **7** to display `Shade(`.
2. Type in the remaining items and press [ENTER].

Note 1: Replace 1 by 2 for a striped shading.

Note 2: To remove the shading, press [2nd] [DRAW] **1**.

How to Find the Best Least-Squares Approximation

A. Obtain the equation of the line. [Assume the points are $(x_1, y_1), \ldots, (x_n, y_n)$.]

1. Press STAT **1** to obtain a table used for entering the data.
2. If there is no data in columns labeled L_1 and L_2, proceed to Step 4.
3. Move the cursor up to L_1 and press CLEAR ENTER to delete all data in L_1's column. Move the cursor right and up to L_2 and press CLEAR ENTER to delete all data in L_2's column.
4. If necessary, move the cursor left to the first blank row of the L_1 column. Type in the value of x_1 and press ENTER, ..., type in the value of x_2 and press ENTER to place the x-coordinates of the points into the $mathrmL_1$ column.
5. Move the cursor right to the L_2 column and place the y-coordinates of the points into the L_2 column.
6. Press STAT ► **4** ENTER to execute `LinReg(ax+b)` and obtain the slope and y-intercept of the least-squares line.

B. Assign the least-squares line to a function.

1. Press Y=, move the cursor to the function, and press CLEAR to erase the current expression for the function.
2. Press VARS **5** ► ► **1** to assign the equation to the function.

C. Display the points from Part A.

1. Press Y= and deselect all functions.
2. Press 2nd [STAT PLOT] ENTER ENTER GRAPH.

Note 1: Make sure the current window is large enough to display the points.

Note 2: To turn off the point-plotting feature, press 2nd [STAT PLOT] ENTER ► ENTER.

D. Display the line and the points from Part A.

1. Press Y= and deselect all functions except for the function containing the equation of the least-squares line.
2. Press 2nd [STAT PLOT] ENTER ENTER GRAPH.

Note 1: To turn off the point-plotting feature, press 2nd [STAT PLOT] ENTER ► ENTER.

Note 2: Make sure the current window setting is large enough to display the points.

APPENDIX E USING THE TI-85 GRAPHING CALCULATOR: A SUMMARY

HOW TO USE FUNCTIONS

A. Define a function

1. Press [GRAPH] [F1] to obtain the screen for defining functions.
2. Type in expression for the function. (*Note*: Press [F1] or [x-VAR] to display x.
3. To delete the expression for an existing function, cursor down to the function and press [CLEAR].
4. To create an additional function, cursor down to the last function and press [ENTER].

B. Select or deselect a function

(Functions with highlighted equal signs are said to be *selected*. Giving the graph command instructs the calculator to graph all selected functions.)

1. Press [GRAPH] [F1] to obtain a list of functions.
2. Cursor down to the function to be selected or deselected.
3. Press [F5] to toggle the state of the function.

C. Display a function name, that is y1, y2, y3, ...

1. Press [2nd] [VARS] [MORE] [F3] to invoke the list of function names.
2. Cursor down to the desired function.
3. Press [ENTER] to display the selected function name.

D. Combining functions.

Suppose y1 is $f(x)$ and y2 is $g(x)$.

1. If y3 = y1 + y2, then y3 is $f(x) + g(x)$. (Similarly for $-$, $\times$, and $\div$.)
2. If y3 = evalF(y1,x,y2), then y3 is $f(g(x))$. (To display evalF(, press [2nd] [CALC] [F1]).

HOW TO SPECIFY WINDOW SETTINGS

A. Customize a window

1. Press [WINDOW] to invoke the window-setting screen.
2. Set xmin to the leftmost value on the x-axis.
3. Set xmax to be the rightmost value on the x-axis.
4. Set xscl to the distance between tick marks on the x-axis.
5. Set ymin to be the bottom value on the y-axis.
6. Set ymax to be the top value on the y-axis.

7. Set yscl to the distance between tick marks on the y-axis.

Note 1: The notation $[a, b]$ *by* $[c, d]$ stands for the window settings xmin $= a$, xmax $= b$, ymin $= c$, ymax $= d$.

Note 2: The default values of xscl and yscl are 1. The value of xscl should be made large (small) if the difference between xmax and xmin is large (small). For instance, with the window setting [0, 100] *by* [−1, 1], good scale settings are xscl = 10 and yscl = .1.

B. Use a predefined window setting

1. Press GRAPH F3 to invoke the ZOOM menu.
2. Press ZSTD to obtain [−10, 10] *by* [−10, 10], xScl = yScl = 1.
3. Press MORE ZDECM to obtain [−6.3, 6.3] *by* [−3.1, 3.1], xScl = yScl = 1.

C. Some nice window setting

With these settings, one unit on the x-axis has the same length as one unit on the y-axis, and tracing progresses over simple values.

1. [−6.3, 6.3] *by* [−3.7, 3.7] (ZDecimal)
2. [−3.15, 3.15] *by* [−1.85, 1.85]
3. [−9.45, 9.45] *by* [−5.56, 5.56]
4. [−12.6, 12.6] *by* [−7.4, 7.4]
5. [0, 12.6] *by* [0, 7.4]
6. [0, 25.2] *by* [0, 14.8]
7. [0, 63] *by* [0, 37]
8. [0, 126] *by* [0, 74]

General principle: (xmax − xmin) should be a number of the form $k \cdot 12.6$, where k is a whole number or $\frac{1}{2}, \frac{3}{2}, \frac{5}{2}, \ldots$. Then (ymax − ymin) should be .5882454(xmax − xmin).

How to Find Tangent Lines and Slopes for a Function

A. Compute $f'(a)$ from the Home screen.

Evaluate `der1(f(x),x,a)` as follows

1. Press 2nd [CALC] F3 to display der1.
2. Enter either y1, y2, ... or an expression for $f(x)$.
3. Type in the remaining items and press ENTER.

B. Compute the slope of a point on a graph.

1. Press GRAPH F5 to display the graph of the function.
2. Press MORE F1 F4.
3. Use the arrow keys to move to the point of the graph.
4. Press ENTER.

Note: This process usually works best with a nice window setting.

C. Draw the tangent line to a point on a graph.

1. Press [GRAPH] [F5] to display the graph of the function.
2. Press [MORE] [F1] [MORE] [MORE] [F3] to select TANLN from the MATH menu.
3. Move the cursor to any point on the graph.
4. Press [ENTER] to draw the tangent line through the point and display the slope of the curve at that point. The slope is displayed at the bottom of the screen as $dy/dx = slope$.
5. To draw another tangent line, press [GRAPH] and then repeat Steps 2–4.

Note: To remove all tangent lines, press [GRAPH] [MORE] [F2] [MORE] [F5].

HOW TO FIND THE COORDINATES OF SPECIAL POINTS ON THE GRAPH OF y1

A. Find a point of intersection with the graph of another function, say y2.

(From the Home screen)

1. Press [2nd] [SOLVER].
2. To the right of `"eqn1:"` enter `y1-y2=0` and press [ENTER]. (*Note*: y1 and y2 can be entered via F keys and the equal sign is entered with [ALPHA] [=].
3. To the right of `"X="` type in a guess for the x-coordinate of the point of intersection, and then press [F5].
4. After a little delay, a value of x for which y1 = y2 will be displayed in place of your guess.

(With the graphs displayed.)

1. Press [GRAPH] [F5] to display the graphs of all selected functions.
2. Press [MORE] [F1] [MORE] [F5] to select ISECT.
3. If necessary, use the up- or down-arrow keys to place the cursor on one of the two curves.
4. Move the cursor close to the point of intersection and then press [ENTER].
5. If necessary, use the up- or down-arrow key to place the cursor on the other curve.
6. Press [ENTER] to display the coordinates of the point of intersection.

B. Find the second coordinate of point with first coordinate specified, call it a.

From the Home screen, compute `evalF(y1,x,a)` as follows:

1. Press [2nd] [CALC] [F1] to display `evalF(`.
2. Enter either y1 or an expression for the function.
3. Type in the remaining items and press [ENTER].
   or

1. Press a STO ▶ x-VAR ENTER to assign the value a to the variable x.
2. Display y1 and press ENTER.

(From the Home screen or with the graph displayed)

1. Press GRAPH MORE MORE F1.
2. Type in the value of a and press ENTER.
3. If desired, press the up-arrow key, ▲, to move to points on other selected functions.

(With the graph displayed)

1. Press TRACE.
2. Move cursor with ▶ and/or ◀ until the x-coordinate of cursor is as close as possible to a. (*Note*: Usually works best if one of the nice window settings discussed above is used.)

C. Find the first coordinate of point with second coordinate specified, call it b.

(From the Home screen)

1. Press 2nd [SOLVER].
2. To the right of "`eqn1:`" enter y1 $=b$ and press ENTER. (*Note*: y1 can be entered via an F key and the equal sign is entered with ALPHA [=].
3. To the right of "`X=`" type in a guess for the x-coordinate of the point of intersection, and then press F5.
4. After a little delay, a value of x for which y1 $=b$ will be displayed in place of your guess.

(With the graph displayed)

1. Set y2 $=b$
2. Find the point of intersection of y1 and y2 as in Part A above.

D. Find an x-intercept

1. Press GRAPH MORE F1 F3 to select ROOT from the MATH menu.
2. Move the cursor along the graph of y1 close to an x-intercept, and press ENTER.

E. Find a relative extreme point

1. Set `y2=der1(y1,x,x)` or set y2 equal to the exact expression for the derivative of y1. (To display der1, press 2nd [CALC] F3.)
2. Select y2 and deselect all other functions.
3. Graph y2.
4. Find an x-intercept of y2, call it r, at which the graph of y2 crosses the x-axis.

5. The point $(r, \text{y1}(r))$ will be a possible relative extreme point of y1.

F. Find an inflection point

1. Set `y2=der2(y1,x,x)` or set y2 equal to the exact expression for the second derivative of y1. (To display der2, press [2nd] [CALC] [F4].)
2. Select y2 and deselect all other functions.
3. Graph y2.
4. Find an x-intercept of y2, call it r, at which the graph of y2 crosses the x-axis.
5. The point $(r, \text{y1}(r))$ will be a possible inflection point of y1.

HOW TO OBTAIN RIEMANN SUMS, DEFINITE INTEGRALS, AND ANTIDERIVATIVES

A. Compute $[f(x_1) + f(x_2) + \cdots + f(x_n)] \cdot \Delta x$.

On the Home screen, evaluate sum(seq($f(x)$, x, x_1, x_n, Δx)) $*$ Δx as follows:

1. Press [2nd] [LIST] [F5] [MORE] [F1] [(] [F3] to display sum(seq(.
2. Enter either a function, such as y1, or an expression for $f(x)$.
3. Type in the remaining items and press [ENTER].

B. Compute $\int_a^b f(x)\,dx$.

On the Home screen, evaluate `fnInt(`$f(x)$`,X,`a`,`b`)` as follows:

1. Press [2nd] [CALC] [F5] to display `fnInt`.
2. Enter either y1 or an expression for $f(x)$.
3. Type in the remaining items and press [ENTER].

C. Find the area of a region under the graph of $\text{y1} = f(x)$.

1. Press [GRAPH] [MORE] [F1] [F5] to select $\int f(x)$ from the MATH menu.
2. Move the cursor to the left endpoint of the region and press [ENTER].
3. Move the cursor to the right endpoint of the region and press [ENTER].

Note: This process usually works best with a nice window setting.

D. Obtain the solution to the differential equation $y' = g(x)$, $y(a) = b$.

1. Set `y1=g(X)`
2. Set `y2=fnInt(y1,x,a,x)+b`. (To display `fnInt`, press [2nd] [CALC] [F5].) The function y2 is an antiderivative of $g(x)$ and can be evaluated and graphed.

How to Find the Best Least-Squares Approximation

A. Obtain the equation of the line. [Assume the points are $(x_1, y_1), \ldots, (x_n, y_n)$.]

1. Press STAT F2 ENTER ENTER to obtain a list used for entering the data.
2. Press F5 to clear all previous data.
3. Enter the data for the points by pressing x_1 ENTER y_1 ENTER x_2 ENTER y_2 ENTER ... x_n ENTER y_n.
4. Press STAT F1 ENTER ENTER F2 to obtain the values of a and b where the least-squares line has equation $y = bx + a$.
5. If desired, the straight line (and the points) can be graphed with the following steps:
   (a) First press GRAPH F1 and deselect all functions.
   (b) Press STAT F3 to invoke the statistical DRAW menu. (If any graphs appear, press F3 to clear them.)
   (c) Press F4 to draw the least-squares line and press F2 to draw the three points.

B. Assign the least-squares line to a function.

1. Press GRAPH F1, move the cursor to the function, and press CLEAR to erase the current expression for the function.
2. Press STAT F5 MORE MORE F2 to assign the equation (known as RegEq) to the function.

ANSWERS TO ODD-NUMBERED EXERCISES

CHAPTER 1

EXERCISES 1.1, PAGE 6

1. $f(x) = 5(x + 3)$ **3.** f multiplies any number t by 6 and then adds the cube of t to the result. **5.** Compute $f(-2)$ **7.** Solve $f(x) = 20$ **9.** 0, 10, 70 **11.** $-\frac{2}{3}, -\frac{6}{5}$ **13.** To determine the value of $f(5)$, start at 5 on the x-axis and move vertically until you reach a point on the graph. Then move horizontally to the y-axis to find the value of $f(5)$. To find a solution to the equation $f(x) = 5$, start at the 5 on the y-axis and move horizontally until you reach a point on the graph. Then move vertically to the x-axis to obtain the value of x. **15.** (a) The number of fax machines sold in 1990. (b) 60 **17.** 1 **19.** 3 **21.** Positive **23.** Positive **25.** $-1, 5, 9$ **27.** 1 unit **29.** After 1 hour **31.** 4; After 5 hours the concentration of the drug in the blood is 4 units. **33.** No **35.** Yes

37. $$f(x) = \begin{cases} .06x & \text{for } 50 \le x \le 300 \\ .02x + 12 & \text{for } 300 < x \le 600 \\ .015x + 15 & \text{for } 600 < x \end{cases}$$

39. $1, \frac{1}{2}, 9$ **41.** $x \neq 1, 2$ **43.** $x < 3$ **45.** Function **47.** Not a function **49.** Not a function

EXERCISES 1.2, PAGE 18

1. 1/(2X) **3.** √ (1 – X^3) or (1 – X^3)^.5 **5.** (X^2)^X or X^(2X) **7.** –9, –4, 4 **9.** –2; 3, 4.3820, 6.6180 **11.** –.21875; .2355, 4.4880 **13.** (3, 3) **15.** (–.6458, –1.6458), (4.6458, 3.6458) **17.** (–3.8708, –4.9354), (–2, –4), (–.1292, –3.0646) **19.** [–2, 2] *by* [–1, 3] **21.** [–7, 7] *by* [–65, 2] **23.** [–1, 3] *by* [–10, 10] **25.** The tick marks are too close together. To correct this, set Xscl = 10 and Yscl = 10. **27.** ZDecimal **29.** 3 **31.** 3.8851 **33.** (TI-82) **23** [STO ▶] [X, T, θ] [ENTER]; (TI-83) **23** [STO ▶] [X, T, θ, n] [ENTER]; (TI-85) **23** [STO ▶] [x-VAR] [ENTER] **35.** If you are not in the Home screen, press [2nd] [QUIT]. Then press [CLEAR] [CLEAR].

EXERCISES 1.3, PAGE 26

1. **3.** **5.**

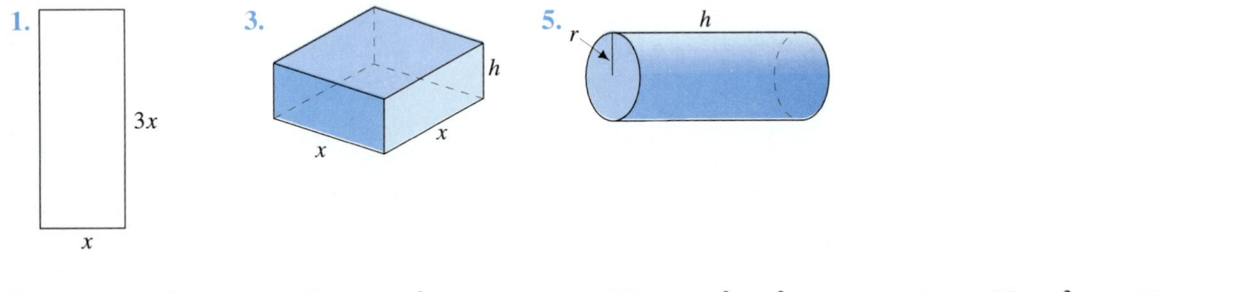

7. $P = 8x$; $3x^2 = 25$ **9.** $A = \pi r^2$; $2\pi r = 15$ **11.** $V = x^2h$; $x^2 + 4xh = 65$ **13.** $\pi r^2 h = 100$; $C = 11\pi r^2 + 14\pi rh$ **15.** $2x + 3h = 5000$; $A = xh$ **17.** $C = 36x + 20h$ **19.** 75 cm^2

21. (a) 38 (b) \$40 **23.** (a) 200 (b) 275 (c) 25 **25.** 270 cents **27.** A 100 in^3 cylinder of radius 3 in. costs \$1.62 to construct. **29.** \$1.08 **31.** The greatest profit, \$52,500, occurs when 2500 units of goods are produced. **33.** Find the point on the graph whose second coordinate is 30,000. **35.** Find $h(3)$. Find the y-coordinate of the point on the graph whose t-coordinate is 3. **37.** Find the maximum value of $h(t)$. Find the second coordinate of the highest point of the graph. **39.** Solve $h(t) = 100$. Find the t-coordinate of the point whose y-coordinate is 100. **41.** (a) \$1050 (b) \$22.11 (c) 10 units **43.** (a) 350 bicycles (b) \$6,000

EXERCISES 1.4, PAGE 39

1. –5 **3.** 0 **5.** $\frac{2}{7}$ **7.** $-\frac{2}{3}$ **9.** $y = 3x - 1$ **11.** $y = x + 1$ **13.** $y = 35 - 7x$ **15.** $y = 4$ **17.** $y = \dfrac{x}{2}$ **19.** $y = -2x$ **21.** $y = 6 - 2x$ **23.** $y = \frac{1}{2}x - \frac{1}{2}$ **25.** (a) C (b) B (c) D (d) A **27.** 2 **29.** –.75 **31.** (2, 5); (3, 7); (0, 1) **33.** $(0, -\frac{5}{4})$; $(1, -\frac{3}{2})$; $(-2, -\frac{3}{4})$ **35.** l_1

37. **39.**

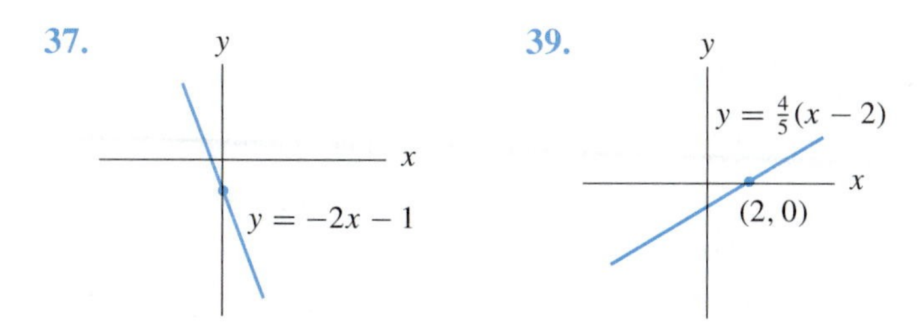

41. $y = 5$ **43.** $4x + 5y = 7$ **45.** $y = -2x$ **47.** Each unit sold increases the pay by \$5. The pay is \$60 if no units are sold. **49.** At the time the piece of equipment is purchased, its price is \$500,000. Each year, the equipment depreciates by \$50,000. **51.** (a) $K = \frac{1}{250}$, $V = \frac{1}{50}$ (b) $(-\frac{1}{k}, 0)$, $(0, \frac{1}{V})$ **53.** Let x be the number of cubic ft of gas (in thousands) and let P represent the payment. Then $P(x) = (.10)x + 5000$ dollars.

55. 65 miles per hour **57.** 13.22 feet

59. (answers may vary)

(a)–(c)

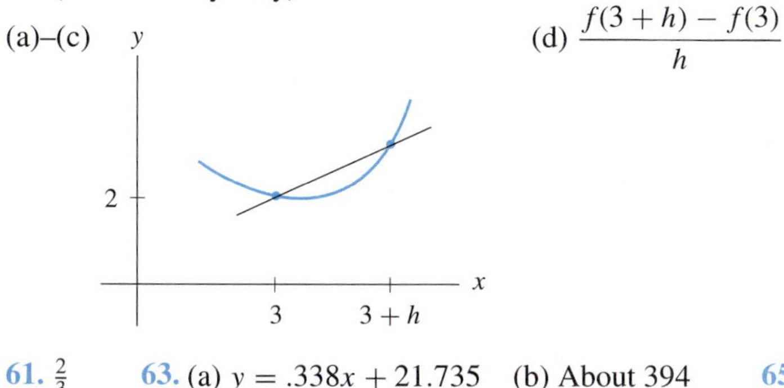

(d) $\dfrac{f(3+h) - f(3)}{h}$

61. $\frac{2}{3}$ **63.** (a) $y = .338x + 21.735$ (b) About 394 **65.** (a) $y = -327.2x + 1591.4$ (b) 610

EXERCISES 1.5, PAGE 49

1. $a = 3, b = -4, c = 0$ **3.** $a = -2, b = 3, c = 1$ **5.** $a = \frac{1}{2}, b = \frac{1}{6}, c = 8$ **7.** $x = -.2, x = 1$ **9.** No solution **11.** $x = 1.1835, x = 2.8165$ **13.** $(x + 5)(x + 3)$ **15.** $(x + \sqrt{13})(x - \sqrt{13})$ **17.** $-2(x + 5)(x - 3)$ **19.** The cost for another 5% is \$25 million. The cost for the final 5% is 21 times as much. **21.** 1 **23.** $\frac{1}{100}$ **25.** $2^{2x}, 3^{\frac{1}{2}x}, 3^{-2x}$ **27.** $2^{2x}, 3^{3x}, 2^{-3x}$ **29.** $2^{-4x}, 2^{9x}, 3^{-2x}$ **31.** $2^{x/2}, 3^{4x/3}$ **33.** $2^{-2x}, 3^x$ **35.** $3^{2x}, 2^{6x}, 3^{-x}$ **37.** $2^x, 3^x, 3^x$ **39.** $2^1 = 2$, $2^{1.4} = 2.639015822$, $2^{1.41} = 2.657371628$, $2^{1.414} = 2.664749650$, $2^{1.4142} = 2.665119089$, $2^{1.41421} = 2.665137562$, $2^{1.414213} = 2.665143104$, $2^{\sqrt{2}} = 2.665144143$; 5 decimal places. **41.** 1 **43.** 2 **45.** -1 **47.** $\frac{1}{5}$ **49.** $\frac{5}{2}$ **51.** -1 **53.** 4 **55.** 2^h **57.** $2^h - 1$

EXERCISES 1.6, PAGE 55

1. $x^2 + 9x + 1$ **3.** $9x^3 + 9x$ **5.** $\dfrac{t}{9} + \dfrac{1}{9t}$ **7.** $\dfrac{3x+1}{x^2 - x - 6}$ **9.** $\dfrac{4x}{x^2 - 12x + 32}$ **11.** $\dfrac{2x^2 + 5x + 50}{x^2 - 100}$ **13.** $\dfrac{2x^2 - 2x + 10}{x^2 + 3x - 10}$ **15.** $\dfrac{-x^2 + 5x}{x^2 + 3x - 10}$ **17.** $\dfrac{x^2 + 5x}{-x^2 + 7x - 10}$ **19.** $\dfrac{-x^2 + 3x + 4}{x^2 + 5x - 6}$ **21.** $\dfrac{-x^2 - 3x}{x^2 + 15x + 50}$

23. $\dfrac{5u-1}{5u+1}, u \neq 0$ **25.** $\left(\dfrac{x}{1-x}\right)^6$ **27.** $\left(\dfrac{x}{1-x}\right)^3 - 5\left(\dfrac{x}{1-x}\right)^2 + 1$ **29.** $\dfrac{t^3 - 5t^2 + 1}{-t^3 + 5t^2}$ **31.** $2xh + h^2$ **33.** $4 - 2t - h$ **35.** (a) $C(A(t)) = 3000 + 1600t - 40t^2$ (b) \$6040 **37.** $h(x) = x + \frac{1}{8}$; $h(x)$ converts British sizes to U.S. sizes. **39.** The graph is reflected about the y-axis. **41.** (a) Even (b) Odd (c) Odd (d) Even (e) Odd (f) Even **43.** $h(x) = x$, domain is the set of all nonnegative real numbers. **45.** $a = \frac{5}{2}$, $b = 0$ (b can be any number other than $\frac{5}{2}$).

CHAPTER 2

EXERCISES 2.1, PAGE 66

1. **3.** **5.**

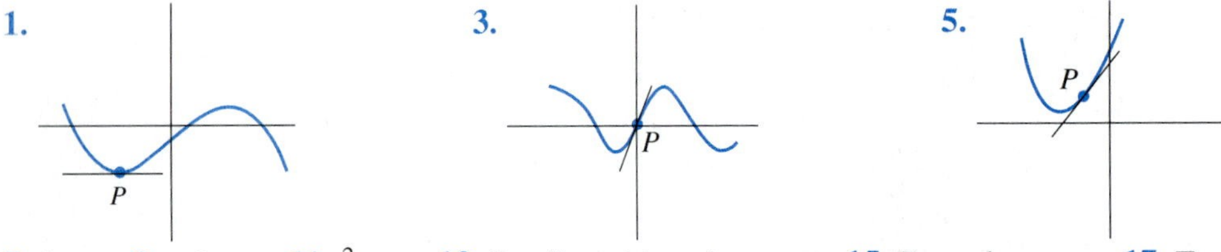

7. 1 **9.** -3 **11.** $\frac{2}{3}$ **13.** Small positive slope **15.** Zero slope **17.** Zero slope **19.** 2 **21.** $\frac{1}{4}$ **23.** 3, $y = 3x - 5$ **25.** $-\frac{1}{4}$, $y = -\frac{1}{4}x + \frac{3}{4}$ **27.** $[.9875, 1.0125]$ *by* $[1.9875, 2.0125]$ **29.** $[.9875, 1.0125]$ *by* $[1.9875, 2.0125]$ **31.** $f(6) = 4$, $f'(x) = \frac{1}{3}$ **33.** $f(-1) = 1$, $f'(-1) = 3$ **35.** $f(4) = 14$, $f'(4) = 3$ **37.** $>$ **39.** $=$ **41.** $<$ **43.** $>$ **45.** $<$ **47.** $\frac{1}{4}$ **49.** -2.4 **51.** $a = 1$, $f(a) = 1.5$, $f'(a) \approx 2$ **53.** (a) slope $= 3$, $d = \sqrt{90}$ (b) increase **55.** (a) 10.5 (b) 10.05 (c) 10

EXERCISES 2.2, PAGE 78

1. (a) 14% (b) $\frac{5}{6}$% per year (c) January 1, 1982 (d) January 1, 1980 **3.** Positive; negative **5.** Negative **7.** On January 1, 1996, the house was worth \$185,000 and was appreciating at the rate of \$3,000 per year. On July 1, 1997, the house will be worth about \$189,500. **9.** On January 1, 1953, there were 5.4 million farms in the United States, and the number of farms was dropping at the rate of 100,000 farms per year. On January 1, 1954, there were about 5.3 million farms in the United States. **11.** Four minutes after it has been poured, the coffee has a temperature of 120°F and the temperature is dropping at a rate of 5°F per minute. After 4.1 minutes, the temperature of the coffee is about 119.5°F. **13.** Manufacturing 50 bicycles each day costs \$5,000. At a production level of 50 bicycles per day, the production of an additional bicycle would increase the daily cost by about \$45. Manufacturing 52 bicycles per day costs about \$5,090. **15.** If 200 units of goods are sold, the revenue is \$30,000, and the additional revenue obtained from selling 1 additional unit of goods is about \$120. If 199 units of goods are sold, the revenue is about \$29,880. **17.** (a) The temperature of the water is increasing steadily at the rate of 8°C per minute. (b) The temperature of the water is rising slowly at first, then more rapidly as time progresses. (c) The temperature of the water is rising quickly at first, then more slowly as time progresses. **19.** a **21.** 16 feet per second **23.** After 7 seconds **25.** After 3.5 seconds **27.** $t = 1$ **29.** 11 feet **31.** $\frac{5}{3}$ feet per second **33.** 48 mph, 40 mph **35.** True **37.** True **39.** $A - d$, $B - b$, $C - a$, $D - c$ **41.** .9 grams; .8 grams per week **43.** 63 units per hour

EXERCISES 2.3, PAGE 87

1. 9 **3.** 2.5, 3.5 **5.** $-\frac{1}{2}$ unit (that is, $f(x)$ *decreases* by one-half unit) **7.** $x = .5$ **9.** (5, .5), [also $\approx$ (.15, 5.5)] **11.** 5.5 **13.** $a = 2.25$ **15.** Less **17.** D **19.** B **21.** 5 **23.** $x = 1$, [also ≈ 4.25] **25.** 1 **27.** 1 **29.** 2 **31.** $y = -x + 3$ **33.** .25, 3.75 **35.** 47,500 **37.** After 2 days **39.** 400°F **41.** After 17 seconds **43.** $1,000 **45.** 50 units **47.** 32 units **49.** 60 feet **51.** 10 feet per sec^2 **53.** After 7 seconds **55.** (a) .5; .25 (b) $f'(x) = \dfrac{1}{2\sqrt{x}}$ (c) $f'(16) = \frac{1}{8}$ (d) $y = \frac{1}{6}x + \frac{3}{2}$ **57.** (a) 3 (b) $x \approx 5$ (c) $x = 6$ (d) $x \approx 8$ (e) $x = 2$ (f) $x = 3$ **59.** (a) $\frac{1}{3}$ (b) $x = 1, x = -1$ (c) $x \approx \pm.25$ (d) $y = \frac{1}{3}x - \frac{3}{2}$ **61.** (a) 960 feet (b) After 7.5 seconds (c) -192 feet per second (d) After 2.25 seconds (e) -32 feet per sec^2 **63.** (a) 1.875 cm (b) About 1.75 cm per week (c) After about 4.94 weeks (d) After about 2.97 weeks and after about 9.63 weeks **65.** (a) 0, 3, 12, 27 (b) $f'(x) = 3x^2$ (c) 75 (d) $y = 48x - 128$

EXERCISES 2.4, PAGE 101

1. (a), (e), (f) **3.** (b), (c), (d) **5.** Decreasing for $x < -2$, relative minimum point at $x = -2$, minimum value $= -2$, increasing for $x > -2$, concave up, y-intercept (0, 0), x-intercepts (0, 0) and (-3.6, 0). **7.** Decreasing for $x < 0$, relative minimum point at $x = 0$, increasing for $0 < x < 2$, relative maximum point at $x = 2$, decreasing for $x > 2$, concave up for $x < 1$, concave down for $x > 1$, inflection point at (1, 3), y-intercept (0, 2), x-intercept (3.4, 0). **9.** Decreasing for $x < 2$, relative minimum at $x = 2$, minimum value $= 3$, increasing for $x > 2$, concave up for all x, no inflection points, defined for $x > 0$, the line $y = x$ is an asymptote, the y-axis is an asymptote. **11.** Decreasing for $1 \le x < 3$, relative minimum point at $x = 3$, increasing for $x > 3$, maximum value $= 6$ (at $x = 1$), minimum value $= 1$ (at $x = 3$), inflection point at $x = 4$, concave up for $1 \le x < 4$, concave down for $x > 4$, the line $y = 4$ is an asymptote. **13.** Slope increases for all x. **15.** Slope decreases for $x < 3$, increases for $x > 3$. Minimum slope occurs at $x = 3$. **17.** (a) C, F (b) A, B, F (c) C

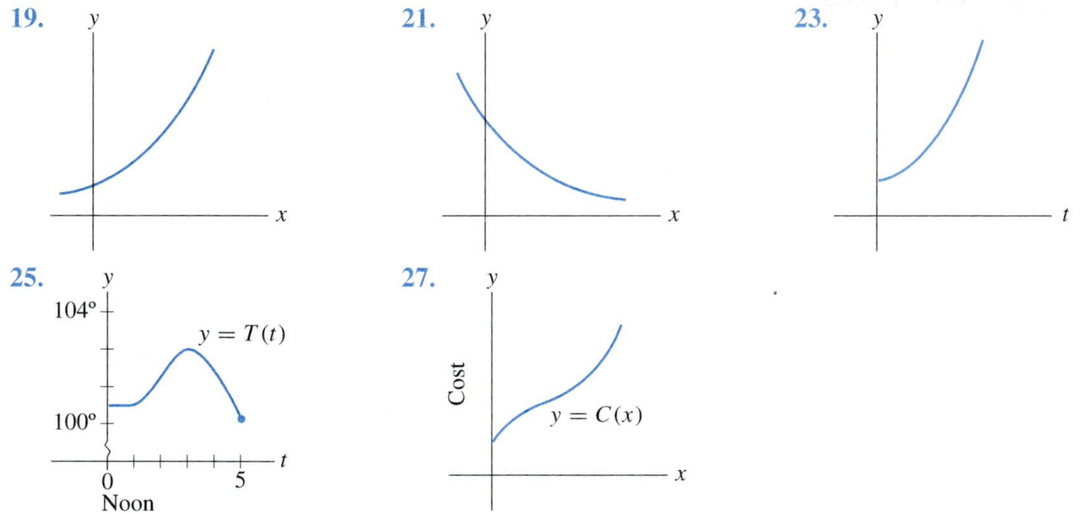

29. Oxygen content decreases until time a, at which time it reaches a minimum. After a, oxygen content steadily increases. The rate of increase increases until b, and then decreases. Time b is the time when oxygen content is increasing fastest. **31.** 1960 **33.** Tap water temperature decreases for the first 25 days of the year, at which time it reaches a minimum value. Then tap water temperature increases until the 208th day, at which time it reaches a maximum value. After the 208th day, it decreases for the remainder of the year. Tap water temperature is increasing at the greatest rate on the 116th day and decreasing at the greatest rate on the 300th day of the year. **35.** 10 million **37.** .9 million people per year **39.** 1940, 150 million, 2.1 million people per year. **41.** After 120 days **43.** After 50 days **45.** After about 193 days **47.** The parachutist's speed levels off to 15 feet per second

49.

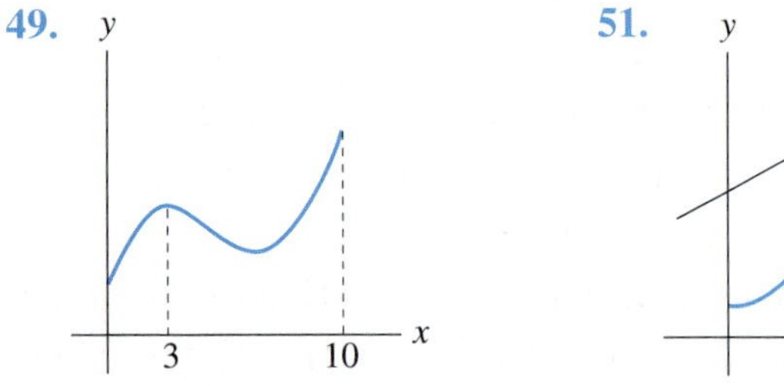

51.

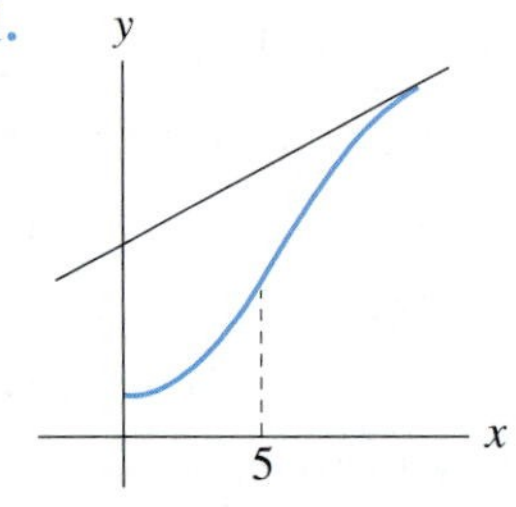

53. (a) Yes (b) Yes **55.** Relatively low **57.** (a) 10; −2 (b) −2; 2 **59.** (a) 7.8125; 2.5 (b) −1.5625; −.5 **61.** (a) 9; 0, 4 (b) 0; 1, 3

EXERCISES 2.5, PAGE 112

1. (b), (c), (f) **3.** (d), (e), (f) **5.** (d)

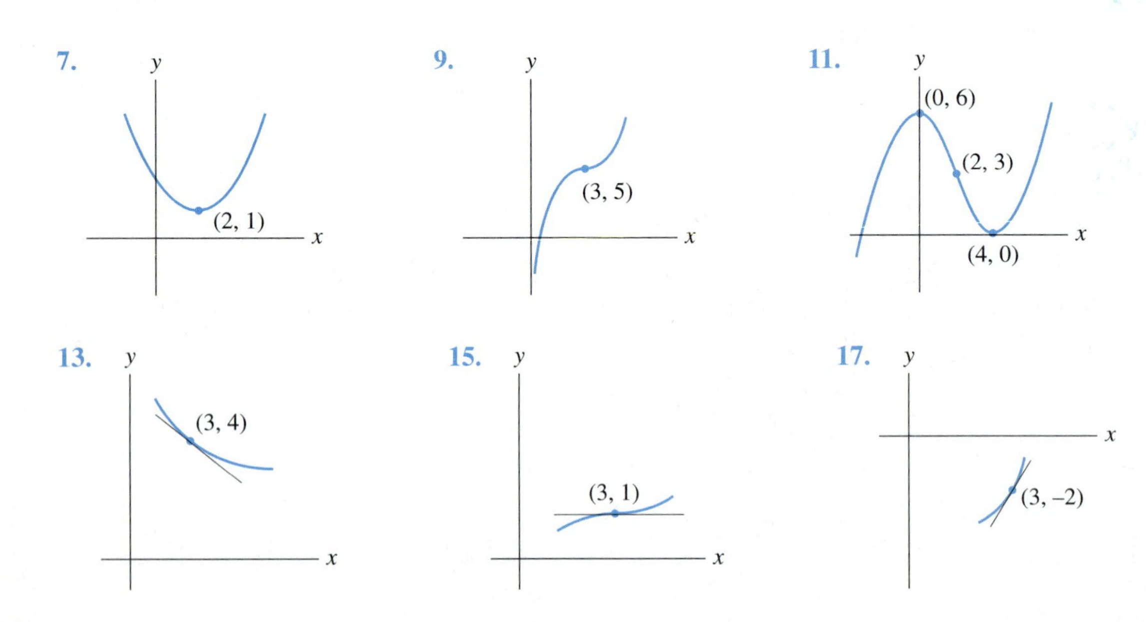

19.

	f	f'	f''
A	Pos.	Pos.	Neg.
B	0	Neg.	0
C	Neg.	0	Pos.

21. The slope is positive because $f'(6) = 2$, a positive number. 23. The slope is 0 because $f'(3) = 0$. Also, $f'(x)$ is positive for x slightly less than 3, and $f'(x)$ is negative for x slightly greater than 3. Hence $f(x)$ changes from increasing to decreasing at $x = 3$. 25. $f'(x)$ is increasing at $x = 0$, so the graph of $f(x)$ is concave up. 27. At $x = 1$, $f'(x)$ changes from increasing to decreasing, so the slope of the graph of $f(x)$ changes from increasing to decreasing. 29. $y - 3 = 2(x - 6)$ 31. 3.25 33. (a) $\frac{1}{6}$ in. (b) ii, because the water level is falling. 35. 4 37. 3 39. Positive 41. At $x = 2$, $f(x)$ has a relative maximum, so the graph is concave down. 43. $y = -8x + 33$ 45. At $x = 6$ 47. $f''(8)$ is positive 49. $f''(x)$ crosses the x-axis at $x = 12$ 51. $f(x)$ has a relative minimum at $x = 8$ 53. $f(x) = g'(x)$ 55. $g(x) = f'(x)$ 57. 30 59. After 9 hours 61. 65 units after 2 hours 63. 69° 65. At 6 weeks and at 44 weeks 67. At about 25 weeks; At about 51 weeks 69. 1984 71. 1987; 40% 73. Relative min. at $x = 1$, relative max. at $x = 5$, inflection point at $x = 3$ 75. Relative min. at $x = 0$, relative maximum at $x = 2.89$, infl. points at $x = .85$ and $x = 4.92$ 77. x-intercepts at 1 and 5, relative min. at $x = 3$ 79. x-intercepts at 0 and ± 1, relative max. at $x = 0$, relative min. at $x = \pm .71$ 81. 12 83. -2.5 85. (a) 6 milliliters (b) After 5.97 weeks (c) 3.16 milliliters per week (d) After 7.24 weeks (e) After 1.12 weeks

EXERCISES 2.6, PAGE 127

1. No limit 3. 1 5. No limit 7. -5 9. 5 11. No limit 13. 288 15. 0 17. 3 19. -4 21. -8 23. $\frac{6}{7}$ 25. No limit 27. $-\frac{2}{11}$ 29. 6 31. 3 33. $-\frac{2}{121}$ 35. $\frac{-\sqrt{3}}{6}$ 37. 0 39. 3 41. $f(x) = \sqrt{x}$; $a = 9$ 43. $f(x) = \frac{1}{x}$; $a = 10$ 45. $f(x) = 3x^2 + 4$; $a = 1$ 47. 0 49. 0 51. 2 53. 3 55. 1 57. 2 59. 3 61. 0

EXERCISES 2.7, PAGE 134

1. No 3. Yes 5. No 7. No 9. Yes 11. No 13. Continuous, differentiable 15. Continuous, not differentiable 17. Continuous, not differentiable 19. Not continuous, not differentiable 21. $f(5) = 3$ 23. Not possible 25. $f(0) = 12$

CHAPTER 3

EXERCISES 3.1, PAGE 143

1. $f(x) = 2x^{-1} + x^{\frac{1}{2}}$ 3. $f(x) = x^{\frac{3}{2}} + 5x^3$ 5. $f'(x) = 4x^3$ 7. $f'(x) = \frac{3}{2}x^{\frac{1}{2}}$ 9. $f'(x) = \frac{-2}{x^3}$

11. $f'(x) = -18x^{-7}$ 13. $f'(x) = -\frac{1}{5}$ 15. $f'(x) = 0$ 17. $f'(x) = x^3 - 6x$ 19. $f'(x) = \dfrac{1}{3} - \dfrac{3}{x^2}$ 21. $f'(x) = \dfrac{-3}{4\sqrt{x^3}} + \dfrac{1}{\sqrt{x}}$ 23. $f''(x) = 12x$ 25. $f''(x) = \dfrac{1}{4\sqrt{x^3}}$ 27. $f''(x) = 0$ 29. $f''(x) = -\frac{2}{3}x^{-\frac{4}{3}}$ 31. -3 33. $y = \frac{5}{2}$ 35. $-2, 3$ 37. (a) \$.80 per unit (b) 450 units 39. (a) \$16.10 (b) \$16 per unit 41. $f(2000) = 50{,}000$, $f'(2000) = 10$. The manufacture and sale of 2000 units of goods yields a profit of \$50,000. At that production level, the additional profit from the manufacture and sale of 1 additional unit of goods is about \$10. 43. Increasing; concave down

EXERCISES 3.2, PAGE 150

1. $10(6x+1)(3x^2+x)^9$ 3. $\frac{3}{2}(5-x)^{\frac{1}{2}}$ 5. $\dfrac{-(2x+1)}{(x^2+x+1)^2}$ 7. $\dfrac{-1}{\sqrt{-2x+1}}$ 9. $3(4t^3+10t)(t^4+5t^2)^2$ 11. $\dfrac{-1}{2\sqrt{1-t}}$ 13. $4\pi r^2$ 15. $4\pi hs$ 17. $6x - \frac{1}{2}$ 19. $3(4-6t)(4t-3t^2)^2$ 21. 20 23. 54 25. 34 27. $8k(2P-1)^{-3}$ 29. $f'(3) = -\frac{1}{2}$, $f''(3) = -\frac{1}{8}$ 31. $\left.\dfrac{dr}{dx}\right|_{x=1500} = 20$. If the company is spending \$1500 on advertising, then by spending one additional dollar, the company's revenue will increase by approximately \$20. 33. $\dfrac{d}{dx}(k \cdot f(x)) = k \cdot \dfrac{d}{dx}f(x)$ 35. (a) .9 grams (b) .6 grams per week (c) After 5.66 weeks (d) After 16 weeks 37. (a) 160 feet per second (b) 336 feet per second (c) 64 feet per second (d) After 8 seconds (e) After 10 seconds (f) -160 feet per second (g) -32 feet per second^2 39. (a) 1 unit (b) Decreasing; Increasing (c) .025 units

41. (a)

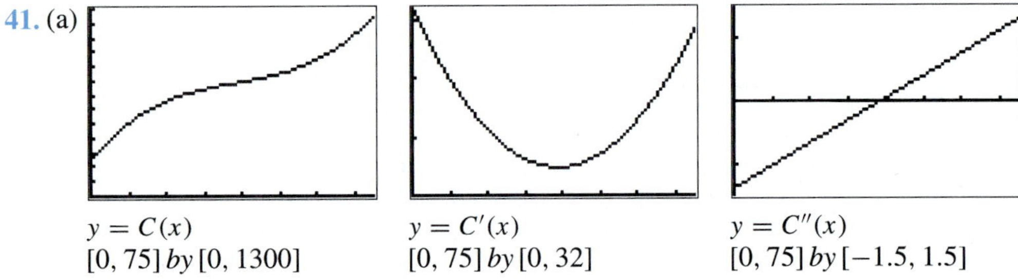

$y = C(x)$ $[0, 75]$ *by* $[0, 1300]$ $y = C'(x)$ $[0, 75]$ *by* $[0, 32]$ $y = C''(x)$ $[0, 75]$ *by* $[-1.5, 1.5]$

(b) \$506 (c) 30 units (d) \$24.51 per unit (e) $27\frac{7}{9}$ units and 50 units (f) Concave up; the marginal cost is increasing at a production level of 50 units (g) Approximately $38\frac{8}{9}$ units (h) .364

EXERCISES 3.3, PAGE 157

1.

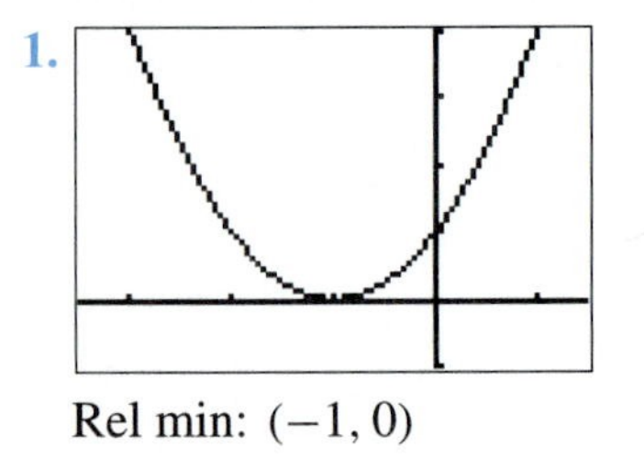

Rel min: $(-1, 0)$

3.

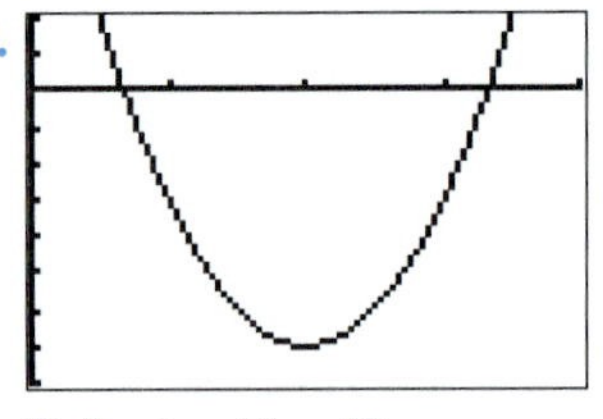

Rel min: $(2, -7)$

5.

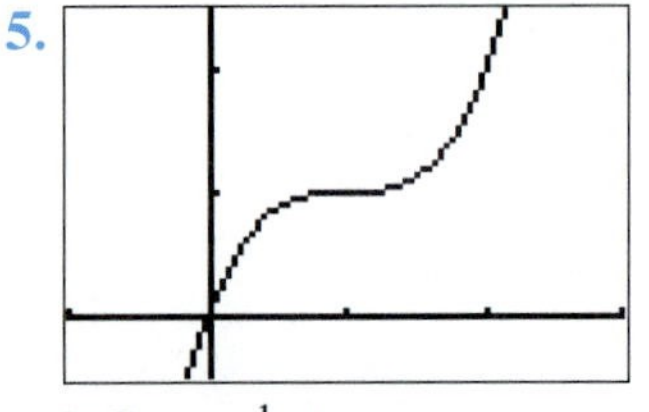

Infl pt: $(\frac{1}{2}, 1)$
Xscl = .5 Yscl = 1

7. 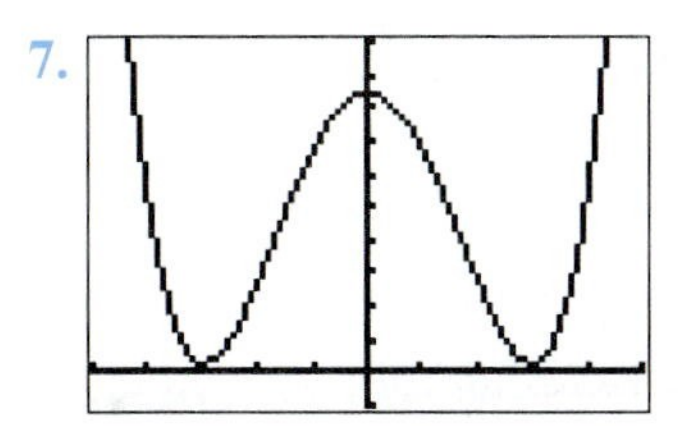

Rel min: $(-3, 3)$, $(3, 3)$
Rel max: $(0, 84)$
Infl pts: $(-\sqrt{3}, 39)$, $(\sqrt{3}, 39)$
Xscl = 1 Yscl = 10

9. 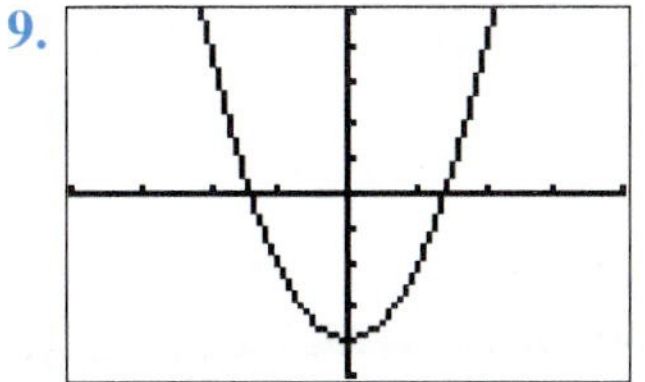

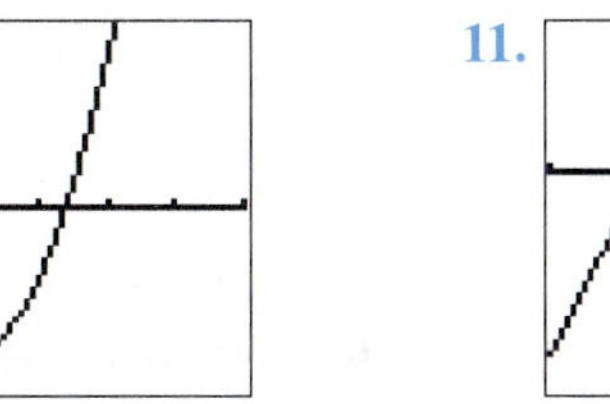

Rel min: $(0, -4)$;
$[-4, 4]$ *by* $[-5, 5]$

11. 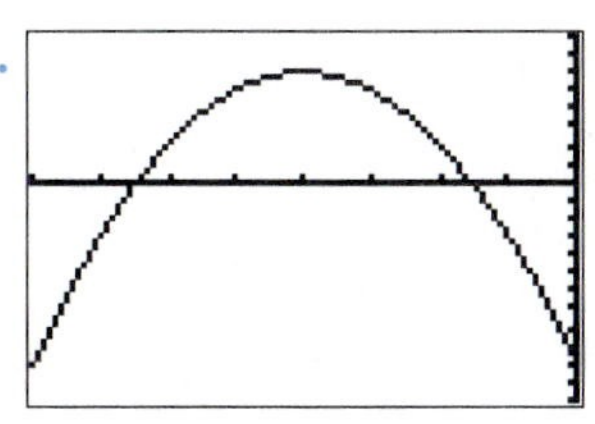

Rel max: $(-4, 6)$
$[-8, 0]$ *by* $[-12, 8]$

13. 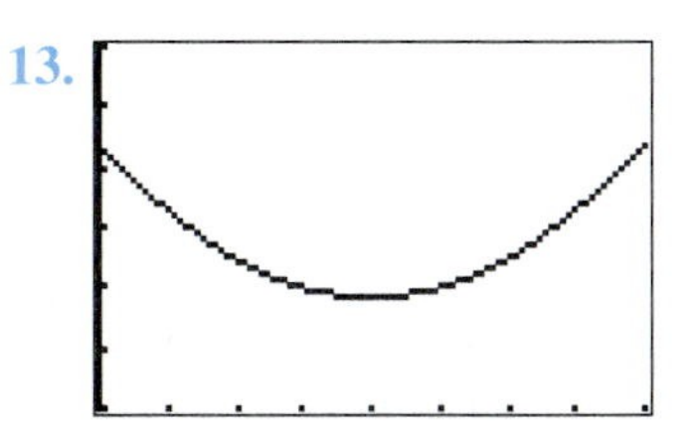

Rel min: $(8, 196)$
$[0, 16]$ *by* $[150, 300]$
Xscl = 2 Yscl = 25

15. 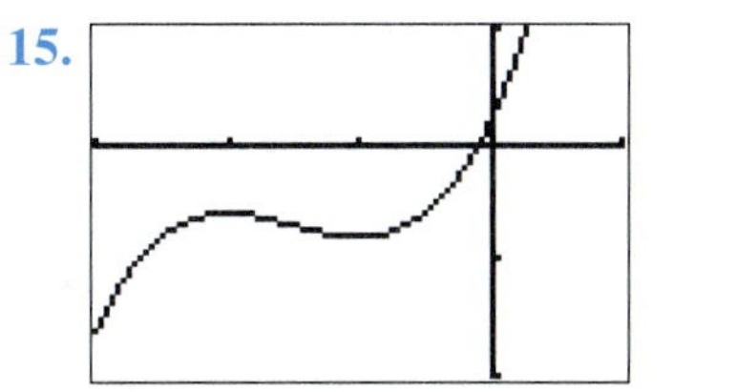

Rel max: $(-2, -3)$
Rel min: $(-1, -4)$
Infl pt: $(-1.5, -3.5)$
$[-3, 1]$ *by* $[-10, 5]$
Xscl = 1 Yscl = 5

17. 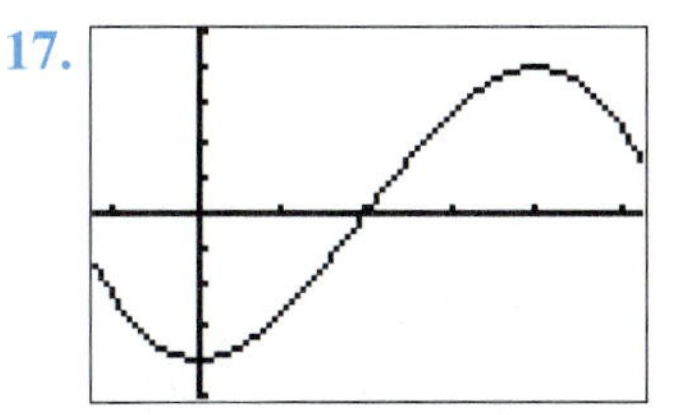

Rel max: $(20, 2000)$
Rel min: $(0, -2000)$
Infl pt: $(10, 0)$
$[-6, 26]$ *by* $[-2500, 2500]$
Xscl = 5 Yscl = 500

19. 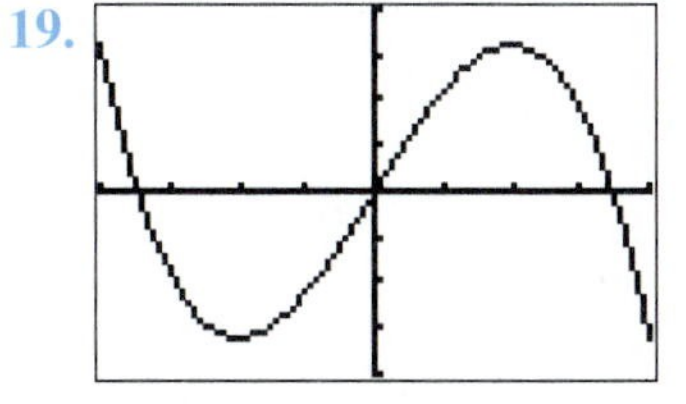

Rel min: $(-2, -16)$
Rel max: $(2, 16)$
Infl pt: $(0, 0)$
$[-4, 4]$ *by* $[-20, 20]$
Xscl = 1 Yscl = 5

21.

Infl pt: $(-1, -5)$
$[-3, 1]$ *by* $[-15, 5]$
Xscl = 1 Yscl = 5

23. 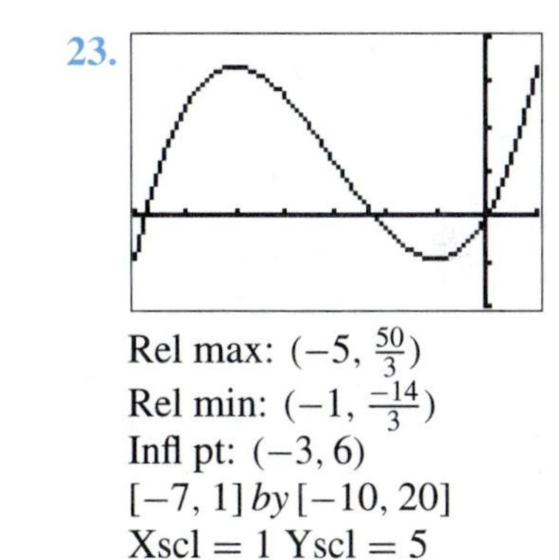

Rel max: $(-5, \frac{50}{3})$
Rel min: $(-1, \frac{-14}{3})$
Infl pt: $(-3, 6)$
$[-7, 1]$ *by* $[-10, 20]$
Xscl = 1 Yscl = 5

25. 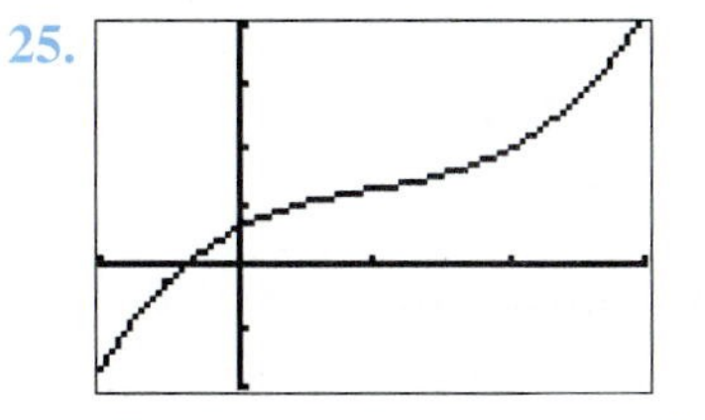

$f'(x)$ is nonnegative for all x
Xscl = 2 Yscl = 5

EXERCISES 3.4, PAGE 163

1.

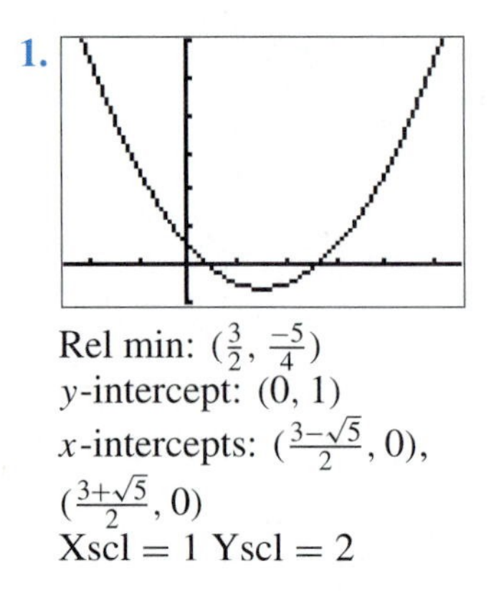

Rel min: $(\frac{3}{2}, \frac{-5}{4})$
y-intercept: (0, 1)
x-intercepts: $(\frac{3-\sqrt{5}}{2}, 0)$, $(\frac{3+\sqrt{5}}{2}, 0)$
Xscl = 1 Yscl = 2

3.

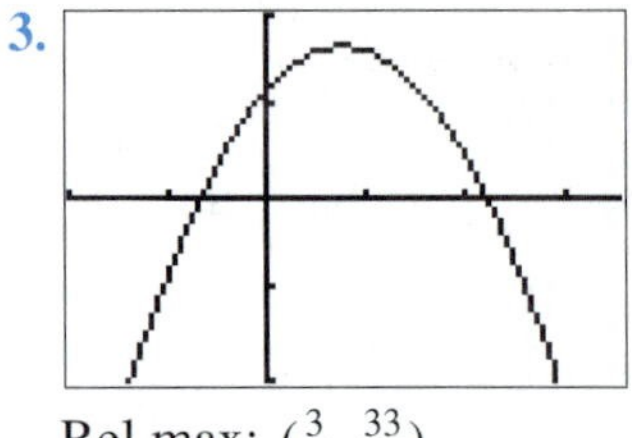

Rel max: $(\frac{3}{2}, \frac{33}{4})$
y-intercept: (0, 6)
x-intercepts: $(\frac{3-\sqrt{33}}{2}, 0)$, $(\frac{3+\sqrt{33}}{2}, 0)$
Xscl = 2 Yscl = 5

5.

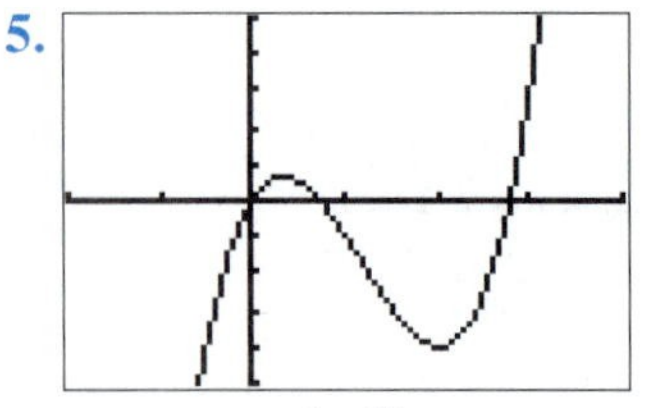

Rel max: $(\frac{1}{3}, \frac{17}{27})$
Rel min: (2, −4)
Infl pt: $(\frac{7}{6}, \frac{-91}{54})$
y-intercept: (0, 0)
x-intercepts: (0, 0), $(\frac{7-\sqrt{17}}{4}, 0)$, $(\frac{7+\sqrt{17}}{4}, 0)$

7.

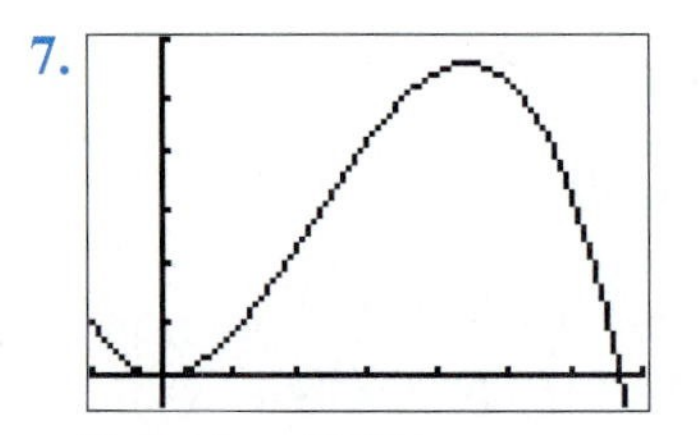

Rel min: (0, 0)
Rel max: $(\frac{28}{\pi}, \frac{10976}{\pi^2})$
Infl pt: $(\frac{14}{\pi}, \frac{5488}{\pi^2})$
y-intercept: (0, 0)
x-intercepts: (0, 0), $(\frac{42}{\pi}, 0)$
Xscl = 2 Yscl = 200

9.

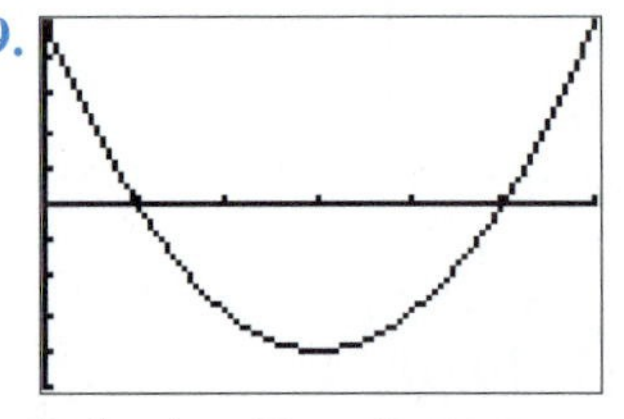

Rel min: (3, −4)
y-intercept: (0, 5)
x-intercepts: (1, 0), (5, 0)
[0, 6] *by* [−5, 5]

11.

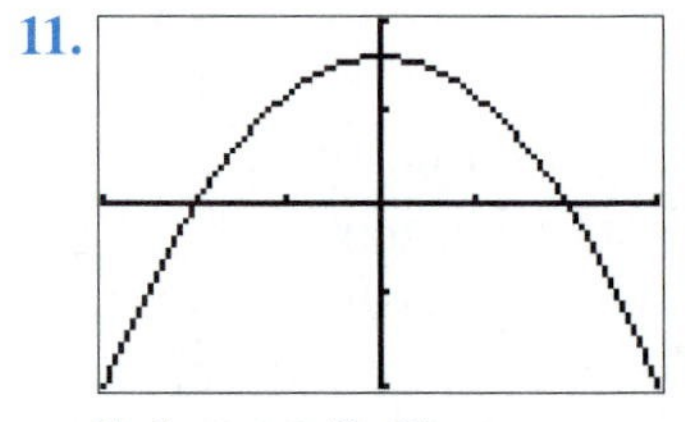

Rel max: (0, 8)
y-intercept: (0, 8)
x-intercepts: (−2, 0), (2, 0)
[−3, 3] *by* [−10, 10]
Xscl = 1 Yscl = 5

13.

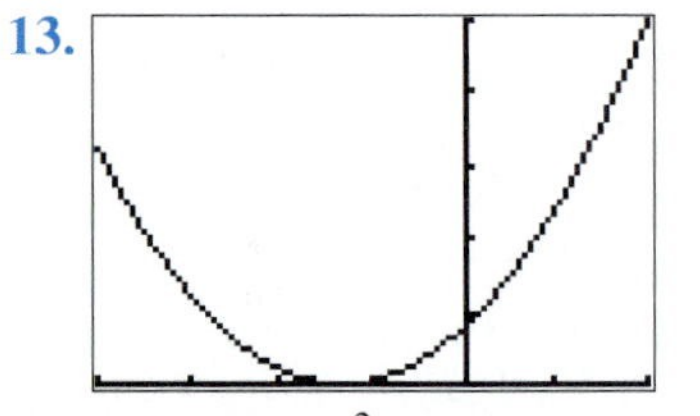

Rel min: $(\frac{-2}{3}, 0)$
y-intercept: (0, 4)
x-intercept: $(\frac{-2}{3}, 0)$
[−2, 1] *by* [0, 25]
Xscl = .5 Yscl = 5

15.

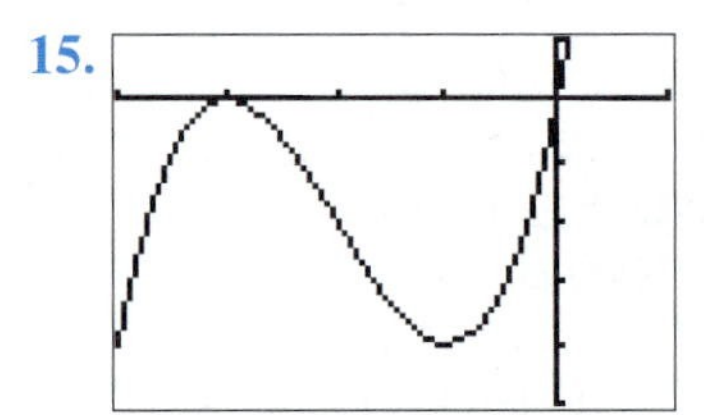

Rel max: (−3, 0)
Rel min: (−1, −4)
Infl pt: (−2, −2)
y-intercept: (0, 0)
x-intercepts: (−3, 0), (0, 0)
[−4, 1] *by* [−5, 1]

17.

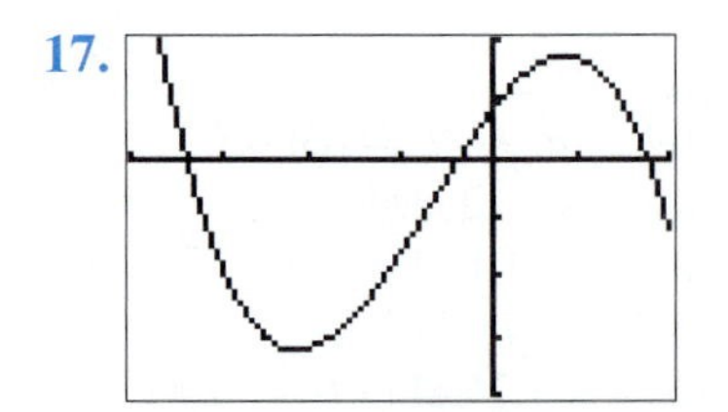

Rel min: (−53.93, −81.00)
Rel max: (20.60, 43.22)
Infl pt: (−16.67, −18.89)
y-intercept: (0, 20)
x-intercepts: (−84.71, 0), (−9.00, 0), (43.71, 0)
[−100, 50] *by* [−100, 50]
Xscl = 25 Yscl = 25

19.

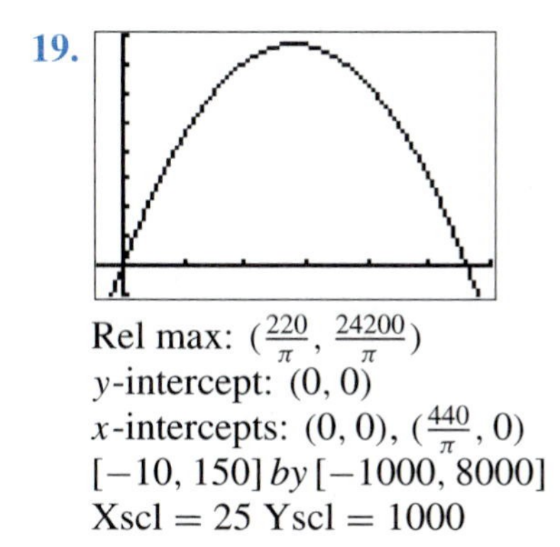

Rel max: $(\frac{220}{\pi}, \frac{24200}{\pi})$
y-intercept: $(0, 0)$
x-intercepts: $(0, 0)$, $(\frac{440}{\pi}, 0)$
$[-10, 150]$ *by* $[-1000, 8000]$
Xscl = 25 Yscl = 1000

21.

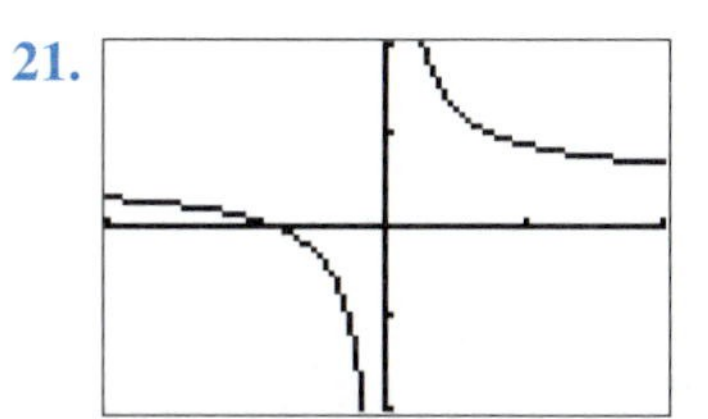

x-intercept: $(-400, 0)$
Vertical asymptote: $x = 0$
Horizontal asymptote: $y = 2500$
$[-1000, 1000]$ *by* $[-10000, 10000]$
Xscl = 500 Yscl = 5000

23.

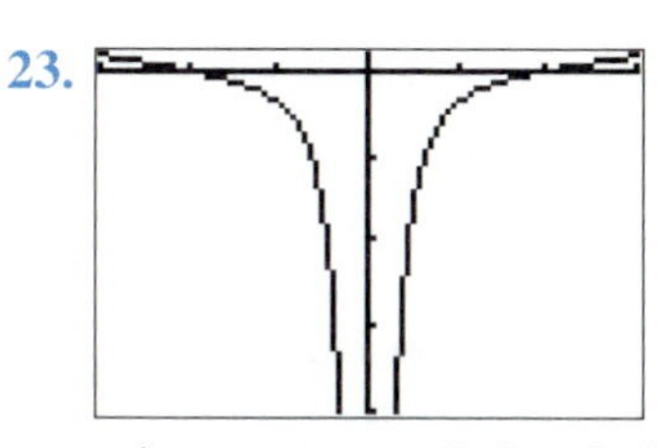

x-intercepts: $(-2, 0)$, $(2, 0)$
Vertical asymptote: $x = 0$
$[-3, 3]$ *by* $[-1000, 50]$
Xscl = 1 Yscl = 250

25. Polynomials of degree 4 have either one or three relative extreme points

27. Polynomials of degree $n + 1$ have at most n relative extreme points

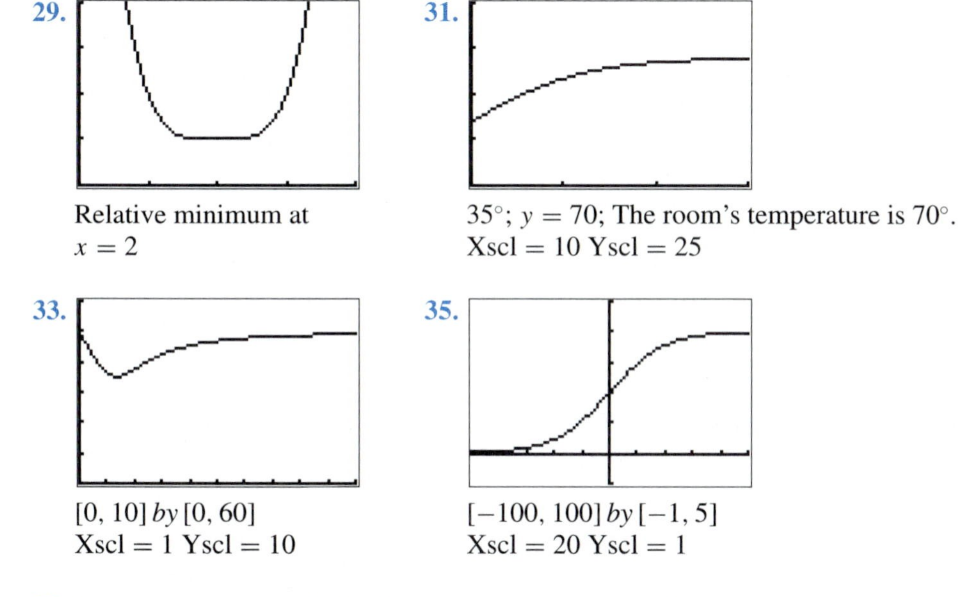

29. Relative minimum at $x = 2$

31. 35°; $y = 70$; The room's temperature is 70°.
Xscl = 10 Yscl = 25

33. $[0, 10]$ *by* $[0, 60]$
Xscl = 1 Yscl = 10

35. $[-100, 100]$ *by* $[-1, 5]$
Xscl = 20 Yscl = 1

37. Yes; $y = 0$; $f''(x)$ will also have $y = 0$ as a horizontal asymptote.

39.

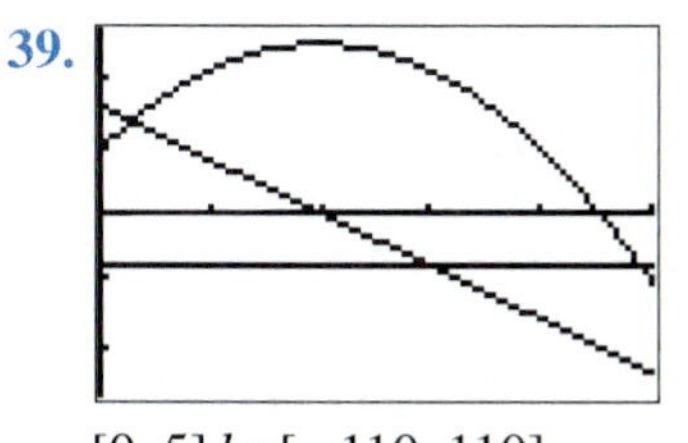

$[0, 5]$ *by* $[-110, 110]$
Xscl = 1 Yscl = 40

(a) $s(t)$: $(0, 38)$, The initial height of the grapefruit is 38 feet.
$s'(t)$: $(0, 64)$, The initial velocity of the grapefruit is 64 feet per second.
$s''(t)$: $(0, -32)$, The grapefruit is initially decelerating at the rate of 32 feet per sec^2
(b) 102 feet (c) After 1 second and after 3 seconds (d) 32 feet per second
(e) After 4 seconds (f) $s(t)$: t-intercept is approximately $(4.5, 0)$, the grapefruit hits the ground after about 4.5 seconds; $s'(t)$: t-intercept is $(2, 0)$, the grapefruit reaches its maximum height after 2 seconds. (g) 102 feet (h) Approximately 80.8 feet per second

EXERCISES 3.5, PAGE 172

1. 20 **3.** $t = 4$, $f(4) = 8$ **5.** (a) Objective equation: $A = xy$, constraint equation: $8x + 4y = 320$ (b) $A = -2x^2 + 80x$ (c) $x = 20$ ft, $y = 40$ ft

7. (a) (b) $h + 4x$ (c) Obj.: $V = x^2h$; con.: $h + 4x = 84$ (d) $V = -4x^3 + 84x^2$ (e) $x = 14$ in., $h = 28$ in.

9. Let x be the length of the fence and y the other dimension. Obj.: $C = 15x + 20y$; con.: $xy = 75$; $x = 10$ ft, $y = 7.5$ ft **11.** Let x be the length of each edge of the base and h the height. Obj.: $A = 2x^2 + 4xh$; con.: $x^2h = 8000$; 20 cm by 20 cm by 20 cm **13.** Let x be the length of the fence parallel to the river and y the length of each section perpendicular to the river. Obj.: $A = xy$; con.: $6x + 15y = 1500$; $x = 125$ ft, $y = 50$ ft **15.** Obj.: $P = xy$; con.: $x + y = 100$; $x = 50$, $y = 50$ **17.** Obj.: $A = \frac{\pi x^2}{2} + 2xh$; con.: $(2 + \pi)x + 2h = 14$; $x = \frac{14}{4 + \pi}$ ft **21.** $(3/2, \sqrt{3/2})$

EXERCISES 3.6, PAGE 181

1. Let x be the number of cases per order and r the number of orders per year. (a) \$4100 (b) Obj.: $C = 80r + 5x$; con.: $rx = 10{,}000$; 400 cases **3.** Let r be the number of production runs and x the number of microscopes manufactured per run. Obj.: $C = 2500r + 25x$; con.: $rx = 1600$; 4 runs **7.** Obj.: $A = (100 + x)w$; con.: $2x + 2w = 300$; $x = 25$ ft, $w = 125$ ft **9.** Obj.: $F = 2x + 3w$; con.: $xw = 54$; $x = 9$ m, $w = 6$ m **11.** Let x be the length of each side of a square end and y the remaining dimension. Obj.: $A = 2x^2 + 3xy$, con.: $x^2y = 36$; $x = 3$ in., $y = 4$ in. **13.** Let x be the length of the front side and y the length of the other side. Obj.: $C = 120x + 100y$, con.: $xy = 12{,}000$; $x = 100$ ft, $y = 120$ ft **15.** Obj.: $A = xh$, con.: $2h + \pi x = 440$; $x = 220/\pi$ yd **17.** Let x be the width of the base and h the height of the box. Obj.: $V = 2x^2h$, con.: $4x^2 + 6xh = 27$; $\frac{3}{2} \times 3 \times 2$ **19.** After 4 weeks **21.** 0 **23.** 19

EXERCISES 3.7, PAGE 192

1. \$1 **3.** 32 **5.** 5 **7.** $x = 20$ units, $p = \$133.33$ **9.** 2 million tons, \$156 per ton **11.** (a) \$2.00 (b) \$2.30 **13.** (a) $x = 15 \cdot 10^5$, $p = \$45$. (b) No. Profit is maximized when price is increased to \$50. **15.** 15%

CHAPTER 4

EXERCISES 4.1, PAGE 201

1. 1.1612, 1.105, 1.10 **3.** 1.005, 1.002, 1 **5.** e^{2x} **7.** e^{2x} **9.** e^{-8x} **11.** e^{1+x} **13.** $-e^{-x}$ **15.** $5e^x$ **17.** $-e^{-.2x}$ **19.** $\frac{-3}{e^{3x}}$ **21.** $-\frac{1}{3}e^{-\frac{x}{3}}$ **23.** $3(e^x - e^{-x})^2(e^x + e^{-x})$ **25.** $20e^x(1 + 5e^x)^3$

27. $4e^{4t}$ 29. $\dfrac{50e^{-.5t}}{(1+4e^{-.5t})^2}$ 31. \$493.19; \$172.62 per year 33. 35.85 units; 6.93 units per hour 35. $f(3) \approx 93$; $f'(3) \approx 47$; after 3 days, about 93 people will have heard the information, and the information will be spreading at the rate of about 47 people per day. 37. 159.62 feet per second; 6.46 feet per sec^2 39. $y' = 4y$ 41. $y' = -.15y$ 43. 1 45. $y = x + 1$ 47. $y = -.18x + .66$

EXERCISES 4.2, PAGE 208

1. $xe^x + e^x$ 3. $\dfrac{(1+2x)e^{2x}}{(1+x)^2}$ 5. $\dfrac{-e^x(x+1)}{(xe^x-1)^2}$ 7. $24x^3 + 6x^2 - 30x - 5$ 9. $\dfrac{4}{(x+2)^2}$ 11. $e^{5x}(15x - 17)$ 13. $y = 2x + 1$ 15. $y = -\frac{1}{30}x + \frac{121}{300}$ 17. 21,361 bacteria; the size of the culture is decreasing by 90 bacteria per day. 19. \$2,000; \$175 per unit 21. (2, 10) 23. Maximum point at $x = 2$; inflection point at $x = \frac{3}{2}$ 25. $y = Ce^{-4x}$ 27. $y = Ce^{\frac{x}{2}}$ 29. 106.066 tons 31. AR is maximized where $0 = \dfrac{d}{dx}(AR) = \dfrac{x \cdot R'(x) - R(x) \cdot 1}{x^2}$. This happens when the production level x satisfies $xR'(x) - R(x) = 0$, and hence $R'(x) = R(x)/x = AR$. 33. 38 in.^{2}/s 37. $\dfrac{1 - 2x \cdot f(x)}{(1+x^2)^2}$ 39. $\frac{1}{8}$ 45. .1448 pints per person per year 47. $f'(x) = e^{-x} - xe^{-x}$ 49. $f'(x) = \dfrac{e^x + e^{-x}}{2}$

51. (a)

Xscl = 20 Yscl = 20

(b) \$262,500,000 (c) 10% and 75% (d) Revenue is decreasing by \$2,500,000 per percent increase in tax rate (e) 17.42% (f) 36.6%

EXERCISES 4.3, PAGE 216

1. $f(g(x)) = e^{5x}$, $g(f(x)) = 5e^x$ 3. $f(g(x)) = e^{2x}\sqrt{e^{2x}+1}$, $g(f(x)) = e^{2x\sqrt{x+1}}$ 5. $f(x) = (1+x)^3$, $g(x) = e^x$ 7. $f(x) = \dfrac{x+1}{x-1}$, $g(x) = e^{3x}$ 9. $f(x) = x^{-\frac{1}{2}}$, $g(x) = 5e^x + 1$ 11. $(2x+3)e^{x^2+3x}$ 13. $2x^2e^{x^2} + e^{x^2}$ 15. $\dfrac{-10}{x^2}e^{3+\frac{2}{x}}$ 17. $\dfrac{2}{x^2}e^{\frac{x-1}{x}}$ 19. $2e^3x$ 21. e^{1+e^x+x} 23. Maximum point: (4, 6); inflection points: $(3, 6e^{-\frac{1}{2}})$, $(5, 6e^{-\frac{1}{2}})$ 25. $y = 15x - 40$ 27. $.02e^{-.01x}e^{-2e^{-.01x}}$ 29. $\dfrac{-2x}{(1+x^2)^2}e^{\left(\frac{1}{1+x^2}\right)}$ 31. $\dfrac{e^{-x^2}(1-4x^2)}{2\sqrt{x}}$

33. (a)

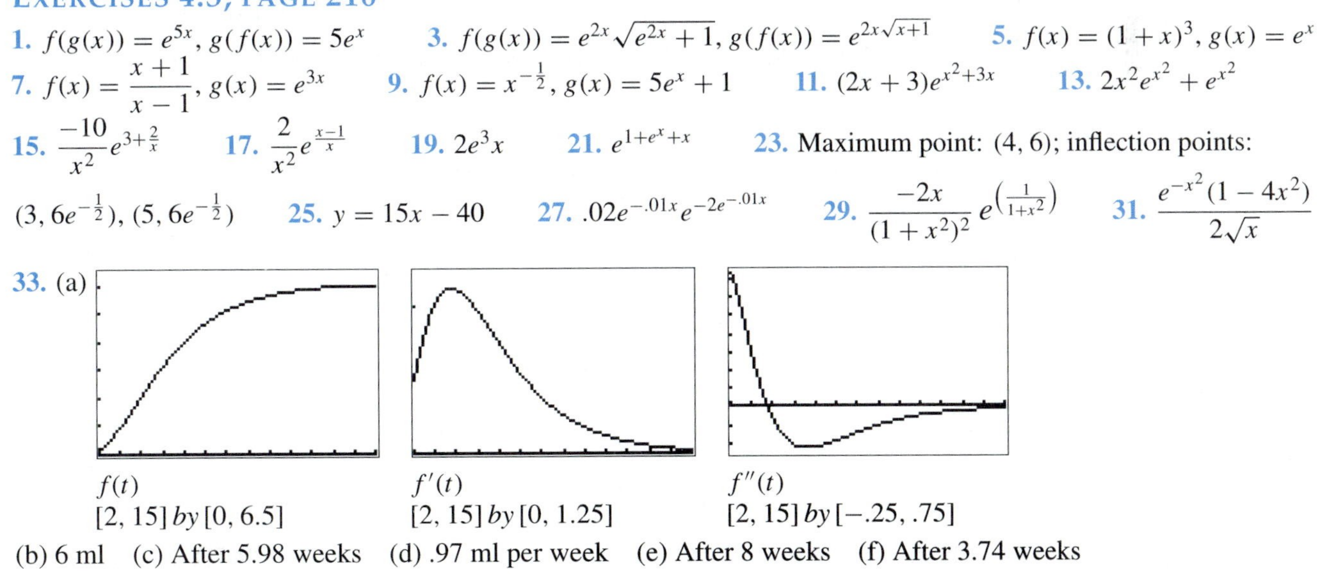

$f(t)$ [2, 15] *by* [0, 6.5] $f'(t)$ [2, 15] *by* [0, 1.25] $f''(t)$ [2, 15] *by* [−.25, .75]

(b) 6 ml (c) After 5.98 weeks (d) .97 ml per week (e) After 8 weeks (f) After 3.74 weeks

EXERCISES 4.4, PAGE 221

1. -1 **3.** $-\ln 1.7$ **5.** $e^{2.2}$ **7.** 2 **9.** e **11.** 1 **13.** $\frac{1}{2}\ln 5$ **15.** $4 - e^{1/2}$ **17.** $\pm e^3$

19. $\dfrac{\ln(.5)}{-.00012}$ **21.** $\frac{3}{5}$ **23.** $\dfrac{e}{2}$ **25.** $3\ln\frac{9}{2}$ **27.** $5\ln 6$ **29.** $\frac{1}{5}\ln\frac{2}{5}$ **31.** $-\ln\frac{3}{2}$

33. $(-\ln 3, 3 - 3\ln 3)$, minimum **35.** $(\frac{1}{2}\ln\frac{3}{2}, \frac{1}{2})$, minimum **37.** $2\ln 51$ **39.** $\ln 2$ **41.** 1.6094; $\ln 5$

43. 7.3891; e^2

45. (a)

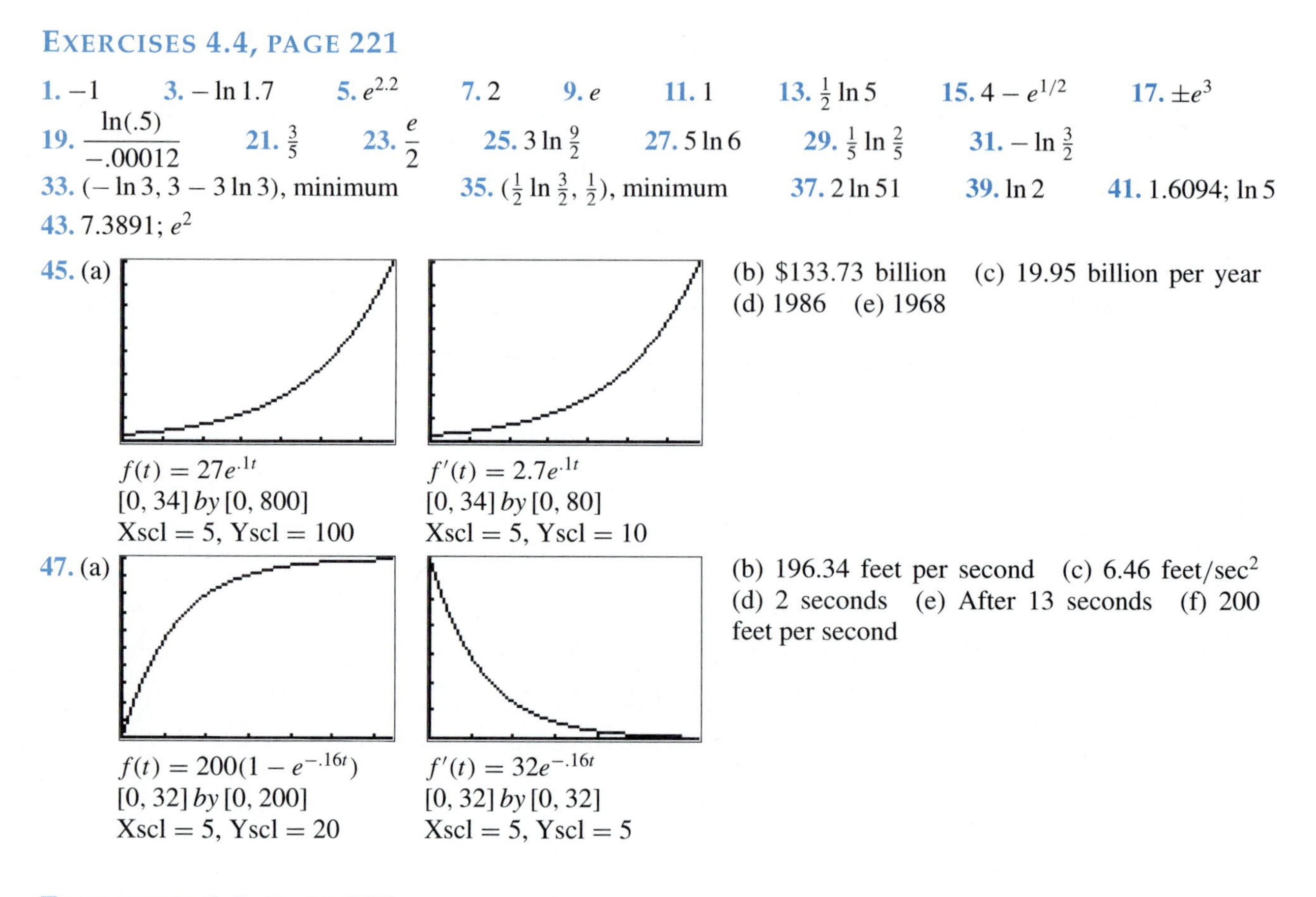

$f(t) = 27e^{.1t}$
$[0, 34]$ *by* $[0, 800]$
Xscl = 5, Yscl = 100

$f'(t) = 2.7e^{.1t}$
$[0, 34]$ *by* $[0, 80]$
Xscl = 5, Yscl = 10

(b) \$133.73 billion (c) 19.95 billion per year (d) 1986 (e) 1968

47. (a)

$f(t) = 200(1 - e^{-.16t})$
$[0, 32]$ *by* $[0, 200]$
Xscl = 5, Yscl = 20

$f'(t) = 32e^{-.16t}$
$[0, 32]$ *by* $[0, 32]$
Xscl = 5, Yscl = 5

(b) 196.34 feet per second (c) 6.46 feet/sec^2 (d) 2 seconds (e) After 13 seconds (f) 200 feet per second

EXERCISES 4.5, PAGE 226

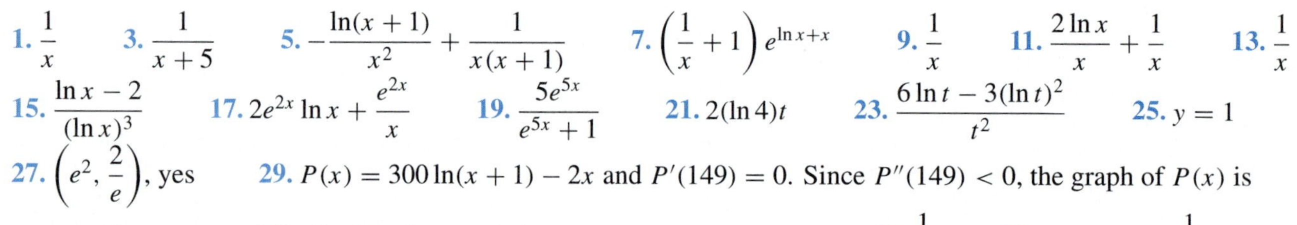

1. $\dfrac{1}{x}$ **3.** $\dfrac{1}{x+5}$ **5.** $-\dfrac{\ln(x+1)}{x^2} + \dfrac{1}{x(x+1)}$ **7.** $\left(\dfrac{1}{x} + 1\right)e^{\ln x + x}$ **9.** $\dfrac{1}{x}$ **11.** $\dfrac{2\ln x}{x} + \dfrac{1}{x}$ **13.** $\dfrac{1}{x}$

15. $\dfrac{\ln x - 2}{(\ln x)^3}$ **17.** $2e^{2x}\ln x + \dfrac{e^{2x}}{x}$ **19.** $\dfrac{5e^{5x}}{e^{5x}+1}$ **21.** $2(\ln 4)t$ **23.** $\dfrac{6\ln t - 3(\ln t)^2}{t^2}$ **25.** $y = 1$

27. $\left(e^2, \dfrac{2}{e}\right)$, yes **29.** $P(x) = 300\ln(x+1) - 2x$ and $P'(149) = 0$. Since $P''(149) < 0$, the graph of $P(x)$ is concave down at $x = 149$. So $P(x)$ has a relative maximum point there. **31.** $\dfrac{1}{e}$ **33.** $f'(x) = 1 - \dfrac{1}{x}$

35. $x + 2x\ln x$

37.

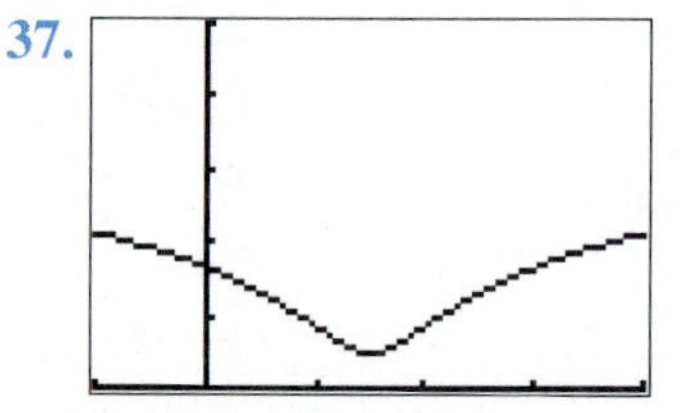

Rel min: (3, 1)
Infl pts: $(2, 1 + \ln 2)$, $(4, 1 + \ln 2)$

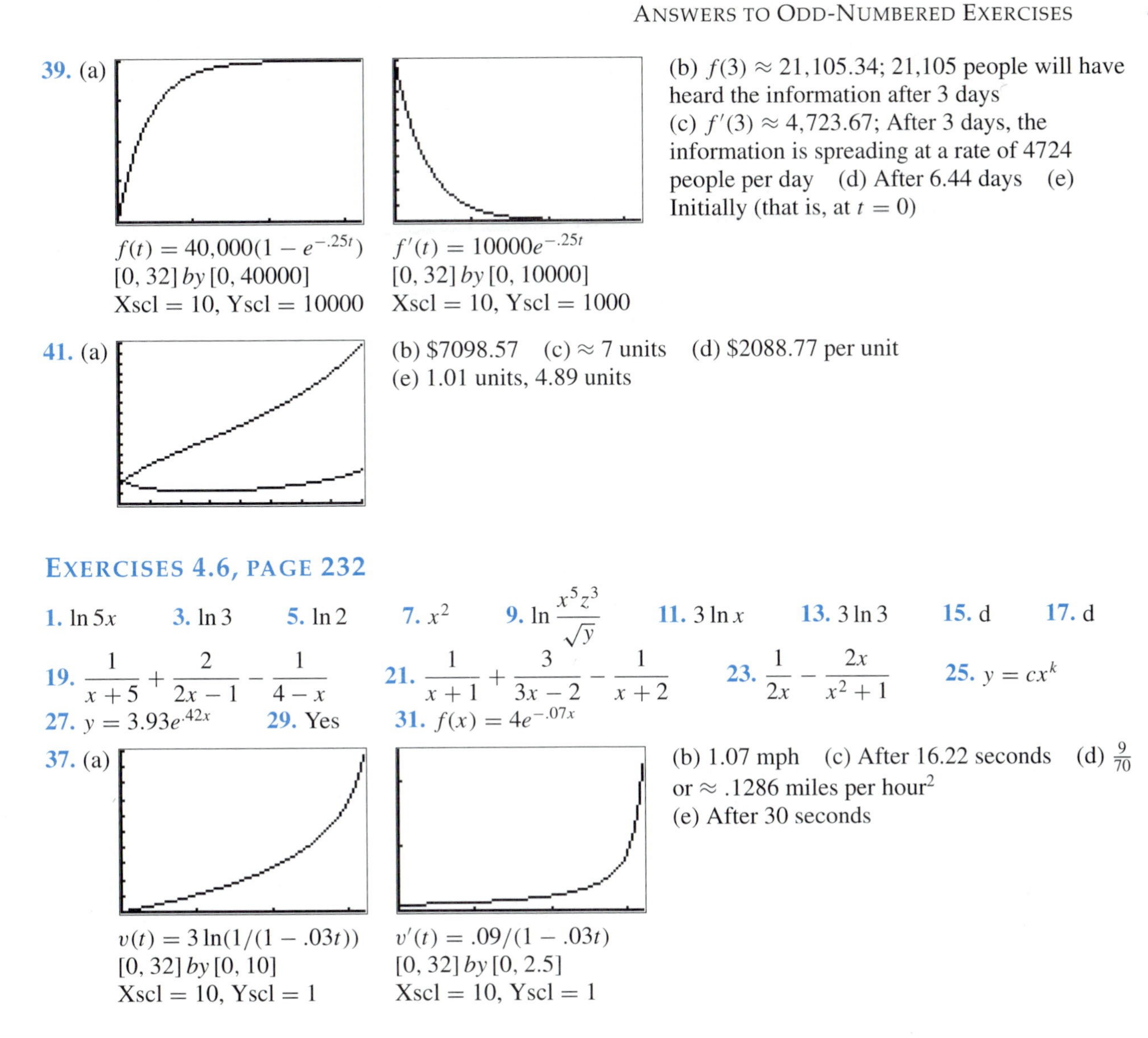

39. (a)

$f(t) = 40{,}000(1 - e^{-.25t})$
[0, 32] *by* [0, 40000]
Xscl = 10, Yscl = 10000

$f'(t) = 10000e^{-.25t}$
[0, 32] *by* [0, 10000]
Xscl = 10, Yscl = 1000

(b) $f(3) \approx 21{,}105.34$; 21,105 people will have heard the information after 3 days (c) $f'(3) \approx 4{,}723.67$; After 3 days, the information is spreading at a rate of 4724 people per day (d) After 6.44 days (e) Initially (that is, at $t = 0$)

41. (a) (b) \$7098.57 (c) ≈ 7 units (d) \$2088.77 per unit (e) 1.01 units, 4.89 units

EXERCISES 4.6, PAGE 232

1. $\ln 5x$ **3.** $\ln 3$ **5.** $\ln 2$ **7.** x^2 **9.** $\ln \dfrac{x^5 z^3}{\sqrt{y}}$ **11.** $3 \ln x$ **13.** $3 \ln 3$ **15.** d **17.** d

19. $\dfrac{1}{x+5} + \dfrac{2}{2x-1} - \dfrac{1}{4-x}$ **21.** $\dfrac{1}{x+1} + \dfrac{3}{3x-2} - \dfrac{1}{x+2}$ **23.** $\dfrac{1}{2x} - \dfrac{2x}{x^2+1}$ **25.** $y = cx^k$

27. $y = 3.93e^{.42x}$ **29.** Yes **31.** $f(x) = 4e^{-.07x}$

37. (a)

$v(t) = 3\ln(1/(1 - .03t))$
[0, 32] *by* [0, 10]
Xscl = 10, Yscl = 1

$v'(t) = .09/(1 - .03t)$
[0, 32] *by* [0, 2.5]
Xscl = 10, Yscl = 1

(b) 1.07 mph (c) After 16.22 seconds (d) $\frac{9}{70}$ or $\approx .1286$ miles per hour^2 (e) After 30 seconds

CHAPTER 5

EXERCISES 5.1, PAGE 244

1. (a) $P(t) = 3e^{.02t}$ (b) 3 million (c) 3.3 million (d) 80,000 people per year (e) 2004 (f) .02 (g) 60,000 people per year (h) 1997 (i) 34.66 **3.** .017 **5.** About 22 **7.** 27 million cells **9.** (a) After 3.8 hours (b) After 3.6 hours **11.** .02; 60,000 people per year; 5 million **13.** \$161,100 **15.** 66.4 **17.** $f'(t) = .0952f(t)$, $f(0) = 1954.52$; \$6737.88 **19.** (a) 9 million (b) 1978 (c) .225 million people per year (d) 1982 **21.** (a) $P(t) = 8e^{-.05t}$ (b) 8 (c) 4.9 g (d) .05 grams per year (e) After 41.6 years (f) .05 (g) .4 grams per year (h) After 21 years (i) 13.9 years (j) 27.7 **23.** 30 years **25.** .0347; 40 days

27. 184 years 29. $f(t) = e^{-.081t}$ 31. 3 grams per day; 8 g 33. No 35. 8990 years 39. a–D, b–E, c–A, d–B, e–C, f–F

EXERCISES 5.2, PAGE 258

1. \$1210 3. $10000(1.02)^{12}$ 5. \$1,521.96 7. \$616.84 9. 14.65% 11. 39% interest compounded continuously 13. After 22 years 15. 2006 17. \$898.66 19. 7.25% 21. \$7,985.16 23. 29% 25. 6% 27. (a) \$600 (b) \$54 per year (c) 9% 29. a–B, b–D, c–G, d–A, e–F, f–E, g–H, h–C 31. .06; .55

EXERCISES 5.3, PAGE 267

1. 20%, 4% 3. 30%, 30% 5. 60%, 300% 7. −25%, −10% 9. 12.5% 11. 5.8 yr 13. $p/(140 - p)$, elastic 15. $2p^2/(116 - p^2)$, inelastic 17. $p - 2$, elastic 19. (a) Inelastic (b) raised 21. (a) Elastic (b) increase 23. (a) 2 (b) yes 29. $p > 2$

CHAPTER 6

EXERCISES 6.1, PAGE 277

1. $\frac{1}{2}x^2 + C$ 3. $\frac{1}{3}e^{3x} + C$ 5. $3x + C$ 7. $-\frac{1}{4}$ 9. $\frac{2}{3}$ 11. -2 13. $-\frac{5}{2}$ 15. $\frac{1}{2}$ 17. -1 19. 1 21. $\frac{1}{15}$ 23. $\frac{x^3}{3} - \frac{x^2}{2} - x + C$ 25. $4\sqrt{x} - 2x^{3/2} + C$ 27. $4t + e^{-5t} + \frac{e^{2t}}{6} + C$ 29. $\frac{2}{5}t^{5/2} + C$ 31. C 33. $\frac{x^2}{2} + 3$ 35. $\frac{2}{3}x^{3/2} + x - \frac{28}{3}$ 37. $2\ln x + 2$ 39. Testing all three functions reveals that (b) is the only one that works.

41.

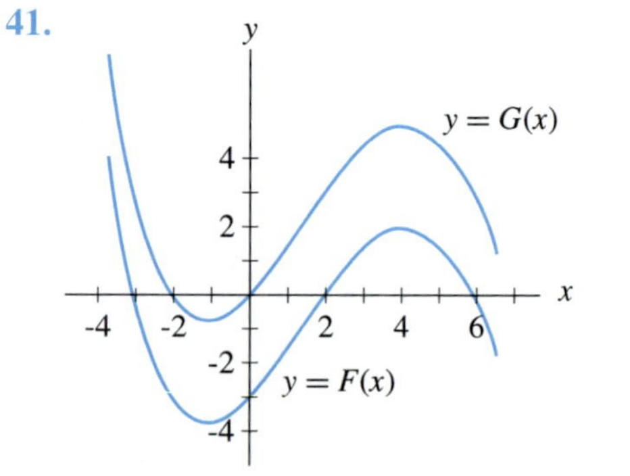

43. $\frac{1}{4}$ 45. $x^2 + 50e^{-.02x}$ 47. $\frac{8}{3}x^3 + 3e^{-x} + \frac{1}{4}x^4$

EXERCISES 6.2, PAGE 284

1. $\frac{2}{5}x^5 - 5x - \frac{3}{x} + C$ 3. $e^3x + C$ 5. $4e^{\frac{x}{4}} - 3\ln|x| + C$ 7. $x^3 - \frac{5}{2}x^2 + 6x + 5$ 9. $2e^{\frac{1}{2}x} - 4$ 11. $-4\ln|x| + 8$ 13. $6e^{\frac{1}{4}x}$ 15. $\frac{1}{4}x^4 - \frac{13}{4}$ 17. (c)

19.

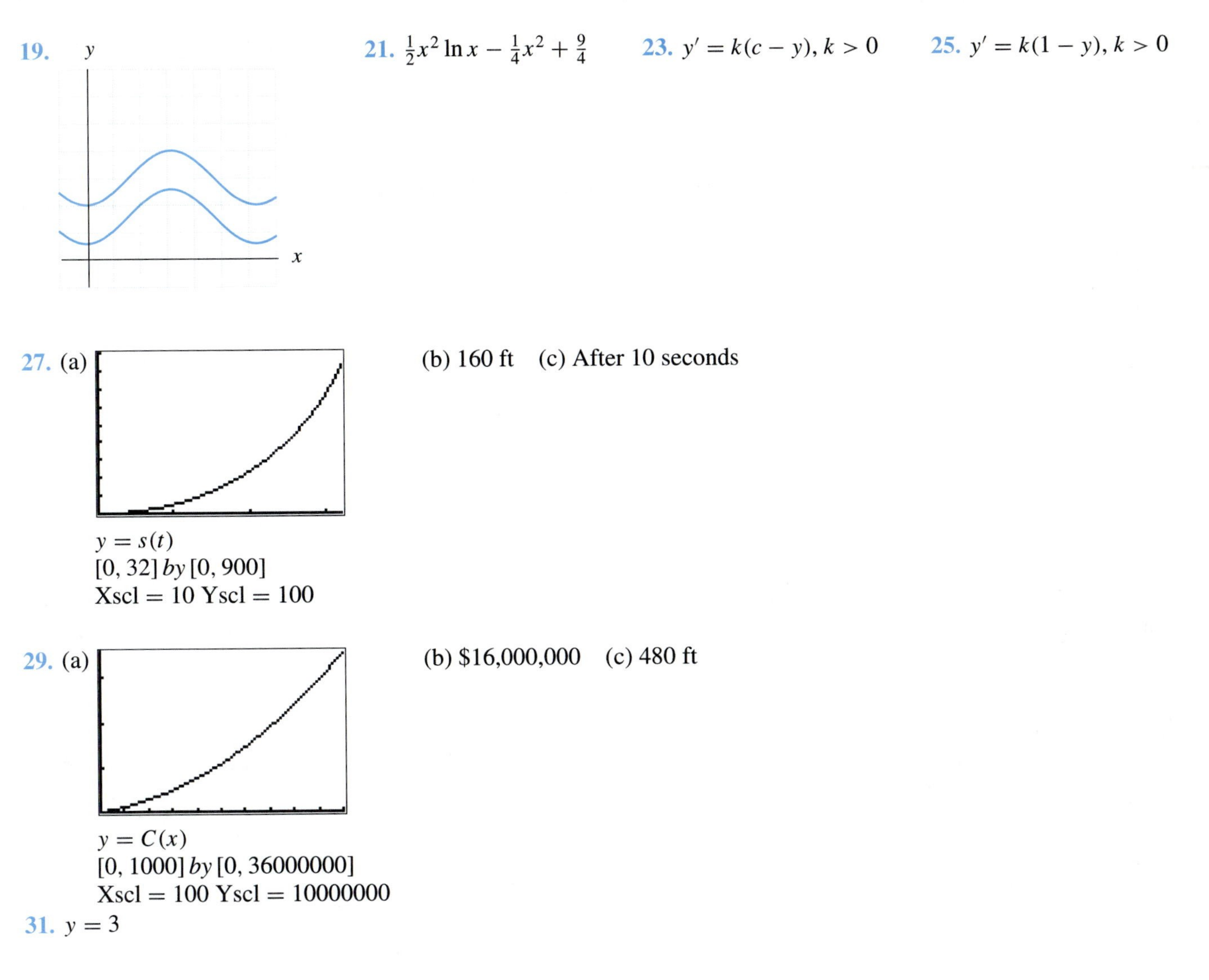

21. $\frac{1}{2}x^2 \ln x - \frac{1}{4}x^2 + \frac{9}{4}$

23. $y' = k(c - y), k > 0$

25. $y' = k(1 - y), k > 0$

27. (a)

$y = s(t)$
[0, 32] *by* [0, 900]
Xscl = 10 Yscl = 100

(b) 160 ft (c) After 10 seconds

29. (a)

$y = C(x)$
[0, 1000] *by* [0, 36000000]
Xscl = 100 Yscl = 10000000

(b) $16,000,000 (c) 480 ft

31. $y = 3$

EXERCISES 6.3, PAGE 294

1.

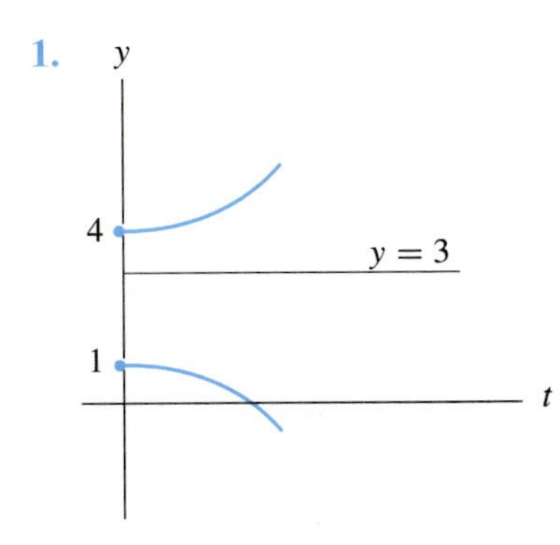

3.

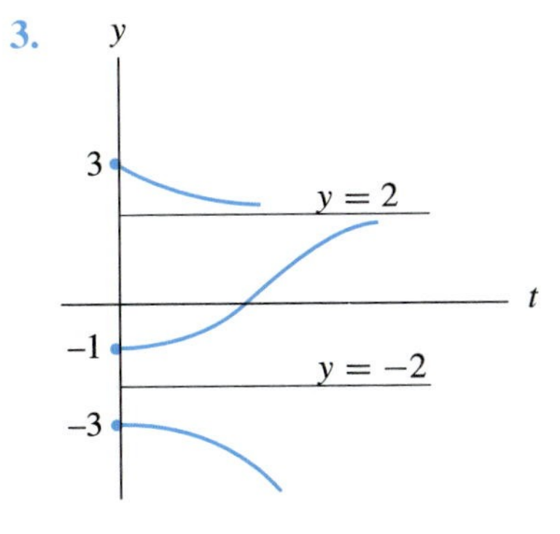

5.

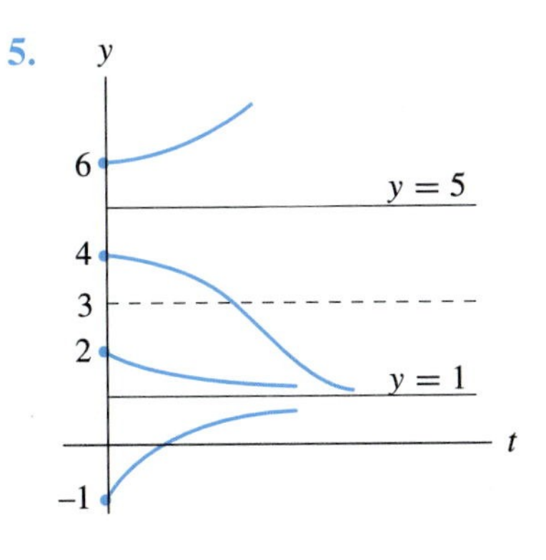

7.

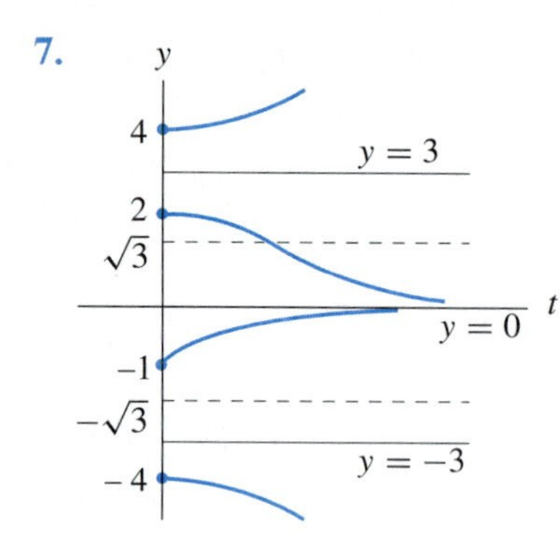

9.

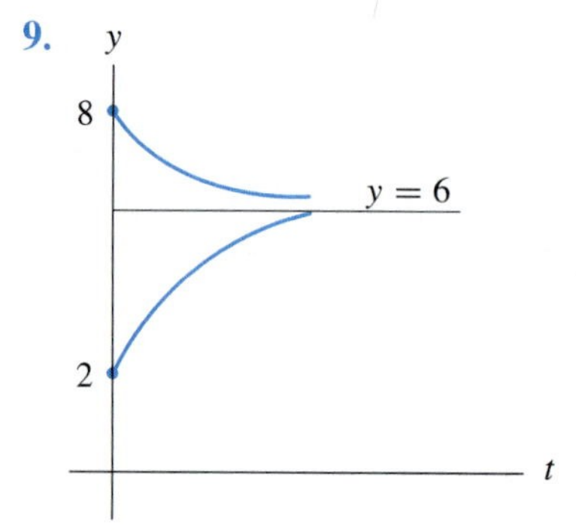

11.

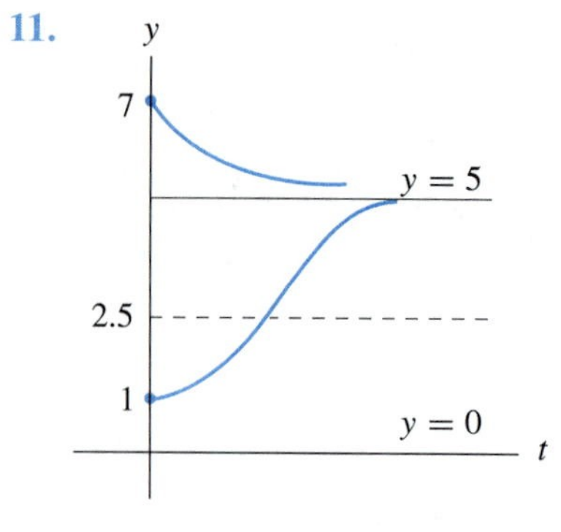

13.

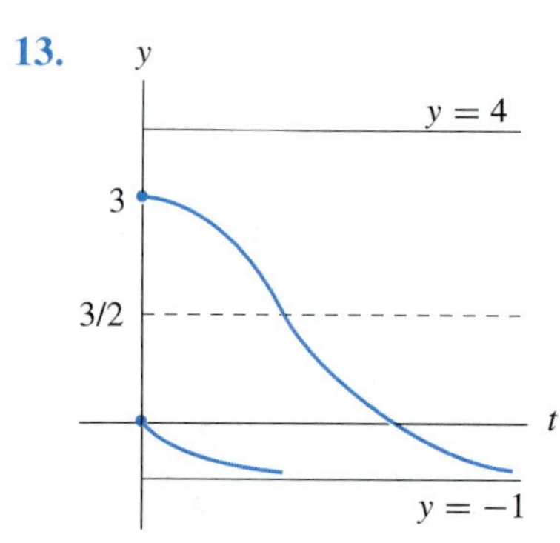

15.

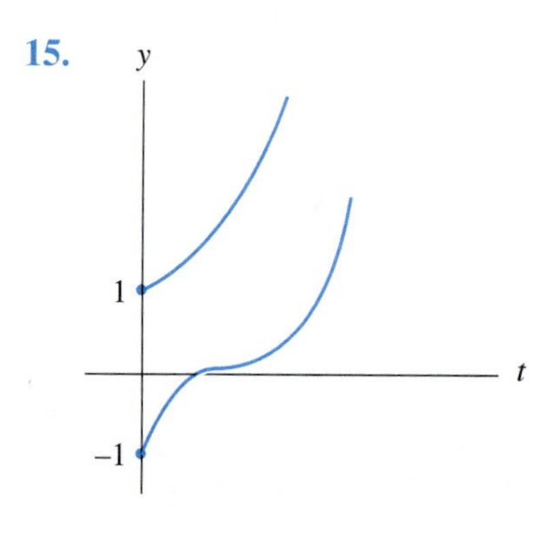

17.

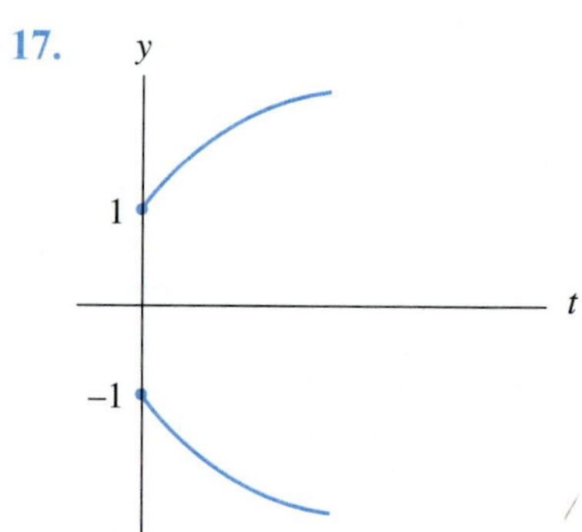

19.

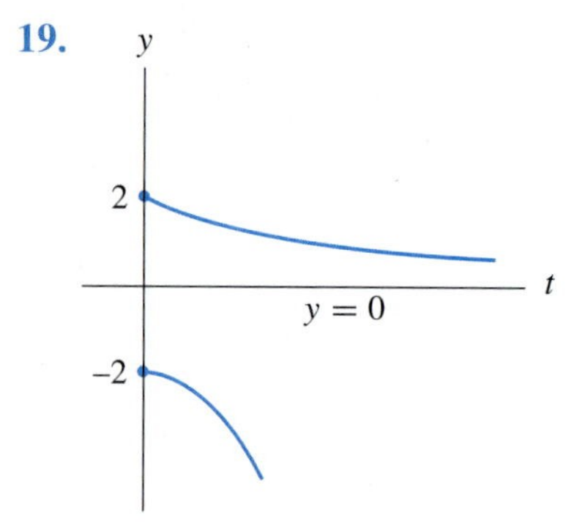

21.

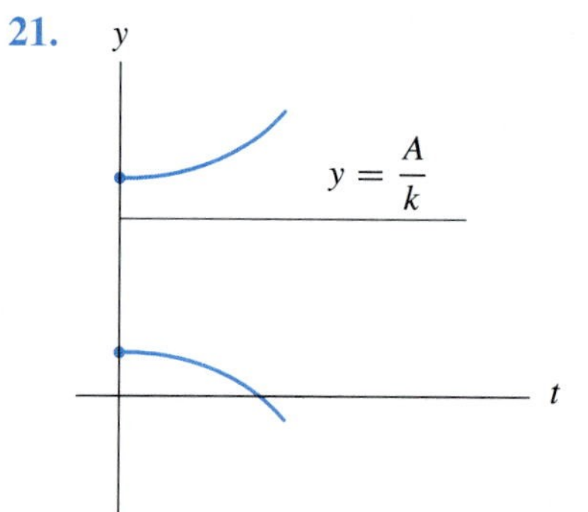

23. $y' = ky(H - y)$, $k > 0$
H = height at maturity
$y = f(t)$

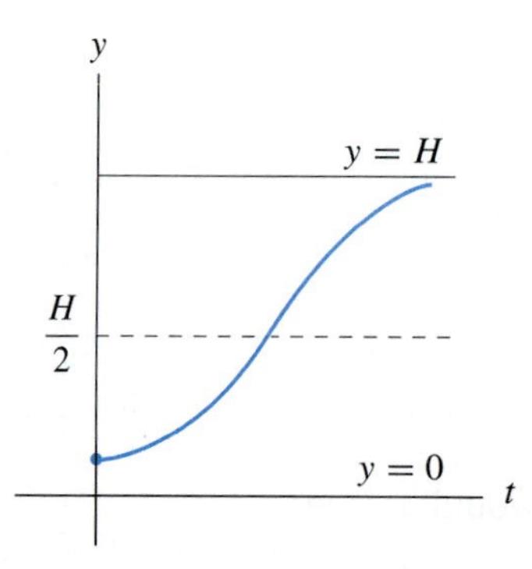

25.

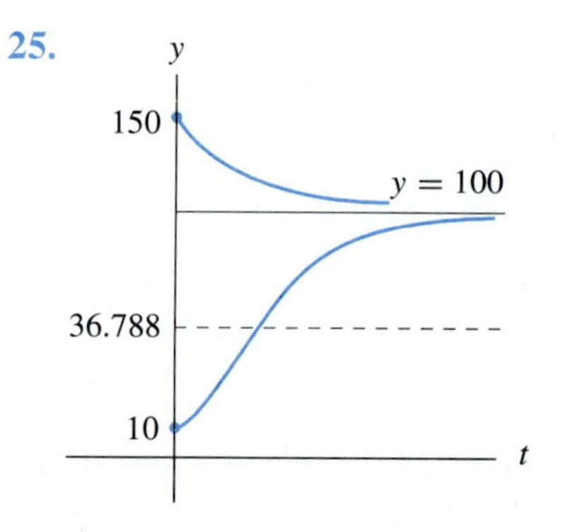

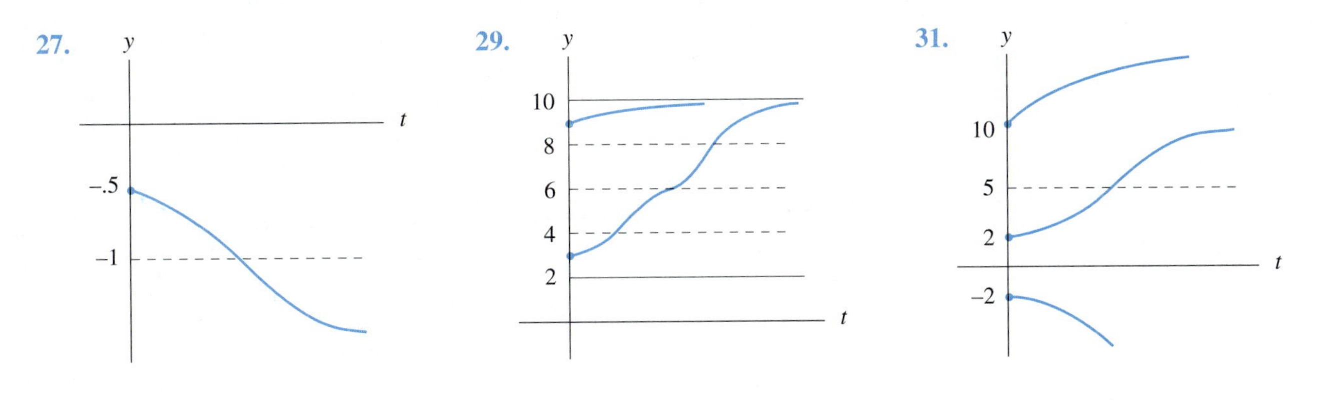

EXERCISES 6.4, PAGE 304

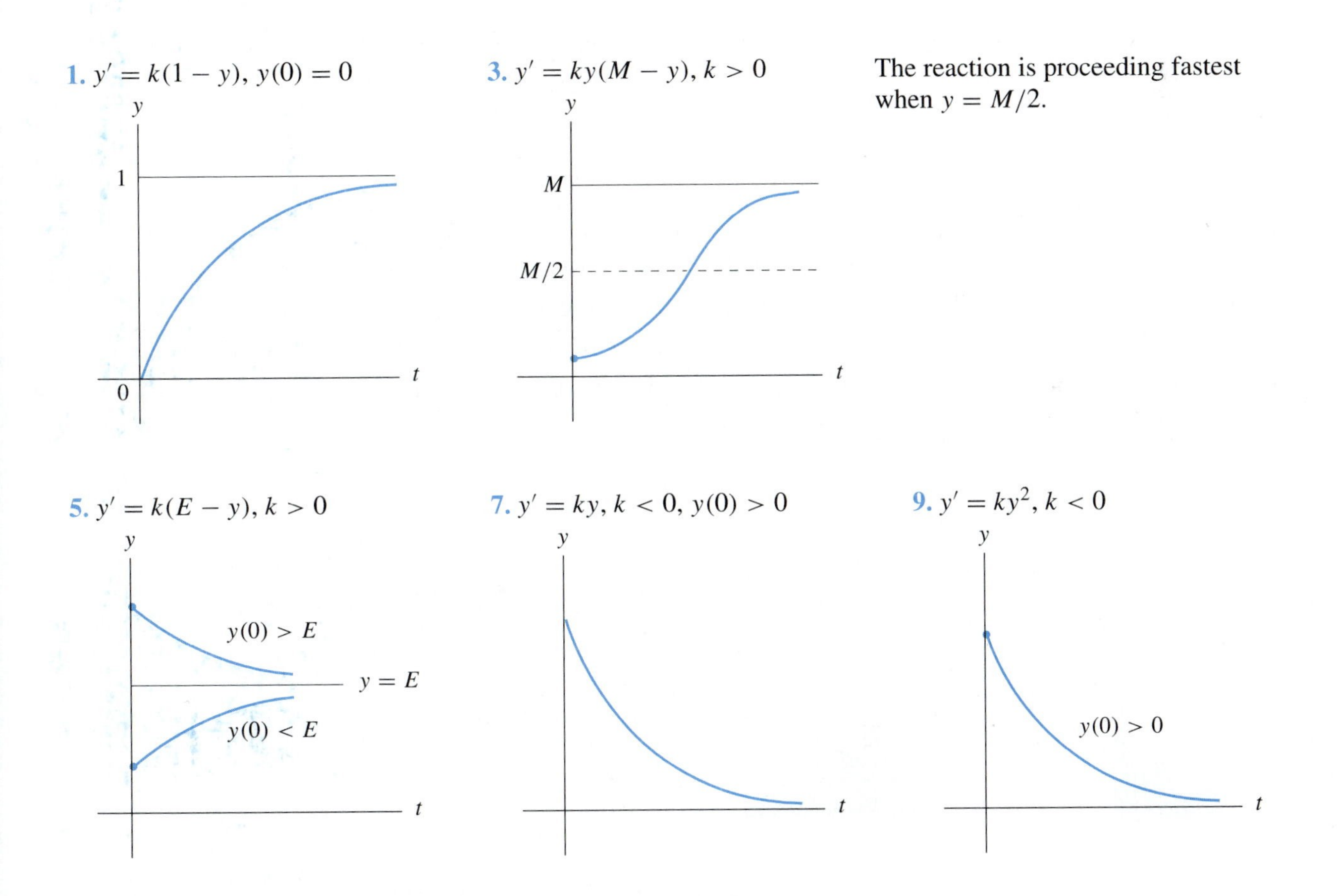

11. $A(t) = \mu(1 - e^{-\lambda t})$; The amount of excess glucose would be doubled.

13. (a) $y' = .2(160 - y)$; $y = 160(1 - e^{.2t})$
(b)

15. (a) $y' = .0042y(55 - y)$, $f(t) = \dfrac{55}{1 + 46.98e^{-.231t}}$
(b)

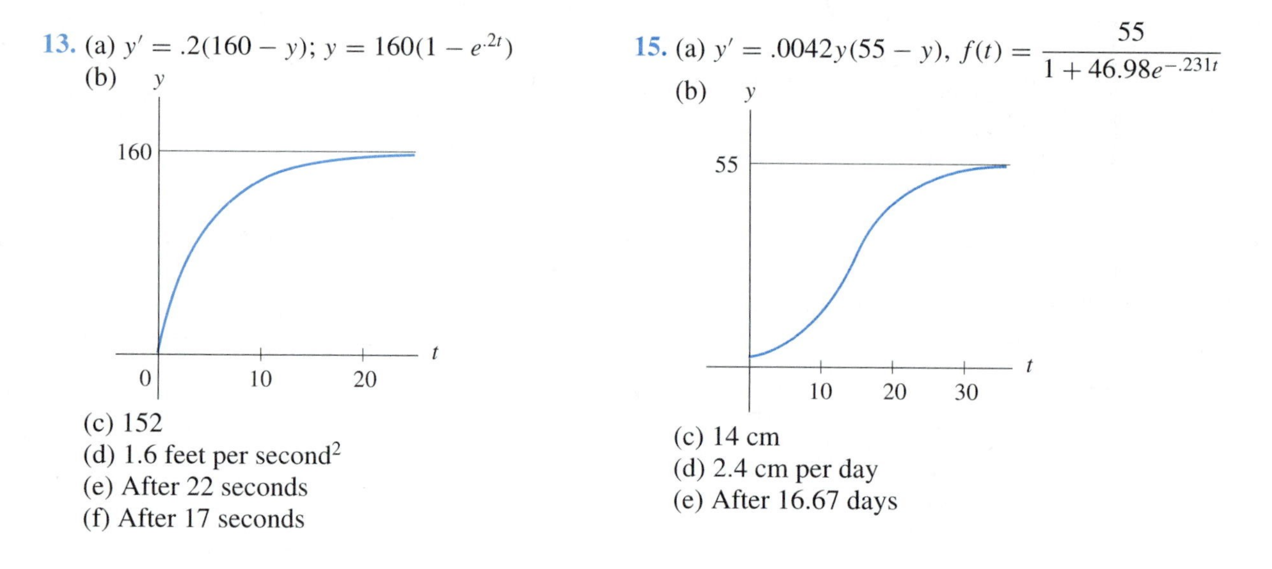

13. (c) 152
(d) 1.6 feet per second^2
(e) After 22 seconds
(f) After 17 seconds

15. (c) 14 cm
(d) 2.4 cm per day
(e) After 16.67 days

CHAPTER 7

EXERCISES 7.1, PAGE 316

1. .5; .25, .75, 1.25, 1.75 **3.** .6; 1.3, 1.9, 2.5, 3.1, 3.7 **5.** 8.625 **7.** 15.12 **9.** .077278 **11.** 40 **13.** 15 **15.** 5.625; 4.5 **17.** 1.61321; error = .04241 **19.** 1.08 L **21.** 2800 ft **23.** Increase in the population (in millions) from 1910 to 1950; rate of cigarette consumption t years after 1985; 20 to 50 **25.** Tons of soil eroded during a 5-day period **27.** The first area is 5 times the second. **29.** The area under the graph of $f(x) + g(x)$ is the sum of the area under the graph of $f(x)$ and the area under the graph of $g(x)$. **33.** 9.59164 **35.** 1.76226 **37.** .877213

EXERCISES 7.2, PAGE 326

1. $\int_1^3 (x^3 - 7x^2 + 18x - 9)\,dx = 13\frac{1}{3}$ **3.** $\int_{.5}^{2} \frac{1}{x}\,dx = 1.3862944$ **5.** $\int_1^3 (1 - x)(x - 3)\,dx = 1\frac{1}{3}$
7. $\int_0^3 (.5x^2 - 2x + 3.5)\,dx + \int_3^5 (5 - x)\,dx = 6 + 2 = 8$

9. **11.**

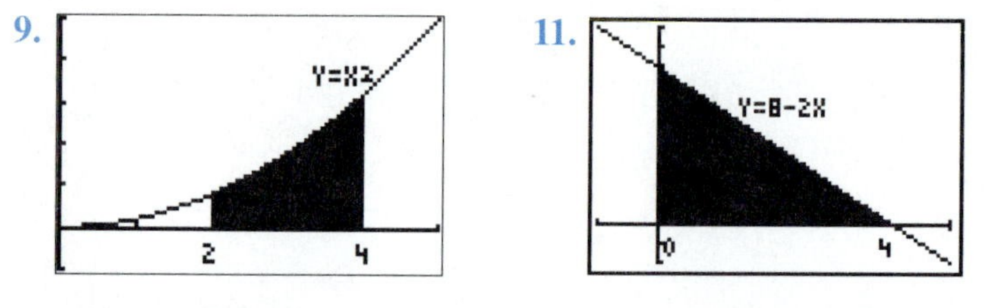

13. 20 **15.** 9 **17.** .68 **19.** .75 **21.** $4 + \frac{9}{4}\pi$ **27.** Positive **29.** $\int_{.2}^{9.8} f(x)\,dx \approx 40.2$
31. $\int_1^{2.8} f(x)\,dx \approx 2.5$ **33.** $\int_8^{10} x^3\,dx$ **35.** $\int_1^2 (x + e^{-x})\,dx$ **37.** $\int_0^3 (3 - x)^2\,dx$ **39.** 49 trillion

41. (a) 30 feet (b)

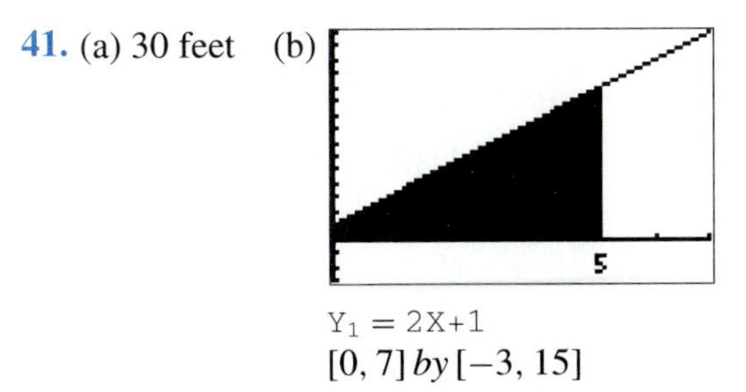

Y1 = 2X+1
[0, 7] *by* [−3, 15]

43. (a) \$1185.75 (b) The area under the marginal cost curve from $x = 2$ to $x = 8$. **45.** The increase in profits resulting from increasing the production level from 44 to 48 units **47.** (a) ≈ 24.5 (b) The amount the temperature falls during the first 2 hours **49.** 2088 million m^3

EXERCISES 7.3, PAGE 336

1. $\int_1^2 f(x)\,dx + \int_3^4 -f(x)\,dx$

3.

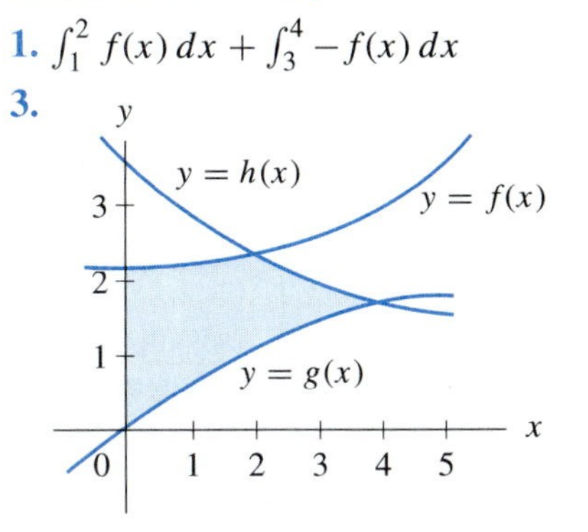

5. $21\frac{1}{3}$ **7.** $17\frac{1}{3}$ **9.** 18 **11.** $10\frac{2}{3}$ **13.** $10\frac{2}{3}$ **15.** 72 **17.** 108 **19.** (a) 4.5 (b) $6\frac{1}{3}$ (c) $13\frac{1}{6}$ **21.** 1.5 **23.** $2 + 12\ln(\frac{3}{2})$ **25.** $\int_0^{20}(76.2e^{.03t} - 50 + 6.03e^{.09t})\,dt$ **27.** No; 20; the additional profit from using the original plan **29.** (a) The distance between the two cars after 1 hour (b) after 2 hours

EXERCISES 7.4, PAGE 345

1. 3 **3.** 0 **5.** 8 **7.** 55° **9.** ≈ 82 g **11.** \$20 **13.** \$404.72 **15.** \$200 **17.** \$25 **19.** Intersection (100, 10), consumers' surplus = \$100, producers' surplus = \$250 **21.** \$3236.68 **23.** \$75,426 **25.** 5.63% compounded continuously **27.** (b) $1000e^{-.06t_1}\Delta t + 1000e^{-.06t_2}\Delta t + \cdots + 1000e^{-.06t_n}\Delta t$ (c) $f(t) = 1000e^{-.06t}$; $0 \le t \le 5$ (d) $\int_0^5 1000e^{-.06t}\,dt$ (e) \$4319.70 **29.** Receive funds steadily **31.** 4188.79 **33.** 19.4779 **35.** 25.1327 **37.** 1.3582

EXERCISES 7.5, PAGE 353

1. $20\frac{5}{6}$ **3.** $5(e^{20} - 1)$ **5.** $10\frac{2}{3}$ **7.** $29\frac{1}{3}$ **9.** $\frac{15}{4}$ **11.** $\frac{4}{3}(1 - e^{-3})$ **13.** −4 **15.** 10 **17.** $\frac{9}{2}$ **19.** $\frac{8}{3}$ **21.** \$1107.01 **23.** In 26 years **25.** \$20 **27.** \$404.72 **29.** 4.5% **31.** \$4066.49 **33.** 4% **35.** $\dfrac{\pi r^2 h}{3}$

EXERCISES 7.6, PAGE 364

1. $\frac{1}{6}(x^2+4)^6+C$ **3.** $\frac{1}{4}(x^2-5x)^4+C$ **5.** $e^{5x-3}+C$ **7.** $\ln\left|x^3-1\right|+C$ **9.** $\frac{2}{3}\sqrt{x^3-1}+C$ **11.** $-e^{1/x}+C$ **13.** $\frac{1}{3}\ln\left|x^3+3x+2\right|+C$ **15.** $\frac{1}{11}(x^5-2x+1)^{11}+C$ **17.** $\frac{3}{2}\ln|2x-4|+C$ **19.** $-\frac{1}{2}e^{-x^2}+C$ **21.** $\frac{1}{5}xe^{5x}-\frac{1}{25}e^{5x}+C$ **23.** $\frac{1}{10}x(2x+1)^5-\frac{1}{120}(2x+1)^6+C$ **25.** $\frac{2}{3}x(x+1)^{3/2}-\frac{4}{15}(x+1)^{5/2}+C$ **27.** $-3xe^{-x}-3e^{-x}+C$ **29.** $x\ln 3x-x+C$ **31.** $\frac{1}{2}xe^{2x}-\frac{1}{4}e^{2x}+C$ **33.** $\frac{1}{2}e^{x^2}+C$ **35.** $\frac{2}{3}(2x-1)(3x-3)^{1/2}-\frac{8}{27}(3x-3)^{3/2}+C$ **37.** $-\frac{1}{18}(6x^2+9x)^{-6}+C$ **39.** $2\sqrt{x}\ln x-4\sqrt{x}+C$ **41.** $\ln|\ln 5x|+C$ **43.** $x\ln x^4-4x+C$ **45.** $(\ln\sqrt{x})^2+C$ **47.** $\sqrt{x^2+9}+3$ **49.** \$4,673.46 **51.** $F(x)=\dfrac{45}{(x^2-x+5)^2}$ **53.** $F(x)=\frac{1}{5}x^5+\frac{15}{4}x^4+25x^3+\frac{125}{2}x^2$

EXERCISES 7.7, PAGE 371

1. 0 **3.** No limit **5.** $\frac{1}{4}$ **7.** 2 **9.** 5 **11.** 6 **13.** 2 **15.** $\frac{1}{2}$ **17.** 2 **19.** 1 **21.** Area under the graph from 1 to b is $\frac{5}{14}(14b+18)^{1/5}-\frac{5}{7}$. This has no limit as $b\to\infty$. **23.** $\frac{1}{2}$ **25.** Divergent **27.** $\frac{1}{6}$ **29.** $\frac{2}{3}$ **31.** 1 **33.** $2e$ **35.** Divergent **37.** $\frac{1}{2}$ **39.** 2 **41.** $\frac{1}{4}$ **43.** 2 **45.** $\frac{1}{6}$ **49.** \$50,000

EXERCISES 7.8, PAGE 378

1. (a) $\frac{11}{32}$ (b) $\frac{7}{27}$ (c) $\frac{27}{32}$ (d) $\frac{27}{128}$ **3.** $\frac{3}{8}$ **5.** (a) .18127 (b) .45085 (c) 13534 (d) .5 **7.** (a) $f(x)=\frac{1}{3}$, $0\le x\le 3$ (b) $\frac{2}{3}$ (c) $\frac{1}{3}$ **9.** (a) .25 (b) 500 acres **11.** .88 **13.** (a) .39347 (b) .36788 **15.** .22313 **17.** $1-e^{-2}$ **19.** .0211 **21.** .08 **23.** .665

CHAPTER 8

EXERCISES 8.1, PAGE 388

1. $f(1,0)=1$, $f(0,1)=8$, $f(3,2)=25$ **3.** $f(0,1)=0$, $f(3,12)=18$, $f(a,b)=3\sqrt{ab}$ **5.** $f(2,3,4)=-2$, $f(7,46,44)=\frac{7}{2}$ **11.** \$50.16; \$50.16 invested at 5% continuously compounded interest will yield \$100 in 13.8 years **13.** (a) \$18,750 (b) \$22,500; yes

15. **17.**

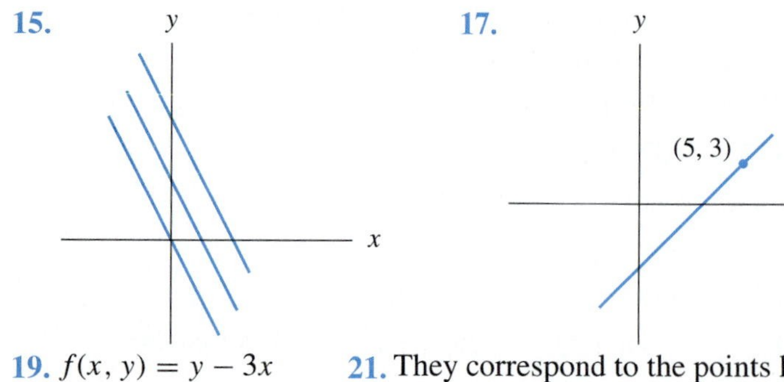

19. $f(x,y)=y-3x$ **21.** They correspond to the points having the same altitude above sea level. **23.** (d) **25.** (c)

EXERCISES 8.2, PAGE 400

1. $5y$, $5x$ 3. $4xe^y$, $2x^2e^y$ 5. $-\dfrac{y^2}{x^2}$, $\dfrac{2y}{x}$ 7. $4(2x-y+5)$, $-2(2x-y+5)$ 9. $(2xe^{3x}+3x^2e^{3x})\ln y$, x^2e^{3x}/y 11. $\dfrac{2y}{(x+y)^2}$, $-\dfrac{2x}{(x+y)^2}$ 13. $\dfrac{3\sqrt{K}}{2\sqrt{L}}$ 15. $\dfrac{2xy}{z}$, $\dfrac{x^2}{z}$, $-\dfrac{1+x^2y}{z^2}$ 17. ze^{yz}, xz^2e^{yz}, $x(yz+1)e^{yz}$
19. 1, 3 21. -12
23. $\dfrac{\partial f}{\partial x}=3x^2y+2y^2$, $\dfrac{\partial^2 f}{\partial x^2}=6xy$, $\dfrac{\partial f}{\partial y}=x^3+4xy$, $\dfrac{\partial^2 f}{\partial y^2}=4x$, $\dfrac{\partial^2 f}{\partial y\partial x}=\dfrac{\partial^2 f}{\partial x\partial y}=3x^2+4y$
25. (a) Marginal productivity of labor = 480; of capital = 40 (b) 240 fewer units produced 27. If the price of a bus ride increases and the price of a train ticket remains constant, fewer people will ride the bus. An increase in train-ticket prices coupled with constant bus fare should cause more people to ride the bus. 29. If the average price of videotapes increases and the average price of a VCR remains constant, people will purchase fewer VCR's. An increase in average VCR prices coupled with constant video tape prices should cause a decline in the number of videotapes purchased.
31. $\dfrac{\partial V}{\partial P}(20, 300)=-.06$, $\dfrac{\partial V}{\partial T}(20, 300)=.004$ 33. $\dfrac{\partial f}{\partial r}>0$, $\dfrac{\partial f}{\partial m}>0$, $\dfrac{\partial f}{\partial p}<0$
35. $\dfrac{\partial^2 f}{\partial x^2}=-\dfrac{45}{4}x^{-5/4}y^{1/4}$. Marginal productivity of labor is decreasing.
39. 3269.95394; .04606; 3209.953464; .046536

EXERCISES 8.3, PAGE 410

1. $(-2, 1)$ 3. $(26, 11)$ 5. $(1, -3)$, $(-1, -3)$ 7. $(\sqrt{5}, 1)$, $(\sqrt{5}, -1)$, $(-\sqrt{5}, 1)$, $(-\sqrt{5}, -1)$ 9. $(\frac{1}{3}, \frac{4}{3})$
11. Relative minimum; neither relative maximum nor relative minimum 13. Relative maximum; neither relative maximum nor relative minimum; relative maximum 15. Neither relative maximum nor relative minimum
17. $(0, 0)$ min 19. $(-1, -4)$ max 21. $(0, -1)$ min 23. $(-1, 2)$ max; $(1, 2)$ neither max nor min
25. $(\frac{1}{4}, 2)$ min; $(\frac{1}{4}, -2)$ neither max nor min 27. $(\frac{1}{2}, \frac{1}{6}, \frac{1}{2})$ 29. 14 in. × 14 in. × 28 in. 31. $x=120$, $y=80$ 35. Relative maximum 37. Relative maximum; relative minimum; saddle point 39. Relative minimum; saddle point 41. Relative minimum; saddle point 43. Relative maximum; saddle point; relative maximum

EXERCISES 8.4, PAGE 422

1. 58 at $x=6$, $y=2$, $\lambda=12$ 3. 13 at $x=8$, $y=-3$, $\lambda=13$ 5. $x=\frac{1}{2}$, $y=2$ 7. 5, 5
9. Base 10 in., height 5 in. 11. $F(x, y, \lambda)=4xy+\lambda(1-x^2-y^2)$; $\sqrt{2}\times\sqrt{2}$
13. $F(x, y, \lambda)=3x+4y+\lambda(18{,}000-9x^2-4y^2)$; $x=20$, $y=60$
15. (a) $F(x, y, \lambda)=96x+162y+\lambda(3456-64x^{3/4}y^{1/4})$; $x=81$, $y=16$ (b) $\lambda=3$ 17. $x=12$, $y=2$, $z=4$
19. $x=2$, $y=3$, $z=1$ 21. $F(x, y, z, \lambda)=3xy+2xz+2yz+\lambda(12-xyz)$; $x=2$, $y=2$, $z=3$
23. $F(x, y, z, \lambda)=xy+2xz+2yz+\lambda(32-xyz)$; $x=y=4$, $z=2$

APPENDIX B

EXERCISES, PAGE 438

1. $2, \frac{3}{2}$ **3.** $\frac{3}{2}$ **5.** No zeros **7.** $1, -\frac{1}{5}$ **9.** $5, 4$ **11.** $2+\sqrt{6}/3, 2-\sqrt{6}/3$ **13.** $(x+5)(x+3)$
15. $(x-4)(x+4)$ **17.** $3(x+2)^2$ **19.** $-2(x-3)(x+5)$ **21.** $x(3-x)$ **23.** $-2x(x-\sqrt{3})(x+\sqrt{3})$
25. $-7, 3$ **27.** $-2, 3$ **29.** -7 **31.** $-2.1, 1.7$ **33.** $-12, 6.5$ **35.** No solution **37.** $-22, 21$

INDEX

O

P

Q

R

S